MANUAL OF THE NEW WORLD GENERA OF THE FAMILY BRACONIDAE (HYMENOPTERA)

Edited By
ROBERT A. WHARTON, PAUL M. MARSH AND MICHAEL J. SHARKEY

SPECIAL PUBLICATION OF
THE INTERNATIONAL SOCIETY OF HYMENOPTERISTS
Number 1

Second edition

SPECIAL PUBLICATION
OF
THE INTERNATIONAL SOCIETY OF HYMENOPTERISTS
Number 1
Second edition

MANUAL OF THE NEW WORLD GENERA OF THE FAMILY BRACONIDAE (HYMENOPTERA)

Edited By
ROBERT A. WHARTON
Department of Entomology, Texas A&M University, College Station, Texas 77843

PAUL M. MARSH
Cooperating Scientist, Systematic Entomology Laboratory, USDA Washington, DC 20560
(present address: P. O. Box 384, North Newton, Kansas 67117)

MICHAEL J. SHARKEY
Department of Entomology, University of Kentucky,
Lexington, Kentucky 40546

Contributing Authors
Paul C. Dangerfield, José L. Fernández-Triana, Rebecca N. Kittel, Robert R. Kula, Paul M. Marsh, Donald L. J. Quicke, Michael J. Sharkey, Scott R. Shaw, Cornelis van Achterberg, Robert A. Wharton, James B. Whitfield, Alejandro Zaldívar-Riverón

Published By
THE INTERNATIONAL SOCIETY OF HYMENOPTERISTS
Washington, DC
1997

EXECUTIVE COMMITTEE
of
THE INTERNATIONAL SOCIETY OF HYMENOPTERISTS

Second edition (1.1): December 23, 2017
(First edition: October 20, 1997)
ISBN 10: 0692944826
ISBN 13: 978-0692944820
Printing and binding: Createspace, Amazon
Typesetting: by the editors and authors

DEDICATION

WILLIAM R. M. MASON

1921-1991

We dedicate this manual to the memory of William (Bill) Richardson Miles Mason who was a mentor and friend to us all until his death from a stroke in1991. Bill was born in Lucknow, India on November 29, 1921. He spent the first years of his life there and then moved back to Canada with his mother and his father who was serving in the Canadian military. Bill spent his formative years in western Canada on a farm outside the town of Olds, Alberta. It was there he first started collecting insects and storing them in a cabinet of shallow drawers bought for him by his parents. He received his B.S. degree from The University of Alberta in 1942. Bill then spent several years as a navigator in the Royal Canadian Air Force and served in Europe during the Second World War before attending Cornell University for a Ph.D. in Entomology.

Bill joined the Systematic Entomology Unit (Canadian National Collection of Insects), Ottawa, in 1948 and served there even after his retirement in 1985. Bill possessed a broad knowledge of the Hymenoptera and was particularly well versed in the systematics of the Ichneumonidae and Braconidae. His earlier works concentrated on the Ichneumonidae including a long series of revisions on the Cteniscini. Believing that the Braconidae were an even more neglected taxon, Bill switched his efforts to the Braconidae and published about 40 papers including some larger revisions such as those on the Muesebeckiini, Brachistini, Braconini, and his influential work on the genera of the Microgastrinae.

Bill was an avid collector and was directly responsible for the world class collection of Braconidae in the Canadian National Collection which he built up over 40 years of intensive collecting. Bill led 17 collecting expeditions to such places as the Arctic, Alaska, Sweden, Mexico, and Nepal. He served as the editor of The Canadian Entomologist in the 1960's and was the head of the Hymenoptera section during that time. His final major contribution to Hymenopterology came in the leading role that he took in the production of the volume "The Hymenoptera of the World."

Bill is sorely missed by us all. Much of what will be considered as a positive contribution in this identification manual is the direct result of his mentoring. We, therefore, take pleasure in dedicating this manual to the memory of our friend and colleague, the late W.R.M. "Bill" Mason.

PROLOGUE

Rebecca Kittel and Michael Sharkey

The Manual, published in 1997, is out of print. A group of us agreed to produce a new manual but realized that it may take a few years to accomplish this. Therefore we are republishing the original Manual. It is dated, and to mitigate this situation we solicited the help of braconologists to add addenda to the subfamily chapters. These will allow any user to find all of the most recent literature, keys, and descriptions of new genera. Although the generic classifications of some of the subfamilies have changed little since 1997, others such as the Agathidinae have been completely revamped. In all cases the addenda will lead readers to the most recent keys and if these do not exist they will warn of potential pitfalls to the present version of the Manual that are due to taxonomic or nomenclatorial changes published since 1997.

An addendum follows each subfamily chapter; however users are cautioned to read the subfamily addendum before the reproduced text of the respective chapter from the original Manual. Addenda to the introductory chapters have not been included; nonetheless the key to subfamilies needs major revision. The subfamily key is still useful because if there is a new subfamily concept, it is explained in the addendum of subfamily the key will lead them to.

The original version of the Manual was published by the International Society of Hymenopterists and all proceeds will still go directly to ISH. One of the advantages of publishing with Amazon is that both electronic and hard copies are available; another benefit is that publication costs are zero and therefore there is no strain on ISH resources. If our community is happy with this republication we will probably use the same venue for the new version(s) of the Manual.

We thank all of the other authors who contributed to the addenda, i.e., Kees van Achterberg, Bob Kula, Donald Quicke, Scott Shaw, José Fernández-Triana, and Alejandro Zaldívar-Riverón. Barb Sharanowski scanned the original Manual. Rebecca Kittel incorporated numerous edits and formatted the text for publication via Amazon. Michael Sharkey edited the preliminary copy of the scanned version, incorporated the figures, and solicited and edited the addenda. All addenda authors edited the error-ridden preliminary copy of the Manual. We apologize for any errors that may have been left undetected.

Table of Contents

MANUAL OF THE NEW WORLD GENERA OF THE FAMILY BRACONIDAE (HYMENOPTERA)

Edited By
ROBERT A. WHARTON, PAUL M. MARSH AND MICHAEL J. SHARKEY

Abstract

This multi authored identification manual presents illustrated keys to the 34 subfamilies and 404 genera of the family Braconidae from the New World. Information about all New World genera described up to 1994 is included which represents an expanded and updated version of the information published in the 1987 Nearctic manual by Marsh, *et. al*. The manual consists of 37 chapters by world authorities plus an index to generic names including subgenera and synonyms, and an appendix listing all the New World genera alphabetically by subfamily. The first chapter provides an introduction including a review of braconid biology, literature, classification, biogeography, collection and curation, and a discussion of the manual format. The second chapter is an extensive illustrated discussion of braconid morphology and terminologies used in the keys. The third chapter is a fully illustrated key to the New World subfamilies of the Braconidae. The New World subfamilies are then presented separately in the remaining 34 chapters. For each, a key to the New World genera is included. Each of these keys is annotated to include estimated number of species, distribution, hosts, and critical references for each genus. Each subfamily chapter is extensively illustrated and the entire manual contains over 750 line drawings and scanning electron micrographs.

INTRODUCTION

R. A. WHARTON

Department of Entomology, Texas A&M University, College Station, Texas 77843, U.S.A.

The Hymenoptera is one of the largest orders of insects. It contains a diverse array of phytophagous insects, social insects, solitary bees and wasps, and parasitoids (= protelean parasites or parasitic wasps). Although the bees, ants, and social wasps are perhaps the most familiar of the Hymenoptera, the parasitoids that develop as immatures on or in the bodies of other arthropods are by far the largest group, and are of considerable importance in the biological control of insect pests.

Among the parasitoids are the two largest families of Hymenoptera, the Ichneumonidae and Braconidae. There are approximately 14,890 valid species of Braconidae worldwide (Marsh, unpublished), somewhat more than recent estimates of the number of described species (Marsh, 1979; M. Shaw and Huddleston, 1991). Published estimates (e.g. van Achterberg, 1984) further suggest that only one-half to one-third of the world's species have been formally described.

Braconids vary tremendously in body length from about 30 mm to less than 1 mm. These figures do not include the ovipositor, which may be equal to or longer than the body in some species. Braconids may be recognized by the following combination of characters (see next chapter for discussion and illustration of terms): absence of a costal cell (costa, subcosta and radius fused or nearly so over basal half of fore wing), fusion of metasomal terga 2 and 3, presence of a trochantellus (i.e. trochanter appears two segmented), and the absence of fore wing 2m-cu (often called the second recurrent vein), except in the rare Chilean genus *Apozyx* Mason. Wingless or brachypterous species usually are either exodont (with non-overlapping, outwardly directly mandibles) or cyclostome (with labrum and ventral portion of clypeus concave). Sharkey and Wahl (1992), in their phylogenetic analysis of the superfamily Ichneumonoidea, defined Braconidae on the basis of four autapomorphs. These are the fusion of metasomal terga 2 and 3, the loss of functional basal hamuli on vein C of the hind wing, loss of an independent stub of vein C of the hind wing basad the distal hamuli, and the basal migration of cross vein Irm of the hind wing to or basal to the separation of veins Rl and RS. In many braconids, the suture line between terga 2 and 3 is still evident even though there is no longer a flexible joint. In aphidiines, however, flexion is still possible, but the suture itself seems to have been lost.

The present work was initially designed to cover the Neotropical fauna, as a companion to the identification manual to braconid genera in America north of Mexico (Marsh et al., 1987). The interest generated by the latter work, together with several subsequent nomenclatural changes and additions to the Nearctic fauna, led to the development of an expanded version covering the entire New World fauna. Because braconids are among the most widely used parasitoids in biological control, we have also included basic biological information under each of the subfamily chapters.

This first chapter covers some of the more general features of the family. Organization and layout of the manual are described at the end of the chapter.

Biology

The vast majority of the braconids are parasitoids of other insects. Only a few species, recently discovered (Macêdo and Monteiro, 1989; Wharton, 1993; Infante et al., 1995), are known to be phytophagous. The most common hosts of braconid parasitoids are the larvae of Lepidoptera, Coleoptera, and Diptera. Unlike members of some of the other families of parasitic Hymenoptera, no braconids have been verifiably reared from non-insect arthropods. As in other parasitoids, the female braconid searches for a suitable host in the appropriate habitat, lays one or more eggs on or in the host, and the immature stages then develop to completion at the expense of a single host, killing it in the process. The family Braconidae contains both ectoparasitoids and endoparasitoids. Though both are well represented amongst the described species, there are more endoparasitoids than ectoparasitoids, Braconid ectoparasitoids are represented largely by members of the subfamilies Braconinae, Hormiinae (in the broad sense as treated in this manual), and Doryctinae. These ectoparasitoids are generally idiobionts, with the immature stages feeding externally on an immobilized host that undergoes no further development after it has been attacked and paralyzed by the female parasitoid. The parasitoid egg may be laid directly on the host or often in the vicinity of the host after the latter has been stung and paralyzed. Braconid endoparasitoids, including all of the species in such large subfamilies as the Alysiinae, Microgastrinae, and Opiinae, are generally koinobionts. Koinobiont braconids only temporarily paralyze their hosts, if at all, and the host continues to develop for a varying length of time after it is parasitized. Koinobiont endoparasitoids generally spend most of their developmental period inside the host, in intimate contact with host tissues, with the host's immune system usually compromised by substances injected by the female during oviposition. Many endoparasitoids emerge from the moribund host near the end of their development, and pupate in a cocoon on, around, or away from the host. This has been referred to as the final ectoparasitic stage of development by some authors (e.g. M. Shaw and Huddleston, 1991). Although most Braconidae are solitary, gregarious parasitism is common, particularly among the Microgastrinae, Braconinae, and Doryctinae.

Most braconids are larval parasitoids, attacking and emerging from the larval stage of holometabolous insects. There are, however, a number of egg-larval parasitoids, a few pupal parasitoids, some that attack nymphs of paurometabolous insects, and one group (Euphorinae) in which many of the species attack adults of either holometabolous or paurometabolous insects. Hyperparasitism is rare. The few recorded instances of hyperparasitism generally involve the attack of Diptera larvae in Lepidoptera and other hosts. Most of these records, however, are insufficiently detailed to determine whether the braconids attacked the maggots before or after the latter had killed the primary host. No species is yet known that develops exclusively in the eggs of other insects, but many species oviposit in the host egg and emerge from a late larval instar or puparium. All known members of the Cheloninae develop in this fashion as egg-larval parasitoids of Lepidoptera. The few available records suggest that members of the helconine tribe Brachistini also develop in this fashion on Coleoptera. Scattered records of egg-larval parasitism are known for the Microgastrinae, Opiinae, and Alysiinae, though emergence is from the puparium in the latter two subfamilies. Host specificity varies in the Braconidae, though most species have relatively narrow host ranges limited either by developmental capabilities or by the microhabitat in which the females search for hosts. As is true of other Hymenoptera, braconids have a haplo-diploid sex determination mechanism. Unfertilized eggs are haploid and fertilized eggs are diploid. Unmated females are fully capable of depositing viable eggs, but these will only produce males.

Considerable additional detail on braconid biology is provided by M. Shaw and Huddleston (1991). Other recent, though less detailed, accounts can be found in Wharton (1993) and S. Shaw (1995). Clausen (1940) is still the best and most detailed review of the earlier literature. Excellent general treatments of parasitoid biology, in which braconids figure prominently, are becoming increasingly available. The works of Beckage, et al. (1993) and Godfray (1994) are typical.

Literature

The literature on Braconidae is enormous. Shenefelt (1965), for example, compiled over 8000 references on the taxonomy and biology of the members of this family and Starý (1987) published a selected bibliography that included 90 pages of references, solely on Aphidiinae, through 1982. Unfortunately, many of the species that have been so intensively studied in recent years by North American workers are not covered by these reviews. These include *Toxoneuron* (as *Cardiochiles*) *nigriceps* (Viereck), *Microplitis croceipes* (Cresson), *Cotesia melanoscelus* (Ratzeburg), *Diachasmimorpha longicaudata* (Ashmead), and numerous aphidiines. Most of the taxonomic research by such prolific authors as S. Belokobyl'skij, M. Fischer, P. Marsh, J. Papp, D. Quicke, P. Starý, V. Tobias, and C. van Achterberg, covering close to 1000 citations, has also been published subsequent to Shenefelt's (1965) review.

The study of Braconidae has been greatly facilitated by the existence of both regional and world catalogs, and these have been extensively referred to in the chapters below. Of special value for the New World are the Catalog of Hymenoptera in America North of Mexico (Marsh, 1979), and the catalogs for Argentina and Brazil (De Santis, 1967, 1980). Gonzalez (unpublished) has also created a specimen-level database for Mexican parasitic Hymenoptera that includes a considerable amount of information on Braconidae. On a world basis, the catalogs and monographs by Mackauer and Starý (1967), Mackauer (1968), Shenefelt (1969, 1970a, b, 1972, 1973a, b, 1974, 1975, 1978, 1980), Fischer (1971, 1972, 1977, 1987), and Shenefelt and Marsh (1976) are invaluable, even though most are now dated because of the wealth of more recent publications and subsequent taxonomic changes.

The present volume, as noted above, builds on the foundation established by the earlier publications of Marsh (1971) and Marsh et al. (1987), and has also benefitted by the recent subfamily keys of van Achterberg (1993) and Sharkey (1993). A relatively large number of genera have been described or redefined since 1987, with especially noteworthy changes in the Aphidiinae, Braconinae, Cardiochilinae, Cenocoeliinae, Cheloninae, Doryctinae, Hormiinae, Ichneutinae, Macrocentrinae, Opiinae, and Rogadinae.

Classification

Hymenoptera. The Hymenoptera are frequently divided into two suborders, the Symphyta (sawflies and horntails) and the Apocrita (bees, wasps, and ants). The Braconidae, as well as nearly all other parasitic Hymenoptera, are placed in the suborder Apocrita. This overly simplified division has come under increasing scrutiny in recent years (for recent discussion, see Gauld and Hanson, 1995). Several researchers, attempting to elucidate the phylogeny of major lineages within the Hymenoptera, have been unable to define the Symphyta on the basis of shared, derived features. The Symphyta has thus been characterized as a grade rather than a clade, raising the question of whether it should be treated in its entirety as a separate suborder. The Apocrita, on the other hand, is widely accepted as a well-defined, monophyletic lineage within the Hymenoptera. The Apocrita contains two major subdivisions, the Aculeata and the Parasitica (=Terebrantia). The Aculeata are clearly defined as monophyletic, based on modifications of the ovipositor, and their recognition as a separate division within the Apocrita has not been disputed. The Parasitica, however, are comprised of a diverse assemblage of superfamilies, and the relationships of these superfamilies, both to each other and to other apocritan and non-apocritan groups, has been the subject of considerable debate since the publications of Koenigsmann in the late 1970's (Rasnitsyn, 1988; Heraty et al, 1994; Dowton and Austin, 1994, and references therein). The Braconidae, grouped together with the Ichneumonidae in the superfamily Ichneumonoidea within the Parasitica, is sometimes considered as one of the most basal lineages within the Apocrita (e.g. Whitfield, 1992b). Understanding the relationship of the Braconidae to other members of the Apocrita is critical to the development of hypotheses concerning evolution of character systems within this family. Our current meager knowledge of these relationships has contributed to the development of several competing hypotheses of braconid phylogeny (van Achterberg, 1984; Quicke and van Achterberg, 1990; Wharton et al., 1992; van Achterberg and Quicke, 1992).

Braconidae. The family Braconidae is routinely subdivided into two informal groups, the cyclostomes and non-cyclostomes, based traditionally on whether or not the labrum and clypeus are indented to form the "cyclostome" condition. Specifically, in the cyclostomes, the ventral portion of the clypeus is recessed, and together with the concave, often glabrous labrum, forms a rounded depression above the mandibles. The cyclostomes originally comprised the genera now included in the Braconinae, Doryctinae, and Rogadinae as well as several genera treated here under Hormiinae. Studies over the last two decades, primarily on internal character sets (e. g. Edson and Vinson, 1979; Maetô, 1987), have resulted in an expanded definition of these two groups, and a consequent shifting of several taxa from the non-cyclostomes to the cyclostomes. Much work still remains, however, as several taxa are questionably included among the cyclostomes. Various aspects of the cyclostome I non-cyclostome question are summarized by Quicke and van Achterberg (1990), Wharton, et al. (1992), and Sharkey (1993).

Non-cyclostomes are endoparasitic. Cyclostomes are predominantly ectoparasitic, with the endoparasitic cyclostomes generally believed to represent separate, multiple forays into an endoparasitic lifestyle from the more primitive ectoparasitic mode (see M. Shaw (1983), Whitfield (1992a), and the excellent summary by M. Shaw and Huddleston (1991)). Thus, endoparasitism is believed to have evolved on at least three separate occasions within the Braconidae: once in the non-cyclostomes, and twice in the cyclostomes (the Rogadinae / Hormiinae and the Opiinae / Alysiinae). Incorporation of the cyclostome I non-cyclostome concepts into a formal classification has met with mixed results. Ectoparasitic cyclostomes generally fall out at the base of the braconid phylogenetic tree, but whether or not they represent a distinct lineage is debatable (Quicke and van Achterberg, 1990; Wharton et al., 1992; van Achterberg and Quicke, 1992). Moreover, the helconines, a group of endoparasitic non-cyclostomes with exceptionally complete wing venation, are often stated to be among the most primitive of the braconids, and should thus be at the base of the non-cyclostome lineage. Yet, the connection between helconines and the cyclostome groups is obscure at best. This sets up the possibility of a biphyletic origin for the Braconidae, with the link between the cyclostome lineage and the non-cyclostome lineage, if indeed the two represent separate, well-defined lineages, lost in evolutionary time. Because of these various problems, formal classifications of the roughly 1000 valid genera of Braconidae have frequently been presented as a large number of weakly connected subfamilies and tribes. The group of non-cyclostome subfamilies clustered around the Microgastrinae represents one of the few well-defined lineages within the Braconidae (Nixon, 1965; Mason, 1981, 1983; Quicke and van Achterberg, 1990; Whitfield and Mason, 1994).

The subfamily classification itself is somewhat controversial and decidedly unstable, with much of the debate centered on whether to recognize a relatively small number of larger subfamilies or a large number of smaller ones. This dichotomy reflects two basic philosophical approaches to classification: emphasizing differences by elevation in rank vs. emphasizing relationships by gathering the components under a single umbrella. Quicke and van Achterberg (1990), for example, recognized 43 subfamilies and van Achterberg (1993) increased the number to 45 in his key to the braconid subfamilies of the world. The differences between Quicke and van Achterberg (1990) and van Achterberg (1993) are the result of a reduction in rank of two of the formerly recognized subfamilies and the elevation in rank of four other groups. Sharkey (1993), on the other hand, recognized only 29 subfamilies in his key to the world fauna, and his classification more closely matches that of Tobias (1986) in this regard. Although Tobias (1986) dealt only with the fauna of the former Soviet Union, and thus did not include several extra-limital subfamilies, he and Sharkey (1993) combined taxa in much the same way. One of the major differences in the way subfamilies are treated by Tobias (1986), Quicke and van Achterberg (1990), van Achterberg (1993), and Sharkey (1993) involves the placement of several of the cyclostome groups treated in this manual as the subfamily Hormiinae. As discussed in the hormiine chapter below, these genera have been variously classified as Rogadinae, Doryctinae, as a separate subfamily (using either the name Hormiinae or Exothecinae), or as a series of smaller subfamilies. Although the object of considerable attention over the last few years, this group of approximately 40 genera worldwide is still one of the most poorly understood of the Braconidae in part because there is so little uniformity of body type and wing venation among the many seemingly related genera.

In this manual, we have treated 34 subfamilies, reflecting the divergent opinions of the various collaborators at the time the project was initiated. Certain subfamilies, such as the Agathidinae, Alysiinae, Braconinae, Cheloninae, Macrocentrinae, and Microgastrinae have long remained stable and will undoubtedly continue to do so because there is broad agreement on their characterization as monophyletic groups. Others, including the Adeliinae, Cardiochilinae, Gnamptodontinae, Meteorideinae, Miracinae, and Sigalphinae represent small subfamilies formerly associated with larger subfamilies but now generally well characterized as separate entities. Several small groups (such as the Betylobraconinae, Cenocoeliinae, Dirrhopinae, Meteorinae, and Microtypinae) have similarly been split off from larger subfamilies, but appear to be less well supported. Thus, several workers (me included) favor maintaining the relevant genera within the confines of larger subfamilies until their relationships are better understood. Unfortunately, this indecision about the smaller subfamilies does nothing to resolve on-going problems associated with the definition of these larger subfamilies, and the Blacinae, Euphorinae, Helconinae, and Orgilinae are particularly affected. The Blacinae, in fact, have been variously placed either in the Helconinae or the Euphorinae by some authors. The problem of defining the Helconinae is highlighted by the way in which this subfamily is treated in the helconine chapter below. The Opiinae, though fairly widely accepted and relatively stable, is another large group that has never been adequately defined, and this may eventually pose problems for its recognition as a separate subfamily distinct from the Alysiinae. The Homolobinae and Ichneutinae, both relatively recently treated on a world basis (van Achterberg, 1979; Sharkey and Wharton, 1994), are presently affected by differences of opinion regarding the inclusion or exclusion of certain elements such as *Charmon* Haliday and *Proterops* Wesmael, respectively. The Ypsistocerinae and Histeromerinae have sometimes been included in the Doryctinae, and the Betylobraconinae do not differ substantially from certain Rogadinae. The exact limits of both the Doryctinae and Rogadinae are also decidedly affected by the way in which the "hormiine" genera are treated. Other putative cyclostome subfamilies, such as the Apozyginae and Aphidiinae, have not always been treated as members of the Braconidae, and some workers may still question their inclusion. However, recent phylogenetic studies on braconids (see previous two paragraphs) unanimously support their inclusion within a monophyletic Braconidae. As a final complication, new taxa are constantly being discovered, as exemplified by the description of two more subfamilies (the Masoninae and Mendesellinae) since the works of van Achterberg (1993) and Sharkey (1993). Thus, with continuing discoveries of new taxa and publication of new data on morphology and biology, braconid classification will remain an exciting and challenging endeavor for the foreseeable future.

No matter which way they are defined, braconid subfamilies differ greatly in size from one another. The Braconinae, with about 2,000 described species (M. Shaw and Huddleston, 1991) is currently the largest subfamily. The Microgastrinae, Opiinae, Alysiinae, and Doryctinae are also very large, with each containing between 1,000 and 1,500 described species. New World collections already contain hundreds of undescribed species in all 5 of these subfamilies and thus their ranking relative to one another (in terms of overall species richness) could change as our knowledge of tropical species, particularly those of small body size, improves over time.

Generic names and why they change. In this manual, we provide keys to 404 genera. Some of these genera are endemic to the New World, but a large number of them are more widespread, occurring in both the New and Old World. Many of the genera are well defined, relatively stable, and their recognition presents few if any difficulties. But many of the problems discussed above for subfamilies also apply to genera, and changes in names are inevitable as new information is analyzed and processed. Thus, the generic names used here may differ somewhat from those used in the key to braconid genera in America north of Mexico (Marsh et al., 1987) despite the relative recency of the latter publication. Of the nearly 3000 generic names that have been proposed since Latreille described *Sigalphus* in 1802 and Fabricius described Bracon in 1804, only about 1000 of these are presently considered valid. The remainder are now treated largely as synonyms or subgenera. Subgenera have been used widely in the subfamilies Braconinae and Opiinae. Within the New World, the largest number of genera are found in the Doryctinae (78 genera), Microgastrinae (44 genera), and Alysiinae (41 genera).

Reasons for name changes are many and varied, and literature pertaining to the examples used here can be found in the chapters below. Most changes are the result of restrictions in the definition of a genus, with the consequent recognition of one or more segregates as separate genera. Perhaps the most familiar example of this is the genus Apanteles Förster. Until the work of Mason (1981), nearly all of the New World species of Microgastrinae with an open second submarginal cell in the fore wing were assigned to the genus *Apanteles*. Mason elevated many of the species groups of Apanteles recognized by Nixon (1965) to the generic rank, described several more of his own, and completely changed the generic classification of the Microgastrinae. *Apanteles* is still a valid genus, with many described species in the New World, including *A. diairaeae* Muesebeck, an effective biological control agent for the sugarcane borer *Diatraea saccharalis* (F.). However, many of the species formerly assigned to this genus are now placed in such genera as *Cotesia* Cameron and *Glyptapanteles* Ashmead (Mason, 1981; Whitfield, 1995). Both *Cotesia* and *Glyptapanteles* are large, with several hundred species, including many that are undescribed. They include some of the most widely used microgastrines in biological control and experimental biology, such as *C. congregata* (Say), *C. flavipes* Cameron, *C. glomerata* (L.), *C. marginiventris* (Cresson), *C. melanoscelus* (Ratzeburg), and *G. militaris* (Walsh). Similar splits have occurred in other commonly encountered genera such as Agathis Latreille, *Aspilota* Förster, *Bracon* F., *Cardiochiles* Nees, *Macrocentrus* Curtis, and *Rogas* Haliday. Recent research on *Agathis, Aspilota,* and *Macrocentrus* has resulted in the recognition of well-defined isolates (*Bassus* F., *Dinotrema* Förster, and *Hymenochaonia* Dalla Torre respectively) that were previously treated as synonyms. These three genera have thus been divided roughly in half, with the addition of three more genera. Such changes represent the formal recognition, by elevation to generic rank, of well-defined and previously recognized species groups. In the case of *Agathis*, a further split has occurred with the recognition of *Alabagrus* Enderlein, a Neotropical element. *Bracon* is one of the largest genera in the New World, and similar attempts have often been made to split it into more manageable units, as a practical matter if for no other reason. Few of these attempts have been successful, but one of the most widely accepted ones is the recognition of a small group of species as the genus *Habrobracon* Ashmead. While relatively few species are involved, at least two of them, *H. gelechiae* (Ashmead) and *H. hebetor* (Say), are extremely important. A slightly different situation has occurred with *Cardiochiles*, at least in the New World. An extensive world revision of the genera has resulted in the shift of a large portion of the New World species from *Cardiochiles* to at least two other genera: one newly described and another whose name has not been used in over 100 years. This change affects one of the most widely studied braconids, for which the valid name is now *Toxoneuron nigriceps* (Viereck). For *Rogas*, a problem in nomenclature regarding misidentification of species over a hundred years ago led to a shift of most of the species formerly placed in Rogas to the genus Aleiodes Wesmael. *Rogas* still exists, and contains a large number of Neotropical species, but the name is now applied to a different group of insects than it has been for most of this century. Though these examples pertain to large genera, affecting many species, small genera are by no means exempt. A typical example is the recognition of *Dinocampus* Förster as a genus distinct from *Perilitus* Nees. This resulted in a change in the name of the well-known coccinellid parasitoid, now called *Dinocampus* coccinellae (Schrank).

Restrictions in the application of names in one part of the world can have a major impact elsewhere, as in the case of *Digonogastra* Viereck. Most species now placed in *Digonogastra* were formerly included in either *Iphiaulax* Förster or *Ipobracon* Dalla Torre. Quicke (1987) restricted the definition of *Iphiaulax* to Old World species, resulting in the transfer of New World species to *Digonogastra*, a name that had previously been treated as a synonym of *Iphiaulax. Digonogastra* is one of the most commonly encountered members of the subfamily Braconinae in the Neotropical and southern Nearctic Regions. *Ipobracon* was placed as a synonym of *Cyanopterus* Haliday, with its definition now restricted to relatively few New World species.

There are other reasons for name changes. Sometimes, especially when intermediate species are discovered, two or more genera are lumped together (synonymy). This is currently an uncommon phenomenon in the Braconidae, but does occur. A recent example is the treatment of *Zelomorpha* Ashmead as a synonym of *Coccygidium* de Saussure. New World species that were formerly placed in *Zelomorpha* are now members of the genus *Coccygidium*. Occasionally, an older name is discovered for one that is currently being used. Thus, the species formerly placed in *Leurinion* Muesebeck were transferred to *Allobracon* Gahan when the true identity of the latter was discovered. Changes of this nature, involving poorly known species, are straight-forward. When such changes involve well known genera, such as the discovery of an older name for *Chelonus* Panzer, it has been possible to conserve usage through formal petition to the International Commission of Zoological Nomenclature. *Homolobus* Förster is an excellent example of some of the more complicated issues involving application of names, including complex synonymies, misidentifications, and forgotten names (van Achterberg, 1979).

Nearly all decisions regarding name changes at these levels are subjective (though by no means arbitrary), involving weighting of characters following detailed studies, often after years of research. Most workers have accepted the above changes, because they were carefully researched, and because they facilitate the handling of large taxa. Nevertheless, there is often disagreement, sometimes based on methodology (concern regarding recognition of paraphyletic taxa), but often over the weight assigned to certain character states. Examples include *Leptodrepana* Shaw and several of the small microgastrine genera.

Biogeographic Considerations

Ranges of genera and individual species. Our knowledge of the ranges of New World braconids, especially in the Neotropical and South Temperate regions, is still rather meager, and this applies equally to both genera and individual species. Little attention has been given to this topic as most revisions have been of a regional nature, and several areas critical to our understanding of distribution patterns have not been intensively sampled until recently. Much of the work on the Nearctic species, for example, has treated the taxa only in terms of America North of Mexico: an artificial, political division used only for convenience. Based on examinations we have made of material in various museums, many of the species recorded in the literature only from the United States and Canada not surprisingly extend well into Mexico. The Mexican fauna, however, has received little attention to date. Fortunately this situation is changing. Extensive collections made in Mexico and Costa Rica over the last 10 years, together with development of local expertise, will enable us to improve our understanding considerably regarding the transition between North Temperate and Neotropical faunas. Similarly, the South American fauna has been largely neglected, with most taxonomic works limited to fairly isolated descriptions of large, showy species and genera. Several species of interest to biological control workers have also been described, but not as part of larger revisions. A few researchers have now initiated revisions focused on the South American fauna, and several South American students are currently being trained.

The extreme richness of the New World braconid fauna, together with the large number of undescribed species, has limited our ability to develop meaningful assessments of biogeographical relationships for most taxa. Now that large collections are available for study, however, this should change. Perhaps for the first time, we are in a position to address a number of fascinating questions. Examples include: What are the relationships of the South Temperate fauna, which in at least some cases appears disjunct from the North Temperate fauna? How much of the Chilean fauna is endemic, and are its relationships closer to the Australian fauna than to the rest of South America? Do Andean and Sierran faunas extend along the full length of these chains, or are they partitioned latitudinally? How are they partitioned vertically? Do South and North American desert faunas grade seasonally into more mesic habitats? How extensively distributed are the mesoamerican dry forest braconids? Is the Costa Rican fauna unique in this regard, or simply a small sample of a much larger fauna that extends beyond mesoamerica? How closely related are the boreal species found in the Nearctic and Palaearctic Regions? Are most of the species actually shared? Is the fauna of the Amazon Basin truly the most species-rich, as random samples seem to indicate? Are braconids unusual in their patterns of distribution, or do they closely match other insect taxa?

Diversity: anomalous or otherwise. Janzen and Pond (1975) noted that ichneumonids (unlike most other plants and animals) seemed to be less diverse in the Neotropics than they were in the Nearctic, much as Owen and Owen (1974) had found for similar comparisons in the Old World. Several publications have subsequently addressed overall species diversity of New World parasitic Hymenoptera, with particular attention to observed differences in species richness with latitude (Hespenheide, 1979; Janzen, 1981; Gauld et al., 1992; Hawkins et al., 1992; Quicke and Kruft, 1995). Although ichneumonids have received the greatest attention in this regard, the braconids have not been entirely neglected (Hawkins et al., 1992; Wharton, 1993; Quicke and Kruft, 1995). The initial claims of anomalous patterns in ichneumonid species richness have been modified over the years to exclude certain groups (e. g. ectoparasitoids, as discussed in Gauld et al., 1992). But even if the pattern holds up for ichneumonids in general, there is an increasing amount of evidence to suggest that braconids simply do not appear to have the same constraints as ichneumonids. While the topic of anomalous diversity has certainly generated some interest, we are greatly limited in our ability to address this issue in a meaningful fashion, especially in tropical regions, by our poor understanding of most large genera. Genera such as *Apanteles, Bracon*, *Chelonus*, *Digonogastra*, *Dinotrema*, and *Heterospilus* Haliday, each containing hundreds of Neotropical species, have never been revised for any Neotropical area. Even the separation of existing collections to morphospecies is a challenging task for these taxa because intraspecific variation has never been adequately assessed. Thus much of the discussion, and in particular the search for mechanisms to explain anomalous patterns in species richness, is premature. It is hoped that this volume will contribute to studies that will eventually enable us to answer such questions unequivocally.

Collection and Curation

Braconids occur in all habitats, and although some of the richest samples we have seen and collected come from wet, tropical forests, there is also a diverse but completely different fauna found in arid regions. Braconid wasps are readily collected in Malaise traps, and in general we have found this to be the most effective method for collecting large numbers of species. Trap size and shape can be varied to fit individual needs (Figs 1, 2), but traps that are about 2 meters long by 2 meters high (at the head end) are most commonly employed. Malaise traps are often vandalized when set in exposed situations, and are often difficult to set up and maintain in some habitats, such as rocky deserts and wind-blown terrains. Braconids may also be taken-in large numbers from low vegetation by using a sweep net or a sweep net fitted with a coarse screen to exclude large leaves and twigs. Screen sweep samples, though excellent, are labor-intensive to process for small species. Yellow pan traps (Figs 3-5) are useful as they sometimes capture species not or only rarely obtained with other methods. Pan traps and Berlese funnels are especially good for apterous or brachypterous species, and fully winged species that spend most of their time searching on the ground. In most cases, however, pan traps collect fewer species and individuals per unit effort than Malaise traps and sweep nets. A number of species are nocturnal, and can be taken at lights or with light traps. These include certain Cheloninae (Phanerotomini), some Hormiinae and

Meteorinae, and most Rogadinae, Macrocentrinae and Homolobinae. See Huddleston and Gauld (1988) for a more detailed account.

In addition to mass collecting methods, braconids are readily collected individually, which has the added advantage of providing useful biological information. They are commonly collected off flowers, and a number of species (e.g. in the Agathidinae, Braconinae, Cardiochilinae, and Microgastrinae) have specialized mouthparts for obtaining nectar. Many of these species also attack hosts in or on flower heads. Similarly, various members of the Braconinae and Doryctinae are frequently collected from logs, where they are either searching for or emerging from their hosts. Braconids in general may be reared from a wide variety of hosts, though larvae of Lepidoptera, Coleoptera, and cyclorrhaphous Diptera are by far the most common. Care should always be exercised when associating the host with a reared parasitoid. Many erroneous host records have been reported in the literature, either because the host or parasitoid was misidentified, or the host was incorrectly associated. Flower heads and woody stems, for example, may contain several potential host species, and identification of the true host requires careful dissection. Where possible, hosts should be individually isolated to avoid such confusion, and it is always preferable to keep the host remains with the parasitoid (if possible, on a separate point or in a gelatin capsule on the same pin). Mummified host remains (Rogadinae and Aphidiinae), puparia, and cocoons are especially valuable because useful information (such as parasitoid larval skins) can be retrieved from them.

Curation. Proper preparation of specimens is essential for their identification, especially by the non-specialist. Specimens should be clean, with appendages extending out from the body and away from each other in a way that affords maximum visibility of all body regions. Many braconids, especially those of small body size, often collapse or shrivel to some extent when air-dried. Collection and preservation in 70-80% ethanol, followed by overnight storage in 95-100% ethanol prior to chemical or critical-point drying, can solve this problem. Other problems are occasionally encountered with these techniques, such as coating of the body with white deposits that represent either glandular residues or contaminants in the preservation fluid. Various chemical preservatives have been discussed in the newsletter Chalcid Forum. Dillon et al. (1996) have noted that specimens killed in 100% ethanol and stored at -4 C give the most consistent results for DNA extraction. Collecting directly into 95-100% ethanol, however, does cause some shriveling and is thus not as good for specimens destined to be pinned or point-mounted.

The arrangement of veins in the fore and hind wings is one of the most important character systems used in identification of Braconidae. Wings should be separated and at least partially spread to enable a clear view of the venation. The best way to view wings is to shine the light beneath the specimen (thus reflecting light back up through the wing) rather than directly on the specimen. This is particularly important for viewing weakly sclerotized veins lacking pigmentation (as is commonly the case with the r-m cross vein of the fore wing). Variation in surface features (e.g. propodeal sculpture) is also difficult to see without proper lighting, but the ability to see such details is essential for identification of some of the most commonly encountered genera of Braconidae. To better visualize sculpture, use fluorescent light sources or objects that diffuse light, such as thin (0.02 mm) pieces of mylar (= Albanene) or acetate. The mylar should be placed between the specimen and the light source, as close as possible to the specimen (without actually breaking the latter).

Format for the Manual

The bulk of this manual consists of 34 separate chapters treating each of the recognized braconid subfamilies occurring in the New World (see comments on subfamily classification above). Within each of these chapters, keys are provided to the known or recognized New World genera. These chapters are preceded by a section on morphology and terminology, and a separate chapter for a key to braconid subfamilies found in the New World. Specimens of interest may be identified directly to genus using the appropriate chapter if the subfamily is already known. Otherwise, the subfamily key may be used to locate the appropriate generic key.

Chapters on individual subfamilies are separately authored by various experts and although edited for consistency of format and terminology, each retains a certain amount of individual style. The keys cover at least all genera described prior to 1994, with a few of the chapters including more recently described taxa. Wings are illustrated for as many taxa as possible, even when features of the wing venation are not used in the keys because the venation pattern is often diagnostic for genera or groups of genera. Many of the SEMs were taken from Marsh et al. (1987). Keys are preceded by a short introduction providing biological information and other relevant data. This preliminary information has been kept relatively short in most cases because of the excellent and detailed treatment provided by M. Shaw and Huddleston (1991), as well as the general discussions provided by Gauld and Bolton (1988), Wharton (1993), Sharkey (1993), Godfray (1994), S. Shaw (1995), and others.

Footnotes have been embedded within the keys to provide information, for each genus, on known distribution, approximate number of species, and references to hosts and/ or taxonomic treatments. Presentation of distribution data differs somewhat from one chapter to the next because the amount of published and unpublished information varies tremendously. In most chapters ranges for the various genera are given specifically for the New World (see Map, Figs 6-7), based on both published sources and museum specimens. Information on extralimital distributions is sometimes provided as well, but for most widespread genera, this can be more readily obtained from the various catalogs noted above. In a few cases (particularly Microgastrinae), much of the described Neotropical fauna remains to be reclassified following extensive generic revisions; data specific to the New World are thus largely unavailable. For many of the Neotropical genera of Braconidae, the only published locality information is from the original description, often limited to one or two countries, and their true ranges are undoubtedly much larger.

The number of described species is given for each genus, except where extensive changes in generic concepts since publication of the last catalogs (as in Microgastrinae) have made this difficult. In the few cases where there have been detailed revisions covering all of the New World (e.g. *Alabagrus* Enderlein and *Homolobus* Förster), the number of described species may be a fairly close approximation of the total number of species. For most of the commonly encountered genera, however, large numbers of species remain to be described, even in the comparatively well-studied Nearctic Region.

The references cited in the footnotes are not meant to be inclusive, but are intended, together with the catalogs by Shenefelt and others, to provide an entry to the literature. For some genera, especially among the Aphidiinae, Braconinae, Microgastrinae, and Opiinae, the amount of published literature, even since the most recent catalogs, is enormous.

Acknowledgements

This manual could not have been completed without the able assistance of many individuals. In particular, Ken Wilks assembled most of the plates and prepared all of the original SEM work for Wharton's chapters, proof read initial drafts of all of the chapters, and provided much additional technical assistance throughout the project. Susan Rigby drew nearly all of the wing illustrations, and Barry Flahey, Imelda Mercado, and Heather Tracy provided some last-minute additions. Jason Hurdis provided original artwork for several chapters, and redrew most of the non-wing illustrations for standard format. Eric Rickey provided technical support for Mike Sharkey, including organizing illustrations. Eric and Rachel Donovan captured all of the artwork and SEM on disk. Michael Newman, Anna Rodriguez, and Gloria Harvey typed seemingly endless drafts of the manuscript, and Michael's skill in handling multiple formats and preparing the manuscript for final submission is especially appreciated.

Mark Shaw, Tom Huddleston, Fred Petitt, John Sivinski, and Kaoru Maetô reviewed various chapters, and Brian Scaccia, Bob Matlock, Ron Cave, and Dan Janzen all provided useful comments on the microgastrine key. We are particularly grateful to Andy Austin who reviewed the bulk of the manuscript and provided many useful comments and suggestions. Four workshops were held during the course of the project to provide input from potential users regarding format and use of keys. Alejandro Gonzalez Hernandez, Enrique Ruiz Cancino, Paul Hanson, Felipe Noguera Martinez, José Luis Carcia R., and Alphonso Pescador provided considerable assistance in arranging the workshops, and the following attendees supplied many useful suggestions for improvement of the keys: Quintin Adolfo Arias Celis, Antonio M. Aranguren, Juan A. Bernal Vega, Rosa Alcira Briceño G., Ronald

D. Cave, Rubén Candia Castro, Gabriela Castaño Meneses, Mario Cerrneli, Doris Chirinos Torres, Ricardina Colmenares Medina, James Coronado R., Juana María Coronado Blanco, Carlos A. Covarrubias Dimas, Samin Dadelahi, Hugo Delfín González, Rosa Isabel Fernández B., Silvia Diaz Figueroa, Carmen Flores, Maríano Flores Davila, Griselda Gaona García, Francis Geraud Povey, William José Goitia Loyo, Randall Alberto Gomez Gutierrez, Sonia Guadalupe Hernández Aguilar, Refugio Lomelí Flores, Celestino Martínez Bernal, Robert B. Matlock, Jr., Heriberto Mejía González, Imelda Mercado Uribe, Santiago Niño Maldonado, José Gregorio Oroño Gomez, Adriana Livier Orozco Cordero, Elsa Ovalles Villegas, Ramón Paz León, Miguel A. Perales Gutierrez, Enrique Piedra Rodriguez, Luis Piñango A., Leslie D. Price, Oscar Rafael Renaud Sotillo, Marco Antonio Reyes Rosas, Alicia Rodriquez-Palafox, Gladys del Carmen Rodríguez González, José Antonio Sánchez Garcia, Jesus Armando Ugalde Gomez, Mercedes Velasquez de Ríos, Alejandro Valerio C., Yanet Zabala Gómez, Nina Zitani, Ronald Zuniga Ramirez. Independent tests of the keys were provided by Carlie Miller, Marcia Trostle, and Jeri Wright.

Funding for preparation and publication of the manual was provided by NSF (DEB 9300517 plus REU supplements to Wharton and S. Shaw). Donald Quicke was supported by the Natural Environment Initiative in Taxonomy. Agriculture Canada is gratefully acknowledged for providing support to Mike Sharkey during the initial phase of the project as is the University of Kentucky through the Kentucky Agricultural Experimental Station Project No. 97-08-68. The Electron Microscopy Center at Texas A&M University provided access to equipment and facilities essential to the completion of the manual.

References

Achterberg, C. van 1979. A revision of the subfamily Zelinae auct. (Hymenoptera: Braconidae). Tijdschrift voor Entomologie 122: 241-479.

Achterberg, C. van 1984. Essay on the phylogeny of Braconidae (Hymenoptera: Ichneumonoidea). Entomologisk Tidskrift 105: 41-58.

Achterberg, C. vart 1993. Illustrated key to the subfamilies of the Braconidae (Hymenoptera: Ichneumonoidea). Zoologische Verhandelingen Leiden 283: 1-189.

Achterberg, C. van and Quicke, D. L. J. 1992. Phylogeny of the subfamilies· of the family Braconidae: A reassessment assessed. Cladistics 8: 237-264.

Beckage, N. E., Thompson, S. N., and Federici, B. A. (Eds.). 1993. Parasites and Pathogens of Insects. Academic Press, San Diego.

Clausen, C. P. 1940. Entomophagous Insects. McGraw-Hill, New York and London. 688 pp.

De Santis, L. 1967. Catálogo de los Himenopteros Argentinos de la Serie Parasitica, incluyendo Bethyloidea. Provincia de Buenos Aires Gobernacion, Comision de Investigacion Cientifica, La Plata. 337 pp.

De Santis, L. 1980. Catálogo de los himenópteros brasileños de la serie parasitica incluyendo Bethyloidea. Publicación de la Universidad Federal de Paraná. 395 pp.

Dillon, N., Austin, A.D., and Bartowsky, E. 1996. Comparison of preservation techniques for DNA extraction from hymenopterous insects. Insect Molecular Biology 5: 21-24.

Dowton, M. and Austin, A. D. 1994. Molecular phylogeny of the insect order Hymenoptera: apocritan relationships. Proceedings of the National Academy of Sciences 91: 9911-9915.

Edson, K.M. and Vinson, S. B. 1979. A comparative morphology of the venom apparatus of female braconids (Hymenoptera: Braconidae). The Canadian Entomologist 111: 1013-1024.

Fischer, M. 1971. Hym. Braconidae. World Opiinae. Index of Entomophagous Insects. Le Francois, Paris. 189 pp.

Fischer, M. 1972. Hymenoptera Braconidae (Opiinae I). Das Tierreich 91: 1-621.

Fischer, M. 1977. Hymenoptera Braconidae (Opiinae II-Amerika). Das Tierreich 96: 1-1001.

Fischer, M. 1987. Hymenoptera: Opiinae III-äthiopische, orientalische, australische und ozeanische Region. Das Tierreich 104: 1-734.

Gauld, I.D. and Bolton, B. (Eds.) 1988. The Hymenoptera. British Museum (Natural History), London, Oxford University Press, Oxford. 332 pp.

Gauld, I.D. and Hanson, P. 1995. The evolution, classification and identification of the Hymenoptera. Pp. 138-156. In: Hanson, P. and Gauld, I. D. (Eds.). The Hymenoptera of Costa Rica. Oxford University Press. Oxford.

Gauld, I.D., Gaston, K.J., and Janzen, D.H. 1992. Plant allelochemicals, tritrophic interactions and the anomalous diversity of tropical parasitoids: the "nasty" host hypothesis. Oikos 65: 353-357.

Godfray, H.C. J. 1994. Parasitoids. Princeton University Press, Princeton. 473pp.

Hawkins, B.A., Shaw, M.R., and Askew, R.R. 1992. Relations among assemblage size, host specialization and climate variability in North American parasitoid communities. American Naturalist 139: 58-79.

Heraty, J.M., Woolley, J.B., and Darling, D.C. 1994. Phylogenetic implications of the mesofurca and mesopostnotum in Hymenoptera. Journal of Hymenoptera Research 3: 241-277.

Hespenheide, H. A. 1979. Are there fewer parasitoids in the tropics? American Naturalist 113: 766-769.

Huddleston, T. and Gauld, I.D. 1988. Parasitic wasps. (Ichneumonoidea) in British light-traps. Entomologist 107: 134-154.

Infante, F., Hanson, P., and Wharton, R. 1995. Phytophagy in the genus *Monitoriella* (Hymenoptera: Braconidae) with description of new species. Annals of the Entomological Society of America 88: 406-415.

Janzen, D.H. 1981. The peak in North American ichneumonid species richness lies between 30° and 42°N. Ecology 62: 532-537.

Janzen, D.H. and Pond, C.M. 1975. A comparison, by sweep sampling, of the arthropod fauna of secondary vegetation in Michigan, England and Costa Rica. Transactions of the Royal Entomological Society of London 127: 33-50.

Macêdo, M.V. de and Monteiro, R.F. 1989. Seed predation by a braconid wasp, *Allorhogas* sp. (Hymenoptera). Journal of the New York Entomological Society 97: 358-363.

Mackauer, M. 1968. Pars 3. Aphidiidae. Pp. 1-103. In: Ferrière, C. and Vecht, J. van der (Eds.). Hymenopterorum Catalogue (nova editio). Dr. W. Junk, The Hague.

Mackauer, M. and Starý, P. 1967. Hymenoptera Ichneumonoidea. World Aphidiidae. Index of Entomophagous Insects. Le Francois, Paris. 195 pp.

Maetô, K. 1987. A comparative morphology of the male internal reproductive organs of the family Braconidae. Kontyu 55: 32-42.

Marsh, P.M. 1971. Keys to the genera of the families Braconidae, Aphidiidae, and Hybrizontidae. Annals of the Entomological Society of America. 64:841-850.

Marsh, P.M. 1979. Family Braconidae. Pp. 144-295. In: Krombein, K. V., Hurd, P. D., Smith, D. R., and Burks, B. D. (Eds.). Catalog of Hymenoptera in America North of Mexico. Smithsonian Institution Press, Washington, D.C.

Marsh, P.M., Shaw, S.R., and Wharton, R. 1987. An identification manual for the North American genera of the family Braconidae (Hymenoptera). Memoirs of the Entomological Society of Washington 13: 1-98.

Mason, W.R.M. 1981. The polyphyletic nature of *Apanteles* Foerster (Hymenoptera: Braconidae): a phylogeny and reclassification of Microgastrinae. Memoirs of the Entomological Society of Canada 115: 1-147.

Mason, W.R.M. 1983. A new South African subfamily related to Cardiochilinae (Hymenoptera: Braconidae). Contributions to the American Entomological Institute 20: 49-62.

Nixon, G.E.J. 1965. A reclassification of the tribe Microgasterini. Bulletin of the British Museum (Natural History) Entomology Supplement 2: 1-284

Owen, D.F. and Owen, J. 1974. Species diversity in temperate and tropical Ichneumonidae. Nature 249: 583-584.

Quicke, D.L.J. 1987. The Old World genera of braconine wasps (Hymenoptera: Braconidae). Journal of Natural History 21: 43-157.

Quicke, D.L.J. and Achterberg, C. van 1990. Phylogeny of the subfamilies of the family Braconidae (Hymenoptera: Ichneumonoidea). Zoologische Verhandelingen Leiden 258: 1-95.

Quicke, D.L.J. and Kruft, R.A. 1995. Latitudinal gradients in North American braconid wasp species richness and biology. Journal of Hymenoptera Research 4: 194-203.

Rasnitsyn, A. P. 1988. An outline of evolution of the hymenopterous insects (Order Vespida), Oriental Insects 22: 115-145.

Sharkey, M.J. 1993. Family Braconidae. Pp. 362-395. In: Goulet, H. and Huber, J. T. (Eds.). Hymenoptera of the World: An Identification Guide to Families. Agriculture Canada, Ottawa.

Sharkey, M.J. and Wahl, D.B. 1992. Cladistics of the Ichneumonoidea (Hymenoptera). Journal of Hymenoptera Research 1:15-24.

Sharkey, M.J. and Wharton, R.A. 1994. A revision of the genera of the world Ichneutinae (Hymenoptera: Braconidae). Journal of Natural History 28: 873-912.

Shaw, M.R. 1983. On evolution of endoparasitism: the biology of some genera of Rogadinae (Braconidae). Contributions of the American Entomological Institute 20: 307-328.

Shaw, M.R. and Huddleston, T. 1991. Classification and biology of braconid wasps (Hymenoptera: Braconidae). Handbooks for the Identification of British Insects 7(11): 1-126.

Shaw, S.R. 1995. The Braconidae. Pp. 431-464. In: Hanson, P. and Gauld, I. (Eds.). The Hymenoptera of Costa Rica. Oxford University Press, Oxford.

Shenefelt, R.D. 1965. A contribution towards knowledge of the world literature regarding Braconidae (Hymenoptera: Braconidae). Beitraege zur Entomologie 15: 243-500.

Shenefelt, R.D. 1969. Pars 4. Braconidae 1, Hybrizoninae, Euphorinae, Cosmophorinae, Neoneurinae, Macrocentrinae. Pp. 1-176. In: Ferrière, Ch. and Vecht, J. van der (Eds.). Hymenopterorum Catalogus (nova editio). Dr. W. Junk, The Hague.

Shenefelt, R.D. 1970a. Pars 5. Braconidae 2, Heliconinae (sic), Calyptinae, Mimagathinae, Triaspinae. Pp. 177-306. In: Ferrière, Ch. and Vecht, J. van der (Eds.), Hymenopterorum Catalogus (nova editio). Dr. W. Junk, The Hague.

Shenefelt, R.D. 1970b. Pars 6. Braconidae 3, Agathidinae. Pp. 307-428. In: Ferrière, Ch. and Vecht, J. van der (Eds.). Hymenopterorum Catalogus (nova editio). Dr. W. Junk, The Hague.

Shenefelt, R.D. 1972. Pars 8. Braconidae 4, Microgasterinae *Apanteles*. Pp. 429-668. In: Vecht, J. van der and Shenefelt, R.D. (Eds.). Hymenopterorum Catalogus (nova editio). Dr. W. Junk, The Hague. Shenefelt, R.D. 1973a. Pars 9. Braconidae 5, Microgasterinae and Ichneutinae. Pp. 669-812. In:

Vecht, J. van der and Shenefelt, R. D. (Eds.), Hymenopterorum Catalogus (nova editio). Dr. W. Junk, The Hague.

Shenefelt, R.D. 1973b. Pars 10. Braconidae 6, Cheloninae. Pp. 813-936. In: Vecht, J. van der and Shenefelt, R.D. (Eds.). Hymenopterorum Catalogus (nova editio). Dr. W. Junk, The Hague.

Shenefelt, R. D. 1974. Pars 11. Braconidae 7, Alysiinae. Pp. 937-1113. In: Vecht, J. van der and Shenefelt, R.D. (Eds.). Hymenopterorum Catalogus (nova editio). Dr. W. Junk, The Hague.

Shenefelt, R.D. 1975. Pars 12. Braconidae 8, Exothecinae Rogadinae. Pp. 1115-1262. In: Vecht, J. van der and Shenefelt, R. D. (Eds.). Hymenopterorum Catalogus (nova edito). Dr. W. Junk, The Hague.

Shenefelt, R.D. 1978. Pars 15. Braconidae 10, Braconinae, Gnathobraconinae, Mesostoinae, Pseudodicrogeniinae, Telengainae, Ypsistocerinae plus Braconidae in general, major groups, unplaced genera and species. Pp. 1425-1872. In: Achterberg, C. van and Shenefelt, R. D. (Eds.), Hymenopterorum Catalogus (nova editio). Dr W. Junk, The Hague.

Shenefelt, R.D. 1980. Pars 16. Braconidae 11, Introduction, Guide to host names, Index to braconid names. Pp. 1-384. In: Achterberg, C. van and Shenefelt, R. D. (Eds.). Hymenopterorum Catalogus (nova editio). Dr W. Junk, The Hague.

Shenefelt, R. D. and Marsh, P. M. 1976. Pars 13. Braconidae 9, Doryctinae. Pp. 1263-1424. In: van der Vecht, J. and Shenefelt, R. D. (Eds.). Hymenopterorum Catalogus (nova edito). Dr. W. Junk, The Hague.

Starý, P. 1987. Subject bibliography of aphid parasitoids (Hymenoptera: Aphidiidae) of the world 1758-1982. Monographs to Applied Entomology 25: 1-101.

Tobias, V.I. 1986. Identification of the Insects of European USSR. Volume III, Part V. Hymenoptera, Braconidae. Akademia Nauk, Leningrad. (In Russian.)

Wharton, R.A. 1993. Bionomics of the Braconidae. Annual Review of Entomology 38: 121-143.

Wharton, R.A., Shaw, S.R., Sharkey, M.J., Wahl, D.B., Woolley, J.B., Whitfield, J.B., Marsh, P.M., and Johnson J.W. 1992. Phylogeny of the subfamilies of the family Braconidae (Hymenoptera: Braconidae): a reassessment. Cladistics 8: 199-235.

Whitfield, J.B. 1992a. The polyphyletic origin of endoparasitism in the cyclostome lineages of Braconidae (Hymenoptera). Systematic Entomology 17: 273'-286.

Whitfield, J.B. 1992b. Phylogeny of the non-aculeate Apocrita and the evolution of parasitism in the Hymenoptera. Journal of Hymenoptera Research 1: 3-14.

Whitfield, J.B. and Mason, W.R.M. 1994. Mendesellinae, a new subfamily of braconid wasps (Hymenoptera: Braconidae) with a review of relationships within the microgastroid assemblage. Systematic Entomology 19: 61-76.

Whitfield, J.B. 1995. Checklist of the Microgastrinae (Hymenoptera: Braconidae) in America north of Mexico. Journal of the Kansas Entomological Society 68: 245-262.

Figs 1-5: Collecting techniques. 1, Malaise trap showing preferred placement at edge of dense vegetation; 2, same, showing readily serviced collecting jar; 3, same showing placement of yellow pan traps to increase capture of insects that fall to the ground after flying into the center panel of the trap; 4-5, detail of yellow pan traps in two different habitats. Large numbers of pan traps can be set out fairly quickly in almost any habitat and require only water and a drop of detergent (to break surface tension).

Fig. 6: Simplified map of North America.

Fig. 7: Simplified map of South America.

MORPHOLOGY AND TERMINOLOGY

M.J. SHARKEY

Department of Entomology, University of Kentucky, S-225 Agricultural Science Building North, Lexington, Kentucky 40546, U.S.A.

R. A. WHARTON

Department of Entomology, Texas A&M University, College Station, Texas 77843, U.S.A.

OVERVIEW. This chapter has six parts. The first is a general section that defines non-morphological terms as well as morphological terms that do not apply to any specific section of the body. For example, most definitions for surface sculpture will be found here. The five remaining sections, head, mesosoma, wings, legs, and metasoma, contain definitions for terms that refer exclusively to a particular area of the body. The mesosoma is the middle section of the body in the parasitic and aculeate Hymenoptera, and is composed of the three thoracic segments plus the first abdominal segment (the propodeum) that is fused to the metathorax. The remaining abdominal segments are referred to as the metasoma (the posterior section of the body). The metasoma is equivalent to the petiole plus gaster. In all of these sections there may be expanded discussions on the more complex or important characteristics. For more detailed information refer to Huber and Sharkey (1993) and Torre-Bueno (1989).

GENERAL

aciculate. Appearing as if scratched with a fine needle, referring to surface sculpture.
aciculo-rugose. Irregularly, longitudinally wrinkled, referring to surface sculpture.
anteriad. Toward the anterior.
apicad. Toward the apex of a structure.
areola (plural = areolae). Central cell or margined area sometimes present on the propodeum (Fig. 8).
areolate. Surface that is divided into a number of cells or areolae, referring to surface sculpture.
basad. Toward the base of a structure.
bifurcate. Cleft, divided into two parts.
carina. Ridge.
carinate. Ridged, referring to surface sculpture.
caudal. Referring to the posterior or anal end of the body.
coriaceous. Leather-like, referring to surface sculpture.
costate. With raised ridges, coarser than striate, referring to surface sculpture.
crenulate. Referring to grooves that are broken by a number of transverse ridges.
deltoid. Elongate triangular shape resembling the Greek letter b. but with the apex extended.
distad. Toward the distal end of a structure.
ectoparasitoid. Parasitoid that, while in the larval stage, feeds externally on its host.
egg-larval. A koinobiont strategy in which the parasitoid egg is deposited inside the egg of the host, but development proceeds slowly and the host is not killed until it becomes a larger instar larva.
emarginate. Notched (indented) on the margin.
endoparasitoid. Parasitoid that, while in the larval stage, feeds on its host internally.

flange. Projecting rim or edge. In this publication flange also refers to a condition of the propleuron where the postero-ventral edge projects over the ventral extremity of the pronotum (Fig. 9).
foveate. Pitted with numerous, uniform depressions that are somewhat larger than punctures.
glabrous. Lacking setae, hairless. granulate. Referring to sculpture that appears granular under high magnification, as in sandpaper.
Holarctic. Composite biotic realm comprising the Nearctic and Palaearctic realms.
hyaline. Transparent and not, or indistinctly, pigmented.
idiobiont. Parasitoid that does not allow further development of the host once oviposition takes place. This is the usual strategy of ectoparasitoids.
infuscate. Darkened, often used in this publication in reference to wings.
koinobiont. Parasitoid that allows the host to develop after oviposition takes place. This is a common strategy for endoparasitoids.
lamella. Thin plate- or leaf-like process.
lamelliform. Leaf-like, in this text sometimes referring to highly raised thin ridges.
laterad. Toward the lateral side of a structure.
maculate. Spotted or mottled, usually referring to wing color patterns.
membranous. Integument that is thin, flexible and weakly sclerotized.
mesad. Toward the mid-line or middle of a structure.
Nearctic. Biotic realm, roughly comprised of the temperate regions of Mexico and all of Canada, Greenland, and Continental United States.
Neotropical. Biotic realm, roughly comprised of the tropical regions of Mexico and all of central and South America.
Palaearctic. Biotic realm roughly comprising Europe, and the northern temperate regions of Asia and Africa.
parasitoid. An insect that has a free-living adult, but whose immature stages develop at the expense of another invertebrate (usually an arthropod), killing it in the process. Differs from a predator in that only a single host individual is required to complete development.
paurometabolous. Those exopterygote insects in which the nymphs are terrestrial and resemble the adults in general form except for the absence of functional wings and reproductive organs (e.g. Psocoptera, Heteroptera). The term hemimetabolous is restricted in this manual to the Ephemeroptera, Odonata, and Plecoptera.
pectinate. Comb-like, with a series of sharp teeth.
petiolate. With a slender (petiole-like) connection. In this text the metasoma is referred to as petiolate when the first segment (or petiole) is very narrow. Wing cells may also be petiolate if they are not in direct contact with a cell but rather connected to it indirectly by way of a vein (see expanded definition in section on wings).
posteriad. Toward the posterior.
punctate. With punctures (fine, small depressions), referring to surface sculpture.
punctuation. Punctures (fine, small depressions).
rugose. Wrinkled, referring to surface sculpture.
rugosities. Wrinkles, referring to surface sculpture.
rugulose. Minutely wrinkled, referring to surface sculpture.
setiferous. Bearing setae, not necessarily with dense setae.
setose. With setae, usually referring to an area with dense setae.
sinuate. Wavy, often referring to wing veins or the margins of sclerites.
stria (plural = striae). Fine groove, finer that costate.
striate. Composed of a number of striae or fine grooves, referring to surface sculpture.
sub-, Latin prefix meaning slightly less than, e.g., subapical, slightly less than apical, or almost apical.
trifid. Divided into three parts. truncate. Cut off squarely at the tip.
tubercle. Small knob-like or rounded projection.
xylophagous. Wood eating.

HEAD

(Figs 1-7)

annellus (= anellus). Weakly sclerotized basal portion of first flagellomere of antenna. Usually distinctly narrower than the remainder of the first flagellomere.

antenna (plural = antennae). Paired segmented appendage located between the compound eyes on the anterior or dorsal surface of the head. These are sensory structures composed of three primary segments, i.e., scape, pedicel, and flagellum. The flagellum is always multi-segmented in the Braconidae (Fig. 3).

anterior tentorial pits. Pair of pits, demarcating the dorso-lateral limits of the clypeus and the insertions of the anterior arms of the tentorium (internal support structure of the head) (Fig. 1).

clubbed. In this text, referring to antennae that are swollen apically.

clypeus. Medial sclerite of the head lying immediately above the labrum and delimited dorso-laterally by the anterior tentorial pits, and dorsally by the face (Fig. 1).

compound eye. Paired organs of vision composed of many ommatidia (facets). Usually positioned laterally on the head, and usually referred to simply as eye (Fig. 1). There are usually microscopic setae (interommatidial hairs) between the facets and these may be enlarged and elongated such that they are readily observable under a dissecting microscope.

cyclostome. Condition describing the oral region of the head in which there is a circular to elliptical opening or gap between the clypeus and mandibles (when the mandibles are in the closed position) and a concave labrum (Fig. 7). The opening is referred to in some of the recent literature as a hypoclypeal depression, but the more traditional terms, cyclostome opening or oral cavity, are used in this publication. Cyclostome also refers to a group of braconid subfamilies, most of which have this condition.

endodont. Usual condition of mandibulate insects in which the teeth of the mandibles are facing towards the midline of the body when the mandibles are closed (Fig. 5).

exodont. Condition in which the teeth of the mandibles are facing anteriorly and do not overlap when the mandibles are closed. (Fig. 6).

face. Part of the anterior surface of the head ventral to the antennal sockets and dorsal to the clypeus (Fig. 1)

filiform. In this text, referring to antennae that have slender, cylindrical flagellomeres making the antennae appear thread-like.

flagellomere. Segment or article of the flagellum (Fig. 3). The first flagellomere refers to the most basal flagellomere, the second flagellomere the next most basal, etc.

flagellum. Third and most distal primary segment of the antenna (Fig. 3). In the Braconidae, each flagellum is subdivided into a number of flagellomeres (or articles). The shape, size, and especially the number of these constitute important characteristics. In most braconid subfamilies, and primitively within the taxon, the number of flagellomeres is variable. This is true even at the species level, with many species showing variability from specimen to specimen. It is not uncommon to see individuals with differing numbers of flagellomeres on the left and right antennae. In some subfamilies, such as the Microgastrinae, the number of flagellomeres is fixed. At least some, and usually most of the flagellomeres of braconids possess sensory structures known as placodes.

frons. Part of the anterior or dorsal part of the head, between the median ocellus and the antennal sockets (Fig. 1). The frons may be variously sculptured. In some Agathidinae for example, lateral carinae extend from the lateral margins of the antennal sockets toward the ocelli. In some Helconinae there are deep grooves posteriad each antennal base.

fronto-clypeal suture. Transverse suture between the anterior tentorial pits, separating the face from the clypeus (Fig. 1).

galea (plural = galeae). Outer apical lobe of the maxilla, sometimes modified into relatively large sheaths to cover the anterior surface of the labium (Fig. 4).

gena (plural = genae). Lateral part of the head, bordered anteriorly by the compound eye, posteriorly by the occipital carina, ventrally by the base of the mandible, and dorsally by the vertex (Fig. 2). The dorsal half of the gena is the temple. The length of the gena is the shortest distance between the posterior margin of the compound eye and the occipital carina.

glossa. Median, apical segment of the labium, often modified into a tongue (Fig. 4).

hypoclypeal depression. Depression in the oral cavity formed by a concave apical margin of the clypeus and a concave labrum (Fig. 7) (see cyclostome).

hypognathous. Orientation of the head when the mouth is facing ventrally (see prognathous).

hypostomal carina. Ridge on the posterior surface of the head, lateral to the maxilla and labium, mesal to the occiput (Fig. 4).

labiomaxillary complex. Portion of the mouthparts composed of the maxillae and labium.

labium. Posterior median appendage of the mouth (Fig. 4). The number of segments of the labial palps is diagnostic for a number of groups of Braconidae. The maximum number of labial palp segments in the Braconidae is four, but reductions have often occurred. The labial palps lie posterior to the maxillary palps and they are almost always shorter and usually have fewer segments.

labrum. Anterior, unpaired, median appendage of the mouth, attached to the posterior side of the clypeus (Figs 1, 7). The shape of the labrum varies considerably in the Braconidae. Most members of the "cyclostome" braconids have the labrum exposed, concave, and glabrous (lacking setae). The labrum of other braconids is mostly concealed by the clypeus and mandibles (Fig. 5).

malar space. Area between the ventral margin of the compound eye and the base of the mandible (Fig. 1).

malar suture. Suture between ventral margin of compound eye and the base of the mandible (Fig. 2).

mandible. Paired, heavily sclerotized appendage of the mouth situated between the labrum and maxilla. In the Braconidae these usually bear two teeth (denticles) apically and the teeth point mesally and somewhat posteriorly when the mandibles are closed (endodont) (Fig. 5). In braconids with exodont mouthparts there are usually 3 or 4 teeth and these face anteriorly when the mandibles are closed (Fig. 6).

maxilla (plural = maxillae). Paired appendage of the mouth lying between the mandible and the labium (Figs 2, 4). The maxillary palps and the galea are parts of the maxilla with diagnostic value.

maxillary palp. Paired palps of the maxillae (Figs 2, 4). These vary in length and number of segments with a maximum of six. They are longer than, and anterior to, the labial palps.

occipital carina. Ridge on the posterior surface of the head that separates the genae and vertex from the occiput (Figs 2, 4). The completeness and the absolute presence or absence of this carina, as well as its point of intersection ventrally (meeting the subgenal carina or the hypostomal carina) are all important diagnostic characters.

occiput. Posterior surface of the head mesal to the occipital carina and lateral to the hypostomal carina (Fig. 4).

ocellus (plural = ocelli). Simple eye composed of one facet. In the Braconidae there are 3 ocelli located on the vertex, one median ocellus and a pair of lateral ocelli (Fig. 1).

oral cavity. Depression between clypeus and mandibles visible when the mandibles are closed (Fig. 7) (see cyclostome).

palp. Tactile, paired, and usually segmented structures borne by the maxillae and the labium (Figs 2, 4). Segments of the labial and maxillary palps are often referred to as palpomeres, but this terminology is not used in the text.

pedicel. Second primary segment of the antenna (Fig. 3).

placode (=longitudinal placode). Sensory structures on the flagellum, tube-like in shape and lying on the longitudinal axis of a flagellomere. The configuration and number of placodes can be important diagnostic features, e.g. Ichneutinae and Microgastrinae.

prognathous. Orientation of the head in which the mouthparts are facing anteriad.

scape. Basal segment of the antenna (Fig. 3).

serrate. Saw-like, in this text referring to the shape of the antenna.

subgenal carina. Carina immediately above the base of the mandible delimiting ventral extent of the gena.

teeth. Denticles of the mandible, usually at the apex and usually two in number (Fig. 2). Teeth or tooth-like projections may also be found at the apex of the clypeus or the apex of the metasoma in certain Braconidae.

temple. Upper part of gena posteriad the compound eye (Fig. 2).
vertex. Dorsal surface of the head, bordered anteriorly by the frons, laterally by the compound eyes, and posteriorly by the occiput (Figs 1, 2, 3).

MESOSOMA

(Figs 8, 9)

antescutal depression. A median, usually triangular depression immediately in front of the anterior margin of the mesoscutum. It appears to be formed by an anterior deflection of the posterior, carinate margin of the pronotum.
coxal cavity. Opening on the body of the thorax into which the coxa is articulated. A coxal cavity is closed when it is surrounded by sclerotized tissue and is not contiguous with another opening in the body wall.
epicnemial carina. Ridge on the anterolateral and antero-ventral surface of the mesopleuron (Fig. 9). The presence or absence of this ridge is a very important diagnostic character. Rarely the fore coxa must be lifted to see this structure. In many older publications, this structure is referred to as the prepectal carina. This term is no longer used because the true prepectus is not visible externally in the Braconidae.
lateral mesonotal lobe. Converging longitudinal grooves (notauli) may divide the mesonotum into a median and two lateral lobes (Fig. 8).
median carina (of propodeum.) Median longitudinal ridge on the propodeum, often present only anteriorly (Fig. 8).
median mesonotal lobe. Converging longitudinal grooves (notauli) may divide the mesonotum into a median and two lateral lobes (Fig. 8).
mesonotum. Tergum or dorsal surface of the mesothorax, composed of the mesoscutum and scutellum.
mesopleuron. Lateral sclerite of the mesothorax (Fig. 9).
mesoscutum. Anterior sclerite of the mesonotum (Fig. 8).
metanotum. Tergum or dorsal surface of the metathorax. In the Braconidae this is a rather narrow band between the scutellum and the propodeum (Figs 8, 9).
metapleuron. Lateral sclerite of the metathorax (Fig. 9). In lateral view, most of the metapleuron lies ventral to the propodeum.
midpit. Pit situated in the postero-median area of the mesoscutum. If notauli are present the midpit lies at or directly posterior to the point at which they converge (Fig. 8).
notaulus (plural = notauli). Pair of longitudinal, posteriorly converging, grooves on the mesoscutum (Fig. 8). These may be short, barely extending beyond the antero-lateral margin, or long, reaching the transscutal articulation. They may be represented by weak, shallow impressions (as in some Microgastrinae) or deep, heavily sculptured grooves.
posterior scutellar depression. Small, median, usually transverse depression or differentiated patch of sculpture along the posterior margin of the scutellum (Fig. 8).
postpectal carina. Transverse carina on the ventral surface of the mesopleuron immediately anterior to the middle coxa (Fig. 9).
pronope. Dorso-median pit on the pronotum (Fig. 8).
pronotum. Tergum of the prothorax. In the Braconidae this is usually a narrow dorsal band anterior to the mesoscutum that extends ventrally on each side and widens on the anterolateral surface of the mesosoma (Figs 8, 9).
propleural carina. Carina on the posterodorsal surface of the propleuron running parallel to the antero-ventral margin of the pronotum.
propleural flange. A flap-like extension of the postero-lateral corner of the propleuron that curves up over the ventral apex of the pronotum (Fig. 9).
propleuron. Lateral sclerite of the prothorax. In the Braconidae this constitutes the antero-ventral surface of the mesosoma, below the head and above the fore coxae (Fig. 9). The prosternum in Braconidae is not visible externally.
propodeal areola. Central cell or area sometimes present on the propodeum (Fig. 8).

propodeum. Most posterior of the four segments of the mesosoma. It is the true first abdominal segment which is fused with the metathorax (Figs 8, 9) and separated from the remainder of the abdomen by a constriction.

scutellar disc. Area of the scutellum posterior to the scutellar sulcus, usually triangular in shape (Fig. 8). scutellar sulcus. Transverse depression of the scutellum directly posterior to the mesoscutum (Fig. 8), often appearing as two large pits.

scutellum. Posterior sclerite of the mesonotum. The area between the anterior margin of the scutellar sulcus and the metanotum (Figs 8, 9). The term scutellum is often used somewhat loosely to refer solely to the scutellar disc.

sternaulus. Groove running from near the base of the middle coxa towards the anterior margin of the mesopleuron (Fig. 9). The sternaulus is frequently sculptured and, depending on the shape of the mesopleuron, may vary in position from horizontal to nearly vertical.

subpronope. One of a pair of pits just off of the midline of the pronotum (Fig. 8).

tegula. Sclerite near the extreme base of the costa of the fore wing that overlaps the wing base (Figs 8, 9).

transscutal articulation. Transverse line or groove that lies directly anterior to the scutellar sulcus (Fig. 8). This structure is absent in some Braconidae.

WINGS

(Figs 13-21)

VENATION

Here we use a modified Comstock-Needham system for naming wing veins and cross veins. Some modifications used in this text are new; these modifications, as well as an overview of the entire nomenclatorial system, are presented here. When discussing wing veins the orientation of the wing is important. The anterior margin of the wing is that margin that is towards the anterior when the wings are extended laterally, perpendicular to the long axis of the body and parallel with the substrate. The costa, or costal vein, thus extends along the anterior margin of the wing.

The ground plan of both the fore and hind wings of the Braconidae, as in other insects, consists of six primary longitudinal veins that run, to a great extent, along the long axis of the wing. These are the costa (C), subcosta (SC), radius (R), media (M), cubitus (CU), and anal veins (A).

Conventions. (Refer to Fig. 13 to better understand the following text). Wing veins may be of three basic morphologies, tubular, nebulous, or spectral (Mason 1986). Tubular veins (and cross veins) are hollow, and when seen under a microscope with backlighting, the veins are darker along their margins. In the illustrations of this text tubular veins are denoted by parallel solid lines or, more rarely, by a single solid line. Non-tubular veins are of two types, if they are pigmented, they are referred to as nebulous veins and if they are not pigmented they are referred to as spectral veins. Unlike tubular veins, nebulous veins are darker medially, fading towards the margins. The margins are thus ill-defined. Spectral veins are best seen with strong light directed from below the specimen so that it reflects up through the wing. In this text, both nebulous and spectral veins are drawn with lines of broken dots. These may be in the form of two parallel lines, a single line, or a mass of dots in no particular pattern. When the margin of a wing is not formed by a vein it is drawn with a dashed line, otherwise it has the appropriate venational convention.

Primary longitudinal veins are abbreviated with capital letters, e.g., costa = C. The convention in the Comstock-Needham system is to name branches with numerical suffixes, using successively larger numbers for more posterior branches. The anal veins, however, are numbered separately (1A, 2A, 3A, and 4A), and thus are considered unbranched. In the Braconidae there are very few veins that branch. In the fore wing, the radial vein (R) usually branches at the base of the stigma; the anterior branch is R1 (referred to as the metacarp in some publications) and the posterior branch is the radial sector (RS). In the Agathidinae and very rarely in several other subfamilies (e.g., Cardiochilinae, Fig. 20), RS of the fore wing may also branch. In this text the anterior branch retains the name RS (to facilitate recognition of homologous veins across Braconidae) and the posterior branch is termed RS2 (Figs 15, 20).

Longitudinal veins may fuse along their long axes as in veins C, SC and R of the fore wing (Fig. 13). Such composite veins are named using the composite vein abbreviations joined by a "+" sign. Far more rarely, composite veins are formed by joining end to end. For example, in Fig. 21 veins r and RS fuse end to end. In this type of fusion the components of the composite vein are joined using "&", and in this example the name is r&RS.

Major veins may be intercepted from the anterior by cross veins. When this occurs the two segments of the intercepted vein are termed abscissae (singular = abscissa). The abscissae of a vein are numbered using numerical prefixes, with the lowest number being the most basal and with more distal abscissae having successively higher numbers. Referring to Fig. 15 observe the hind wing vein M. The basal abscissa is 1M, the middle abscissa is 2M, and the most distal abscissa is 3M.

Any one abscissa of a longitudinal vein may be intercepted from the posterior by cross veins. When this occurs, the abscissa that is intercepted takes on alphabetical suffixes. The more basal section of a subsegmented abscissa takes the suffix "a" and the more distal section takes the suffix "b". Refer to the first abscissa of the radial sector vein (1RS) of the hind wing in Fig. 15. In this case the more basal of the two sections of IRS becomes 1RSa and the more distal becomes 1RSb. This is a modification of the Comstock-Needham system introduced here to facilitate the recognition of homologous wing abscissae across the Braconidae. Some rationalization seems appropriate, and we will here try to demonstrate the usefulness of the system. Refer to vein 1CU of the fore wings of Figs 15 and 20. In Fig. 20 we see that 1cu-a intercepts vein 1CU, resulting in two subsections of 1CU, i.e., 1CUa and 1CUb. Now, referring to Fig. 15, note that 1cu-a intercepts 1CU at its point of origin (more or less) and therefore 1CU is not subdivided. In the system developed for this text, 1CU in Fig. 15 becomes 1CUa and 1CUb in Fig. 20. The homologous 1CU, i.e., that section of CU between 1M and 1m-cu, is still recognizable and comparable across the two taxa. In older systems that regard every section of the wing as an abscissa, regardless of whether the intercepting cross vein is anterior or posterior, the two homologies of the 1CU abscissa would be lost. Using this older system, the abscissa would remain 1CU in Fig. 15, but in Fig. 20, 1CU (1CUa plus 1CUb) would become 1CU and 2CU. Naturally this would inflate all of the more distal abscissae of CU as well, and therefore none of the abscissae of CU could be used, in a comparative sense, when discussing the venation of the two taxa.

Cross veins are veins that join two primary longitudinal veins. They are written in small case and usually have a composite name. The first part of a cross vein name is derived from the more anterior of the two veins being connected, and the second part of the name is derived from the more posterior. The two components of a cross vein name are joined by "-." If a cross vein connects two veins, one of which is a composite vein, the situation is somewhat more complicated. If the composite vein is the more anterior of the two veins being connected by a cross vein, then the name of the hypothetically more posterior of the two veins composing the composite vein lends its name to the cross vein. For example, in Fig. 15, vein RS+M is connected to vein CU by cross vein 1m-cu, not by cross vein rs-cu. The opposite is true if the composite vein forms the posterior vein of two veins being connected; i.e., it is the hypothetically more anterior of the two veins that lends its name.

Fore wing veins. The costa (C) runs along the anterior margin of the wing and extends to the base of the stigma where it ends abruptly. There is usually a wing fold, indicated by a break, on the anterior wing margin between C and the stigma. C is fused to various degrees with the subcosta+radius (SC+R). This fusion varies from complete, such that it is impossible to discern C from SC+ R, to a rather incomplete fusion where C is distinct from, though in contact with, SC+ R. Although C and SC+ R are contiguous for most of their lengths in all Braconidae, there may be a small separation between them directly basal to the stigma. The subcosta and radius are completely fused and it is not possible to distinguish the two components in the fused vein. This composite vein is either completely fused with C or runs directly posterior to, and is contiguous with, C. SC+ R thus extends along the anterior margin of the wing from the wing base to just anterior to the base of the stigma. At this distal end, SC terminates, and R splits into two branches. The anterior branch, Rl, continues along the anterior margin of the fore wing, through the stigma, and usually terminates near the apex of the wing. Just basal to the stigma, R1 may swell forming the parastigma. The posterior branch or R is the radial sector (RS). There may be a small, basal, free, abscissa of RS, 1RS, or the radial sector may fuse immediately with the media (M) to form a composite vein, RS+M. Typically, RS+M has one or two abscissae after which point the two veins diverge. RS migrates· anteriorly toward the stigma, and a small cross vein usually connects RS and the stigma. This cross vein is simply referred to as r. Rarely, as in some Cardiochilinae (Fig. 20), a more basal cross vein is present between the stigma and RS. In this case the more basal cross vein is 1r and the more distal becomes 2r. After the intersection with r, RS usually continues to a point distad the stigma, often near the distal apex of the wing. In some members of the Agathidinae and Cardiochilinae (Fig. 20), the RS branches in the distal third of the fore wing. The anterior branch retains the name RS, although it could also be called RS1. The posterior branch is referred to as RS2. When present, RS2 bends abruptly posteriad along the path of the r-m cross vein and then turns sharply towards the apex of the wing. The RS2 vein is thought to be a secondary and independent development within both the Agathidinae and the Cardiochilinae and not part of the ground plan of the Braconidae. The so-called RS2 in the Braconidae is not homologous with RS2 of the sawfly family Xyelidae. The real origin of the vein is unknown. It could simply be an extension of the r-m cross vein. If RS2 is an element of RS, it is not known if it fuses with the r-m cross vein basally or if it is independent. Our decision to refer to the basal section of RS2 as RS2a (Fig. 15) is arbitrary, but perhaps simpler than referring to it is as RS2+r-m.

The position of 2RS may shift relative to m-cu, and this is an important character for distinguishing genera in several of the larger subfamilies (e.g. Doryctinae and Alysiinae). The shifting of veins affects the naming of RS+M. When the anterior end of m-cu is basad 2RS, RS+M is subdivided into (RS+M)a and (RS+M)b. This condition is shown in Figs 13, 15, 16, 17, 19, 20. Commonly, however, the anterior end of m-cu is either directly in line with 2RS or is posteriad 2RS (see frontispiece for Doryctinae chapter). When this condition occurs, RS+M is undivided but 2M may become divided into 2Ma and 2Mb. The changes in this region of the wing may be due to the shifting of RS+M anteriorly or posteriorly along m-cu and 2RS or they may be due to a basal or distal displacement of either m-cu or 2RS.

Basally, the media (M) is fused with the cubitus (CU) to form the composite vein M +CU. This vein runs along the long axis of the wing and it is usually closer to the median line of the long axis of the wing than any other vein. At some point in the basal half of the wing, M splits from CU and migrates towards the anterior margin of the wing as the first independent abscissa of the medius, 1M. Near the anterior margin of the wing and basal to the stigma, M fuses with RS to form RS+M. (Note that when two veins fuse it is the vein that is most anterior that forms the prefix of the composite name.) Somewhere posterior to the stigma, M splits from RS and continues in a relatively direct line to a point posterior to the wing apex.

After CU diverges from M, CU continues toward the distal margin of the wing. Usually, and as part of the ground plan of the Braconidae, CU turns abruptly posteriad for a short distance until it meets the second cubital-anal cross vein, 2cu-a. At this point the cubitus regains its former path by bending again, this time at a sharp angle toward the postero-distal margin of the wing. In the ground plan of the Braconidae there are two cross veins connecting CU with M and two cross veins connecting CU with A. Except for the one known species of Apozyx, 2m-cu cross vein is absent. On the other hand, many braconids have retained both cu-a cross veins, and the loss of 2cu-a is an important character in the higher classification of the Braconidae.

There are two anal veins in the Braconidae, 1A and 2A. Unlike other veins, the numerical prefixes do not refer to abscissae but rather to different longitudinal veins, 1A being the more anterior of the two. 1A runs more or less parallel to the posterior margin of the wing and weakens and eventually terminates distal to 2cu-a. The second anal vein, 2A, is usually absent and at most represented by a relatively short stub. Emanating from the posterior side of 1A, there may be one or two short stubs. Although they are referred to here as anal cross veins, la and 2a (la is basal to 2a), the real homologies are not known. Since these two stubs are absent in the Ichneumonidae, it is possible that they are secondary veins. When referring to abscissae of 1A, a numerical prefix is added with a hyphen, for example the first abscissae of 1A is 1- 1A.

Hind wing veins. C of the hind wing is a short stub that is fused with SC+ R for a short distance along the base of the anterior margin of the wing (Fig. 15). SC+ R then migrates away from the anterior margin. C may terminate at this point (Fig. 15) or continue for a very short distance along the anterior margin of the wing as a free abscissa (Fig. 16). In any case, C always terminates well before reaching the distal hamuli. This results in an open costal cell along the basal, anterior margin of the hind wing.

As in the fore wing, SC is completely fused to R, forming composite vein SC+ R. After SC+ R separates from C near the base of the wing, SC+ R continues toward the apex of the wing parallel and close -to the fore margin of the wing. At the point where the r-m cross vein intercepts SC+ R from the posterior side, SC is considered to terminate and all elements distal to this point are those of the radius (Fig. 15).

After SC terminates there may be a short section of R before it branches, or R may branch immediately into an anterior R1 vein and a more posterior RS vein. R1 migrates to the anterior margin of the wing where it continues toward the wing apex. Near, but somewhat distal to, the point where R1 meets the anterior margin are the hamuli. These are modified setae that are heavily sclerotized and hooked. The hamuli serve to couple the fore and hind wings.

As in the fore wing, veins M and CU are fused basally to form the composite vein M +CU. M and CU separate in the basal third of the wing and M continues toward the apex of the wing, or migrates somewhat anteriad and then bends again towards the apex (Fig. 15). There is usually no free CU in the hind wing and CU often terminates at the point where cu-a intercepts M+CU (Fig. 14). In a relatively few Braconidae, e.g., many Agathidinae, CU is present as an independent vein in the hind wing. In these cases, CU bends sharply posteriad. After traveling towards the posterior margin for a short distance it bends sharply again along the long axis of the wing where it very quickly fades into a weak vein before disappearing (Fig. 15). CU is intercepted posteriorly by cu-a, thus creating two segments, CUa and CUb.

There is usually only one anal vein in the hind wing, 1A. This is a relatively short vein roughly paralleling the hind margin of the wing. Rarely, a second anal vein, 2A, is also present as a short stub at the base of and posterior to 1A. 1A is usually intercepted near its distal end by cu-a, resulting in a basal abscissa, 1-1A and a short apical abscissa, 2-1A. Since both 2-1A and CUb arise from the distal border of the subbasal cell of the hind wing, it may be challenging to determine which vein it is when only one is present. The task is made easier by the presence of a wing fold between the junction of 1A and cu-a. If the vein segment in question arises posteriad the wing fold (as shown in Fig. 20), it is 2-1A. If it arises anteriad the wing fold (see frontispiece for Agathidinae chapter), it is CUb.

In some taxa, extra cross veins are present on the hind wing (Fig. 15). These are veins that are not present in the sister group of the Braconidae (Ichneumonidae) nor in other Apocrita, and M.J.S. therefore considers them to be features derived within the Braconidae. An alternative viewpoint, held by R.A.W., considers these cross veins to be derived directly from symphytan ancestors. Examples of three of the most common of these cross veins in the hind wing follow. A radial cross vein (r) may join RS with R1 in the antero-distal area of the hind wing. A second r-m cross vein (2r-m) may join RS with M, and an m-cu cross vein may be present posterior to M and directed towards the posterior margin of the wing.

CELLS

For simplicity and ease of comparison across taxa, we use a system for naming cells that is not tied to the names of the veins. This system is illustrated in Fig. 16. The loss of veins and/or cross veins is common in Braconidae, resulting in adjacent cells becoming confluent. Thus, in Fig. 17, fore wing r-m is lost and the 2nd and 3rd submarginal cells are confluent (the marginal cell is also greatly reduced in size).

ADDITIONAL TERMS

areolet. The second submarginal cell is sometimes referred to as the areolet, especially when it is small, e.g. Microgastrinae (Fig. 19).

brachypterous. Very short wings (not extending past first metasomal tergum) with venation greatly reduced or absent. Most often these wings are in the form of short stubs.

bulla. Break in a wing vein, usually indicating an area on a vein through which a fold traverses.

cell height. Height of a cell is measured from the anterior margin of the cell to the posterior margin of the cell when the wings are extended laterally, i.e., the distance along the short axis of the wing.

cell length. Length of a cell is measured from the basal margin of a cell to the distal (apical) margin of the cell, i.e., the distance along the long axis of the wing.

closed cell. Wing cell that is surrounded by veins on all sides. These veins need not be tubular. In Fig. 16), the2nd submarginal cell of the fore wing is closed.

cross vein. Vein connecting two of the major (longitudinal) veins. Cross veins are usually relatively short and run parallel to the short axis of the wing.

hamulus (plural; hamuli). One of a series (often 3) of hooked, bristle-like setae on the anterior margin of the hind wing. These couple with the hind margin of the fore wing (Fig 15).

hyaline. Transparent and not or very weakly pigmented.

infuscate. Referring to dark wings with melanic pigmentation; wings with a blackish tinge.

interstitial. A vein is said to be interstitial when it meets another vein end to end; that is, the vein intercepts a point between two cells.

longitudinal vein. Veins that run primarily along the long axis of the wing. In the Braconidae, these are, from anterior to posterior, costa (C), subcosta (SC), radius (R), radial sector (RS), media (M), cubitus (CU), first anal vein (1A), and second anal vein (2A).

maculate. Wings spotted or mottled with dark markings.

nebulous vein. Non-tubular vein indicated by pigmentation that is darker medially than along its edges.

open cell. Wing cell that is not surrounded by veins. In Fig. 16, the 3rd submarginal cell of the fore wing is open. pentagonal. Used in the subfamily key to refer to the 2nd submarginal cell when m-cu divides 2M into a short 2Ma and a long 2Mb, resulting in a cell with five sides (though one of those sides is usually very short).

petiolate cell. Used in this text to refer to a triangular second submarginal cell with a short stalk anteriad the junction of 2RS with r-m. In the Agathidinae, this stalk appears to be formed by the basal migration of r-m.

quadrangular. Used to refer to a 4-sided 2nd submarginal cell, as in Figs 13, 16, and 19.

spectral vein. Weakened vein that is not tubular and not pigmented; rather one that is indicated only by a slight depression or crease on the dorsal surface of the wing. Spectral veins are best viewed with strong light reflected from the dorsal surface of the wing.

stigma (plural = stigmata). (= pterostigma) Sclerotized, pigmented area on the distal half of the fore margin of the fore wing. Rarely, sclerotized spots also occur in the hind wing, e.g., some Doryctinae such as *Heterospilus*. These are also termed stigmata, though they are not homologous with the fore wing structure.

triangular. Used in reference to the 2nd submarginal cell, when the anterior end of 2RS intersects the anterior end of r-m, resulting in the loss of 3RSa.

tubular vein. Vein that is hollow. Tubular veins, and cross veins, appear darker on their edges than they do medially.

vannal lobe. In Braconidae, referring to a posterior basal lobe of the hind wing (barely indicated in Fig. 19).

wing length. Refers to the length from base to apex, that is, the long axis of the wing.

wing width. Refers to the distance from anterior margin to posterior margin when the wings are extended laterally, that is, the short axis of the wing.

LEGS

(Fig. 10)

arolium. Pad-like median lobe between tarsal claws.

basitarsus. Most basal of the tarsal segments (Fig. 10).

coxa. Most basal of the primary leg segments (Fig. 10).

femur. Third major segment of the leg, lying between the trochanter and tibia (Fig. 10). The femur is subdivided and the basal portion is known as the trochantellus.

pectin. Comb, referring to a comb-like structure on the tarsus.

tarsal claws. Pair of claws borne by the telotarsus (apical tarsomere) (Fig. 10).

tarsomere. Segment of the tarsus. The most basal segment is tarsomere 1 (or basitarsus), the next most basal segment is tarsomere 2, etc. Most members of the Braconidae have 5 tarsomeres on all legs. The apical tarsomere (tarsomere 5 or telotarsus) bears the tarsal claws (Fig. 10).

tarsus. Most apical of the primary leg segments. In the Braconidae, the tarsus is subdivided into five tarsomeres (Fig. 10).

telotarsus. Apical tarsomere (Fig. 10). tibia. Fourth primary leg segment lying between the femur and tarsus (Fig. 10).

tibial spur. Spine-like structure at the apex of the tibiae. In Braconidae, the fore tibia has only one tibial spur and this is modified for grooming the antennae. Each of the mid- and hind tibiae have two tibial spurs, one median and one lateral (Fig. 10). The median spur is usually the longest of the pair.

trochantellus. Basal division of the femur, lying between the trochanter and the femur proper (Fig. 10).

trochanter. The second major segment of the leg, lying between the coxa and the trochantellus (Fig. 10).

METASOMA (Figs 11, 12)

carapace. Refers to the dorsal surface of the metasoma when it is composed of no more than 4 visible terga that are more or less rigid and expanded laterally and posteriorly so as to cover the entire metasoma.

cercus (plural = cerci). In the Braconidae, these are small patches of setae or an unsegmented tubercle on the apical abdominal tergum (Fig. 11).

dorsal carinae. Pair of prominent longitudinal carinae on the first metasomal tergum (Fig. 12).

dorsope. Pit, usually at the base of the dorsal carinae of the first metasomal tergum (Figs 11, 12).

hypopygium. Last visible abdominal sternum (Fig. 11).

lateral tergite. Metasomal terga are often divided by a pair of lateral longitudinal ridges, creases, or folds, into a median tergite and two lateral tergites (Fig. 11).

laterope. Deep invagination or pit on the lateral, basal, surface of the first metasomal tergum. If a dorsope is present the laterope is ventral to the dorsope (Figs 11, 12).

median tergite. Metasomal terga are often divided, by a pair of lateral longitudinal ridges, creases, or folds, into a median tergite and two lateral tergites (Fig. 11).

ovipositor. In the Braconidae, an egg laying apparatus composed of two pairs of interlocking styles that form a hollow tube (Fig. 11).

ovipositor sheath. One of a pair of setiferous sheaths that envelop the ovipositor (Fig. 11).

petiole. First metasomal segment (Fig.11).

second valvifers (= second gonocoxa). The basal appendage to which the ovipositor sheaths are attached. This is often concealed by the hypopygium.

sternum (plural = sterna). Ventral sclerite of a metasomal segment. Sterna are numbered from the most basal to the most apical as sternum 1, sternum 2, etc. (Fig. 11).

tergite. Subplate of a tergum. In the Braconidae the terga of the metasoma, especially the anterior terga are divided into one median and two lateral tergites (Fig. 11).

tergum (pl. terga). Dorsal sclerite of a metasomal segment. Often divided into one median and two lateral tergites. Terga are numbered from the most basal to the most apical as tergum 1, tergum 2, etc. (Fig. 11).

REFERENCES

Huber, J. T. and Sharkey, M. J. 1993. Structure. Pp. 18-59. In: Goulet, H. and Huber, J. T. (Eds.). Hymenoptera of the world: an identification guide to the families. Agriculture Canada Publication 1894/E, Ottawa.

Mason, W. R. M. 1986. Standard drawing conventions and definitions for venational and other features of wings of Hymenoptera. Proceedings of the Entomological Society of Washington 88: 1-7

Torre-Bueno, J. R. de la, 1989. The Torre-Bueno glossary of Entomology compiled by Stephen W. Nichols; including supplement A by George S. Tulloch. Revised edition, New York Entomological Society, New York. 840 pp.

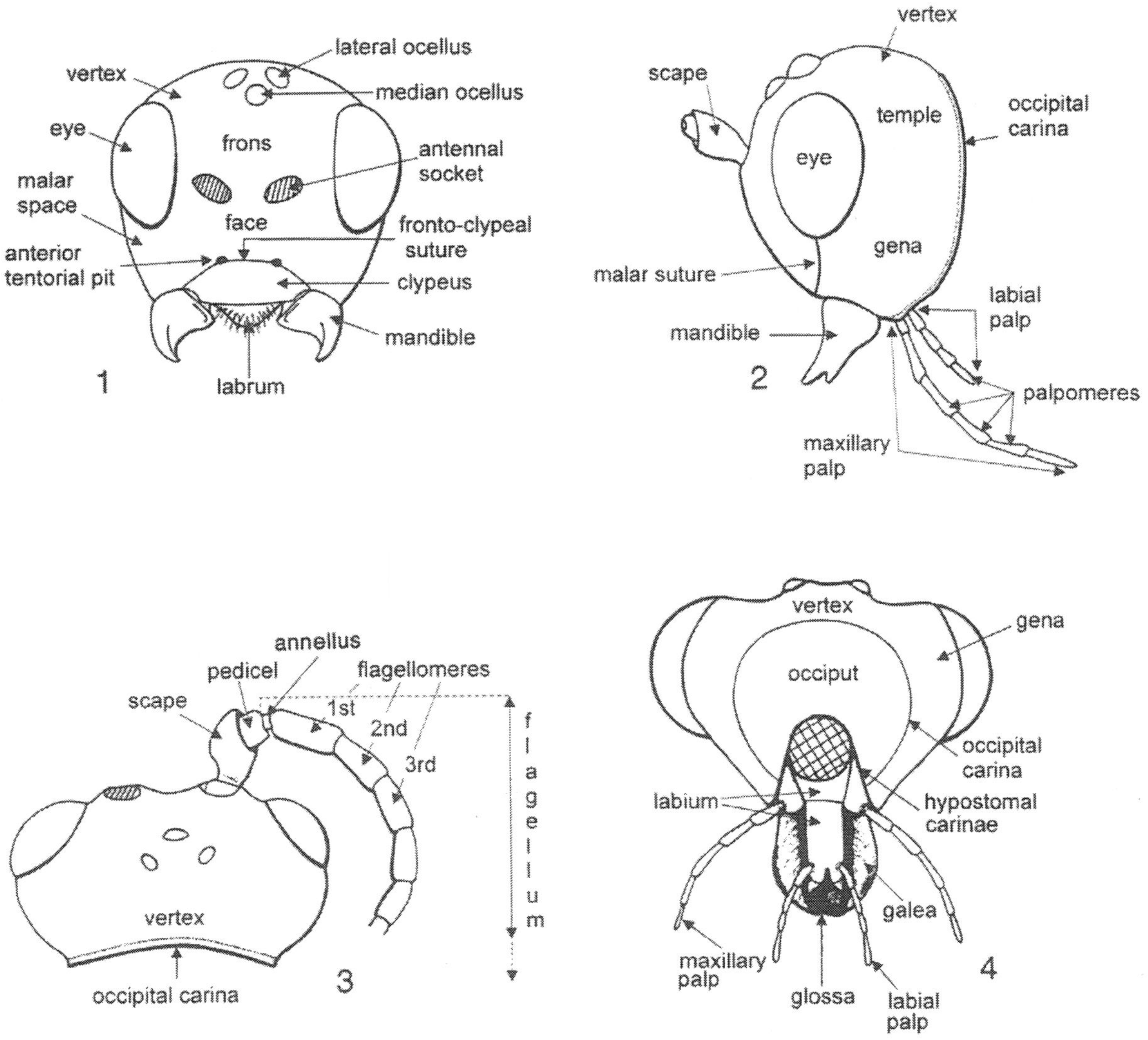

Figs 1-4: Head. 1, anterior view; 2, lateral view; 3, dorsal view; 4, posterior view.

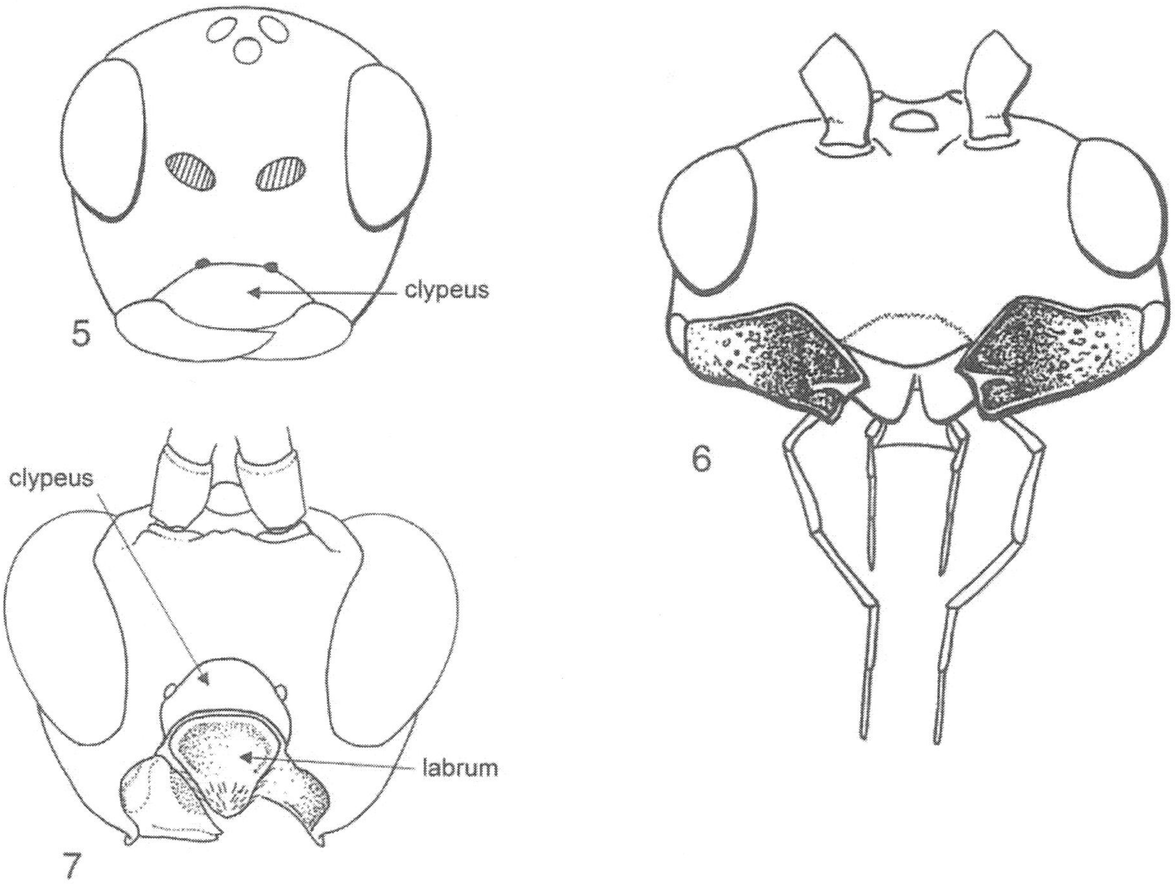

Figs 5-7: Head, anterior view. 5, non-cyclostome condition of clypeus and typical, endodont mandibles; 6, exodont mandibles; 7, cyclostome condition of clypeus and concave labrum (left mandible removed).

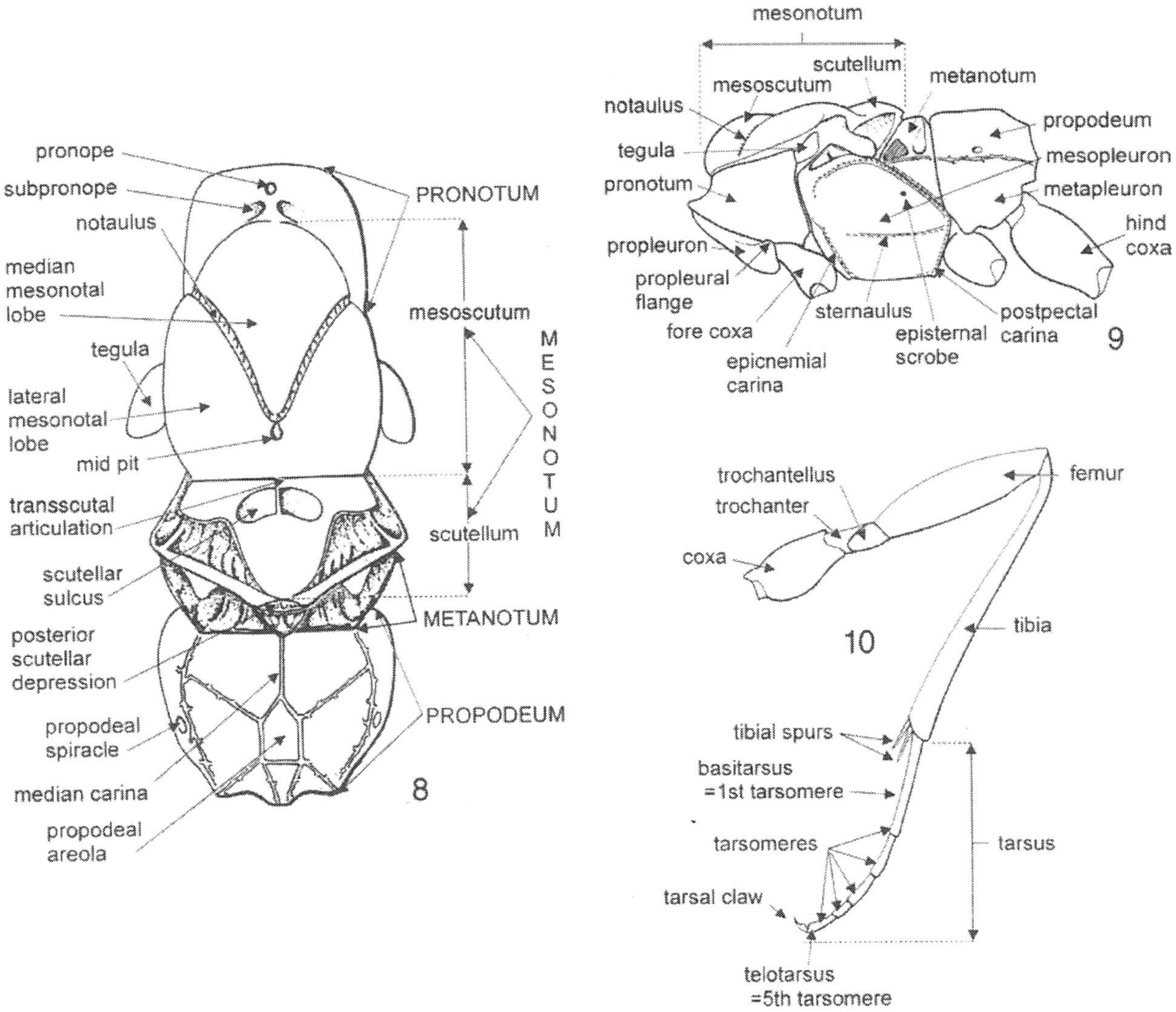

Figs 8-10: 8, mesosoma, dorsal view; 9, mesosoma, lateral view; 10, hind leg, lateral view.

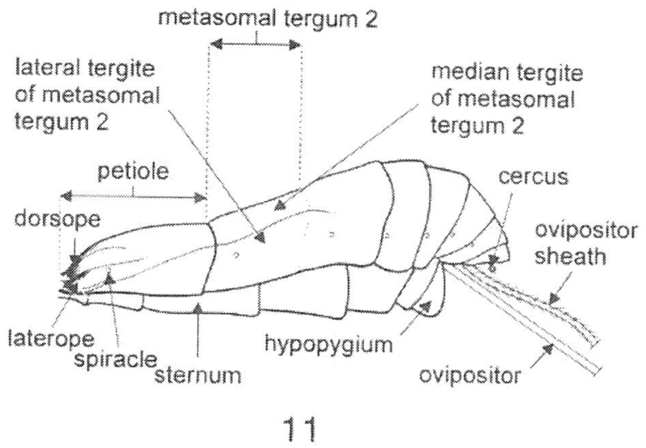

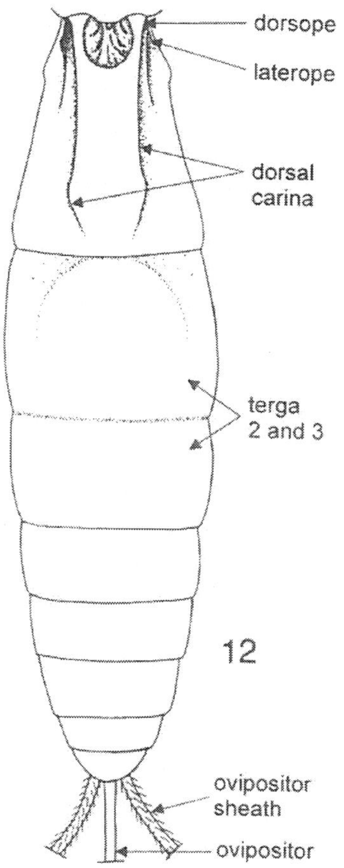

Figs 11-12: Metasoma. 11, lateral view; 12, dorsal view.

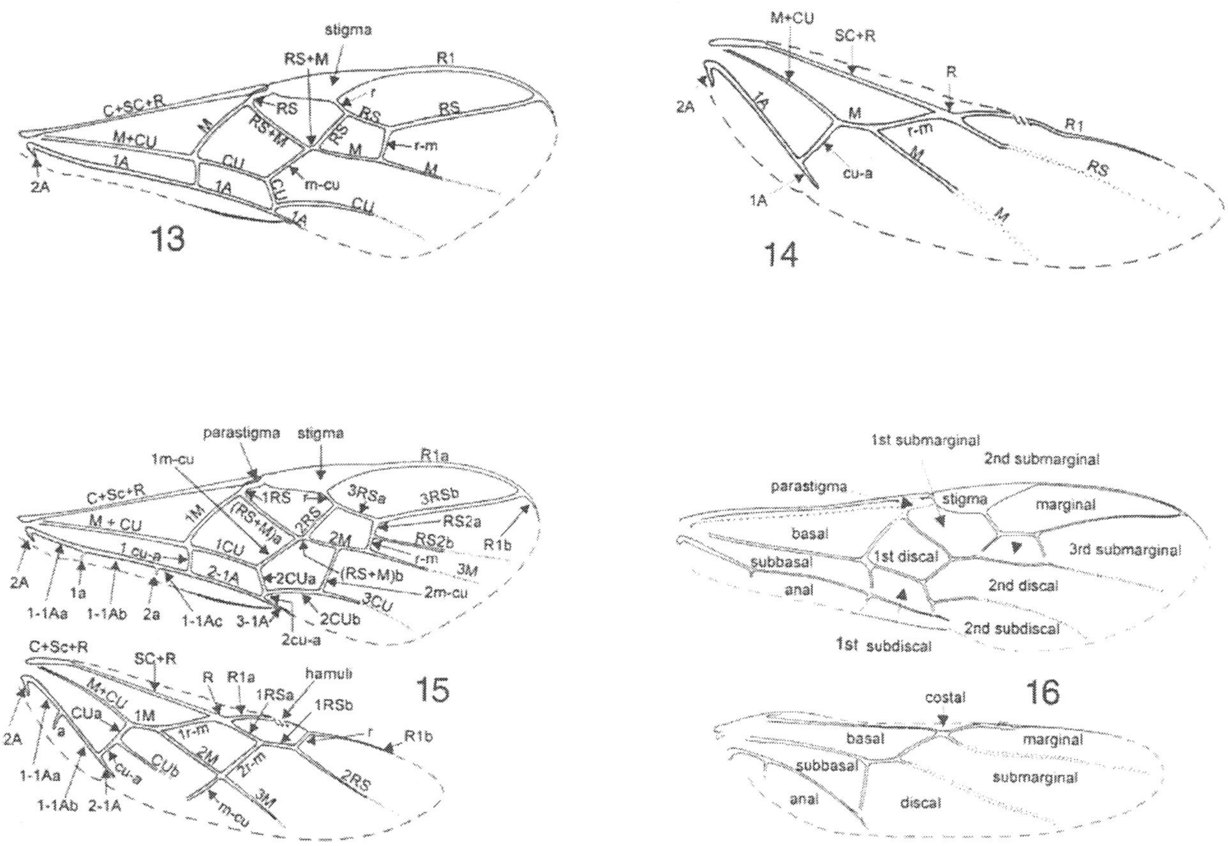

Figs 13-16: Wings. 13, fore wing of generalized Braconidae (abscissae not numbered); 14, hind wing of generalized Braconidae (abscissae not numbered); 15, fore and hind wing of hypothetical braconid showing all common primary and secondary veins and cross veins; 16, fore and hind wing of *Zele reticulatus* (Muesebeck) showing wing cells.

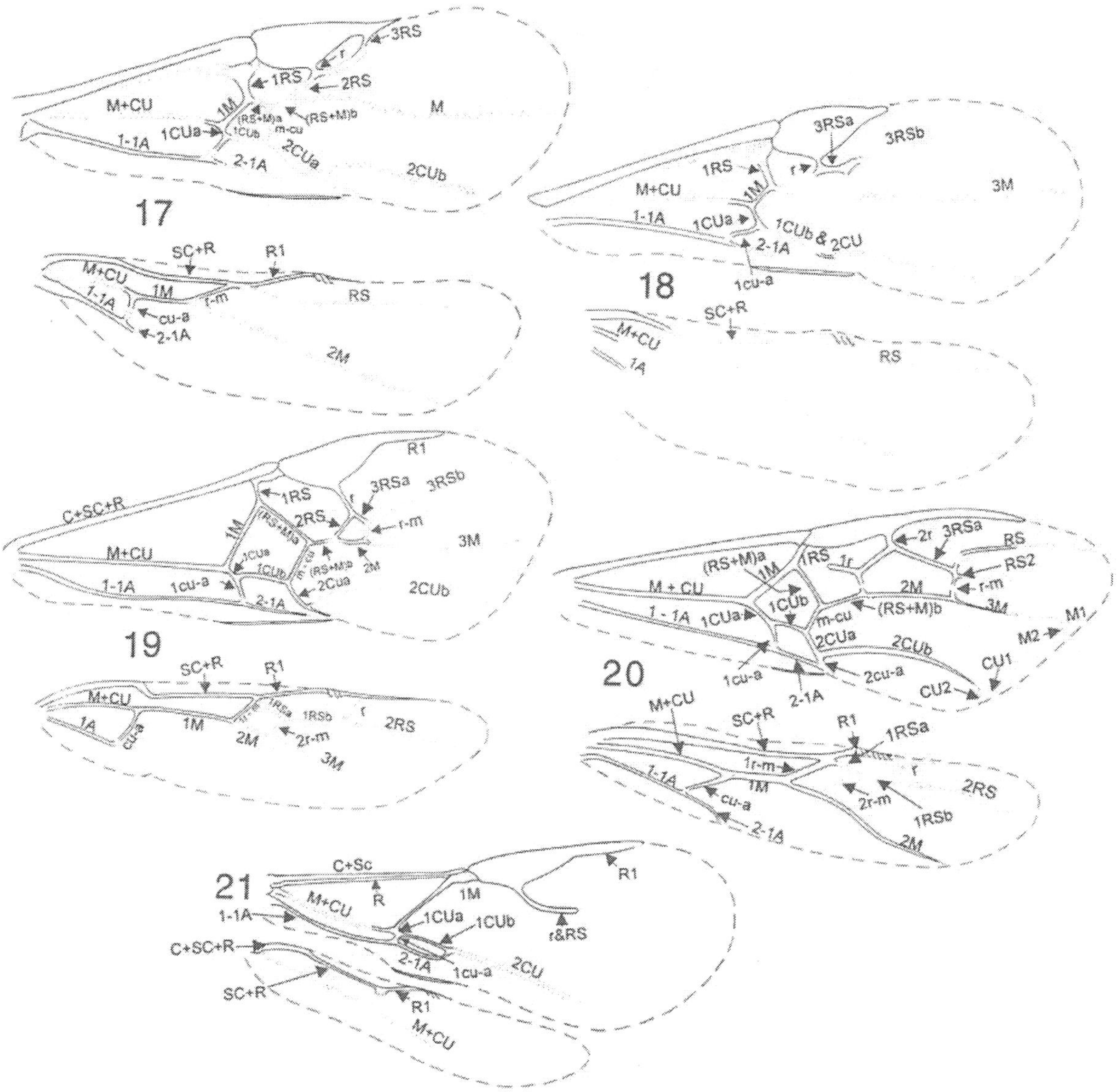

Figs 17-21: Fore and hind wing of species with different venation patterns to illustrate nomenclature for wing venation. 17, *Neoneurus* sp.; 18, *Elasmosoma* sp.; 19, *Snellenius* sp.; 20, *Wesmaelella* sp.; 21, *Diaeretiella rapae* (M'Intosh).

KEY TO NEW WORLD SUBFAMILIES OF THE FAMILY BRACONIDAE

M. J. SHARKEY

Department of Entomology, University of Kentucky, S-225 Agricultural Science Building North, Lexington, Kentucky 40546, U.S.A.

Most subfamilies appear more than once in the key because they are morphologically diverse. In some cases (e.g. couplets 8, 11, 16) a single genus is sufficiently distinct to be keyed separately. I use the terms "a few," "some," "most" or" all" to indicate the frequency with which members of a subfamily key to a particular couplet. Additionally, taxa that are seldom encountered are indicated by the term [Rare] and those frequently collected are indicated by the term [Common]. Where percentages are used (as in couplet 3), these refer to the expected proportion of specimens keying to that half of the couplet that are likely to possess the characteristic in question.

1. Wings absent or greatly reduced, not extending beyond petiole when directed 79
- Wings present and extending well beyond petiole when folded over body and directed 2

2(1). Exodont (Figs 1, 105): mandibles not touching or overlapping when closed, and with 3 or more teeth facing outwardly. [Common] .... most **ALYSIINAE**
Exodontiella (Opiinae) and some *Paroligoneurus* (Ichneutinae) also key here, and can be identified using the key to genera of Alysiinae.
- Endodont (Figs 2, 3, 106): mandibles touching and usually overlapping when closed, and usually with two teeth facing inwardly 3

3(2). Labrum exposed and concave and/or ventral margin of clypeus and dorsal surface of mandibles forming an ovoid cavity (Figs 2, 106). Spiracle of tergum 2 usually (95%) on median tergite (Fig. 12). Hind wing m-cu often (40%) present (Fig. 42) 59
- Labrum completely concealed by clypeus, or if visible not concave (Fig. 3); ventral margin of clypeus and dorsal surface of mandibles not forming an ovoid cavity (Fig. 3). Spiracle of tergum 2 usually (95%) on lateral tergite (Fig. 13). Hind wing m-cu a1Most always (99.9%) absent (Fig. 43).. 4

4(3). Occipital carina always completely absent (Fig. 9) 5
- Occipital carina present and either complete (Figs 7, 8) or incomplete (Figs 4-6) 9

5(4). Fore wing RS not reaching wing margin as a tubular vein (Figs 59, 61, 67, 90, 94) 6
- Fore wing RS reaching wing margin as a tubular vein (Figs 60, 63) 9

6(5). Fore wing m-cu tubular (Fig. 90) 54
- Fore wing m-cu absent (Fig. 61) or if present, not tubular (Fig. 59) 7

7(6). Mid tibia with preapical lateral spines (Figs 31, 110) .... a few **AGATHIDINAE**
- Mid tibia without preapical lateral spines (Fig. 32) 8

8(7). Fore wing M+CU tubular (Fig. 59); fore wing venation as in Fig. 59. [Rare] .... some **ICHNEUTINAE**
- Fore wing M+CU not tubular (Fig. 61); fore wing venation as in Fig. 61. [Rare]..
.... *Elasmosoma* **NEONEURINAE**

9(4, 5). First three metasomal terga forming carapace covering all other terga (Figs 14, 15,117, 118)....................
.... 10
- First three metasomal terga not forming carapace covering all other terga (Figs 12, 13, 16, 17) 12

10(9). Fore wing r-m present (though not always tubular and complete), thus second submarginal cell closed distally (Fig. 62) 11

- Fore wing r-m absent, thus second submarginal cell open distally (Fig. 63) 12

11(10). Metasomal terga 1 and 2 fused (Figs 14, 117). [Common] ....all **CHELONINAE**

- Metasomal terga 1 and 2 not fused, capable of articulation (Fig, 15; also as in Fig. 118). [Rare] .... *Sigalphus* **SIGALPHINAE**

12(9,10). Epicnemial carina absent (Fig. 33) 13

- Epicnemial carina present (Fig. 34, 107) 17

13(12). Metasomal tergum 2 with basal, transverse area that is differentiated from the remainder of the tergum by a groove or ridge (Fig. 16); the posterior margin of this area is straight or evenly rounded. [Rare] .... most **GNAMPTODONTINAE**

- Metasomal tergum 2 usually (90%) without basal, transverse area (Fig. 17), but if present, posterior margin neither straight nor evenly rounded 14

14(13). Fore wing 3RS not reaching wing margin as a tubular vein (Fig. 65). [Rare] ....all **ADELIINAE**

- Fore wing 3RS; reaching wing margin as a tubular vein (Figs 64, 66) 15

15(14). Fore wing with anal cross vein (a) present. (Fig. 64) .... some **ICHNEUTINAE**

- Fore wing with anal cross vein absent (Figs 60, 66) 16

16(15). Fore wing 3RS meeting wing margin near stigma (Fig. 66). [Rare] .... *Neoneurus* **NEONEURINAE**

- Fore wing 3RS meeting wing margin near apex of wing, far from stigma (Fig. 60)....................some **OPIINAE**

17(12). Hind wing m-cu present (as in Fig. 42). Metasomal tergum 2 with spiracle on median tergite (Fig. 12) .... [Rare] .... all **BETYLOBRACONINAE**

- Hind wing m-cu usually (99%) absent (Fig, 43). Metasomal tergum 2 usually (98%) with spiracle on lateral tergite (Fig. 13). Never with combination of hind wing m-cu present and spiracles on median tergite .. 18

18(17). Distance from posterior margin of eye to occipital carina (in dorsal view) greater than greatest diameter of eye (from any angle) (Fig. 11). [Rare]. ....a few **MASONINAE**

- Distance from posterior margin of eye to occipital carina (in dorsal view) less than greatest diameter of eye (from any angle) (Fig. 10), or occipital carina absent 19

19(18). Fore wing r-m present (though not always tubular and complete), thus second submarginal cell closed distally (Figs 62, 71, 72), though sometimes open basally (Fig. 67) 20

- Fore wing r-m absent, thus second submarginal cell open distally (Fig. 63) 41

20(19). Fore wing with both (RS+M)a and 2RS absent, creating a single confluent cell in middle of wing (Fig. 67); this cell (1st submarginal + 2nd submarginal +1st discal) may be completely closed, as in Fig. 67, or partially open posteriorly (Fig. 68) ....some **APHIDIINAE**

- Fore wing without a large confluent cell as described above; 2R$ always present, thus second submarginal cell closed basally (Figs 72-74, 77, 99, 103); (RS+M)a absent in some Agathidinae (as in Fig. 87) but otherwise present 21

21(20). Fore wing 2cu-a tubular (Fig. 69) 22

- Fore wing 2cu-a absent (Fig. 70) or if present, not tubular 36

22(21). Hind wing CUb present (Figs 44, 45). [Rare] 23

- Hind wing CUb absent (Fig. 46). [Common] 24

This character is sometimes difficult to interpret because vein CUb may be very posterior in position, as in Fig. 45, and therefore appear to be vein 2-1A. A typical position for vein 2-1A is illustrated in Fig. 46. The best way to distinguish the two veins is as follows: if there is a break (bulla) directly basad or posterior to the base of the vein in question it is CUb, and if there is a break (bulla) directly anterior to the base of the vein in question it is 2-1A.

23(22). Metasomal terga 2 and 3 smooth, unsculptured or nearly so, never striate. [Rare] ....all **METEORIDIINAE**

- Metasomal terga 2 and 3 striate (Fig. 18). [Rare] ....*Pselaphanus* **SIGALPHINAE**

24(22). Metasoma attached to propodeum well above hind coxal cavities (Fig. 37)....................all **CENOCOELIINAE**

- Metasoma attached to propodeum directly above hind coxal cavities (Fig. 38)....................25

25(24). Second submarginal cell of fore wing pentagonal, with origin of m-cu distad base of 2RS (Fig. 76) .... some **APHIDIINAE**

- Second submarginal cell of fore wing triangular (Figs 71, 72) or quadrangular (Fig. 73) 26

26(25). Second submarginal cell of fore wing triangular (Fig. 71, 72)... 27

- Second submarginal cell of fore wing quadrangular, with origin of m-cu basad or directly in line with 2RS (Fig. 73) 29

27(26). Fore wing r-m very short, almost absent, thus 2M and 3RSa meeting distally or nearly so (Fig. 71)............ ....some *Ichneutes* **ICHNEUTINAE**

- Fore wing r-m distinct, veins 2M and 3RSa not meeting distally (Fig. 72) 28

28(27). Posterior scutellar depression present (Figs 35, 112). Hind wing 1M less than 2 times longer than M+CU (Fig. 47) .... all **MICROTYPINAE**

- Posterior scutellar depression absent (Figs 36, 113). Hind wing 1M more than 2 times longer than M+CU (Fig. 48) ....some **ORGILINAE**

29(26). Occipital carina completely absent (Fig. 9) 30

- Occipital carina present, at least in part (Fig. 4-8) 31

30(29). Trochantellus of hind leg with spines (Fig. 26). Hind wing RS always present (Fig. 49)............................... ....all **MACROCENTRINAE**

- Trochantellus of hind leg without spines. Hind wing RS often absent (Fig. 50)... .... some **ICHNEUTINAE**

31(29). Petiole at least 2.5 times wider at posterior margin than at narrowest point (Figs 20, 111); dorsope of petiole often (60%) present (Fig. 111) ....some **METEORINAE**

- Petiole less than 2.5 times wider at posterior margin than at narrowest point (Figs 19, 21); dorsope absent (Fig. 109) 32

32(31). Basal half of fore wing M+CU not tubular (Fig. 76); hind wing cu-a absent (Fig. 52). Scutellar sulcus without longitudinal carinae (Fig. 113)..a few **APHIDIINAE**

- Basal half of fore wing M+CU mostly tubular (Fig. 75); hind wing cu-a present (Fig. 51). Scutellar sulcus nearly always (95%) with longitudinal carinae (Figs 112, 114) 33

33(32). Occipital carina incomplete, absent mid-dorsally. [Rare] ....*Antestrix* **ORGILINAE**

- Occipital carina complete, present mid-dorsally (Fig. 7) 34

34(33). Longest spur of hind tibia as long as or longer than half length of basitarsus (Fig. 28) ....most **HOMOLOBINAE**

- Longest spur of hind tibia shorter than half length of basitarsus. (Fig. 27) 35

35(34). Fore wing (RS+M)b present, origin of m-cu distinctly separated from base of 2RS (Fig. 103) ....some **HELCONINAE**

- Fore wing (RS+M)b absent or nearly so, m-cu directly in line with 2RS. [Rare].. ....*Helconichia* **ICHNEUTINAE**

36(21). Occipital carina completely absent (Fig. 9) ....most **AGATHIDINAE**

- Occipital carina present, at least in part (Fig. 4-8) 37

37(35). Petiole more than 4 times longer than apical width (Fig. 19) ....some **EUPHORINAE**

- Petiole less than 4 times longer than apical width (Figs 20, 21) 38

38(37). Metasomal tergum 2 granulate. Fore wing 3RSb not reaching wing margin as a tubular vein. [Rare] ....*Ademon* **OPIINAE**

- Metasomal tergum 2 not granulate. Fore wing 3RSb always reaching wing margin as a tubular vein 39

39(38). Second submarginal cell of fore wing pentagonal, with origin of m-cu distad base of 2RS (Fig. 74). Petiole with a pair of dorsal carinae (Fig. 116); spiracle of petiole in basal half (Fig. 116). [Rare] .... *Dyscoletes* **BLACINAE**

- Lacking one or more of the above characteristics: second submarginal cell usually (90%) quadrangular (Fig. 77); if pentagonal, then petiole without carinae (Fig. 111) and spiracle in apical half (Fig. 111) 40

40(39). Petiole strongly narrowed basally, at least 2.5 times wider at posterior margin than at narrowest point (Figs 20, 111)...some **METEORINAE**

- Petiole not strongly harrowed basally, less than 2.5 times wider at posterior margin than at narrowest point (Figs 21, 116). [Rare] .... a few **HELCONINAE**

41(19). Hind wing r-m absent (Fig. 53) or if present not tubular (Fig. 54) 42
- Hind wing r-m present and tubular (Fig. 55).......... 43
42(41). Fore wing RS curved sharply towards anterior margin of wing (Fig. 78); (RS+M)a often (70%) present, at least in part (Fig. 78) ....some **EUPHORINAE**
- Fore wing RS not curved sharply towards anterior margin of wing (Fig. 79); (RS+M)a completely absent (Fig. 79)..........some **APHIDIINAE**
43(41). Fore wing 2RS absent (Figs 80, 81) 44
- Fore wing 2RS present (Figs 82, 83).......... 45
44(43). Fore wing as in Fig. 80; fore wing RS represented by a small stub on wing margin distad stigma (Fig. 80)... .... *Euphoriella* **EUPHORINAE**
- Fore wing otherwise (Figs 79, 81); fore wing RS arising from stigma and not reaching wing margin (Figs 79, 81)..........some **APHIDIINAE**
45(43). Fore wing 2cu-a present (Fig. 69) 46
- Fore wing 2cu-a absent (Fig. 70).......... 48
46(45). Hind wing anal cross vein (a) present (Fig. 56).......... *Charmon* **HOMOLOBINAE**
- Hind wing anal cross vein absent (Fig. 55).......... 47
47(46). Median tibial spur of hind leg more than one third as long as basitarsus (Fig. 28); hind tibia usually (90%) with apical pegs or spines (Fig. 28). Occipital carina usually (85%) absent mid-dorsally (Fig. 4).......... some **ORGILINAE**
- Median tibial spur of hind leg usually (80%) less than or equal to one third length of basitarsus (Fig. 27); hind tibia without apical pegs or spines (Fig. 27). Occipital carina complete dorsally (Fig. 7) ....some Brachistini **HELCONINAE**
48(45). Metasomal tergum 1 (petiole) fused with tergum 2 (Figs 14, 117)..........some Brachistini **HELCONINAE**
- Metasomal tergum 1 articulating with tergum 2 (Figs 15, 118 49
49(48). Occipital carina present, at least laterally (Figs 4, 5, 7, 8) 50
- Occipital carina completely absent (Fig. 9).......... 53
50(49). Fore wing with one or more of the following veins absent or not tubular: (RS+M)a, m-cu, at least basal three-fourths of M +CU (Fig. 84)..........some **EUPHORINAE**
- Fore wing with all of the following veins tubular: (RS+M)a, m-cu, and at least apical three-fourths of M+CU (Fig. 85) 51
51(50). Fore wing 1RS present (Fig. 86)..........some **EUPHORINAE**
- Fore wing 1RS absent (Fig. 85).......... 52
52(51). Scutellar disc usually (90%) margined by carinae (Fig. 114). First 3 metasomal terga not forming carapace (Fig. 115); dorsope deep (Fig. 116) some BLACINAE Scutellar disc not margined by carinae (Fig. 112). First 3 metasomal terga sometimes (30%) forming carapace (Figs 117, 118); dorsope not so deep (Fig. 118)..... ....some Brachistini **HELCONINAE**
53(49). Petiole more than 4 times longer than its posterior width (Fig. 19)..........some **EUPHORINAE**
- Petiole less than 4 times longer than its posterior width (Figs 16, 17). [Rare] ....some **AGATHIDINAE**
54(6). Antenna with 12 flagellomeres. Fore wing (RS+M)a usually (90%) not tubular basally (Fig. 89) ....all **MIRACINAE**
- Antenna with 16 or more flagellomeres. Fore wing (RS+M)a tubular throughout its length (Fig. 88).......55
55(54). Fore wing r-m absent (Figs 88, 90) 56
- Fore wing r-m present (Figs 91-94), though usually unpigmented.......... 57
56(55). Fore wing RS curving anteriad (Fig. 88). Antenna usually with more than 16 flagellomeres. [Rare].......... ....all **DIRRHOPINAE**
- Fore wing RS straight or apparently absent (Fig. 90). Antenna with 16 flagellomeres (see footnote for couplet 57). [Common] ....most **MICROGASTRINAE**
57(55). Antenna with 16 flagellomeres. Second submarginal cell of fore wing usually small, not longer than high, often triangular..........some **MICROGASTRINAE**

In many species of Microgastrinae each flagellomere has longitudinal placodes arranged into two sets, one distal and one basal. This often results in an apparent constriction at the midlength of each flagellomere. The constrictions are often sharp enough to make each flagellomere appear as two. Any flagellum with 32 articles should be examined carefully with this in mind.

- Antenna with more than 16 flagellomeres. Second submarginal cell of fore wing distinctly longer than high (Figs 91-93)........58

58(57). Spiracle of petiole on lateral tergite (Fig. 24). Fore wing 3RSb usually strongly curved (Fig. 92), rarely weakly curved ....all **CARDIOCHILINAE**

- Spiracle of petiole on median tergite (Fig. 25). Fore wing 3RSb (if visible) straight (Fig. 91) or very weakly curved (Fig. 93). [Rare] ....all **MENDESELLINAE**

59(3). Hind wing SC+R largely obliterated (Fig. 104). Malar space much greater than eye height (Fig. 39). [Rare] . ...some **YPSISTOCERINAE**

- Hind wing SC+R complete (Figs 46, 49) or with stigma-like swelling (Fig. 58). Malar space usually (90%) distinctly less than eye height........60

60(59). Epicnemial carina absent (Fig. 33) 61

- Epicnemial carina present (Figs 34, 107) 71

61(60). Occipital carina present, at least in part (Figs 4-7) 62

- Occipital carina completely absent (Fig. 9)........68

62(61). Hind leg with basitarsus longer than tarsomeres 2-5 combined (Fig. 40). [Rare] ....all **HISTEROMERINAE**

- Hind leg with basitarsus shorter than tarsomeres 2-5 combined (as in Figs 27, 28, 30) 63

63(62). Fore wing as in Fig. 98: (RS+M)a partially or completely absent, r-m always absent, RS not reaching wing margin as a tubular vein. [Rare]........*Cosmophorus* **EUPHORINAE**

- Fore wing otherwise. RS+M and r-m nearly always (99%) present 64

64(63). Second submarginal cell of fore wing pentagonal, m-cu arising slightly but distinctly distad base of 2RS (Figs 74, 95); rarely with fore wing r-m absent........65

- Second submarginal cell of fore wing quadrangular, m-cu arising basad (Figs 73, 102) or directly in line with base of 2RS; never with fore wing r-m absent........67

65(64). Metasomal terga 2+3 and often terga 4-6, desclerotized and membranous medially (as in Fig. 22) Fore wing 2RS equal to or longer than 3RSa, or r-m absent........ *Parahormius* **HORMIINAE**

- Metasomal terga normally sclerotized (Fig. 23). Fore wing 2RS usually (90%) shorter than 3RSa; r-m always present........66

66(65). Notauli deep, well-developed over anterior half of mesoscutum, terminating postero-medially in patch of coarse sculpture (as in Fig. 112). Occipital carina meeting hypostomal carina above base of mandible (Fig. 7). [Rare]........*Monitoriella* **HORMIINAE**

- Notauli usually poorly developed, rarely extending beyond middle of mesoscutum, and never terminating in patch of coarse sculpture. Occipital carina usually (95%) distinctly separated from hypostomal carina ventrally (Fig. 4) ....some **OPIINAE**

67(64). Fore wing length, from base of stigma to fore wing apex, more than 1.4 times as long as length from base of stigma to apex of tegula (Fig. 96) ....some **OPIINAE**

- Fore wing length, from base of stigma to fore wing apex, less than 1.3 times as long as length from base of stigma to apex of tegula (Fig. 97)........some **HORMIINAE**

68(61). Metasomal tergum 2 with basal, transverse area that is differentiated from the remainder of the tergum by a groove or ridge (Fig. 16); the posterior margin of this area is straight or evenly rounded. [Rare] ....most **GNAMPTODONTINAE**

- Metasomal tergum 2 usually (90%) without basal, transverse area (Fig. 17), but if present, posterior margin neither straight nor evenly rounded 69

69(68). Hind wing M+CU less than half as long as 1M (Fig. 57). [Common] ....all **BRACONINAE**

- Hind wing M +CU more than half as long as 1M (Fig. 47)........70

70(69). Metasomal terga 1-3 completely sclerotized (Fig. 23) ....some **OPIINAE**

- Posterior portion of petiole and most of terga 2 and 3 membranous (Fig. 22) ....some **HORMIINAE**

71(60). Fore wing 2m-cu present (Fig. 100). [Rare, known only from Chile] ....all **APOZYGINAE**

- Fore wing 2m-cu absent 72

72(71). Fore leg tarsomere 5 much wider than tarsomere 4 and as long as tarsomeres 2, 3 and 4 combined (Fig. 29); fore leg tarsomeres 2, 3 and 4 each as wide, or wider than, long (Fig. 29) ....some **ROGADINAE**

- Fore leg tarsomere 5 subequal in width to tarsomere 4 or not longer than tarsomeres 2, 3 and 4 combined; tarsomeres 2, 3 and 4 of fore leg mostly longer than wide (Fig. 30) 73

73(72). Metasomal terga in the form of a 3-segmented carapace (as in Figs 15, 117), heavily sculptured throughout..74

- Metasomal terga heavily sculptured or not, but always with at least 4 terga visible in dorsal view (Figs 119, 120) 75

74(73). Fore wing r-m absent (Fig. 63) ....*Choreborogas* **ROGADINAE**

- Fore wing r-m present (Figs 99, 101) ....a few **HORMIINAE**

75(73). Fore tibia with pegs or spines present on anterior surface (may be small and difficult to see) (Fig. 31) Fore wing 2RS often (30%) absent or not tubular (Fig. 101). Metasomal terga 1 and 2 without median longitudinal carina (Fig. 120). Males of some species with stigma-like swelling in hind wing (Fig. 58) ....most **DORYCTINAE**

- Fore tibia without pegs or spines on anterior surface (Fig. 32). Fore wing 2RS present and mostly tubular (Fig. 102). Metasomal terga 1 and 2 often (30%) with median longitudinal carina (Fig. 119). Hind wing never with stigma-like swelling 76

76(75). Metasomal tergum 2 with median longitudinal carina (Fig. 119) and/or with longitudinal striae over most of surface (as in Fig. 18). Fore wing m-cu distinctly basad 2RS, thus (RS+M)b present (Fig. 102) ....most **ROGADINAE**

- Metasomal tergum 2 without median longitudinal carina; usually (98%) either un- sculptured, coriaceous, or finely granular. If largely striate, then (RS+M)b absent 77

77(76). Fore wing r-m absent (Fig. 63)..78

- Fore wing r-m present (Figs 99, 103) ....some **HORMIINAE**

78(77). Fore wing RS not reaching wing margin as a tubular vein (as in Fig. 88). [Rare].............. a few **HORMIINAE**

- Fore wing RS reaching wing margin as a tubular vein (Fig. 63).......................................a few **ROGADINAE**

79(1). Exodont (Figs 1, 105): mandibles not touching or overlapping when closed, and with 3 or more teeth facing outwardly ....some **ALYSIINAE**

- Endodont (Figs 2, 3, 106): mandibles touching and usually overlapping when closed, and usually with two teeth facing 80

80(79). Fore leg tarsomere 5 as long as or longer than tarsomeres 2 and 4 combined (Fig. 29)........................... 81

- Fore leg tarsomere 5 shorter than tarsomeres 2 and 4 combined (Fig. 30) 82

81(80). Occipital carina present, at least dorsally (Figs 11, 41). [Rare] ....some **MASONINAE**

- Occipital carina completely absent (Figs 9, 39). [Rare] .... some **YPSISTOCERINAE**

82(80). Anterior surface of fore tibia with spines (Fig. 31) ....some **DORYCTINAE**

- Anterior surface of fore tibia without spines (Fig. 32) ....some **HORMIINAE**

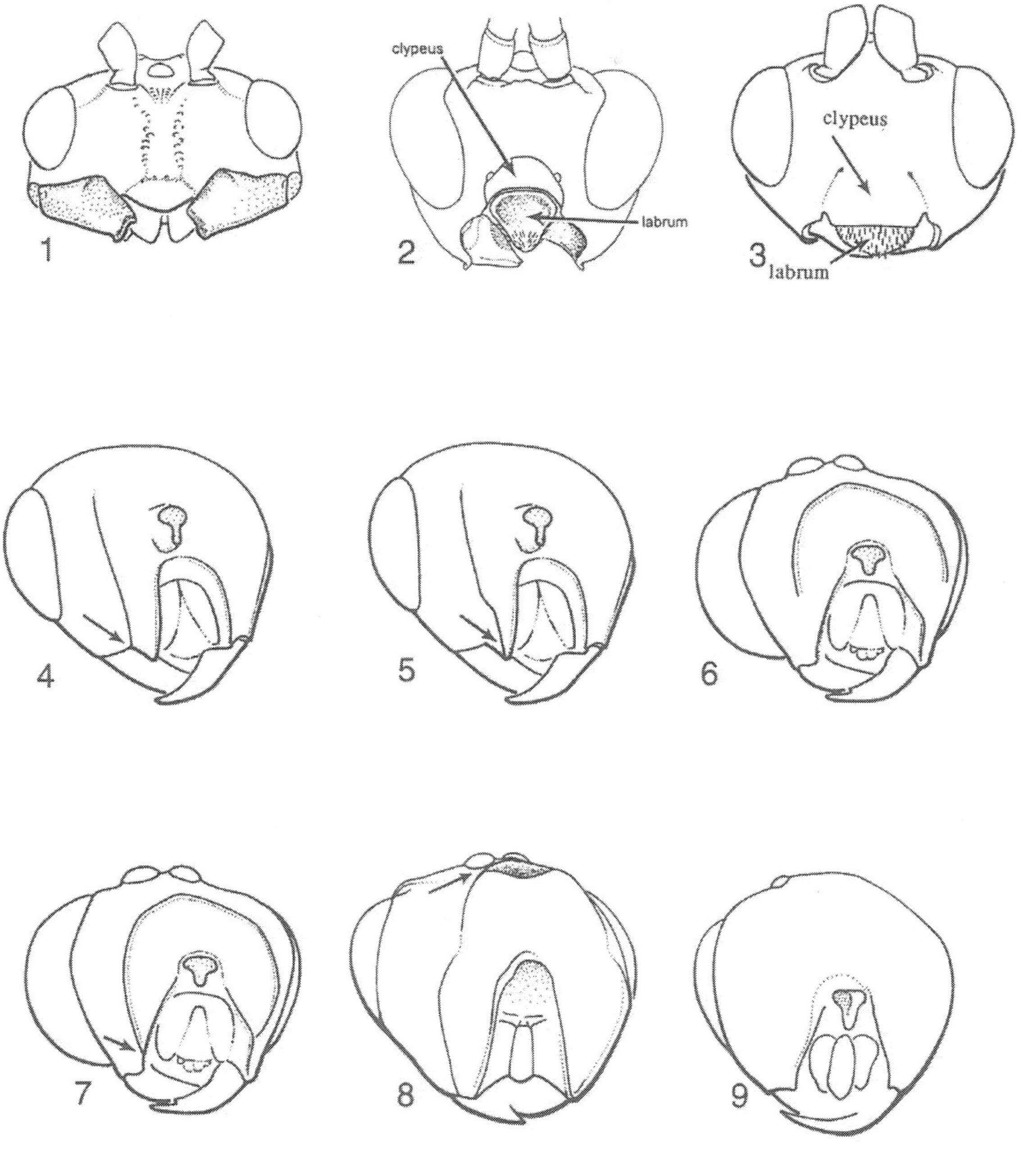

Figs 1-9: Heads. 1, Alysiinae, showing exodont mandibles; 2, cyclostome braconid, left mandible removed, showing exposed, concave labrum and short clypeus; 3, non-cyclostome braconid with flat, exposed labrum; 4-9, back of head: 4, occipital carinal absent dorsally, widely separated from hypostomal carina ventrally; 5, occipital carina absent dorsally, narrowly separated from hypostomal carina ventrally; 6, occipital carina absent ventrally, present dorsally; 7, occipital carina complete, meeting hypostomal carina ventrally at arrow; 8, occipital carina complete, extending to ocelli dorsally; 9, occipital carina completely absent.

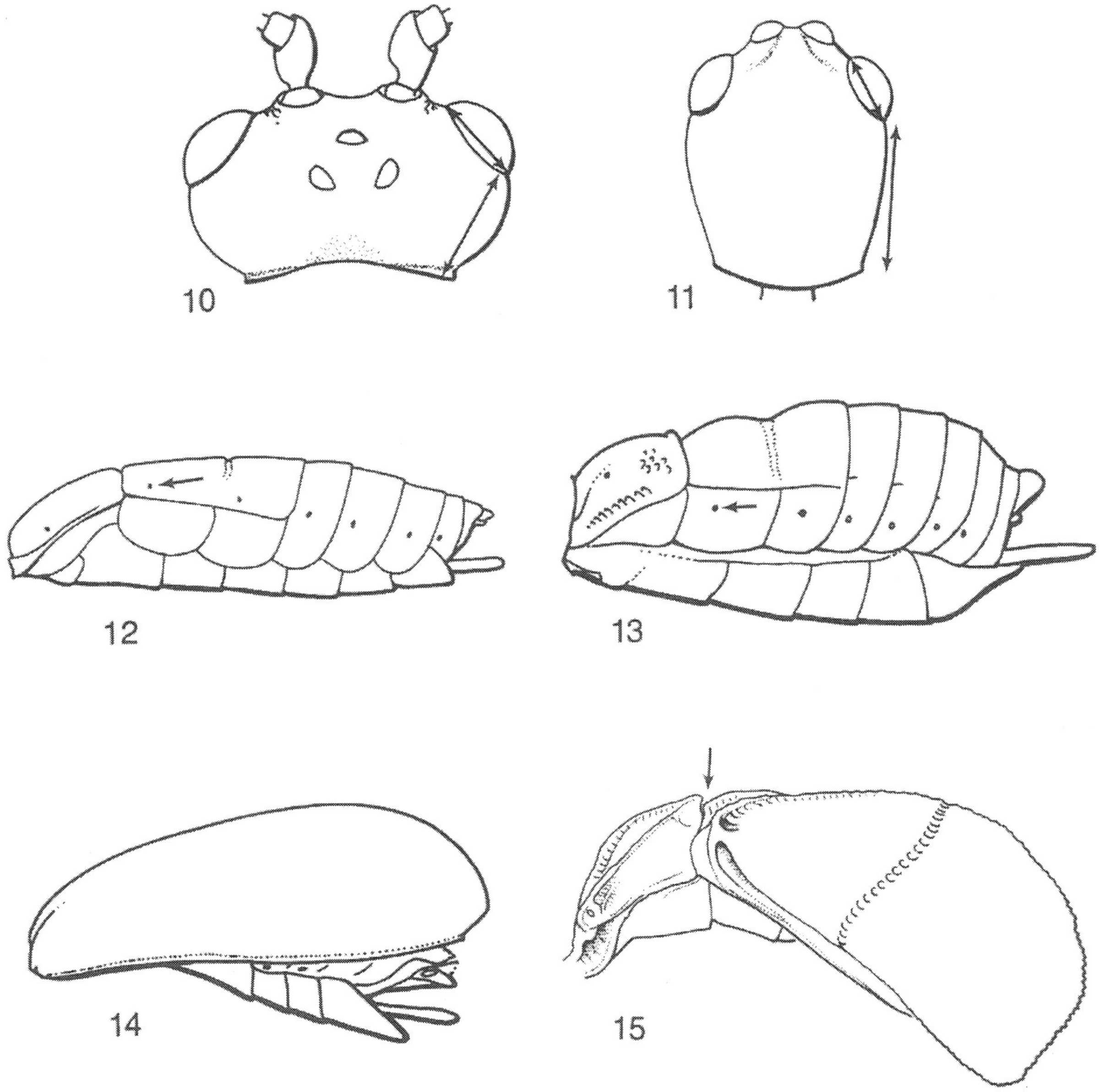

Figs 10-15: Head and body. 10-11, head, dorsal view, showing relative sizes of eye and temple. 12-15, metasoma, lateral view: 12, arrow = spiracle of second metasomal tergum in. median tergite; 13, arrow = spiracle of second metasomal tergum in lateral tergite; 14, *Chelonus*, showing carapace-like metasoma without articulation between segments; 15, *Sigalphus*, arrow = articulation between first two terga of carapace.

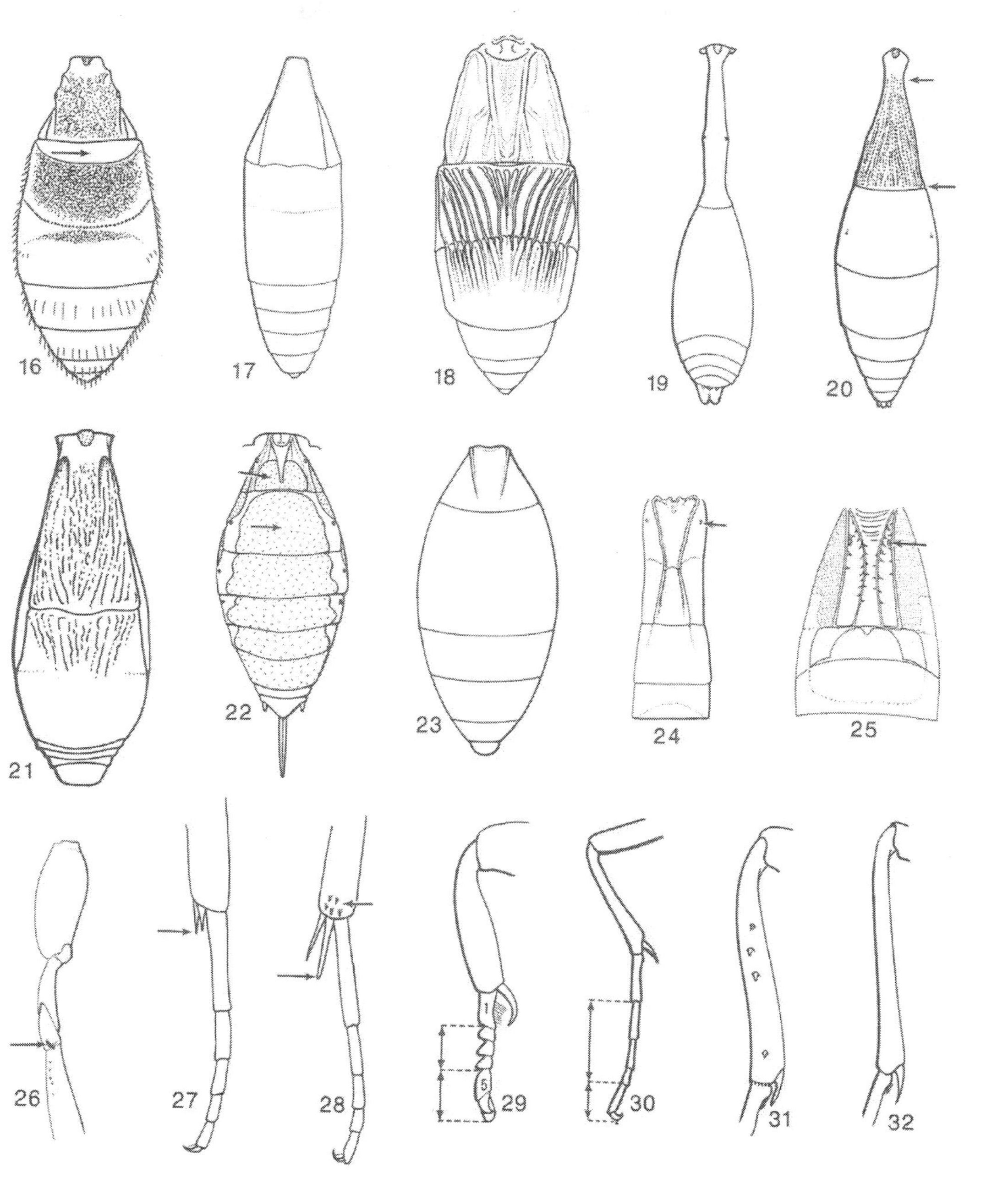

Figs 16-32: Metasoma and legs. 16-25, metasoma, dorsal view: 16, *Gnamptodon*, arrow = basal, transverse area of tergum 2; 17, Agathidinae; 18, *Pselaphanus*; 19, Euphorinae showing long, narrow petiole; 20, *Meteorus*, showing petiole that is much wider at apex than at base; 21, Helconinae: 22, *Allobracon,* arrows = membranous, desclerotized regions of petiole and tergum 2; 23, Hormiinae without membranous terga; 24, Microgastrinae, arrow = spiracle of petiole in lateral tergite; 25, Mendesellinae, arrow = spiracle of petiole in median tergite. 26-32, legs: 26, hind leg, Macrocentrinae, arrow = spines on trochantellus; 27, hind leg, arrow = short median tibial spur; 28, hind leg, arrows = long median tibial spur and apical tibial spines; 29, fore leg, *Yelicones*, showing short, broad tarsomeres; 30, fore leg showing normal tarsomeres; 31, inner surface of fore tibia, Doryctinae, showing spines; 32, inner surface of tibia without spines.

Figs 33-41: Bodies. 33-34, mesosoma, lateral view: 33, arrow = epicnemial carina absent; 34, arrow = epicnemial carina present; 35-36, mesosoma, dorsal view: 35, arrow = posterior scutellar depression present; 36, posterior scutellar depression absent; 37-38, lateral views: 37, Cenocoeliinae, arrows = metasoma arising from mesosoma well above base of hind coxa; 38, Meteorinae, arrow = metasoma arising in normal position at or between bases of hind coxae; 39, Ypsistocerinae, arrows = long malar space and notauli absent; 40, Histeromerinae, arrows = long malar space and long hind basitarsus; 41, Masoninae, arrow = occipital carina present dorsally.

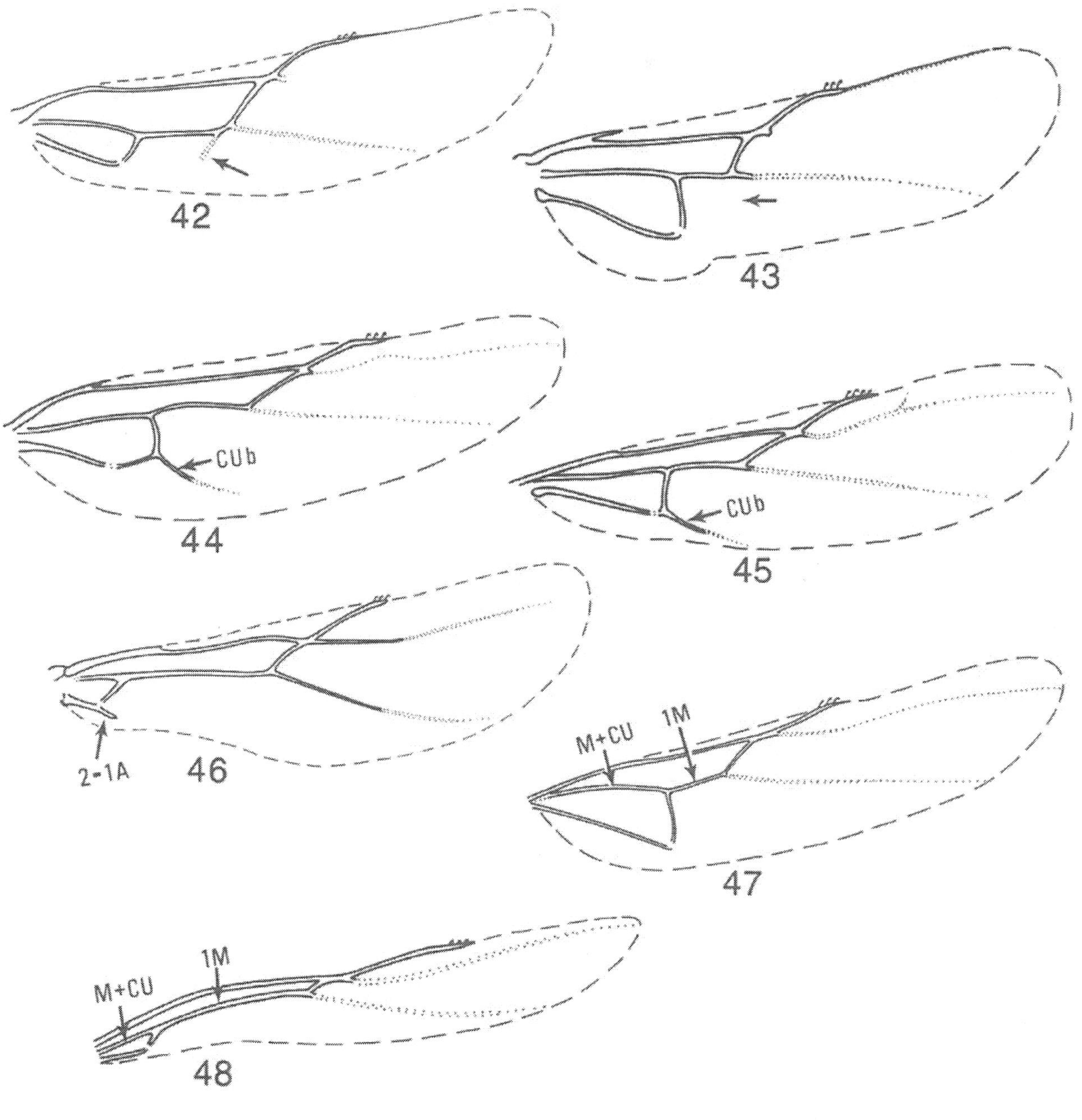

Figs 42-48: Hind wings. 42, arrow = m-cu present; 43, arrow = m-cu absent; 44-45, arrows = CUb present; 46, CUb absent, arrow = 2-1A; 47-48, arrows showing relative differences in length of M+CU and 1M.

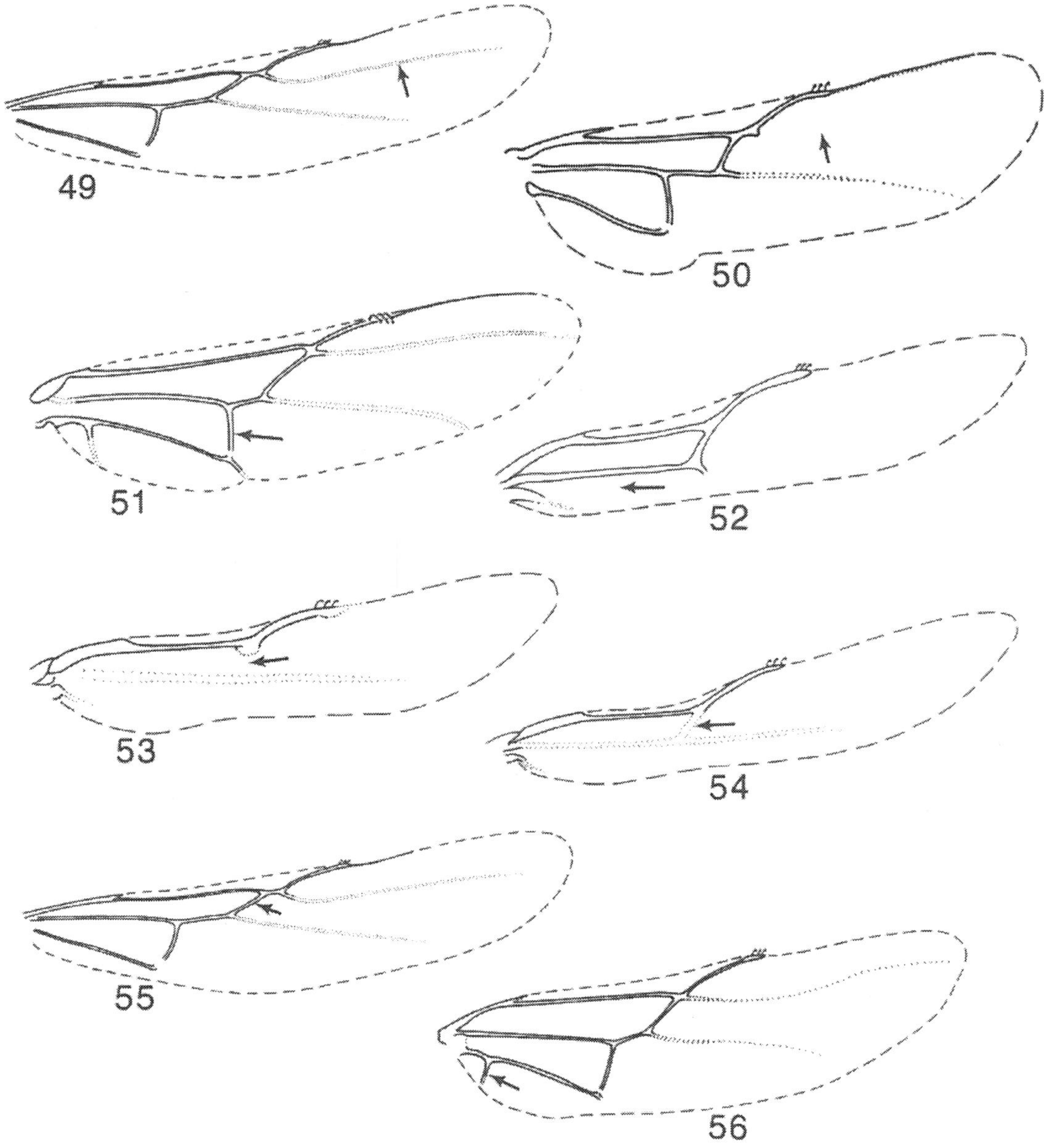

Figs 49-56: Hind wings. 49, arrow = RS present; 50, arrow = RS absent: 51, arrow = cu-a present; 52, arrow = cu-a absent; 53, arrow = r-m absent; 54, arrow = r-m present but not tubular; 55, arrow = r-m present and tubular; 56, arrow = anal cross vein.

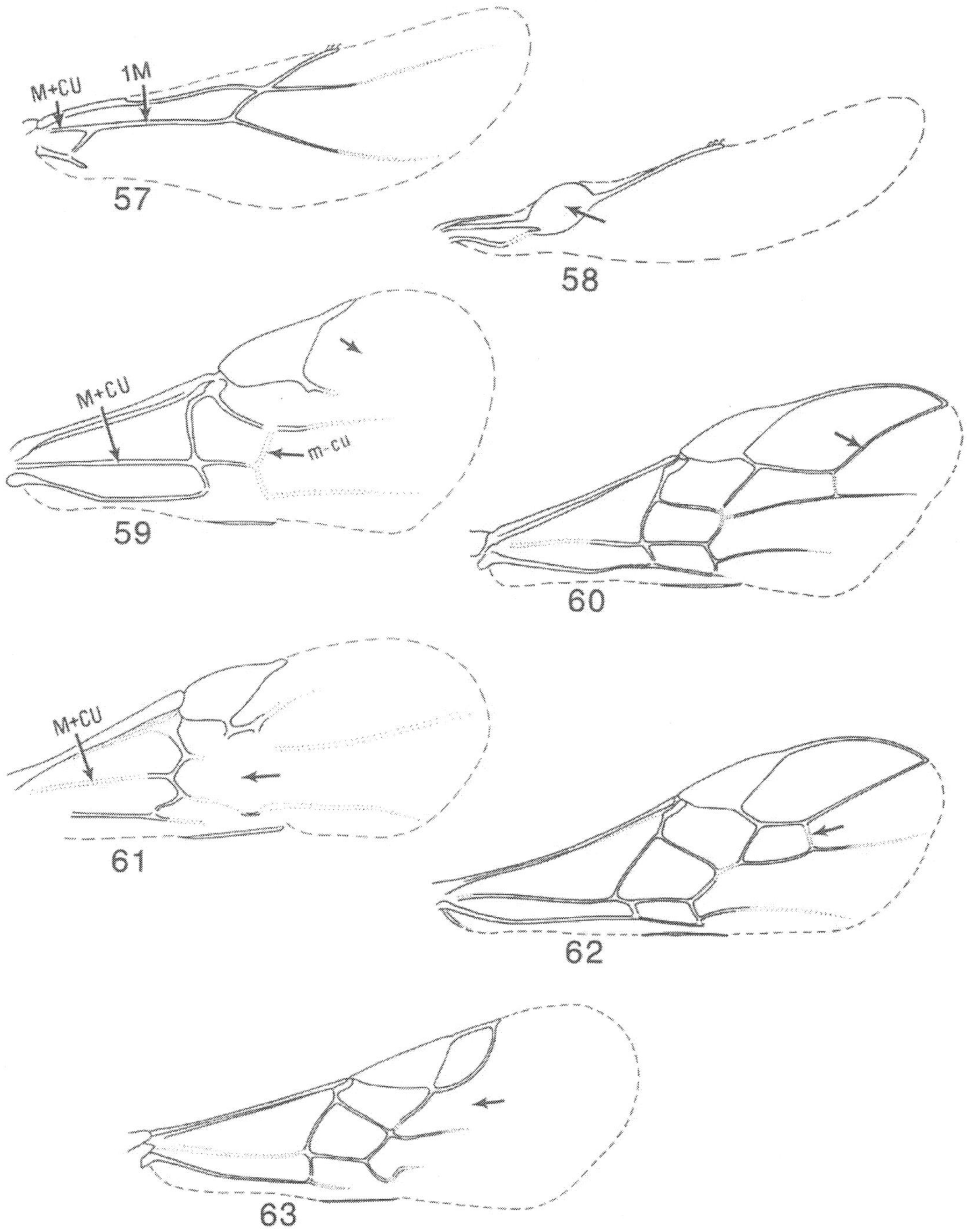

Figs 57-:63: Wings. 57-58, hind wings: 57, arrows = relative lengths of M+CU and 1M; 58, Doryctinae, arrow = enlarged, stigma-like swelling. 59-63, fore wings: 59, Ichneutinae, arrows = RS absent distally, m-cu present but not tubular, and M+CU present and tubular throughout; 60, Opiinae, arrow = 3RSb extending to wing margin as a tubular vein; 61, *Elasmosoma*, arrows = m-cu absent and M+CU present but not tubular basally; 62, arrow = r-m present: 63, Euphorinae, arrow = r-m absent.

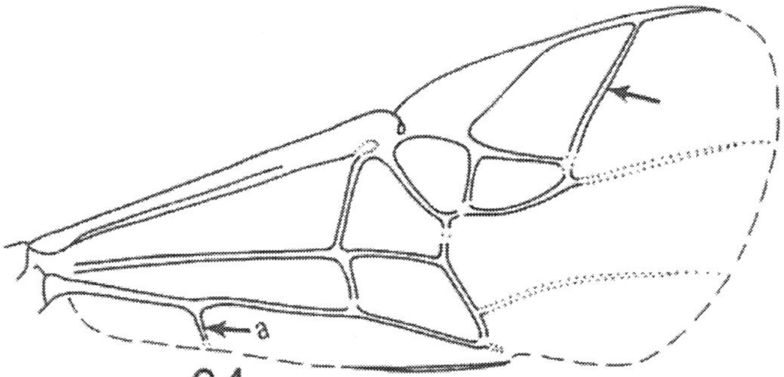

64

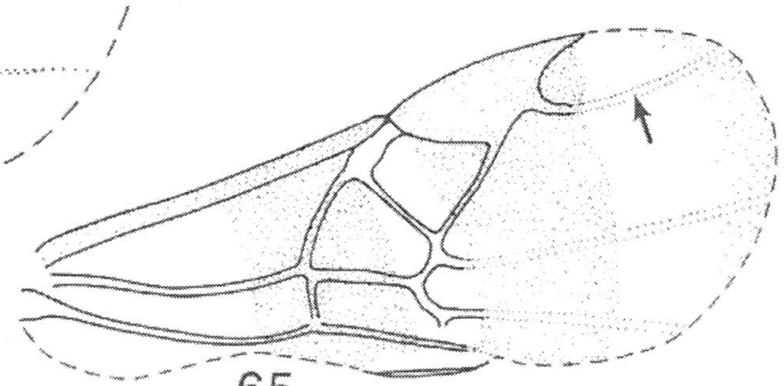

65

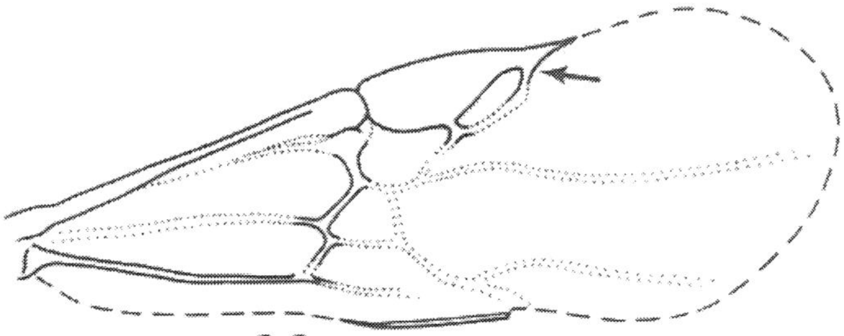

66

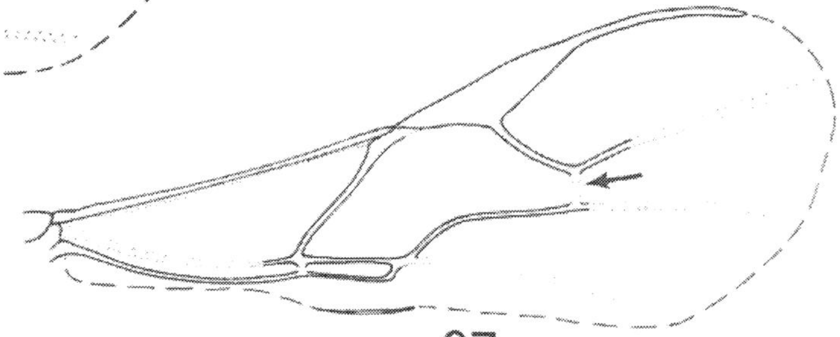

67

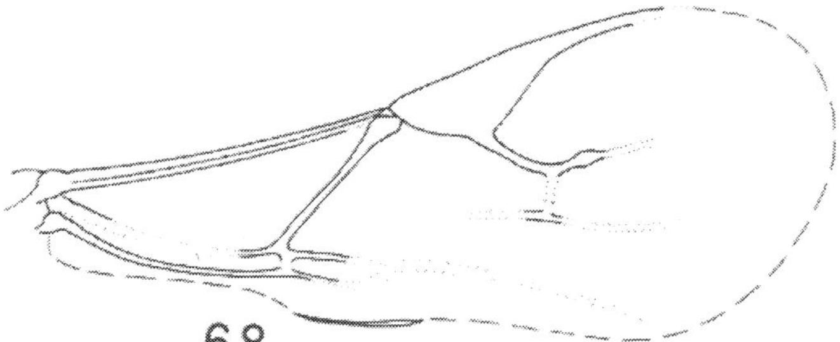

68

Figs 64-68: Fore wings. 64, Ichneutinae, arrow = 3RSb extending to wing margin as a tubular vein, anal cross vein present; 65, Adeliinae, arrow = RS extending to wing margin, but not as a tubular vein; 66, *Neoneurus*, arrow = RS, represented by a tubular vein at wing margin even though it is not tubular posteriorly; 67, *Aphidius*, arrow = r-m present; 68, *Lysiphlebus*, showing r-m present, though weak.

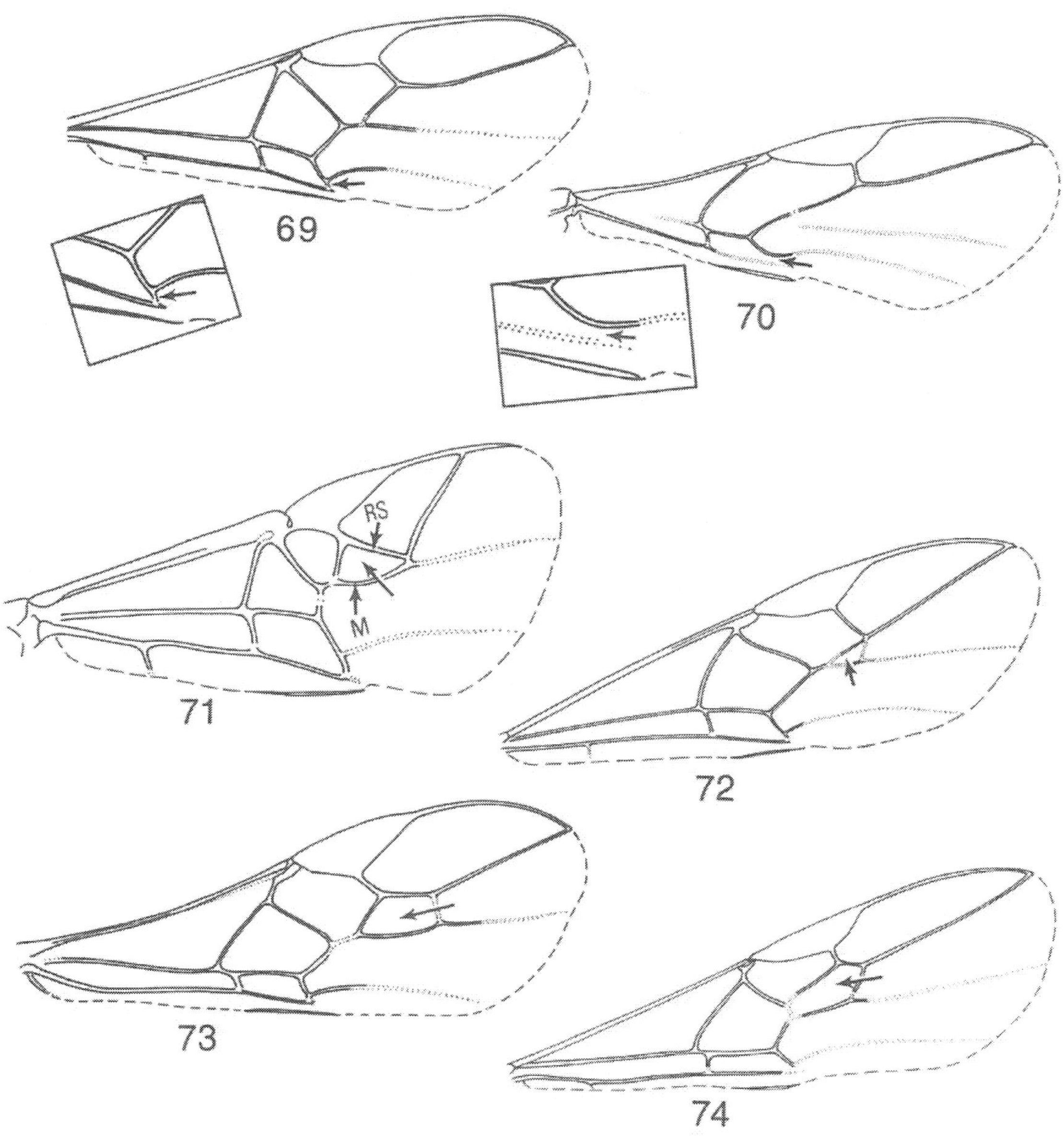

Figs, 69-74: Fore wings. 69, arrow and inset = 2cu-a present; 70, arrow and inset = 2cu-a absent; 71, Ichneutinae, arrows = triangular second submarginal cell; 72, Orgilinae, arrow = triangular second submarginal cell; 73, arrow = quadrangular second submarginal cell; 74, arrow = pentagonal second submarginal cell.

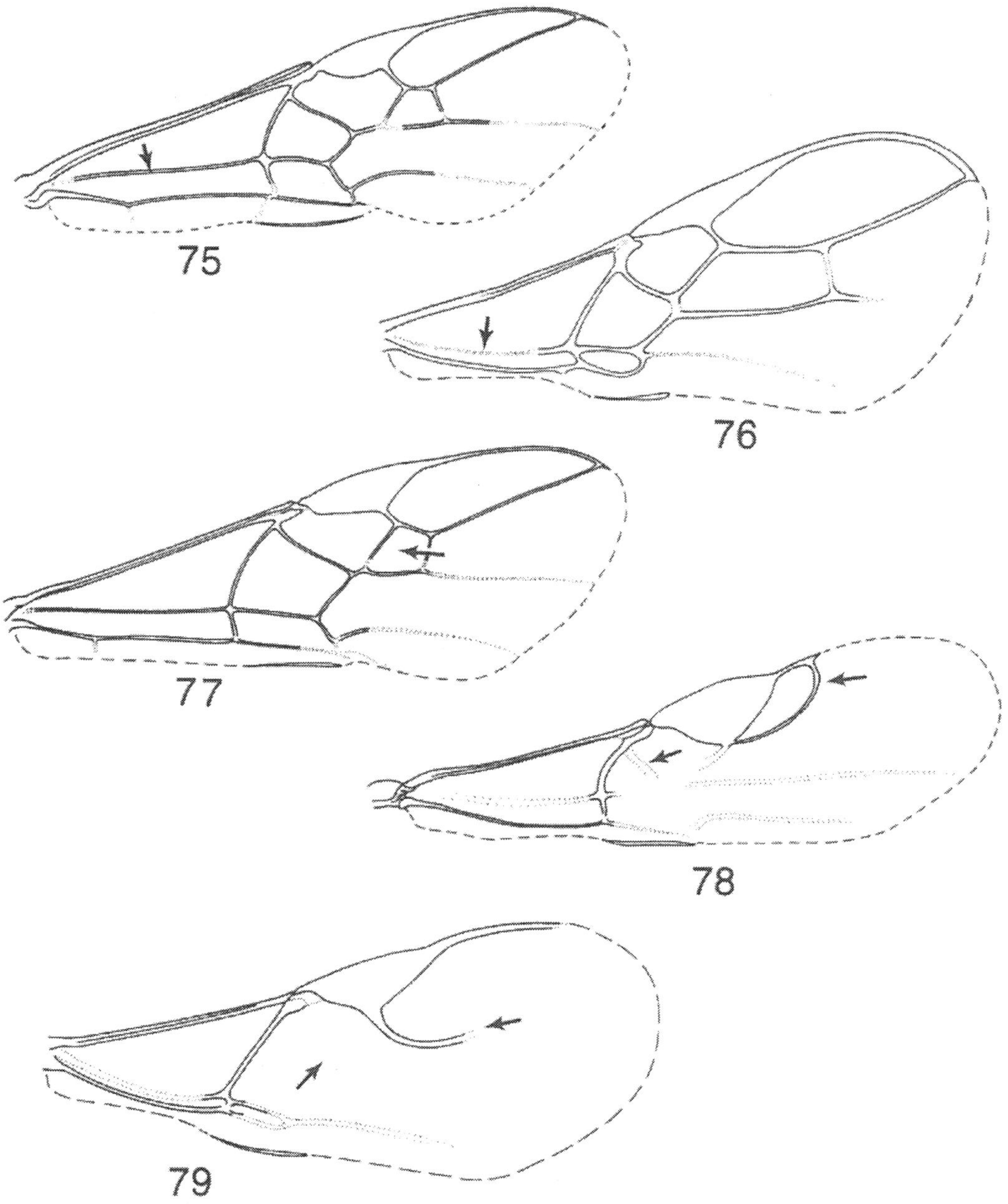

Figs 75-79: Fore wings. 75, Helconinae, arrow = M +CU present and tubular throughout; 76, Aphidiinae, arrow = M+CU not represented by a tubular vein basally; 77, arrow = second submarginal cell quadrangular; 78, Euphorinae, arrows = (RS+M)a present but not tubular and RS curved sharply towards anterior margin of wing; 79, Aphidiinae, arrows = (RS+M)a completely absent and RS not extending to wing margin.

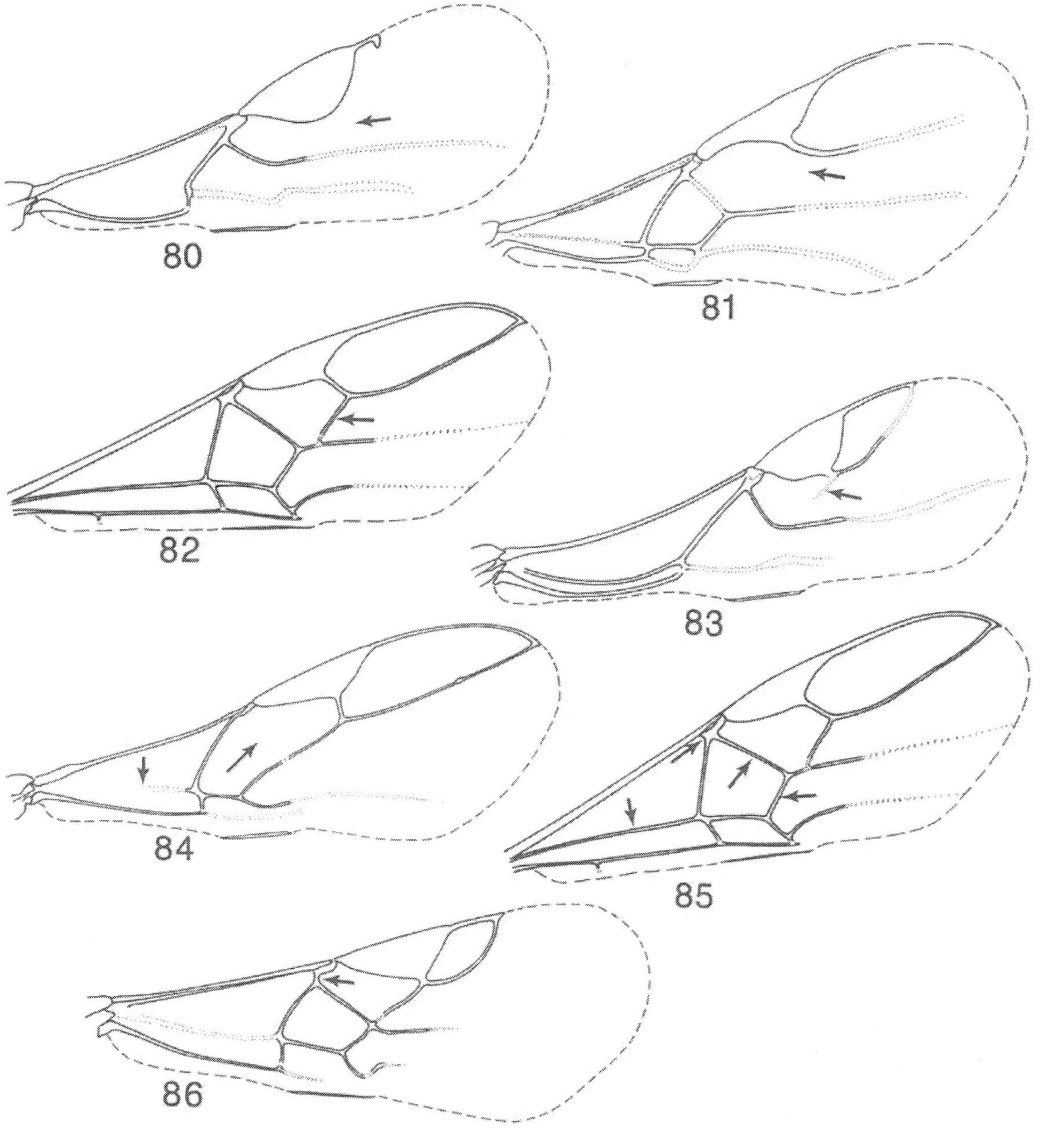

Figs 80-86: Fore wings. 80, *Euphoriella*, arrow = 2RS and r-m absent; 81, *Praon*, arrow = 2RS and r-m absent; 82, r-m absent, arrow = 2RS present; 83 = Euphorinae, arrow = 2RS present but incomplete and not tubular; 84, Euphorinae, arrows = M+CU absent basally and present but not tubular distally, (RS+M)a absent; 85, Helconinae, arrows = M+CU present and tubular, 1RS absent, (RS+M)a present, m-cu present and tubular; 86, Euphorinae. arrow = 1RS present.

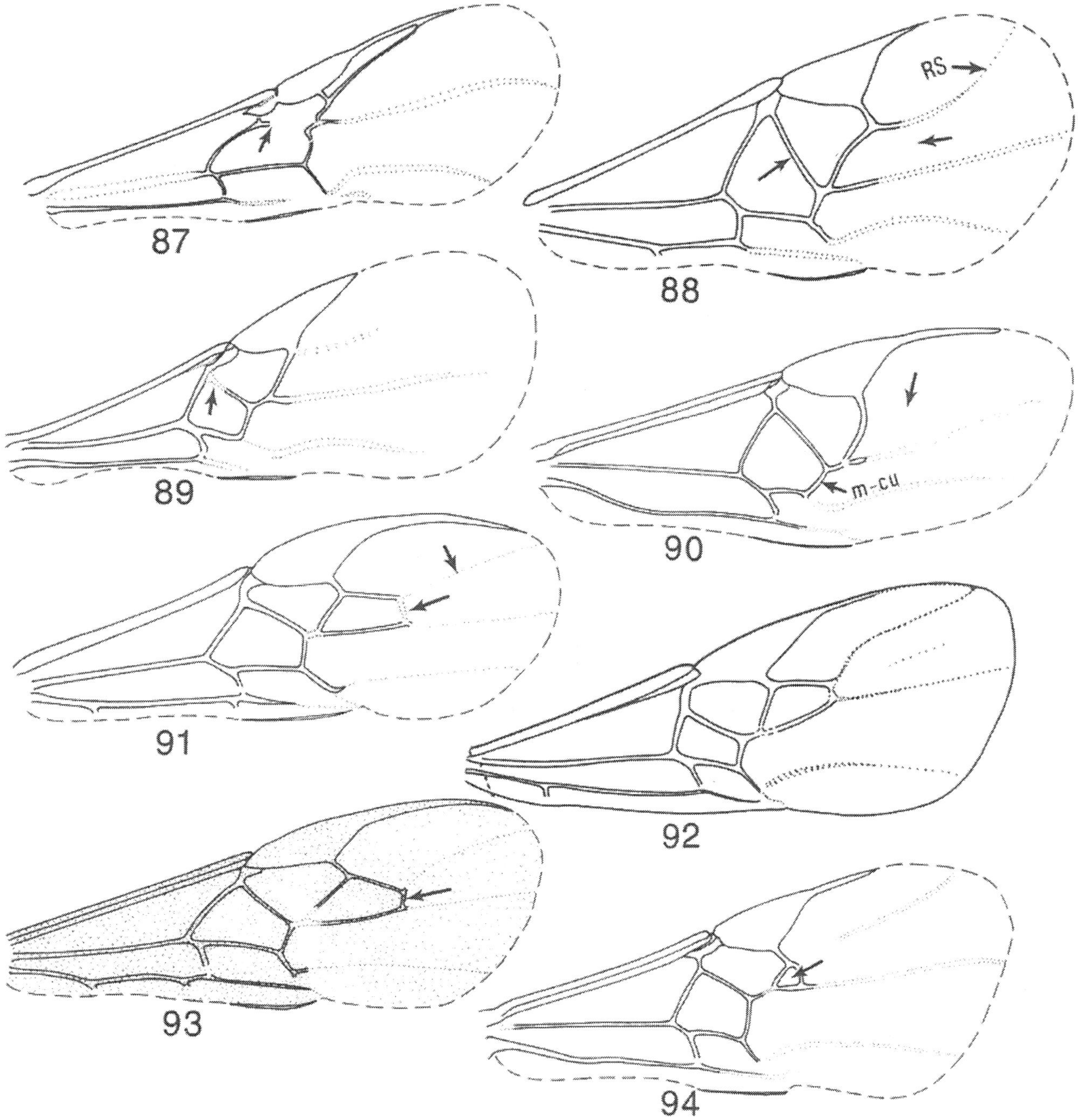

Figs 87-94: Fore wings. 87, Agathidinae, arrow = (RS+M)a absent; 88, Dirrhopinae, arrows = (RS+M)a present, r-m absent, RS not extending to wing margin as a tubular vein; 89, Miracinae, arrow = (RS+M)a not tubular at base; 90, Microgastrinae, arrows = m-cu tubular and RS absent; 91, arrows = r-m present and RS not reaching wing margin as a tubular vein; 92, Cardiochilinae, showing recurved 3RSb; 93, Mendesellinae, arrow =r-m present; 94, Microgastrinae, arrow = small second submarginal cell.

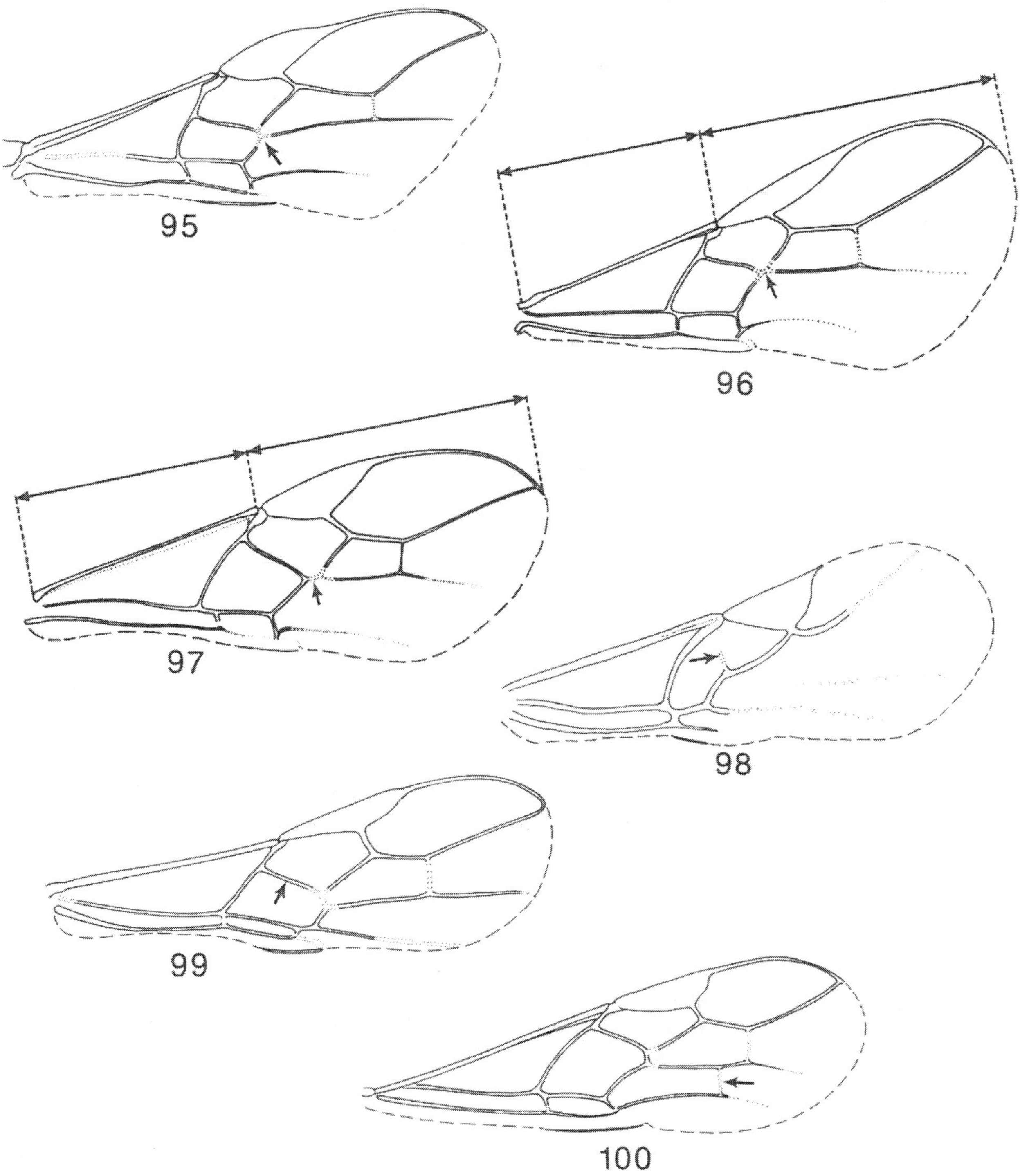

Figs 95-100: Fore wings. 95, Opiinae, arrow = m-cu arising distad base of 2RS, creating a pentagonal second submarginal cell; 96, Opiinae, showing portion of wing distad base of stigma longer than basal portion of wing; 97, Hormiinae, showing basal and distal portions of wing more nearly equal in length; 98, *Cosmophorus*, arrow = (RS+M)a present but incomplete; 99, Hormiinae, (RS+M)a complete and largely tubular; 100, *Apozyx*, arrow = 2m-cu.

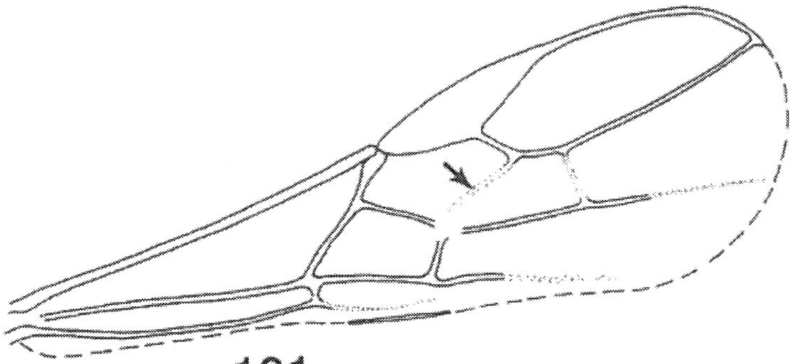

101

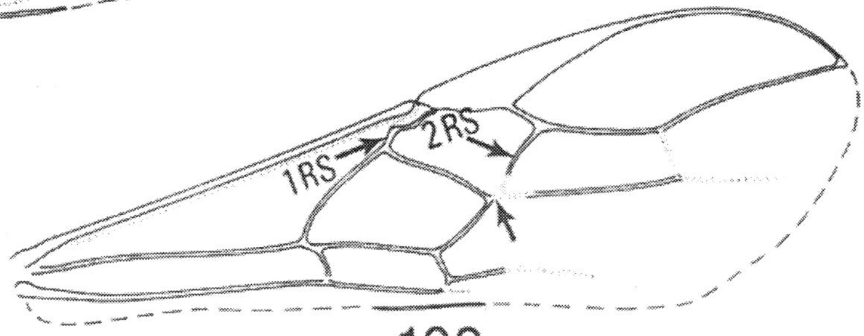

102

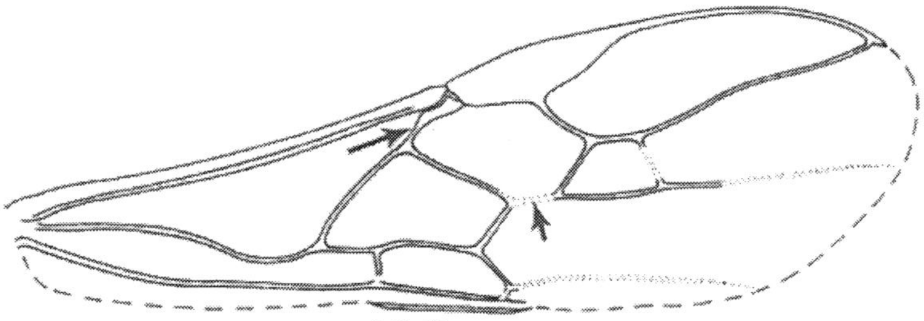

103

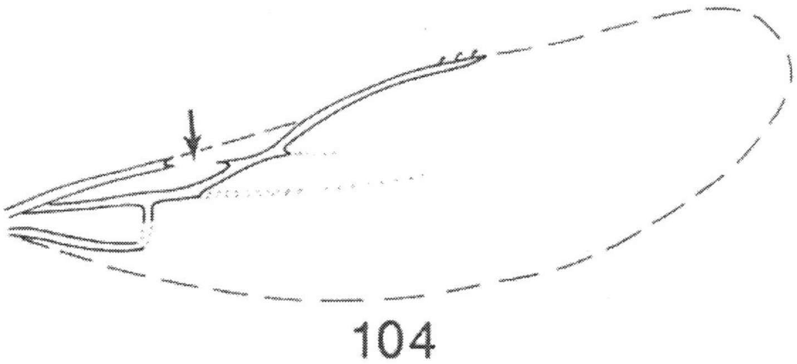

104

Figs 101-104: Wings. 101, fore wing of *Heterospilus*, arrow = incomplete, non-tubular 2RS; 102, fore wing showing m-cu narrowly separated from 2RS and 1RS present; 103, fore wing showing m-cu widely separated from 2RS and 1RS present; 104, hind wing of Ypsistocerinae, arrow = SC+R incomplete.

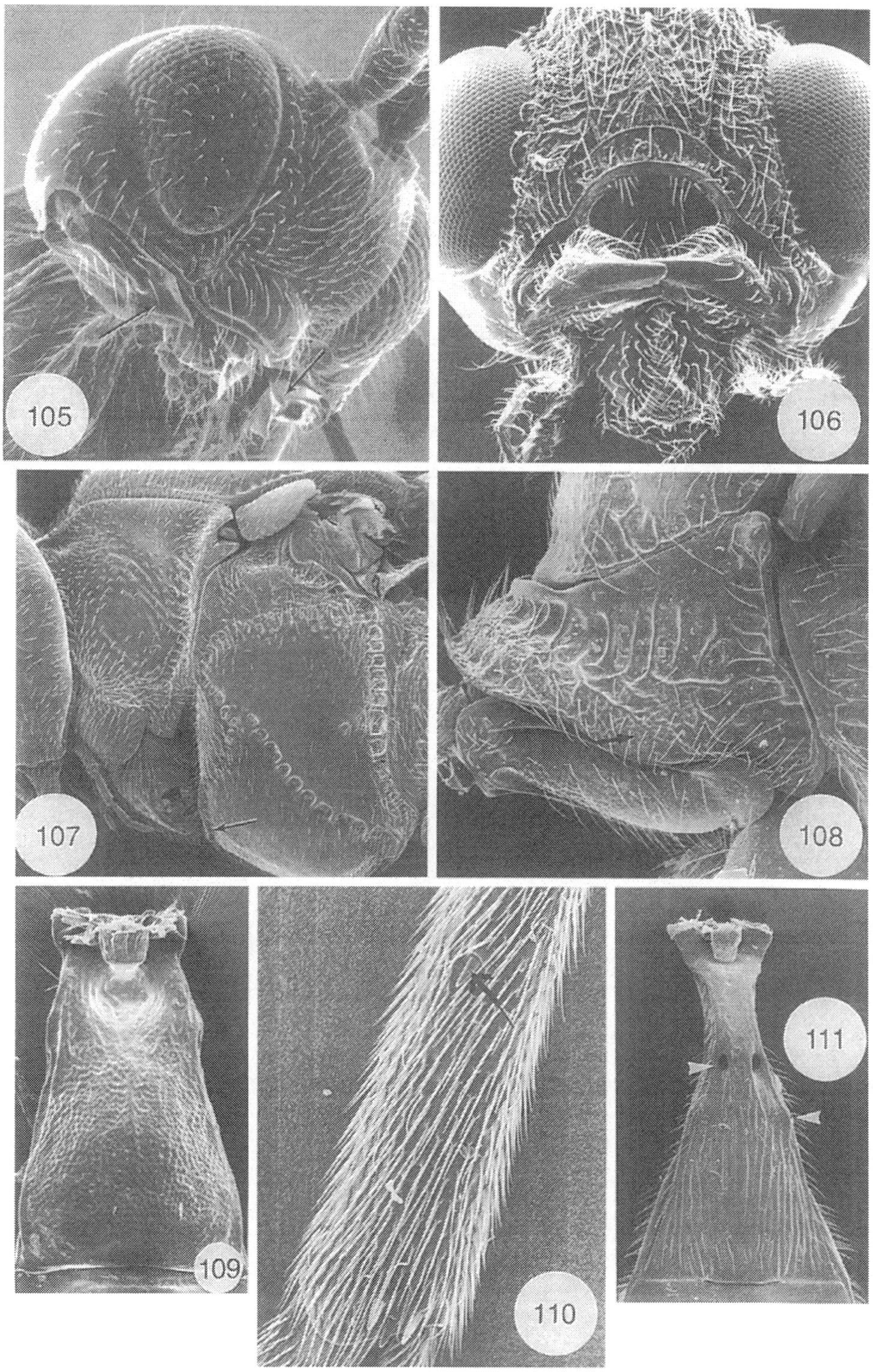

Figs 105-111: SEM. 105, Alysiinae, arrows = exodont mandibles; 106, Braconinae, showing cyclostome opening above mandibles created by deeply concave labrum; 107, lateral view of thorax, arrow = epicnemial carina; 108, lateral view of prothorax of Rogadinae, arrow = granulate sculpture; 109, petiole, dorsal view, showing absence of dorsope and dorsal carinae: 110, inner side of fore leg, arrow = spine; 111, petiole of Meteorinae, dorsal view, arrows = dorsope (anterior) and spiracle (posterior).

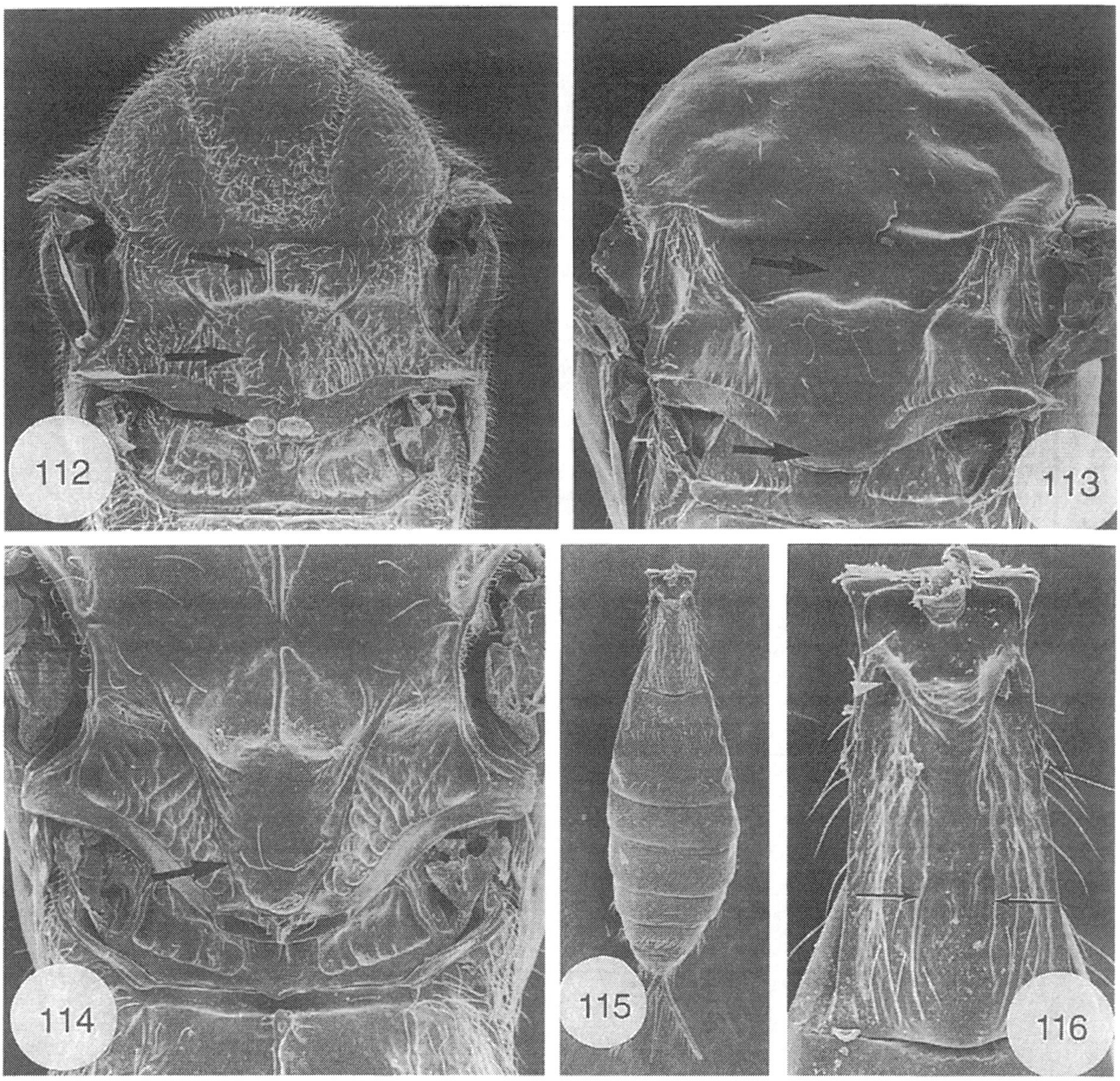

Figs 112-116: SEM. 112, thorax of Meteorinae, dorsal view, arrows = scutellar sulcus with carina, lateral margin of scutellum without carina, and postscutellar depression present and represented by two small pits; 113, thorax of Aphidiinae, dorsal view, arrows = scutellar sulcus without carina, postscutellar depression absent; 114, thorax, dorsal view, arrow = lateral carina of scutellum present; 115, *Dyscoletes* metasoma, dorsal view; 116, *Dyscoletes* petiole, dorsal view, white arrow = large dorsope present basally, black arrows = dorsal carinae present.

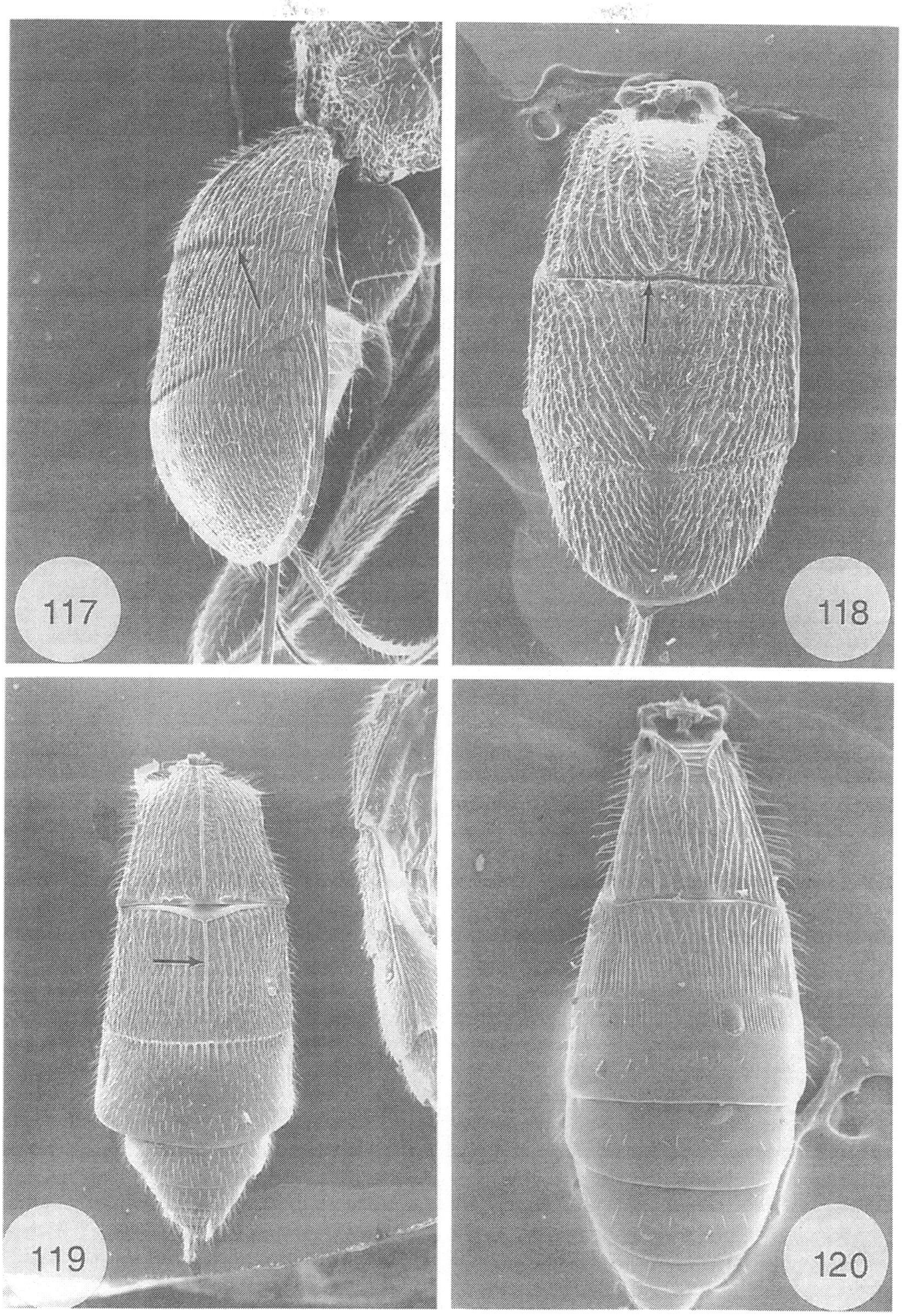

Figs 117-120: SEM, metasomata. 117, *Triaspis*, lateral view, showing first 3 terga carapace-like, arrow = suture present but no articulation between tergum 1 and 2; 118, *Aliolus*, dorsal view, showing first 3 terga carapace-like, arrow = suture present and tergum 1 and 2 capable of articulation; 119, *Aleiodes*, dorsal view, arrow = median longitudinal carina on tergum 2; 120, *Heterospilus*, showing striate sculpture without median longitudinal carina.

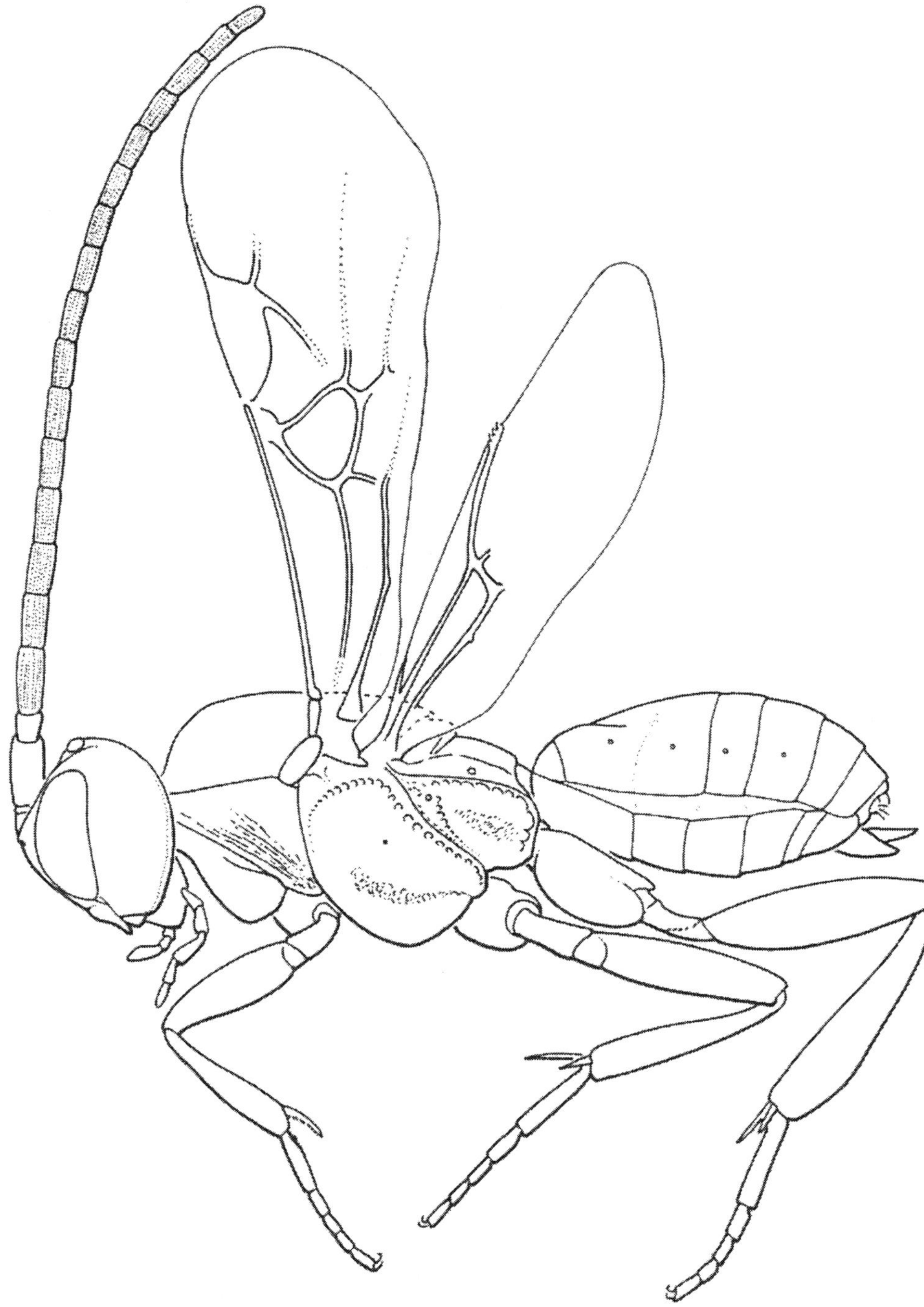

SUBFAMILY ADELIINAE

J.B. WHITFIELD

Department of Entomology, 320 Morrill Hall, University of Illinois, 505 S. Goodwin Ave, Urbana, IL 61801, U.S.A.

INTRODUCTION. Adeliines are small, stout-bodied braconids that parasitize the larvae of leaf-mining moths. They tend to be rare in collections but are distributed widely and may be locally common. Two genera are known in the New World (Marsh et al., 1987; Whitfield, 1989). The North American species have been treated by Muesebeck (1922, *Adelius* Haliday) and Whitfield (1989, *Paradelius* de Saeger).

BIOLOGY. All of the species so far known are solitary endoparasitoids of leaf-mining moths, especially (perhaps exclusively) of the family Nepticulidae (Whitfield and Wagner, 1991), and they emerge from the host cocoon. At least some of the species resemble ants when they run about on plants, and one species is known to produce a formic acid-like odor when handled (Whitfield, 1989).

BIOLOGICAL CONTROL POTENTIAL. Conceivably some of the species could prove useful against leaf-miners on coffee or on ornamental shrubs and trees, but no species have been used in this manner.

COMMON GENERA *Adelius* and *Paradelius* are generally recognized. Belokobyl'skij (1988) has described another genus, *Sculptomyriola* Belokobylski], from the eastern Palaearctic Region, for those species possessing more longitudinally sculptured anterior metasomal tergites and a fully sculptured third metasomal tergite, but this appears to represent only an extreme of trends found within *Paradelius*. The New World fauna is probably best treated as being composed of two genera until the fauna is better known. *Adelius* has sometimes been referred to as *Acoelius*, due to inconsistencies in the original description (Haliday, 1833), but this has been cleared up recently (Mason, 1985) and the accepted name is now *Adelius*.

DISTRIBUTION. Both *Adelius* and *Paradelius* are present in North America south through Mexico. Though very poorly surveyed, they probably occur more or less throughout the region. In the Neotropics only *Paradelius* has so far been collected, in Costa Rica. The same two genera are also present in Africa, Australia and Eurasia.

DISTINGUISHING FEATURES. Adeliines share with the Cheloninae the fusion of the first two (sometimes the first three or four) metasomal terga, and the absence of spiracles on both metasomal terga 6 and 7. Relative to chelonines the wing venation is reduced apically (Fig. 1) and the metasomal terga are less carapace-like, but the two subfamilies are clearly closely related (Whitfield and Mason, 1994) and might eventually be merged into one. Both appear to be related to the microgastroid complex of subfamilies (Quicke and van Achterberg, 1990; Wharton et al., 1992; Whitfield and Mason, 1994).

KEY TO THE NEW WORLD GENERA OF THE SUBFAMILY ADELIINAE

1. First 2 metasomal terga sculptured, and usually separated by a crenulate groove (although immovably fused) (Fig. 2) .... ***Paradelius*** de Saeger
Distribution: U.S.A. to Costa Rica; 2 described New World species. Biology: reared from leaf-mining Lepidoptera. References: de Saeger (1942), Whitfield (1989).

\- First 2 metasomal terga not or only weakly sculptured; division between them obscure.... ***Adelius*** Haliday
Distribution: Canada to Mexico; 3 described New World species. Biology: reared from leaf-mining Lepidoptera. References: Muesebeck (1922), Whitfield and Wagner (1991).

REFERENCES

Belokobyl'skij, S.A. 1988. Subfamily Adeliinae (Hymenoptera) in the Far East of the U.S.S.R. Sistematika Nasekomiya i Klezhei 70: 144-152. (In Russian.)

Haliday, A.H. 1833. An essay on the classification of the parasitic Hymenoptera of Britain, which correspond with the *Ichneumones minuti* of Linnaeus. Entomological Magazine 1: 259-276.

Marsh, P.M., Shaw, S.R., and Wharton, R.A. 1987. An identification manual to the North American genera of the family Braconidae (Hymenoptera). Memoirs of the Entomological Society of Washington 13: 1-98.

Mason, W.R.M. 1985. The correct spelling for *Adelius* Haliday (Hymenoptera: Braconidae). Proceedings of the Entomological Society of Washington 87: 896.

Muesebeck, C.F.W. 1922. A revision of the North American ichneumon-flies, belonging to the subfamilies Neoneurinae and Microgasterinae. Proceeedings of the United States National Museum 61: 1-76.

Quicke, D.L.J. and Achterberg, C. van 1990. Phylogeny of the subfamilies of the family Braconidae (Hymenoptera: Ichneumonoidea). Zoologische Verhandelingen Leiden 258; 1-95.

Saeger, H. de 1942. Paradelius; genre nouveau de Microgasterinae (Hymen. Braconidae). Revue Zoologique et Botanique Africaines 36: 313-316.

Wharton, R.A., Shaw, S.R., Sharkey, M.J., Wahl, D.B., Woolley, J.B., Whitfield, J.B., Marsh, P.M., and Johnson, J.W. 1992. Phylogeny of the sub families of the Braconidae (Hymenoptera: Braconidae): a reassessment. Cladistics 8: 199-235.

Whitfield, J.B. (1988) 1989. Two new species of *Paradelius* (Hymenoptera: Braconidae) from North America, with biological notes. Pan-Pacific Entomologist 64: 313-319.

Whitfield, J.B. and Mason, W.R.M. 1994. Mendesellinae, a new subfamily of braconid wasps (Hymenoptera: Braconidae) with a review of relationships within the microgastroid assemblage. Systematic Entomology 19: 61-76.

Whitfield, J.B. and Wagner, D.L. 1991. Annotated key to the genera of Braconidae (Hymenoptera) attacking leaf-mining Lepidoptera in the Holarctic Region. Journal of Natural History 25: 733-754.

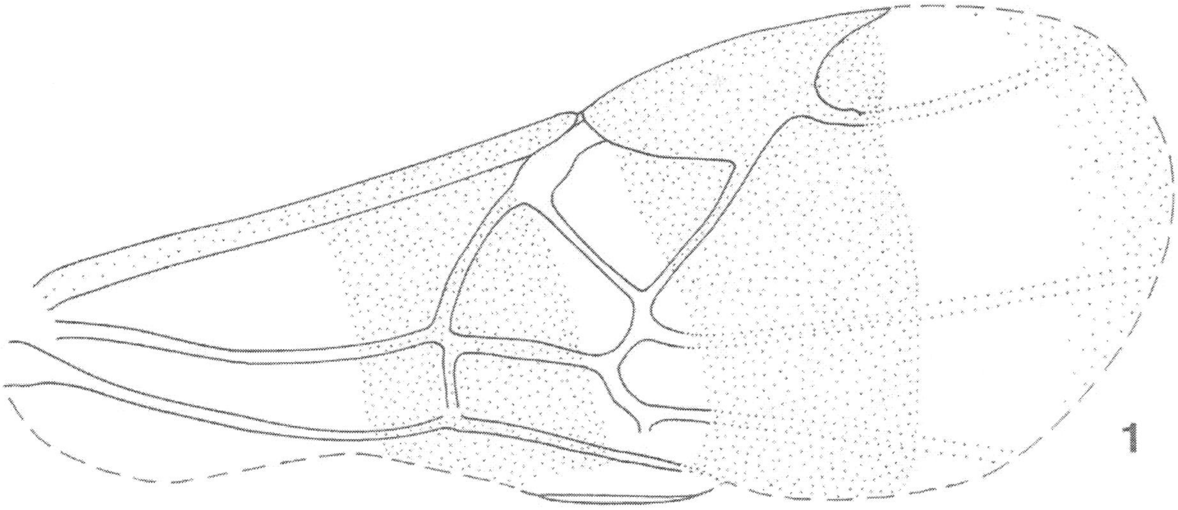

1

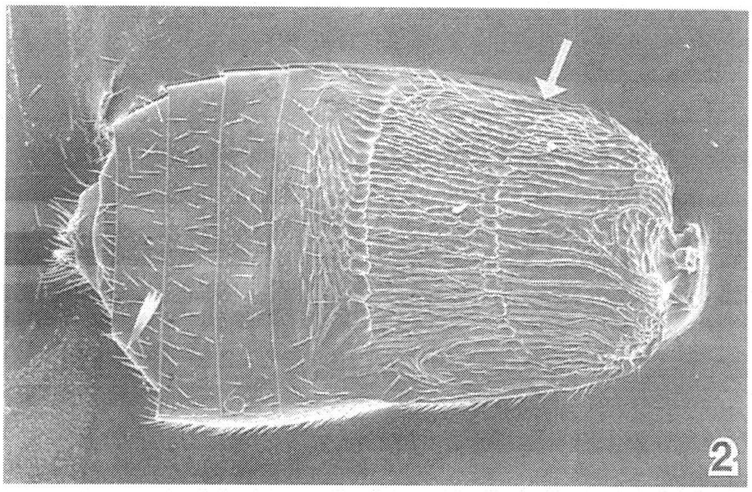

Figs 1-2: Adeliinae. 1, wing of *Paradelius* sp.; 2, metasomal terga 1 and 2 of *Paradelius* sp., arrow = crenulate groove.

ADELIINAE UPDATE

S.R. Shaw
Department of Ecosystem Science and Management, University of Wyoming, Laramie, Wyoming 82071, U.S.A.
braconid@uwyo.edu

The Adeliinae chapter of the manual continues to work perfectly well for the identification of New World genera; however, according to Kittel *et al.* (2016) the adeliines should be classified as members of the subfamily Cheloninae.

Annotated list of recent papers on New World Adeliinae

Kittel RN, Austin AD, Klopfstein S. 2015. Molecular and morphological phylogenetics of chelonine parasitoid wasps (Hymenoptera: Braconidae), with a critical assessment of divergence time estimations. *Molecular Phylogenetics and Evolution* 101: 224-241.

A molecular phylogenetic analysis of the Cheloninae was conducted, concluding that the genera comprising the former subfamily Adeliinae (including *Adelius* and *Paradelius*) are nested within the Cheloninae. Therefore, the adeliines are best reclassified as belonging in the subfamily Cheloninae, where they can be treated as tribe Adeliini.

Shimbori EM, Bortoni MA, da Silva Souza C, de Campos Muradas P, Haimowitz L, Penteado-Diaz AM, Shaw SR. 2017. Revision of the New World genera *Adelius* Haliday and *Paradelius* de Saeger (Hymenoptera: Braconidae: Cheloninae: Adeliini). (manuscript).

At the time of this writing, this manuscript is in an advanced stage of preparation, and publication is anticipated later in 2017.

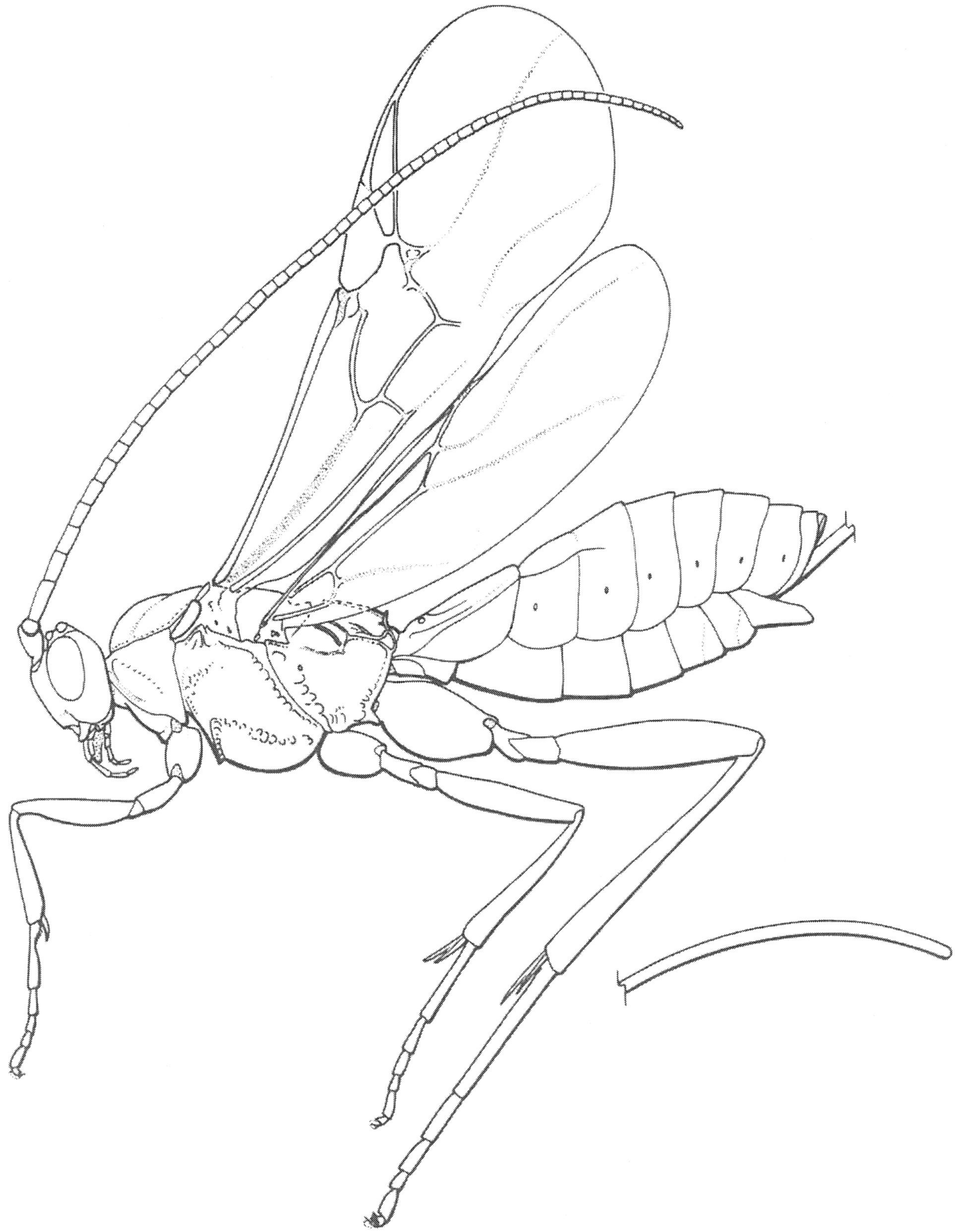

SUBFAMILY AGATHIDINAE

M. J. SHARKEY

Department of Entomology, University of Kentucky, S-225 Agricultural Science Building North, Lexington, Kentucky 40546, U.S.A.

INTRODUCTION. The subfamily Agathidinae contains 20 genera in the New World and 52 worldwide. Species revisions for the New World include those by Muesebeck (1927) on Agathidinae north of Mexico, Marsh (1961) on *Cremnops* Förster north of Mexico, Sharkey (1983, 1986, 1988, 1990) on *Marjoriella* Sharkey, *Pharpa* Sharkey, *Alabagrus* Enderlein, and *Zacremnops* Sharkey and Wharton, and van Achterberg (1990) on *Plesiocoelus* van Achterberg and *Mesocoelus* Schulz. Sharkey (1992) revised the Agathidinae of the world at the tribal level and is preparing a checklist of the Nearctic species north of Mexico. Shenefelt (1970) cataloged the world species, but there have been significant changes in generic concepts (e.g. Sharkey, 1985, 1988, 1992) since then.

BIOLOGY. Members of the Agathidinae are solitary koinobiont endoparasitoids of concealed lepidopterous larvae except for members of the tribe Disophrini which attack free-living larvae. Members of the tribes Agathidini, Earinini, and Eumicrodini attack the first or second instar larvae of their hosts; those of the Disophrini attack later instars, and members of Cremnoptini appear to be able to parasitize any larval stage. In all cases the adult parasitoid emerges after the final instar of the host has spun its cocoon (Nickels et al., 1950; Dondale, 1954; Odebiyi and Oatman, 1972, 1977). Most species are diurnal but many members of the Disophrini are nocturnal with typical pale coloration and enlarged ocelli. A few detailed studies of the biology of species of Agathidinae have been conducted. Most notable among the New World species are the following: Simmonds (1947) on *Cremnops vulgaris* (Cresson) (as *Agathis vulgaris*), a parasitoid of the sugarbeet web-worm (*Loxostege sticticalis* L.); Dondale (1954) on *Bassus dimidiator* Nees (as *Agathis laticinctus* Cresson), a parasitoid of the eye-spotted bud moth *Spilonota ocellana* (Denis and Schiffermueller), an orchard pest; Needham (1955) on *Alabagrus texanus* (Cresson) (as *Bassus texanus*), a parasitoid of *Sylepta silicalis* (Guenée); Balduf (1966) on *Bassus calcaratus* (Cresson) (as *Agathis calcarata*), a parasitoid of *Acrobasis rubrifasciella* Packard; Quednau (1970) on *Agathis pumila* (Ratzeburg), a parasitoid of the larch case-bearer *Coleophora laricella* (Huebner): Odebiyi and Oatman (1972, 1977) on *Agathis gibbosa* (Say) and *Bassus unicoloratus* (Shenefelt) (as *Agathis unicolor*), parasitoids of the potato tuberworm *Phthorimaea operculella* (Zeller); and Hummelen (1974) on *Alabagrus stigma* Brullé (as *Agathis stigmaterus* (Cresson)), a parasitoid of the sugarcane borer *Diatraea saccharalis* (F.). Each of these works gives a detailed account of the immature stages and assesses the value of the parasitoids as natural control agents. Shaw and Huddleston (1991) provide a more thorough review of the biology of the Agathidinae.

BIOLOGICAL CONTROL POTENTIAL. Many members of the Agathidinae are important in the natural control of pest species of Lepidoptera. Some have been used in classical biological control programs. These include *A. pumila* for the larch case-bearer, *C. laricella* (Ryan, 1990), *A. stigma* against the sugarcane borer, *D. saccharalis* (Hummelen, 1974) and *B. unicoloratus* for the potato tuberworm, *P. operculella* (Odebiyi and Oatman, 1977).

COMMON GENERA. The genera *Agathis* Latreille, *Alabagrus*, *Bassus* F., *Coccygidium* de Saussure and *Cremnops* all contain dozens of species in the New World. The genera *Coccygidium* and *Bassus* contain large Neotropical elements that have not been revised. *Dichelosis* Szépligeti and *Liopisa* Enderlein are separated from *Coccygidium* on the basis of characters involving superficial losses. I believe that they will be considered as derived lineages within *Coccygidium* once a cladistic analysis is completed.

DISTRIBUTION. Worldwide. Generally, agathidines are more diverse in humid tropical regions but some genera such as Agathis and *Earinus* Wesmael are exceptional in that they are more speciose in temperate regions. The genera *Coccygidium*, *Agathirsia* Westwood, *Agathis* and *Crassomicrodus* Ashmead have members that prefer desert habitats.

DISTINGUISHING FEATURES. Agathidines can be recognized by the following combination of characters: fore wing M +CU not tubular in basal third or more (Figs 1-18); fore wing RS complete to wing margin and terminating far basad the apex of the wing (Figs 1-12, 14, 16-18), except *Mesocoelus* and *Plesiocoelus* (Figs 13, 15); occipital carina absent (Fig. 24); second submarginal cell of fore wing usually present (90%), when present it is small and quadrate (20%) (Fig. 8) or triangular (80%) (Figs 2-4).

KEY TO THE NEW WORLD GENERA OF THE SUBFAMILY AGATHIDINAE

1. Fore tarsal claw cleft, i.e. with 2 sharp teeth (Figs 19, 20) 2

\- Fore tarsal claw simple or with basal lobe (Figs 21, 22, 23) 7

2(1). Ovipositor sheaths longer than half length of metasoma. Base of fore tarsal claw usually (85%) pectinate, with a comb-like row of teeth (Fig. 20) (Tribe Cremnoptini) 3

\- Ovipositor sheaths shorter than half length of metasoma (Fig. 25). Base of fore tarsal claw not pectinate (Fig. 19) (Tribe Disophrini) 5

3(2). Lateral carina of frons present (Fig. 26) .... ***Cremnops*** Förster

Distribution: widespread in temperate and tropical regions; about 40 New World species described and undescribed. Biology: reared from Pyralidae (Simmonds, 1947). Reference: Marsh (1961).

\- Lateral carina of frons absent (Fig. 27) 4

4(3). Propodeum areolate with fine granulate microsculpture in posterior areolae (Fig. 28).Petiole twice as wide at apex as at base (Fig. 28) .... ***Labagathis*** Enderlein

Distribution: northern South America; one described species, *L. rufoaira* Enderlein. Biology: unknown. Reference: Enderlein (1920).

\- Propodeum areolate without granulate microsculpture (Fig. 29). Petiole about as wide at apex as at base (Fig. 29) .... ***Zacremnops*** Sharkey and Wharton

Distribution: southern U.S.A. to northern Argentina; 4 described species. Biology: reared from Pyralidae. Reference: Sharkey (1990).

5(2). Notauli absent (Fig. 30) .... ***Dichelosus*** Szepligeti

Distribution: Tropical Central and South America; 7 described species. Biology: unknown. References: Shenefelt (1970), Sharkey (1992). This genus is separated from *Coccygidium* only with difficulty.

\- Notauli present, at least weakly (Fig. 31) 6

6(5). Propodeum smooth, lacking areolae .... ***Liopisa*** Enderlein

Distribution: Tropical Central and South America; only one described species. Biology: unknown. Reference: Sharkey (1992). This genus is separated from Coccygidium only with difficulty.

\- Propodeum areolate, at least weakly (Fig. 31) .... ***Coccygidium*** Saussure

Distribution: warm temperate and tropical regions, worldwide; about 100 New World species, mostly undescribed. Biology: reared from Noctuidae and Arctiidae. References: Shenefelt (1970, as *Zelomorpha*), Marsh (1979, as *Zelomorpha*.

7(1). Fore wing (RS+M)a complete (Fig. 9). Notauli absent (Tribe Earinini) 8

\- Fore wing (RS+M)a incomplete (Fig. 3). Notauli variable 9

8(7). Tarsal claw simple (as in Fig. 23) .... ***Sesioctonus*** Viereck

Distribution: Tropical Central and South America; one described species, about 30 undescribed. Biology: unknown. Reference: Sharkey (1992).

\- Tarsal claw with basal lobe (as in Fig. 22) .... ***Earinus*** Wesmael (in part)

Distribution: cool temperate, Holarctic, southern South America; about 40 New World species, 2 in Canada and the U.S.A.; most of those in the southern hemisphere are undescribed. Biology: reared from spring species of Gracillariidae, Tineidae, Noctuidae, Geometridae, Coleophoridae, Tortricidae and Oecophoridae. References: Muesebeck (1927), Sharkey (1992).

9(7). Labio-maxillary complex elongate (galea longer than wide and usually longer than mandible) (Fig. 32) .. 10

\- Labio-maxillary complex of normal dimensions (galea at most as long as wide and usually shorter than mandible) (Fig. 33) 17

10(9). Petiole with granulate microsculpture (Fig. 34) 11

\- Petiole lacking granulate microsculpture 12

11(10). Frons bordered laterally with carina (Fig. 35) .... ***Trachagathis*** Viereck

Distribution: northern South America; 2 described species. Biology: unknown. Reference: Sharkey (1992).

\- Frons not bordered laterally with carina (Fig. 36) 12

12(11). Strong transverse carina between hind coxae present (Fig. 38); hind coxa and metasoma separated by wide sclerite (Fig. 38) .... ***Bassus*** Fabricius (in part)

Distribution: throughout New World; several hundred species, most undescribed. Biology: reared from many families of Lepidoptera, common hosts are Gelechiidae, Noctuidae, Pyralidae, Tortricidae, Tineidae and Coleophoridae (Dondale, 1954; Balduf, 1966; Odebiyi and Oatman, 1977). References: Muesebeck (:l927, key to species. in America north of Mexico), Sharkey (1985). The genus is a diverse polyphyletic assemblage.

14(13). Petiole smooth, lacking sculpture (as in Fig. 39) .... ***Crassomicrodus*** Ashmead (in part)

Distribution: southern Canada to Costa Rica, small genus of about 10 species. Biology: one species has been reared from *Autographa californica* (Speyer) (Noctuidae). Reference: Muesebeck (1927).

\- Petiole with sculpture (Fig. 40) .... ***Agathis*** Latreille (in part)

Distribution: widely distributed in temperate regions worldwide; about 20 New World species, mostly Nearctic. Biology: reared from wide range of lepidopterous families including Coleophoridae, Gelechiidae and Pyralidae (Odebiyi and Oatman, 1972, 1977). Reference: Muesebeck (1927, key to species in America north of Mexico).

15(13). Petiole with sculpture (Fig. 40) .... ***Agathis*** Latreille (in part)

See footnote under couplet 14.

\- Petiole smooth, lacking sculpture (Fig. 39) 16

16(15). Mandible thick; teeth of mandible round in cross section, not flattened (Fig. 41) .....***Agathirsia*** Westwood

Distribution: southern Canada to Mexico; small genus of about 10 species. Biology: unknown. Reference: Muesebeck (192.7).

\- Mandible thin; teeth of mandibles flat in cross section (Fig. 42) .... ***Agathis*** Latreille (in part)

See footnote under couplet 14.

17(9). Fore wing RS2 long and tubular (Fig. 11). Ovipositor barely exerted.......... ***Marjoriella*** Sharkey

Distribution: wet tropical regions of South America; small genus with 4 described species. Biology: unknown. Reference: Sharkey (1983).

\- Fore wing RS2 absent or weak (Figs 3, 4); if present then ovipositor at least half as long as metasoma...18

18(17). Fore wing venation much reduced, second submarginal cell open or absent; vein RS not reaching wing margin (Figs 13, 15). Propodeum granulate (Fig. 44) 19

\- Fore wing venation not greatly reduced, second submarginal cell open or closed, never absent; vein RS usually complete, when incomplete there remains an apical portion meeting wing margin (Figs 1-8). Propodeum variable, usually (90%) not granulate 20

19(18). Tarsal claw with basal lobe (as in Fig. 22). Combined discal + first submarginal cell of fore wing present (Fig.15) .... ***Plesiocoelus*** van Achterberg

Distribution: Tropical Central and South America; one described species, *P. bassicornis* van Achterberg. Biology: unknown. Reference: van Achterberg (1990).

\- Tarsal claw simple, without basal lobe (as in Fig. 23). Combined discal + first submarginal cell of fore wing absent (as in Fig. 13) .... ***Mesocoelus*** Schulz

Distribution: Tropical Central and South America; 2 described species. Biology: reared from Gracillariidae. Reference: van Achterberg (1990).

20(18). Frons margined laterally with carina (Fig. 45) 21

\- Frons not margined laterally with carina (Fig. 36) 22

21(20). Petiole with 2 dorsal carinae that do not meet medially (Fig. 46) .... ***Pharpa*** Sharkey

Distribution: Tropical Central and South America; 3 described species. Biology: unknown. Reference: Sharkey (1986).

\- Petiole usually (90%) without 2 dorsal carinae but if present then these meet medially (Figs 47, 48) .... ***Alabagrus*** Enderlein

Distribution: southern Canada to Northern Argentina, primarily tropical; about 100 species. Biology: Needham (1955), Hummelen (1974). Reared from Pyralidae. Reference: Sharkey (1988, revision, key).

22(20). Gena greatly expanded postero-ventrally (Fig. 49) 23

\- Gena not greatly expanded postero-ventrally (Fig. 33) 24

23(22). Fore wing 2RS and r-m partly fused, thus second submarginal cell appearing petiolate (Fig. 18). Propleuron with distinct bump (Fig. 50) .... ***Zamicrodus*** Viereck

Distribution: Tropical Central and South America; about 15 species, mostly undescribed. Biology: unknown. Reference: Sharkey (1992).

\- Fore wing 2RS and r-m separated by short but distinct 3RSa, thus second submarginal cell not petiolate (Fig. 1). Propleuron lacking distinct bump (as in Fig. 43) .... ***Aerophilus*** Szépligeti

Distribution: Peru (Amazon area); one species, *A. brullei* Szepligeti. Biology: unknown. Reference: Sharkey (1992).

24(22). Notauli present (Fig. 51) 25

\- Notauli absent (Fig. 53) 27

25(24). Petiole sculptured, at least in part (as in Fig. 40) .... ***Bassus*** Fabricius (in part)

Distribution: throughout New World; several hundred New World species, most undescribed. Biology: reared from many families of Lepidoptera, common hosts are Gelechiidae, Noctuidae, Pyralidae, Tortricidae, Tineidae and Coleophoridae (Dondale, 1954; Balduf, 1966;

Odebiyi and Oatman, 1977). References: Muesebeck (1927, key to species in America north of Mexico), Sharkey (1985). The genus is a diverse polyphyletic assemblage.

- Petiole smooth (as in Fig. 39) 26

26(25). Tarsal claw simple, lacking basal lobe (Fig. 23) .... ***Crassomicrodus*** Ashmead (in part)

Distribution: southern Canada to Costa Rica, small genus of about 10 species. Biology: one species has been reared from *Autographa californica* (Speyer) (Noctuidae). Reference: Muesebeck (1927).

- Tarsal claw with basal lobe (in part Figs 21, 22) .... ***Bassus*** Fabricius (in part)

See footnote under couplet 25.

27(24). Specimens from North America .... ***Bassus*** Fabricius (in part)

See footnote under couplet 25.

- Specimens from South America 28

28(27). Hind coxal cavities separated from metasomal foramen by wide sclerite (Fig. 52)***Bassus*** Fabricius (in part)

See footnote under couplet 25.

- Hind coxal cavities not separated from metasomal foramen (Fig. 54) or separated by a very narrow sclerite (as in Fig. 37) .... ***Earinus*** Wesmael (in part)

Distribution: cool temperate, Holarctic, southern South America; about 40 New World species, 2 in Canada and the U.S.A.; most of those in the southern hemisphere are undescribed. Biology: Hosts are members of spring species of Gracillariidae, Tineidae, Noctuidae, Geometridae, Coleophoridae, Tortricidae and Oecophoridae. References: Muesebeck (1927), Sharkey (1992).

REFERENCES

Achterberg, C. van 1990. Revision of the subtribe Mesocoelina Viereck (Hymenoptera: Braconidae). Zoologische Mededelingen Leiden 64: 31-57.

Balduf, W. V. 1966. Life of *Acrobassis rubrifasciella* (Lepidoptera: Phycitidae), its main parasite, *Agathis calcarata* (Hymenoptera: Braconidae), and three hyperparasites (Chalcidoidea). Annals of the Entomological Society of America 59: 1038-1049.

Dondale, C. D. 1954. Biology of *Agathis laticinctus* (Cress.) (Hymenoptera: Braconidae) a parasite of the eye-spotted bud moth in Nova Scotia. The Canadian Entomologist 86: 40-44.

Enderlein, G. 1920 (1918). Zur Kenntnis auβereuropäischer Braconiden. Archive für Naturgeschichte 84A(II): 51-224.

Hummelen, P. J. 1974. Relations between two rice borers in Surinam, *Rupela albinella* (Cr.) and *Diatraea saccharalis* (F.), and their hymenopterous larval parasites. Mededelingen Landbouwhogeschool, Wageningen 74(1): 1-88.

Marsh, P. M. 1961. A taxonomic study of the genus *Cremnops* in America north of Mexico (Hymenoptera, Braconidae). Annals of the Entomological Society of America 54: 851-861.

Muesebeck, C. F. W. 1927. A revision of the parasitic wasps of the subfamily Braconinae occurring in America north of Mexico. Proceedings of the United States National Museum 69: 1-73. Needham, J. G., Jr. 1955. Notes on a leaf-rolling caterpillar and some of its associates. Ecology 36:346-352.

Nickels, C. B., Pierce, W. C., and Pinkney, C. C. 1950. Parasites of the pecan nut casebearer in Texas. Technical Bulletin, United States Department of Agriculture 1011: 1-21.

Odebiyi, J. and Oatman. E. R. 1972. Biology of *Agathis gibbosa* (Hymenoptera: Braconidae), a primary parasite of the potato tuberworm. Annals of the Entomological Society of America 65: 1104-1114.

Odebiyi, J. and Oatman E. R. 1977. Biology of Agathis unicolor (Schrottky) and *Agathis gibbosa* (Say) (Hymenoptera: Braconidae), primary parasites of the potato tuberworm. Hilgardia 45: 123-151.

Quednau, F. W. 1970. Notes on life-history, fecundity, longevity and attack pattern of *Agathis pumila* (Hymenoptera: Braconidae), a parasite of the larch casebearer. The Canadian Entomologist 102: 736-745.

Ryan, R. B. 1990. Evaluation of biological control: Introduced parasites of larch casebearer (Lepidoptera: Coleophoridae) in Oregon. Environmental Entomology 19: 1873-1881.

Sharkey, M. J. 1983. *Marjoriella*, a new neotropical genus of Agathidinae (Braconidae, Hymenoptera). Contributions of the American Entomological Institute 20; 94-100.

Sharkey, M. J. 1985. Notes on the genera *Bassus* Fabricius and *Agathis* Latrielle with a description of *Bassus arthurellus* n. sp. (Hymenoptera: Braconidae). The Canadian Entomologist 117: 1497-1502.

Sharkey, M. J 1986. *Pharpa*, a new genus of neotropical Agathidinae (Hymenoptera: Braconidae) with a discussion of phylogenetic relationships. The Canadian Entomologist 118: 1231-1239.

Sharkey, M. J. 1988. A taxonomic revision of *Alabagrus* (Hymenoptera: Braconidae). Bulletin of the British Museum (Natural History), Entomology 57: 311-437.

Sharkey, M. J. 1990. A revision of *Zacremnops* Sharkey and Wharton (Hymenoptera: Braconidae: Agathidinae). Proceedings of the Entomological Society of Washington. 92: 561-570.

Sharkey, M. J. 1992. Cladistics and tribal classification of the Agathidinae (Hymenoptera: Braconidae). Journal of Natural History 26: 425-447.

Shaw, M. R. and Huddleston, T. 1991. Classification and biology of braconid wasps (Hymenoptera: Braconidae). Handbooks for the Identification of British Insects 7(11): 1-126.

Shenefelt, R. D. 1970. Pars 6. Braconidae 3, Agathidinae. Pp. 307-428. In: Ferrière, Ch. and Vecht, J. van der (Eds.). Hymenopterorum Catalogus (nova editio). Dr. W. Junk, The Hague.

Simmonds, F. J. 1947. The biology of the parasites of *Loxostege sticticalis* L., in North America-*Bracon vulgaris* (Cress.) (Braconidae, Agathinae [sic]). Bulletin of Entomological Research 38: 145-155.

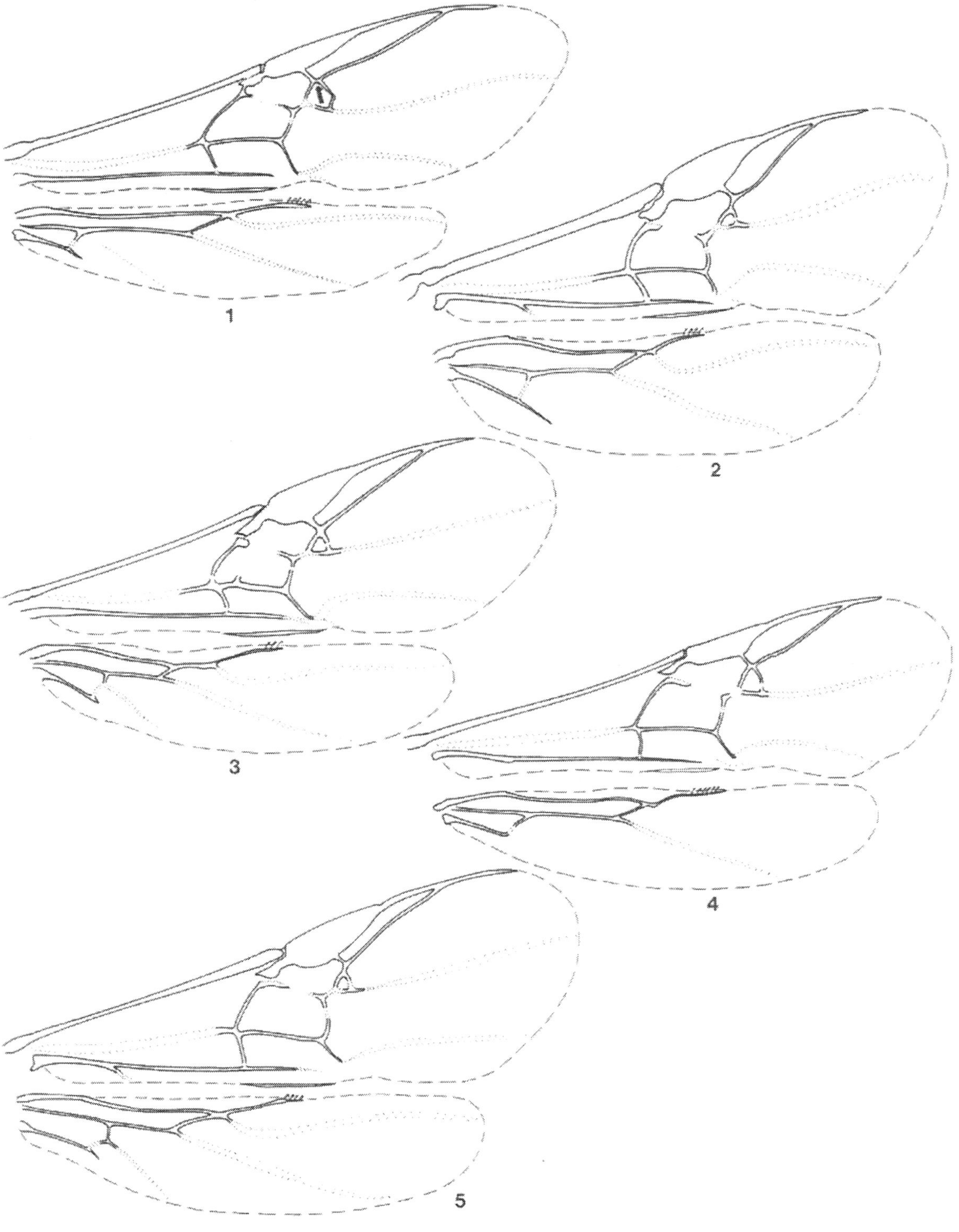

Figs 1-5: Wings of Agathidinae. 1, *Aerophilus brullei* Szepligeti, arrow = 2nd submarginal cell not petiolate; 2, *Agathirsia testacea* Muesebeck, 2nd submarginal cell petiolate: 3, *Agathis malvacearum* Latrielle: 4, *Alabagrus stigma* (Brullé): 5, *Bassus simillimus* (Cresson).

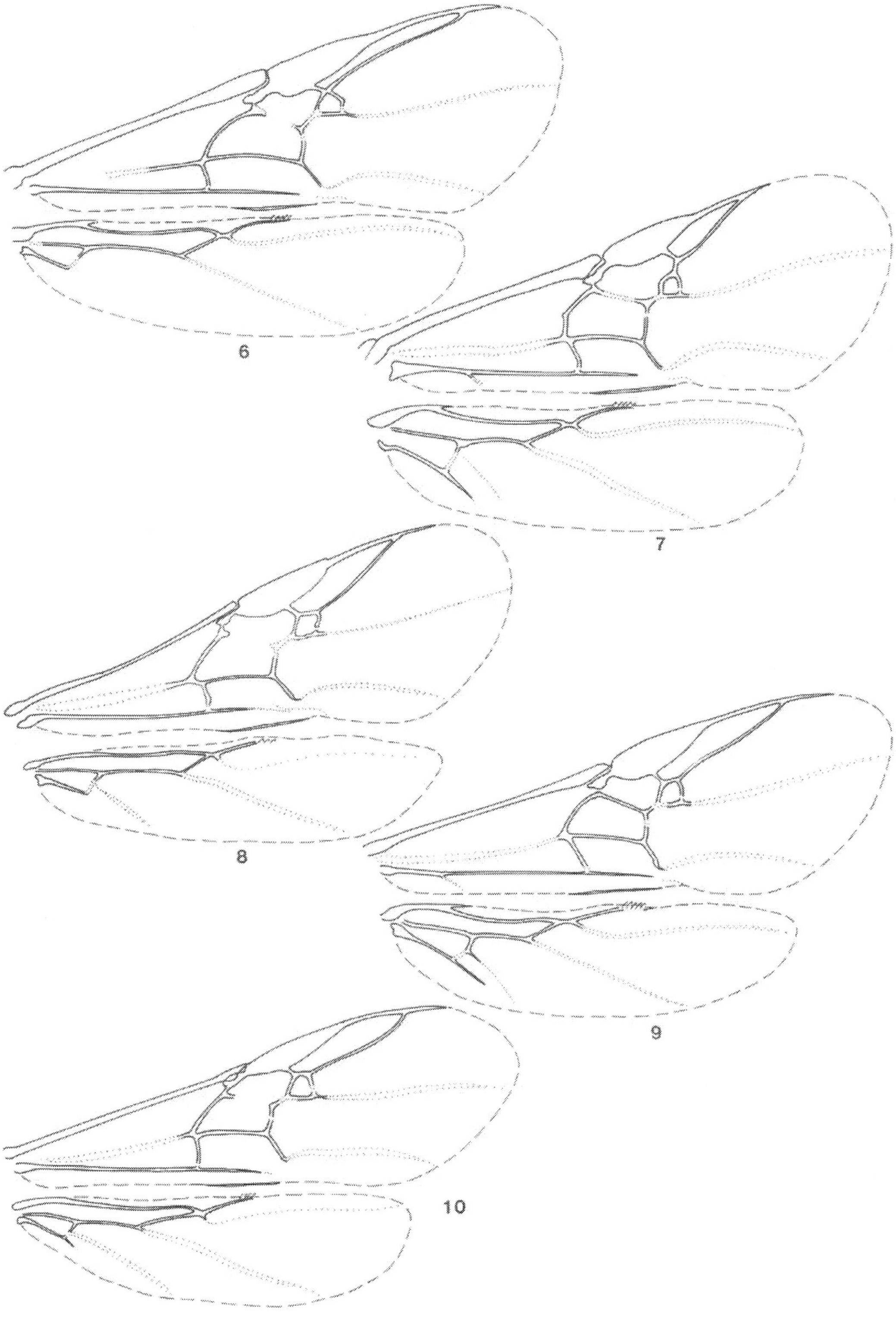

Figs 6-10: Wings of Agathidinae. 6, *Coccygidium* sp.; 7, *Crassomicrodus divisus* (Cresson); 8, *Cremnops desertor* (L.); 9, *Earinus limitaris* (Say); 10, *Labagathis rufoatra* Enderlein.

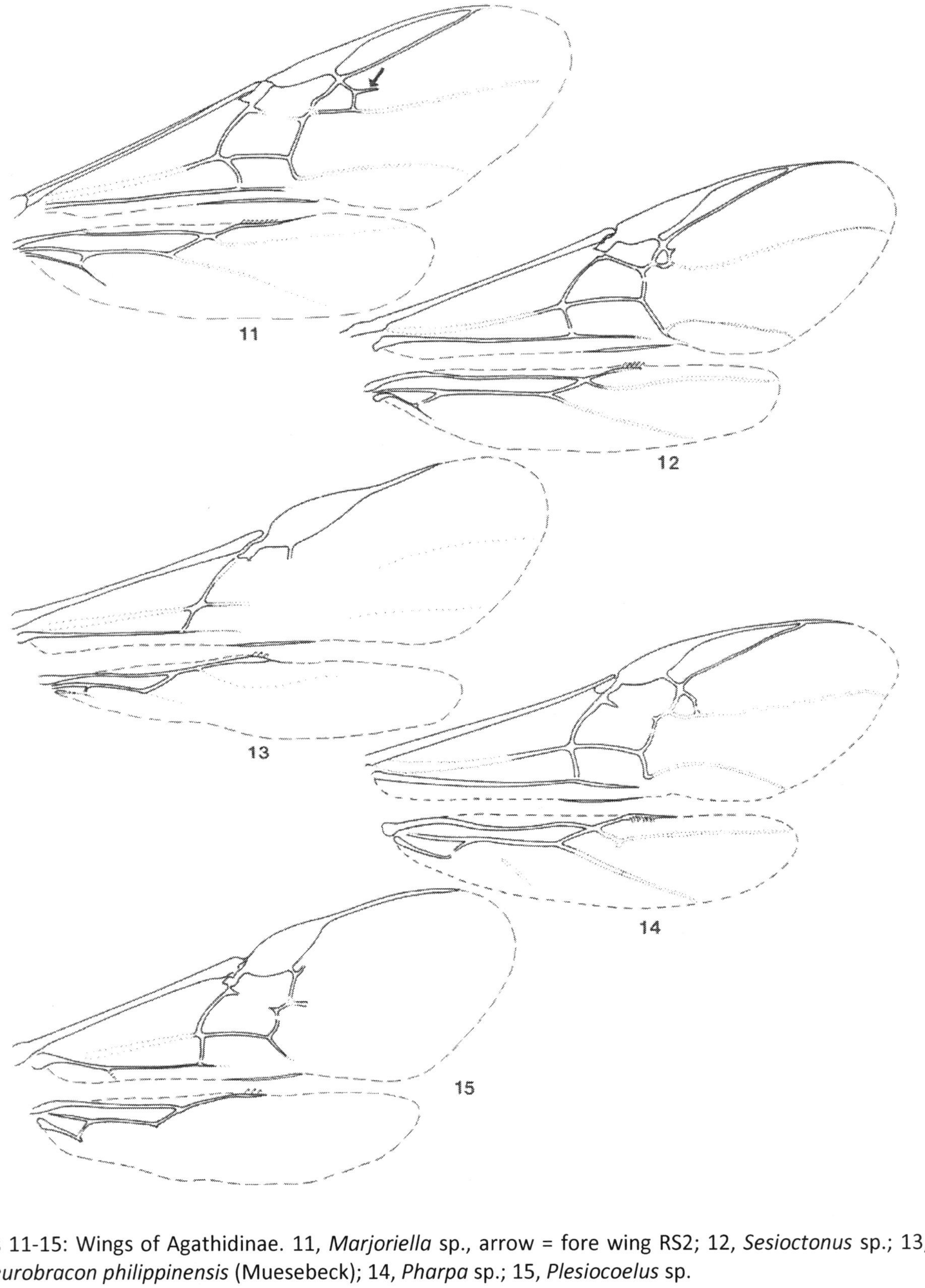

Figs 11-15: Wings of Agathidinae. 11, *Marjoriella* sp., arrow = fore wing RS2; 12, *Sesioctonus* sp.; 13, *Aneurobracon philippinensis* (Muesebeck); 14, *Pharpa* sp.; 15, *Plesiocoelus* sp.

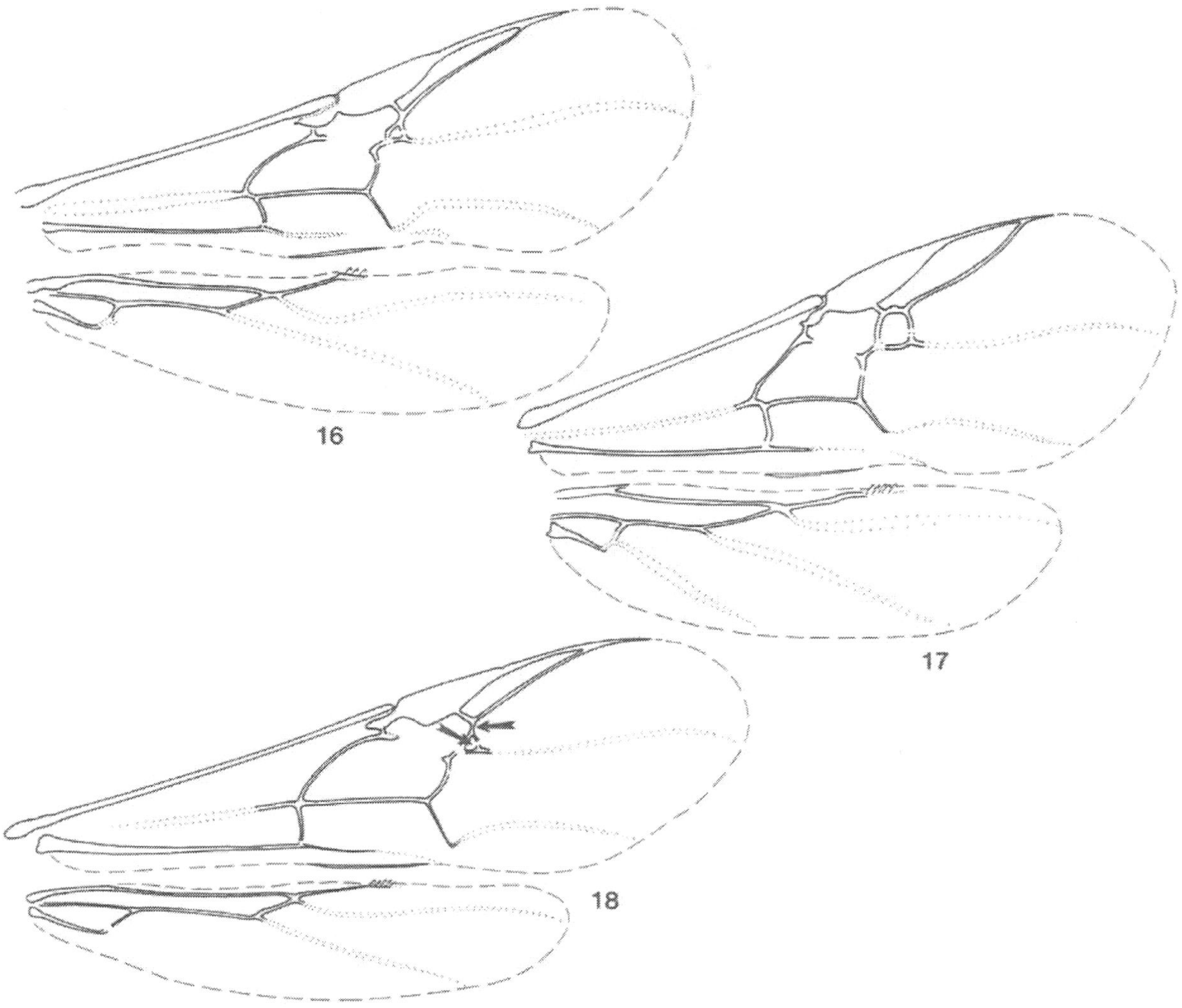

Figs 16-18: Wings of Agathidinae. 16, *Trachagathis* sp.; 17, *Zacremnops* sp.; 18, *Zamicrodus* sp., arrows = petiolate 2nd submarginal cell.

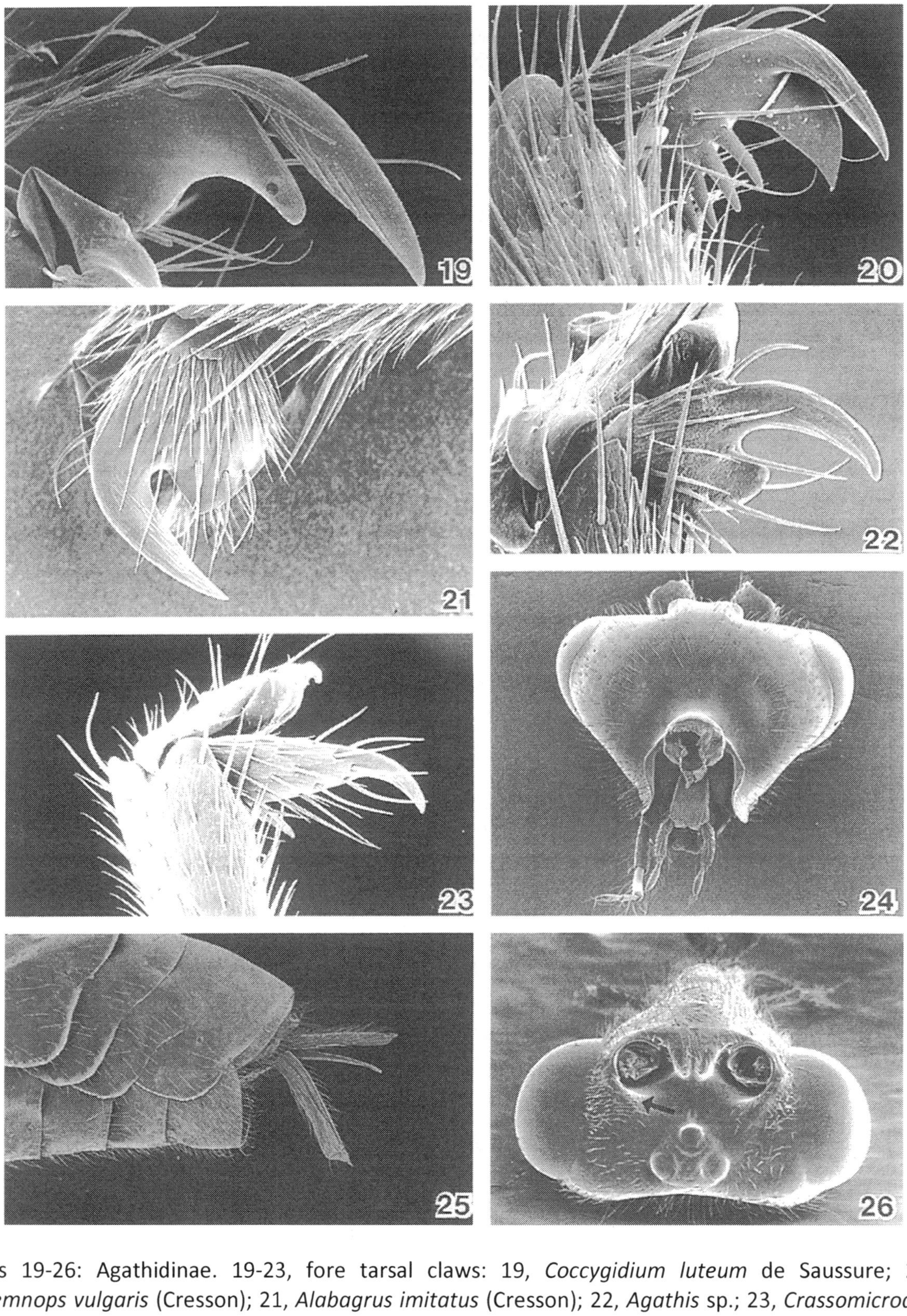

Figs 19-26: Agathidinae. 19-23, fore tarsal claws: 19, *Coccygidium luteum* de Saussure; 20, *Cremnops vulgaris* (Cresson); 21, *Alabagrus imitatus* (Cresson); 22, *Agathis* sp.; 23, *Crassomicrodus fulvescens* (Cresson); 24, head of *Alabagrus texanus* (Cresson), posterior view; 25, apex of metasoma of *Coccygidium luteum* de Saussure, lateral view; 26, head of *Cremnops* sp., dorsal view, arrow = lateral carina of frons.

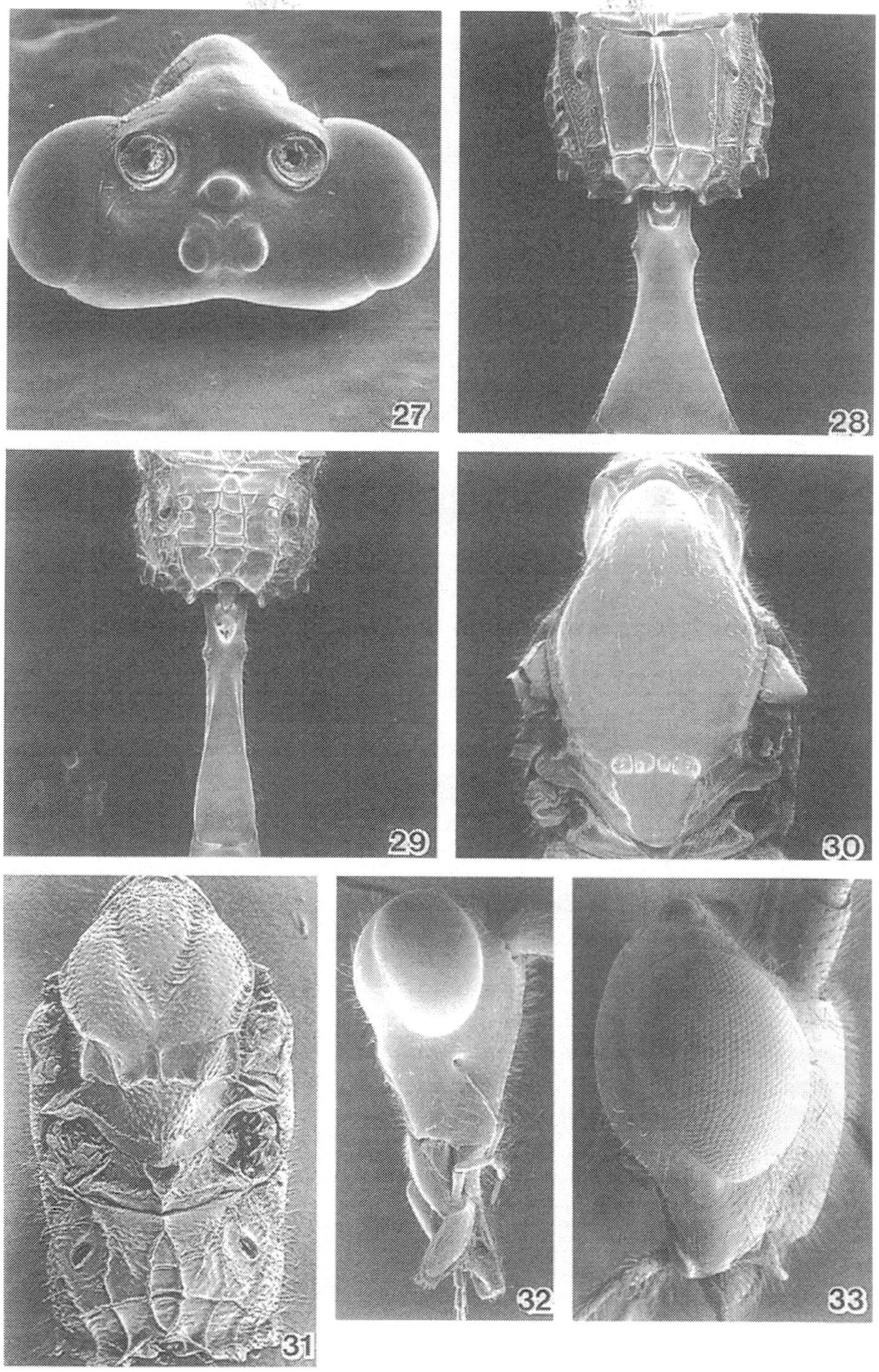

Figs 27-33: Agathidinae. 27, head of *Zacremnops* sp., dorsal view; 28, propodeum and metasomal tergum 1 of *Labagathis rufoatra* Enderlein, dorsal view; 29, propodeum and metasomal tergum 1 of *Zacremnops cressoni* (Cameron), dorsal view; 30, mesosoma of *Dichelosus* sp., dorsal view; 31, mesosoma of *Coccygidium* sp., dorsal view; 32, head of *Agathis* sp., lateral view; 33, head of *Bassus* sp., lateral view.

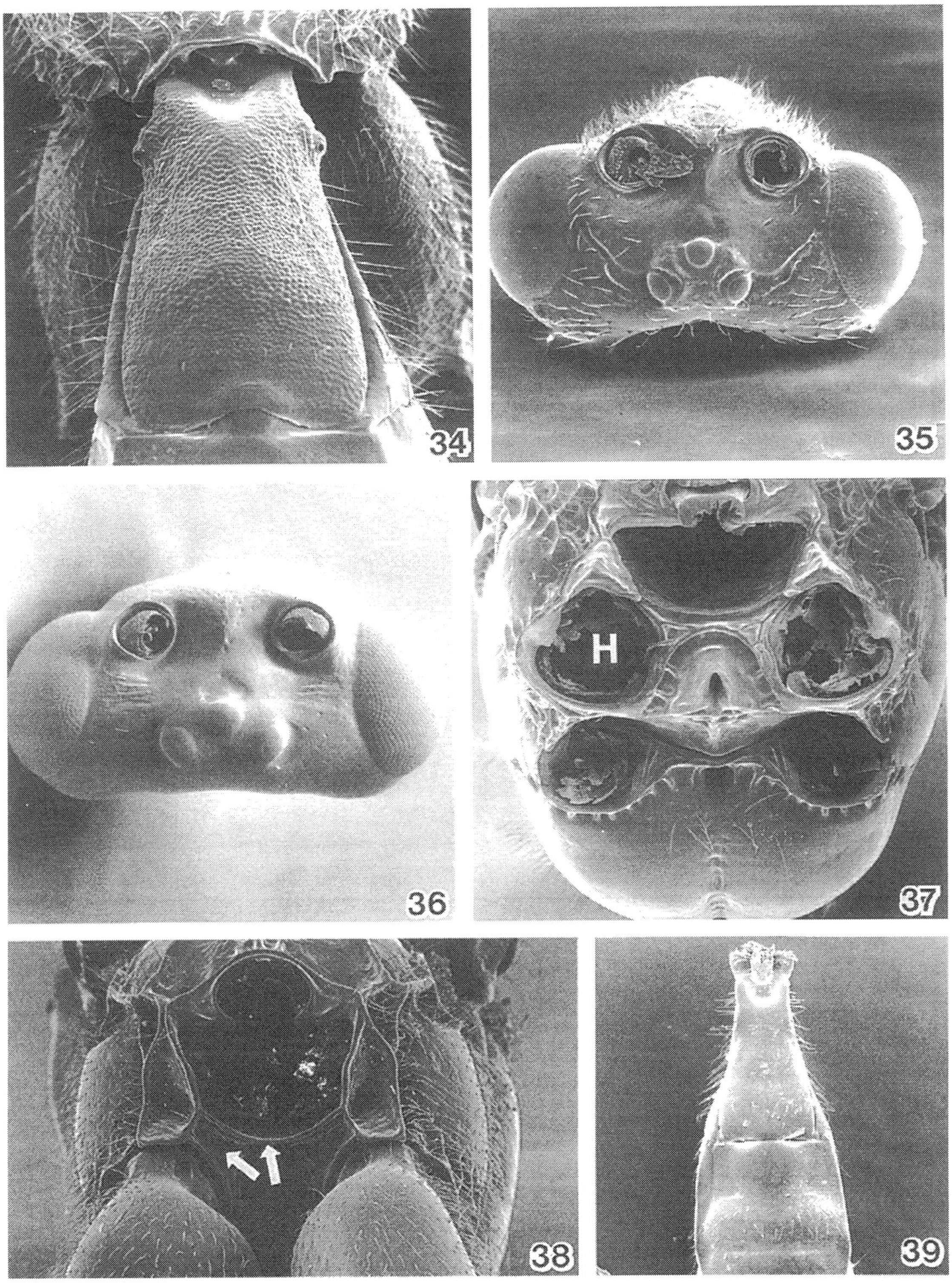

Figs 34-39: Agathidinae. 34, anterior metasoma of *Trachagathis* sp., dorsal view; 35, head of *Trachagathis* sp., dorsal view; 36, head of *Bassus* sp., dorsal view; 37, mesosoma of *Agathis* sp. with legs and metasoma removed, postero-ventral view, H = hind coxal cavity; 38, mesosoma of *Bassus* sp. with metasoma removed, posterior view, arrows = transverse carina between hind coxa and wide sclerite between metasoma and hind coxa: 39, petiole of *Cremnops* sp., dorsal view.

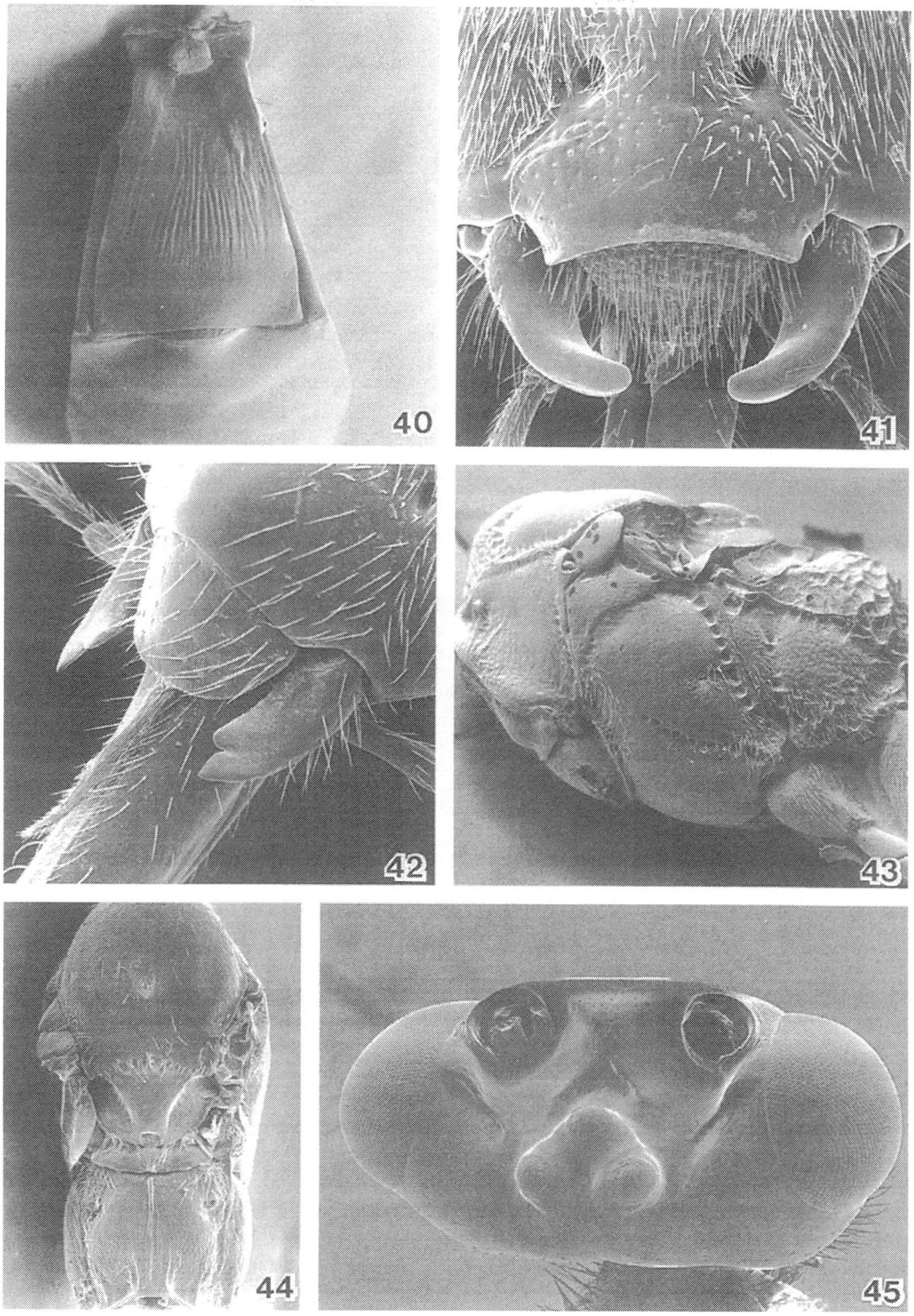

Figs 40-45: Agathidinae. 40, metasoma of *Agathis* sp., dorsal view; 41, head of *Agathirsia nigricauda* (Viereck), anterior view; 42, head of *Agathis* sp., anterior view; 43, mesosoma of *Bassus cingulipes* (Nees), lateral view; 44, mesosoma of *Mesocoelus* sp., dorsal view; 45, head of *Alabagrus texanus* (Cresson), dorsal view.

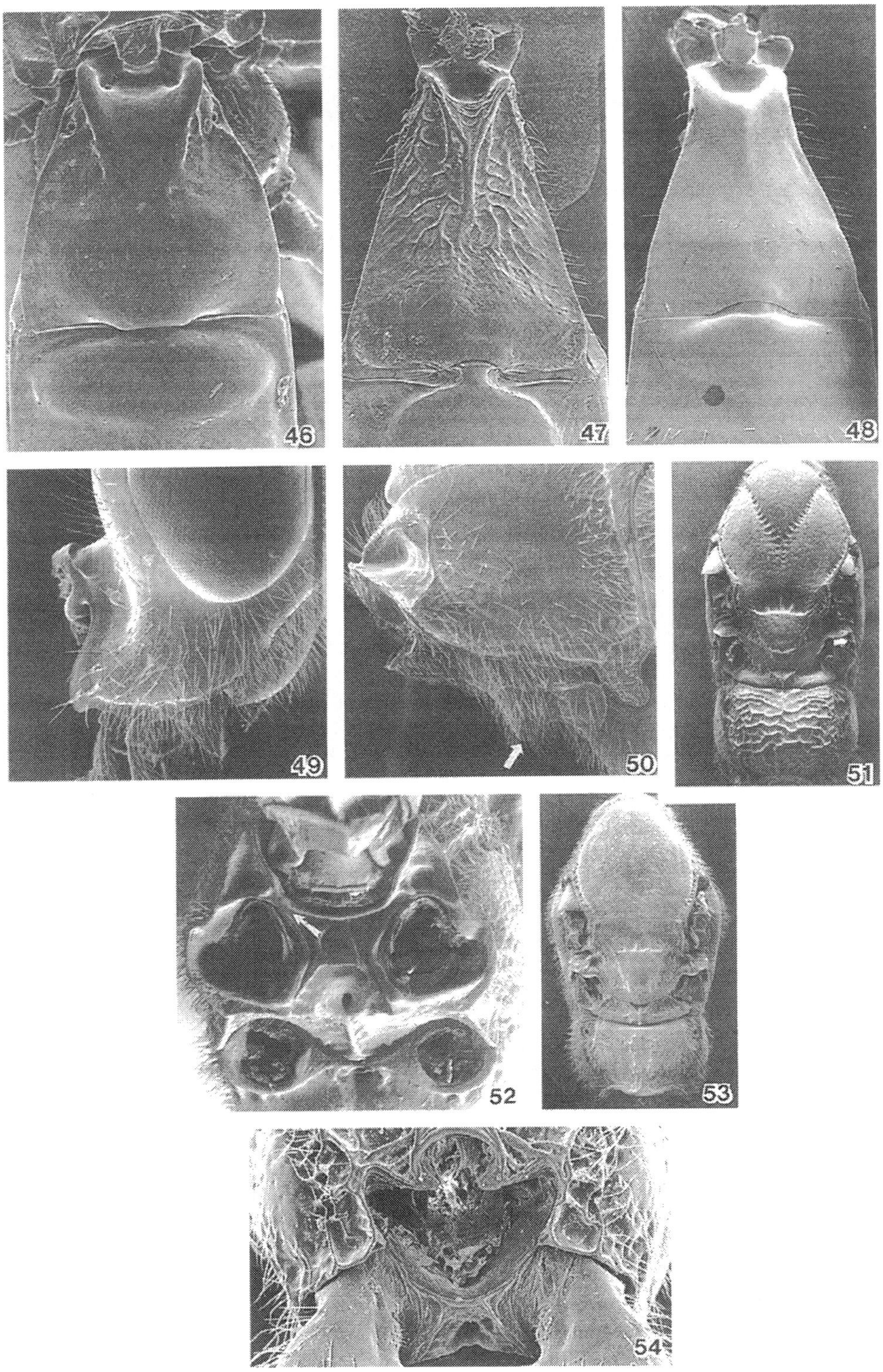

Figs 46-54: Agathidinae. 46-48, anterior portion of metasoma: 46, *Pharpa*. sp., dorsal view; 47, *Alabagrus stigma* (Brullé), dorsal view; 48, *Alabagrus texanus* (Cresson), dorsal view; 49, head of *Zamicrodus* sp., lateral view; 50, anterior mesosoma of *Zamicrodus* sp., lateral view, arrow = bump on propleuron; 51-54, mesosomas:51, *Bassus festivus* (Muesebeck), dorsal view; 52, *Bassus festivus* with legs removed, postero-ventral view, arrow = sclerite between metasoma and hind coxa; 53, *Earinus limitaris* (Say), dorsal view; 54, *Earinus* sp. with metasoma removed, posterior view.

AGATHIDINAE UPDATE

M.J. Sharkey
Department of Entomology, University of Kentucky, Lexington, Kentucky 40546-0091 U.S.A. msharkey@uky.edu

The Agathidinae chapter of the manual is outdated. Sharkey and Chapman (2015) provide an updated key to the Nearctic genera. There are two generic names that are misapplied in their key as explained in Sharkey and Chapman (2016), i.e., *Lytopylus* should be named *Aerophilus*, and *Austroearinus* should be named *Lytopylus* (a senior synonym). The biggest changes at the generic level since the publication of the Manual in 1997 are the redefinition of *Bassus* to refer to a small genus confined to the Old World (Sharkey *et al.,* 2009), and the addition of six new or resurrected genera. Sharkey (2006) published a key to the New World genera of Agathidinae; however, this too is now outdated so there is no reliable key available for Neotropical specimens. Specimens that keyed to *Bassus* in the Manual are now in the following genera: *Aerophilus* Szépligeti, *Amputoearinus* Sharkey, *Aphelagathis* Sharkey, *Gelastagathis* Sharkey, *Neothlipsis* Sharkey, *Pneumagathis* Sharkey, *Smithagathis* Sharkey, and *Therophilus* Wesmael. *Dichelosus* Szépligeti is a junior synonym of *Zelomorpha* (Sharkey and Chapman, in press) and *Liopisa* Enderlein is here synonymized with *Zelomorpha* **new synonym**.

Annotated list of recent papers on New World Agathidinae

Briceño GR. 2003. Taxonomic revision of the Genus *Sesioctonus* Viereck (Hymenoptera: Braconidae: Agathidinae). *Journal of Hymenoptera Research* 12 (2): 238-271.
Twenty-six species are treated, and a key is included; however, the most inclusive key is that of Sulca and Sharkey (2012).

Briceño R, Sharkey MJ. 2000. *Bassus macadamiae* sp. n. (Hymenoptera: Braconidae: Agathidinae), parasitoid of *Ecdytolopha torticornis* (Lepidoptera: Tortricidae) in Macademia nut crops in Central and South America. *Journal of Hymenoptera Research* 9: 99-103.
Bassus macadamiae was later transferred to *Aerophilus* (Sharkey and Chapman, 2016).

Figueroa-de la Rosa JI, Sharkey MJ, Lopez Martinez V. 2005. First records of males and new distribution records for two species of *Agathirsia* Westwood (Hymenoptera: Braconidae). *Entomological News* 116(2): 113-114.

Figueroa JI, Sharkey MJ, Nápoles JR, García JAS, Martinez AM, López-Martínez V, Pineda S. 2011. Revision of the new world genus *Crassomicrodus* Ashmead (Hymenoptera, Braconidae, Agathidinae), with an identification key to species. *Zookeys* 142: 27-75.

Janzen DH, Sharkey MJ, Burns JM. 1998. Parasitization biology of a new species of Braconidae (Hymenoptera) feeding on larvae of Costa Rican dry forest skippers (Lepidoptera: Hesperiidae: Pyrginae). *Journal of Neotropical Lepidoptera* 9: 33-41.
This publication discusses the biology of *Bassus brooksi*, which was transferred to the genus *Pneumagathis* by Sharkey and Chapman (2015).

Leathers J, Sharkey MJ. 2003. Taxonomy and Life History of Costa Rican *Alabagrus* (Hymenoptera: Braconidae), with a key to world species. *Contributions in Science* 497: 1-82.
Several new species of *Alabagrus* are described, and a key to all species is provided. Sharkey and Chapman are currently revising the Costa Rican species of *Alabagrus* with the aid of molecular data, and many of the species included in Leathers and Sharkey (2003) will be redefined.

Lindsay C, Sharkey MJ. 2006. Revision of the genus *Amputoearinus* (Hymenoptera: Braconidae: Agathidinae) with fourteen new species. *Zootaxa* 1329: 1-27.

Fourteen new species are described for this genus, which was proposed by Sharkey (2006) who included two species at that time. A key to all species is given.

Newton B, Sharkey MJ. 2000. A new species of *Bassus* (Hymenoptera: Braconidae: Agathidinae) parasitic on *Sameamultiplicalus*, a natural control agent of waterlettuce. *Florida Entomologist* 83: 284-289.

The biology of a new species *Bassus agathoides* is detailed. It was transferred to the new genus *Neothlipsis* in Sharkey *et al.* (2011).

Pucci T, Sharkey MJ. 2004. A revision of *Agathirsia* Westwood (Hymenoptera: Braconidae: Agathidinae) with notes on mouthpart morphology. *Journal of Hymenoptera Research* 13: 64-107.

Thirty one species of *Agathirsia* are keyed including 23 new species.

Sarmiento CE, Sharkey MJ. 2005. On the status of some species of Braconidae (Hymenoptera) described by J. C. Fabricius and the synonymy of *Dichelosus* Szépligeti with *Coccygidium* De Saussure. *Zootaxa* 1067: 59-68.

Sharkey *et al.* (2006) transferred all New World members of *Coccygidium* to *Zelomorpha*. *Dichelosus* is now a junior synonym of *Zelomorpha*.

Sarmiento CE, Sharkey MJ, Janzen DH. 2004. The first gregarious species of the Agathidinae (Hymenoptera: Braconidae). *Journal of Hymenoptera Research* 13: 295-301.

This species was transferred to *Zelomorpha* (see Sharkey *et al.*, 2006); it was referred to as *Coccygidium gregarium* in the above publication.

Sharkey MJ. 2004. Synopsis of the Agathidinae (Hymenoptera: Braconidae) of North America, north of Mexico. *Proceedings of the Russian Entomological Society* 75: 134-152.

This is a checklist of all Canadian and USA species of Agathidinae with several new combinations. Due to the redefinition of *Bassus* and the erection of many new genera since 2004, this publication is largely outdated.

Sharkey MJ. 2005. A new species of *Agathirsia* Westwood (Hymenoptera: Braconidae) from Mexico. *Zootaxa* 1070: 43-47.

Sharkey MJ. 2006. Two new genera of Agathidinae (Hymenoptera: Braconidae) with a key to the genera of the New World. *Zootaxa* 1185: 37-51.

Due to the redefinition of *Bassus* and the erection of many new genera since 2004, the key is largely outdated. Two new genera are described, *Amputoearinus* Sharkey and *Austroearinus* Sharkey. Recently, Sharkey and Chapman (2016) placed *Austroearinus* in synonymy with *Lytopylus*. Biological information is given for members of both genera.

Sharkey MJ. 2007. Revision of the Neotropical genus *Trachagathis* Viereck (Hymenoptera: Braconidae: Agathidinae). *Journal of the Entomological Society of Ontario* 137: 51-61.

Two new species are described, and the three species belonging to the genus are included in a key.

Sharkey MJ, Chapman EG. 2015. The Nearctic genera of Agathidinae (Hymenoptera: Braconidae) with a phylogenetic analysis, illustrated generic key and the description of three new genera. *Zootaxa* 4000(1): 49-72.

Three new genera are proposed, i.e., *Aphelagathis, Gelastagathis, and Pneumagathis*. A key to the Nearctic genera is included. There are two generic names that are misapplied in this key as explained in Sharkey and Chapman (2016), i.e., *Lytopylus* should be named *Aerophilus*, and *Austroearinus* should be named *Lytopylus* (a senior synonym).

Sharkey MJ, Briceño R. 2005. Review of *Sesioctonus* Viereck (Hymenoptera: Braconidae: Agathidinae) from Colombia with descriptions of five new species. *Zootaxa* 1002: 45-58.
Five new species of *Sesioctonus* are described, and a key for all species of *Sesioctonus* is included.

Sharkey MJ, Laurenne NM, Sharanowski B, Quicke DLJ, Murray D. 2006. Revision of the Agathidinae (Hymenoptera: Braconidae) with comparisons of static and dynamic alignments. *Cladistics* 22: 546-567.
Zelomorpha is resurrected to refer to all New World species formerly referred to as *Coccygidium*, and the tribal level classification of Agathidinae is discussed.

Sharkey MJ, Yu DS, van Noort S, Seltmann K, Penev L. 2009. Revision of the Oriental genera of Agathidinae (Hymenoptera, Braconidae) with an emphasis on Thailand including interactive keys to genera published in three different formats. *ZooKeys*. 21: 19-54.
The genus *Bassus* is redefined to refer to a small group of species found only in the Old World.

Sharkey MJ, Chapman EG. 2016. Revision of *Aerophilus* Szépligeti (Hymenoptera, Braconidae, Agathidinae) from eastern North America, with a key to the Nearctic species. *Contributions in Science* 524: 51-110.
Here the generic concepts *Austroearinus* and *Lytopylus* are changed to *Lytopylus* and *Aerophilus*, respectively. More specifically, *Aerophilus* is shown to be the oldest name for the genus referred to as *Lytopylus* by Sharkey and Chapman (2015). *Lytopylus* is shown to be a senior synonym of *Austroearinus*. All 35 Canadian and USA species of *Aerophilus* are treated in the key; however, only eastern USA species were intensively surveyed, and these were almost exclusively from three localities. Sixteen new species are described.

Sharkey MJ, Chapman EG. In press. Phylogeny of the Agathidinae (Hymenoptera: Braconidae) with a revised tribal classification and the description of a new genus. *Proceedings of the Entomological Society of Washington*.

Sharkey MJ, Chapman EG, Janzen DH, Hallwachs W, Smith MA. 2015. Revision of *Aphelagathis* (Hymenoptera, Braconidae, Agathidinae, Agathidini). *Zootaxa* 4000(1): 73-89.
The eleven species of *Aphelagathis*, including ten new species, are revised, and a key presented. Host data are presented for two species. *Aphelagathis* was first proposed by Sharkey (Sharkey and Chapman 2015).

Sharkey MJ, Parys KA, Clutts SA. 2011. A new genus of Agathidinae with the description of a new species parasitic on *Samea multiplicalis* (Guenee). *Journal of Hymenoptera Research* 23: 43-53.
The genus *Neothlipsis* is described, and ten species are included only one of which is new. The remaining nine species were transferred from the polyphyletic genus *Bassus*.

Sulca L, Sharkey, MJ. 2012. Three new species of *Sesioctonus* Viereck (Hymenoptera: Braconidae: Agathidinae) from Peru. *Zookeys* 196: 1-10.
This includes the most recent and inclusive key to the species of *Sesioctonus*.

Tucker EM, Chapman EG, Sharkey MJ. 2015. A revision of the New World species of *Cremnops* Förster (Hymenoptera: Braconidae: Agathidinae). *Zootaxa* 3916(1): 001-083.
Thirty-three species, including five new species, are treated; an identification key is presented.

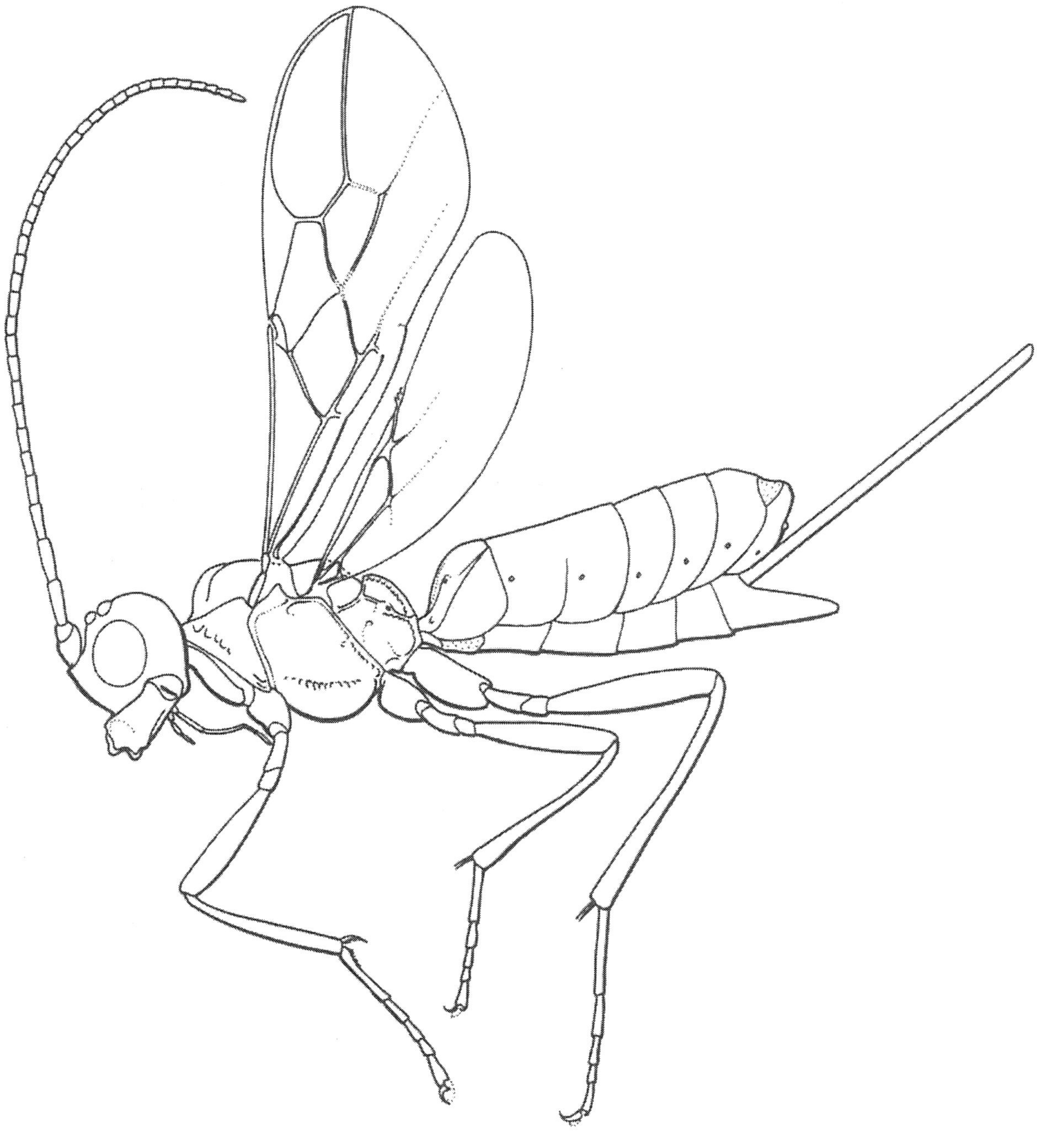

SUBFAMILY ALYSIINAE

R. A. WHARTON

Department of Entomology, Texas A&M University, College Station, Texas 77843, U.S.A.

INTRODUCTION. The Alysiinae is a very large subfamily containing over 1000 described species worldwide. Shaw and Huddleston (1991) have presented an excellent summary of the biology and taxonomic literature, with emphasis on the British fauna. Keys to the Palaearctic species can be found in Tobias (1986) and in the important contributions of Fischer and others cited by Shenefelt (1974) and Shaw and Huddleston (1991). A more recent key to the Alysiini of China is provided by Chen and Wu (1994). Citations given in the key below are therefore limited primarily to the New World fauna.

There are two currently recognized tribes: the Alysiini and the Dacnusini (Shenefelt, 1974). The New World genera of Alysiini were reviewed by Wharton (1980), and the Dacnusini by Riegel (1982) and Wharton (1994). Most of the genera of the New World Alysiini have been revised (Papp, 1969; Shenefelt, 1974; Fischer, 1974b, 1975a, b; Wharton, 1977b, 1980, 1986, 1988b; van Achterberg, 1986), but new species are constantly being discovered, especially in *Dinotrema* Förster (formerly treated as a synonym of *Aspilota* Förster) and *Phaenocarpa* Förster. The New World species of the Dacnusini are badly in need of revision.

Griffiths (1964, 1966a, b, 1967, 1968a, b) completely reclassified the dacnusine parasitoids of the Agromyzidae using biological data to evaluate species concepts and an assessment of shared, derived characters to establish his tribal and generic concepts. Many of the genera of the Alysiini, however, are still based on traditional concepts (Wharton, 1980), and the tribe itself has never been defined as monophyletic. Attempts to redefine larger or more complex groups such as *Dinotrema* / *Aspilota* (van Achterberg and Bin, 1981; Wharton, 1985; van Achterberg, 1988), *Dapsilarthra* Förster (Wharton, 1980; van Achterberg, 1983), *Phaenocarpa* (Wharton, 1980; Fischer, 1990), and *Coelinius* Nees (Wharton and Austin, 1991) have met with only partial success, and much work is still needed to delineate more adequately the genera in these large complexes.

A close relationship between the Alysiinae and Opiinae, the two braconid subfamilies that are exclusively endoparasitic on cyclorrhaphous Diptera, is firmly established (Fischer, in Griffiths, 1964; Čapek, 1970; van Achterberg, 1983; Buckingham and Sharkey, 1988; Quicke and van Achterberg, 1990; Wharton et al., 1992). While the Alysiinae are readily characterized as monophyletic on the basis of the exodont mandibles and complete loss of the occipital carina, opiines are much more difficult to define relative to Alysiinae. For this reason, recognition of Alysiinae renders Opiinae paraphyletic (defined solely by homoplasies or underlying synapomorphies). Available evidence (Čapek, 1970; Edson and Vinson, 1979; Quicke and van Achterberg, 1990) suggests that opiines and alysiines are derived members of the cyclostome lineage within the Braconidae.

BIOLOGY. All alysiines are koinobiont endoparasitoids of cyclorrhaphous Diptera (Griffiths, 1964; Wharton, 1984). They oviposit in host larvae or eggs and emerge from the puparium. As in the Opiinae, the available evidence (limited to a few species) suggests that the molt from first to second instar of the parasitoid coincides with pupariation of the host. Most species are solitary, but several of the species of *Aphaereta* Förster are gregarious.

The Alysiini use a wide variety of cyclorrhaphous hosts, often in moist habitats and decaying, ephemeral substrates. While a few species (e.g. *Aphaereta pallipes* (Say)) appear to have a moderately broad host range, attacking Diptera from several different families, most appear to be fairly specific. For example, nearly all members of the genera *Aspilota* and *Dinotrema* are parasitic on Phoridae, and almost all of the species of *Asobara* Förster attack Drosophilidae. Additional information on hosts is given by Wharton (1984), van Achterberg (1988), and in the excellent paper on comparative searching behavior by Vet and van Alphen (1985). Our ability to generalize about alysiine biology is limited because detailed information is available for relatively few species of Alysiini. Noteworthy among these are *Alysia manducator* (Panzer), *Aphaereta minuta* (Nees), *A. aotea* Hughes and Woolcock, and *A. pallipes* (Say) (see Reznik et al. (1992), Visser (1994), and literature reviews by Shenefelt (1974) and Wharton (1984)). Several species of *Asobara* (Prince, 1976; Vet and van Alphen, 1985; Kraaijeveld, 1994) and two species of *Dapsilarthra* (Keilin and Tate, 1943; Sugimoto et al., 1986, 1987, 1988a-d) are also well known. In the New World, the most intensively studied species are *Aphaereta excavata* McComb, *A. pallipes*, *Alysia alticola* (Say), *Alysia fossulata* Provancher, *Gnathopleura cariosa* (Marsh), *G. ridibunda* (Say), *Alysiasta* caltagironei Wharton, *Phaenocarpa ruficeps* (Nees), *P. raptatrix* Fischer, and *P. brachyptera* Fischer (Wharton, 1984).

Dacnusines, on the other hand, are confined almost exclusively to leaf and stem-mining Agromyzidae, Ephydridae, and Chloropidae, and are relatively better known biologically (at least in the Old World). Griffiths (1964, 1968a) provided extensive data on the hosts of European species, and hypothesized that most dacnusines exhibit a high degree of host specificity. A detailed study on the introduced species *Dacnusa dryas* (Nixon) was published by Guppy et al. (1988).

The subfamily is characterized by the outwardly-directed, non-overlapping, "exodont" mandibles (Fig. 87). These mandibles are used to escape from the host puparium (by pushing outwardly to mimic the function of the ptilinum) and from the substrate in which the puparium may be buried. They are also used by at least a few species to push through the host habitat to locate concealed hosts (Hackman and Meinander, 1979; Wharton, 1984).

BIOLOGICAL CONTROL POTENTIAL. Alysiini in the genera *Gnathopleura* Wharton, *Aphaereta* and Alysia Latreille have been reared and released for the biological control of calliphorids and muscids, but without much success (Legner, 1978; Wharton, 1984). Some dacnusines, however, have been successful in the control of leaf-mining agromyzids (Hendrikse et al., 1980; Drea and Hendrickson, 1986; Guppy et al., 1988), and *Dacnusa sibirica* Telenga is routinely mass reared for this purpose in Europe. The ephydrid *Hydrellia* Robineau-Desvoidy has been used in biological control programs directed against aquatic weeds. Dacnusines in the genera *Chaenusa* Haliday and *Chorebus* Haliday attack *Hydrellia* and must therefore be carefully excluded during the quarantine phase of an importation program.

COMMON GENERA. By far the largest and most commonly encountered genus is *Dinotrema*, with several hundred species in the New World. Less than half of the Nearctic fauna has been described, and virtually nothing has been published on the rich Neotropical *Aspilota /Dinotrema* fauna. Van Achterberg (1988) made an important contribution to our understanding of this group by using the shape of the anterior tentorial pits, previously noted by Fischer (1969a),to divide *Aspilota s. l.* into two large component genera (*Dinotrema* and *Aspilota s. s.*) plus several smaller genera. Even though *Dinotrema* is not clearly defined by apomorphic states (see Wharton, 1985), I treat them here as separate genera to facilitate identification of this large complex. Other commonly encountered genera of small body size are *Asobara*, *Aphaereta*, and *Chorebus*. Of those that are somewhat larger (5-10 mm), *Phaenocarpa* and Alysia are common in the Nearctic Region, while *Gnathopleura* and *Microcrasis* Fischer are frequently collected in the Neotropics.

DISTRIBUTION. The Dacnusini are known primarily from temperate and boreal regions of the Northern Hemisphere. A few, undescribed species of *Coelinius s.l.*, *Dacnusa*, and *Chorebus* have been collected in Mexico, but the distribution of dacnusines in other Latin American countries is essentially unknown. Numerous samples of braconids from several different countries suggest a depauperate Latin American fauna, but this pattern needs verification. The Alysiini are more evenly distributed with numerous species in all biogeographic regions. In the New World, *Aspilota*, *Dinotrema*, and *Aphaereta* are widely distributed throughout both temperate and tropical regions. The next two largest genera, Alysia and *Phaenocarpa*, are essentially temperate and boreal, but have representatives in the high mountains at least throughout Mexico. This pattern is undoubtedly common for most Nearctic groups, but has not yet been documented for the more rarely encountered genera. *Alysiasta*, *Gnathopleura*, and *Microcrasis* are primarily Neotropical, with a few species extending into the Nearctic Region. *Phaenocarpa* also has lowland tropical representatives which fall into distinctive species groups (Wharton, 1994). *Dapsildiastema* Wharton may be endemic to Chile and Argentina, *Vachterbergia* Wharton is South American, and both *Pneumosema* Fischer and *Dinostigma* Fischer are Nearctic. The last four are the only New World endemics.

Of the 41 alysiine genera recorded for the New World, only 19 are known south of Mexico. Further collecting is warranted before much can be said about alysiine distribution patterns in the Neotropics. The paucity of our knowledge to date is reflected by the fact that 15 genera are recorded from Mexico for the first time in this chapter (see key below). Thirteen of these are either southern or northern range extensions.

DISTINGUISHING FEATURES. The exodont condition of outwardly directed, non-overlapping mandibles is found only in a few other, rarely encountered genera outside the Alysiinae. *Exodontiella* Wharton (of the Opiinae) and some species of *Paroligoneurus* Muesebeck (Ichneutinae) are exodont, but readily separated from New World Alysiinae by their distinctive venation patterns (Figs 1 and 2) and the absence of a broad, crenulate suture midventrally on the mesothorax. Some undescribed Hormiinae (= Exothecinae) also have broad, outwardly directed mandibles, but the ones I have studied all overlap slightly when closed. Alysiine mandibles are generally 3-toothed, but several have 4 teeth and a few appear to have either 2 or 5 teeth. The mandibles are used extensively for the identification of alysiines, and it is therefore important to prepare specimens so that these can be seen clearly. All alysiines lack both the occipital and epicnemial carinae, and the longitudinal suture running mid-ventrally along the mesothorax is often broad and crenulate.

KEY TO THE NEW WORLD GENERA OF THE SUBFAMILY ALYSIINAE AND OTHER EXODONT BRACONIDAE

Many couplets use features of the mandible, which are often difficult to see and interpret. Generally, when counting teeth, tooth #2 is considered the longest of the 3 major teeth, as in Figs 72, 74. When additional teeth are present, they may be located either between the first 2 teeth (as in Fig. 73, arrow), between tooth #2 and tooth #3 (as in Fig. 56, arrow), or along the ventral margin basad tooth #3 (as in Fig. 63). Also, when comparing the lengths of the first 2 flagellomeres (e.g. couplet 26), the annellus is not included (see Fig. 67).

1. Wings absent or wings replaced by short stubs (and veins thus absent or irregular, Fig. 3). Rarely collected 64

\- Wings present and fully developed, though venation pattern variable (Figs 4, 5, 11, 14, 18, 25, 26, 32). Species such as *Idiasta hispida* Wharton, in which wings may be greatly reduced but veins are present, can be keyed through both halves of this couplet 2

2(1). Fore wing with marginal cell open distally, RS largely absent, not extending to wing margin as a tubular vein (Fig. 1); m-cu unpigmented, represented only by a crease .... ***Paroligoneurus*** Muesebeck
Distribution: southern U.S.A. (Texas to Florida); 2 described species, both rare. Biology: unknown for exodont species. References: Mason (1991, as *Anaprixia* Mason), Sharkey and Wharton (1994). Belongs in Ichneutinae.

\- Marginal cell completely enclosed, RS extending to wing margin (Figs 2, 4-45) or rarely partially fused to stigma; m-cu distinct, at least partially pigmented 3

3(2). Marginal cell in fore wing exceptionally short, at least twice as high as long (Fig. 2), RS sometimes partially fused with stigma .... ***Exodontiella*** Wharton
Distribution: western U.S.A. and Canada; very rare, 2 described species, 7 specimens known. Biology: unknown. References: Wharton (1977a, description, 1988a, 1994), Quicke and van Achterberg (1990, as Alysiinae). An exodont member of the Opiinae, though there is some disagreement on its subfamilial relationships.

\- Marginal cell normal, maximum length usually greater than height (Figs 4, 5, 25-27), rarely with height subequal to length (Fig. 6) 4

4(3). Fore wing r-m absent; RS and M widely separated (Figs 4-13) (Tribe Dacnusini and one species of Alysiini) 5

\- Fore wing r-m present (Figs 14-46), though very short when RS and M converge and nearly touch each other in some smaller species (compare Figs 14 and 17) (Tribe Alysiini) 17

5(4). Metasoma short and broad, terga 2+3 carapace-like, longitudinally striate and covering most of abdomen (Fig. 47). Wings: Fig. 11 .... ***Symphya*** Förster
Distribution: Alaska and Canada to Mexico; 6 described New World species. Biology: parasitoids of *Phytobia* Lioy (Agromyzidae) boring in the cambium of trees (Griffiths, 1964). References: Rohwer (1914), Griffiths (1964).

- Metasomal terga 2+3 not completely carapace-like, unsculptured in most species; if both terga are sculptured, then metasoma is long and narrow (Figs 49, 50) 6

6(5). Eyes setose (Fig. 48). Fore wing RS+M absent in some species (Fig. 4) .... ***Chaenusa*** Haliday

Distribution: U.S.A. and Canada, also Argentina and Chile; 10 described New World species. Biology: parasitoids of *Hydrellia* (Ephydridae). References: Griffiths (1964), Riegel (1982, including *Chorebidella* Riegel and *Chorebidea* Viereck), situated between original tooth 2 and 3, or basad tooth 3 (Figs 55, 56).

Eyes virtually bare (Figs 58). Fore wing RS+M present (Figs 5, 6) 7

7(6) Pubescence on metapleuron usually in form of dense rosette of radiating hairs around a raised swelling (Figs 51, 52). Sternaulus always present on mesopleuron, usually sculptured (Fig. 51) but represented by a long, narrow, deeply impressed line (Fig. 52) in some species groups. Mandibles generally with 4 teeth, the additional tooth situated between original teeth 2 and 4, or basad tooth 3 (Figs 55, 56). Fore wing with r arising basad midpoint of stigma (Fig. 5). .... ***Chorebus*** Haliday (in part)

Distribution: Canada to Argentina and Chile, but not yet recorded from Caribbean; the largest and most commonly encountered genus in the Dacnusini, but only 18 of the New World species have been described. Biology: primarily parasitoids of Agromyzidae, with one group of species diversifying onto *Hydrellia* (Ephydridae) and at least one species on Psilidae. References: Griffiths (1964), Marsh (1979).

- Pubescence on metapleuron sparse (Fig. 53), or if dense then evenly distributed and directed downwards towards coxa (Fig. 54). Sternaulus present or absent. Mandibles with 3 teeth, or with additional tooth usually situated between original teeth 1 and 2 (Figs 57, 58). Only rarely (*Epimicta* and *Synelix*) with mandibles as above. Origin of r variable, but arising from or distad mid-stigma in one commonly encountered genus (Fig. 6) 8

8(7). Fore wing m-cu arising distad 2RS (Fig. 7) .... ***Exotela*** Forester (in part)

Distribution: Alaska and Canada to Mexico; all New World species are undescribed. Biology: parasitoids of Agromyzidae, especially *Agromyza* Fallén and *Cerodontha* Rondani. References: Griffiths (1964, 1966a, b, 1968b), Marsh (1979).

- Fore wing m-cu arising basad 2RS (Figs 6, 8-10) 9

9(8). First subdiscal cell of fore wing open at apex, 2cu-a absent (Fig. 10). Propodeal spiracle enormously enlarged, its diameter equal to distance between spiracle and anterior margin of propodeum (Fig. 59) .... ***Pneumosema*** Fischer (in part)

Distribution: U.S.A. (Louisiana); known from one specimen. Biology: unknown. Reference: Wharton (1994).

- First subdiscal cell usually closed (2cu-a present). If cell open, spiracle much smaller, often difficult to locate (as in Fig. 54) 10

10(9). Fore wing r arising from or distad middle of stigma; stigma short, thick, usually less than 5 times as long as wide (Fig. 6). Mandible long and narrow, with 4th tooth usually present between tooth 1 and 2 (Figs 57, 58). Female metasoma laterally compressed, at least apically (Figs 49, 50) 11

- Fore wing r usually arising near base of stigma, stigma elongate, often narrow and more than 6 times as long as wide (Figs 8, 9, 12). Mandible of most species 3-toothed; female metasoma dorso-ventrally depressed 12

11(10). Mandible with 4th tooth, when present, located on dorsal side of elongate tooth 2 (Fig. 58) .... ***Coelinius*** Nees

Distribution: Canada to Mexico; 40 described New World species. Biology: parasitoids of Chloropidae. References: Riegel (1982, as *Coelinidea* Viereck), Maetô (1983), Wharton & Austin (1991), Wharton (1994).

- Mandible with 4th tooth located on ventral side of elongate tooth 2 (Fig. 56) ...***Chorebus*** Haliday (in part)

Distribution: Canada to Argentina and Chile, but not yet recorded from Caribbean; very few species key here, most go to couplet 7. Biology: primarily parasitoids of Agromyzidae, with one group of species diversifying onto *Hydrellia* (Ephydridae) and at least one species on Psilidae, References: Griffiths (1964), Marsh (1979).

12(10). First subdiscal cell of fore wing open at apex, 2cu-a absent (Fig. 12). Rarely collected 13

- First subdiscal cell of fore wing closed (2cu-a present) (Figs 8, 9) 14

13(12). Metasomal tergum 2 striate (Fig. 62). Mandible: Fig. 65 .... ***Synelix*** Förster

Distribution: Alaska and British Columbia east to Labrador; Michigan; known from one Holarctic species. Biology: reared from *Nanna* Becker (Scathophagidae). References: Griffiths (1964), Fischer (1976), Wharton (1994).

- Metasomal tergum 2 smooth, polished .... ***Coloneura*** Förster

Distribution: Canada to Mexico; New World species are undescribed. Biology: parasitoids of Agromyzidae. References: Griffiths (1964), van Achterberg (1976).

14(12). Metasomal tergum 2 striate (Fig. 61). Sternaulus sculptured (as in Fig. 51) 15

- Metasomal tergum 2 smooth, polished; if striate (very rare), then sternaulus unsculptured (Fig. 54)......16

15(14). Mandible broad, expanded distally, with 4th tooth along ventral margin (Fig. 63) ....

....**Epimicta** Förster

Distribution: U.S.A. (Texas); only one described species in New World. Biology: unknown. References: Griffiths (1964), Wharton (1994).

- Mandible parallel-sided, with 4th tooth between tooth 1 and 2 (Fig. 64) ....**Laotris** Nixon

Distribution: U.S.A. and Canada; 2 species recorded from New World, neither described, Biology: European species reared from *Cerodontha* (Agromyzidae). References: Griffiths (1964), Wharton (1994).

16(14). Sternaulus of mesopleuron sculptured (as in Fig. 51). Metapleural pubescence sparse. Stigma of fore wing narrow, about 10 times as long as wide (as in Figs 7, 8), never enlarged or wedge shaped..**Exotela** Förster (in part)

Distribution: Alaska and Canada to Mexico; all New World species are undescribed. Biology: parasitoids of Agromyzidae, especially *Agromyza* and *Cerodontha*. References: Griffiths (1964, 1966a, b, 1968b), Marsh (1979).

- Sternaulus usually unsculptured. Metapleural pubescence often dense (Fig. 54). If sternaulus sculptured, then stigma enlarged or wedge shaped..**Dacnusa** Haliday

Distribution: Canada to Mexico; 6 described species recorded from New World, but generic placement of the one species described from Chile needs confirmation. Biology: parasitoids of Agromyzidae and leaf-mining Drosophilidae. References: Griffiths (1964), Marsh (1979).

17(4). First submarginal cell of fore wing confluent with second submarginal cell and/or discal cell (Figs 14-18).18

- First submarginal cell separated from both second submarginal cell and discal cell by distinct veins (Figs 19-46)...22

18(17). Both first and second submarginal cells of fore wing confluent with discal cell, thus forming one large cell similar in appearance to that of ichneumonids (Fig. 14)...................................**Aphaereta** Förster (in part)

Distribution: Mexico to Brazil; one described New World species, other species key to couplet 20. Biology: unknown. Reference: Wharton (1994, as species group of *Aphaereta*).

- Not as above; either 2RS absent (Figs 15, 16) or RS+M absent (Figs 17, 18), but never with both absent 19

19(18). First and second submarginal cells of fore wing confluent, 2RS absent; first submarginal and discal cells separated, RS+M present (Figs 15, 16) 20

- Discal and first submarginal cells confluent, RS+M absent; first and second submarginal cells separated, 2RS present (Figs 17, 18)...21

20(19). First subdiscal cell of fore wing absent; fore wing with r indistinct (Fig. 15). Sternaulus unsculptured ....**Dinostigma** Fischer

Distribution: northeastern U.S.A.; known only from one specimen. Biology: unknown. References: Fischer (1966a), van Achterberg (1988), Wharton (1994).

- First subdiscal cell present; fore wing r distinct (Fig, 16). Sternaulus nearly always sculptured 57

Distribution: Alaska and Canada to Argentina and Chile, including Caribbean; 22 described New World species in addition to those listed under couplet 57. Biology: parasitoids primarily on Phoridae. References: Fischer (1967, as *Synaldis*, 1974a).

21(19). First flagellomere longer than second (Fig. 67). Second submarginal cell of fore wing very short, nearly square, with 2RS and 3RSa subequal (Fig. 17)..**Syncrasis** Förster

Distribution: U.S.A.; one species recorded from New World, rarely collected. Biology: unknown. Reference: Wharton (1980).

- First flagellomere shorter than second (Fig. 68). Second submarginal cell elongate, with 2RS much shorter than 3RSa (Fig. 18)..**Aphaereta** Förster (in part)

Distribution: Cosmopolitan; 16 described New World species. Biology: reared from at least 15 families of Diptera, though most rearing records are from calypterate muscoid flies; includes *A. pallipes*, the most frequently reared and intensively studied alysiine in. North America. References: Wharton (1977b, 1984), Visser (1994).

22(17). Metasomal terga 2 and l or 3 variously sculptured, the sculpture covering at least basal half of tergum 2 (Figs 47, 60-62)..23

- Metasomal terga 2 and 3 unsculptured, or very rarely with individuals of some species with faint strigose sculpture at extreme base of tergum 2 30

23(22). First subdiscal cell of fore wing open, at least 2cu-a absent; fore wing 2CUb arising above middle of first subdiscal cell. Ovipositor not, or only barely, extending beyond apex of metasoma (Fig. 83)...................24

- First subdiscal cell closed, 2cu-a present; vein 2CUb arising from or below middle of first subdiscal cell. Ovipositor distinctly longer, usually extending well beyond apex of metasoma......................................26

24(23). Metasoma usually carapace-like, with only 3 segments visible in dorsal view (Fig. 60); suture separating tergum 2 from tergum 3 crenulate or otherwise sculptured. Wings: Fig. 19 ....**Oenonogastra** Ashmead

Distribution: U.S.A. to Brazil, Caribbean; one described species. Biology: parasitoids of leaf-mining Agromyzidae. References: Stegmaier (1972), Wharton (1980).

- Metasoma not carapace-like, with more than 3 segments visible in dorsal view; suture separating tergum 2 from tergum 3 unsculptured, barely visible 25

25(24). Petiole broad, gradually widening posteriorly (as in Fig. 69), not striate..

.... ***Glyphogaster*** Muesebeck and Walkley

Distribution: northeastern U.S.A. and Canada; 2 species known, one of these undescribed. Biology: originally described as "bred from a tachinid puparium from an unknown geometrid larva" on *Ambrosia artemisiifolia* L.; record needs verification. References: Muesebeck and Walkley (1951), Wharton (1980, redescription).

- Petiole narrow, parallel-sided, striate (Fig. 70) .... ***Dapsildiastema*** Wharton (in part)

Distribution: Chile; 2 described species. Biology: unknown. Reference: Wharton (1994).

26(23). First flagellomere longer than second (Fig. 67) 27

- First flagellomere shorter than second (Figs 68, 84) 28

27(26). Fore wing stigma nearly linear, r arising basad midpoint of stigma (Fig. 21) ....
.... ***Apronopa*** van Achterberg

Distribution: U.S.A. and Canada: 2 species worldwide, North American species undescribed, rarely collected. Biology: unknown. References: van Achterberg (1980), Fischer (1991), Wharton (1994).

- Fore wing stigma short, broad, discrete, r arising at or distad its midpoint (as in Fig. 19) ....
.... ***Symphanes*** Förster

Distribution: southeastern Canada, northeastern U.S.A.; one species described from New World, but its placement in *Symphanes* needs to be re-evaluated (Wharton, 1994). Biology: reared from elm tree slime flux, host unknown. Reference: Wharton (1980).

28(26). Mandible with 4th tooth along ventral border basad tooth 3 (as in Figs 63, 71) ....
.... ***Microcrasis*** Fischer (in part)

Distribution: Costa Rica and Peru; uncommon species group keys here, all species undescribed. Biology: other species have been reared from Tephritidae and Lonchaeidae in fruit. Reference: Wharton (1980).

- Mandible with only 3 teeth (Figs 66, 72) 29

29(28). Mandible strongly excavated apically, tooth I and 2 angled outwardly nearly 90° from surface (Fig. 66). Ovipositor less than twice length of hind tibia .... ***Phasmalysia*** Tobias

Distribution: northern U.S.A. and Canada; 2 described species, one Nearctic, one Palaearctic, both rare. Biology: Palaearctic species reared from Lonchaeidae in a fir log. References: Tobias (1971), Wharton (1980), Fischer (1991).

- Mandible normal; not resembling above. Ovipositor more than twice length of hind tibia. Wings: Fig. 23..
.... ***Phasmidiasta*** Wharton

Distribution: northern U.S.A. and Canada; one described species, rarely collected. Biology: reared from Lonchaeidae. References: Wharton (1980), Fischer (1991).

30(22). First flagellomere shorter than second (Figs 68, 84). Fore wing never with combination of linear stigma and venation pattern as in Fig. 36 31

- First flagellomere equal to or longer than second (Fig. 67). Many of the smaller species with first and second flagellomeres approximately equal in length have fore wing venation as in Fig. 36, and should be taken through this half of the couplet 42

31(30). Scutellum conically elevated (Fig. 75). Wings maculate .... ***Hoplitalysia*** Ashmead

Distribution: southern U.S.A. and Mexico; one described species, Biology: unknown, but one specimen teared from bovine dung. References: Wharton (1980), Fischer (1991).

- Scutellum at most weakly convex. Wings of most species not maculate 32

32(31). First subdiscal cell of fore wing open, 2cu-a and most of 2-1A absent (Fig. 24) 33

- First subdiscal cell closed, 2cu-a present and meeting 2-1A 34

33(32). Either hind wing 1M much longer than M +CU or hind wing cu-a absent (Fig. 24). Ovipositor longer than hind tibia .... ***Asobara*** Förster

Distribution: Cosmopolitan; commonly collected; 8 described New World species. Biology: many species associated with Drosophilidae (Vet and van Alphen, 1985), but Sepsidae and Tephritidae also attacked (Wharton, 1984, 1994). Reference: Papp (1969).

- Hind wing 1M shorter than M+CU, cu-a always present (Fig. 25). Ovipositor shorter than hind tibia ....
.... ***Dapsilarthra*** Förster (in part)

Distribution: U.S.A. and Canada; common, though most specimens will key to couplet 48; one described species in New World keys here. Biology: parasitoids of Agromyzidae. References: Griffiths (1968a, b), Wharton (1980), van Achterberg (1983, as subgenus *Heterolexis* Förster).

34(32). Mandible with 4th tooth along ventral border basad of tooth 3 (as in Figs 63, 71).Fore wing vein 2RS longer than 3RSa (Fig. 22) .... ***Microcrasis*** Fischer (in part)

Distribution: U.S.A. to Brazil and Peru; only 2 described species. Biology: reared from fruit-infesting Lonchaeidae and Tephritidae. References: Costa Lima (1938), Fischer (1975a), Wharton (1980).

- Mandible without 4th tooth along lower border basad of tooth 3 (Figs 72, 74, 80, 84). Relative lengths of 2RS and 3RSa variable 35

35(34). Fore wing stigma long and narrow, with r arising distinctly basad its midpoint (Fig. 27), or stigma so narrow that its midpoint is not discernible 36

- ....F
ore wing stigma broad, discrete, with r arising at or distad its midpoint (Fig. 26) 37

36(35). Fore wing m-cu arising distad 2RS (as in Fig. 27)..***Dapsilarthra*** Förster (in part)

Distribution: Canada to Brazil; only a few species key here, none described. Biology: parasitoids of leaf-mining Tephritidae and Anthomyiidae. References: Griffiths (1968a, b), Wharton (1980), van Achterberg (1983, as *Adelurola* Strand).

- Fore wing m-cu arising basad 2RS (as in Fig. 30)..***Phaenocarpa*** Förster (in part)

Distribution: U.S.A. to Brazil; only 2 described and a few undescribed species key here, most species key to couplet 41. Biology: unknown for the species that key here, but at least one species associated with mushrooms. Reference: Fischer (1974b).

37(35). Border of mandible between tooth 1 and tooth 2 swollen to form a 4th tooth (Fig. 73). Wings entirely dark or dark with yellow spot around stigma. Fore wing 2RS equal to, or longer than, 3RSa (Fig. 28).........
....***Gnathopleura*** Wharton (in part)

Distribution: central and eastern U.S.A. to Argentina, including Caribbean, commonly encountered; 12 described species. Biology: reared from Calliphoridae, Muscidae, and Sarcophagidae usually in association with dung and carrion (Marsh, 1968; Wharton, 1984). References: Fischer (1975a), Wharton (1980).

- Lacking above combination of characters. Mandible often with deep cleft between tooth 1 and 2 (Figs 72, 80, 84), only rarely with a flange between tooth 1 and 2 swollen to form a weak, knob-like 4th tooth (as in Fig. 74). Wings of most species hyaline or very weakly infuscate. Fore wing 3RSa often longer than 2RS (Figs 29, 30) 38

38(37). Fore wing 2CUb arising near or below middle of distal border of first subdiscal cell; 2CUa strongly deflected posteriorly (Fig. 31)...39

- Fore wing 2CUb arising well above middle of first subdiscal cell, often directly in line with 1CU; 2CUa absent, or present and at most weakly deflected posteriorly (Figs 29, 30, 42)...41

39(38). Anterior tentorial pit greatly enlarged, extending nearly to compound eye (Figs 87, 88)............................
....***Idiasta delicata*** (Papp)

Distribution: Colombia to Peru and Argentina. Generic placement uncertain (Wharton, 1980). Biology: unknown. References: Papp (1969), Fischer (1975a).

- Anterior tentorial pit relatively small, extending less than half distance to eye (Fig. 86)..........................40

40(39). Ovipositor sheath of female very sparsely setose, the setae separated from each other by a distance equal to or greater than their length. Dorsal carinae of petiole widely separated, parallel or slightly diverging from each other at mid-length. Wings: Fig. 26 ....***Alysiasta*** Wharton (in part)

Distribution: U.S.A. to Panama; 4 described species. Biology: 2 species reared from dung-breeding flies, one of these from Sarcophagidae (Wharton, 1984). References: Wharton (1980, description and key), Fischer (1991).

- Ovipositor sheath of female densely setose, the setae separated from each other by less than half their length. Dorsal carinae of petiole strongly converging basally, often uniting to form a median longitudinal carina. Wings: Fig. 31...***Idiasta*** Förster (in part)

Distribution: Alaska and Canada to Mexico; 8 described species in New World. Biology: largely unknown; two species have been reared from *Lispe* (Muscidae) (Wharton, 1984). Reference: Wharton (1980, key to Nearctic species).

41(38). Hind wing m-cu extending more than half way to wing margin as a heavily pigmented vein. Mandible with deep cleft between tooth 1 and tooth 2, without dorsal knob on tooth 2 (Fig. 84). Fore wing 3RSb ending in Rl distinctly basad wing tip (Fig. 26) ....***Alysiasta*** Wharton (in part)

Distribution: U.S.A. to Panama; 4 described species. Biology: 2 species reared from dung-breeding flies, one of these from Sarcophagidae (Wharton, 1984). References: Wharton (1980, description and key), Fischer (1991).

- Hind wing m-cu weak to absent, rarely extending more than half way to wing margin. Fore wing vein 3RSb of most species ending at wing tip (Fig. 29). Mandible in some species with flange or tooth between tooth 1 and 2 (Fig. 74), thus lacking deep cleft..............................***Phaenocarpa*** Förster (in part)

Distribution: Cosmopolitan; about 80 described species in New World. After *Dinotrema*, this is the largest genus in the New World. Biology: reared from hosts representing at least 9 different families of cyclorrhaphous Diptera, most frequently recorded from Anthomyiidae and Scathophagidae (Shenefelt, 1974; Wharton, 1984); often associated with dung or fungi (Vet and van Alphen, 1985). References: Papp (1969), Fischer (1974b, 1975b, 1990). Males of the *Idiasta hispida* will also key here but are readily identified by the short wings with unusual venation (Wharton, 1980).

42(30). Mandible with 4 or 5 teeth, the extra teeth located along the lower border basad tooth 3 (Figs 63, 71). Sternaulus unsculptured (as in Fig. 54) 43

- Mandible usually with 3 teeth (Figs 80-82); 4th tooth, if present, located between tooth 1 and tooth 2 (as in Figs 73, 74), never along lower border. Sternaulus usually sculptured (the presence of a sculptured sternaulus is a particularly useful character for small species whose mandibles are difficult to see)........44

43(42). First subdiscal cell of fore wing closed; r arising basad middle of long, narrow stigma (Fig. 32) ....

....***Anisocyrta*** Förster

Distribution: Alaska and Canada to Mexico; 4 described species in New World. Biology: collected from mushrooms, hosts unknown. References: Wharton (1980), van Achterberg (1986).

- First subdiscal of fore wing cell open at apex; r of most species arising at or distad middle of stigma (Fig. 33) ....***Pentapleura*** Förster

Distribution: Alaska and Canada to Mexico, also Ecuador and Peru; commonly encountered; 4 described species in New World. Biology: reared from Sphaeroceridae (Wharton, 1984). Reference: Wharton (1980).

44(42). Border of mandible between tooth 1 and tooth 2 swollen to form a 4th tooth, never deeply indented (Fig. 73). Fore wing 2CU arising near upper edge of first subdiscal cell (Fig. 28). Large, brightly colored species with darkened wings ....***Gnathopleura*** Wharton (in part)

Distribution: central and eastern U.S.A. to Argentina, including Caribbean; commonly encountered; 12 described New World species. Biology: reared from Calliphoridae, Muscidae, and Sarcophagidae usually in association with dung and carrion (Marsh, 1968; Wharton, 1984). References: Fischer (1975a), Wharton (1980).

- Border of mandible between tooth 1 and tooth 2 deeply indented; if apparently swollen to form a 4th tooth (Fig. 74), then 2CU with distal abscissa (2CUb) arising from or below middle of first subdiscal cell. Body and wing color variable, but rarely as above, except in *Ilatha* 45

45(44). First subdiscal cell of fore wing open; at least at apex (2cu-a absent) (Figs 20, 25, 45) 46. First subdiscal cell closed; 2cu-a present 50

46(45). Propodeal spiracle greatly enlarged, diameter of spiracle greater than half distance between spiracle and anterior border of propodeum (Fig. 59) 47

- Propodeal spiracle normal (small); diameter of spiracle much less than half distance between spiracle and anterior border of propodeum (as in Fig. 54) 48

47(46). Mandible small, less than half length of head (Fig. 77) ....***Dinotrema*** Förster (in part)

Distribution: Canada to Mexico; only 2-3 described species key here. Biology: unknown, though probably on Phoridae. References: Fischer (1969a), Wharton (1980).

- Mandible larger, more than half length of head (Fig. 78). Wings: Fig. 45..
....***Pneumosema*** Fischer (in part)

Distribution: eastern U.S.A. and 'southeastern Canada; 2 species known, only one of which keys here. Biology: unknown. References: Fischer (1966b), Wharton (1994).

48(46). Fore wing 2CU arising directly in line with 1CU, thus 2CU not divided into 2CUa and 2CUb by 2cu-a (Fig. 25) ....***Dapsilarthra*** Förster (in part)

Distribution: Canada to Costa Rica; common in Nearctic Region, though only one described species recorded from New World. Biology: parasitoids of leaf-mining Agromyzidae and possibly Anthomyiidae. References: Griffiths (1968a, b) and Wharton (1980, as *balteaia* group); van Achterberg (1983, as subgenus *Heterolexis*).

- Fore wing 2CUa present and posteriorly-directed, forming a distinct angle with the distally-directed 2CUb (Fig. 20)..49

49(48). Petiole nearly parallel-sided, about twice longer than apical width (Fig. 70). Fore wing m-cu arising basad 2RS, (RS+M)b present (Fig. 20) ....***Dapsildiastema*** Wharton (in part)

Distribution: Chile and Argentina (Andean); 2 described species, rarely collected. Biology: unknown. Reference: Wharton (1994).

- Petiole short, broad, not parallel-sided, length about equal to apical width, at most 1.2 times longer than apical width (about as in Fig. 69). Fore wing m-cu arising distad 2RS (as in Fig. 23), rarely interstitial, with (RS+M)b absent ....***Alysia*** Latreille (in part)

Distribution: Holarctic; 2 described species from western and northern U.S.A. and Canada key here. Biology: unknown. Reference: Wharton (1988b).

50(45). Mandibular tooth 3 short, broad, scoop-like; mandibular teeth separated from base by well-defined, distally curving keel (Fig. 79). Fore wing m-cu arising distad RS, (RS+M)b absent; 3RSa usually much longer than 2RS (Fig. 37) ....***Orthostigma*** Ratzeburg

Distribution: Alaska and Canada to Costa Rica, Caribbean; 9 described species in New World. Biology: parasitoids of Phoridae often associated with fungi (van Achterberg, 1988). References: Fischer (19.69a), Wharton (1980).

- Mandible never shaped as in Fig. 79; mandibular teeth rarely separated from base by well-developed, distally curving keel; mandibular tooth 3 only rarely expanded and scoop-like. Fore wing variable, though often as above (e.g. *Dinotrema* and *Aspilota*, Fig. 36) 51

51(50). Fore wing 2CU arising directly in line with 1CU or nearly so; 2CUa absent or, if present, distinctly shorter than 2cu-a (Figs 27, 38, 41) 52

- Fore wing vein 2CUa present, posteriorly-directed and equal to or longer than 2cu-a; 2CUb thus arising at or below middle of distal border of first subdiscal cell (Figs 34, 39, 40, 44) 58

52(51). Head strongly flattened antero-posteriorly (Fig. 76). Wings: Fig. 41........................***Vachterbergia*** Wharton

Distribution: Venezuela and Ecuador; 2 known species, rarely collected. Biology: unknown. Reference: Wharton (1994).

\- Head rounded (Figs 77, 84)..53

53(52). Mandibular tooth 2 long, as long as apical width of mandible (Fig. 81) ....***Alloea*** Haliday

Distribution: U.S.A. and Canada; only one species described from New World. Biology: parasitoids of Lonchopteridae. References: Fischer (1966c), Wharton (1980), Wharton and Chou (1985), van Achterberg (1988).

\- Mandibular tooth 2 small to minute, about half as long as apical width of mandible (Figs 72, 82) 54

54(53). Fore wing r arising from or distad middle of broad stigma (Figs 38, 42)..55

\- Fore wing r arising distinctly basad narrowly elongate (or sometimes indistinct) stigma (Fig. 36)...........56

55(54). Fore wing 2RS equal to or longer than 3RSa; stigma of normal thickness throughout (Fig. 38) ....***Cratospila*** Förster

Distribution: Alaska and Canada to Brazil, including Caribbean; common, but only one species described from New World. Biology: unknown. Reference: Wharton (1980).

\- Fore wing 2RS distinctly shorter than 3RSa; portion of stigma basad r usually distinctly narrowed; stigma of male often split by hyaline streak (Fig. 42) ....***Phaenocarpa*** (in part)

Distribution: Mexico to Brazil; a moderately common species group keys here, but only 2 of the species are described. Biology: unknown. Reference: Wharton (1994, description as *cratomorpha* species group).

56(54). Fore wing m-cu nearly always arising basad 2RS, (RS+M)b present (Fig. 43). If m-cu distad 2RS, antenna twice as long as body ....***Dapsilarthra*** Förster (in part)

Distribution: Canada to Mexico; commonly collected; 2 described species in New World. Biology: primarily parasitoids of leaf-mining Agromyzidae. References: Griffiths (1968a, b), Wharton (1980, 1984), van Achterberg (1983).

\- Fore wing m-cu arising distad 2RS, (RS+M)b absent (Fig. 36). Antenna never twice as long as body. Extremely common 57

57(20, 56, 62). Anterior tentorial pit extending to eye (Fig. 85)..***Aspilota*** Förster *s.s.*

Distribution: Cosmopolitan; common, about 15 described New World species. Biology: primarily parasitoids of Phoridae and Platypezidae, often associated with fungi (Fischer, 1974a; van Achterberg and Bin, 1981, Wharton, 1984). References to restricted concept: Fischer (1969e, 1970, keys to Nearctic species), Wharton (1985), van Achterberg (1988).

\- Anterior tentorial pit not extending to eye (Fig. 86) ....***Dinotrema*** Förster (in part)

Distribution: Cosmopolitan; the most commonly collected genus of Alysiinae; about 40 described New World species, most species undescribed. Biology: primarily parasitoids of Phoridae and Platypezidae often associated with fungi (Fischer, 1974a; van Achterberg and Bin, 1981, Wharton, 1984). References: Fischer (1969a-d, key to Nearctic species as *Aspilota*), van Achterberg and Bin (1981, as valid genus), Wharton (1985), van Achterberg (1988, including key to genera of *Aspilota* group).

58(51). Sternaulus unsculptured (as in Fig. 54). Fore wing stigma discrete, fairly broad 59

\- Sternaulus sculptured (as in Fig. 51); if nearly smooth, then stigma narrow, linear, not easily distinguished distally from vein Rl (Fig. 36) 60

59(58). More than 70 flagellomeres, Wing membrane black, yellow or both. Tropical species, often brightly colored ....***Ilatha*** Fischer

Distribution: Mexico to northern Argentina and Peru, not recorded from Chile or Caribbean; 5 described New World species. Biology: unknown. References: Fischer (1975a), Wharton (1980).

\- Less than 50 flagellomeres. Wing membrane hyaline. Temperate species, never brightly patterned.......... ....***Mesocrina*** Förster

Distribution: Canada to Mexico: relatively uncommon, only one species described from New World. Biology: reared from mushrooms, host unknown. References: Wharton (1980, as *Pseudomesocrina*, van Achterberg (1983, as subgenus of *Dapsilarthra*).

60(58). Fore wing m-cu arising basad 2RS, (RS+M)b present (as in Figs 31, 43)..61

\- Fore wing m-cu arising distad 2RS, (RS+M)b absent (Figs 36, 44) 62

61(60). Fore wing vein r arising basad middle of long, narrow stigma (Fig. 43) ....***Dapsilarthra*** Förster (in part)

Distribution: Canada to Mexico; commonly collected; 2 described species in New World. Biology: primarily parasitoids of leaf-mining Agromyzidae. References: Griffiths (1968a, b), Wharton (1980, 1984), van Achterberg (1983).

\- Fore wing with r arising at or distad middle of short, broad stigma (Fig. 34) ....***Alysia*** Latreille (in part)

Distribution: Canada to Mexico; commonly collected, about 15 described species from New World key here. Biology: reared from leaf-mining Anthomyiidae and Tephritidae, carrion-feeding Calliphoridae, Muscidae and Sarcophagidae, mushroom-infesting Anthomyiidae, Heleomyzidae, and Muscidae, and Scathophagidae attacking flower heads, stems, bulbs, roots, and rhizomes. References: Hackman and Meinander (1979), Wharton (1984, 1986, 1988b).

62(60). Fore wing stigma narrow, linear, not easily distinguished from R1 distally (Fig. 36); 2RS shorter than 3RSa. Mandible usually unsculptured and sparsely setose medially (Fig. 82). Males of the few species with exceptionally enlarged stigma (as in Fig. 46) can be most easily recognized by the sparsely setose mandibles..57

- Stigma discrete, usually fairly broad; 2RS usually equal to or longer than 3RSa. Mandible usually sculptured and moderately setose medially (as in Figs 74, 80) 63

63(62). Fore wing r arising from or basad middle of stigma (Fig. 44) ....***Tanycarpa*** Förster

Distribution: Alaska and Canada to Brazil and Peru; 4 described species in New World. Biology: 2 species have been reared on Drosophilidae (Vet and van Alphen, 1985). Reference: Wharton (1980).

- Fore wing r arising distad middle of stigma (Fig. 34)..***Alysia*** Latreille (in part)

Distribution: Canada to Mexico; commonly collected; about 15 described species from New World key here. Biology: reared from leaf-mining Anthomyiidae and Tephritidae, carrion-feeding Calliphoridae, Muscidae and Sarcophagidae, mushroom-infesting Anthomyiidae, Heleomyzidae, and Muscidae, and Scathophagidae attacking flower heads, stems, bulbs, roots, and rhizomes. References: Hackman and Meinander (1979), Wharton (1984, 1986, 19S8b).

64(1). Second fagellomere longer than first (Figs 68, 84) 65

- Second flagellomere equal to or shorter than first (Fig. 67)...67

65(64). Scutellum conically elevated (Fig. 75). Wings of male reduced to petiolate knobs (Fig. 3) ....
....***Hoplitalysia*** Ashmead

Distribution; southern U.S.A. and Mexico, one described species. Biology: unknown, but one specimen reared from bovine dung. Reference: Wharton (1980).

- Scutellum at most convex. Male brachypterous, but some venation discernible 66

66(65). Mesosoma and metasoma usually yellow. Temple smooth...***Aphaereta*** Förster

Distribution: southeastern U.S.A.: one described species keys here. Biology: parasitoid of Sarcophagidae in pitcher plants. References: McComb, (1960), Wharton (1977b).

- Mesosoma and metasoma black. Temple rugose ....***Idiasta*** Förster

Distribution: northern Canada; only one species described from New World keys here. Biology: unknown. References: Tobias (1975, as *Euphaenocarpa)*, Wharton (1980).

67(64). Mandible with tooth 2 long, narrow, about as long as apical width of mandible (Fig. 81). Head in dorsal view narrower at temples than eyes ....***Alloea*** Haliday

Distribution: U.S.A. and Canada; only one species described from New World. Biology: parasitoids of Lonchopteridae. References: Fischer (1966c), Wharton (1980).

- Mandible with tooth 2 shorter and broader, only half as long as apical width of mandible. Head in dorsal view distinctly broader at temples than at eyes..68

68(67). Mandible less than half length of head (Fig. 77). Width of propodeal spiracles less than half distance between spiracle and anterior border of propodeum...***Dinotrema*** Förster

Distribution: U.S.A. (New York), only one described species keys here. Biology: unknown. Reference: Wharton (1980).

- Mandible more than half length of head (Fig. 78). Width of propodeal spiracles greater than half distance between spiracle and anterior border of propodeum (Fig. 59) ....***Pneumosema*** Fischer

Distribution: eastern U.S.A. and southeastern Canada; 2 species known, only one of which keys here. Biology: unknown. References: Fischer (1966b), Wharton (1994).

REFERENCES

Achterberg, C. van 1976. Revisionary notes on the genus *Coloneura* Foerster with description of a new subgenus *Coloneurella*, from the Netherlands (Hym., Braconidae, Alysiinae), Entomologische Betichten 36: 186-192.

Achterberg, C. van 1980. Three new Palearctic genera of Braconidae (Hymenoptera). Entomologische Berichten 40: 72-80.

Achterberg, C. van 1983. Revisionary notes on the genera *Dapsilarthra* auct. and *Mesocrina* Foerster (Hymenoptera, Braconidae, Alysiinae). Tijdschrift voor Entomologie 126: 1-24.

Achterberg, C. van 1986. The Holarctic genus *Anisocyrta* Foerster (Hymenoptera: Braconidae: Alysiinae). Zoologische Mededelingen Leiden 60: 285-297.

Achterberg, C. van 1988. The genera of the *Aspilota* group and some descriptions of fungicolous Alysiini from the Netherlands (Hymenoptera: Braconidae: Alysiinae). Zoologische Verhandelingen Leiden 247: 1-88.

Achterberg, C. van and Bin, F. 1981. Notes on two species of *Dinotrema* Foerster (Hym., Braconidae, Alysiinae) with observations on the hymenopterous parasite-complex of *Spiniphora dorsalis* Becker (Dipt., Phoridae) in dead *Helix* spp. (Mollusca). Entomologische Berichten 41: 104-112.

Buckingham, G. R. and Sharkey, M. J. 1988. Abdominal exocrine glands in Braconidae (Hymenoptera). Pp. 199-242. In: Gupta, V. K. (Ed.). Advances in Parasitic Hymenoptera Research. Brill, Leiden.

Čapek, M. 1970. A new classification of the Braconidae (Hymenoptera) based on the cephalic structures of the final instar larva and biological evidence. The Canadian Entomologist 102: 846-875.

Chen, J. and Wu, Z. 1994. The Alysiini of China (Hymenoptera; Braconidae: Alysiinae), China Agriculture Press, Fuzhou. 178 pp. (In Chinese.)

Costa Lima, A. da 1938. Vespas parasitas de moscas de frutas (Hymenoptera: Braconidae). O Campo 8: 69-72.

Drea, J. J. and Hendrickson, R. M., Jr. 1986. Analysis of a successful classical biological control project: The alfalfa blotch leafminer (Diptera: Agromyzidae) in the northeastern United States. Environmental Entomology 15: 448-455.

Edson, K. M. and Vinson, S. B. 1979. A comparative morphology of the venom apparatus of female braconids (Hymenoptera: Braconidae). The Canadian Entomologist 111: 1013-1024.

Fischer, M. 1966a. Studien über Alysiinae (Hymenoptera, Braconidae). Annalen Naturhistoriches Museum Wien 69: 177-205.

Fischer, M. 1966b. Zwei neue Alysiinen-Gattungen aus Nordamerika (Hymenoptera, Braconidae). Annalen des Naturhistorichen Museums Wiens 69: 207-212.

Fischer, M. 1966c. Über gezüchtete Braconiden aus Europa (Hymenoptera). Zeitschrift für angewandte Entomologie 58: 323-339.

Fischer, M. 1967. Die nearktischen Arten der Gattung *Synaldis* Foerster (Hymenoptera, Braconidae, Alysiinae). Polskie Pismo Entomologiczne 37: 431-478.

Fischer, M. 1969a. Über nordamerikanische Arten der Gattungen *Orthostigma* Ratzeburg und *Aspilota* Foerster (Hymenoptera, Braconidae, Alysiinae). Acta Entomologica Musei Nationalis Prague 38: 81-114.

Fischer, M. 1969b. Die nearktischen *Aspilota*-Arten der *petiolata*-Gruppe (Hymenoptera, Braconidae, Alysiinae). Bollettino del Laboratorio di Entomologia Agraria "Filippo Silvestri" di Portici 27: 55-78.

Fischer, M. 1969c. Revision der nearktischen *Aspilota*-Arten der *signifrons*-Gruppe (Hymenoptera, Braconidae, Alysiinae). Sitzungberichten Österreichischen Akademie der Wissenschaften, Mathernatisch-naturwissenschafliche Klasse I (178): 243-259.

Fischer, M. 1969d. Die nearktischen Arten der *Aspilota*-Gruppe und der *Aspilota smithi* Gruppe. Redia 51: 187-209.

Fischer, M. 1969e. Die nearktischen *Aspilota*-Arten der Sektion B (Hymenoptera: Braconidae). Proceedings of the Entomological Society of Washington 71: 361-368.

Fischer, M. 1970. Revision der nearktischen *Aspilota*-Arten der Sektion D und Ergänzungen zu anderen Artengruppen (Hymenoptera, Braconidae, Alysiinae), Annalen Naturhistoriches Museum Wien 74: 91-127.

Fischer, M. 1974a. *Aspilota*-Arten, gezogen aus Phoriden (Hymenoptera, Braconidae, Alysiinae), Bollettino de] Laboratorio di Entomologia Agraria "Filippo Silvestri" di Portici 31: 253-267.

Fischer, M. 1974b. Die nearktischen *Phaenocarpa*-Arten. Revision der Gruppe B (Hymenoptera, Braconidae, Alysiinae). Polskie Pismo Entomologiczne 44: 103-230.

Fischer, M. 1975a. Taxonomische Untersuchungen an Neotropischen Alysiini, besonders solchen aus dem British Museum (Natural History) in London (Hym. Braconidae: Alysiinae). Studia Entomologica 18: 127-152.

Fischer, M. 1975b. Die nearktischen Phaenocarpu-Xsten. Revision der Gruppe A (Hymenoptera, Braconidae, Alysiinae). Polskie Pismo Entomologiczne 45: 279-356.

Fischer, M. 1976. Durchsicht der Gattung *Synelix* Foerster (Hymenoptera, Braconidae, Alysiinae). Bollettino del laboratorio di Entomologia Agraria "Filippo Silvestri" di Portici 36: 3-13.

Fischer, M. 1990. Westpalaarktische *Phaenocarpa*-Arten: Vorläufiger Bestimmungsschlüssel, Deskriptionen und Redeskriptionen (Hymenoptera, Braconidae, Alysiinae). Annalen Naturhistoriches Museum Wien 91: 105-135.

Fischer, M. 1991. Integration einiger Gattungen der Alysiini in ein diagnostisches System (Hymenoptera: Braconidae, Alysiinae). Linzer Biologisches Beiträge 23: 3-35.

Griffiths, G. C. D. 1964. The Alysiinae (Hym. Braconidae) parasites of the Agromyzidae (Diptera). I. General questions of taxonomy, biology and evolution. Beiträge zur Entomologie 14: 823-914.

Griffiths, G. C. D. 1966a. The AJysiinae (Hym, Braconidae) parasites of the Agromyzidae (Diptera). IL The parasites of *Agromyza* Fallen. Beiträge zur Entomologie 15: 551-605.

Griffiths, G. C. D. 1966b. The Alysiinae (Hym. Braconidae) parasites of the Agromyzidae (Diptera). III. The parasites of *Paraphytomyza* Enderlein, *Phytagromyza* Hendel and *Phytomyza* Fallen. Beiträge zur Entomologie 16: 775-951.

Griffiths, G. C. D. 1967. The Alysiinae (Hym. Braconidae) parasites of the Agromyzidae (Diptera) IV. The parasites of *Hexomyza* Enderlein, Melan *Agromyza* Hendel, *Opiomyia* Braschnikov and *Napemyza* Westwood. Beiträgezur Entomologie 17: 653-696.

Griffiths, G. C. D.1968a. The Alysiinae (Hym. Braconidae) parasites of the Agromyzidae (Diptera). V. The parasites of *Liriomyza* Mik and certain small genera of Phytomyzinae. Beiträge zur Entomologie 18: 5-62.

Griffiths, G. C. D. 1968b. The Alysiinae (Hym. Braconidae) parasites of the Agromyzidae (Diptera). VI. The parasites of *Cerodontha* Rondani *s.l.* Beiträge zur Entomologie 18: 63-152.

Guppy, J.C., Meloche, F., and Harcourt, D. G. 1988. Seasonal development, behavior, and host synchrony of *Dacnusa dryas* (Nixon) (Hymenoptera: Braconidae) parasitizing the alfalfa blotch leafminer, *Agromyza frontella* (Rondani) (Diptera: Agromyzidae). The Canadian Entomologist 120: 145-152.

Hackman, W. and Meinander, M. 1979. Diptera feeding as larvae on macrofungi in Finland. Annales Zoologici Fennici 16: 50-83.

Hendrikse, A., Zucchi, R., Lenteren, J. C. van, and Woets, J. 1980. *Dacnusa sibirica* Telenga and *Opius pullipes* Wesmael (Hym., Braconidae) in the control of the tomato leafminer *Liriomijza bryoniae* Kalt. Bulletin IOBC/WPRS 3: 83-98.

Keilin, D. and Tate, P. 1943. The larval stages of the celery fly (*Acidia heraclei* L.) and of the braconid *Adelura apii* (Curtis), with notes upon an associated parasitic yeast-like fungus. Parasitology 35: 27-36.

Kraaijeveld, A. R. 1994. Local adaptations in a parasitoid-host system. A coevolutionary arms race? Ph.D. Dissertation, University of Leiden, The Netherlands.

Legner, E. F. 1978·. Calliphoridae, Muscidae. Pp. 335-338, 346-355. In: Clausen, C. P. (Ed.). Introduced Parasites and Predators of Arthropod Pests and Weeds: A World Review. United States Department of Agriculture, Agriculture Handbook No. 480.

Maetô, K. 1983. Discovery of two genera, *Sarops* Nixon and *Protodacnusa* Griffiths, from Japan (Hymenoptera, Braconidae). Kontyu 51: 247-253.

Marsh, P. M. 196S. A new species of *Alysia* parasitic on carrion-feeding flies (Hymenoptera: Braconidae). Proceedings of the Entomological Society of Washington 70: 208-210.

Marsh, P. M. 1979. Family Braconidae. Pp. 144-295. In: Krombein, K. V., Hurd, P. D., Smith, D. R., and Burks, B. D. (Eds.). Catalog of Hymenoptera hi America North of Mexico. Smithsonian Institution Press, Washington, D.C.

Mason, W.R. M. 1991. A new genus of exodont Ichneutinae (Hymenoptera, Braconidae). Insecta Mundi 5: 95-99.

McComb, C. W. 1960. Two new species of *Aphaereta*, with a key to the Nearctic species (Hymenoptera: Braconidae). Entomological News 71: 175-178.

Muesebeck, C. F. W. and Walkley, L. M. 1951. Family Braconidae, Pp. 90-184. In: Muesebeck, C. F. W., Krombein, K. V., and Townes, H. K. (Eds.). Hymenoptera of America North of Mexico, Synoptic Catalog. United States Department of Agriculture, Agriculture Monograph 2.

Papp, J. 1969. A synopsis of the *Phaenocarpa* Foerster species of the Neotropical Region (Hym., Braconidae: Alysiinae). Acta Zoologica Academiae Scientiarum Hungaricae 15: 379-389.

Prince, G. J. 1976. Laboratory biology of *Phaenocarpa persimilis* Papp (Braconidae: Alysiinae), a parasitoid of *Drosophila*. Australian Journal of Zoology 24: 249-264.

Quicke, D. L J. and Achterberg, C. van 1990. Phylogeny of the subfamilies of the family Braconidae (Hymenoptera: Ichneumonoidea). Zoologische Verhandelingen Leiden 258: 1-95.

Reznik, S. Ya., Chernoguz, D. G., and Zinovjeva, K.B. 1992. Host searching, oviposition preferences and optimal synchronization in *Alysia manducator* (Hymenoptera: Braconidae), a parasitoid of the blowfly, *Calliphora uicina*. Oikos 65: 81-88.

Riegel, G. T. 1982. The American species of Dacnusinae, excluding certain Dacnusini (Hymenoptera: Braconidae). Novitates Arthropodae 1: 1-185.

Rohwer, S. A. 1914. The Nearctic species of the hymenopterous genus *Symphya* Foerster. Entomological News 25: 158-172.

Shaw, M. R. and Huddleston, T. 1991. Classification and biology of braconid wasps (Hymenoptera: Braconidae). Handbooks for the Identification of British Insects 7(11): 1-126.

Sharkey, M. J. and Wharton, R. A. 1994. A revision of the genera of the world Ichneutinae (Hymenoptera: Braconidae). Journal of Natural History 28: 873-912.

Shenefelt, R D. 1974. Pars 11. Braconidae 7, Alysiinae. Pp. 937-1113. In: Vecht, J. van der and Shenefelt, R. D. (Eds.). Hymenopterorum Catalogus (nova editio). Dr. W. Junk, The Hague.

Stegmaier, C. E. Jr. 1972. Parasitic Hymenoptera bred from the family Agromyzidae (Diptera) with special reference to South Florida. Florida Entomologist 55: 273-282.

Sugimoto, T., Ichikawa, T., Mitomi, M., and Sakuratani, Y. 1988b. Foraging for patchily distributed leaf-miners by the braconid *Dapsilarthra rufiventris* (Hymenoptera: Braconidae). IV. Analysis of sounds emitted by a feeding host. Applied Entomology and Zoology 23: 209-211.

Sugimoto, T., Kameoka, H., Kusatani, S., Inui, 0., and Otsuka, K. 1988c. Foraging for patchily distributed leaf-miners by the parasitoid, *Dapsilartha rufiuentris* (Hymenoptera: Braconidae), V. Plant odour as a cue to long range patch-location, Applied Entomology and Zoology 23: 135-143.

Sugimoto, T., Kawado, K., and Tadera, K. 1988d. Responses to the host mine by two parasitic wasps. *Dapsilarthra rufiventris* (Hymenoptera: Braconidae) and *Chrysocharis pentheus* (Hymenoptera: Eulophidae). Journal of Ethology 6: 55-58.

Sugimoto, T., Murakami, H., and Yamazaki, R. 1987. Foraging for patchily-distributed leaf miners by the parasitoid, *Dapsilarthra rufiventris* (Hymenoptera: Braconidae). II. Stopping rule for host search. Journal of Ethology 5: 95-103.

Sugimoto, T., Shimano, Y., Hata, Y., Nakai, A., and Yahara, M. 1988a. Foraging for patchily distributed leaf-miners by the parasitoid, *Dapsilarthra rufiventris* (Hymenoptera: Braconidae). III. Visual and acoustic cues to a close range patch-location. Applied Entomology and Zoology 23: 113-121.

Sugimoto, T., Uenishi, M., and Machida, F. 1986. Foraging for patchily-distributed leaf-miners by the parasitoid, *Dapsilartha rufiventris* (Hymenoptera: Braconidae). I. Discrimination of previously searched leaflets. Applied Entomology and Zoology 21: 500-508.

Tobias, V. I. 1971. Review of the Braconidae (Hymenoptera) of the USSR. Trudy Vsesoyuznogo Entomologicheskogo Obshchestva 54: 156-268. (In Russian.)

Tobias, V. I. 1975. Two new species and a new genus of braconids (Hymenoptera, Braconidae, Alysiinae) from Mongolia. Insects of Mongolia 3: 306-309. (In Russian.)

Tobias, V. I. 1986. Subfamily Alysiinae. Pp. 100-231. In: Tobias, V. I. (Ed.). Identification of the Insects of European USSR. Volume UI, Part V. Hymenoptera, Braconidae. Akademia Nauk, Leningrad. (In Russian.)

Vet, L. E. M. and van Alphen, J. J.M. 1985. A comparative functional approach to the host detection behaviour of parasitic wasps. I. A qualitative study on Eucoilidae and Alysiinae. Oikos 44: 478-486.

Visser, M. E. 1994. The importance of being large: the relationship between size and fitness in females of the parasitoid *Aphaereta minuta* (Hymenoptera: Braconidae). Journal of Animal Ecology 63: 963-978,

Wharton, R. A. 1977a. Exodontiellini, a new tribe of Opiinae with exodont mandibles (Hymenoptera: Braconidae). Pan-Pacific Entomologist 53: 297-303.

Wharton, R. A. 1977b. New world *Aphaereta* species (Hymenoptera: Braconidae), with a discussion of terminology used in the tribe Alysiini. Annals of the Entomological Society of America 70: 782-803.

Wharton, R. A. 1980. Review of New World Alysiini (Hymenoptera: Braconidae) with discussion of generic relationships within the tribe. University of California Publications in Entomology 88: 1-104.

Wharton, R. A. 1984. Biology of the Alysiini (Hymenoptera: Braconidae), parasitoids of cyclorrhaphous Diptera. Texas Agricultural Experiment Station Technical Monograph 11: 1-39.

Wharton, R.A. 1985. Characterization of the genus *Aspilota* (Hymenoptera: Braconidae). Systematic Entomology 10: 227-237.

Wharton, R. A. 1986. The braconid genus *Alysia* (Hymenoptera): a description of the subgenera and a revision of the subgenus *Alysia*. Systematic Entomology 11: 453-504.

Wharton, R. A. 1988a. Classification of the braconid subfamily Opiinae (Hymenoptera). The Canadian Entomologist 120: 333-360.

Wharton, R. A. 1988b. The braconid genus *Alysia* (Hymenoptera): a revision of the subgenus *Anarcha*. Contributions of the American Entomological Institute 24: 1-69.

Wharton, R. A. 1994. New genera, species, and records of New World Alysiinae (Hymenoptera: Braconidae). Proceedings of the Entomological Society of Washington 96: 630-664.

Wharton, R. A. and Austin, A. D. 1991. Revision of Australian Dacnusini (Hymenoptera: Braconidae: Alysiinae), parasitoids of cyclorrhaphous Diptera. Journal of the Australian Entomological Society 30: 193-206.

Wharton, R. A. and Chou, L.-y. 1985. Revision of the Taiwanese species of *Alloea* Haliday (Hymenoptera: Braconidae, Alysiinae). Journal of Agricultural Research of China 34: 352-367.

Wharton, R. A., Shaw, S. R., Sharkey, M. J., Wahl, D. B., Woolley, J. B., Whitfield, J. B., Marsh, P. M., and Johnson, J. W. 1992. Phylogeny of the subfamilies of the family Braconidae (Hymenoptera: Ichneumonoidea): a reassessment. Cladistics 8: 199-235.

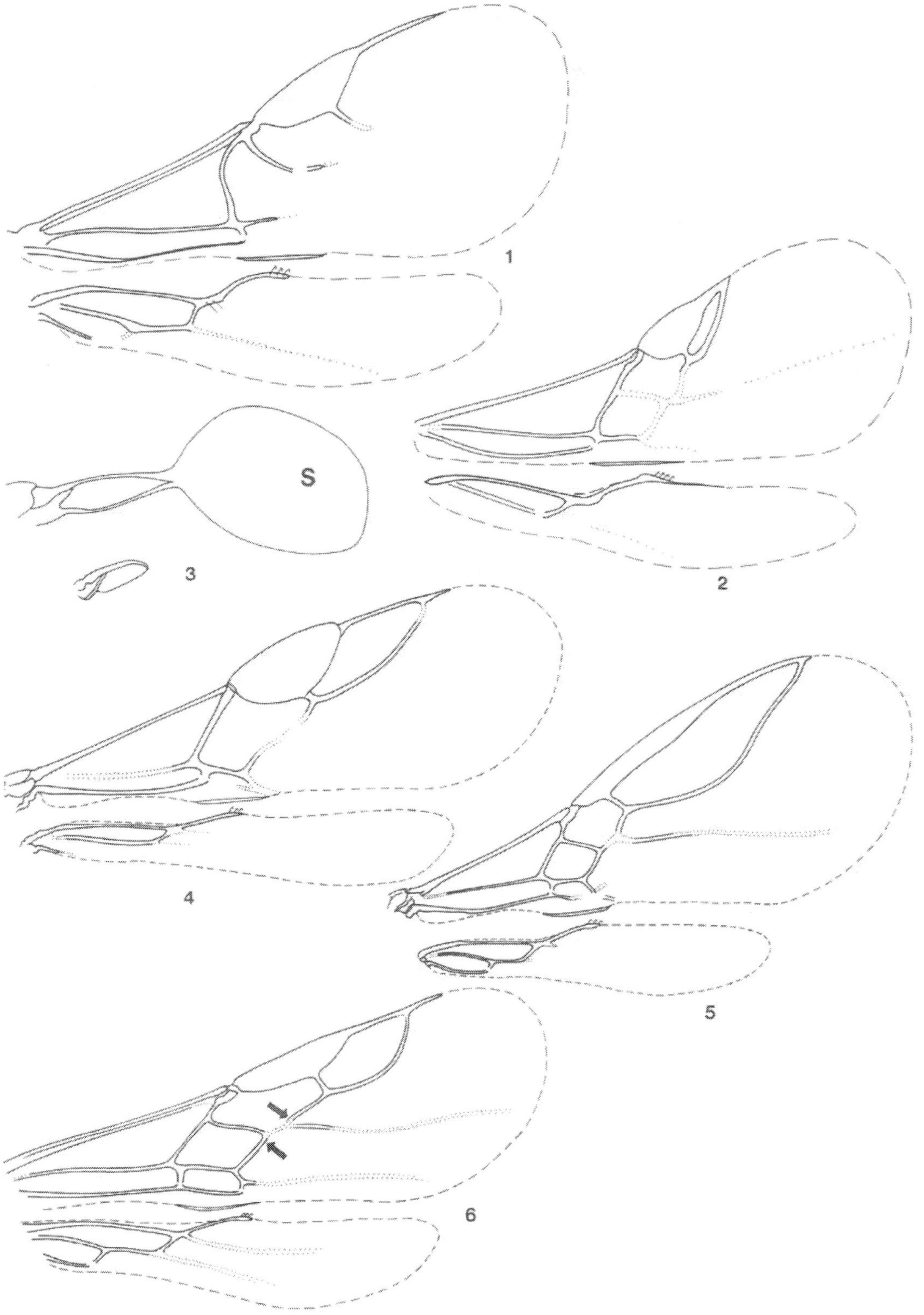

Figs 1-6: Wings of Alysiinae and other exodont Braconidae. l, *Paroligoneurus* sp.; 2, *Exodondiella* sp.; 3, *Hoplitalysia slossonae* Ashmead, male, s = enlarged stigma; 4, *Chaenusa* sp.; 5, *Chorebus* sp.; 6, *Coelinius* sp., arrows = m-cu origin relative to 2RS.

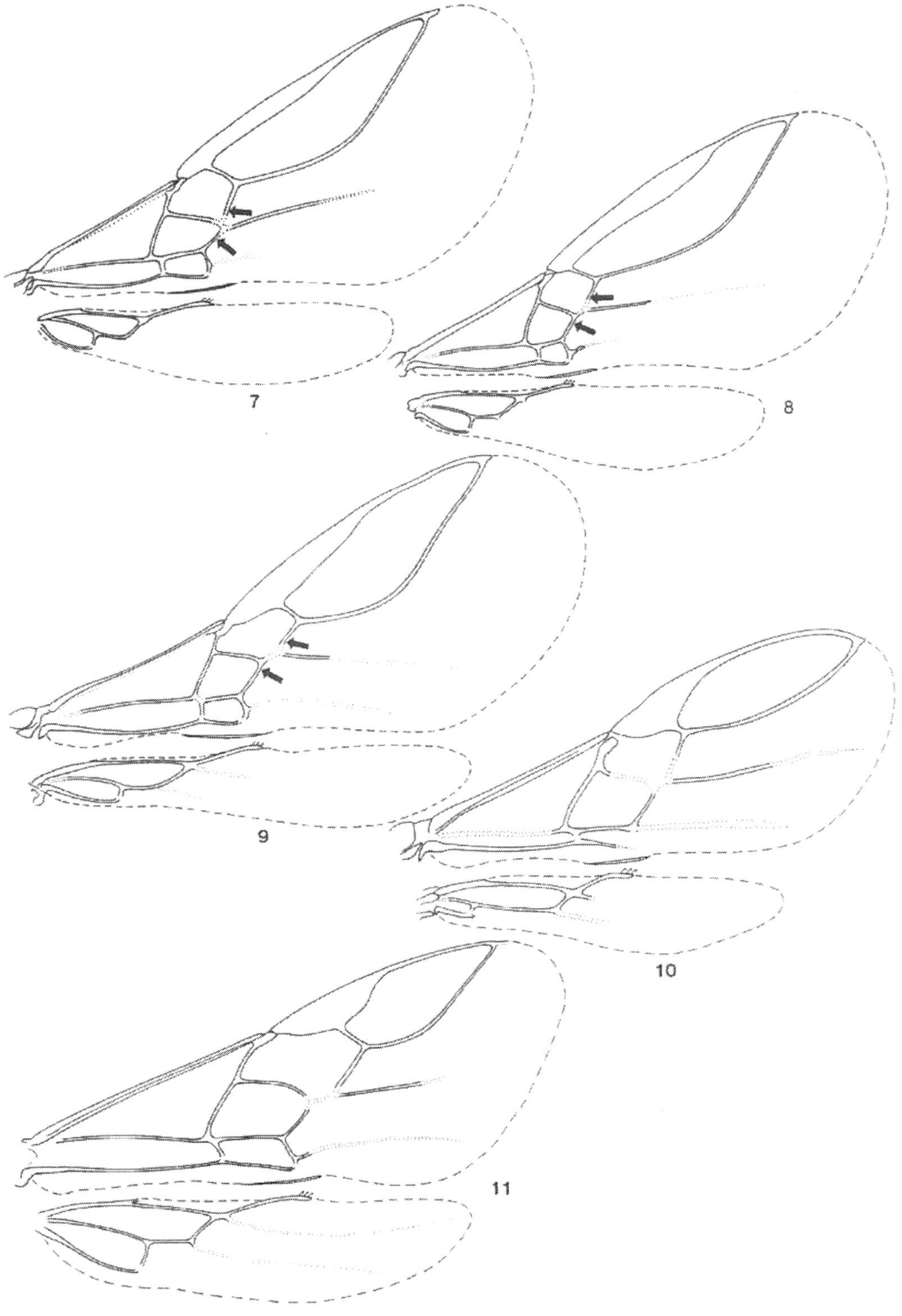

Figs 7-11: Wings of Alysiinae. 7, *Exotela* sp., arrows = m-cu origin relative to 2RS; 8, *Dacnusa* sp., arrows = m-cu origin relative to 2RS; 9, *Dacnusa siberica* Telenga, arrows = m-cu origin relative to 2RS; 10, *Pneumosema perditum* Wharton; 11, *Symphya* sp.

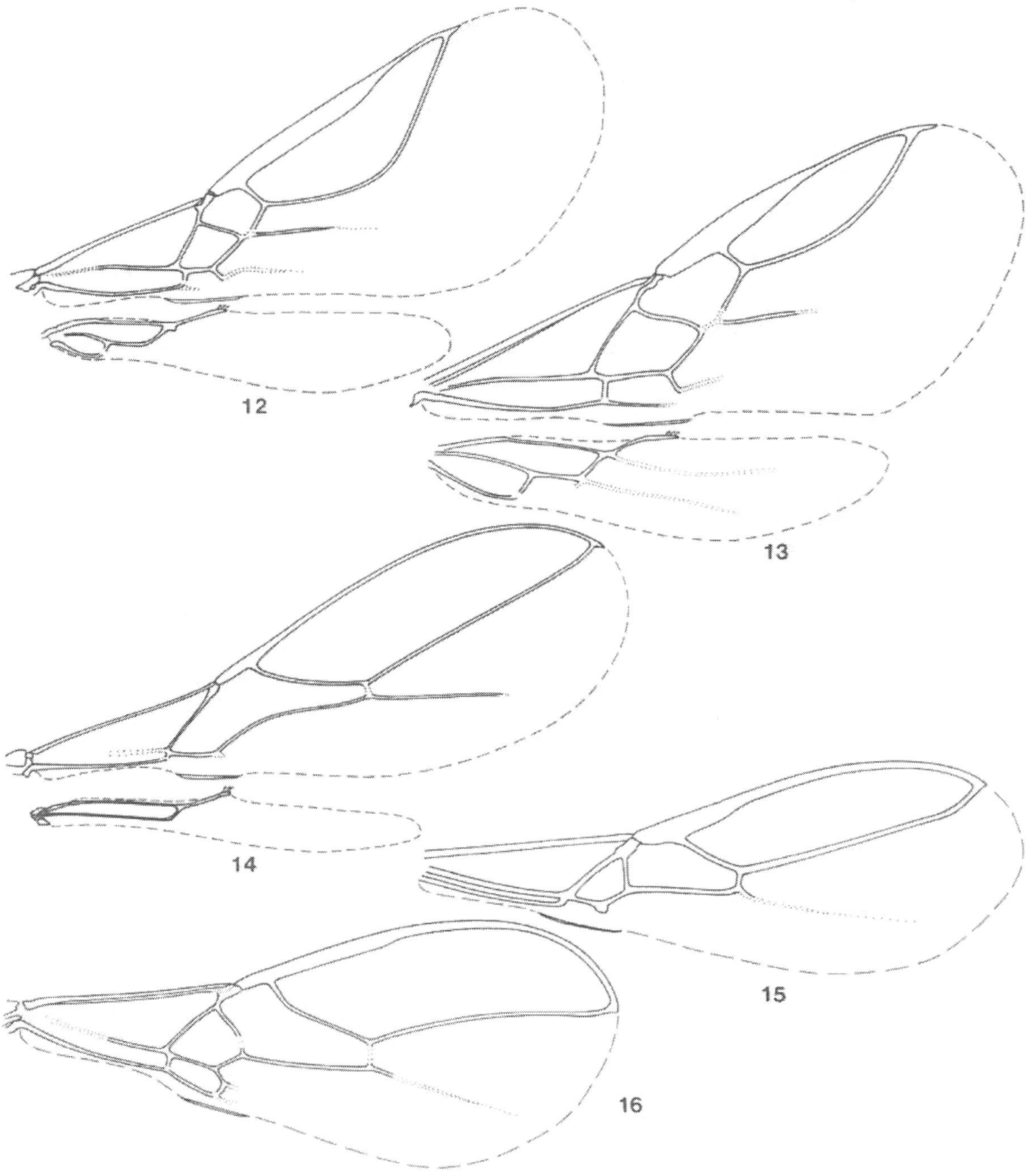

Figs 12-16: Wings of Alysiinae. 12, *Coloneura* sp.; 13, *Synelix semirugosa* (Haliday), 14, *Aphaereta confusa* Wharton; 15, *Dinostigma* sp.; 16, *Dinotrema* sp.

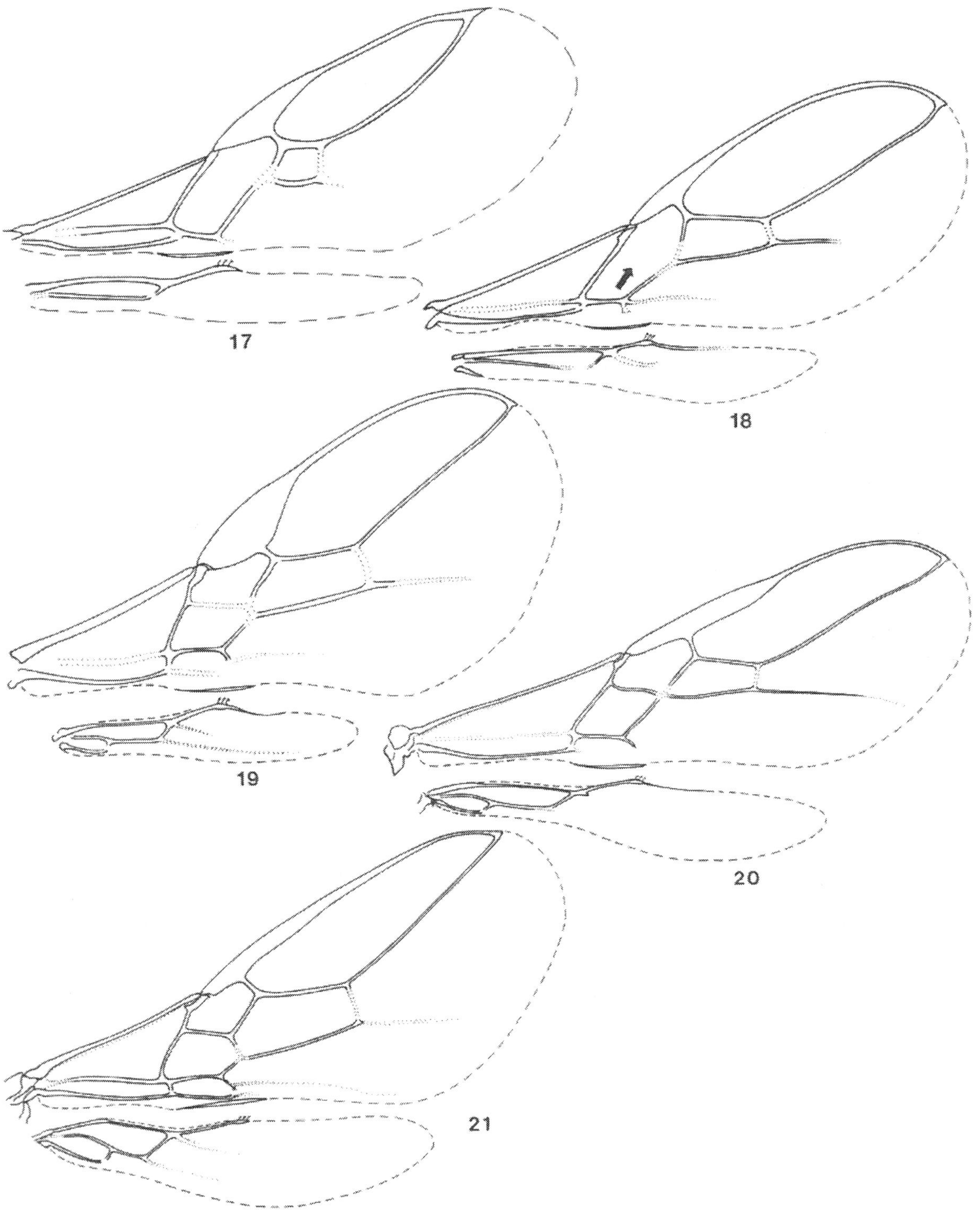

Figs 17-21: Wings of Alysiinae. 17, *Syncrasis* sp.; 18, *Aphaereta pallipes* Say, arrow = (RS+M)a absent; 19, *Oenonogastra* sp.; 20, *Dapsildiastema angustum* Wharton; 21, *Apronopa* sp.

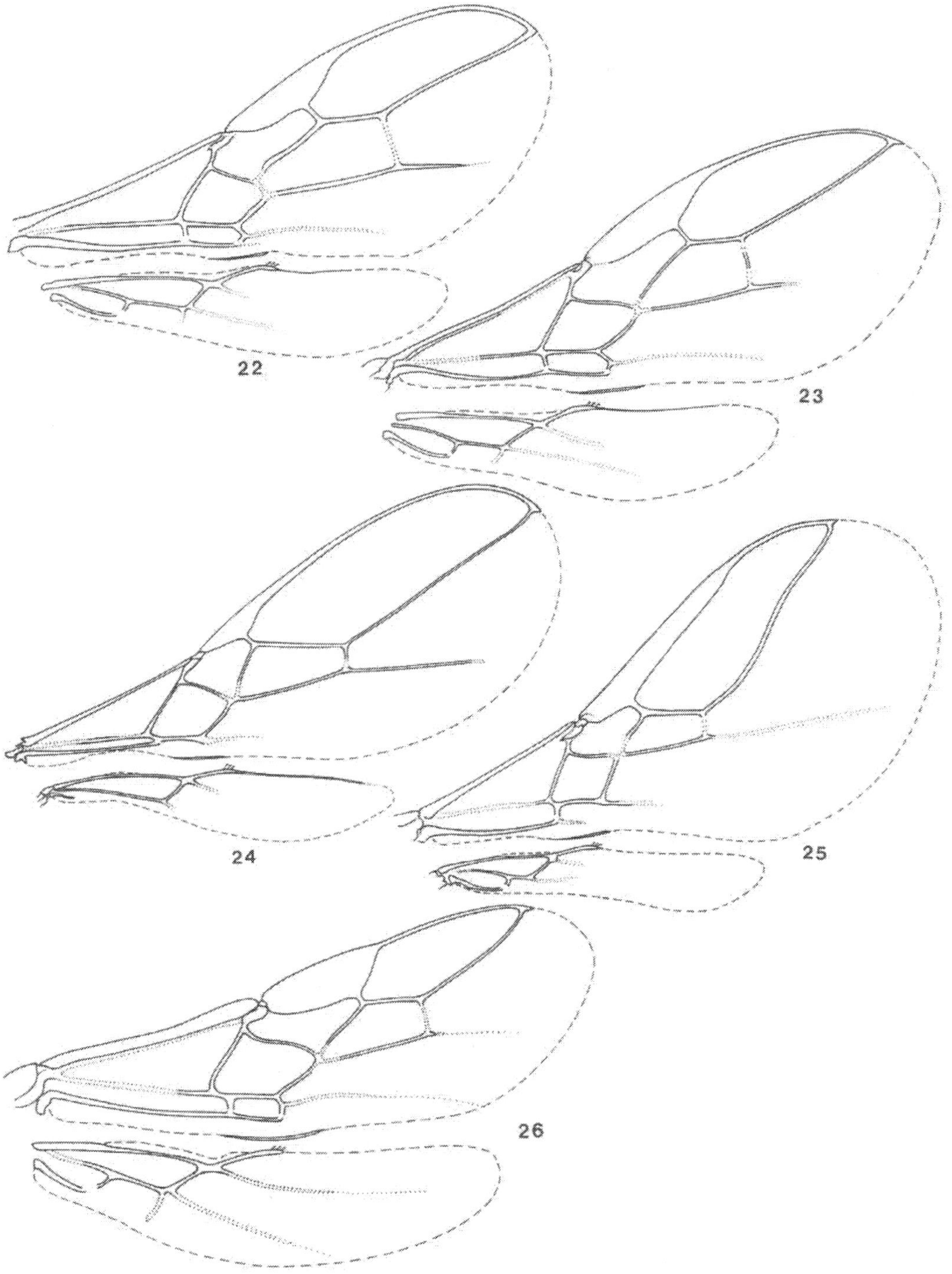

Figs 22-26: Wings of Alysiinae. 22, *Microcrasis* sp.; 23, *Phasmidiasia lia* Wharton; 24, *Asobara* sp.; 25, *Dapsilarthra* sp.; 26, *Alysiasta caltagironei* Wharton.

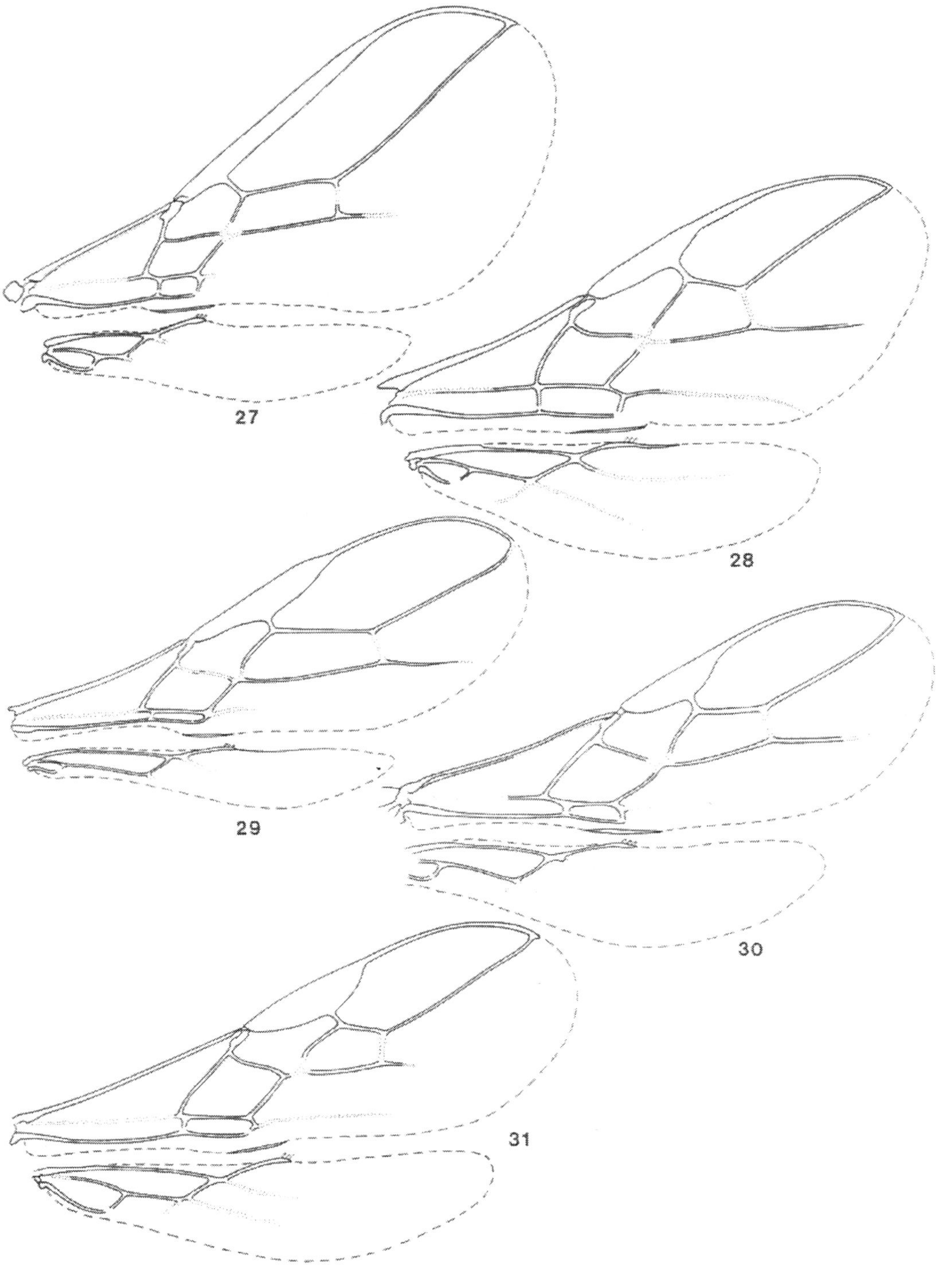

Figs 27-31: Wings of Alysiinae. 27, *Dapsilarthra* sp.; 28, *Gnathopleura cariosa* (Marsh); 29, *Phaenocarpa sharkeyi* Wharton; 30, *Phaenocarpa brachyptera* Fischer; 31, *Idiasta longicornis* (Provancher).

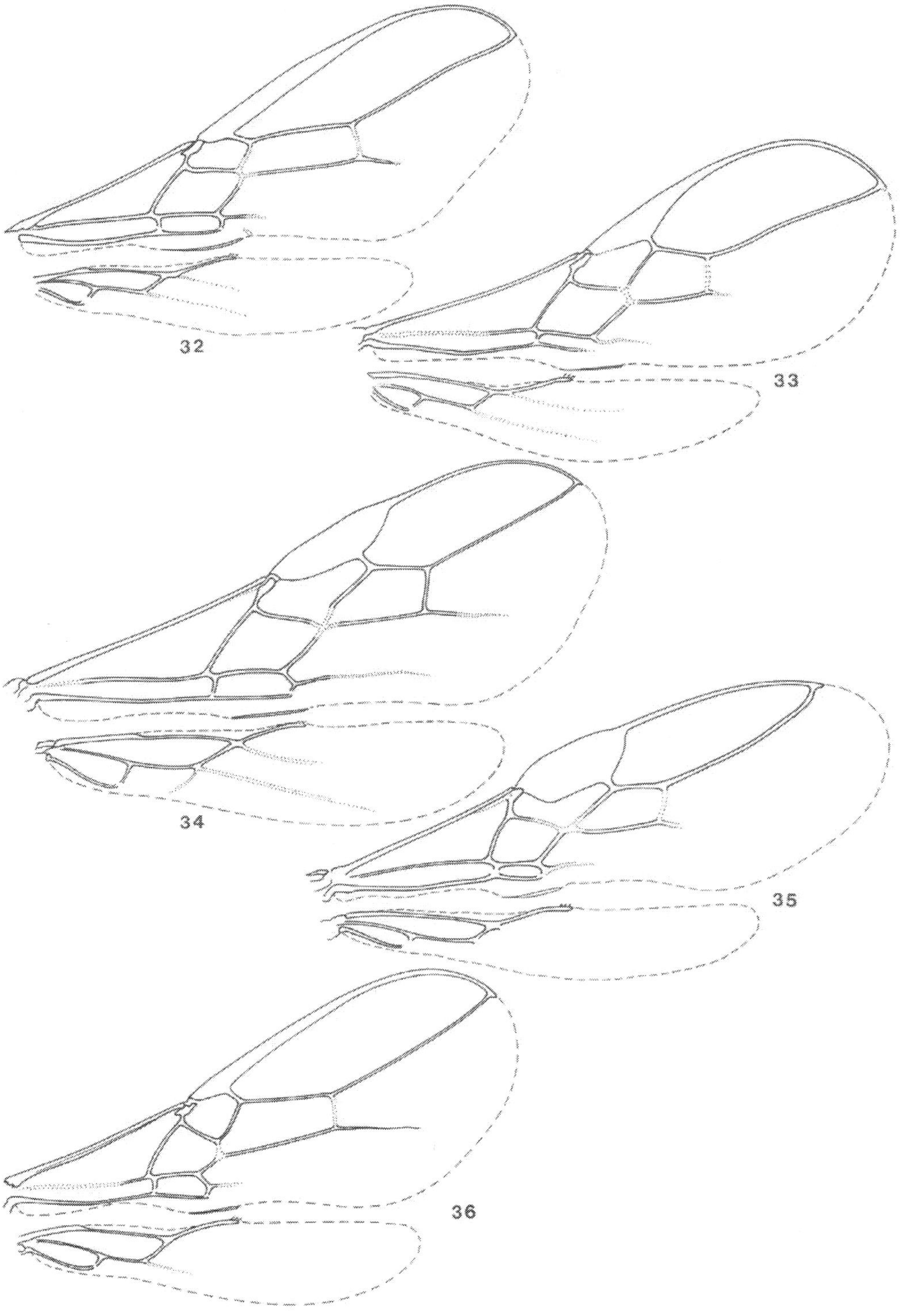

Figs 32-36: Wings of Alysiinae. 32, *Anisocyrta* sp.; 33, *Peniapleura* sp.; 34, Alysia sp.; 35, *Alloea* sp.; 36, *Aspilota* sp.

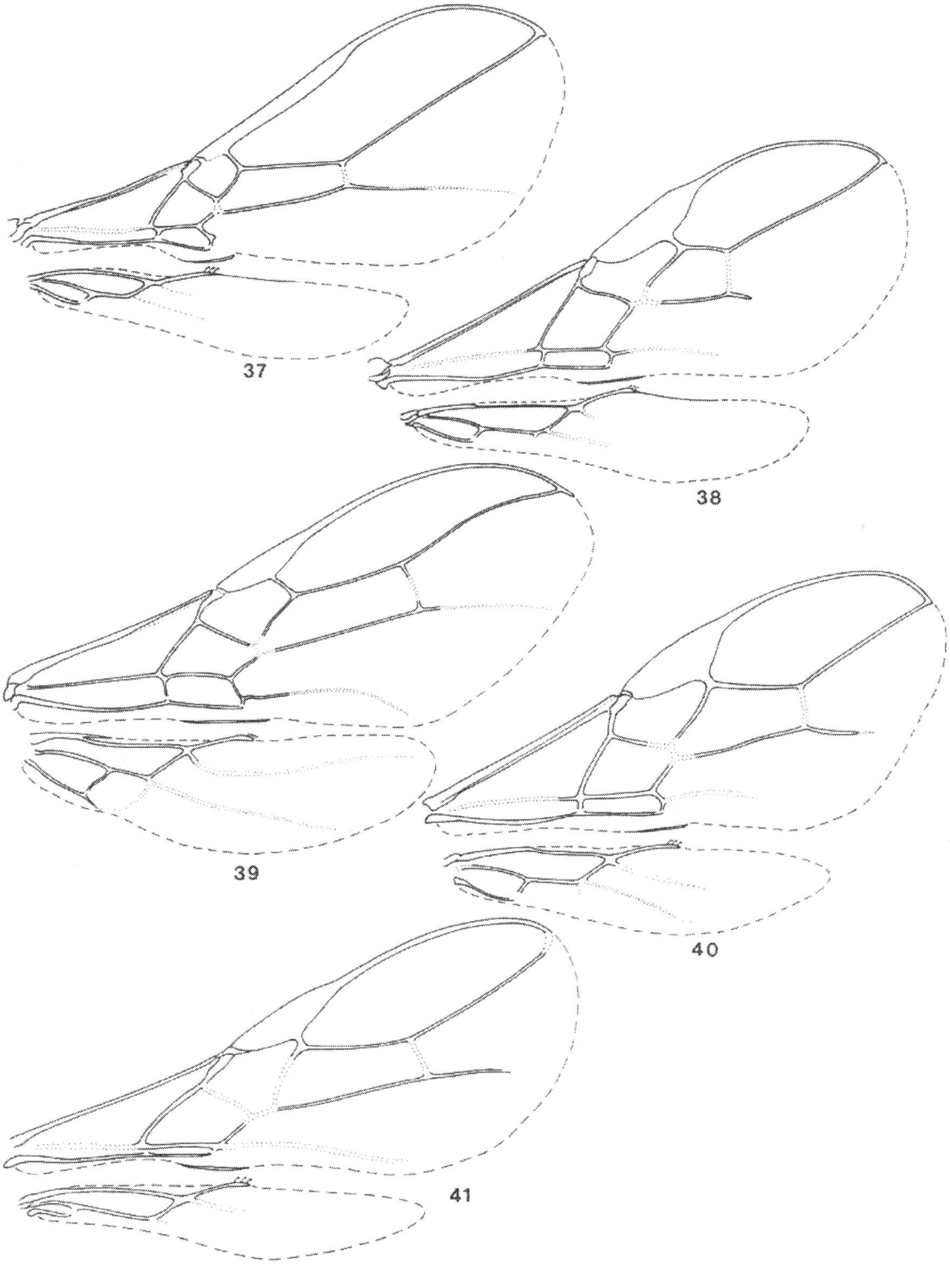

Figs 37-41: Wings of Alysiinae. 37, *Orthostigma* sp.; 38, *Cratospila neocirce* Wharton; 39, *Ilatha* sp.; 40, *Mesocrina* sp.; 41, *Vachterbergia grayae* Wharton.

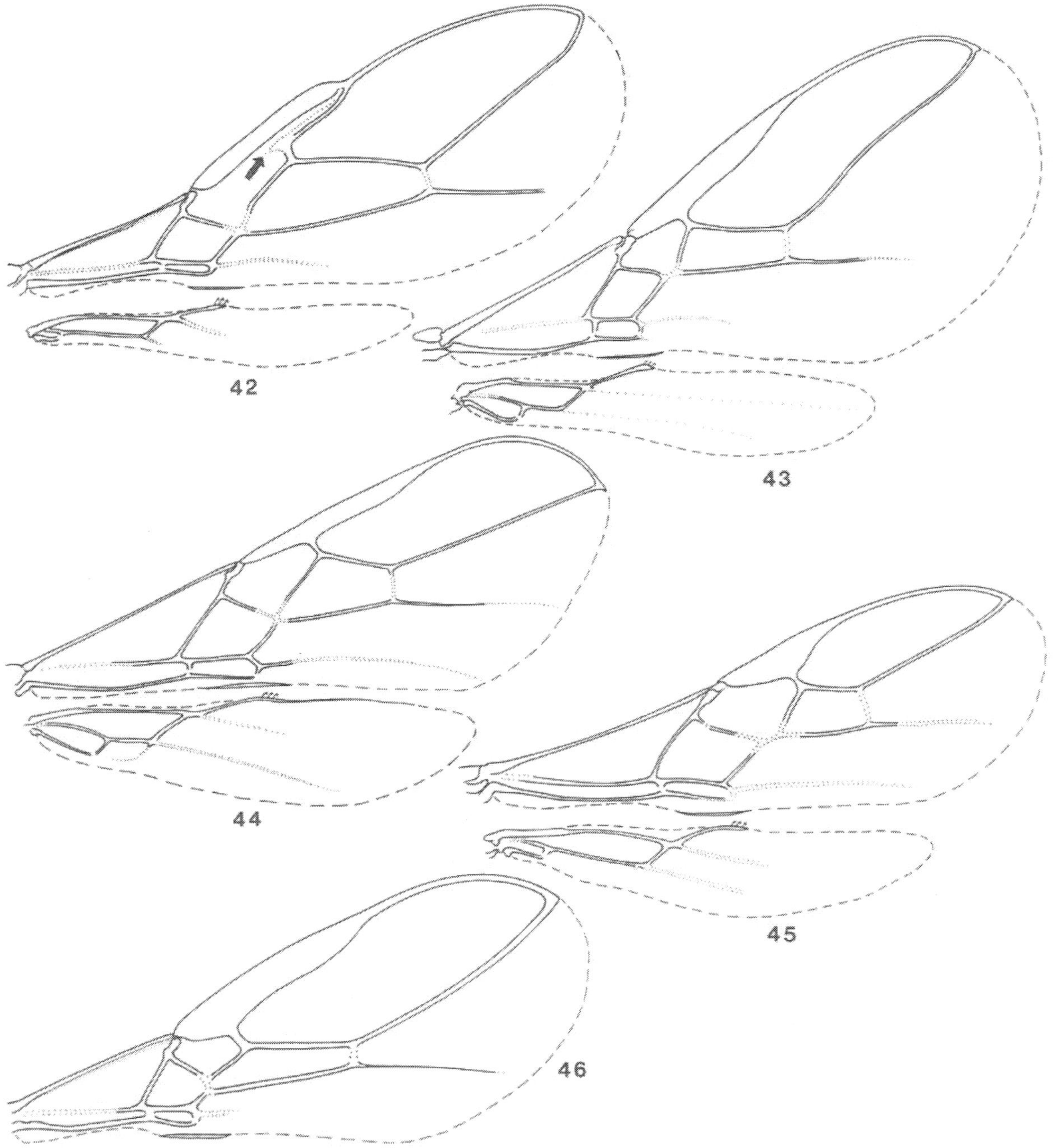

Figs 42-46: Wings of Alysiinae. 42, *Phaenocarpa crutomorpha* Wharton, arrow = split in stigma; 43, *Dapsilarthra* sp.; 44, *Tanicarpa* sp.; 45, *Pneumosema siigmaticuni* Fischer; 46, *Dinotrema* sp. (= *Prosapha*).

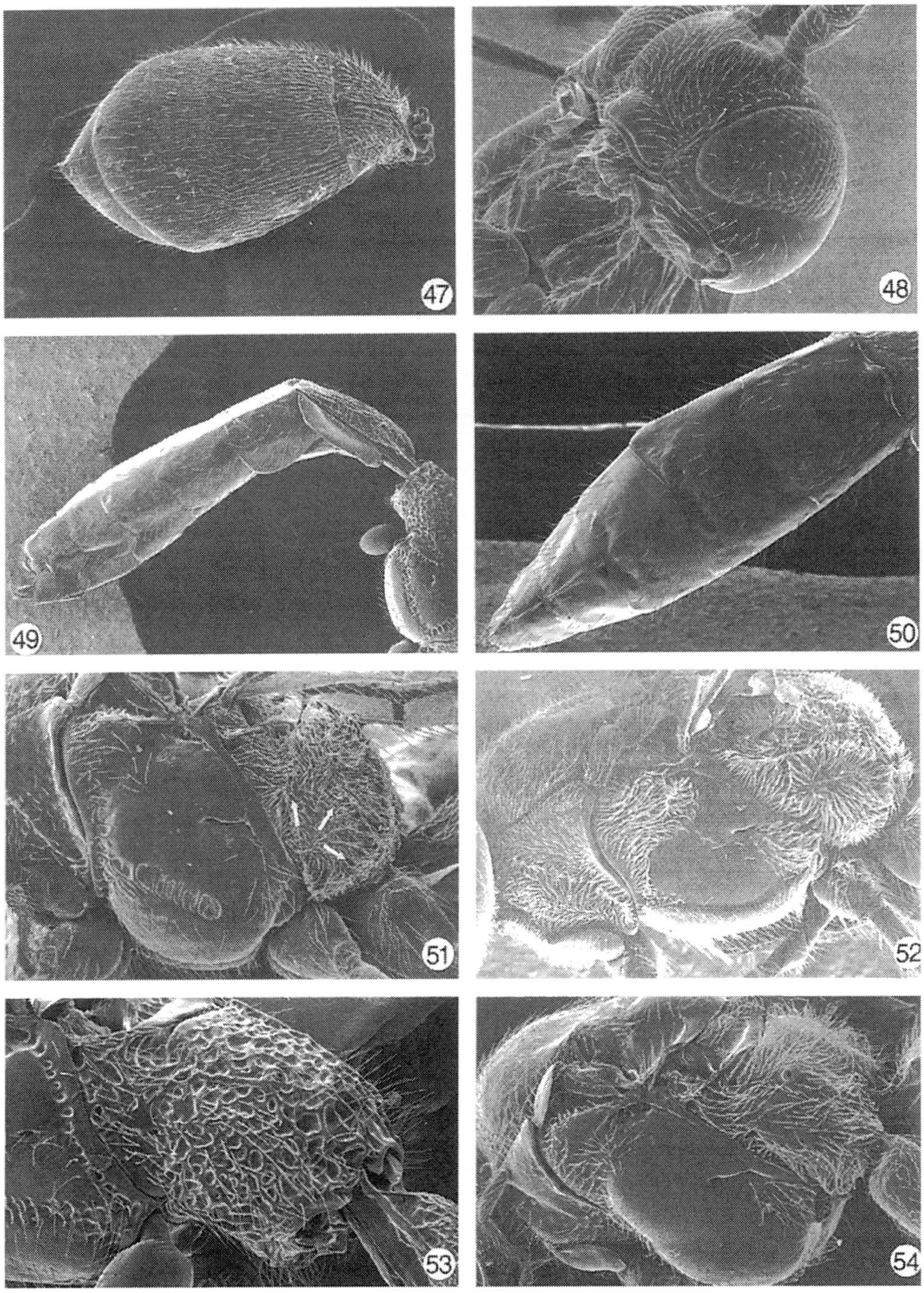

Figs 47-54: Alysiinae. 47, carapace-like metasoma of *Symphya* sp., dorsal view; 48, head of *Chaenusa* sp. showing setose eyes; 49, metasoma of *Coelinius* sp., lateral view; 50, metasoma of *Coelinius* sp., dorsal view; 51-54, mesosomas: 51, *Chorebus* sp., lateral view, arrows = rosette of metapleural pubescence; 52, *Chorebus* sp., lateral view showing denser rosette of metapleural pubescence; 53, *Coelinius* sp., lateral view, showing sparse metapleural pubescence; 54, *Dacnusa* sp., lateral view showing dense metapleural pubescence not in a rosette.

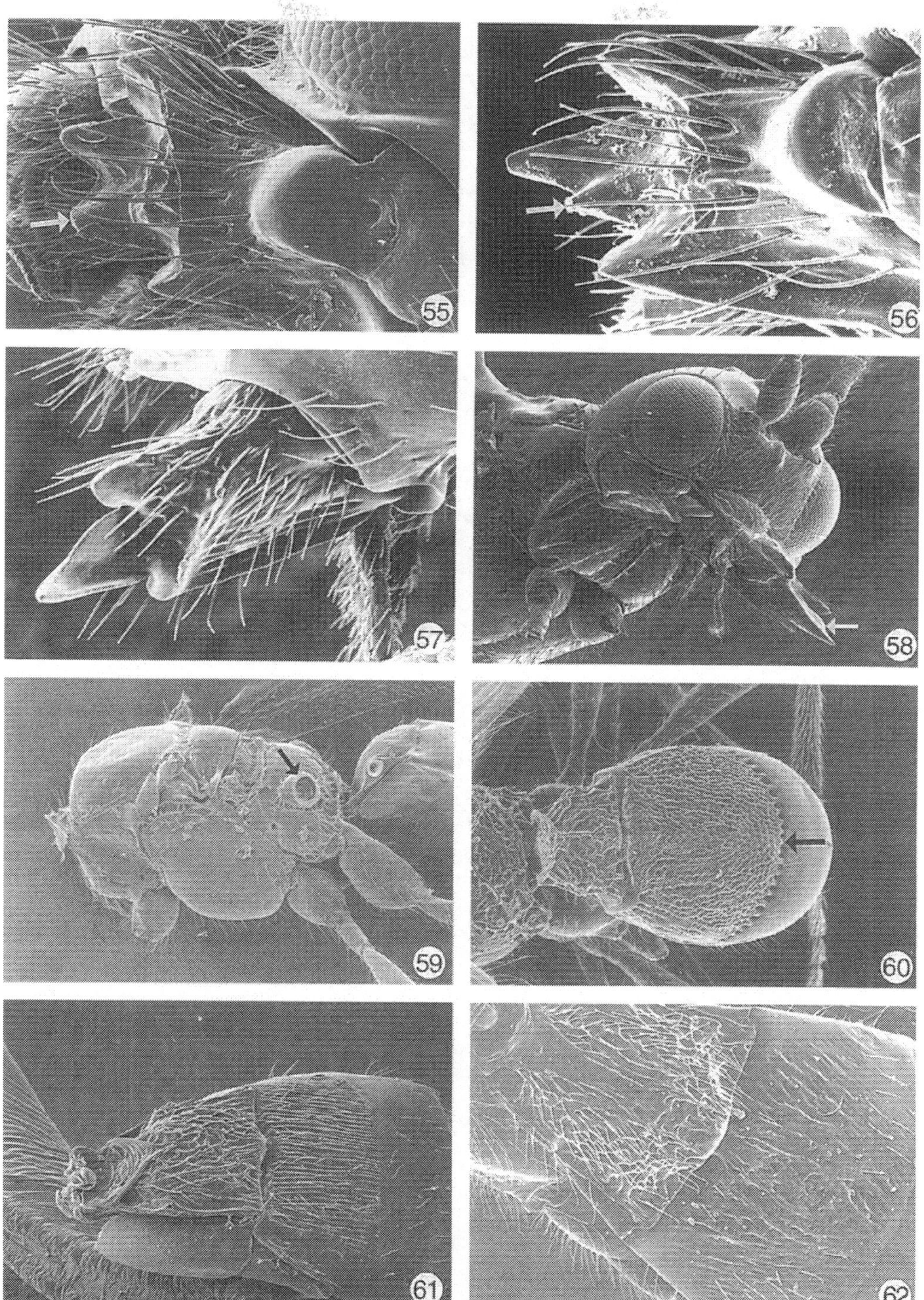

Figs 55-62: Alysiinae. 55-57, left mandibles: 55, *Chorebus* sp., arrow = large 4th tooth between tooth 2 and tooth 3; 56, *Chorebus* sp., arrow = small 4th tooth between tooth 2 and tooth 3; 57, *Coelinius* sp.; 58, head of *Coelinius* sp., frontal view, arrow = weakly developed 4th tooth between tooth 1 and tooth 2; 59, mesosoma of *Pneumosema* stigmaticum Fischer, arrow = propodeal spiracle; 60, metasoma of *Oenonogastra microrhopalae* Ashmead, dorsal view, arrow = crenulate suture between tergum 2 and tergum 3; 61, metasoma of *Epimicia griffithsi* Wharton, dorsal view (after Wharton, 1994); 62, first 2 metasomal terga of *Synelix semirugosa* (Haliday).

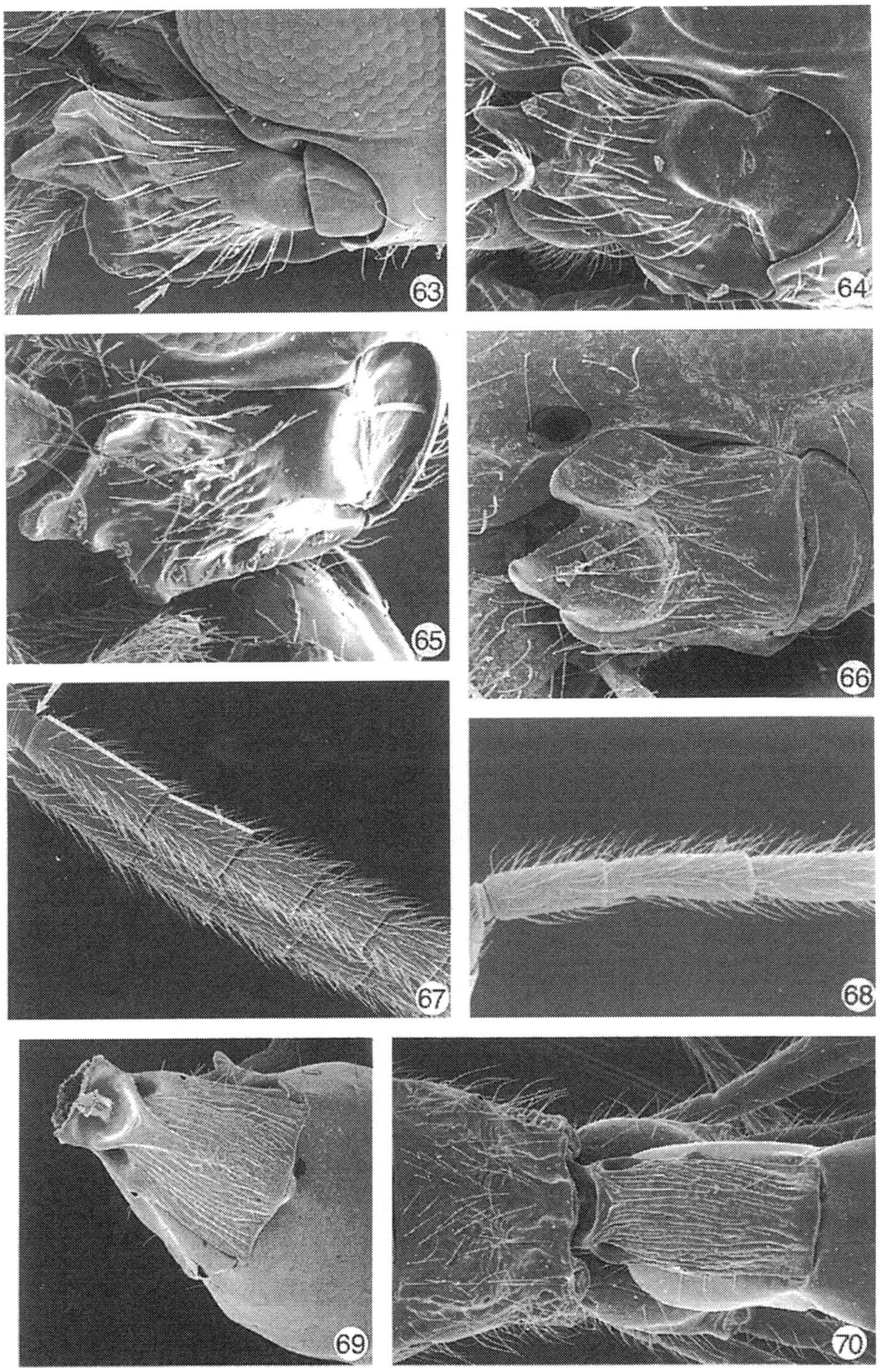

Figs 63-70: Alysiinae. 63-66, left mandibles: 63, *Epimicta griffithsi* Wharton, arrow = 4th tooth along ventral margin basad tooth 3; 64, *Laoiris* sp.; 65, *Synelix semirugosa* (Haliday); 66, *Phasmalusia* sp.; 67, basal 5 flagellomeres of antennae of *Alysia glabra* Wharton, white bars showing measurements for relative length of 1st 2 flagellomeres, arrow = annellus; 68, basal 3 flagellomeres of antenna of *Phaenocarpa anomala* Wharton; 69, petiole of *Alysia alticola* (Ashmead), dorsal view (after Wharton, 1986); 70, petiole of *Dapsildiastema angustum* Wharton, dorsal view. Figures 63, 68, and 70 after Wharton (1994).

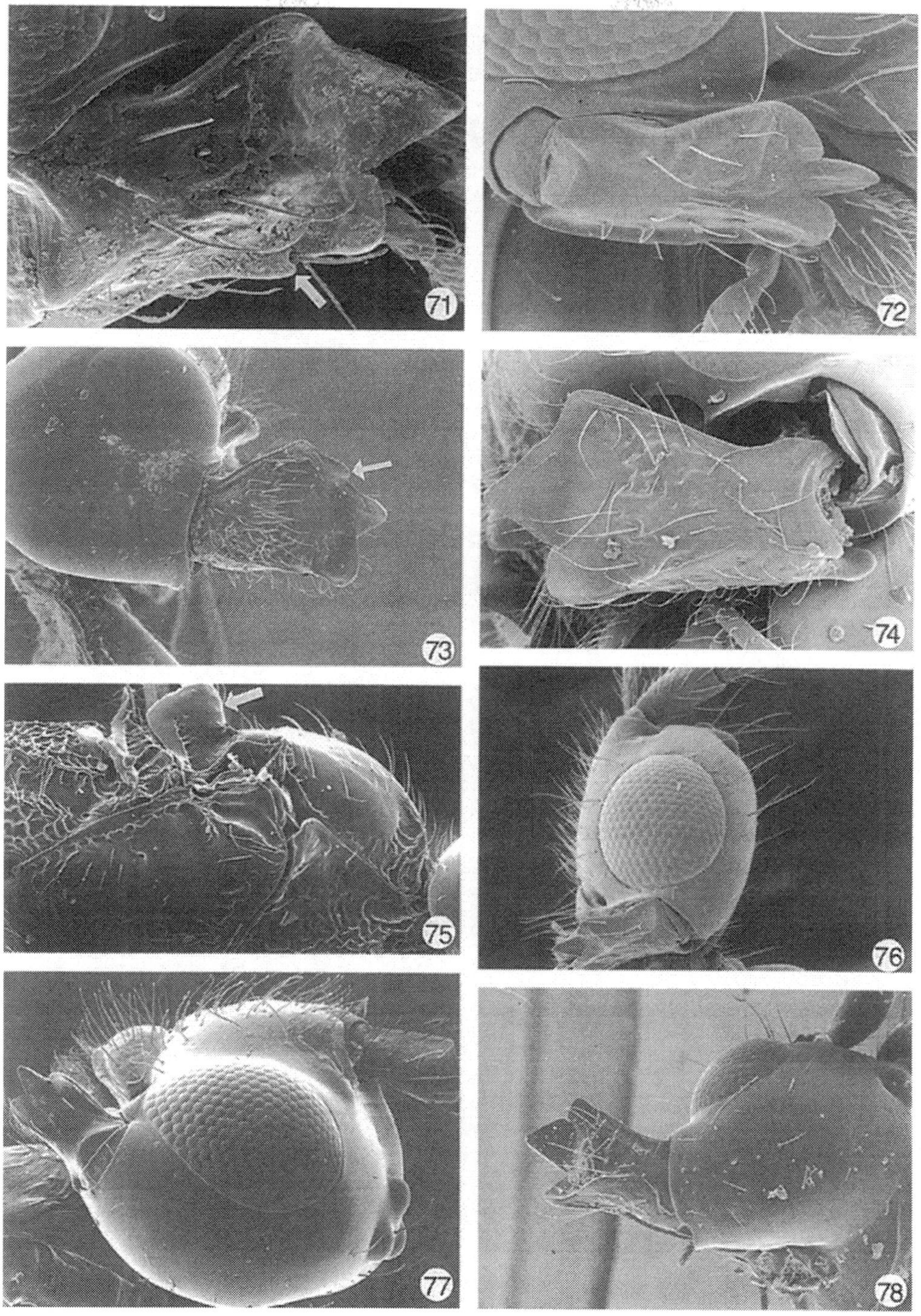

Figs 71-78: Alysiinae. 71-73, right mandibles: 71, *Pentapleura pumilio* (Nees), arrow =4th tooth along ventral margin basad tooth 3; 72, *Phaenocarpa cratomorpha* Wharton (after Wharton, 1994); 73, *Gnathopleura* sp., arrow = 4th tooth between tooth 1 and tooth 2; 74, left mandible of *Idiasta delicata* (Papp); 75, mesosoma of *Hoplitalysia slossonae* Ashmead, lateral view, arrow = conical scutellum; 76, head and mandible of *Vachterbergia* sp. (after Wharton, 1994); 77, head and mandible of *Dinotrema* sp.; 78, head and mandible of *Pneumosema stigmaticum* Fischer.

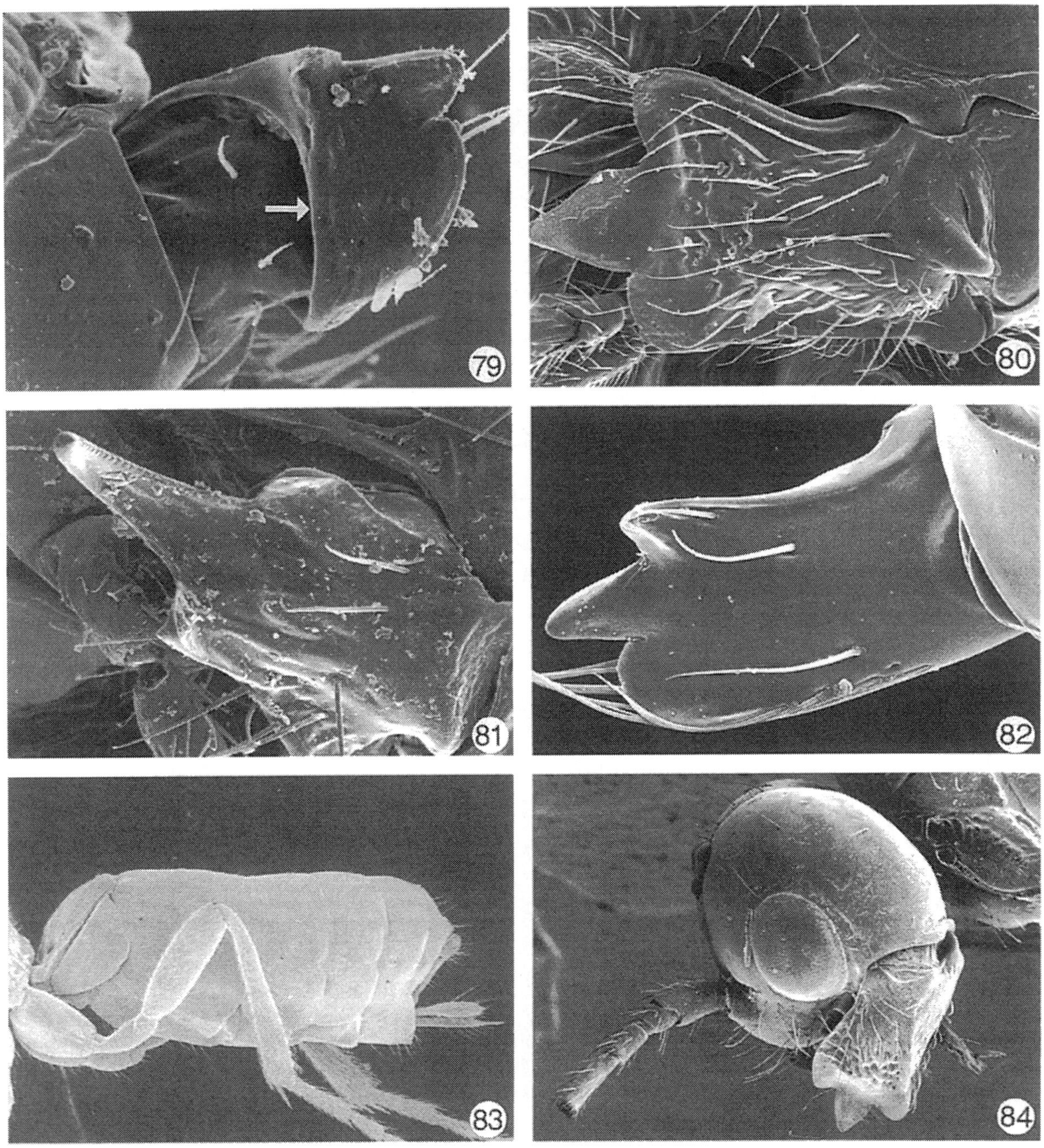

Figs 79-84: Alysiinae. 79, right mandible of *Orthostigma* sp., arrow = transverse keel; 80-82, left mandibles: in, 1988b); 81, *Alloea* sp.; 82, *Dinotrema* sp. (after Wharton, 1985); 83, metasoma of lateral view showing short ovipositor (after Wharton, 1986); 84, head and mandible of *Alysiasta caltagironei* Wharton.

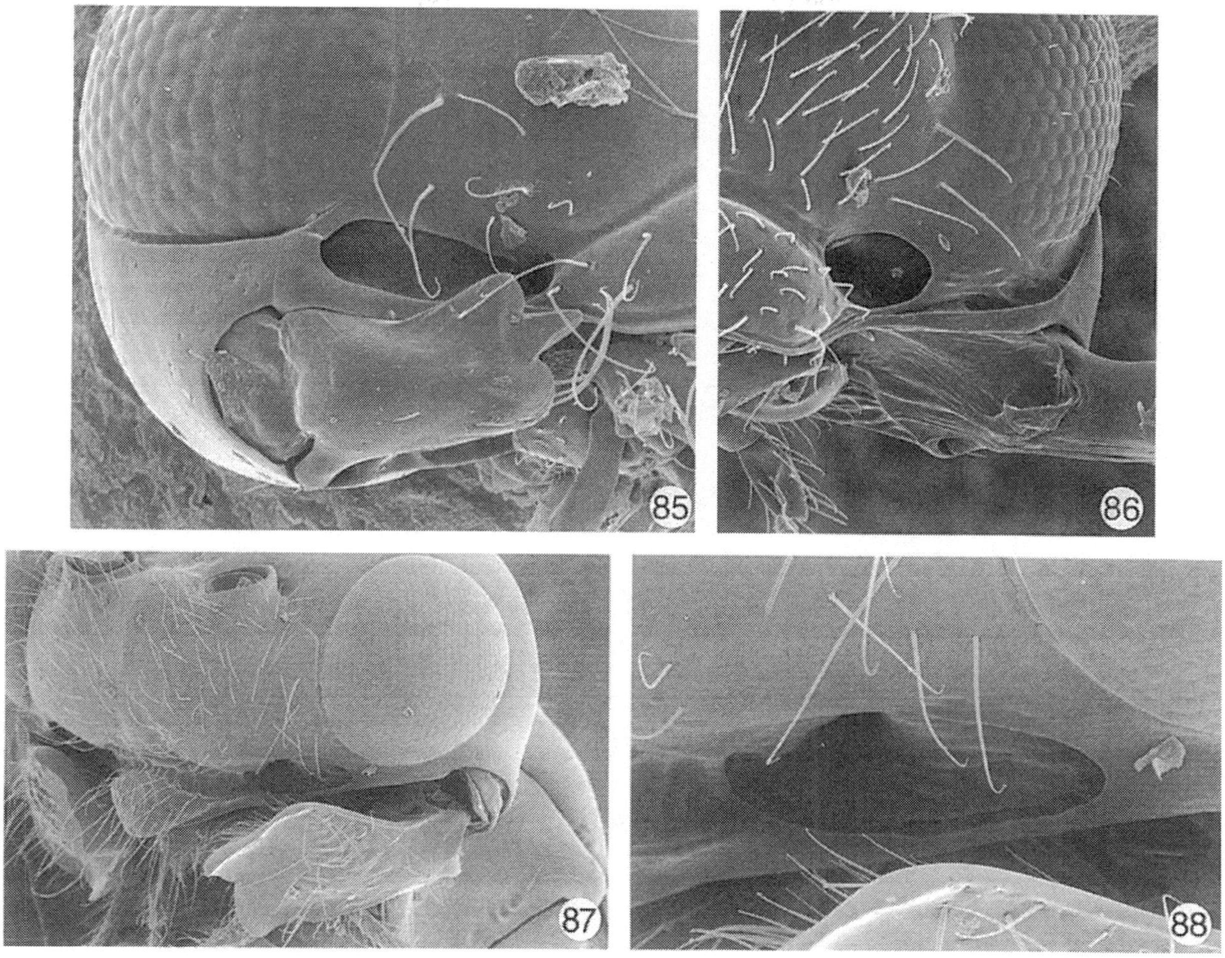

Figs 85-88: Alysiinae. 85, mandible and anterior tentorial pit of *Aspilota* sp.; 86, anterior tentorial pit of *Dinotrema* sp.; 87, face of *Idiasia delicata* (Papp) showing large anterior tentorial pit; 88, close up of anterior tentorial pit of *I. delicata* (Papp). Figures 85 and 86 after Wharton (1985).

ALYSIINAE UPDATE

R.R. Kula
Systematic Entomology Laboratory, Beltsville Agricultural Research Center, Agricultural Research Service, U.S. Department of Agriculture, c/o National Museum of Natural History, Smithsonian Institution, P.O. Box 37012, MRC-168, Washington, DC 20013-7012 U.S.A. Robert.Kula@ars.usda.gov*

Alysiinae currently contains ~403 species in the New World placed in two tribes: Alysiini with 312 species and Dacnusini with 91 species (Yu *et al.*, 2012). Approximately 80% of the described New World species are known from the Nearctic Region, but species richness observed in Malaise trap samples from the Neotropical Region suggests there are many undescribed species, especially for Alysiini (R. Kula, pers. obs.).

Taxonomic works on Alysiini published after Wharton (1997) focused on the Neotropical fauna. Trostle *et al.* (1999) described two new species of *Phaenocarpa* Förster and provided a key to species for the Neotropical Region. Braet and van Achterberg (2003), Arouca *et al.* (2004), and Arouca and Penteado-Dias (2006, 2011) collectively described eight new species of *Phaenocarpa* and provided updated keys incorporating the species described. Dix (2010) described one new species each in *Alysiasta* Wharton and *Idiasta* Förster, as well as two new species in *Tanycarpa* Förster, and provided a key to the New World species of *Alysiasta*. Kula *et al.* (2012) described six new species of *Ilatha* Fischer along with a key to all species. Papp (2012) proposed the genus *Telmogarbus* Papp, described the single included species, and described one new species each in *Aspilota* Förster and *Synaldis* Förster. Peris-Felipo and Belokobylskij (2016) described four new species of *Dinotrema* Förster and provided a key to the New World species. Cerântola *et al.* (2016) described a new species of *Tanycarpa* along with a key to the Neotropical species.

Other notable taxonomic contributions for Alysiini were descriptions of new species in *Alysiasta*, *Gnathopleura* Wharton, *Phasmidiasta* Wharton (Fischer, 2006); *Asobara* Förster (Fischer, 2007); *Aphaereta* Förster (Arouca *et al.*, 2009); *Microcrasis* Fischer and *Phaenocarpa* (Fischer, 2010); and *Aphaereta, Dinotrema*, and *Phaenocarpa* (Fischer, 2014). Fischer (2006) provided keys to some species of *Alysiasta*, *Phasmalysia* Tobias, and *Phasmidiasta* that included New World species. Fischer (2007) provided a key to some species of *Asobara* that included species from the Neotropical Region. Peris-Felipo *et al.* (2012) provided a key to all species of *Adelphenaldis* Fischer, with two from the Nearctic Region.

Relatively recent taxonomic publications on New World dacnusines have mostly treated species from the Nearctic Region. Kula and Zolnerowich (2005) described a new species of *Epimicta* Förster and provided a key to all species. Kula and Zolnerowich (2008) revised the New World species of *Chaenusa* Haliday *sensu lato*, including a key to species; Kula *et al.* (2009) supplemented the revision. Kula (2008) published new combinations and synonymies, as well as clarified the status of primary type specimens, for species of *Coelinius* Nees *sensu lato*. Van Achterberg (2006) and Kula (2013) described one new species of *Dacnusa* Haliday and *Coelinius s.l.*, respectively, and provided information on morphology and diversity for other species in Alysiini and Dacnusini. Kula and Harms (2016) described a new species of *Chaenusa* from the Nearctic Region and amended the key to New World species.

As noted in Wharton (1997), species of Alysiinae have been considered most closely related to species of Opiinae based on koinobiont endoparasitism of cyclorrhaphous Diptera and morphological features. Phylogenetic relationships among alysiines and opiines are unresolved, with results from cladistic analyses varying depending on molecular marker selection and analytical approach (Gimeno *et al.*, 1997; Dowton *et al.*, 1998; Wharton *et al.*, 2006; Zaldívar-Riverón *et al.*, 2006; Pitz *et al.*, 2007). In addition to the aforementioned cladistic analyses, Quicke *et al.* (1997) compared morphological features of the venom gland and reservoir for species of Alysiinae and Opiinae and used the results as a basis for discussing possible relationships.

Contemporary cladistic analyses for inferring relationships within Alysiinae are very limited, with Kula *et al.* (2006) for *Chaenusa* and Yaakop *et al.* (2009) for *Heratemis* Walker.

As noted in Wharton (1997), species of Dacnusini exclusively parasitize plant-mining flies and have been used in biological control. However, dacnusines are also potential biological control antagonists, particularly species of *Chaenusa* and *Chorebus* Haliday parasitic on *Hydrellia* Robineau-Desvoidy flies that feed on aquatic weeds. Kula (2009) described a new species of *Chaenusa* reared from *Hydrellia* species during foreign exploration in India and in a U.S. quarantine facility in a biological control program targeting *Hydrilla*. Kula *et al.* (2009) also described a new species of *Chaenusa* reared from a species of *Hydrellia* during foreign exploration in Argentina for biological control agents of Brazilian waterweed.

Most species of Alysiini have been reported as parasitoids of saprophagous flies, and some species attack herbivorous flies (Yu *et al.*, 2012). However, Kula *et al.* (2012) reported two species of *Ilatha* as parasitoids of tachinid flies through nymphalid and geometrid lepidopterans in Ecuador. Only a few species of Alysiinae have been reported as parasitoids of tachinids, and currently, the two species of *Ilatha* are the only species of Alysiinae known from tachinids in the New World.

Recent papers on New World Alysiinae

Achterberg C van. 2006. The Braconidae (Hymenoptera) of Greenland. *Zoologische Mededelingen, Leiden* 80(1): 13-62.

Arouca RG, Gomes SAG, Yamada, MV, Penteado-Dias AM. 2009. New species of *Aphaereta* Foerster, 1862 (Hymenoptera: Braconidae: Alysiinae) from Brazil. *Zoologische Mededelingen, Leiden* 83: 819-823.

Arouca RG, Penteado-Dias AM. 2006. Contribution to systematics of Neotropical species *Phaenocarpa* Foerster (Hymenoptera: Braconidae: Alysiinae) with description of three new species. *Studies on Neotropical Fauna and Environment* 41: 33-43.

Arouca RG, Penteado-Dias AM. 2011. A new species of *Phaenocarpa* Foerster (Hymenoptera: Braconidae: Alysiinae) from Brazil parasitizing *Neosilba perezi* (Diptera: Lonchaeidae). *Studies on Neotropical Fauna and Environment* 46: 63-67.

Arouca RG, Penteado-Dias AM, van Achterberg C. 2004. A new species of *Phaenocarpa* Foerster (Hymenoptera: Braconidae: Alysiinae) from Brazil. *Zoologische Mededelingen, Leiden* 78: 123-130.

Braet Y, van Achterberg C. 2003. New species of *Pambolus* Haliday and *Phaenocarpa* Foerster (Hymenoptera: Braconidae: Pambolinae, Alysiinae) from French Guyana, Suriname and Panama. *Zoologische Mededelingen, Leiden* 77: 153-179.

Cerântola PCM, Souza-Gessnera CS, Penteado-Dias AM. 2016. A new species of *Tanycarpa* Förster (1862) (Hymenoptera, Braconidae: Alysiinae) from Itatiaia National Park, Rio de Janeiro, Brazil. *Brazilian Journal of Biology* 76: 750-756.

Dix O. 2010. Cuatro especies nuevas de Alysiinae (Hymenoptera: Braconidae) y registro nuevo para Colombia. *Revista Colombiana de Entomología* 36: 304-314.

Dowton M, Austin AD, Antolin MF. 1998. Evolutionary relationships among the Braconidae (Hymenoptera: Ichneumonoidea) inferred from partial 16S rDNA gene sequences. *Insect Molecular Biology* 7: 129-150.

Fischer M. 2006. Neue Kieferwespen aus der Sammlung des Biologiezentrums des Oberösterreichischen Landesmuseums in Linz und Mitteilungen über andere Arten (Hymenoptera, Braconidae, Alysiinae). *Linzer Biologische Beiträge* 38: 605-651.

Fischer M. 2007. Neue Arten der Gattung *Asobara* Foerster aus drei verschiedenen Erdteilen (Hymenoptera, Braconidae, Alysiinae, Alysiini). *Linzer Biologische Beiträge* 39: 857-875.

Fischer M. 2010. Einige neue Taxa der Kieferwespen aus der Sammlung des Biologiezentrums des Oberösterreichischen Landesmuseums in Linz (Hymenoptera, Braconidae, Alysiinae). *Linzer Biologische Beiträge* 42: 635-657.

Fischer M. 2014. Winzige neue Arten der Kiefer- und Madenwespen aus Ecuador (Insecta: Hymenoptera: Braconidae: Alysiinae, Exothecinae, Opiinae). *Annalen des Naturhistorischen Museums in Wien*. Serie B für Botanik und Zoologie 116: 21-39.

Gimeno C, Belshaw R, Quicke DLJ. 1997. Phylogenetic relationships of the Alysiinae/Opiinae (Hymenoptera: Braconidae) and the utility of cytochrome b, 16S and 28S D2 rRNA. *Insect Molecular Biology* 6: 273-284.

Kula RR. 2008. Taxonomic status and location of type specimens for species of *Coelinidea* Viereck and *Sarops* Nixon (Hymenoptera: Braconidae: Alysiinae) described by Garland T. Riegel. *Journal of Hymenoptera Research* 17: 138-156.

Kula RR. 2009. A new species of *Chaenusa* (Hymenoptera: Braconidae) reared from *Hydrellia pakistanae* and *Hydrellia sarahae laticapsula* (Diptera: Ephydridae) infesting *Hydrilla verticillata* (Alismatales: Hydrocharitaceae) in India and Pakistan. *The Florida Entomologist* 92: 139-146.

Kula RR. 2013. A new species and new distribution records for Braconidae from Mountain Lake Biological Station in southwestern Virginia and a redescription of *Pentapleura foveolata* Viereck. *Zootaxa* 3641: 501-523.

Kula RR, Harms NE. 2016. A new species of *Chaenusa* Haliday *sensu lato* (Hymenoptera: Braconidae) from the Nearctic Region. *Proceedings of the Entomological Society of Washington* 118: 101-108.

Kula RR, Martinez JJ, Cabrera Walsh G. 2009. Supplement to revision of New World *Chaenusa* Haliday *sensu lato* (Hymenoptera: Braconidae: Alysiinae). *Proceedings of the Entomological Society of Washington* 111: 641-655.

Kula RR, Zolnerowich G. 2005. A new species of *Epimicta* Förster (Hymenoptera: Braconidae) from North America and new distribution records for *Epimicta griffithsi* Wharton. *Proceedings of the Entomological Society of Washington* 107: 78-83.

Kula RR, Zolnerowich G. 2008. Revision of New World *Chaenusa* Haliday *sensu lato* (Hymenoptera: Braconidae: Alysiinae), with new species, synonymies, hosts, and distribution records. *Proceedings of the Entomological Society of Washington* 110: 1-60.

Kula RR, Zolnerowich G, Ferguson CJ. 2006. Phylogenetic analysis of *Chaenusa sensu lato* (Hymenoptera: Braconidae) using mitochondrial NADH 1 dehydrogenase gene sequences. *Journal of Hymenoptera Research* 15: 251-265.

Kula RR, Dix-Luna O, Shaw, SR. 2012. Review of *Ilatha* Fischer (Hymenoptera: Braconidae: Alysiinae), including descriptions of six new species and a key to species. *Proceedings of the Entomological Society of Washington* 114: 293-328.

Papp J. 2012. Five new braconid species from Colombia (Hymenoptera, Braconidae). *Journal of Hymenoptera Research* 28: 67-84.

Peris-Felipo FJ and Belokobylskij SA. 2016. First record of the genus *Dinotrema* Foerster, 1862 (Hymenoptera, Braconidae, Alysiinae) from the Neotropical region, with description of four new species and a key to the New World taxa. *European Journal of Taxonomy* 179: 1-23

Peris-Felipo FJ. Belokobylskij SA, Jiménez-Peydró R. 2012. Genus *Adelphenaldis* Fischer, 2003 (Hymenoptera: Braconidae: Alysiinae) in Spain, with a key to the world species. *Annales Zoologici* 62: 287-296.

Pitz KM, Dowling APG, Sharanowski BJ, Boring CA, Seltmann KC, Sharkey MJ. 2007. Phylogenetic relationships among the Braconidae (Hymenoptera: Ichneumonoidea): a reassessment of Shi *et al.* (2005). *Molecular Phylogenetics and Evolution* 43: 338-343.

Quicke DLJ, van Achterberg K, Godfray HCJ. 1997. Comparative morphology of the venom gland and reservoir in opiine and alysiine braconid wasps (Insecta, Hymenoptera, Braconidae). *Zoologica Scripta* 26: 23-50.

Trostle M, Carrejo NS, Mercado I, Wharton RA. 1999. Two new species of *Phaenocarpa* Foerster (Hymenoptera: Braconidae: Alysiinae) from South America. *Proceedings of the Entomological Society of Washington* 101: 197-207.

Wharton RA. 1997. Subfamily Alysiinae. Pp. 84-117 In: Wharton, R.A., Marsh, P.M., and Sharkey, M.J. (Eds.). Manual of the New World Genera of the Family Braconidae (Hymenoptera). Special Publication No. 1. International Society of Hymenopterists, Washington, DC.

Wharton RA, Yoder MJ, Gillespie JJ, Patton JC, Honeycutt RL. 2006. Relationships of *Exodontiella*, a non-alysiine, exodont member of the family Braconidae (Insecta, Hymenoptera). *Zoologica Scripta* 35: 323-340.

Yu DS, van Achterberg K, Horstmann K. 2012. World Ichneumonoidea 2011. Taxonomy, Biology, Morphology and Distribution. USB Flash drive. Taxapad, Ottawa, Canada. www.taxapad.com.

Yu DSK, van Achterberg C, Horstmann K. 2016. Taxapad 2016, Ichneumonoidea 2015. Database on flash-drive. Nepean, Ontario, Canada. http://www.taxapad.com.

Zaldívar-Riverón A, Mori M, Quicke, DLJ. 2006. Systematics of the cyclostome subfamilies of braconid parasitic wasps (Hymenoptera: Ichneumonoidea): a simultaneous molecular and morphological Bayesian approach. *Molecular Phylogenetics and Evolution* 38: 130-145.

Yaakop S, van Achterberg C, Ghani IbA. 2009. *Heratemis* Walker (Hymenoptera: Braconidae: Alysiinae: Alysiini): revision and reconstruction of the phylogeny combining molecular data and morphology. *Tijdschrift voor Entomologie* 152: 3-64.

*Mention of trade names or commercial products in this publication is solely for the purpose of providing specific information and does not imply recommendation or endorsement by the USDA. USDA is an equal opportunity provider and employer.

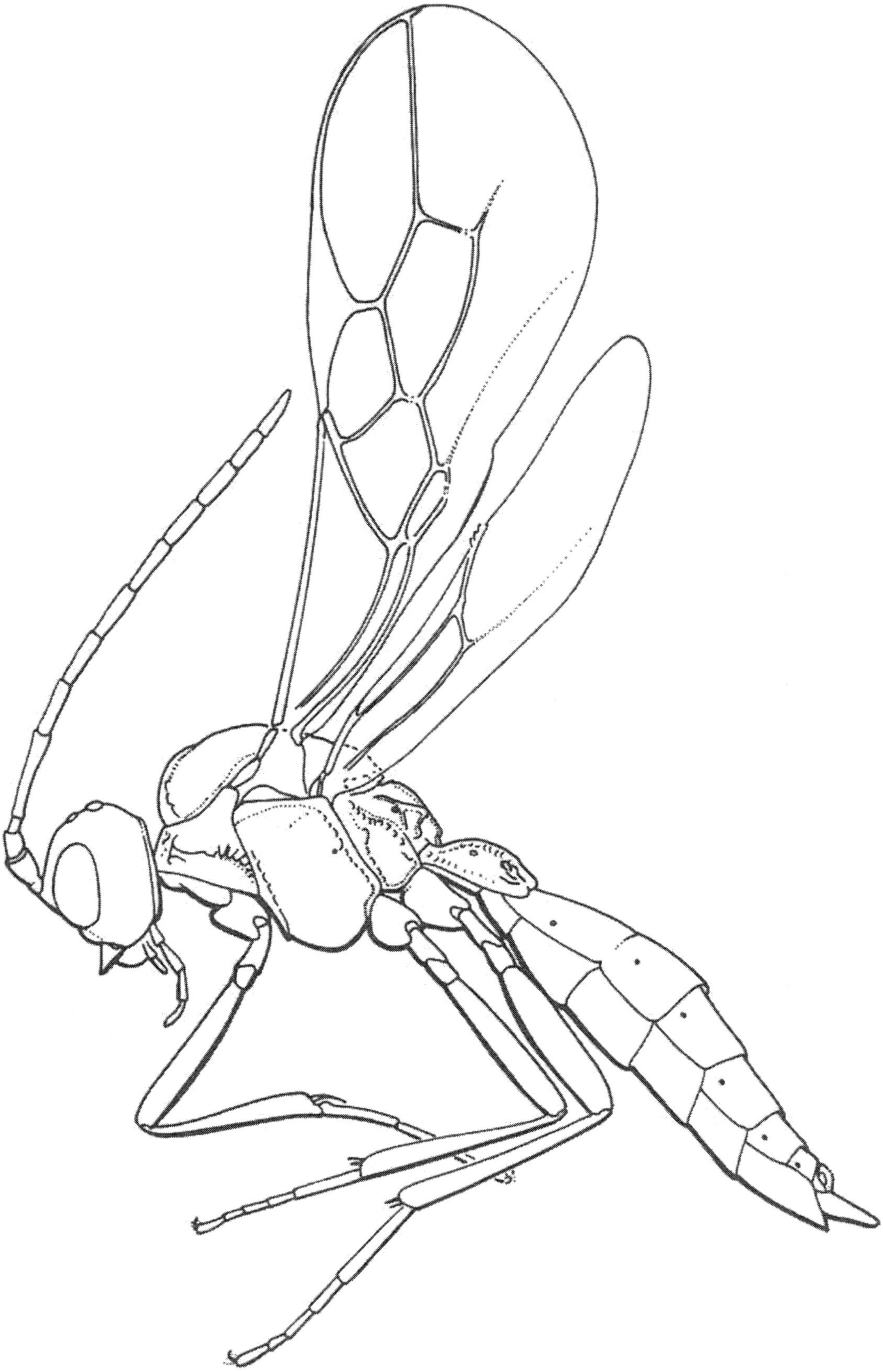

SUBFAMILY APHIDIINAE

C. VAN ACHTERBERG

Afdeling Entomologie (Hymenoptera), Nationaal Natuurhistorisch Museum, Postbus 9517,
2300 RA Leiden, The Netherlands

INTRODUCTION. The subfamily Aphidiinae is a medium-sized and rather uniform group, with species usually 1.5-3.5 mm long. Females have a short ovipositor (Figs 23, 31), and its sheath may be strongly widened, or the hypopygium may possess prongs (Fig. 24). Several genera are known and most are widespread. Four tribes are recognized: the Ephedrini, with fairly complete fore wing venation, the Praini, with a closed cell in the hind wing, and two tribes with reduced venation, the Aphidiini and the Trioxini. The females of the latter usually possess prongs on their hypopygium.

The Nearctic aphidiines were revised by Smith (1944), and more recent treatments are available for several Latin American countries (Starý, 1981b, 1983, 1995; Starý and Remaudière, 1982, 1983). The Nearctic species of the genus *Praon* Haliday were revised by Johnson (1987). Additional taxonomic information on species in the New World can be found in several papers treating specific hosts (e.g. Marsh, 1977,1991; Unruh et al., 1989). Aphidiines were catalogued on a World basis by Mackauer and Starý (1967) and Mackauer (1968) and the fauna of America north of Mexico was catalogued by Marsh (1979). Starý (1987) published a useful bibliography on aphidiines covering the period from 1758-1982, arranged by subject.

BIOLOGY. Aphidiines have been reared extensively; all are solitary, koinobiont endoparasitoids of adult and immature aphids. The biology of Aphidiinae has been summarized by Starý (1970) and Shaw and Huddleston (1991). In addition to extensive publications on hosts, a large number of excellent studies have been conducted on the biology of aphidiines. Recent examples include parameters affecting foraging behavior (Read et al.,1970; Liu et al., 1984, Hofsvang and Hagvar, 1986; Powell and Wright, 1988; Sheehen and Shelton, 1989; Völkl and Mackauer, 1990; Powell, 1991; Bueno et al., 1993; Pennacchio et al., 1994), basic life history (Calvert, 1973; Chow and Sullivan, 1984; Tripathi and Singh, 1990; Reed et al., 1992), immatures (O'Donnell, 1989; Finlayson, 1990), host-induced effects and other factors affecting performance (Singh and Srivastava, 1989; Kambhampati and Mackauer, 1989; Powell and Wright, 1991), biotypes and sibling species (Flint, 1980; Tremblay and Pennacchio, 1988). The following families of aphids are attacked: Anoeciidae, Aphididae, Drepanosiphidae, Greenideidae, Hormaphididae, Lachnidae, Mindaridae, Pemphigidae, and Thelaxidae. Aphid classification follows Heie (1980).

BIOLOGICAL CONTROL POTENTIAL. Aphidiines have been used extensively in biological control and integrated pest management (Clausen, 1978), and their distribution pattern in the New World has been greatly influenced by purposeful and accidental introductions. In the New World, successful classical biological control programs include those for alfalfa aphid (Hagen and van den Bosch, 1968; Flint, 1980), walnut aphid (van den Bosch et al., 1979), and aphids on urban trees (Zuparko, 1983). In Europe, *Aphidius colemani* Viereck is used against aphids in greenhouses with good results.

COMMON GENERA. *Aphidius* Nees, *Lysiphlebus* Förster, and *Diaeretiella rapae* (M'Intosh) have been found in every country intensively surveyed in the New World, and thus are the most commonly encountered genera. *Binodoxys* Mackauer, *Pauesia* Quilis Pérez, *Praon* Haliday, *Trioxys* Haliday, and *Ephedrus* Haliday are also major New World genera, at least in temperate regions.

DISTRIBUTION. The Aphidiinae are Worldwide in distribution with several widespread introduced species. As most species of aphids occur in the northern Hemisphere, the Aphidiinae are primarily a Holarctic group. However, many species have been introduced to South America either accidentally with their hosts or purposefully for biological control (e.g. Starý, 1995), and *Binodoxys* is particularly diverse in Mexico. Five genera are endemic to the New World, and one of these (*Pseudephedrus* Starý) is unknown north of the Tropic of Cancer.

DISTINGUISHING FEATURES. Aphidiinae have a fragile habitus, with metasomal tergites weakly sclerotized, with rather short antennae (usually consisting of 18 or fewer segments) and often with modified ovipositor sheath or hypopygium in females. Unlike all other Braconidae, the suture between the second and third metasomal segments is flexible, allowing the metasoma to bend at this joint (Figs 31, 32). During oviposition, the metasoma usually swings vertically and often beneath the rest of the body (with the ovipositor thus projecting anteriorly). Many of the species have a single, large, median cell in the fore wing (second submarginal cell open basally and united with discal and first submarginal cells as in Figs 12, 14, 15). Of further assistance in recognition of aphidiines are the smooth scutellar sulcus (very rarely sculptured); the absence of cross vein cu-a in the hind wing; and the presence, at least laterally, of an occipital carina.

KEY TO NEW WORLD GENERA OF THE SUBFAMILY APHIDIINAE

1. Stigma of fore wing nearly absent, vein-like (Fig. 1), not easily distinguished from R1 .... ***Pseudephedrus*** Starý
Distribution: Neotropical (originally described from Cuba); 5 described species. Biology: reared from *Neolizerius* sp. on *Nectandra* sp. and *Neuquenaphis* spp. associated with *Nothofagus* spp. References: Schlinger (1974), Starý (1972, 1976, 1994, key to species on *Nothofagus*).

\- Stigma distinct, triangular (Figs 2-15), easily distinguished from R1 (when R1 present) 2

2(1). Venation of fore wing complete, with 3 submarginal cells present and marginal cell closed (3RSb reaching wing margin) (Figs 2, 3) 3

\- Venation of fore wing reduced, without 3 clearly separated submarginal cells; marginal cell open distally (RS not reaching wing margin) (Figs 4-15) 4

3(2). Antenna with 18 (female) or more (male) segments. Ovipositor sheath curved downward, deltoid, and apically trifid (Fig. 16). Wings: Fig. 2 .... ***Toxares*** Haliday
Distribution: U.S.A. (Idaho, Texas) and Canada; rare; only one described species in New World. Biology: reared from Aphididae (Aphidinae). Reference: Marsh (1979).

\- Antenna with 11-12 segments. Ovipositor sheath straight or curved upward, slender and apically narrowed (Fig. 17). Wings: Fig. 3 .... ***Ephedrus*** Haliday
Distribution: Alaska and Canada to Argentina and Chile; common; 8 described species in New World. Biology: Völkl and Mackauer (1990); reared from Aphididae, Drepanosiphidae, Pernphigidae, and Anoeciidae. References: Smith (1944), Marsh (1979), Gärdenfors (1986), Starý (1995).

4(2). At least basal two-fifths of petiole tube-shaped, its sternum fused with the median tergite, at least to level of spiracles (Figs 18, 19) 5

\- Basal half of petiole not distinctly tube-shaped, its sternum at least partly free from median tergite, fused part not reaching level of spiracles (Fig. 20) 9

5(4). Hypopygium of female without prongs (Figs 21, 22). Fore wing m-cu (and usually also r-m) at least partly present (Fig. 4), rarely completely absent 6

\- Hypopygium with prongs (Figs 23, 24). Fore wing m-cu and r-m absent (Fig. 5) 8

6(5). Petiole slender, its length 3-4 times its apical width (Fig. 33). Ventral margin of ovipositor sheath straight or nearly so .... ***Quadrictonus*** Starý and Remaudière
Distribution: Mexico; known only from the type species. Biology: reared from Aphididae (Macrosiphini). References: Starý & Remaudière (1982), van Achterberg (1989).

\- Petiole moderately robust, its length 1.5-2.0 times its apical width (Figs 34, 35). Ventral margin of ovipositor sheath curved (Fig. 21) 7

7(6). Propodeal areola open anteriorly and medially; propodeum largely smooth (Fig. 34) Ovipositor sheath comparatively slender, gradually widened ventrally .... ***Harkeria*** Cameron
Distribution: U.S.A. (Maine, Pennsylvania): uncommon; one described species in New World. Biology: reared from Aphididae (Aphidinae). Reference: van Achterberg (1989).

\- Propodeal areola closed anteriorly (Fig. 35), if carinae are weak then propodeum rugose medially. Ovipositor sheath distinctly ploughshare-shaped (Fig. 21). Wings: Fig. 4 .... ***Monoctonus*** Haliday
Distribution: Alaska and Canada to Mexico, Chile; at least 3 described species in New World. Biology: Calvert (1973); reared from Aphididae (Aphidinae), Thelaxidae and Drepanosiphidae. References: Smith (1944), Starý and Smith (1976), van Achterberg (1989), Starý (1995).

8(5). Accessory prongs of female hypopygium present, spine-like (Fig. 24). Subapical row of pegs or spiny bristles present on fifth, sixth, and seventh terga of female (Fig. 24). Wings: Fig. 5 ..***Acanthocaudus*** Smith

Distribution: across southern Canada to Texas and Florida; uncommon; 3 described species. Biology: apparently restricted to the aphid genus *Dactylonotus* Rafinesque. Reference: Smith (1944).

\- Accessory prongs of female absent (Fig. 23). Fifth, sixth, and seventh terga of female without subapical row of pegs or spiny bristles (Fig. 23).. ***Binodoxys*** Mackauer

Distribution: U.S.A., Canada, Mexico, Cuba, Venezuela, Chile; common; at least 12 described species. in New World; includes *Cristicaudus* Starý, a probable synonym. Biology: reared from Aphididae (primarily Aphidinae). References: Marsh (1979), Starý (1995).

9(4). Hind wing M+CU distinctly sclerotized (tubular) (Fig. 6).. 10

\- Hind wing M+CU not sclerotized, at most pigmented (Figs 9-15), but never tubular 11

10(9). Fore wing RS+M completely absent (Fig. 7). Ovipositor sheath densely setose (Fig. 36). Sternaulus absent .... ***Pseudopraon*** Starý

Distribution: Holarctic to Mexico; rare; one described species in New World. Biology: Starý (1975); reared from *Mindarus* Koch; pupation inside (oviposition in non diapause nymphs) or outside and below aphid (oviposition in diapause nymphs).

\- Fore wing RS+M present, at least as a fold (Fig. 6). Ovipositor sheath sparsely setose as in Fig. 22. Sternaulus shallowly impressed .... ***Praon*** Haliday

Distribution: Alaska and Canada to Argentina and Chile; common; at least 17 described species in New World. Biology: Schlinger and Hall (1960), Chow and Sullivan (1984); reared from Aphididae (Aphidinae), Lachnidae, Thelaxidae, and Drepanosiphidae (Chaitophorinae). Reference: Johnson (1987; key to species).

11(9). Fore wing m-cu and r-m completely absent (Figs 8-11).. 12

\- Fore wing r-m and/or m-cu at least partly present (Figs 12-15) 16

12(11). Scutellum with lateral carina present anteriorly (Fig. 37). First subdiscal cell of fore wing narrow, faintly indicated (Fig. 8) .... ***Boreogalba*** Mackauer

Distribution: Alaska, northeastern Canada; rare; only known species: *B. gladifer* Mackauer. Biology: unknown. Reference: Mackauer (1962).

\- Scutellum with lateral carina absent (Fig. 38). First subdiscal cell of fore wing variable, but usually better developed, at least basally (Figs 9-11) 13

13(12). Hypopygium of female with prongs (Fig. 23). Fore wing r&RS comparatively long and strongly curved (Fig. 9) and arising near middle of stigma .... ***Trioxys*** Haliday

Distribution: Alaska and Canada to Mexico; common; at least 13 described species in New World. Biology: Schlinger and Hall (1961), Singh and Srivastava (1989); reared from Aphididae, Thelaxidae, and Drepanosiphidae. References: Smith (1944), Marsh (1979), Starý (1981a).

\- Hypopygium of female without prongs (Figs 21, 22). Fore wing r&RS usually of medium-length and slightly curved (Fig. 10); if long and distinctly curved then r&RS arising closer to base of stigma 14

14(13). Notauli completely absent (Fig. 39). Propodeum with large, wide areola medio-posteriorly (Fig. 25) .... ***Diaeretus*** Förster

Distribution: Nearctic; rare; only known species: *D. leucopterus* (Haliday). Biology: reared from Lachnidae (*Eulachnus* del Guercio). Reference: Mackauer (1968).

\- Notauli present on anterior, perpendicular part of mesoscutum (Fig. 40). Propodeum with areola absent or with narrow areola (Figs 26, 27) 15

15(14). Propodeal carinae largely absent (Fig. 26). Sternaulus absent or shallow and smooth. Wings: Fig. 11 .... ***Adialytus*** Förster

Distribution: Canada to Mexico; uncommon; 4 described species in New World. Biology: reared from Aphididae, Drepanosiphidae, and Thelaxidae. Reference: Marsh (1979, as subgenus of *Lysiphlebus*).

\- Propodeum with carina or narrow areola present medio-posteriorly (Fig. 27). Sternaulus deep and finely crenulate. Wings: Fig. 10 .... ***Diaeretiella*** Starý

Distribution: throughout New World; common; only known species: *D. rapae* (M'Intosh). Biology: reared From Aphididae, commonly on cabbage aphid. Reference: Read et al. (1970).

16(11). Propodeum deeply impressed medially and smooth (Fig. 28). Petiole slender basally (Fig. 28). Wings: Fig. 12 .... ***Xenostigmus*** Smith

Distribution: U.S.A. and Canada; uncommon; only known species: *X. bifasciatus* (Ashmead). Biology: reared from Lachnidae. References: Smith (1944), Marsh (1979).

\- Propodeum not deeply impressed medially, if apparently impressed then propodeum areolate (Figs 25, 27, 29, 30). Petiole robust basally (Figs 26, 30) 17

17(16). Fore wing m-cu absent or nearly so (Fig. 13); first subdiscal cell of fore wing open distally in most species18

\- Fore wing m-cu present (Figs 14, 15), rarely reduced; first subdiscal cell of fore wing closed, bordered by sclerotized veins (Figs 14, 15) 19

18(17). Propodeum with narrow areola present, bordered by carinae (Fig. 30) .... ***Lysaphidus*** Smith

Distribution: southern Canada to Mexico; rarely collected; at least 4 described species in New World; includes *Diaeretellus* Starý, a probable synonym. Biology: reared from Aphididae (Aphidinae). References: Smith (1944), Marsh (1979).

- Propodeum unsculptured, lacking closed areola (Fig. 26), at most with pair of short carinae at apex. Wings: Fig.13 .... ***Lysiphlebus*** Förster

  Distribution: throughout New World, probably the most commonly encountered aphidiine in southern U.S.A; at least 7 described species in New World. Biology: Rotundo et al. (1988); reared from Aphididae and Anoeciidae. References: Smith (1944), Marsh (1979), Starý (1995).

19(17). Areola of propodeum, when present, very wide posteriorly, its maximum width exceeding its maximum length (Figs 25, 29); if areola absent, then transverse carinae very strong and posterior face of propodeum almost perpendicular to dorsal face. Antennal segments 18-30. Wings: Fig. 14 .... ***Pauesia*** Quilis Pérez

  Distribution: Canada to Mexico; moderately common; at least 17 described species in New World; includes *Chaetopauesia* Mackauer, a probable synonym. Biology: reared from Lachnidae (especially *Cinara* Curtis), and probably also from Hormaphididae. References: Smith (1944), Marsh (1979, 1991), Pike and Starý (1996).

- Areola of propodeum comparatively narrow posteriorly, its maximum width less than or equal to its maximum length (Figs 27, 30), or areola absent and replaced by median, posterior carina. Antennal segments 12-21.. 20

20(19). Pronotum with pair of deep, well-defined depressions positioned as in Fig. 41. Tarsal claws as long as or longer than arolium (Fig. 43).. ***Euaphidius*** Mackauer

  Distribution: U.S.A. and Canada; rare; 2 described species in New World. Biology: reared from Aphididae (*Pterocomma* Buckton and *Periphyllus* van der Hoeven). Reference: Marsh (1979).

- Pronotum with depressions usually absent or poorly defined (Fig. 42); if apparently deep, then not positioned as in Fig. 41. Tarsal claws somewhat shorter than arolium (Fig. 44). Wings: Fig. 15 .... ***Aphidius*** Nees

  Distribution: throughout New World; common; about 30 described species in New World. Biology: Bueno et al. (1993); reared from Aphididae, Thelaxidae, and Drepanosiphidae (Chaitophorinae). References: Pungerl (1983, 1986), Starý (1995, Chile).

REFERENCES

Achterberg, C. van 1989. Revision of the subtribe Monoctonina Mackauer sensu stricto (Hymenoptera: Braconidae: Aphidiinae), Zoologische Mededelingen Leiden 63: 1-22.

Bosch, R. van den, Horn, R., Matteson, P., Frazer, B. D., Messenger, P. S., and Davis, C. S. 1979. Biological control of the walnut aphid in California: impact of the parasite, *Trioxys pallidus*. Hilgardia 47: 1-13.

Bueno, B. H. P., Gutierrez, A. P., and Ruggle, P. 1993. Parasitism by *Aphidius ervi* [Hym.: Aphidiidae]: preference for pea aphid and blue alfalfa aphid [Horn.: Aphididae] and competition with *A. smithi*. Entomophaga 38: 273-284.

Calvert, D. J. 1973. Experimental host preferences of *Monoctonus paulensis* (Hymenoptera: Braconidae), including a hypothetical scheme of host selection. Annals of the Entomological Society of America 66: 28-33.

Chow, F. J. and Sullivan, D. J. 1984. Developmental stages of *Praon pequodorum* Viereck (Hymenoptera: Aphidiidae), a pea aphid parasitoid. Annals of the Entomological Society of America 77: 319- 322.

Clausen, C. P. 1978. Introduced parasites and predators of arthropod pests and weeds: a World review. Agriculture Handbook No. 480. United States Government Press, Washington, D. C. 545 pp.

Finlayson, T. 1990. The systematics and taxonomy of final instar larvae of the family Aphidiidae (Hymenoptera). Memoirs of the Entomological Society of Canada 152: 1-74.

Flint, M. L. 1980. Climatic ecotypes in *Trioxys complanatus*, a parasite of the spotted alfalfa aphid. Environmental Entomology 9: 501-507.

Gärdenfors, U. 1986. Taxonomic and biological revision of Palaearctic *Ephedrus* Haliday (Hymenoptera: Braconidae, Aphidiinae). Entomologica Scandinavica, Supplement 27: 1-95.

Hagen, K. S. and Bosch, R. van den 1968. Impact of pathogens, parasites, and predators on aphids. Annual Review of Entomology 13: 325-384.

Heie, 0. E. 1980. The Aphidoidea (Hemiptera) of Fennoscandia and. Denmark. I. General part. The families Mindaridae, Hormaphididae, Thelaxidae, Anoeciidae, and Pemphigidae. Fauna Entomologica Scandinavica 9: 1-236.

Hofsvang, T. and Hagvar, E. B. 1986. Oviposition behaviour of *Ephedrus cerasicola* [Hym.: Aphidiidae] parasitizing different instars of its aphid host. Entomophaga 31: 261-267.

Johnson, J. W. 1987. A revision of the species of *Praon* Haliday in North America north of Mexico (Hymenoptera: Aphidiidae). The Canadian Entomologist 119: 999-1025.

Kambhampati, S. and Mackauer, M. 1989. Multivariate assessment of inter and intraspecific variation in performance criteria of several pea aphid parasites (Hymenoptera: Aphidiidae). Annals of the Entomological Society of America 82: 314-324.

Liu, S.-s., Morton, R., and Hughes, R. D. 1984. Oviposition preferences of a hymenopterous parasite tor certain instars of its aphid host. Entomologia Experimentalis et Applicata 35: 249-254.

Mackauer, M. 1962. A new genus of Aphidiidae (Hymenoptera) from northern Canada. The Canadian Entomologist 94: 1107-1109.

Mackauer, M. 1968. Pars 3. Aphidiidae. Pp. 1-103. In: Ferrière, C. and Vecht, J. van der (Eds.). Hymenopterorum Catalogus (nova editio). Dr. W. Junk, The Hague.

Mackauer, M. and Starý, P. 1967. Hymenoptera Ichneumonoidea. World Aphidiidae. Index of Entomophagous Insects. Le Francois, Paris. 195 pp.

Marsh, P. M. 1977. Notes on the taxonomy and nomenclature of *Aphidius* species (Hym., Aphidiidae) parasitic on the pea aphid in North America. Entomophaga 22: 365-372.

Marsh, P. M. 1979. Family Aphidiidae, Pp. 295-313. In: Krombein, K. V., Hurd, P. O., Smith, D. R., and Burks, B. D. (Eds.). Catalog of Hymenoptera in America North of Mexico. Smithsonian Institution Press, Washington, D.C.

Marsh, P. M. 1991. A new species of *Pauesia* (Hymenoptera: Braconidae, Aphidiinae) from Georgia and introduced into South Africa against the black pine aphid (Homoptera: Aphididae). Journal of Entomological Sciences 26: 81-84.

O'Donnell, D. J. 1989. A morphological and taxonomic study of first instar larvae of Aphidiinae (Hymenoptera: Braconidae). Systematic Entomology 14: 197-219.

Pennachio, F., Digilio, M. C., Tremblay, E., and Tranfaglia, A. 1994. Host recognition and acceptance behaviour in two aphid parasitoid species: *Aphidius ervi* and *Aphidius microlophii* (Hymenoptera: Braconidae). Bulletin of Entomological Research 84: 57-64.

Pike, K. S. and Starý, P. 1996. New species of *Pauesia* (Hymenoptera: Braconidae: Aphidiinae) parasitoids on *Cinara* (Homoptera: Aphidiinae: Lachninae) associated with conifers in the Pacific Northwest. Proceedings of the Entomological Society of Washington 98: 324-331.

Powell, W. 1991. Tritrophic influences on aphid parasitoids (Hym., Braconidae; Aphidiinae). Redia 74 (Appendice): 129-133.

Powell, W. and Wright, A. F. 1988. The abilities of the aphid parasitoids *Aphidius ervi* Haliday and *A. rhopalosiphi* De Stephani Pérez (Hymenoptera: Braconidae) to transfer between different known host species and the implications for the use of alternative hosts in pest control strategies. Bulletin of Entomological Research 78: 683-693.

Powell, W. and Wright, A. F. 1991. The influence of host food plants on host recognition by four aphidiine parasitoids (Hymenoptera: Braconidae). Bulletin of Entomological Research 81: 449-453.

Pungerl, N. P. 1983. Variability in characters commonly used to distinguish *Aphidius* species (Hymenoptera: Aphidiidae). Systematic Entomology 8: 425-430.

Pungerl, N. P. 1986. Morphometric and electrophoretic study of *Aphidius* species (Hymenoptera: Aphidiidae) reared from a variety of aphid hosts. Systematic Entomology 11: 327-354.

Read, D. P., Feeny, P. P., and Root, R. B. 1970. Habitat selection by the aphid parasite *Diaeretiella rapae* (Hymenoptera: Braconidae) and hyperparasite *Charips brassicae* (Hymenoptera: Cynipidae). The Canadian Entomologist 102: 1567-1578.

Reed, H. C., Reed, D. K., and Elliott, N. C. 1992. Comparative life table statistics of *Diaeretiella rapae* and *Aphidius matricariae* on the Russian wheat aphid. Southwestern Entomologist 17: 307-312.

Rotundo, G., Cavalloro, R., and Tremblay, E. 1988. In vitro rearing of *Lysiphlebus fabarum* [Hym.: Braconidae]. Entomophaga 33: 261-267.

Schlinger, E. I. 1974. Continental drift, *Nothofagus*, and some ecologically associated insects. Annual Review of Entomology 19: 323-343.

Schlinger, E. I. and Hall, J. C. 1960. The biology, behavior and morphology of *Praon palitans* Muesebeck, an internal parasite of the spotted alfalfa aphid, *Therioaphis maculata* (Buckton) (Hymenoptera: Braconidae, Aphdiinae). Annals of the Entomological Society of America 53: 144-160.

Schlinger, E. I. and Hall, J. C. 1961. The biology, behavior, and morphology of *Trioxys (Trioxys) utilis*, an internal parasite of the spotted alfalfa aphid, *Therioaphis maculata* (Buckton) (Hymenoptera: Braconidae, Aphidiinae). Annals of the Entomological Society of America 54: 34-45.

Shaw, M. R. and Huddleston, T. 1991. Classification and biology of braconid wasps (Hymenoptera: Braconidae). Handbook for the Identification of British Insects 7(11): 1-126.

Sheehan, W. and Shelton, A. M. 1989. The role of experience in plant foraging by the aphid parasitoid *Diaeretiella rapae* (Hymenoptera: Aphidiidae). Journal of Insect Behaviour 2: 743-759.

Singh, R. and Srivastava, M. 1989. Bionomics of *Trioxys indicus* [Hym.: Aphidiidae], a parasitoid of *Aphis craccivora* [Hem.: Aphididae]: 31. Effect of host haemolymph on the numerical response of the parasitoid. Entomophaga 34: 581-586.

Smith, C.F. 1944. The Aphidiinae of North America (Braconidae: Hymenoptera). Ohio State University Contributions in Zoology and Entomology 6: 1-154.

Starý, P. 1970. Biology of aphid parasites (Hymenoptera: Aphidiidae) with respect to integrated control. Series Entomologica 6: 1-643.

Starý, P. 1972. New aphid parasites (Hymenoptera: Aphidiidae) from Cuba. Annales Zoologici, Warszawa 29: 317-322.

Starý, P. 1975. *Pseudopraon mindariphagum* gen. n., sp. n. (Hymenoptera, Aphidiidae) description and life history of a parasite of *Mindarus abietinus* (Homoptera, Mindaridae) in Central Europe. Acta Entomologica Bohemoslovaca 72: 249-259.

Starý, P. 1976. Two. new *Pseudephedrus* Starý aphid parasites (Hymenoptera: Aphidiidae) associated with *Nothofagus* in South America. With notes on continental drift. Entomologica Scandinavica 7: 24-30.

Starý, P. 1981a. Biosystematical classification of *Trioxys* Hal. and related genera (Hymenoptera, Aphidiidae). Bollettino del Laboratorio di Entomologia Agraria "Filippo Silvestri", Portici 38: 85-93.

Starý, P. 1981b. Aphid parasitoids (Hymenoptera, Aphidiidae) of Cuba. Acta Entomologica Bohemoslovaca 78: 33-42.

Starý, P. 1983. New species and records of aphid parasitoids from Mexico (Hymenoptera, Aphidiidae). Acta Entomologica Bohemoslovaca 80: 35-48.

Starý, P. 1987. Subject bibliography of aphid parasitoids (Hymenoptera: Aphidiidae) of the World 1758-1982. Monographs to Applied Entomology 25: 1-101.

Starý, P. 1994. Aphid parasitoid fauna (Hymenoptera, Aphidiidae) of the southern beech (*Nothofagus*) forest. Studies on Neotropical Fauna and Environment 29: 87-98.

Starý, P. 1995. The Aphidiidae of Chile (Hymenoptera, Ichneumonoidea, Aphidiidae). Deutsche Entomologische Zeitschrift 42: 113-138.

Starý, P. and Remaudière, G. 1982. New genera, species, and host records of aphid parasitoids (Hymenoptera, Aphidiidae) from Mexico. Annales de la Société Entomologique de France (N.S.) 18: 107-127.

Starý, P. and Remaudière. G. 1983. Complements to the aphid parasitoid fauna of Mexico (Hymenoptera, Aphidiidae). Annales de la Société Entomologique de France 19: 113-116.

Starý, P. and Smith, C. F. 1976. The Nearctic distribution, host plants and nomenclature of *Monoctonus* (Hymenoptera: Aphidiidae). Proceedings of the Entomological Society of Washington 78: 171-175.

Tremblay, E. and Pennachio, F. 1988. Speciation in aphidiine Hymenoptera. Pp. 139-146. In: Gupta, V. K. (Ed.). Advances in Parasitic Hymenoptera Research. Brill, Leiden.

Tripathi, R. N. and Singh, R. 1990. Fecundity, reproductive rate, longevity, and intrinsic rate of increase of an aphidiid parasitoid *Lysiphlebia mirzai*. Entomophaga 35: 601-610.

Unruh, T. R., White W., Gonzalez, D., and Woolley, J. B. 1989. Genetic relationships among seventeen *Aphidius* (Hymenoptera: Aphidiidae) populations, including six species. Annals of the Entomological Society of America 82: 754-768.

Völkl, W. and Mackauer, M. 1990. Age-specific pattern of host discrimination by the aphid parasitoid *Ephedrus californicus* Baker (Hymenoptera: Aphidiidae). The Canadian Entomologist 122: 349-361.

Zuparko, R. 1983. Biological control of *Eucallipterus tiliae* [Hom.: Aphididae] in San Jose, Calif., through establishment of *Trioxus curvicaudus* [Hym.: Aphidiidae]. Entomophaga 28: 325-330.

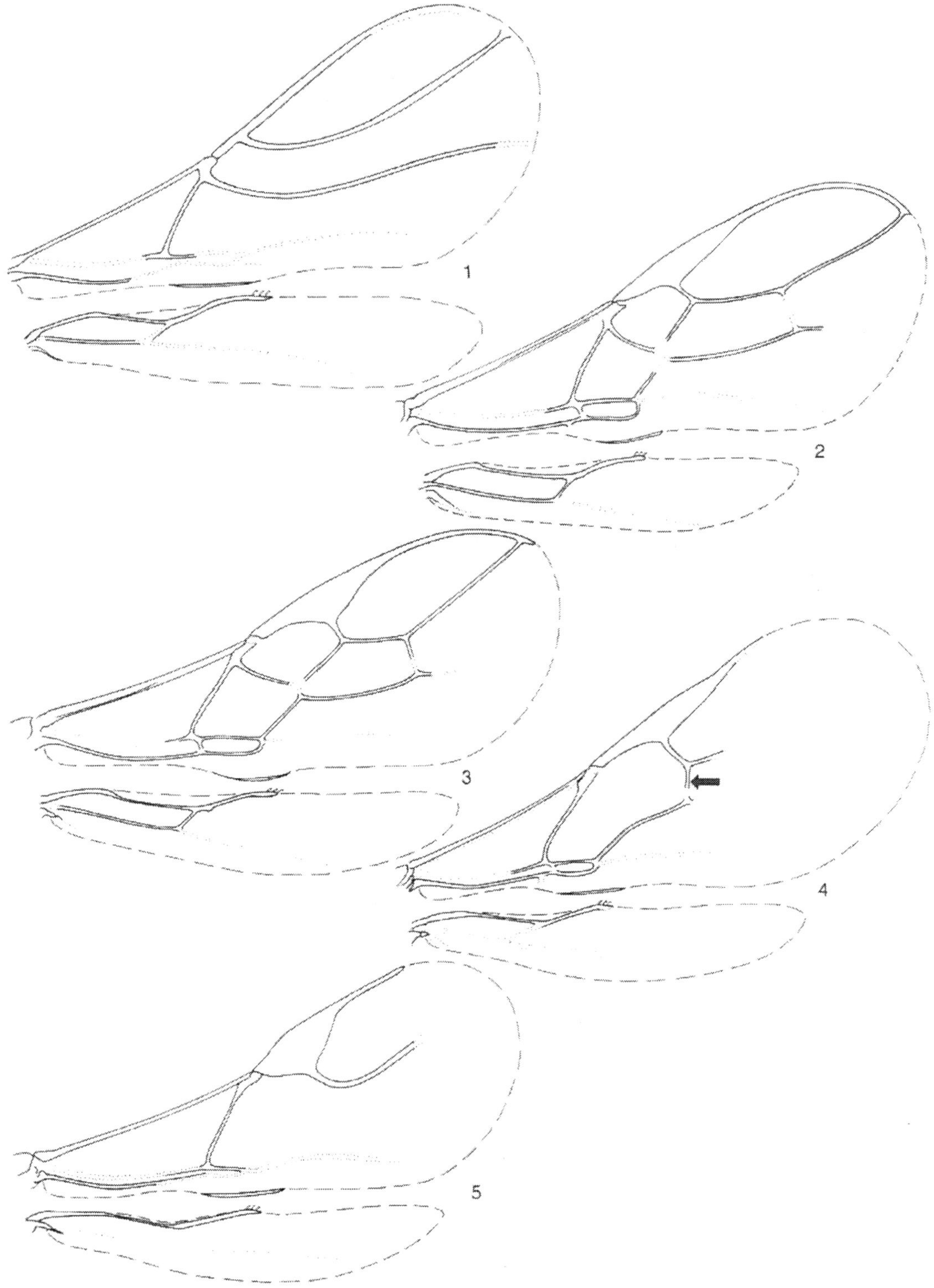

Figs 1-5: Wings of Aphidiinae. 1, *Pseudephedrus* sp.; 2, *Toxares* sp.; 3, *Ephedrus* sp.; 4, *Monoctonus*. sp., arrow = r-m: 5, *Acanthocaudus* sp.

Figs 6-10: Wings of Aphidiinae. 6, *Praon* sp., arrow = RS+M; 7, *Pseudopraon* sp.; 8, *Boreogalba* sp.; 9, *Trioxys* sp.; 10, *Diaeretiella rapae* (M'Intosh).

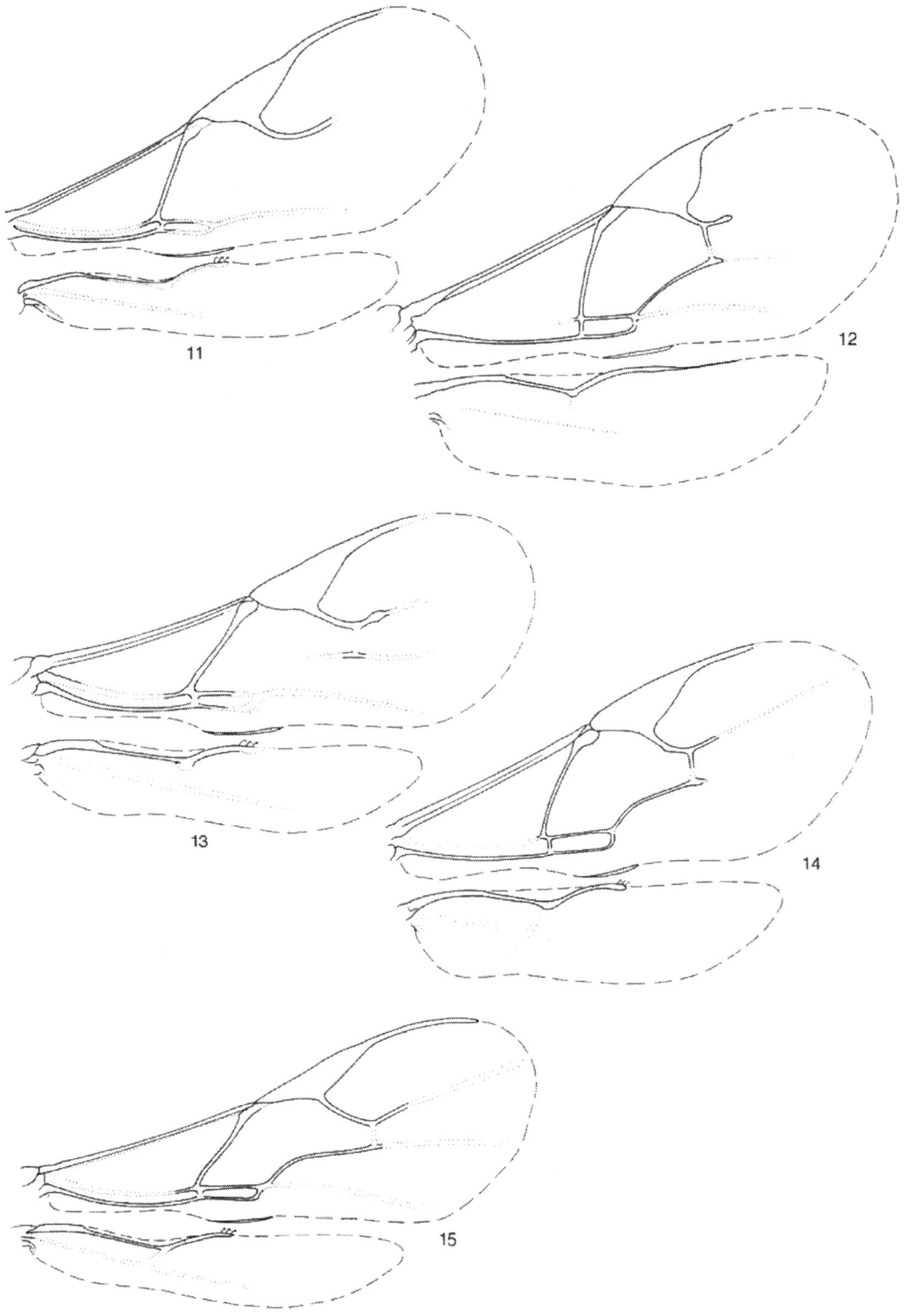

Figs 11-15: Wings of Aphidiinae. 11, *Adialytus* sp.; 12, *Xenostigmus bifasciatus* (Ashmead); 13, *Lysiphlebus testaceipes* (Cresson); 14, *Pauesia* sp.; 15, *Aphidius* sp.

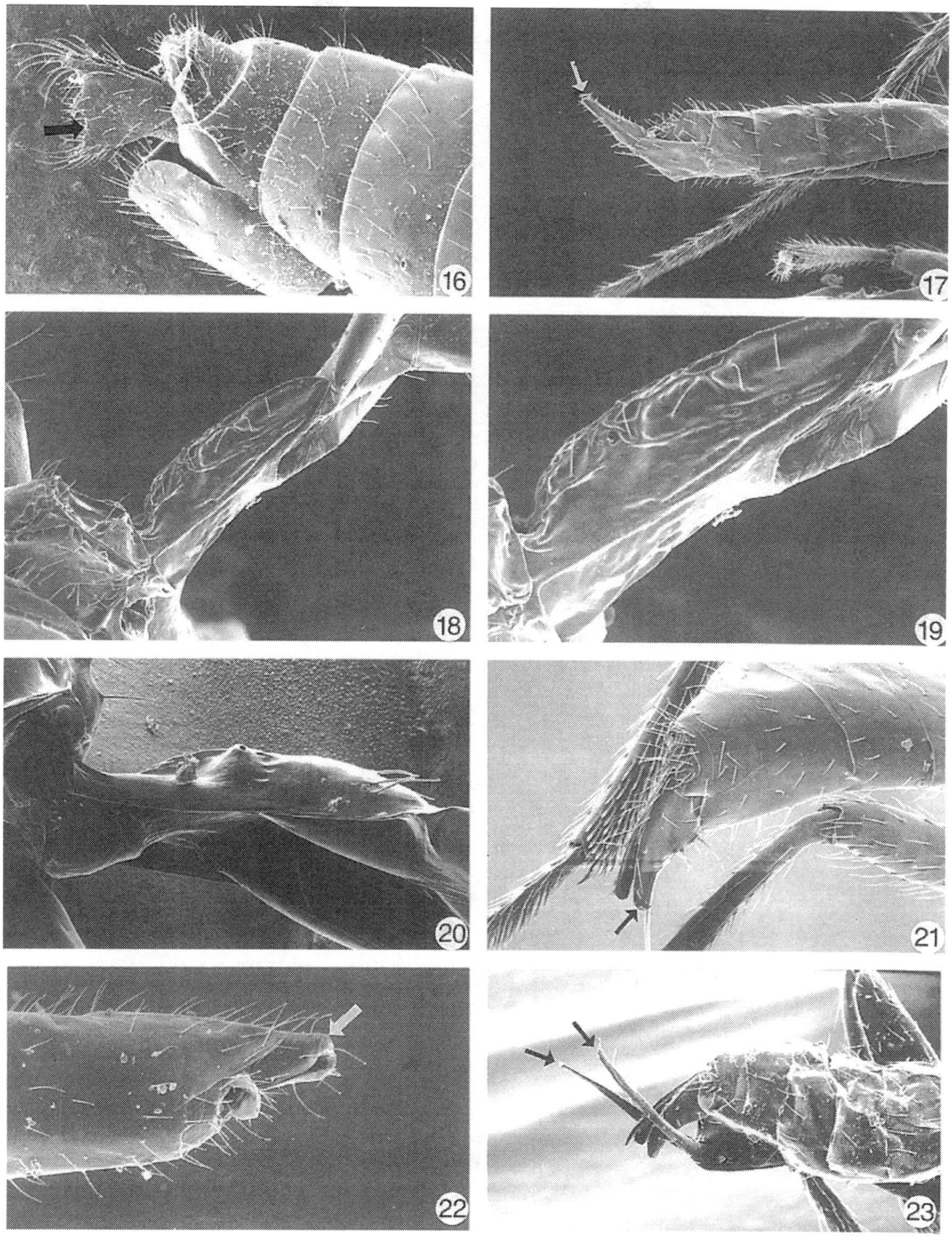

Figs 16-23: Aphidiinae. 16, apex of metasoma of *Toxares* sp., arrow = ovipositor sheath; 17, metasoma of *Ephedrus* sp., arrow = ovipositor sheath; 18-20, petioles: 18, *Monoctonus* sp., ventro-lateral view; 19, *Monoctonus* sp., ventro-lateral view; 20, *Lysiphlebus testaceipes* (Cresson), ventro-lateral view; 21, apex of metasoma of *Monoctonus* sp., arrow = ovipositor sheath; 22, apex of metasoma of Aphidiinae, arrow = ovipositor sheath; 23, metasoma of *Binodoxys* sp., arrows = prongs on hypopygium.

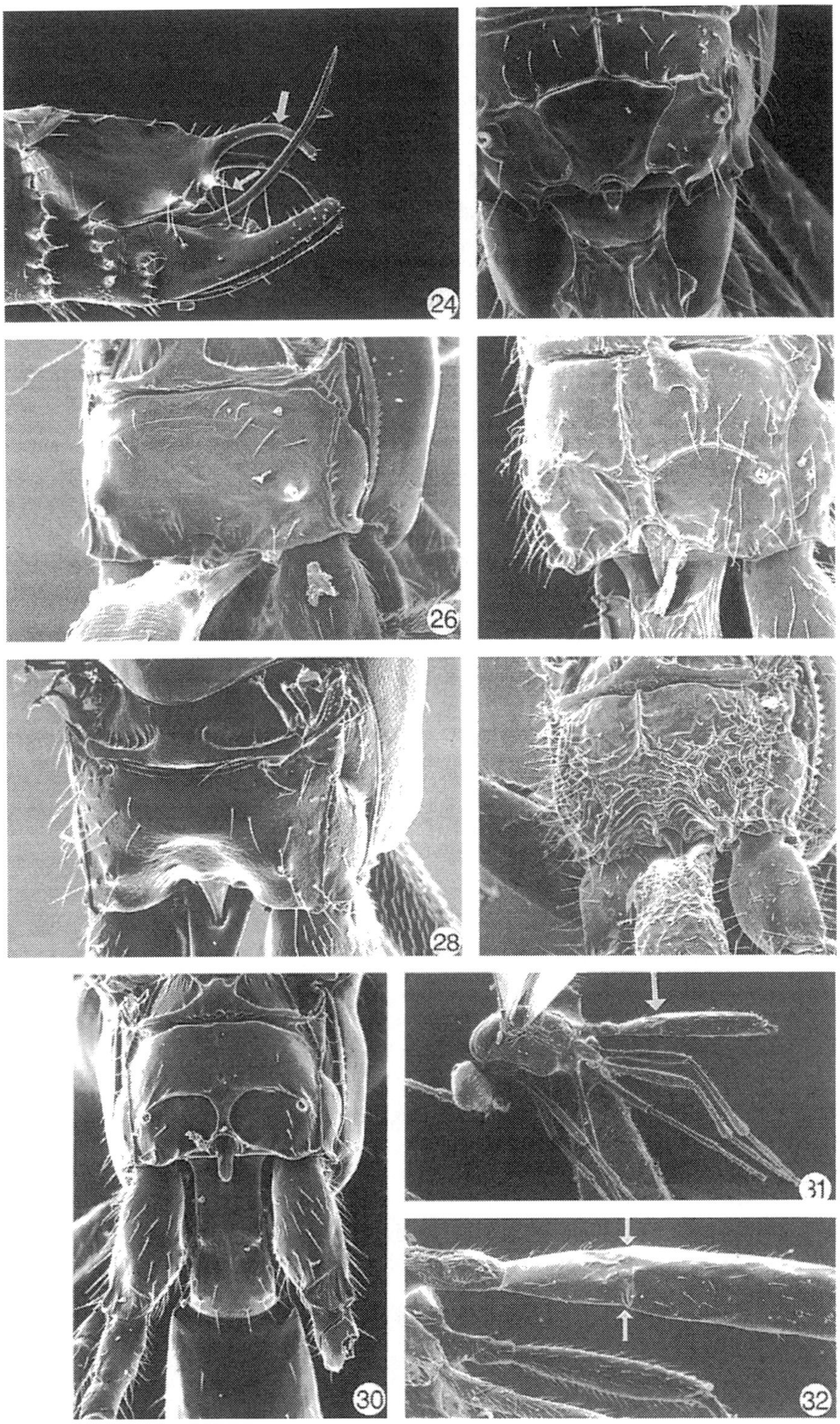

Figs 24-32: Aphidiinae. 24, apex of metasoma of *Acanthocaudus* sp., arrows = prong and accessory prong (both slightly broken) of hypopygium; 25-27, propodeum: 25, *Binodoxys* sp.; 26, *Lysiphlebus* sp.; 27, *Aphidius* sp.; 28, propodeum and base of petiole of *Xenostigmus* sp.; 29, propodeum and petiole of *Pauesia* sp.; 30, propodeum and petiole of *Lysaphidus* sp.; 31, habitus of *Pauesia* sp., arrow = articulation between metasomal terga 2 and 3; 32, metasoma of *Pauesia* sp., arrow = articulation between metasomal terga 2 and 3.

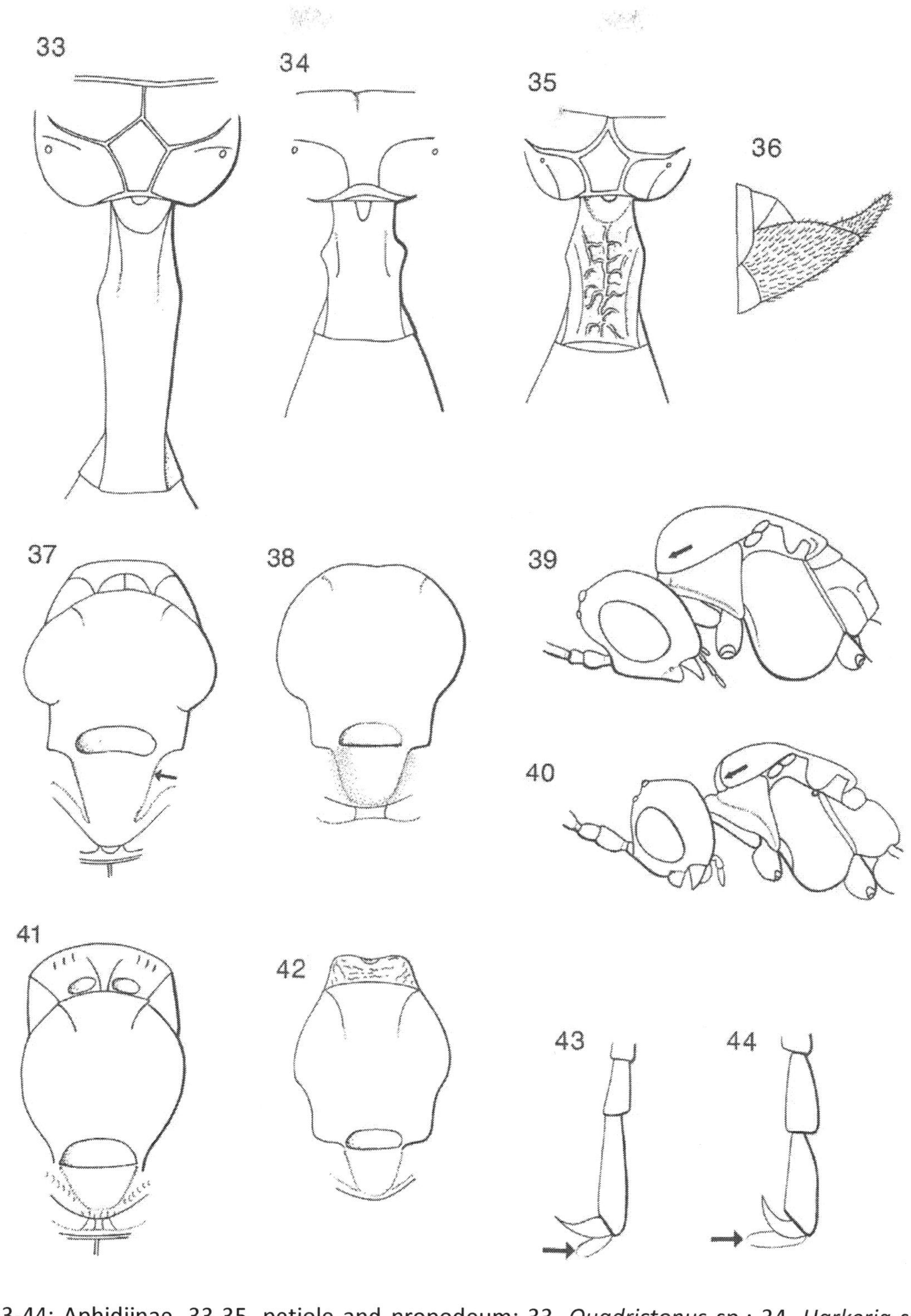

Figs 33-44: Aphidiinae. 33-35, petiole and propodeum: 33, *Quadrictonus* sp.; 34, *Harkeria* sp.; 35, *Monoctonus* sp.; 36, ovipositor sheath and hypopygium of *Pseudopraon* sp.; 37, scutellum of *Boreogalba* sp., arrow = lateral carina; 38, scutellum of *Trioxys* sp.; 39, head and mesosoma of *Diaeretus* sp., arrow = notauli absent; 40, head and mesosoma of *Adialytus* sp., arrow = notauli; 41, pronotum of *Euaphidius* sp. showing pair of pits; 42, pronotum of *Aphidius* sp.; 43, tarsal claws of *Euaphidius* sp., arrow = arolium; 44, tarsal claws of *Aphidius* sp., arrow = arolium.

APHIDIINAE UPDATE

C. van Achterberg
Department of Terrestrial Zoology, Naturalis Biodiversity Center, Postbus 9517, 2300 RA Leiden, The Netherlands.
c.vanachterberg@xs4all.nl

The Aphidiinae chapter of the manual needs a minor update. The species of *Acanthocaudus* Smith are apparently restricted to the aphid genus *Uroleucon* Mordvilko (Yu *et al.*, 2016). According to Völkl and Mackauer (2000) and Kambhampati *et al.* (2000), *Lysaphidus* Smith belongs to a clade within *Aphidius* Nees and was synonymized with *Aphidius* by Tomanović *et al.* (2007). Van Achterberg (2006) and Milosević *et al.* (2015) lowered the status of *Euaphidius* Mackauer to a subgenus of *Aphidius* Nees.

Annotated list of recent papers on New World Aphidiinae

Achterberg C van. 2006. The Braconidae (Hymenoptera) of Greenland. *Zoologische Mededelingen, Leiden* 80(1): 13-62.
The Aphidiinae from Greenland are revised.

Kambhampati S, Völkl W, Mackauer M. 2000. Phylogenetic relationships among genera of Aphidiinae (Hymenoptera: Braconidae) based on DNA sequence of the mitochondrial 16S rRNA gene. *Systematic Entomology* 25: 437-445.
The position of the genus *Lysaphidus* Smith is discussed.

Kula RR, Johnson PJ, Heidel-Baker TT, Boe A. 2017. A new species of *Acanthocaudus* Smith (Braconidae: Aphidiinae), with a key to species and new host and distribution records for aphidiines associated with *Silphium perfoliatum* L. (Asterales: Asteraceae). *Zootaxa* 4236: 543-552.

Milošević MI, Petrović A, Stanković SS, Čkrkić J, Starý P, Zikić V, Tomanović Z. 2015. Taxonomic position and phylogenetic relationships of the genera and species *Euaphidius* and *Remaudierea* (Hymenoptera: Braconidae: Aphidiinae) analyzed using molecular markers and geometric morphometrics. *Annals of the Entomological Society of America* 108(3): 435-445.
The genus *Euaphidius* Mackauer is treated as a subgenus of *Aphidius* Nees.

Pike KS, Starý P. 1996. *Binodoxys conei* sp. n. (Hymenoptera, Braconidae, Aphidiinae), a new parasitoid of hop aphid, *Phorodon humuli* (Schrank) in Washington State. *Journal of the Kansas Entomological Society* 68(4): 444-450.

Pike KS, Starý P. 1996. New species of parasitic wasps attacking cereal aphids in the Pacific Northwest (Hymenoptera: Braconidae: Aphidiinae). *Journal of the Kansas Entomological Society* 68(4): 408-414.

Pike KS, Starý P. 1996. New species of *Pauesia* (Hymenoptera: Braconidae: Aphidiinae) parasitoids on *Cinara* (Homoptera: Aphidiinae: Lachninae) associated with conifers in the Pacific Northwest. *Proceedings of the Entomological Society of Washington* 98(2): 324-331.

Pike KS, Foottit RG, Starý P, Allison DW, Graf G, Maw HEL. 2004. *Aphis boydstoni*, n. sp., and *Aphis ceanothi* Clarke (Hemiptera: Aphididae: Aphidinae) on *Ceanothus velutinus* Douglas, including associated parasitoids. *Proceedings of the Entomological Society of Washington* 106(4): 801-810.

Pike KS, Starý P, Allison D, Graf G, Boydston L, Miller T. 1997. Parasitoids (Hymenoptera: Braconidae: Aphidiinae) of aphids on big sagebrush (*Artemisia tridentata* Nuttall) and prairie sage (*Artemisia ludoviciana* Nuttall) in Washington State. *Proceedings of the Entomological Society of Washington*. 99(1): 143-155.

Pike KS, Starý P, Brewer MJ, Noma T, Langley S, Kaiser M. 2007. A new species of *Binodoxys* (Hymenoptera: Braconidae: Aphidiinae), parasitoid of the soybean aphid, *Aphis glycines* Matsumura, with comments on biocontrol. *Proceedings of the Entomological Society of Washington* 109(2): 359-365.

Pike KS, Starý P, Graf G, Allison D. 2002. *Pauesia columbiana*, n. sp. (Hymenoptera: Braconidae: Aphidiinae) on juniper aphids, and a key to related species. *Proceedings of the Entomological Society of Washington* 104(3): 646-654.

Pike KS, Starý P, Miller R, Allison D, Boydston L, Graf G, Gillespie R. 1997. Small-grain aphid parasitoids (Hymenoptera: Aphelinidae and Aphidiidae) of Washington: Distribution, relative abundance, seasonal occurrence, and key to known North American species. *Environmental Entomology*. 26(6): 1299-1311.

Pike KS, Starý P, Miller R, Allison D, Boydston L, Graf G, Miller T. 1996. New species and host records of aphid parasitoids (Hymenoptera: Braconidae: Aphidiinae) from the Pacific Northwest, U.S.A. *Proceedings of the Entomological Society of Washington* 98(3): 570-591.

Pike KS, Starý P, Miller T, Allison D, Graf G, Boydston L., Miller R, Gillespie R. 1999. Host range and habitats of the aphid parasitoid *Diaeretiella rapae* (Hymenoptera: Aphidiidae) in Washington State. *Environmental Entomology* 28(1): 61-71.

Pike KS, Starý P, Miller T, Graf G, Allison D, Boydston L, Miller R,. 2000. Aphid parasitoids (Hymenoptera: Braconidae: Aphidiinae) of northwest USA. *Proceedings of the Entomological Society of Washington* 102(3): 688-740.

Pike KS, Starý P, Miller T, Miller R, Allison D, Graf G, Boydston L. 1999. *Ephedrus clavicornis* sp. n. (Hymenoptera, Braconidae, Aphidiinae), an aphid parasitoid of the Pacific northwest, and key to Nearctic *Ephedrus*. *Journal of the Kansas Entomological Society* 72(1): 10-16.

Pike KS, Starý P, Graf G. 2003. Description of a new species of Braconidae, *Monoctonus allisoni* (Hymenoptera: Braconidae: Aphidiinae), and discussion of the aphid-parasitoid guild for *Nasonovia* spp. (Hemiptera: Aphididae) in northwestern United States. *Proceedings of the Entomological Society of Washington* 105(2): 380-387.

Pike KS, Starý P, Graf G, Raworth DA, Mathur S, Tanigoshi LK, Murray T. 2011. A new Aphidius Nees (Hymenoptera, Braconidae, Aphidiinae) of *Ericaphis fimbriata* (Richards) (Hemiptera, Aphididae) and key to parasitoids of blueberry aphid in the Pacific Northwest. *Zootaxa* 2802: 58-62.

Tomanović Z, Rakhshani E, Starý P, Kavallieratos NG, Stanisavljević LZ, Zikić V, Athanassiou CG. 2007. Phylogenetic relationships between the genera *Aphidius* and *Lysaphidus* (Hymenoptera: Braconidae: Aphidiinae) with description of *Aphidius iranicus sp. nov. Canadian Entomologist* 139: 297-307.
The genus *Lysaphidus* Smith is synonymized with *Aphidius* Nees.

Völkl W, Mackauer M. 2000. Oviposition behaviour of aphidiine wasps (Hymenoptera: Braconidae, Aphidiinae): Morphological adaptations and evolutionary trends. *Canadian Entomologist* 132(2): 197-212.
The genus *Lysaphidus* Smith belongs to a clade within the genus *Aphidius* Nees.

Yu DS, van Achterberg C, Horstmann K. 2016. Taxapad 2016, Ichneumonoidea 2015. *Database on flash-drive*. www.taxapad.com, Nepean, Ontario, Canada.

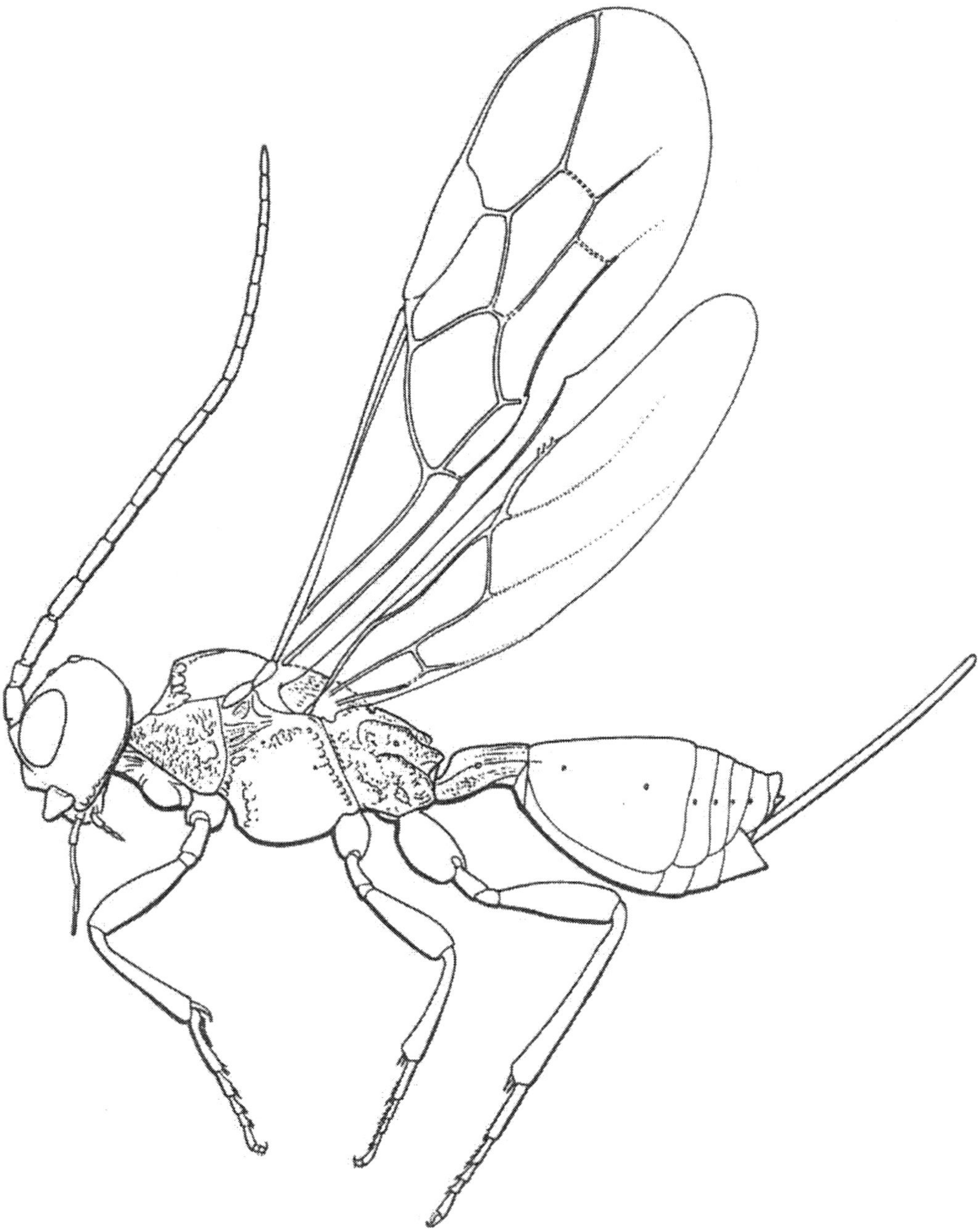

SUBFAMILY APOZYGINAE

M. J. SHARKEY

Department of Entomology, University of Kentucky, S-225 Agricultural Science Building North, Lexington, Kentucky 40546, U.S.A.

INTRODUCTION. The subfamily Apozyginae contains one species, Apozyx *penyai* Mason. Mason (1978) proposed a new family for the species and Mason (1987) described the previously unknown female of the same species. Sharkey and Wahl (1992) discussed the phylogenetic position of the taxon within the Braconidae.

BIOLOGY. The biology of *A. penyai* is unknown but due to its basal phylogenetic position within the Braconidae and its general similarity to some members of the Doryctinae (Sharkey 1993) it may be an idiobiont ectoparasitoid of xylophagous coleopterous larvae.

DISTRIBUTION. Known only from Chile.

DISTINGUISHING FEATURES. This is the only braconid in which 2m-cu (the second recurrent vein) is present (Fig. 1). Additionally, vein 2CUb is present in the hind wing, and the oral cavity is cyclostome (Fig. 2).

REFERENCES

Mason, W. R. M. 1978. A new genus, species and family of Hymenoptera (Ichneumonoidea) from Chile. Proceedings of the Entomological Society of Washington 80: 606-610.

Mason, W. R. M. 1987. Discovery of female *Apozyx* (Hymenoptera: Apozygidae) and comments on its taxonomic position. Proceedings of the Entomological Society of Washington 89: 226-229.

Sharkey, M. and Wahl, D. B. 1992. Cladistics of the Ichneumonoidea (Hymenoptera). Journal of Hymenoptera Research 1:15-24.

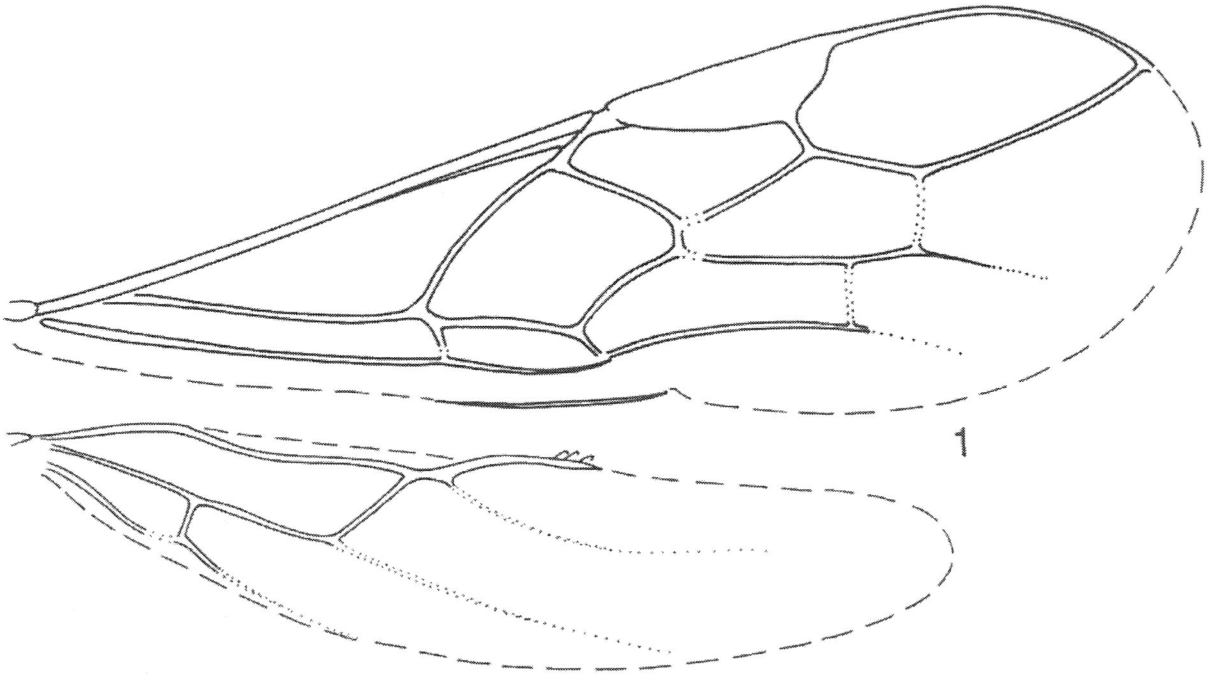

1

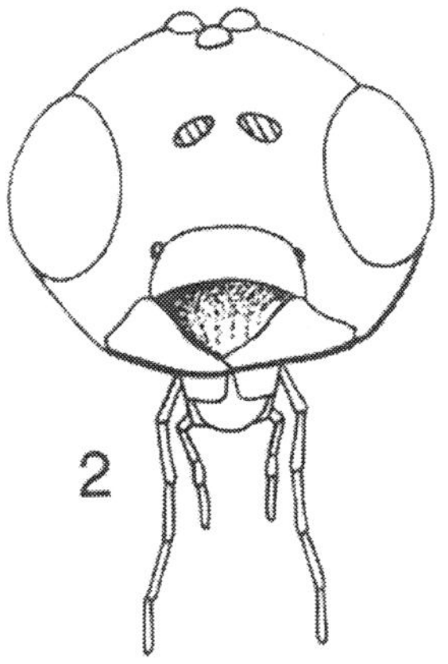

2

Figs, 1-2: Apozyginae, 1, wings of *Apozyx penyai* Mason; 2, face of *A. penyai*; showing exposed labrum.

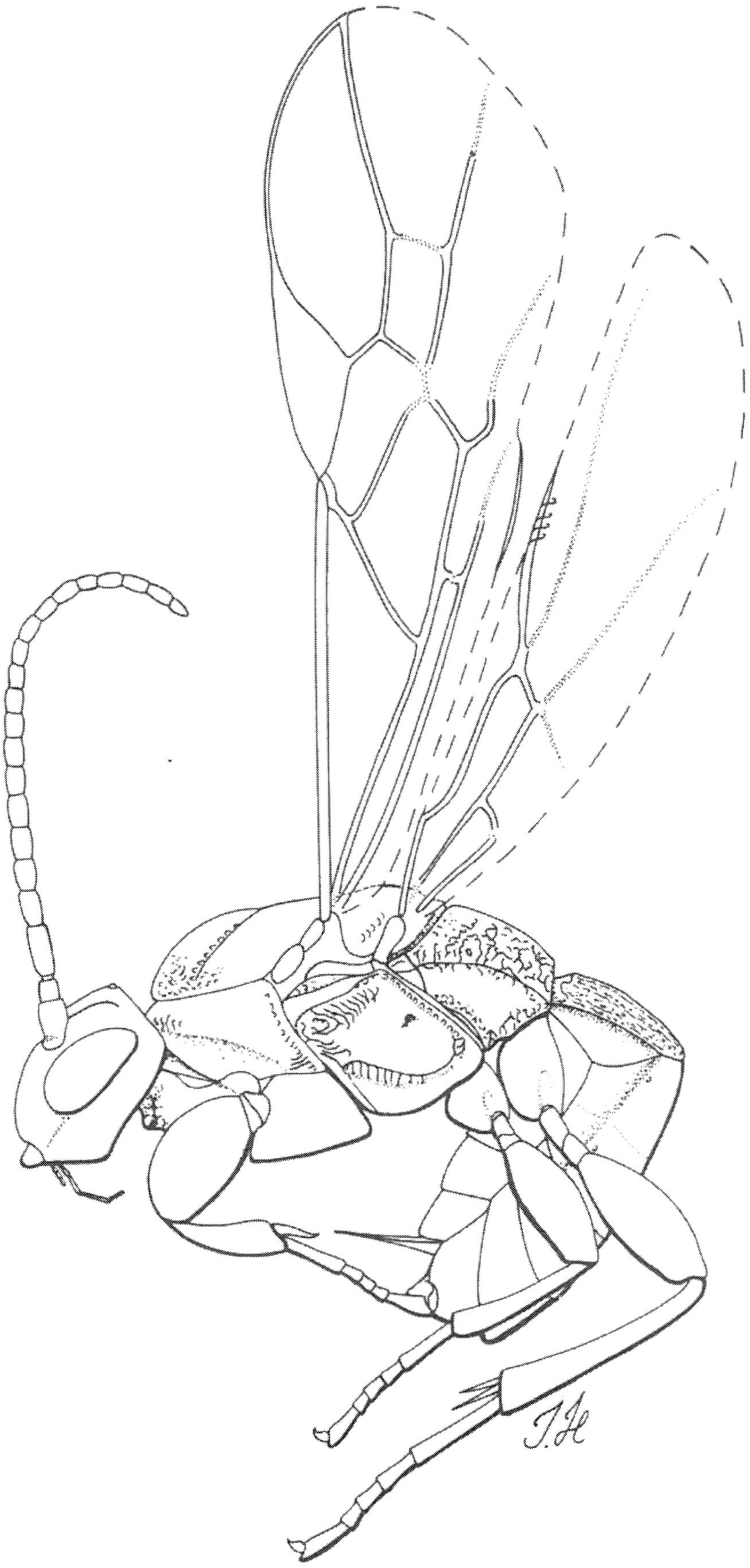
T.H

SUBFAMILY BETYLOBRACONINAE

C. VAN ACHTERBERG

Afdeling Entomologie (Hymenoptera), Nationaal Natuurhistorisch Museum, Postbus 9517,
2300 RA Leiden, The Netherlands

INTRODUCTION. The subfamily Betylobraconinae is a small group, with species 2-4 mm long. Females have a short ovipositor, much shorter than the metasoma. The subfamily is divided into three tribes by van Achterberg (1995): Planitorini (Australian), Betylobraconini (Australian) and Facitorini (widespread). Seven genera are known and two occur in the New World. Both of the New World genera belong to the tribe Facitorini. The two New World genera differ considerably from *Betylobracon* Tobias, the only included genus when the subfamily was originally proposed by Tobias (1979).The New World species are described by van Achterberg (I995).

BIOLOGY. Unknown.

BIOLOGICALCONTROL POTENTIAL. Potential unknown.

COMMON GENERA. Neither of the two New World genera, *Conobregma* van Achterberg and *Jannya* van Achterberg, is commonly collected.

DISTRIBUTION. The Betylobraconinae are almost worldwide in distribution, but are unknown from the northern Holarctic and Afrotropical regions. Four of the seven genera are known only from the Australian region. Of the two New World genera, one is found in Central America and the other in southern United States and the Caribbean.

DISTINGUISHING FEATURES. New World Betylobraconinae are small, moderately slender wasps with an anteriorly protruding head (Fig. 3), short antennae (Figs 4, 5) and short, protruding ovipositor in the females. The venation of the New World genera is diagnostic: vein 2M of the hind wing is situated far above the level of vein 1M (Figs 1, 2), arising near the distal end of SC+R in a position usually occupied by RS; hind wing RS absent or represented by a short stub (Fig. 1); vein M +CU of fore wing distinctly curved (Figs 1, 2). The cyclostome opening is absent or minute (Figs 4, 5), and the occipital carina is complete in the New World species.

KEY TO THE NEW WORLD GENERA OF THE SUBFAMILY BETYLOBRACONINAE

1. Fore wing r-m absent; fore wing r about as long as 2RS or longer (Fig. 2) ....***Jannya*** van Achterberg
   Distribution: Costa Rica to Ecuador; 2 described New World species. Biology: unknown. Reference: van Achterberg (1995).

\- Fore wing r-m present; r shorter than 2RS (Fig. 1)..***Conobregma*** van Achterberg
   Distribution: U.S.A. (Texas and North Carolina to Florida), Dominican Republic; 5 described New World species. Biology: unknown. Reference: van Achterberg (1995).

REFERENCES

Achterberg, C. van 1995. Generic revision of the subfamily Betylobraconinae (Hymenoptera: Braconidae) and other groups with modified fore tarsus. Zoologische Verhandelingen Leiden 298: 1-242.

Tobias, V. I. 1979. Two new sand little known subfamilies of Braconidae (Hymenoptera) from Australia. Entomologicheskoe Obozrenie 58: 128-142. (In Russian.)

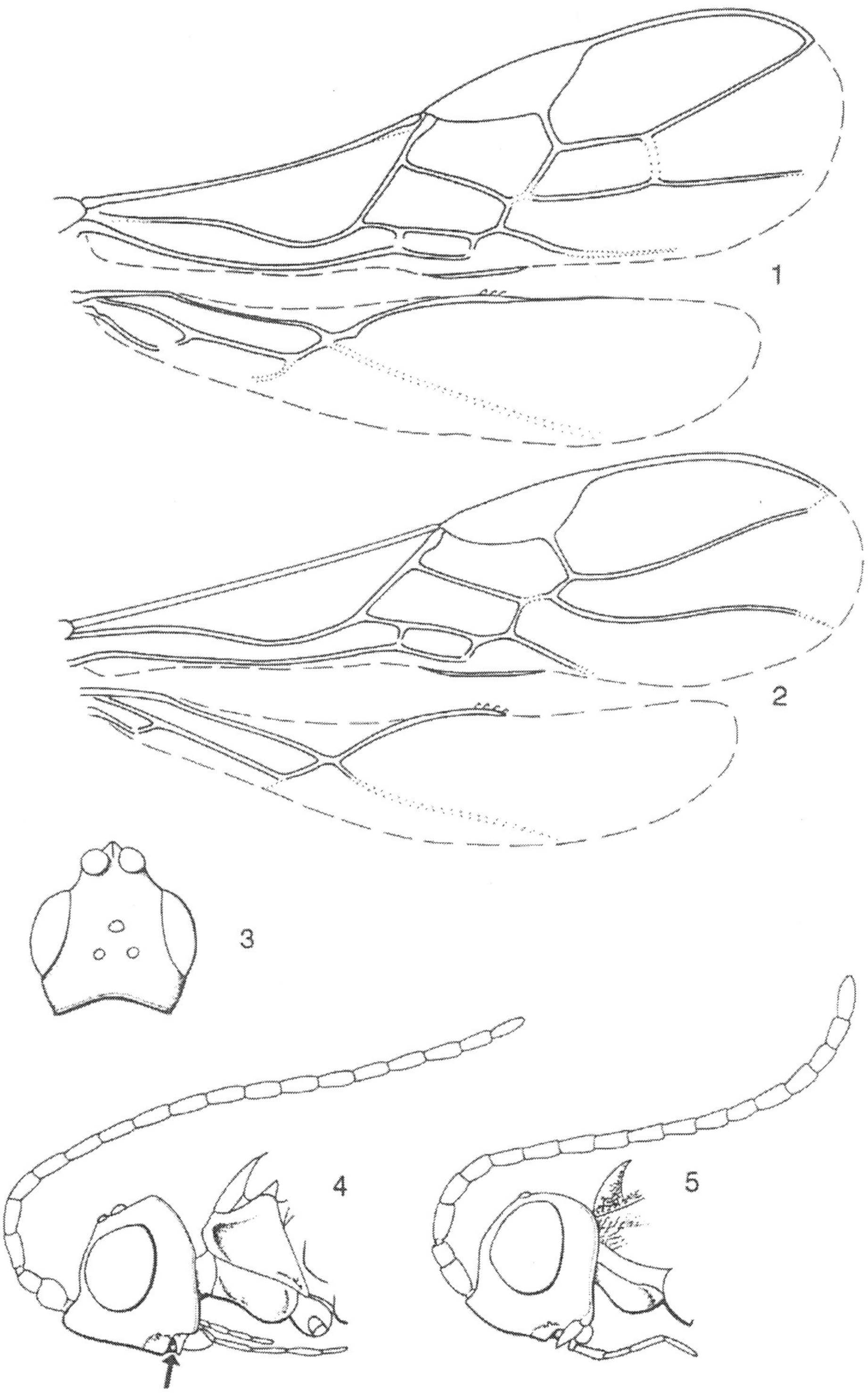

Figs 1-5: Betylobraconinae. 1, wings of *Conobregma brevinervis* van Achterberg; 2, wings of *Jannya* sp.; 3-5, heads: 3, *Jannya brevitarsus* van Achterberg, dorsal view; 4, *Jannya brevitarsus* van Achterberg, lateral view, arrow = cyclostome opening; 5, *Conobregma cometes* van Achterberg, lateral view. Figs 3-5 after van Achterberg (1995).

BETYLOBRACONINAE UPDATE

D.L.J. Quicke
Department of Biology, Faculty of Life Sciences, Chulalongkorn University, Bangkok, Thailand.
d.quicke@email.com

Van Achterberg (1995a) included three tribes in the Betylobraconinae: Betylobraconini, Facitorini, and Planitorini. The first two of these are now considered to belong to the Rogadinae, but morphological characters mislead van Achterberg into treating the Planitorini as a cyclostome braconid (it is not morphologically cyclostome) and has subsequently been shown unambiguously to belong to the non-cyclostome subfamily Euphorinae (Stigenberg *et al.*, 2015).

The Betylobraconinae were recorded from the New World by van Achterberg (1995a), who erected the tribe Facitorini to receive two genera then known only from the New World, *Conobregma* van Achterberg and *Jannya* van Achterberg, though the latter was subsequently recorded from Asia (van Achterberg 1995b). Molecular phylogenetic studies have shown that these two genera, along with the Old World *Facitorus* van Achterberg, are not closely related to *Betylobracon*, the type genus of Betylobraconinae, but rather belong in a clade within the Rogadinae tribe Yeliconini to where they were transferred. Further research also led to the conclusion that remnant Betylobraconinae, *Betylobracon* and *Mesocentrus* were also Rogadinae, and Butcher and Quicke (2015) synonymized the two subfamilies but retained Betylobraconini as a separate tribe.

Penteado-Dias has found a new species of *Jannya* from southern Brazil, which represents the first occurence of the genus to Brazil (soon to be published).

The Asian species of *Conobregma* were transferred to a new genus, *Asiabregma*, by Belokobylskij *et al.* (2008) based on four characters. However, discovery of an intermediate species from South Africa led Butcher *et al.* (2016) to synonymize the two, thus *Conobregma* (Rogadinae: Yeliconini) occurs in both the Old and New World.

Up to this point, the Betylobraconini were unknown from the New World, but its presence there was reported by Quicke and Butcher (2015) based on the discovery and description of a new genus, *Gondwanocentrus* Quicke & Butcher, from Chile. It is effectively impossible to assign most Betylobraconinae *s.l.* to either Rogadinae: Yeliconini: Facitorini or Rogadinae: Betylobraconini based on morphology alone (Butcher and Quicke, 2015), but importantly, they are readily separated using DNA sequence data.

Annotated list of the recent papers on New World 'betylobraconines' = (Rogadinae: Yeliconini: Facitorini) and Betylobraconini *s.s.*

Achterberg C van. 1995a. Generic revision of the sub family Betylobraconinae (Hymenoptera: Braconidae) and other groups with modified fore tarsus. *Zoologische Verhandelingen Leiden* 298: 1-242.
Descriptions of *Conobregma* and *Jannya*.

Achterberg C van. 1995b. New taxa of the subfamilies Betylobraconinae, Cenocoeliinae, Ecnomiinae, Homolobinae, and Sigalphinae (Hymenoptera: Braconidae) from East Indonesia. *Zoologische Mededelingen, Leiden* 69**:** 307-328.
First record of *Conobregma* from Asia.

Belokobylskij SA, Zaldívar-Riverón A, Maetô K, Sáez AG. 2008. Asian Betylobraconinae (Hymenoptera, Braconidae), with description of a new genus and phylogenetic affinities of the tribe Facitorini. *Insect Systematics and Evolution* 39: 133-154.
Transfer of Facitorini to Rogadinae based on molecular evidence.

Butcher BA, Quicke DLJ. 2015. A remarkable new genus and species of Rogadinae (Hymenoptera: Braconidae) of uncertain tribal placement, from Papua New Guinea, resembling Betylobraconini *stat. nov. Journal of Natural History* 49: 2045-2054.
Synonymy of Betylobraconinae with Rogadinae and treatment of it as a separate tribe within Rogadinae. Description of an extralimital new genus of uncertain relationship.

Butcher BA, Quicke DLJ, Shreevihar S, Ranjith AP. 2016. Major range extensions for two genera of the parasitoid subtribe Facitorina, with a new generic synonymy (Braconidae, Rogadinae, Yeliconini). *ZooKeys* 584: 109-120.
Synonymy of *Asiabregma* with *Conobregma*.

Quicke DLJ, Butcher BA. 2015a. Description of a new Betylobraconini-like parasitoid wasp genus and species (Hymenoptera: Braconidae: Rogadinae) from Chile. *Zootaxa* 4021: 459-466.
A new genus, *Gondwanocentrus*, from Chile, the only genuine member of the Betylobraconini from the New World, is described and illustrated.

Stigenberg J, Boring CA, Ronquist F. 2015. Phylogeny of the parasitic wasp subfamily Euphorinae (Braconidae) and evolution of its host preferences. *Systematic Entomology* 40: 570-591.

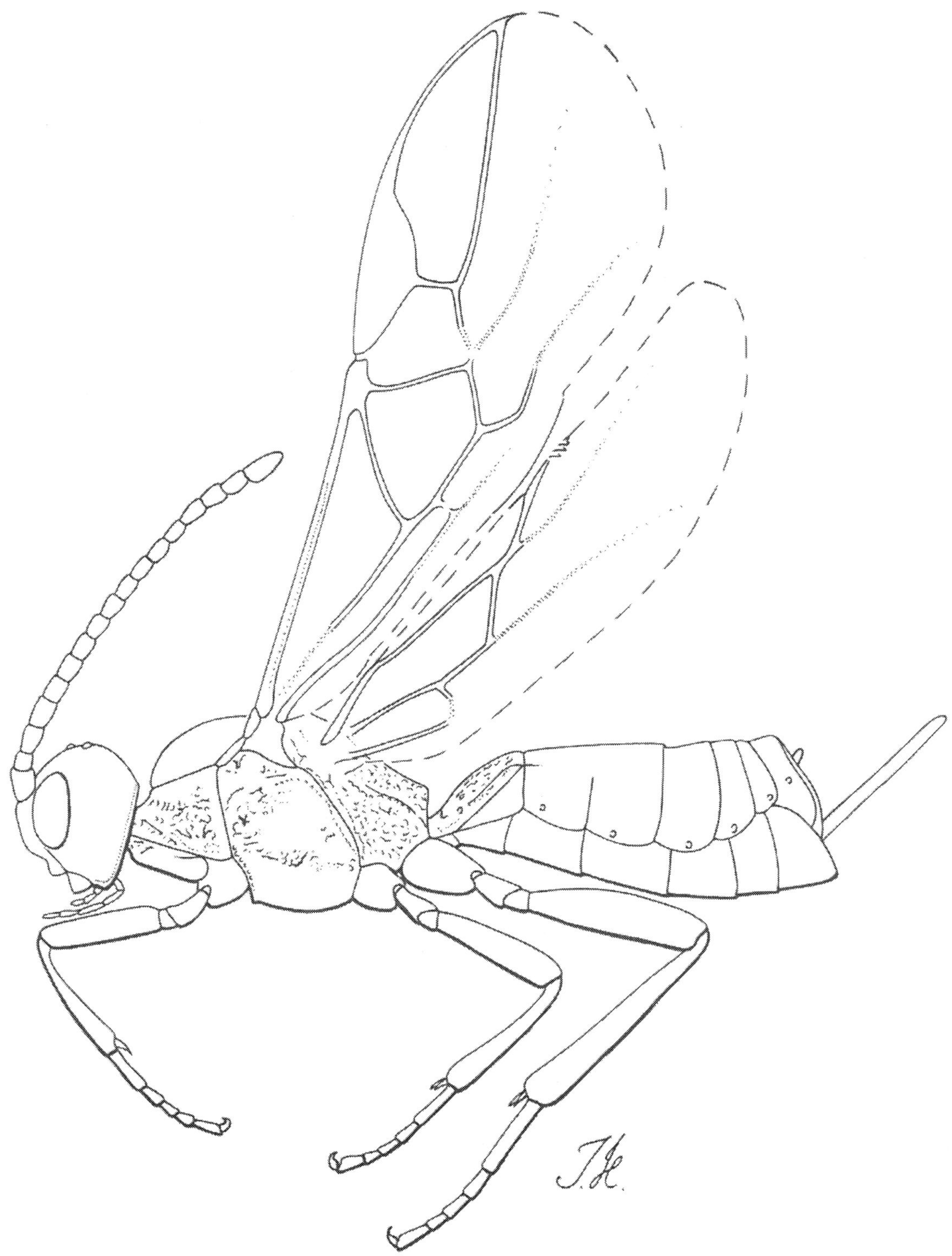
J.H.

SUBFAMILY BLACINAE

M. J. SHARKEY

Department of Entomology, University of Kentucky, S-225 Agricultural Science Building North, Lexington, Kentucky 40546, U.S.A.

INTRODUCTION. The subfamily Blacinae is worldwide in distribution and contains 11 genera, nine of which are found in the New World. Shenefelt (1969) cataloged the World species of Blacinae. Van Achterberg (1988) presented a key to the World genera and subgenera and presented keys to the species of the World of which there are about 100. Most of the keys to species are divided into geographical realm. Van Achterberg (1976, 1988) described many species and discussed phylogenetic relationships among genera, tribes, and subgenera. Van Achterberg (1988) included five tribes in the Blacinae. The monophyly of the Blacinae is supported by three character states, according to van Achterberg (1988:17), "strong tendency to lose veins CU1B (= 2cu-a) and r-m of fore wing" and "larval epistoma reduced". The first two characters are tendencies, not synapomorphies, and the larval character is known for only a few members of two of the five tribes. Despite the dubious foundation of this subfamily no alternative classification based on cladistic arguments has been proposed.

BIOLOGY. The biology of all but two genera is unknown. What is known is summarized in van Achterberg (1988). *Blacus* Nees are parasitoids of larval Coleoptera. Records of species being parasitic on dipterous larvae are almost certainly wrong according to Shaw and Huddleston (1991). The two described species of *Dyscoletes* Haliday are parasites of larval Boreidae of the genus *Boreus* Latrielle (Mecoptera). Some species of *Blacus* have male mating swarms (Haeselbarth, 1973).

BIOLOGICAL CONTROL POTENTIAL. No attempts have been made to use members of the Blacini in classical biological control programs, perhaps because they are so poorly known biologically. Because of their diversity of hosts, including members of Cerambycidae and Curculionidae, members of the Blacinae may be involved in the natural control of potential pest species. However, this remains to be demonstrated.

COMMON GENERA. *Blacus* is common, with about 40 described New World species and many more undescribed. All other genera are represented by one or a few species.

DISTRIBUTION. *Blacus* is cosmopolitan but all other genera have restricted distributions.

DISTINGUISHING FEATURES. This is a rather diverse group all of which have an occipital carinae and non-cyclostome mouthparts. *Stegnocella* van Achterberg, *Mesoxiphium* van Achterberg, *Apoblacus* van Achterberg and *Blacus* can be distinguished from other braconids by the following combination of characters: occipital carina present; r-m and 2cu-a of fore wing absent; petiole with pair of deep dorso-lateral pits near base (dorsope); subalar depression without transverse carina. *Glyptoblacus* van Achterberg is similar to *Blacus* but with 2cu-a of the fore wing present. *Dyscoletes* may be distinguished by the following combination: r-m of fore wing present; 2cu-a of fore wing absent; m-cu arising distad 2RS, thus (RS+M)b absent and second submarginal cell five sided (Fig. 1); metasomal terga not forming a carapace. *Chalarope* van Achterberg may be differentiated by the following combination: M+CU of fore wing tubular; r-m of fore wing absent; 2cu-a of fore wing absent; petiole lacking carinae and with pair of deep dorso-lateral pits near base (dorsope). *Blacozona* van Achterberg and *Grypokeros* van Achterberg may be distinguished from other Braconidae by the following combination of characters: frons concave; r-m of fore wing absent; 2cu-a of fore wing present; metasomal terga not forming a carapace.

KEY TO THE NEW WORLD GENERA OF THE SUBFAMILY BLACINAE

1. Second submarginal cell of fore wing closed, r-m present (Fig. 1) .... ***Dyscoletes*** Haliday
Distribution: Holarctic; 2 described species, one Nearctic and one Palaearctic. Bi- ology: both species reared from *Boreus* (Mecoptera). References: Mason (1976), van Achterberg (1988).

\- Second submarginal cell of fore wing open, r-m absent (Figs 2-8) 2

2(1). Fore wing 1RS present (Fig. 2). Petiole without longitudinal striae and without a pair of longitudinal carinae (Fig. 9) .... ***Chalarope*** van Achterberg
Distribution: Chile; one described species. Biology: unknown. Reference: van Achterberg (1988).

\- Fore wing 1RS absent (Figs 3, 4) or, if present as in Fig. 5 (1 %), then petiole with longitudinal striae or a pair of longitudinal carinae (Fig. 10) 3

3(2). Fore wing 2cu-a present (Figs 3, 4) 4

\- Fore wing 2cu-a absent (Figs 5-8) 6

4(3). Frons with median horn (Fig. 11). Wings: Fig. 3 .... ***Grypokeros*** van Achterberg
Distribution: Chile; 2 described species. Biology: unknown. Reference: van Achterberg (1988).

\- Frons lacking median horn (Figs 12, 13) 5

5(4). Pronotum punctate (Fig. 14). Frons excavated (Fig. 12)..................................... ***Blacozona*** van Achterberg
Distribution: Argentina; one described species. Biology: unknown. References: van Achterberg (1976, 1988).

\- Pronotum mostly rugose-striate (Fig. 15). Frons not excavated (Fig. 13). Wings: Fig. 4...............................
.... ***Glyptoblacus*** van Achterberg
Distribution: Honduras; one described species. Biology unknown. Reference: van Achterberg (1995).

6(3). Metasomal terga enlarged and enveloping sterna (Fig. 16); no crease between median tergite 2 and lateral tergite 2. Wings: Fig. 6 .... ***Apoblacus*** van Achterberg
Distribution: Chile; one described species. Biology: unknown. References: van Achterberg (1976, 1988).

\- Metasomal terga not enlarged and hot enveloping sterna (Fig. 18); crease usually (90%) present between median tergite 2 and lateral tergite 2..7

7(6). Hind wing M+CU shorter than 1M or subequal (Fig. 5). Tarsal claws of fore leg of female often with black bristles (Fig. 19). Scutellum usually with lateral carinae (Fig. 17)... ***Blacus*** Nees
Distribution: large cosmopolitan genus; about 40 described species in New World. Biology: endoparasitoids of Coleoptera larvae. References: van Achterberg (1976, 1988, phylogeny and revision of world species).

\- Hind wing M +CU longer than 1M or subequal (Fig. 7). Tarsal claws of fore leg of female without black bristles (Fig. 20). Scutellum without lateral carinae 8

8(7). Fore wing 2a present (Fig. 7) .... ***Stegnocella*** van Achterberg
Distribution: Chile; one described species. Biology: unknown. References: Achterberg (1976, 1988).

\- Fore wing 2a absent (Fig. 8) .... ***Mesoxiphium*** van Achterberg
Distribution: Chile; 2 described species. Biology: unknown. References: van Achterberg (1976, 1988).

REFERENCES

Achterberg, C. van 1976. A revision of the tribus Blacini (Hymenoptera, Braconidae, Helconinae). Tijdschrift voor Entomologie 118: 159-322.

Achterberg, C. van 1988. Revision of the subfamily Blacinae Foerster (Hymenoptera, Braconidae). Zoologische Verhandelingen Leiden 249: 1-324. Achterberg, C. van 1995. *Glyptoblacus* gen. nov. (Hymenoptera, Braconidae, Blacinae) from Honduras. Zoologische Mededelingen Leiden 69: 303- 306.

Haeselbarth, E. 1973. Die *Blacus*-Arten Europas und Zentral-Asiens (Hymenoptera, Braconidae). Veröffentlichungen der zoologischen Staatssammlung München 16: 69-170.

Mason; W.R. M. 1976. A revision of *Dyscoletes* Haliday (Hym.: Braconidae). The Canadian Entomologist 108: 855-858.

Shaw, M. R. and Huddleston, T. 1991. Classification and biology of braconid wasps (Hymenoptera: Braconidae), Handbooks for the Identification of British Insects 7(11): 1-126.

Shenefelt, R. D. 1969. Pars 4. Braconidae 1, Hybrizoninae, Euphorinae, Cosmophorinae, Neoneurinae, Macrocentrinae. Pp. 1-176. In: Ferrière, Ch. and Vecht, J. van der (Eds.). Hymenopterorum Catalogus (nova editio). Dr. W. Junk, The Hague.

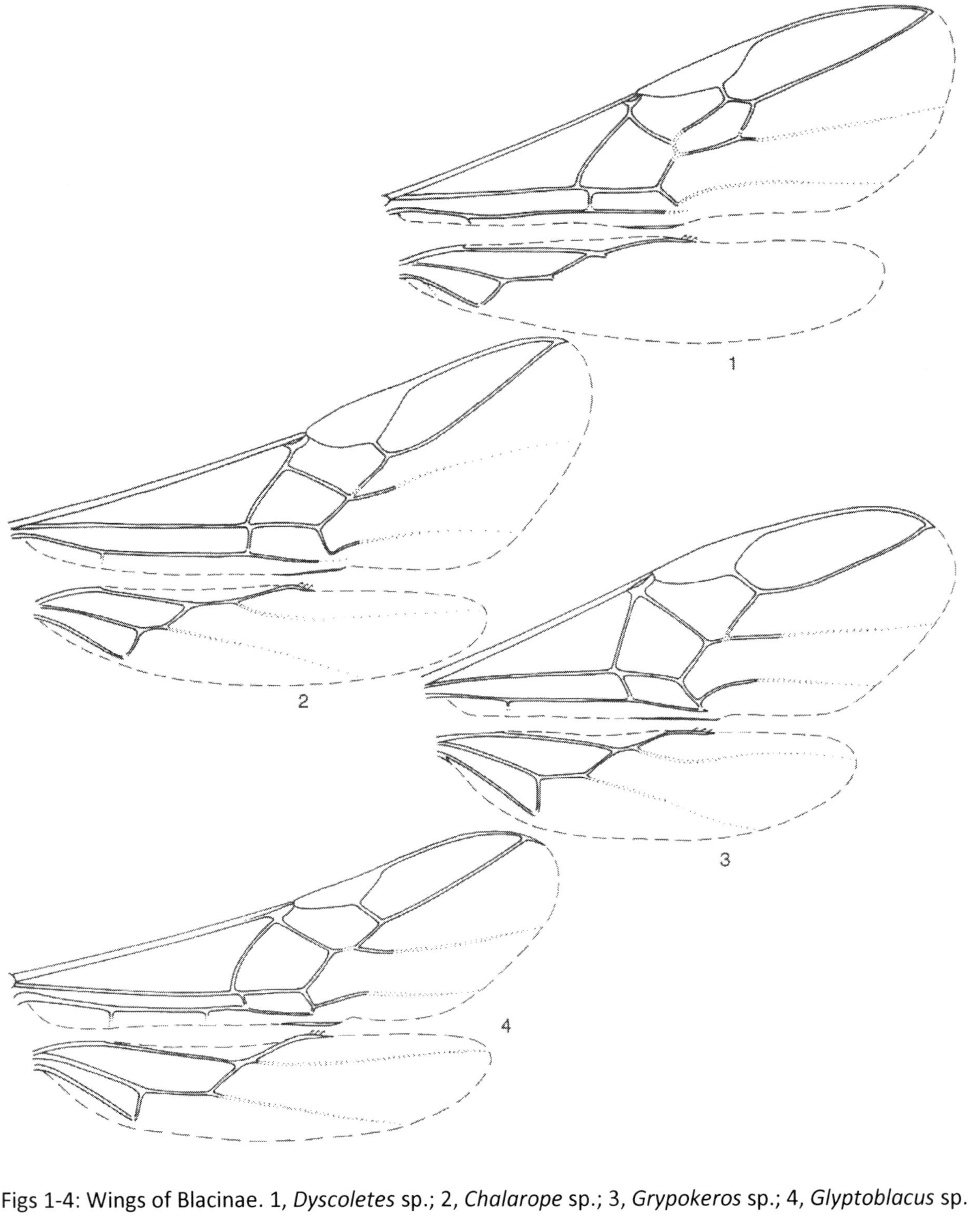

Figs 1-4: Wings of Blacinae. 1, *Dyscoletes* sp.; 2, *Chalarope* sp.; 3, *Grypokeros* sp.; 4, *Glyptoblacus* sp.

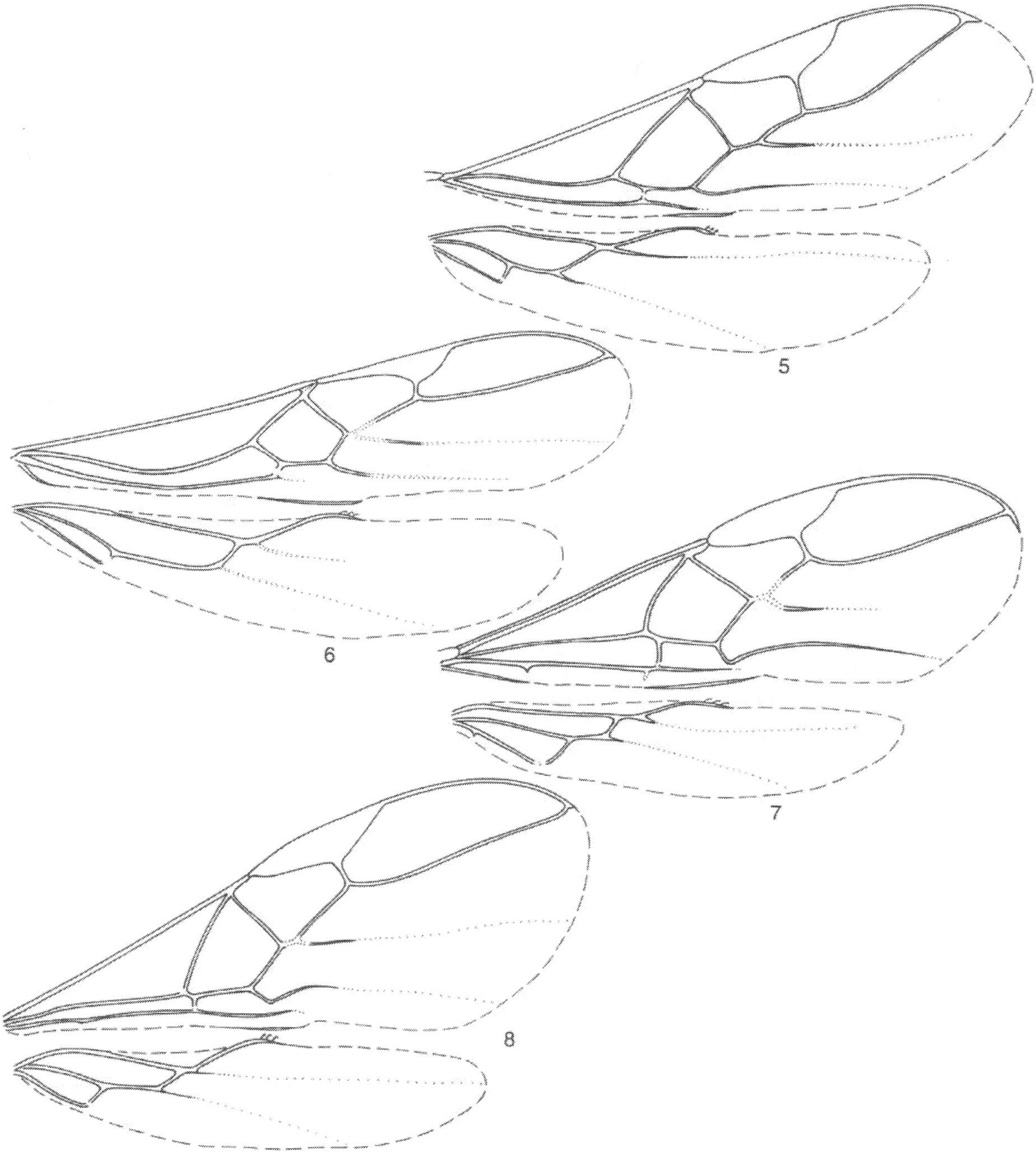

Figs 5-8: Wings of Blacinae. 5, *Blacus* sp.; 6, *Apoblacus* sp.; 7, *Stegnoceila* sp.; 8, *Mesoxiphium* sp.

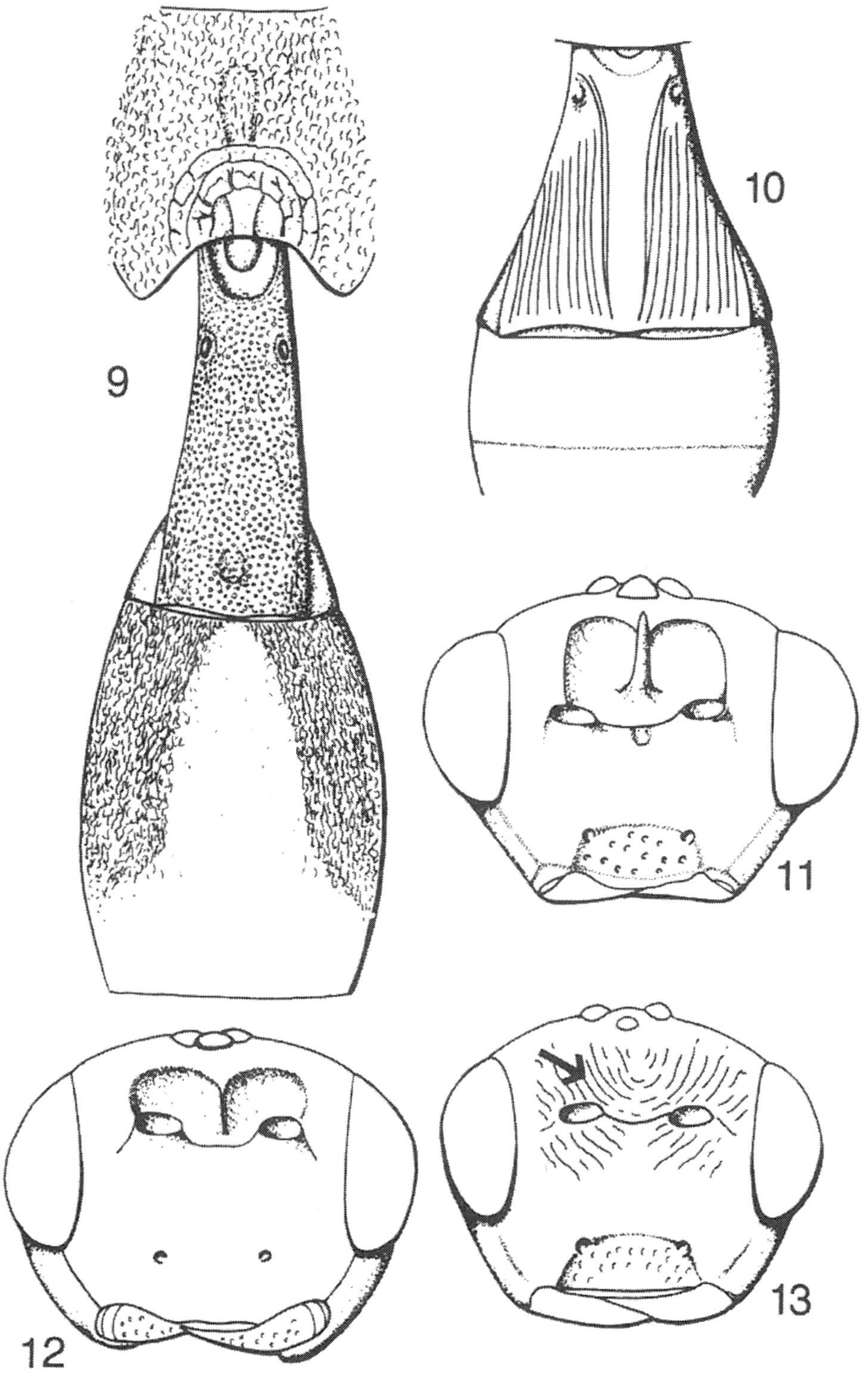

Figs 9-13: Blacinae. 9, petiole of *Chalarope* sp. showing absence of both carinae and striate sculpture; 10, petiole of *Blacus* sp. showing longitudinal carinae and striate sculpture; 11-13, head, frontal view: 11, *Grypokeros* sp. showing median horn on frons; 12, *Blacozona* sp. showing excavated frons: 13, *Glyptoblacus*, sp., arrow = frons. All figures after van Achterberg (1988, 1995).

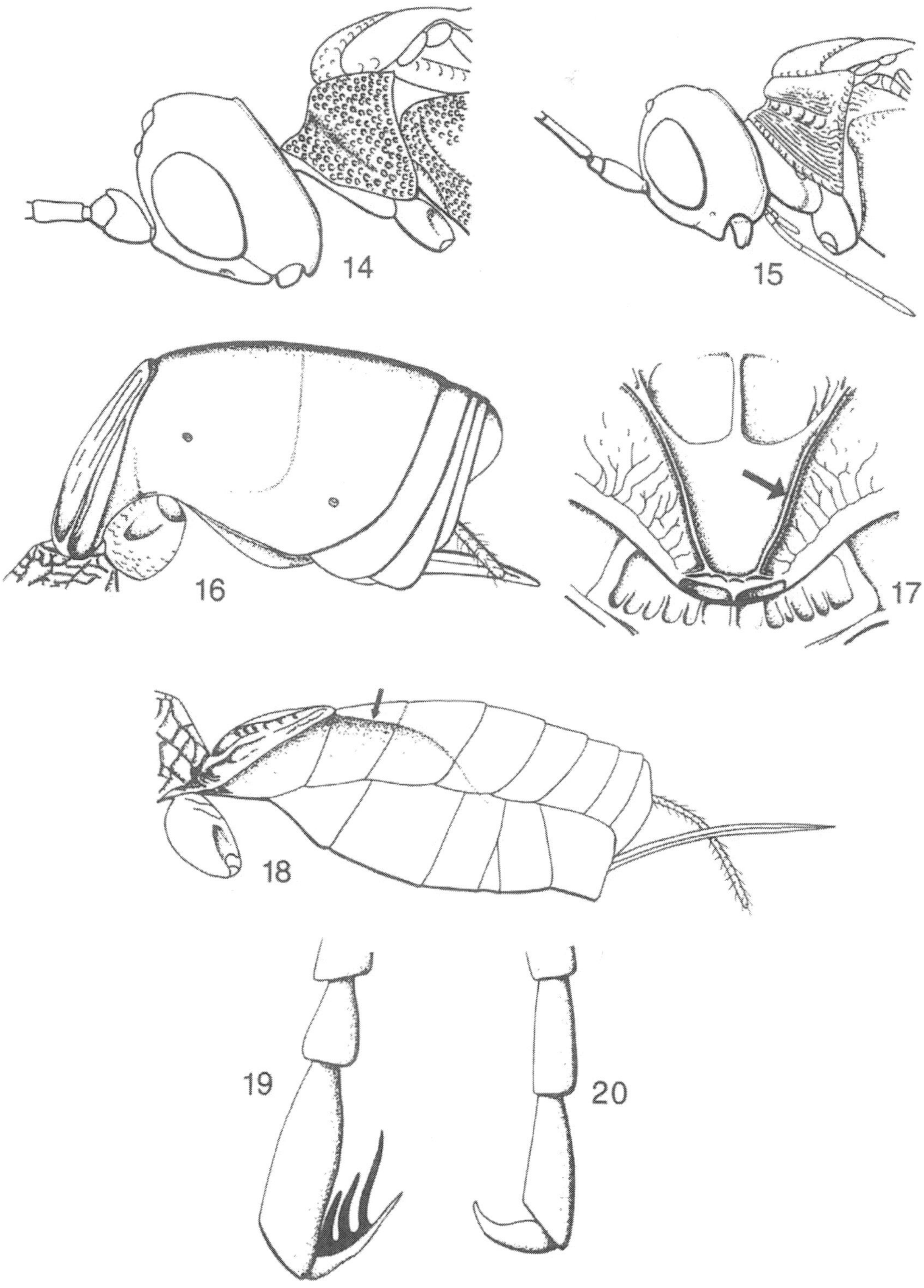

Figs 14-20: Blacinae. 14, pronotum of *Blacozona* sp.; 15, pronotum of *Glyptoblacus* sp.; 16, metasoma of *Apoblacus* sp., lateral view showing enlarged terga; 17, scutellum of *Blacus* sp., dorsal view, arrow = lateral carina; 18, metasoma of *Blacus* sp., arrow = crease between median and lateral tergites; 19, fore tarsal claw of female *Blacus* sp.; 20, fore tarsal claw of female *Mesoxiphium* sp. All figures after van Achterberg (1988; 1995).

BLACINAE UPDATE

M.J. Sharkey
Department of Entomology, University of Kentucky, Lexington, Kentucky 40546-0091, U.S.A.
msharkey@uky.edu

Sharanowski *et al.* (2011) proposed that Blacinae should be changed to Blacini and included with Brachistinae (separate from Helconinae) with the following four tribes Brachistini, Diospilini, Blacini and Brulleiini.

Four new species from Mexico were described by Sanchez-Garcia *et al.*, 2010. There are no generic changes since publication of the Manual.

Recent papers on New World Blacini

Achterberg C van. 2006. The Braconidae (Hymenoptera) of Greenland. *Zoologische Mededelingen, Leiden* 80(1): 13-62.

Sharanowski BJ, Dowling APG, Sharkey MJ. 2011. Molecular phylogenetics of Braconidae (Hymenoptera: Ichneumonoidea), based on multiple nuclear genes, and implications for classification. *Systematic Entomology* 36: 549-572.

Sanchez-Garcia J, Wharton RA, Romero-Napoles J, Figueroa JI, Lopez-Martinez V, Martinez AM, Perez-Pacheco R, Pineda S. 2010. Four new species of *Blacus* Nees (Hymenoptera, Braconidae, Blacinae) from Mexico. *Zootaxa* 2641: 37-46.

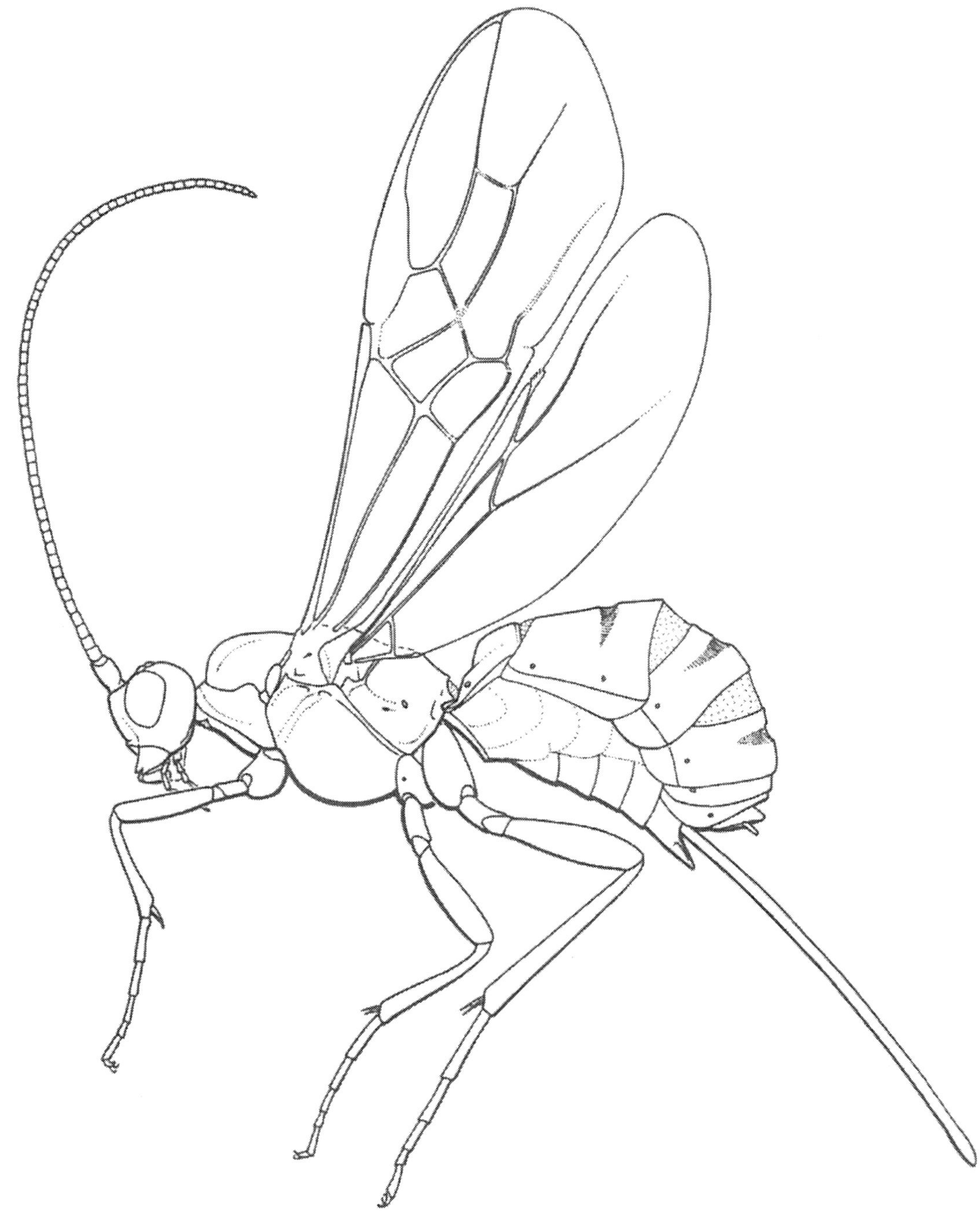

SUBFAMILY BRACONINAE

D. L. J. Quicke

Department of Biology, Faculty of Life Sciences, Chulalongkorn University, Bangkok, Thailand.

INTRODUCTION. The Braconinae is a large subfamily of moderately small to large wasps with more than 2000 described species worldwide (Shaw and Huddleston, 1991). They are most diverse in the Old World tropics but are also well represented in the New World. Comparatively few studies have been made on the New World fauna since the first two decades of this century. The species described up until 1976 have been cataloged by Shenefelt (1978) and the North American fauna have been cataloged by Marsh (1979). In both catalogs most of the New World species were included under the generic names *Iphiaulax* Förster (a genus that in reality has an entirely Old World distribution), *Ipobracon* Thomson (a junior synonym of *Cyanopterus* Haliday), or *Bracon* Fabricius, with *Iphiaulax* and *Bracon* used in a far wider sense than they are today. Many of these species have subsequently been reclassified by Quicke (1988a, 1989a, 1991) and Quicke and Genise (1994) mostly into *Digonogastra* Ashmead. The aberrant Neotropical genus *Gnathobracon* Costa was recently redescribed and transferred formally from the Rogadinae to the Braconinae (Quicke and Huddleston, 1991). Several new genera have been described in the last few years by Shenefelt (in Mason, 1978), Quicke (1988b, 1989b, 1994, 1995), van Achterberg (1989), Quicke and Sharkey (1989), and Quicke and Delobel (1995).

BIOLOGY. As far as is known, all New World braconines are idiobiont ectoparasitoids of concealed holometabolous insect larvae, especially of the Lepidoptera and Coleoptera, though a few species, mostly in the genus *Bracon*, attack concealed dipterous or sawfly larvae. Indeed, this is true for the whole subfamily with the exception of the small, entirely Old World tropical subtribe Aspidobraconina which are endoparasitic in butterfly pupae. Braconines are synovigenic and their large eggs are usually laid on the host which was previously paralyzed by injection of venom (Shaw and Huddleston,1991). Host feeding appears to be quite common among the smaller species and sometimes involves the construction of feeding tubes. Both solitary and gregarious parasitism occur, sometimes with closely related species displaying different strategies (Wharton et al., 1989). Some genera appear to be somewhat specialized in their host range though most species are at least oligophagous. Thus, the species of *Myosoma* Brullé, for example, are parasitoids of stem-boring Lepidoptera (Pyralidae and Sesiidae) while species of *Coeloides* Wesmael are specialized on bark-boring beetle larvae. Unfortunately the biologies of many of the Neotropical genera are still unknown, but members of one recently described genus, *Cyclaulacidea* Quicke and Delobel, have been reared from bruchid beetle larvae.

BIOLOGICAL CONTROL POTENTIAL. Several species have been utilized in biocontrol programs in the New World. Various species of *Bracon*, *Habrobracon* Ashmead and *Pigeria* van Achterberg have been released to control a range of insects including stored product and stem borer pests. For example, *H. brevicornis* (Wesmael) has been released in the U.S.A. on numerous occasions against *Heliothis virescens* (F.), *Helicoverpa zea* (Boddie), *Ostrinia nubilalis* (Hubner) and *Pectinophora gossypiella* (Saunders) while *Pigeria piger* (Wesmael) has been released against *Etiella zinckenella* (Treit) and *Cnephasia* sp. *Bracon mellitor* Say is one of the most important natural parasitoids of the boll weevil, *Anthonomus grandis* Boheman, in the United States (Tillman, 1985) while various other braconines are important in the natural regulation of bark beetle pests in North American forests. Members of several endemic New World genera are at least locally important parasitoids of pyralid cane borers, *Diatraea* spp., in Central and South America, including *Digonogastra kimballi* Kirkland, *Palabracon diairaeaphilus* Quicke and *Myosomatoides myersi* Quicke (= *pennipes* (Myers)). The biocontrol potential of all but the first of these remains to be assessed. Undoubtedly, difficulties in identifying New World braconines in general and

Neotropical species in particular have hampered and even discouraged work on their biology and hence possible use in biological or integrated control programs.

COMMON GENERA. The majority of New World genera are only infrequently collected but members of two speciose genera, *Bracon* and the generally larger and more brightly colored *Digonogastra*, are abundant virtually everywhere. For the tropical region there is no reliable way to identify members of either genus to species level, and the same is true of North American *Digonogastra*. Muesebeck (1925) has provided a key to the species of *Bracon* (as *Microbracon* Ashmead) in America north of Mexico, though it is probable that many species still await description. More recently, Shenefelt (1943) and Mason (1978) provided keys to the North American species of three smaller genera, viz. *Atanycolus* Förster, *Coeloides* and *Myosoma* respectively. Inayatullah (1992) has recently revised the moderately common, but principally North American genus *Vipio* Latreille. Members of the genera *Atanycolus* and *Coeloides* are both common parasitoids of wood or bark-inhabiting beetle larvae.

DISTRIBUTION. The Braconinae have a cosmopolitan distribution, with the greatest generic and probably specific diversity being in the Old World tropical regions. The Nearctic fauna is similar in its generic composition to that of the Palaearctic but in the Neotropics, most genera are endemic. The *Compsobracon* Ashmead group of genera (keying out in couplets 8-12 below) shows a possible biogeographic link with the Australian genera such as *Virgulibracon* Quicke and *Calcaribracon* Quicke (Quicke, 1988c). These genera share the possession of a strongly curved fore wing vein (RS+M)a, lack of a midbasal area on metasomal tergum 2 and possess a mid-ventral gland between metasomal sterna 2 and 3. Alternatively, Neotropical genera such as *Cervellus* Szépligeti, *Gnathobracon* Costa, Lasiophorus Haliday, *Leptobracon* Szépligeti, *Megabracon* Szépligeti, *Palabracon* Quicke and *Pheloura* van Achterberg appear to be most closely related to Holarctic elements, in particular to *Cyanopterus* and *Digonogastra*. The genus *Vipiomorpha* Tobias, an arguably ancient member of the Glyptomorphini, is recorded here from the New World for the first time.

DISTINGUISHING FEATURES. Most, if not all Braconinae, while alive produce a distinctive smell when handled. All New World species are distinctly cyclostome (i.e. have a large, dorsally rounded depression above the mandibles formed by a ventrally recessed clypeus and an exposed, concave and usually glabrous labrum) though in both *Alienoclypeus* Shenefelt and *Gnathobracon*, the lower face and clypeal configuration are somewhat modified (Figs 40, 49, 50). The occipital carina is always absent (except in one aberrant Old World genus), likewise the epicnemial carina, and the propleuron never has a longitudinal lateral carina just "below its junction with the pronotum. The median tergite of the petiole is usually distinctly flattened laterally with a domed median area, the hind wing cross vein m-cu is always absent and the hind wing vein M +CU is less than 0.5 times the length of vein 1M (the latter nearly always being markedly thicker proximally than distally).

KEY TO THE NEW WORLD GENERA OF THE SUBFAMILY BRACONINAE

The shape of the antennal scape is important for identification of genera. For determination of the relative lengths of the dorsal and ventral sides (couplets 5, 11, and 15), the antenna should be positioned as in Fig. 39.

1 Face with a median projection that is approximately round in cross section (Figs 16-18, 21) and either apically bifurcate (females; Fig. 17) or simple (males; Fig. 18). Wings: Fig. 1 ....***Lasiophorus*** Haliday

Distribution: Mexico to Brazil and Ecuador; few species, infrequently collected. Biology: unknown. References: Fahringer (1930, key), Quicke (1989a, transfer of 2 species to *Palabracon*).

\- Face without a projection (Figs 30-36, 47, 48, 52) or if with a projection, then projection with a very different form (Figs 19, 20, 22, 23) 2

2(1). First flagellomere with thorn-like, pre-apical projection (Figs 19, 24). Face with distinctive and massive protuberance with concave dorsal face and concave frontal aspect (Figs 19, 20). Propleuron with pair of boss-like, medially-orientated projections antero-medially (Fig. 25). Wings: Fig. 2........***Cervellus*** Szépligeti

Distribution: Trinidad, South America, but not recorded from Chile; few species, infrequently collected. Biology: one species has been reared from a coleopteran larva (Quicke, 1989c). Reference: Fahringer (1930, key).

- First flagellomere simple (Figs 38, 39, 41) or occasionally apically flared (Figs 26, 27), without pre-apical, thorn-like projection. Face simple (Figs 30-36, 47, 48, 52) or with plate-like projection (Figs 22, 23). Propleuron simple, without antero-median projections 3

3(2). Face with large, horizontal, apically dentate and dorsally concave plate-like projection (Figs 22, 23) .... ***Palabracon*** Quicke

Distribution: Mexico, Brazil, Ecuador; infrequently collected. Biology: one species has been reared from *Diatraea* Guilding (Lepidoptera: Pyralidae). Reference: Quicke (1988b).

- Face simple, without median, horizontal projection (Figs 30-36, 47, 48, 52) 4

4(3). Basal 1, 2 or 3 flagellomeres distinctly expanded apically (Figs 26, 27); first flagellomere distinctly shorter than third (Fig. 27) 5

- Basal flagellomeres simple, not or hardly expanded apically (Figs 38, 39, 41); first flagellomere longer than third (Figs 41) 6

5(4). Scape shorter ventrally than dorsally (antenna directed anteriorly) (as in Figs 28, 39); flagellum uniformly dark, without white or cream-colored median band. Wings: Fig. 3 .... ***Coeloides*** Wesmael

Distribution: Palaearctic, Canada, U.S.A., and Mexico. Biology: parasitoids of bark-boring beetle larvae, especially Scolytidae, Curculionidae and Buprestidae. References: Mason (1978, key to species of U.S.A. and Canada).

- Scape longer ventrally than dorsally (antenna directed anteriorly) (as in Fig. 29); flagellum with white or cream-colored median band .... ***Megacoeloides*** Quicke

Distribution: Peru; rarely collected; one described and one undescribed species known. Biology: unknown. Reference: Quicke (1989b).

6(4). Middle part of face clearly demarked from lateral parts by a pair of sub-medial, longitudinal carinae running ventrally from the antenna! sockets (Figs 30-34) or by a pair of broad, sculptured grooves (Figs 35, 36). Metasomal tergum 2 without midbasal, posteriorly-narrowing triangular area (Figs 74-77, 79-81, 90, 91). Fore wing (RS+M)a strongly curved (Figs 4, 6) or rarely Strongly angled (Fig. 5) 7

- Middle part of face not clearly demarked from lateral parts by longitudinal grooves or carinae (Figs 42, 45-48, 52), though sometimes middle of face conspicuously glabrous (Figs 42, 45). Metasomal tergum 2 with (Figs 51, 71, 72, 84) or without midbasal, posteriorly-narrowing triangular area. Fore wing (RS+M)a variable but rarely strongly curved (Figs 8-14 13

7(6). Scape highly modified: apically with false margin well-separated from the real margin, giving rise to large pre-apical ledge ventro-medially (Figs 35-37, 43); basally (except in a few *Cyclaulax*) abruptly and concavely narrowing (Fig. 35; see also Figs 44-46). Upper half of face with deep groove extending between antennal sockets (Figs 34, 35) 8

- Scape simple, neither abruptly and concavely narrowing basally nor with wide apical ledge formed by pre-apical false margin (see Figs 21, 28, 29, 40). Upper half of face without deep groove between antennal sockets (Figs 31, 32) 10

8(7). Antennal sockets protruding and distally acutely angulate in profile (Fig. 36); scape very short and distinctive (Figs 35, 36). Fore basitarsus not laterally compressed. Wings: Fig 4. ......***Calobracon.*** Szépligeti

Distribution: Brazil, Peru, Argentina; few species. Biology: unknown. Reference: Shenefelt (1978).

- Antennal sockets not protruding and not acutely angulate in profile. Scape longer (Fig. 37; see also Figs 44-46). Fore basitarsus strongly laterally compressed (Fig. 60) 9

9(8). Metasomal tergum 2 medially more than 0.67 times median length of tergum 3 (Fig. 90); metasomal tergum 2 strongly 'pinched-up' antero-medially (Fig. 90); last metasomal tergite of male with pair of conspicuous projections. Wings: Fig. 5 .... ***Gracilibracon*** Quicke

Distribution: Neotropical; 3 described species, and at least 3 undescribed. Biology: unknown. Reference: Quicke (1995).

- Metasomal tergum 2 medially less than 0.5 times median length of tergum 3 (Figs 76, 91); metasomal tergum 2 not or hardly 'pinched-up' antero-medially (Figs 76, 91). Last metasomal tergite of male unmodified. Wings: Fig. 6 .... ***Cyclaulax*** Cameron

Distribution: Neotropical; frequently collected. Biology: unknown. Reference: Shenefelt (1978).

10(7). Hypopygium truncated in lateral aspect (Figs 57, 87). Hind telotarsus usually with a ventral, pre-apical cluster of especially thickened setae (Fig. 61) .... ***Compsobracon*** Ashmead

Distribution: southern U.S.A. (one record) to Brazil and Paraguay including Bahamas and Trinidad; many are particularly large and brightly colored and have thus been collected quite frequently. Biology: one species has been reared from Pyralidae (Quicke, 1989c). References: Mason (1978), Shenefelt (1978).

- Hypopygium acutely pointed in lateral aspect (Figs 55, 56, 58). Hind telotarsus ventrally without a cluster of especially thickened setae (Figs 62-64) 11

11(10). Scape shorter ventrally than dorsally (antenna directed anteriorly) (as in Figs 28, 39). Face usually with fine coriaceous or rugulose sculpture between longitudinal ridges (Figs 30, 31) .... ***Compsobraconoides*** Quicke

Distribution: southern U.S.A. to South America including the Caribbean; at least 5 described species; a very variable genus as currently defined, some species have well-developed mid-longitudinal grooves on the petiole (Fig. 75), some have claws with rounded basal lobes, some have pale-striped antennae, some have the metasoma distinctly sinuous in dorsal profile, and some have the suture between metasomal terga 2 and 3 strongly buckled medially (Fig. 74). Biology: unknown. Reference: Quicke and Sharkey (1989).

- Scape longer ventrally than dorsally (antenna directed anteriorly) (as in Fig. 29). Face usually coarsely sculptured between longitudinal ridges (Figs 32, 33) 12

12(11). Raised median area of petiole rather square in cross-section, lateral margin more or less cariniform posteriorly (Figs 77, 78). Face with finely sculptured, tear-drop shaped area medially (Fig. 32) .... ***Cyclaulacidea*** Quicke and Delobel

Distribution: Neotropical; one described species. Biology: parasitoids of Bruchidae and Curculionidae in palm fruit and similar situations. Reference: Quicke and Delobel (1995).

- Raised median area of petiole more rounded in cross-section (Fig. 80). Face coarsely carinate-rugose medially without a more-finely sculptured, tear-drop shaped area (Fig, 33) .... ***Sacirema*** Quicke

Distribution: Neotropical; one described species, about 5 undescribed. Biology: unknown. Reference: Quicke (1995).

13(6). Scape highly modified: basally with concave excavation and apically with medioventral false margin well-separated from real margin, giving rise to large pre-apical ledge ventro-medially (Figs 41, 43-46). Apex of hind wing C+SC+R with only one especially thickened seta (Figs 7, 53; sometimes these setae are lost but their large sockets can be seen) 14

- Scape simple, neither abruptly concavely narrowing basally nor with wide apical ledge formed by pre-apical, medic-ventral false margin (Figs 28, 29, 38, 39). Apex of hind wing C+SC+R often with more than one especially thickened seta and nearly always so in wasps with wing span greater than 13 mm (Figs 15, 54; sometimes these setae are lost but their large sockets can be seen) 15

14(13). Raised median area of petiole abruptly and semicircularly emarginate postero-laterally, the resulting concavities being bordered laterally by the dorso-lateral carinae (Fig. 72) though sometimes with these rather weak (Fig. 73); if weak, then metasomal tergum 3 longer than wide; petiole often with a mid-longitudinal carina or 'Y' -shaped ridge. Wings: Fig. 7 .... ***Hemibracon*** Szépligeti

Distribution: Costa Rica to Argentina; commonly collected; many species. Biology: unknown. Reference: Shenefelt (1978).

- Raised median area of petiole not abruptly emarginate postero-laterally, and without mid-longitudinal carina or 'Y'-shaped ridge .... ***Atanycolus*** Förster

Distribution: Palaearctic, Old World tropics (rarely), Canada to Costa Rica; common in northern part of range. Biology: parasitoids of bark-boring beetles especially Cerambycidae and Buprestidae. References: Shenefelt (1943, key to North American species).

15(13). Scape shorter ventrally than dorsally (antenna directed anteriorly) (Figs 28, 39), not or only weakly emarginate apico-laterally and not emarginate apico-medially 16

- Scape longer ventrally than dorsally (antenna directed anteriorly) (Figs 29, 42), apico-laterally emarginate and usually also distinctly emarginate apico-medially 23

16(15). Marginal cell of fore wing short, vein 3RSb reaching wing margin at most 0.7 times distance between apex of stigma and wing tip (Fig. 8). Either clypeus with pair of 'hair-pencils' composed of several long setae arranged in a cluster and becoming fused apically (Fig. 48) and I or metasomal tergum 4 with characteristic pattern of fine striae that curve away from the midline (Fig. 95) .... ***Vipio*** Latreille

Distribution: Palaearctic, Palaeotropics (occasional), Canada to South America though less common in the southern portion of the range. Biology: parasitoids of concealed lepidopterous, coleopterous, and probably symphytan larvae. References: Wharton and Mason (1991), Inayatullah (1992), ICZN (1994).

- Marginal cell of fore wing usually longer, vein 3RSb reaching wing margin at least 0.8 times distance between apex of stigma and wing tip (Figs 1-7); if shorter, then clypeus without pair of 'hair-pencils' (clypeal guard setae evenly spaced and not fused, Figs 47, 52) and metasomal tergum 4 without fine striae that curve away from the midline 17

17(16). Hind tibia strongly laterally compressed, very broad in lateral aspect with especially long setae (Figs 92, 94). Metasomal tergum 2 with a mid-longitudinal ridge (Fig. 93) .... ***Myosomatoides*** Quicke

Distribution: Neotropical, widely distributed but rarely collected; one described species. Biology: one species has been reared from larvae of the stem-boring pyralid moth *Diatraea* sp. (Quicke, 1994). Reference: Quicke (1994).

\- Hind tibia usually not strongly laterally compressed and usually without especially long marginal setae. If rather strongly compressed with long marginal setae (most *Myosoma* species) then metasomal tergum 2 has no mid-longitudinal ridge (Fig. 77) 18

18(17). Body very strongly dorso-ventrally depressed, mesosoma more than 2.8 times longer than maximally high (Fig. 59) .... ***Chartobracon*** van Achterberg

Distribution: known only from Canada. (1 species) and northern Europe. Biology: the European species, *C. huggerti* van Achterberg, has been reared from *Callidium* F. (Cerambycidae) tunnels under spruce bark. Reference: van Achterberg (1983).

\- Body not or only moderately dorso-ventrally depressed, mesosoma less than 2.3 times longer than maximally high (see Figs 65, 66) 19

19(18). Median tergite of petiole very long and narrow, more than 2.5 times longer than maximally wide (Fig. 83). Hind femur and tibia usually markedly flattened, with long marginal setae. Wings: Fig. 9 .... ***Myosoma*** Brullé

Distribution: widely distributed from southern U.S.A. to Brazil. Biology: parasitoids of lepidopterous borers, especially Pyralidae, Reference: Mason (1978, key to species of North America).

\- Petiole with median tergite not especially elongate, less than 2.2 times longer than maximally wide (as in Figs 75-77, SQ). Hind femur and tibia normal 20

20(19). Propodeum and metanotum both with a complete lamelliform mid-longitudinal carina (Fig. 68). Sternaulus deep and crenulate (Fig. 66) .... ***Lapicida*** Quicke

Distribution: Canada through central U.S.A.; one described and one undescribed species known; rarely collected because of its particular habitat. Biology: associated with aquatic vegetation (*Sagittaria*). Reference: Quicke and Sharkey (1989).

\- Propodeum and metanotum both simple (Fig. 67) or at most propodeum with a short mid-longitudinal carina posteriorly and/ or with metanotum forming a short carina mid-anteriorly. Sternaulus not impressed or if weakly impressed then not crenulate (Fig. 65) 21

21(20). Posterior part of propleuron with a longitudinal carina or ridge sub-laterally (Fig. 70) .... ***Pigeria*** van Achterberg

Distribution: Palaearctic; one species, *P. piger* (Wesmael), has been introduced into California and it seems highly likely that this species may also be native to North America (Quicke and Sharkey, 1989). Biology: parasitic on a wide variety of concealed Lepidoptera larvae. References: van Achterberg (1985), Quicke and Sharkey (1989).

\- Posterior part of propleuron smooth, without a longitudinal carina or ridge sub-laterally 22

22(21). Fore wing 3RSa less than 1.5 times length of r (usually less than 1.2 times) (Fig. 10). Fore wing r often distinctly sinuate. Antenna with fewer than 20 flagellomeres (Fig. 38). Wings: Fig. 10 .... ***Habrobracon*** Ashmead

Distribution: cosmopolitan; few species. Biology: extensively studied in the laboratory and attacks concealed larvae of a wide variety of Coleoptera, Lepidoptera, Diptera and Hymenoptera but is best known for its gregarious parasitism of flour and grain moths. Reference: Fischer (1968, key).

\- Fore wing 3RSa more than 1.6 times length of r (usually more than 1.8 times) (Fig. 11). Fore wing r not or hardly sinuate (Fig.11). Antenna usually with more than 20 flagellomeres .... ***Bracon*** Fabricius (part)

Distribution: cosmopolitan; very common and highly speciose, several Neotropical species groups are distinctive and some in particular are larger and/ or more heavily sculptured than is typical of most Nearctic and Old World species; nevertheless, separation of these into distinct genera would almost certainly render the remainder of *Bracon* paraphyletic. Biology: parasitoids of a wide range of lepidopterous, coleopterous and dipterous larvae. Reference: North American species were keyed by Muesebeck (1925) though there are undoubtedly many species awaiting description from that region; at present identification of Neotropical species is virtually impossible.

23(15). Propodeum coarsely sculptured (Fig. 69). Fore wing marginal cell short, 3RSb reaching wing margin less than 0.8 times distance between apex of stigma and wing tip (Fig. 12); fore wing (RS+M)a straight (Fig. 12) .... ***Vipiomorpha*** Tobias

Distribution: principally a Palaearctic and Afrotropical genus, but also occurring in Australia and South America though very seldom collected there. Biology: unknown. Reference: Tobias (1986).

\- Propodeum smooth and shiny, at most with weak punctures at bases of setae (Fig.67). Fore wing marginal cell long, Vein 3RSb reaching wing margin more than 0.8 times distance between apex of stigma and wing tip (Figs 11, 13-15); fore wing (RS+M)a variable, but often distinctly curved or angled posteriorly shortly after arising from 1M (see Figs 4-6) 24

24(23). Ventral border of clypeus sinuate in frontal aspect (Fig. 40). Flagellum with fewer than 50 flagellomeres. Wings: Fig. 13 .... ***Alienoclypeus*** Shenefelt

Distribution: Arid regions of southern. U.S.A. and Mexico; uncommon; one described species. Biology: parasitoid of weevil *Scyphophorus* Schoenherr. Reference: Shenefelt in Mason (1978).

- Ventral border of clypeus arched in frontal aspect (see Figs 31, 52). Flagellum usually with more than 60 flagellomeres 25

25(24). Mandibles massive (Figs 49, 50) .... ***Gnathobracon*** Costa

Distribution: French Guiana, Peru; rarely collected; one described species. Biology: unknown. Reference: Quicke and Huddleston (1991).

- Mandibles of normal size (Figs 30, 31, 52) 26

26(25). With three very long pseudo-ovipositors (more than 5 times longer than body) extending from anal region (true ovipositor and ovipositor sheaths much shorter) (Fig. 96) .... ***Pheloura*** van Achterberg

Distribution: Neotropical (Colombia, Costa Rica); one species; extremely rarely collected, known to date from only three specimens. Biology: unknown. Reference: van Achterberg (1989).

- Without pseudo-ovipositors (Fig. 87) 27

27(26). Scape very long, more than 2.5 times longer than mid-width (Fig. 42). Metasoma smooth and shiny (as in Figs 74, 79) .... ***Megabracon*** Szépligeti

Distribution: Neotropical; few species; rarely collected. Biology: unknown.

- Scape much shorter, less than 2.0 times longer than mid-width (Fig. 29). Metasomal sculpture variable, often with suture between terga 2 and 3 crenulate and/ or otherwise coarsely sculptured (as in Figs 51, 84) 28

28(27). Metasomal terga 2-5 very long and slender, tergum 3 more than 2 times longer than maximally wide: posterior margins of terga 3-5 strongly emarginate and membranous medially, with extremely long internal apodemes (Fig. 88). A long, slender wasp (Fig.97). Wings: Fig. 14 .... ***Leptobracon*** Szépligeti

Distribution: Neotropical; few species; rarely collected. Biology: unknown. Reference: Fahringer (1930).

- Metasomal terga 2-5 never so long and slender, tergum 3 less than 2 times longer than maximally wide (see Figs 51, 84); posterior margins of terga 3-5 at most weakly emarginate and never with a distinct membranous median zone, and with internal apodemes never more than 2 times longer than medially wide (Fig. 89). Habitus otherwise 29

29(28). Posterior margins of metasomal terga 3-5 thickly sclerotized and dorsally convex in lateral aspect (Fig. 85); metasomal terga coarsely sculptured in most species (Figs 51, 84); raised median area of petiole with short mid-longitudinal carina or pair of submedial pits medio-posteriorly in most species (Fig. 84); dorso-lateral carinae of petiole well developed and, in most species, lamelliform. Wings: Fig. 15 .... ***Digonogastra*** Viereck

Distribution: New World possibly extending into the eastern Palaearctic; very speciose, many species previously included in the genus *Iphiaulax* Förster, the latter now restricted to the Old World (Quicke, 1988d). Biology: hosts include larvae of Coleoptera and Lepidoptera. References: Quicke (1988d, 1989a).

- Posterior margins of metasomal terga 3-5 thinly sclerotized or membranous, profile straight in lateral aspect (Fig. 86); metasomal terga smooth (as in Fig. 74); raised median area of petiole without carina or pair of submedial depressions; dorso-lateral carinae of petiole absent or indistinctly defined .... ***Cyanopterus*** Haliday

Distribution: New World and Palaearctic; many species. Biology: hosts include larvae of Coleoptera and Lepidoptera. Note: it is very difficult to draw a clear line between this and the preceding genus, especially where the Neotropical fauna is concerned. Reference: Quicke and Sharkey (1989).

REFERENCES

Achterberg, C. van 1983. Three new Palaearctic genera of Braconinae (Hymenoptera: Braconidae), Entomologia Scandinavica 14: 69-76.

Achterberg, C. van 1985. Notes on Braconidae I-IV. II. *Pigeria* gen. nov., a new palaearctic genus of the Braconinae (Hymenoptera; Braconidae). Zoologische Mededelingen 59: 168-174.

Achterberg, C. van 1989. *Pheloura* gen. nov., a neotropical genus with an extremely long pseudoovipositor (Hymenoptera: Braconidae). Entomologische Berichten 49: 105-108.

Fahringer, J. 1930. Über einige südamerikanischer Braconidengattungen. Wiener entomologische Zeitung 47: 19-31.

Fischer, M. 1968. Über gezüchtete Raupenwespen (Hymenoptera, Braconidae). Pflanzengschutz Berichte 37: 97-140.

ICZN 1994. Opinion 1758. *Vipio* Latrielle, 1804 (Insecta, Hymenoptera): *Agathis longicauda* Boheman, 1853 designated as the type species, Bulletin of Zoological Nomenclature 51: 76-78.

Inayatullah, M. 1992. A systematic study of the genus *Vipio* Latreille (Hymenoptera: Braconidae) of the Nearctic and Neotropical Regions. Unpublished Ph.D. thesis, University of Wyoming, Laramie, Wyoming.

Marsh, P. M. 1979. Family Braconidae. Pp. 144-295. In: Krombein, K. V., Hurd, P. D., Smith, D. R., and Burks, B. D. (Eds.). Catalog of Hymenoptera in America North of Mexico. Smithsonian Institution Press, Washington, D.C.

Mason, W. M. 1978. A synopsis of the Nearctic Braconini, with revisions of Nearctic species of *Coeloides* and *Myosoma* (Hymenoptera: Braconidae). The Canadian Entomologist 110: 721-768.

Muesebeck, C. F. W. 1925. A revision of the parasitic wasps of the genus *Microbracon* occurring in America north of Mexico. Proceedings of the United States National Museum 67: 1-85.

Quicke, D. L. J. 1988a. Reclassification of some Neotropical Braconinae (Hym., Braconidae). Entomologist's Monthly Magazine 124: 195-199.

Quicke, D. L. J. 1988b. A new genus and species of Braconinae (Hymenoptera: Braconidae) parasitic on *Diatraea* sp. (Lepidoptera: Pyralidae) from Ecuador. Bulletin of Entomological Research 78: 15-18.

Quicke, D. L. J. 1988c. The higher classification, zoogeography and biology of the Braconinae, Pp. 117-138. In: Gupta, V. K. (Ed.). Advances in Parasitic Hymenoptera Research. E.J. Brill Publishing Co., Amsterdam.

Quicke, D. L. J. 1988d. *Digonogastra*: the correct name for Nearctic *Iphiaulax* of authors (Hymenoptera, Braconidae). Proceedings of the Entomological Society of Washington 90: 196-200.

Quicke, D. L. J. 1989a. Reclassification of some New World species of Braconinae (Hym., Braconidae). Entomologist's Monthly Magazine 125: 119-121.

Quirke, D. L. J. 1989b. A new Neotropical genus and species of Coeloidini (Hym., Braconidae, Braconinae). Entomologist's Monthly Magazine 125: 9-12.

Quicke, D. L. J. 1989c. Further new host records for genera and species of Braconinae (Hym., Braconidae). Entomologist's Monthly Magazine 125: 199-205.

Quicke, D. L. J. 1991. The Braconinae type specimens of G. Szépligeti in Budapest (Hymenoptera, Braconidae). Annales Historico-Naturales Musei Nationalis Hungarici 83: 169-186.

Quicke, D. L. J. 1994. *Myosomatoides* gen. nov. (Hymenoptera: Braconidae), a Neotropical larval parasitoid of stem-boring pests, *Diatraea*. Journal of Hymenoptera Research 3: 227-231.

Quicke, D. L. J. 1995. Two new Neotropical genera of Braconinae (Hymenoptera, Braconidae). Entomologist's Monthly Magazine 131: 223-228.

Quicke, D. L. J. and Delobel, A. 1995. A new Neotropical braconine (Hym., Braconidae) parasitic on Bruchidae (Col.). Entomologist's Monthly Magazine 131: 215-222.

Quicke, D. L. J. and Genise, J. F. 1994. Reclassification of the South American species of Braconinae (Hym., Braconidae) described by Brèthes. Entomologist's Monthly Magazine 130: 35-45.

Quicke, D. L. J. and Huddleston, T. 1991. The extraordinary genus *Gnathobracon* Costa (Hym., Braconidae), with the description of a new species from Peru. Entomologist's Monthly Magazine 127: 191-195.

Quicke, D. L. J. and Sharkey, M. J. 1989. A key to and notes on the genera of Braconinae (Hymenoptera: Braconidae) from America north of Mexico with descriptions of two new genera and three new species. The Canadian Entomologist 121: 337-361.

Shaw, M. R. and Huddleston, T. 1991. Classification and biology of braconid wasps (Hymenoptera: Braconidae). Handbooks for the Identification of British Insects 7(11): 1-126.

Shenefelt, R. D. 1943. The genus *Atanycolus* Foerster in America north of Mexico. Research Studies of the State College of Washington 11: 51-163.

Shenefelt, R. D. 1978. Pars. 15. Braconidae 10, Braconinae, Gnathobraconinae, Mesostoinae, Pseudodicrogeniinae, Telengainae, Ypsistocerinae plus Braconidae in general, major groups, unplaced genera and species. Pp. 1425-1872. In: Achterberg, C. van and Shenefelt, R. D. (Eds.). Hymenopterorum Catalogus (nova editio). Dr W. Junk, The Hague.

Tillman, P. G. and Cate, J. R. 1989. Six new hosts of *Bracon mellitor* (Hymenoptera: Braconidae), with a review of recorded hosts. Environmental Entomology 18: 328-333.

Tobias, V. I. 1986. Subfamily Braconinae. Pp. 94-149. In: Tobias, V. I. (Ed.). Identification of the Insects of European U.S.S.R. Volume III, Part V. Hymenoptera, Braconidae. Akademia Nauk, Leningrad. (In Russian.)

Wharton, R. A. and Mason, W.R. M. 1991. *Vipio* Latrielle, 1804 (Insecta, Hymenoptera): proposed designation of *Agathis longicauda* Boheman, 1853 as the type species. Bulletin of Zoological Nomenclature 48: 45-49.

Wharton, R. A., Smith, J. W., Quicke, D. L. J., and Browning, H. W. 1989. Two new North American species of *Digonogastra* Viereck (Hym., Braconidae) parasitic on cane boring Lepidoptera. Bulletin of Entomological Research 79: 401-410.

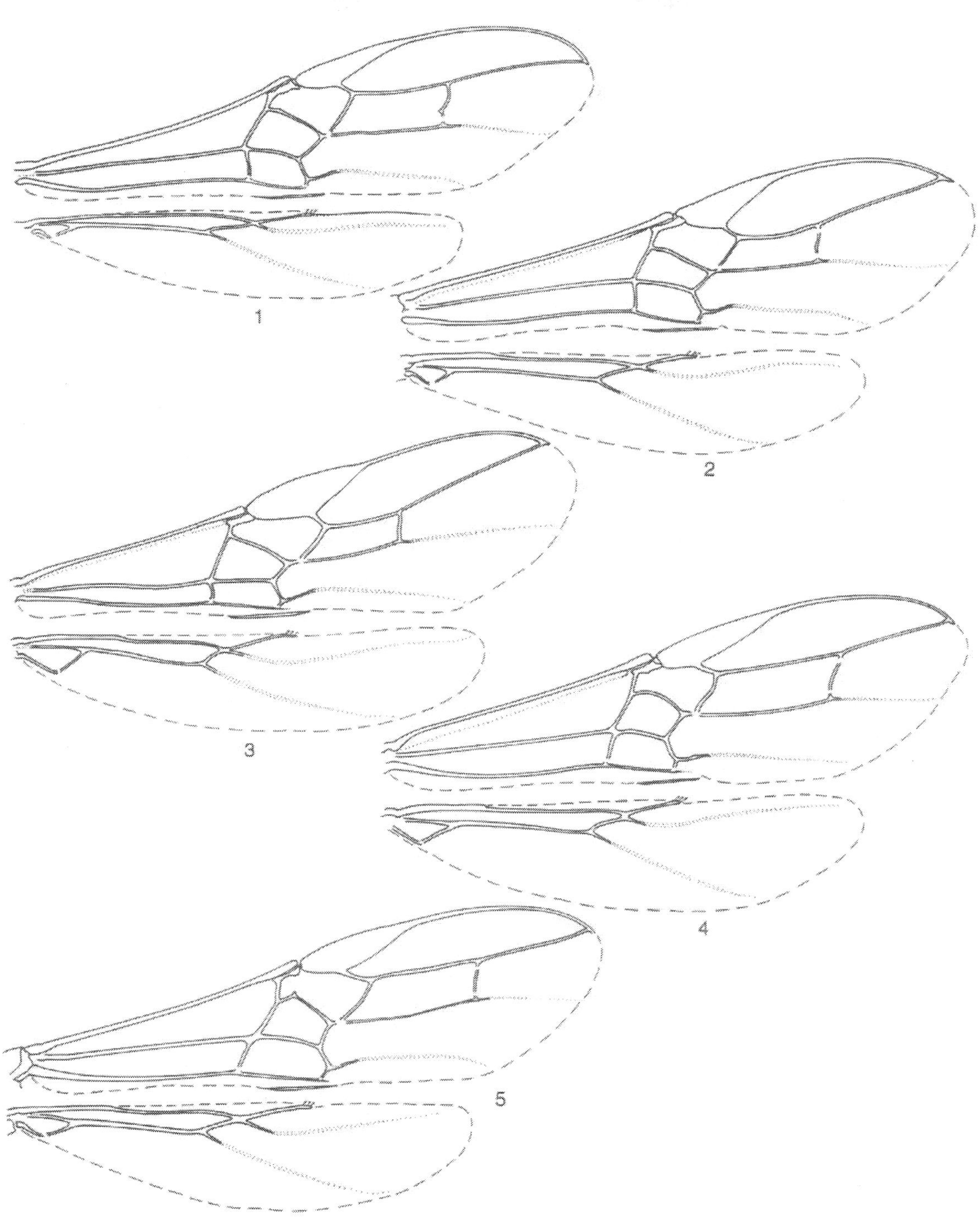

Figs 1-5: Wings of Braconinae. 1, *Lasiophorus* sp.; 2, *Cervellus* sp.; 3, *Coeloides* sp.; 4, *Calobracon* sp.; 5, *Gracilibracon* sp.

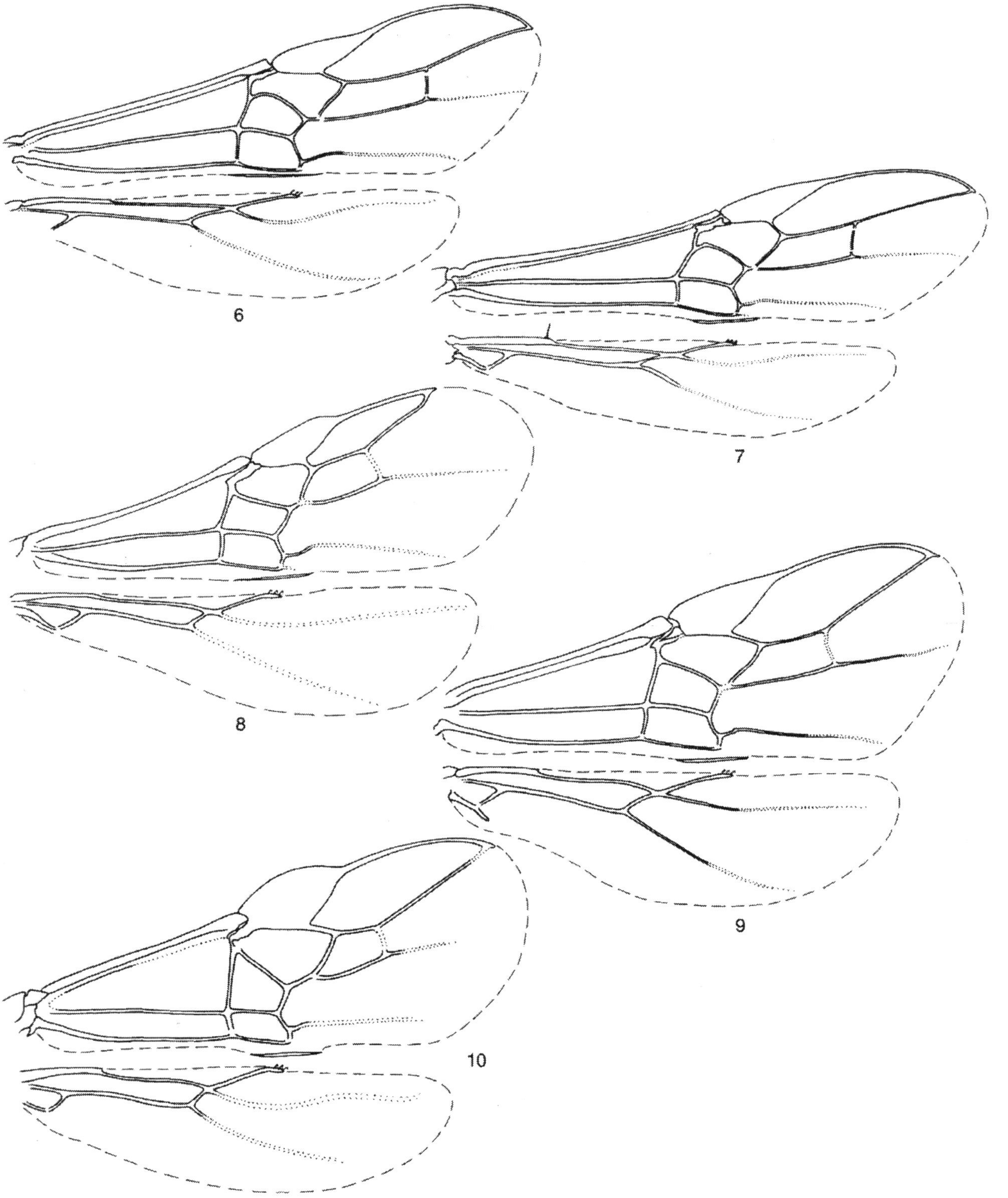

Figs 6-10: Wings of Braconinae. 6, *Cyclaulax* sp.; 7, *Hemibracon* sp.; 8, *Vipio* sp.; 9, *Myosoma* sp.; 10, *Habrobracon* sp.

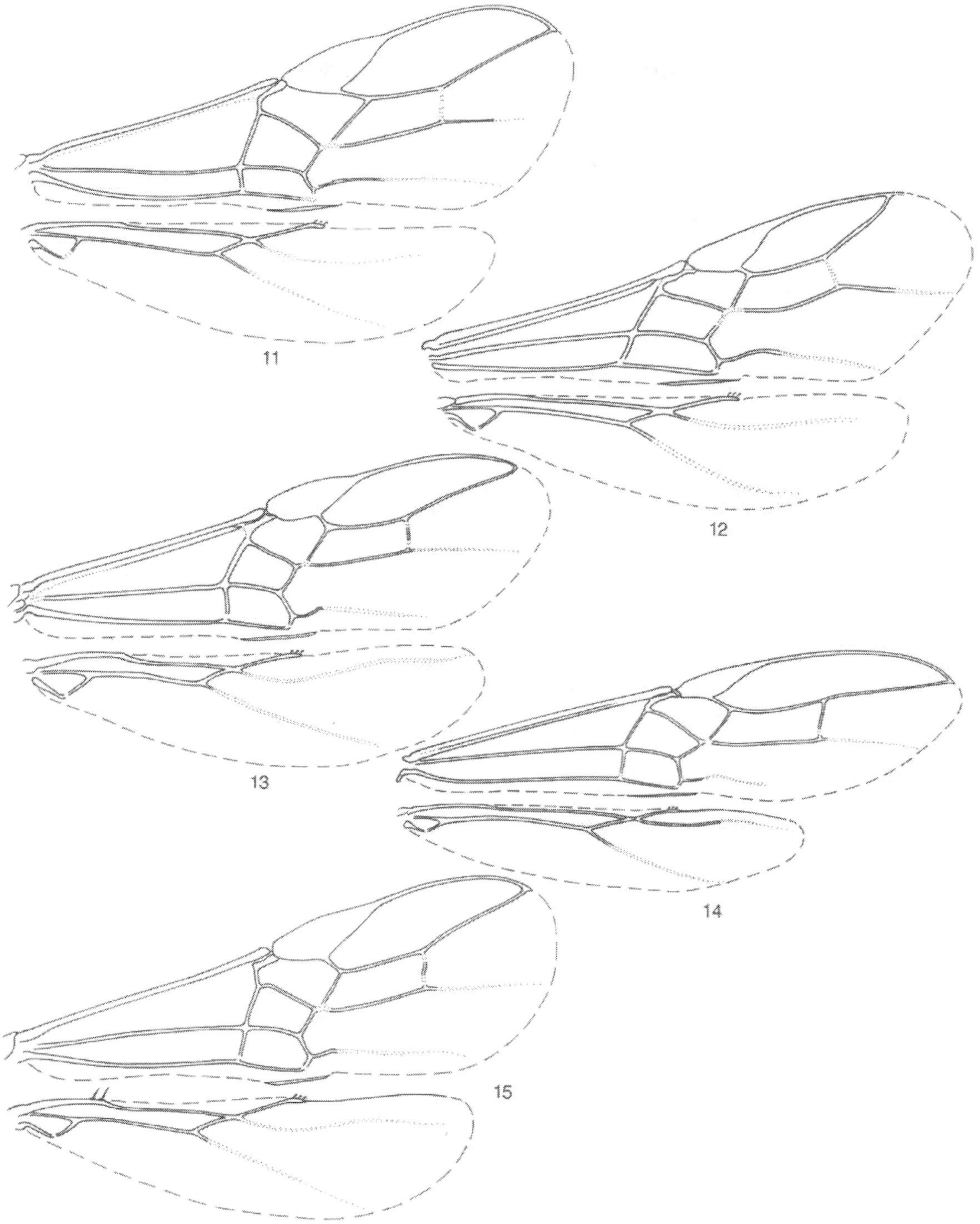

Figs 11-15: Wings of Braconinae. 11, *Bracon* sp.; 12, *Vipiomorpha* sp.; 13, *Alienoclypeus* sp.; 14, *Leptobracon* sp.; 15, *Digonogastra* sp.

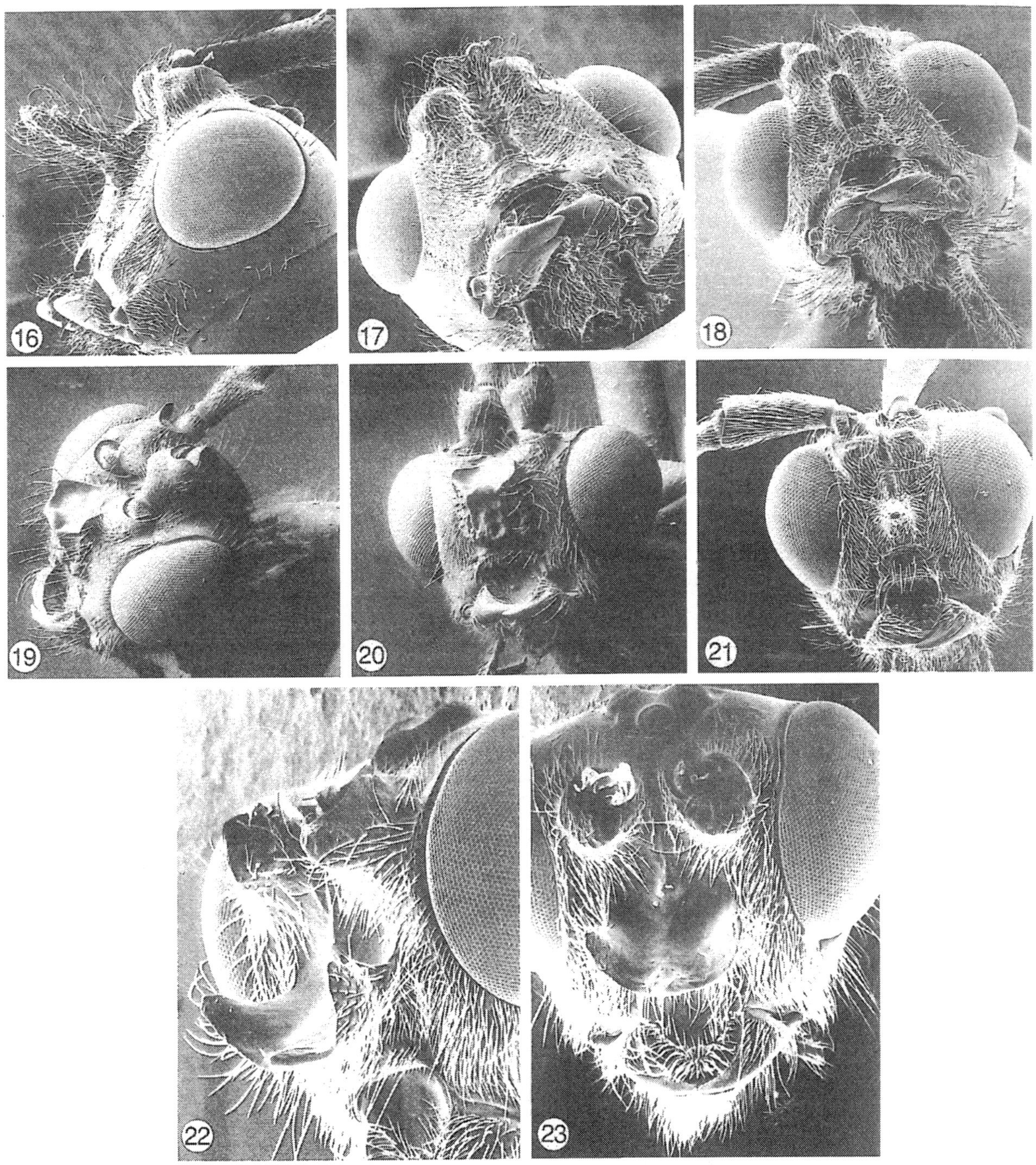

Figs 16-23: Heads of Braconinae. 16, *Lasiophorus* sp. female; 17, *Lasiophorus* sp. female; 18, *Lasiophorus* sp. male; 19, *Cervellus* sp.; 20, *Cervellus* sp.; 21, *Lasiophorus* sp. male; 22, *Palabracon* sp.; 23, *Palabracon* sp.

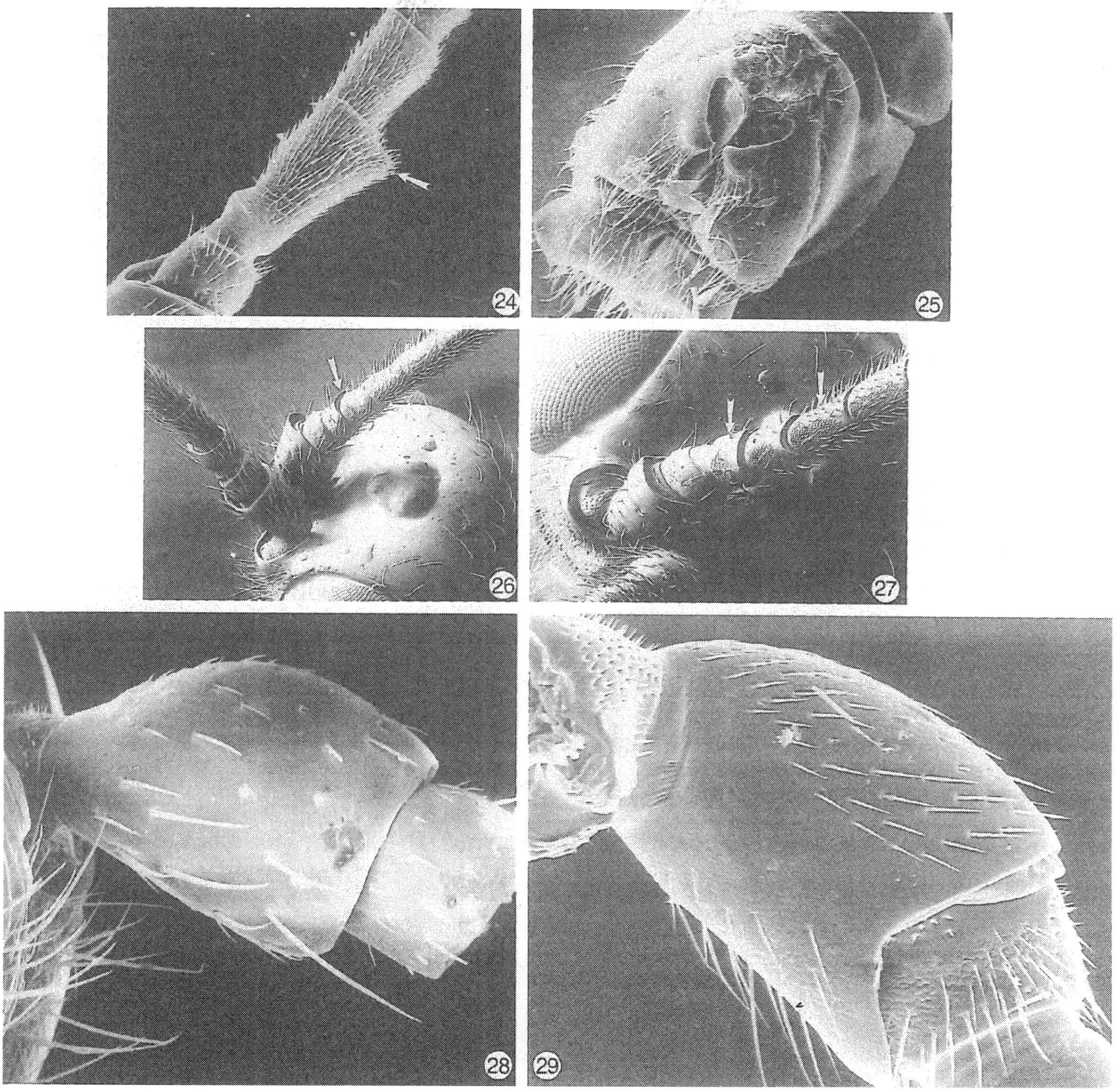

Figs 24-29: Braconinae. 24, antennae of *Cervellus* sp., arrow = thorn-like projection on first flagellomere; 25, propleuron of *Cervellus* sp., antero-ventral view; 26, basal antennal segments of *Megacoeloides* sp., arrow = expanded apex of first flagellomere; 27, basal antennal segments of *Megacoeloides* sp., arrows = first and third flagellomeres; 28, scape and pedicel of *Compsobracon* sp., lateral view; 29, scape and pedicel of *Leptobracon* sp., lateral view.

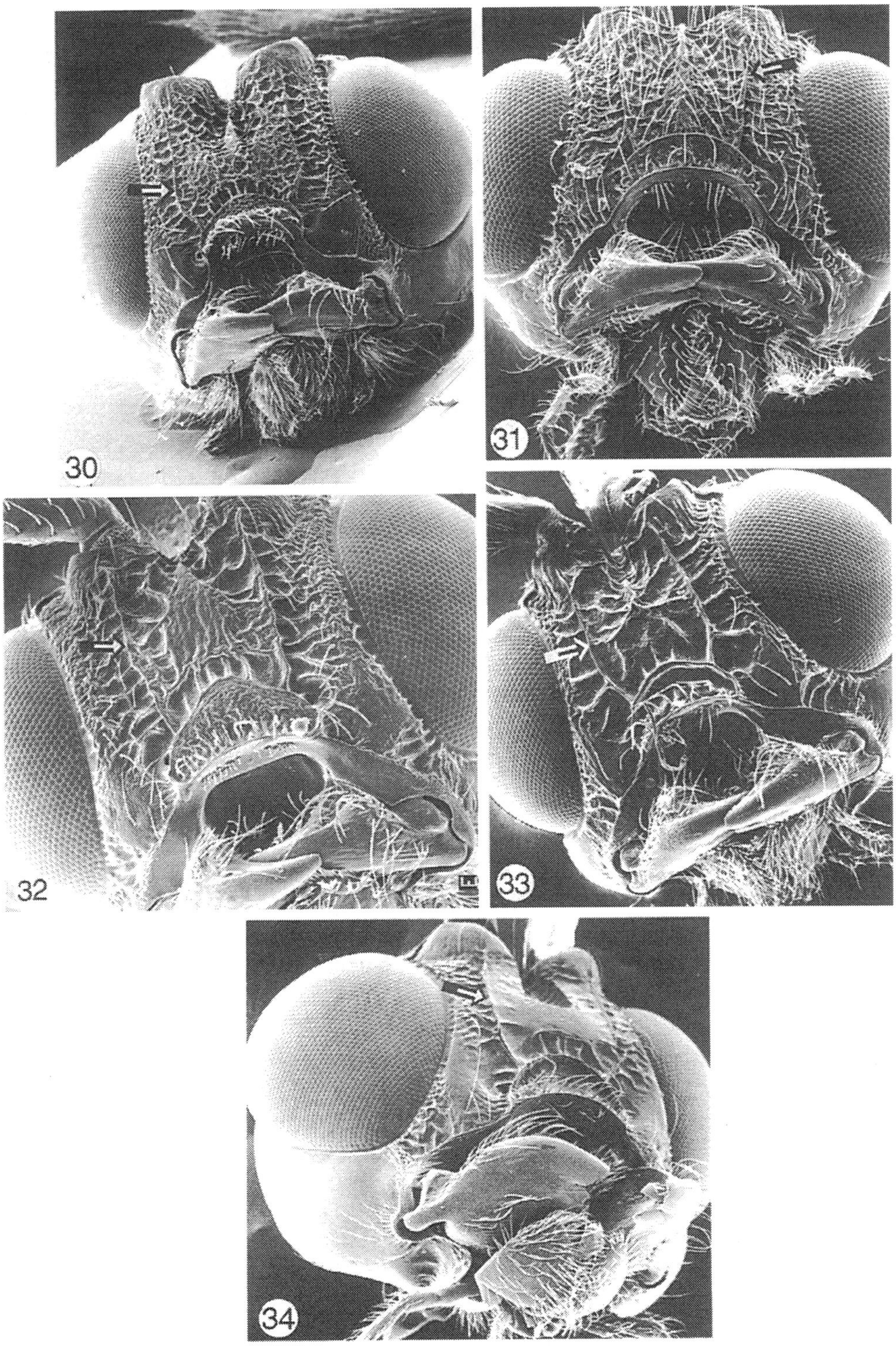

Figs 30-34: Heads of Braconinae, frontal views. 30 and 31, *Compsobraconoides* sp., arrow = sub-medial, longitudinal carina; 32, *Cyclaulacidea* sp., arrow = sub-medial, longitudinal carina (after Quicke and Delobel, 1995); 33, *Sacirema* ·sp., arrow = sub-medial, longitudinal carina (after Quicke, 1995); 34, *Gracilibracon* sp., arrow = sub-medial, longitudinal carina.

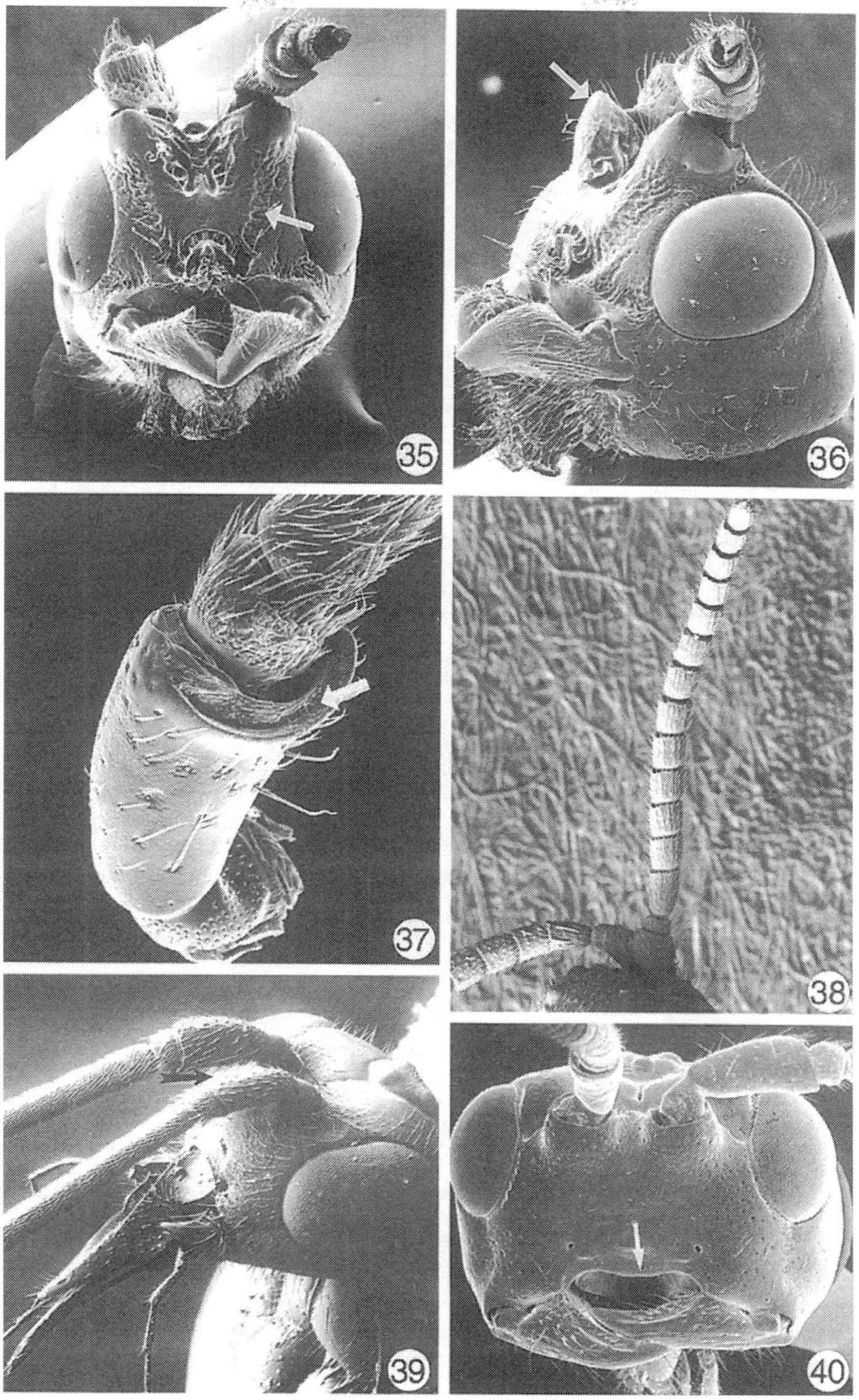

Figs 35-40: Braconinae. 35, face of *Calobracon* sp., arrow = sculptured groove demarking median part of face; 36, head of *Calobracon* sp., profile, arrow = protruding antennal socket; 37, scape and pedicel of *Cyclaulax* sp., arrow = pre-apical ledge of scape; 38, antenna bf *Habrobracon*. sp.; 39, head of *Vipio* sp., arrow = "dorsal" or upper surface of scape; 40, face of *Alienoclypeus* sp., setae removed, arrow = sinuate margin of clypeus (after Quicke and Sharkey, 1989).

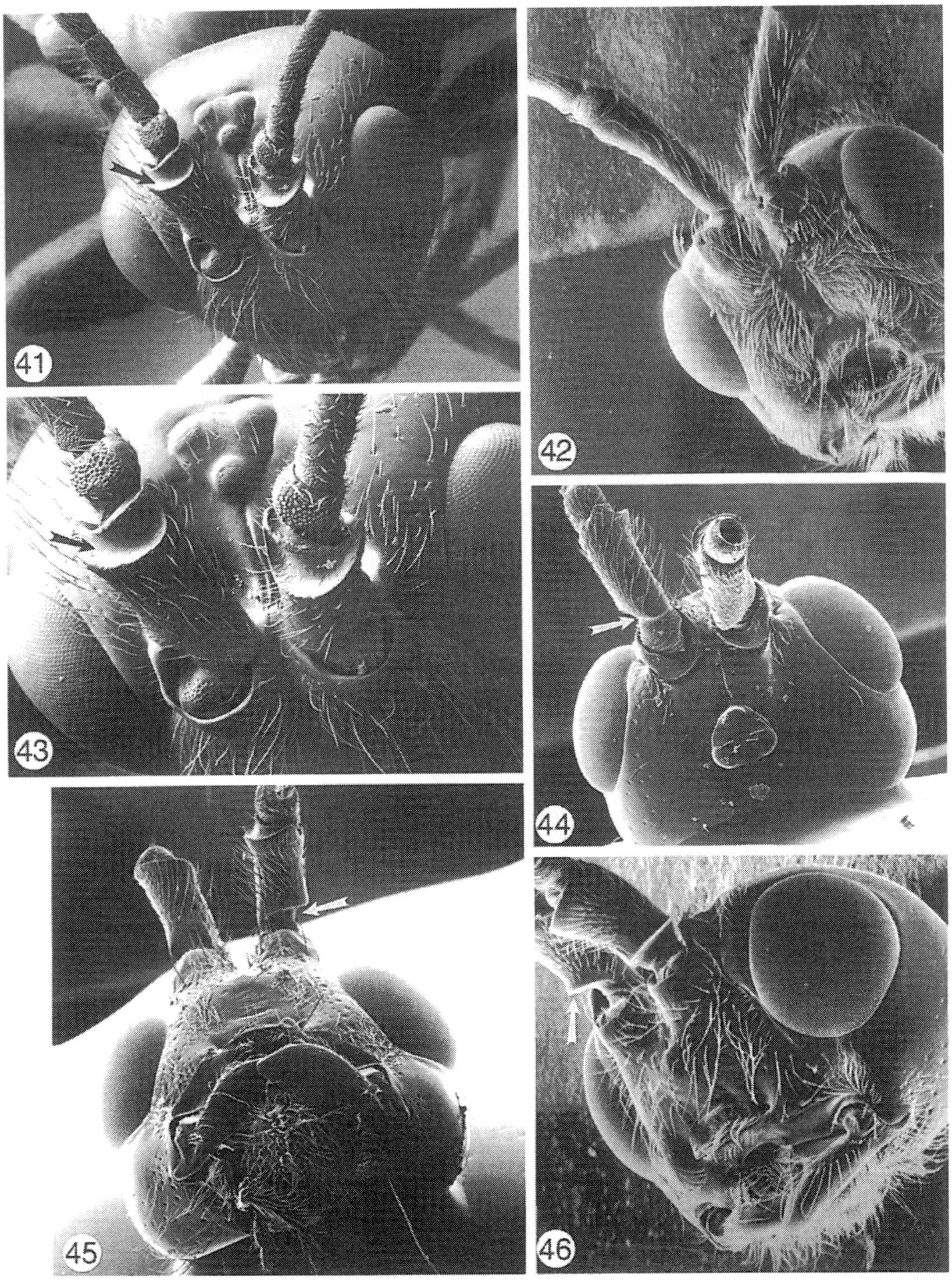

Figs 41-46: Braconinae. 41, head of *Hemibracon* sp., arrow = pre-apical ledge of scape; 42, head of *Megabracon* sp. showing long scape; 43; scape and pedicel of *Hemibracon* sp., arrow = pre-apical ledge of scape; 44-46, heads: 44, *Atanycolus* sp., dorsal view, arrow = deeply excavated basal portion of scape; 4,5, *Atanycolus* sp., antero-ventral view, arrow = deeply excavated basal portion of scape; 46, *Atanycolus* sp., arrow = deeply excavated basal portion of scape.

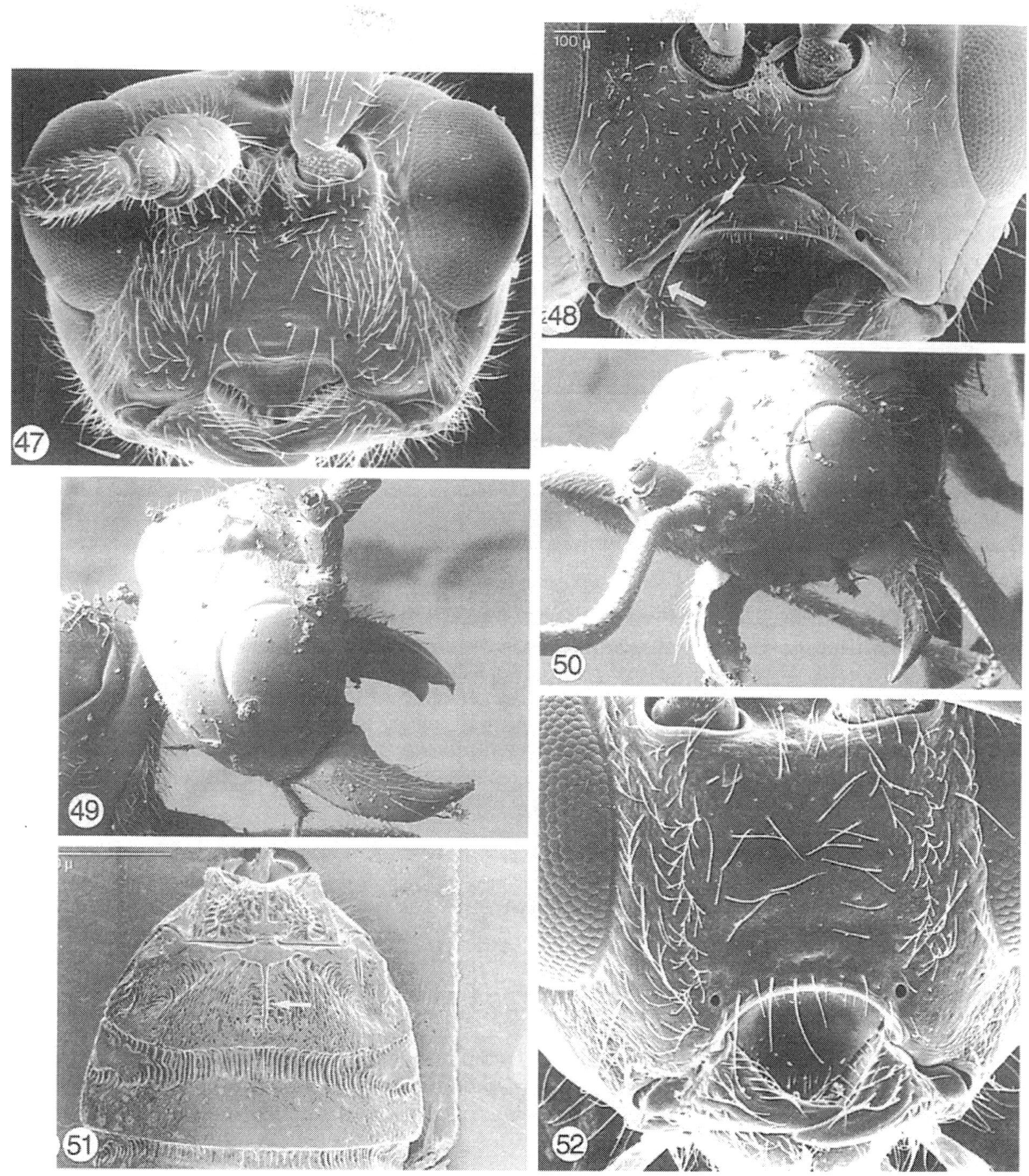

Figs 47-52: Braconinae. 47, face of *Alienoclypeus* sp.; 48, face of *Vipio* sp., arrow = fused hair-pencils on clypeus (after Quicke and Sharkey, 1989); 49 and 50, head of *Gnathobracon* sp.; 51, metasomal terga 1-3 of *Digonogastra* sp. (after Quicke and Sharkey, 1989); 52, face of *Bracon* sp.

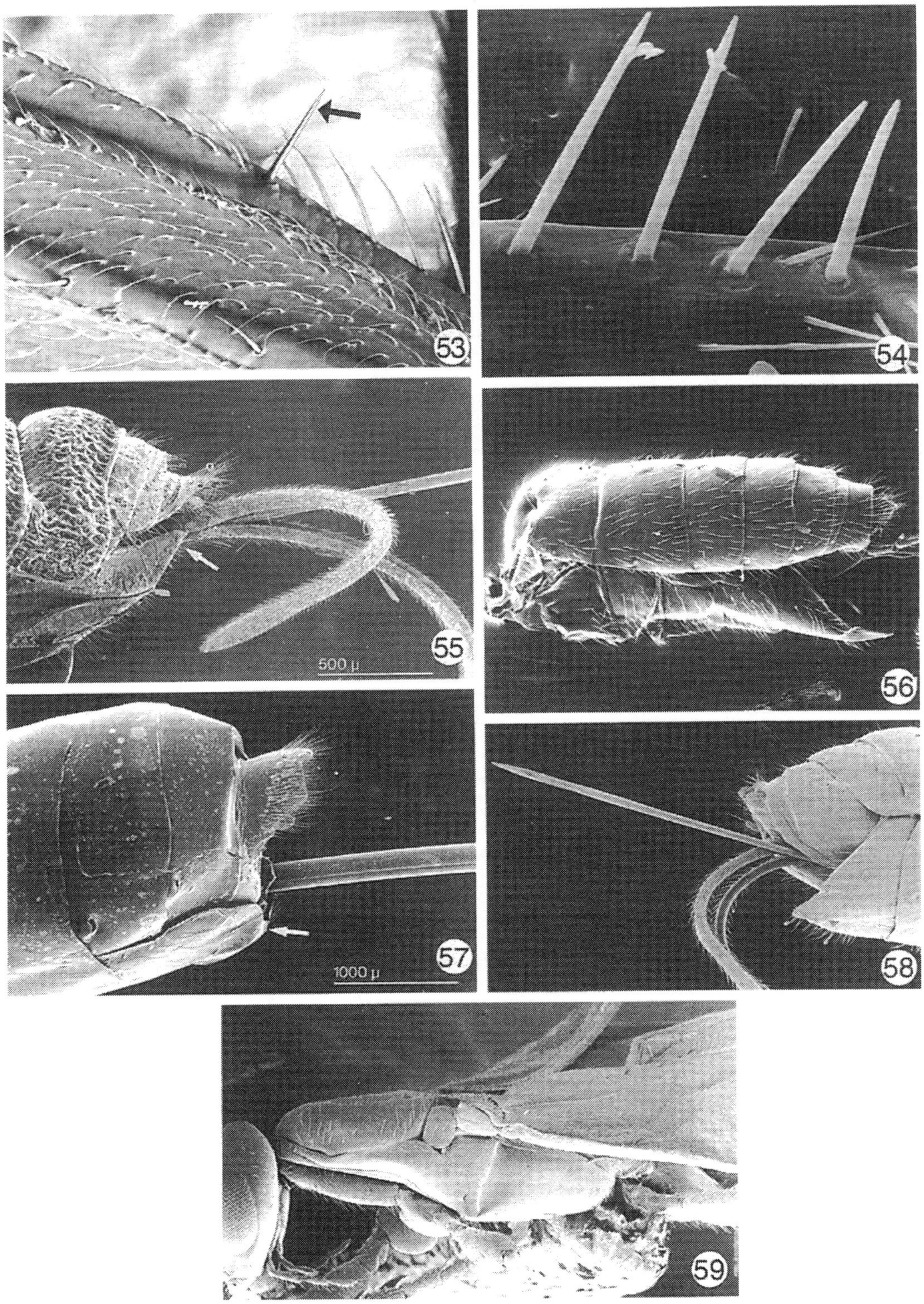

Figs 53-59: Braconinae. 53, hind wing of Bracon sp., arrow = single thickened seta on C+SC+R; 54, hind wing of *Digonogastra* sp. showing 4 thickened setae on C +SC+ R; 55-58, apex of metasoma: 55, *Digonogastra* sp., arrow = pointed hypopygium; 56, *Compsobraconoides* sp.; 57, *Compsobracon* sp., arrow = truncated hypopygium; 58, Bracon sp.; 59, body of *Chartobracon* sp. showing dorso-ventral flattening. Figures 55 and 57-59 after Quicke and Sharkey (1989).

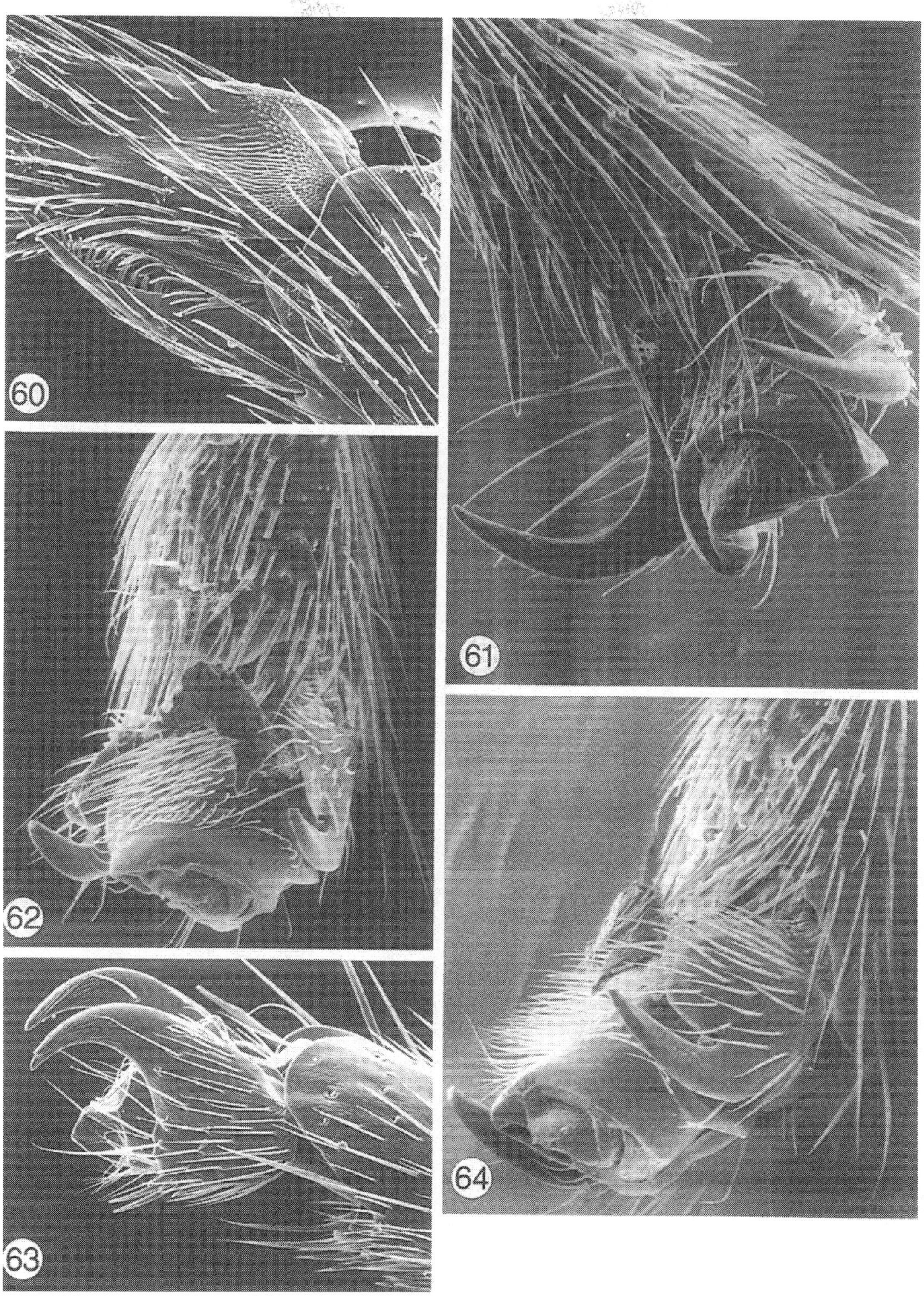

Figs 60-64: Braconinae. 60, strongly flattened fore basitarsus of *Cyclaulax* sp.; 61-64, hind telotarsi and claws: 61, *Compsobracon* sp. showing thickened ventral setae; 62, *Myosoma* sp.; 63, *Bracon* sp.; 64, *Bracon* sp.

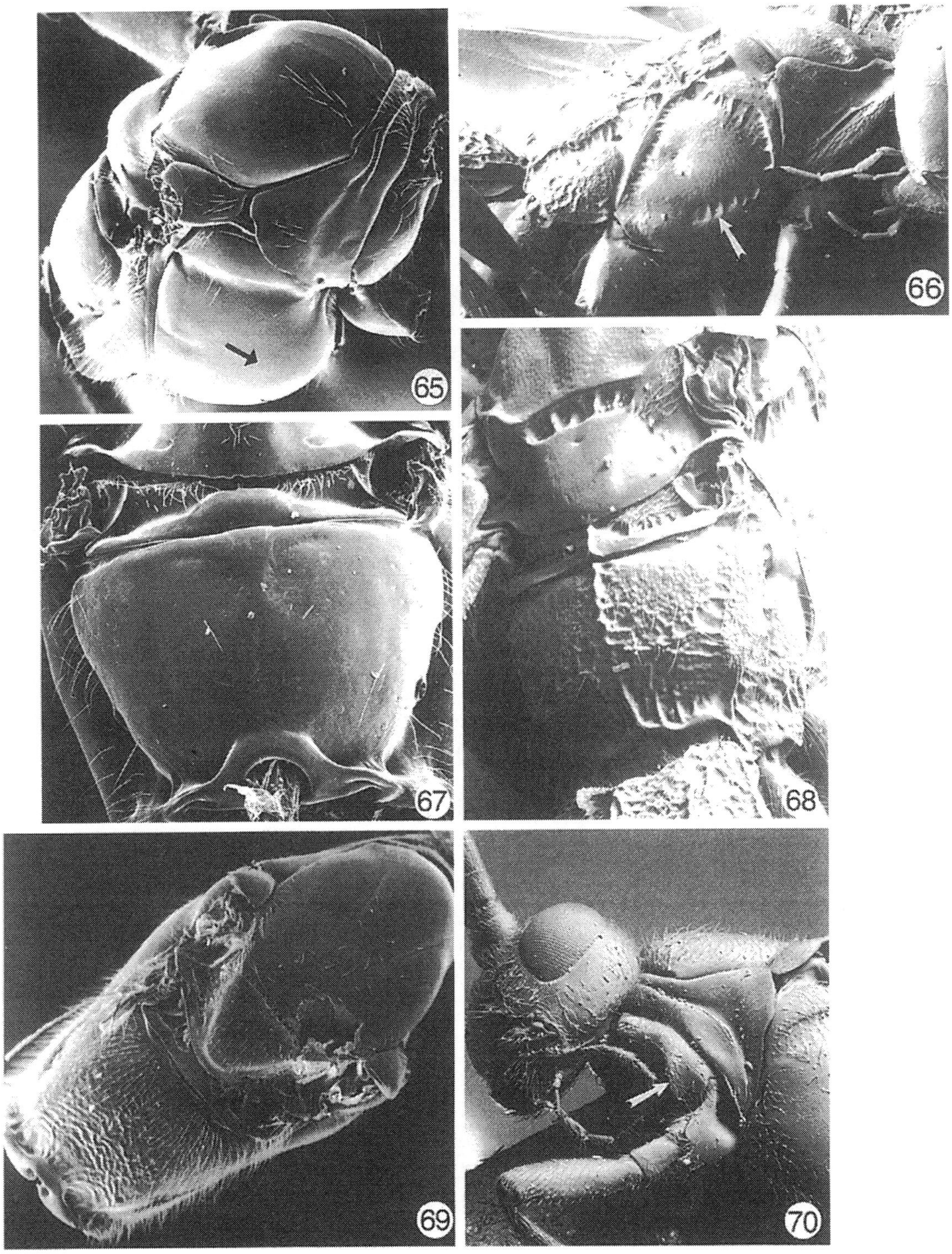

Figs 65-70: Braconinae. 65, mesosoma of *Compsobraconoides* sp., arrow = absence of sternaulus; 66, mesosoma of *Lapicida* sp., lateral view, arrow = sternaulus; 67, propodeum of *Cyclaulacidea* sp., dorsal view; 68, propodeum of *Lapicida* sp., dorsal view; 69, mesosoma of *Vipiomorpha* sp., dorsal view; 70, head and anterior portion of thorax of *Pigeria piger*, arrow = ridge on propleuron.

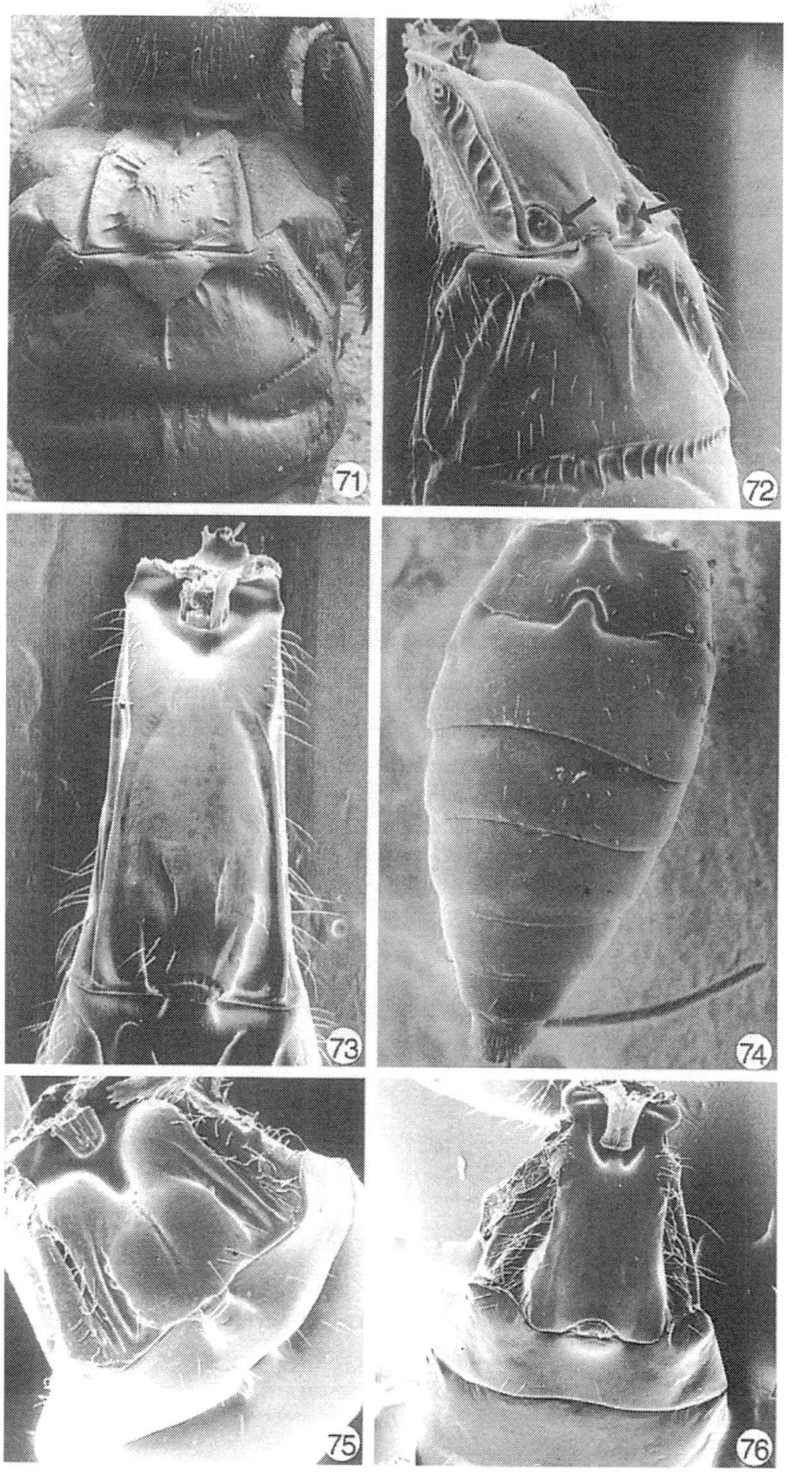

Figs 71-76: Braconinae. 71, metasomal terga 1-3 of *Digonogastra* sp., dorsal view; 72, petiole and metasomal tergum 2 of *Hemibracon*. sp., arrows = posterior-lateral concavities 011 petiole; 73, petiole of *Hemibracon* sp.; 74, metasomal terga 2-8 of *Compsobraconoides* sp.; 75 petiole and metasomal terga 2 + 3 of *Compsobraconoides* sp.; 76, petiole and metasomal terga 2 + 3 of *Cyclaulax* sp.

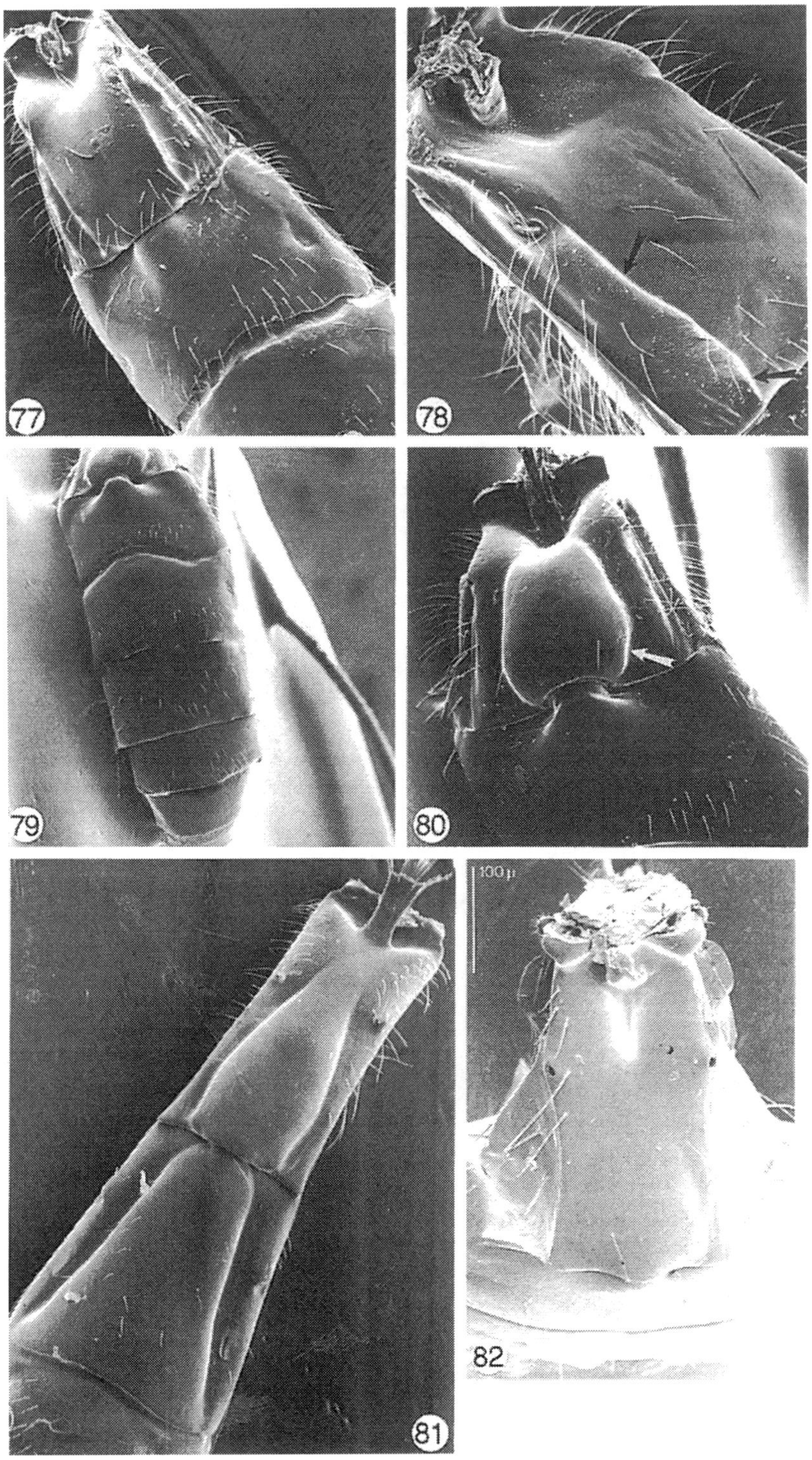

Figs 77-82: Braconinae. 77, petiole and tergum 2 of *Cyclaulacidea* sp.; 78, petiole of *Cyclaulacidea* sp., arrow = carinate margin (after Quicke and Delobel, 1995); 79 metasoma of *Sacirema* sp., dorsal view; 80, petiole of *Sacirema* sp., arrow = rounded margin; 81, petiole and tergum 2 of *Compsobracon* sp.; 82 petiole of *Bracon* sp.

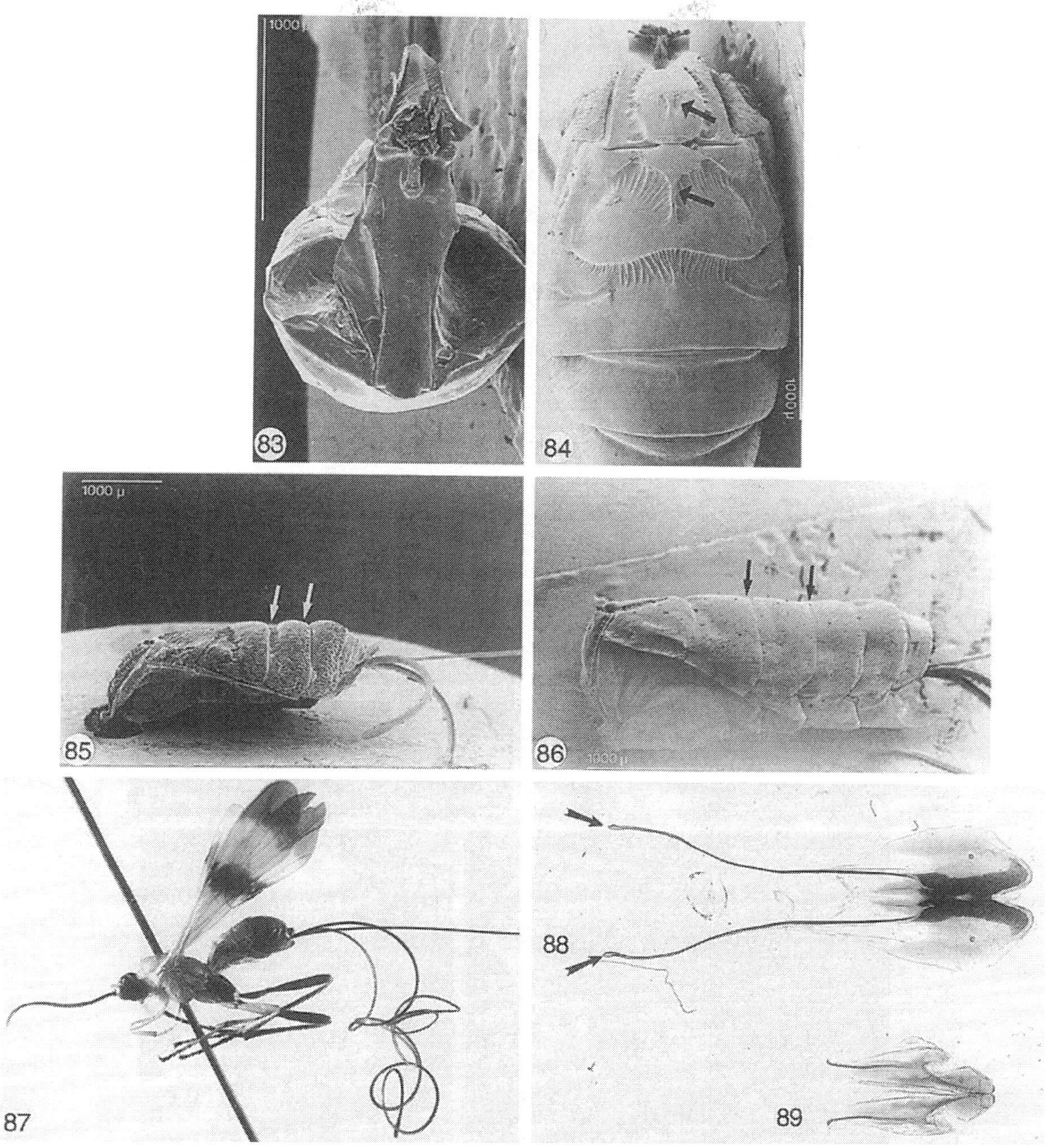

Figs 83-89: Braconinae. 83, petiole of *Myosoma* sp., dorsal view; 84-86, metasomas: 84, *Digonogastra* sp., arrows = median triangular area on tergum 2 and pair of pits postero-medially on the petiole; 85, *Digonogastra* sp.; lateral view, arrows = dorsally convex terga; 86, *Cyanopterus* sp., lateral view, arrows = thin posterior margins of dorsally flattened terga; 87, habitus of *Compsobracon* sp.; 88, metasomal tergum 4 of *Leptobracon* sp., arrows = apodemes; 89, metasomal sternum 4 of *Lepiobracon* sp. Figures 83-86 after Quicke and Sharkey (1989).

Figs 90-95: Braconinae, 90, metasoma of *Gracilibracon* sp., arrow = pinched up antero-median ridge (after Quicke, 1995); 91, metasoma of *Cyclaulax* sp.; 92, hind leg of *Myosomatoides* sp., lateral view; 93, metasoma of *Myosomatoides* sp., arrow = median carina on tergum 2; 94, habitus of *Myosomatoides* sp.; 95, metasomal terga 1-5 of *Vipio* sp. cleared to show sculpture.

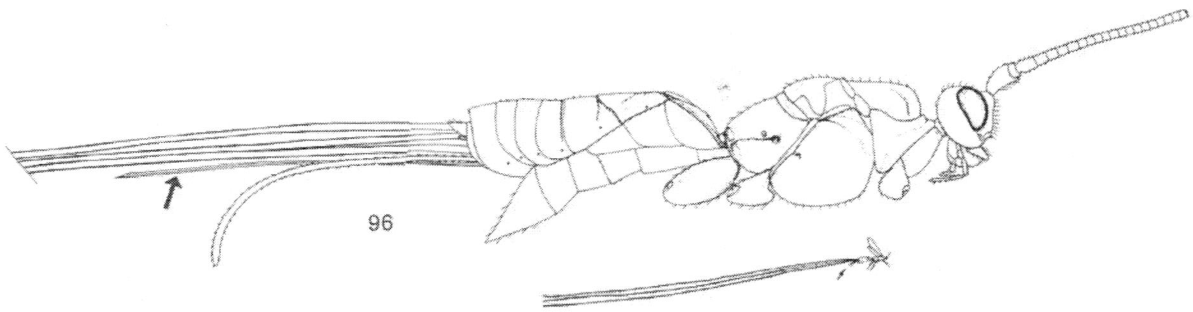

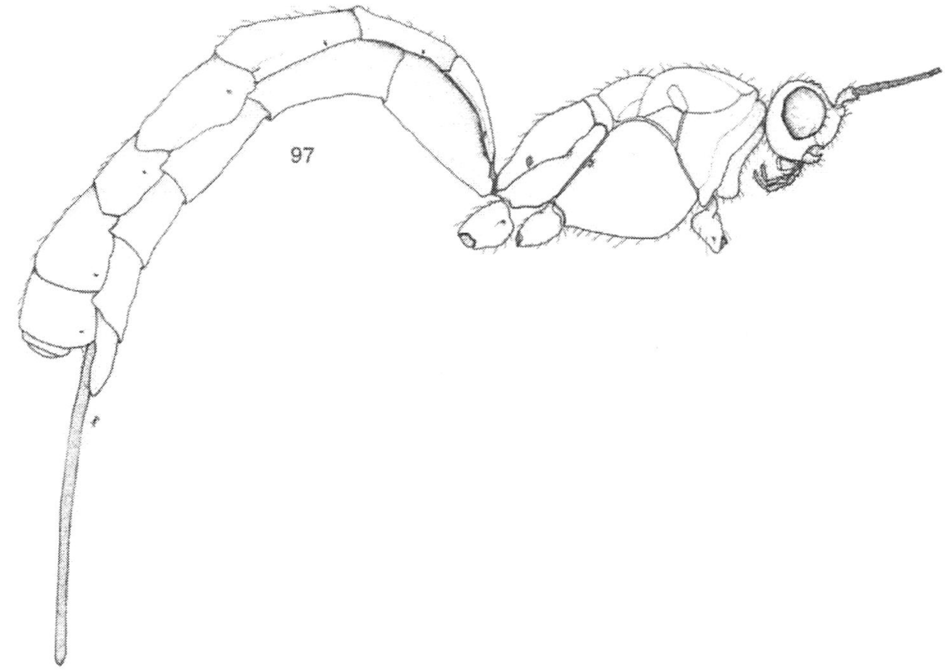

Figs 96-97: Braconinae. 96, *Pheloura* sp., arrows = true ovipositor (after van Achterberg, 1989); 97, habitus of *Leptobracon* sp.

BRACONINAE UPDATE

D.L.J. Quicke
Department of Biology, Faculty of Life Sciences, Chulalongkorn University, Bangkok, Thailand.
d.quicke@email.com

The Braconinae chapter of the manual is slightly out of date. Only one new genus has been described since then and one Palaeotropical genus recorded from the Neotropics for the first time. A few species-level revisions have been published. It should be noted that a number of recent authors have treated *Pigeria* as a synonym of *Bracon* and have sometimes afforded it subgeneric status.

In the 1997 version, the genus *Compsobraconoides* Quicke was broadly circumscribed, and some of the variation was commented upon. Papp (2007b) provided some resolution of this by describing a distinct group of species in his new genus *Gozmanycomp*. The original couplet 11 can me modified thus to allow their separation:

11a(10). Scape shorter ventrally than dorsally (antenna directed anteriorly) (as in Figs. 28, 39). Face usually with fine coriaceous or rugulose sculpture between longitudinal ridges (Figs. 30, 31) 11b
- Scape longer ventrally than dorsally (antenna directed anteriorly) (as in Fig. 29). Face usually coarsely sculptured between longitudinal ridges (Figs. 32, 33)...12

11b(11a). Propodeum simple. Second metasomal tergite not pinched-up medio-anteriorly. Suture between metasomal terga 2 and 3 strongly buckled medially..............................***Compsobraconoides*** Quicke
- Propodeum with a distinct midlongitudinal carina posteriorly. Second metasomal tergite pinched-up medio-anteriorly. Suture between metasomal terga 2 and 3 strongly buckled medially (Fig. 74) ....***Gozmanycomp*** Papp

Papp (2012) described a new species from Honduras that he placed in the Old World genus *Amyosoma* Viereck. On the basis of current morphological understanding this may be correct and the original couplet 11 can me modified thus to allow their separation in accordance with Papp's key:

19a(18). Median tergite of petiole very long and narrow, more than 2.5 times longer than maximally wide (Fig. 83). Hind femur and tibia usually markedly flattened, with long marginal setae. Wings: Fig. 9 straight 19b
- Petiole with median tergite not especially elongate, less than 2.2 times longer than maximally wide (as in Figs. 75-77, SQ). Hind femur and tibia normal...20

19b(19a). Hind femur and tibia with short hairs as usually in Braconinae. Vein 1r-m of hind wing weakly bent to almost straight ....***Amyosoma*** Viereck
- Hind femur and hind tibia with long hairs. Vein 1r-m of hind wing distinctly bent ...***Myosoma*** Brullé

Annotated list of recent papers on New World Braconinae

Cappaert D, McCullough DG. 2009. Occurrence and seasonal abundance of *Atanycolus cappaerti* (Hymenoptera: Braconidae) a native parasitoid of emerald ash borer, *Agrilus planipennis* (Coleoptera: Buprestidae). *The Great Lakes Entomologist* 42: 16-29.
Biological observations on a recently described, potentially important biological control agent against emerald ash-borer.

Crabb BA, Pellmyr O. 2006. Impact of the third trophic level in an obligate mutualism: do *Yucca* plants benefit from parasitoids of *Yucca* moths? *International Journal of Plant Science* 167: 119-124.
Biological study of an undescribed *Digonogastra* species as a parasitoid of Yucca moths.

De Santis L, Esquivel L. 1966. Tercera lista de Himenopteros parasitos y predatores de los insectos de la Republica Argentina. *Revista del Museo de La Plata. Sec. Zoologia. (N.S.)(Zool.)* 9(69): 47-215.
Report of *Megabracon* as a parasitoid of the cerambycid beetle *Mallodon spinibarbis* (Linné) (as *Stenodontes spinibarbis*).

Flores S, Nassar JM, Quicke DLJ. 2006. Reproductive phenology and pre-dispersal seed predation in *Protium tovarense* (Burseraceae), with description of the first known phytophagous *'Bracon'* species (Hymenoptera: Braconidae: Braconinae). *Journal of Natural History* 39: 3663-3685.
Description of a remarkable, completely phytophagous species that keys to, and is currently included in, *Bracon* and represents a distinct species group.

Fortier J, Nishida K. 2004. A new species and host association biology of Neotropical *Compsobraconoides* Quicke (Hymenoptera: Braconidae). *Journal of Hymenoptera Research* 13(2): 228-233.
Description of *Gozmanycomp cinnamoni* (Fortier & Nishida) (= *Compsobraconoides cinnamoni*) reared from galls caused by larvae of the weevil, *Camptocheirus* sp. (Curculionidae), on branches of *Cinnamonum cinnamomifolium* (Lauraceae).

Marsh PM, Strazanac JS, Laurusonis SY. 2009. Description of a new species of *Atanycolus* (Hymenoptera: Braconidae) from Michigan reared from the Emerald Ash Borer, *Agrilus planipennis* (Coleoptera: Buprestidae: Agrilinae). *The Great Lakes Entomologist* 42, 8-15.
Description of a new Nearctic species of *Atanycolus*, potentially important biological control agent against emerald ash-borer.

Papp J. 2007a. Szépligeti's *Cyclaulax* types deposited in the Hungarian Natural History Museum (Hymenoptera: Braconidae: Braconinae). *Journal of Hymenoptera Research* 16: 147-166.
The nine *Cyclaulax* species represented by types and deposited in the Hungarian National Museum, Budapest, as well as the type-species of the genus *Cyclaulax grandiceps* Cameron, housed in the Natural History Museum, London, are redescribed. An identification key to the ten species is presented. (Abstract of paper).

Papp J. 2007b. A new genus and two new species of Braconidae (Hymenoptera) dedicated to Lásló Gozmány. *Acta Zoologica Academiae Scientiarum Hungaricae* 53(2): 295-305.
Description of the new genus *Gozmanycomp* and its differentiation from *Compsobraconoides*.

Papp J. 2007c. First survey of the *Sacirema* Quicke species (Hymenoptera: Braconidae: Braconinae). *Genus (Wroclaw)* 18(3): 507-515.
Describes a new species from the Lesser Antilles, a new combination of a species formerly in *Bracon*, redescribes the type species and provides an illustrated key to the three known species.

Papp J. 2011. Redescriptions of eight species by Gy. Szépligeti belonging to the genera *Calobracon*, *Diamblomera* and *Hemibracon* with a new *Diamblomera* species (Hymenoptera, Braconidae: Braconinae). *Annales Historico-naturales Musei Nationalis Hungarici* 103: 315-348.
Redescribes and illustrates the type species of *Calobracon*. Papp notes that the species runs to *Gracilibracon* and *Cyclaulax* in the 1997 key, but it appears to me to key out correctly at couplet 8. Provides a key to and redescribes the six species of *Hemibracon* described by G. Szépligeti.

Papp J. 2012. A taxonomic study of the *Myosoma* genus-group with description of *Amyosoma cavei* sp. n. from Honduras (Hymenoptera: Braconidae: Braconinae: Braconini). *Acta Zoologica Academiae Scientiarum Hungaricae* 58(1): 1-29.
Papp maintained the separation of *Myosoma* from *Amysoma* Viereck and described a species from Honduras under that name. Transfers *Myosoma ruber* Brullé to *Compsobraconoides* and notes two new species-level synonyms of it.

Santos JC, Viana, LR, Silveira FAO, Penteado-Dias AM, Fernandes GW. 2004. Ovipositing behaviour of *Compsobracon mirabilis* (Szépligeti) (Hymenoptera, Braconidae) in a cerrado habitat, southeastern Brazil. *Revista Brasileira de Entomologia* 48(1): 139-140.

Biological observations of a *Compsobracon* species parasitizing larvae of an unidentified branch-boring Lepidoptera species.

Yu DW, Quicke DLJ. 1997. *Compsobraconoides* (Braconidae: Braconinae), the first hymenopteran ectoparasitoid of adult *Azteca* ants (Hymenoptera: Formicidae). *Journal of Hymenoptera Research* 6: 419-421.
Describes the remarkable biology of an undescribed species of *Compsobraconoides* that attacks adult *Azteca* ants in their domatia.

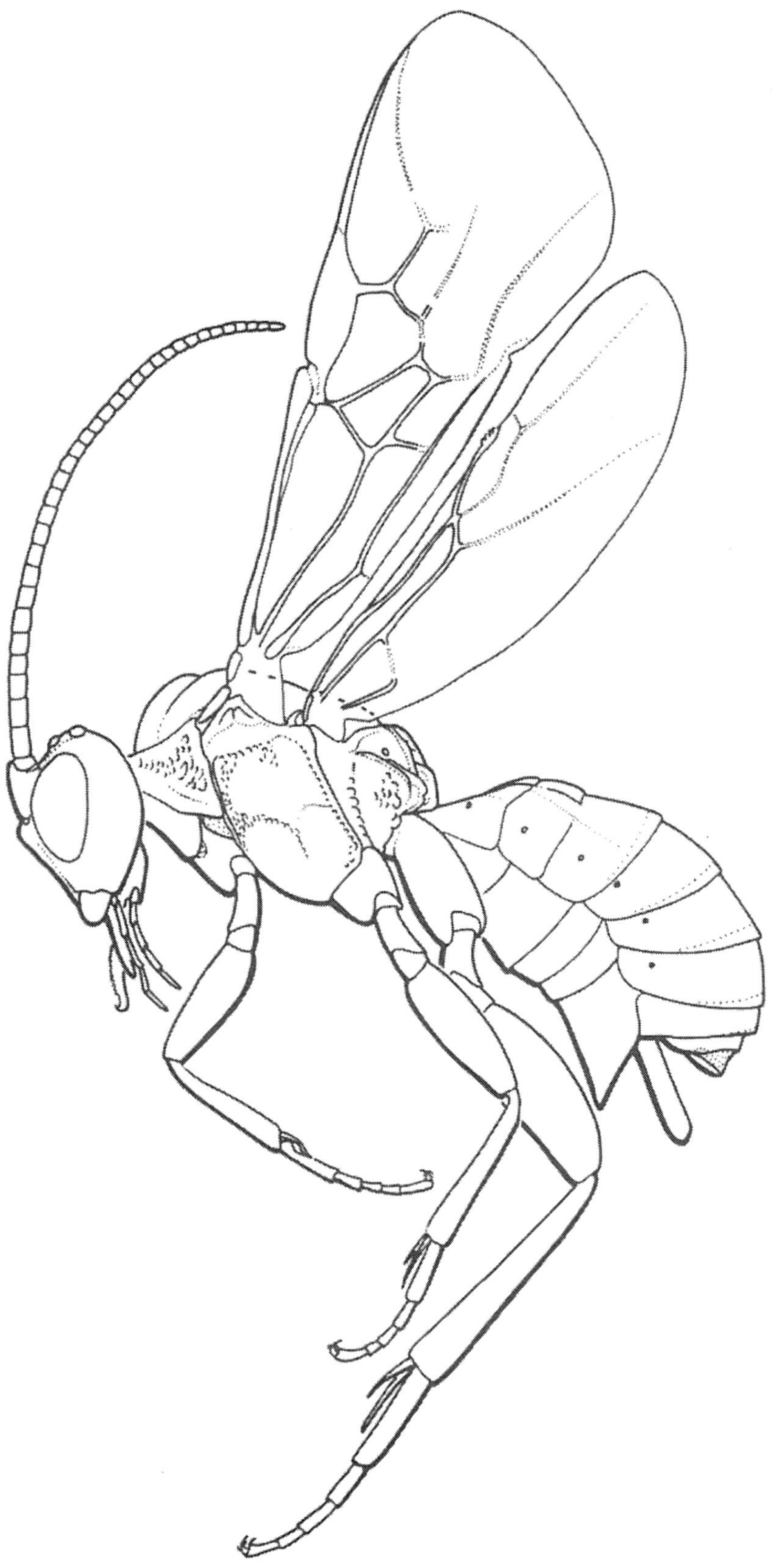

SUBFAMILY CARDIOCHILINAE

J.B. WHITFIELD

Department of Entomology, 320 Morrill Hall, University of Illinois, 505 S. Goodwin Ave, Urbana, IL 61801 U.S.A.

P. C. DANGERFIELD

Waite Campus, University of Adelaide, Glen Osmond, 5064 South Australia, Australia

INTRODUCTION. New World species of the subfamily Cardiochilinae have usually been placed in the genus *Cardiochiles* Nees, and cardiochilines are treated as a tribe of Microgastrinae in many catalogs (Shenefelt, 1973; Marsh, 1979). The generic classification of the Cardiochilinae has recently been revised, however, and nearly 20 genera are now recognized (Dangerfield et al., 1999) about half of which are present in the New World. Mao (1945, 1949) revised both the Mexican and North American species; both revisions need updating, especially the Mexican one. The Central and South American faunas are badly in need of study at the species level. Many of the species are components of mimicry (color pattern) complexes with other braconids and some ichneumonids, vespids, and pompilids.

BIOLOGY. All of the species for which host data are available parasitize the larvae of Lepidoptera. As far as is known, all of the species are solitary, internal parasitoids, attacking the early instar larvae and emerging from the late instar larvae or prepupae (Huddleston and Walker, 1988). The known hosts are most commonly from the Pyralidae and Noctuidae, but a number of other host groups have been recorded, including Apatelodidae (Dangerfield et al., 1996) and leaf-mining moth larvae (Whitfield and Wagner, 1991). The species appear to carry polydnaviruses (Stoltz et al., 1984) with which they compromise the immune system and physiology of their host insects. The most fully studied species is *Toxoneuron* (usually referred to as *Cardiochiles*) *nigriceps* (Viereck), a parasitoid of the tobacco budworm, *Heliothis virescens* (F.). In this case, not only the field biology but morphological, physiological, immunological, and biochemical aspects of the parasitoid's biology have been investigated, largely through the work of S. B. Vinson and associates (e.g. Vinson and Lewis, 1965; Vinson, 1969, 1972, 1978; Vinson and Scott, 1974).

BIOLOGICAL CONTROL POTENTIAL. The most familiar species in North America is *Toxoneuron nigriceps* (Viereck), which attacks the tobacco budworm. A few other species have been released for the control of pyralid pests of cucurbits in Florida (*Cardiochiles diaphaniae* Marsh) and of vegetable crops in India (*C. hymeniae* Fischer and Prasad). For a review of the published host records see Huddleston and Walker (1988). Additional records are accumulating from many parts of the world, including Central and South America (e.g. Dangerfield et al., 1996).

COMMON GENERA. On a worldwide basis, the most commonly collected genera are *Cardiochiles* and *Schoenlandella* Cameron. In dry habitats in Western North America and Mexico, species of *Toxoneuron* Say can be locally quite abundant. Other genera can also be encountered relatively frequently in tropical forests of South and Central America.

DISTRIBUTION. The subfamily is found worldwide, as are the genera *Cardiochiles* and *Schoenlandella*. All of the other genera included in the key are restricted to the New World. *Neocardiochiles* Szépligeti, *Wesmaelella* Spinola and *Heteropteron* Brullé have only been found so far in South America.

DISTINGUISHING FEATURES. The subfamily can be easily recognized on the basis of the structure of the anteriorly arched 3RS vein of the fore wing and the structure of the petiole. The lateral tergites of the petiole are only partially sclerotized, are incompletely separated from the median tergite, and bear the spiracles.

KEY TO NEW WORLD GENERA OF THE SUBFAMILY CARDIOCHILINAE (based primarily on females)

1. Propodeum mostly smooth, without defined areola (Figs 15-17). Stigma of fore wing narrow and elongate (Figs 1-3). Body large, body length > 10 mm long 2

\- Propodeum usually sculptured, often with fully developed areola bordered by carinae (Fig. 18). Stigma of fore wing usually not as narrow (Figs 4-6). Body < 10 mm long 4

2(1). Fore wing 1r present, in part (Fig. 1). Propodeum smooth, without median furrow (Fig. 15) .... ***Wesmaelella*** Spinola

Distribution: Brazil, Peru; relatively rare; 2 described species. Biology: unknown. Reference: Dangerfield et al. (1999)

\- Fore wing 1r absent (Figs 2, 3). Propodeum with median furrow at least anteriorly 3

3(2). Propodeum with complete median longitudinal furrow (Fig. 16). Fore wing 1a present and pigmented (Fig. 2). Hind tarsal claws pectinate .... ***Neocardiochiles*** Szépligeti

Distribution: Suriname, Brazil; relatively rare; one described species. Biology: unknown. Reference: Dangerfield et al. (1999).

\- Propodeum with short, antero-median furrow or pit (Fig. 17). Fore wing 1a absent (Fig. 3). Hind tarsal claws simple .... ***Heteropteron*** Brullé

Distribution: Brazil; relatively rare; one described species. Biology: unknown. Reference: Dangerfield et al. (1999).

4(1). Clypeal margin either convex and evenly rounded, or with a single median point, but without two apical tubercles (Fig. 21) 5

\- Clypeal margin with 2 median tubercles present (Figs 9, 20, 22) 7

5(4). Ovipositor sheath very short, < 0.2 times as long as hind tibia, and downturned along length (Fig. 12). Hind tibia usually without cup-like apical projection. Galea short and broad (Fig. 21) 6

\- Ovipositor sheath variable, but usually > 0.2 times as long as hind tibia; not downturned along its full length. Hind tibia with cup-like projection at apex (Fig. 8); if projection slight then ovipositor sheaths smooth and polished. Galea variable. Wings: Fig. 4 .... ***Toxoneuron*** Say

Distribution: New World, especially western U.S.A. and Mexico; about 20 described species, many undescribed. Biology: attack macrolepidoptera, especially Noctuidae. References: Mao (1945, 1949, as *Cardiochiles*, in part), Dangerfield et al. (1999).

6(5). Hind wing 2A present half way to wing margin (as in Fig. 4); with 6-7 harrtuli. Clypeus broad, > 2.5 times as broad as long. Petiole short, median tergite < 2.0 times longer than broad .... **undescribed genus #1**

Distribution: New World; only a few described species, several undescribed. Biology: apparently attack macrolepidoptera. References: Mao (1945, 1949, as *Cardiochiles*, in part), Dangerfield et al. (i1999).

\- Hind wing 2A absent or present only as a basal stump (as in Fig. 5); with 4-5 hamuli. Clypeus not as broad, < 2.5 times as broad as long. Petiole slightly longer, median tergite > 2.5 times longer than broad . ... **undescribed genus #2**

Distribution: principally Old World, especially Africa and the Philippines; a few undescribed New World species seem to key here, so its distribution needs further study. Biology: one record from Geometridae. Reference: Dangerfield et al. (1999).

7(4). Glossa deeply bilobed at apex; galea elongate, narrow and bladelike, portion that is visible beyond mandibles 1.5-4 times longer than broad (Fig. 20). Spectral trace of 3r present in fore wing, RS angled at junction with 3r (Fig. 6) .... ***Schoenlandella*** Cameron

Distribution: Cosmopolitan; fairly common, especially in dry habitats; nearly all New World species are undescribed. Biology: solitary on microlepidoptera. References: Mao (1945, 1949, as *Cardiochiles*. in part), Dangerfield et al. (1999, for clarification of generic concept).

\- Glossa only weakly to moderately bilobed at apex; galea just visible beneath mandibles, not as long and bladelike (Fig. 22). Fore wing 3r absent, RS usually evenly curved (Fig. 5) 8

8(7). Petiole variable but median tergite relatively broad, usually less than 2.5 times as long as apical width (as in Fig. 18). Metasoma not strongly laterally compressed anteriorly. Hypopygium with median desclerotized area which may be either infolded, but not wholly membranous, or membranous and folded fanlike in repose (Fig. 13). Ovipositor sheaths moderately long and clearly exerted. Wings: Fig. 5 ... . ***Cardiochiles*** Nees

Distribution: Cosmopolitan; common; about 16 described New World species, many more undescribed. Biology: apparently all attack microlepidoptera. References: Mao (1945, 1949, as *Cardiochiles*, in part), Dangerfield et al. (1999).

\- Petiole with median tergite elongate, > 4 times longer than broad; lateral tergites parallel and at right angles to median tergite so that metasomal terga 1-3 are strongly compressed (Fig. 19). Hypopygium evenly sclerotized throughout and relatively short (Fig. 23). Ovipositor sheaths short and weakly decurved .... ***Hansonia*** Dangerfield

Distribution: Mexico to Costa Rica; only 1 described species. Biology: solitary parasitoid of Apatelodidae. Reference: Dangerfield et al. (1996).

REFERENCES

Cameron, P. 1906. On the phytophagous and parasitic Hymenoptera collected by Mr. Ernest Green in Ceylon. Spolia Zeylanica 3: 67-97.

Dangerfield, P. C., Austin, A. D, and Whitfield, J.B. 1999. Systematics of the world genera of Cardiochilinae (Hymenoptera: Braconidae). Invertebrate Taxonomy. 13(6): 917-976.

Dangerfield, P. C., Whitfield, J. B., Sharkey, M. J., Janzen, D. H., and Mercado, I. 1996. *Hansonia*, a new genus of cardiochiline Braconidae (Hymenoptera) from Costa Rica, with notes on its biology. Proceedings of the Entomological Society of Washington 98: 592-596.

Huddleston, T. and Walker, A. K. 1988. *Cardiochiles* (Hymenoptera: Braconidae), a parasitoid of lepidopterous larvae, in the Sahel of Africa, with a review of the biology and host relationships of the genus. Bulletin of Entomological Research 78: 435-461.

Mao, Y.-T. 1945. Synopsis of the Mexican species of *Cardiochiles* Nees (Hymenoptera, Braconidae). Pan-Pacific Entomologist 21: 125-134.

Mao, Y.-T. 1949. The species of ichneumon-flies of the genus *Cardiochiles* occurring in America north of Mexico. Proceedings of the United States National Museum 99: 229-266.

Marsh, P. M. 1979. Family Braconidae. Pp. 144-295. In: Krombein, K. V., Hurd, P. D., Smith, D.R., and Burks, B. D. (Eds.). Catalog of Hymenoptera in America North of Mexico. Smithsonian Institution Press, Washington, D.C.

Shenefelt, R. D. 1973. Pars 9. Braconidae 5, Microgasterinae and Ichneutinae. Pp. 669-812. In: Vecht, J. van der and Shenefelt, R. D. (Eds.). Hymenopterorum Catalogus (nova editio). Dr. W. Junk, The Hague.

Stoltz, D. B., Krell, P., Summers, M. D., and Vinson, S. B. 1984. Polydnavitidae-a proposed family of insect viruses with segmented, double-stranded, circular DNA genomes. Intervirology 21: 1-4.

Vinson, S. B. 1969. General morphology of the digestive and internal reproductive systems of adult *Cardiochiles nigriceps* (Hymenoptera: Braconidae). Annals of the Entomological Society of America 62: 1414-1419.

Vinson, S. B. 1972. Factors in successful attack on *Heliothis virescens* by the parasitoid *Cardiochiles nigriceps*. Journal of Invertebrate Pathology 20: 118-123.

Vinson, S. B. 1978. Courtship behavior and sources of a sexual pheromone from *Cardiochiles nigriceps*. Annals of the Entomological Society of America 71: 832-837.

Vinson, S. B. and Lewis, W. J. 1965. A method of host selection by *Cardiochiles nigriceps*. Journal of Economic Entomology 58: 869-871.

Vinson, S. B. and Scott, J. R. 1974. Ultrastructure of teratocytes of *Cardiochiles nigriceps* Viereck (Hymenoptera: Braconidae). International Journal of Insect Morphology and Embryology 3: 293-304.

Whitfield, J. B. and Wagner, D. L. 1991. Annotated key to the genera of Braconidae (Hymenoptera) attacking leaf-mining Lepidoptera in the Holarctic Region. Journal of Natural History 25: 733-754.

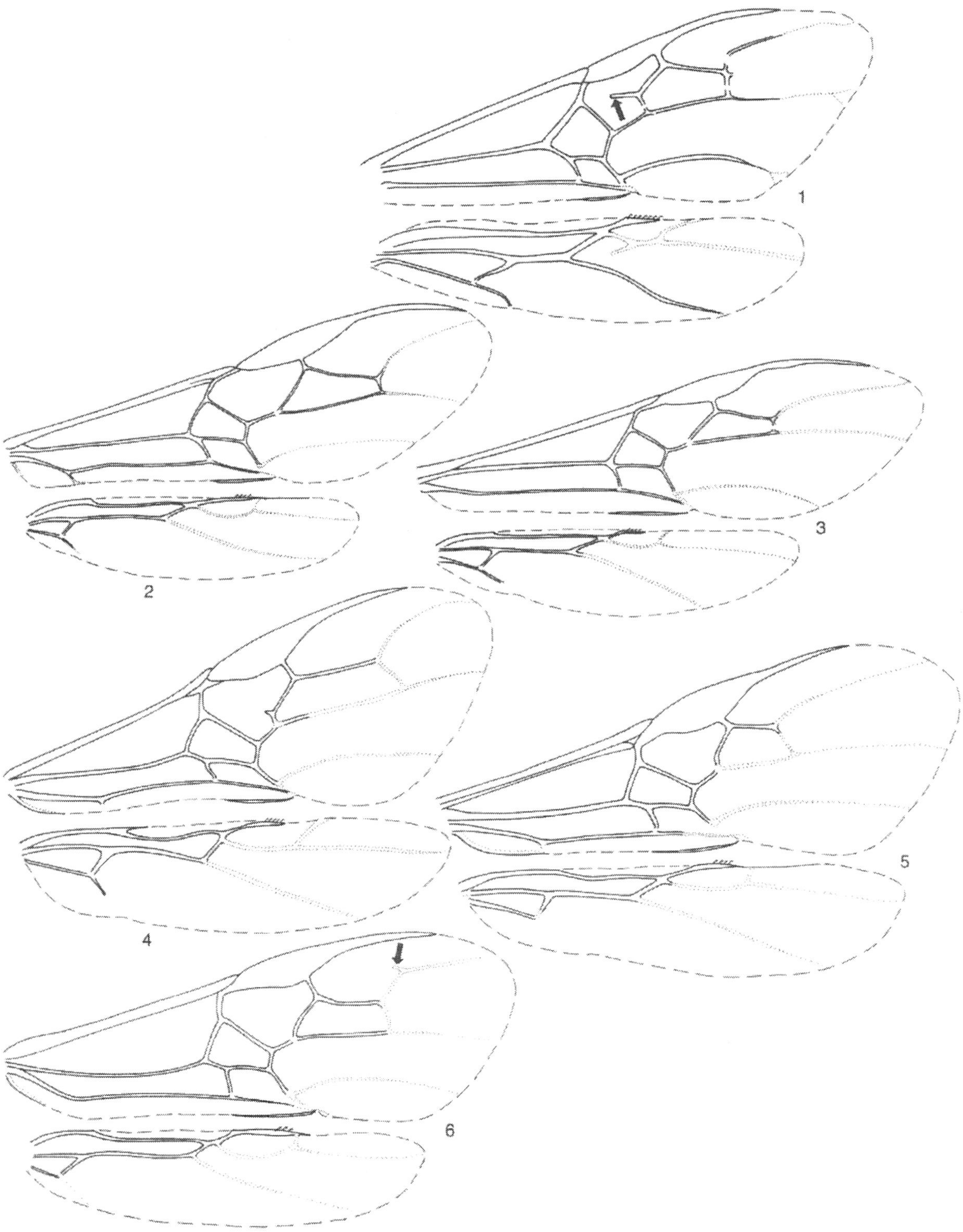

Figs 1-6: Wings of Cardiochilinae. 1, *Wesmaelella* sp., arrow = lr; 2, *Neocardiochiles* sp.; 3, *Heteropteron* sp.; 4, *Toxoneuron* nigriceps (Viereck); 5, *Cardiochiles* saliator (F.); 6, *Schoenlandella* sp., arrow = 3r.

CARDIOCHILINAE UPDATE

Jose Fernández-Triana
Canadian National Collection of Insects, 960 Carling Avenue, Ottawa, Canada.
jose.fernandez@agr.gc.ca

There has been little change to our knowledge of New World Cardiochilinae over the past 20 years. The key in the Manual was outdated by a world revision of the subfamily (Dangerfield *et al.*, 1999) and a new genus, *Retusigaster* Dangerfield, Austin & Whitfield, was described in the same paper. The genus *Bohayella* Belokobylskij, 1987 was recorded for the first time from the region. Since the publication of the Manual in 1997, seven new species have been described (Mercado and Wharton 2003), with the known diversity for New World Cardiochilinae now standing at 54 species (Yu *et al.*, 2016).

Recent papers on New World Cardiochilinae

Belokobylskij SA. 1987. A new genus in the subfamily Cardiochilinae (Hym.: Brac.) from the Soviet Far East. [In Russian with English summary] *Zoologicheskii Zhurnal* 66(2): 302-304.

Dangerfield PC, Austin AD, Whitfield JB. 1999. Systematics of the world genera of Cardiochilinae (Hymenoptera: Braconidae). *Invertebrate Taxonomy* 13(6): 917-976.

Kula RR. 2016. Cardiochilinae and Ichneutinae (Hymenoptera: Braconidae) of Konza Prairie. *Proceedings of the Entomological Society of Washington* 118: 273-288.

Mercado I, Wharton RA. 2003. Mexican cardiochiline genera (Hymenoptera: Braconidae), including a preliminary assessment of species-groups in *Toxoneuron* Say and *Retusigaster* Dangerfield, Austin and Whitfield. *Journal of Natural History* 37: 845-902.

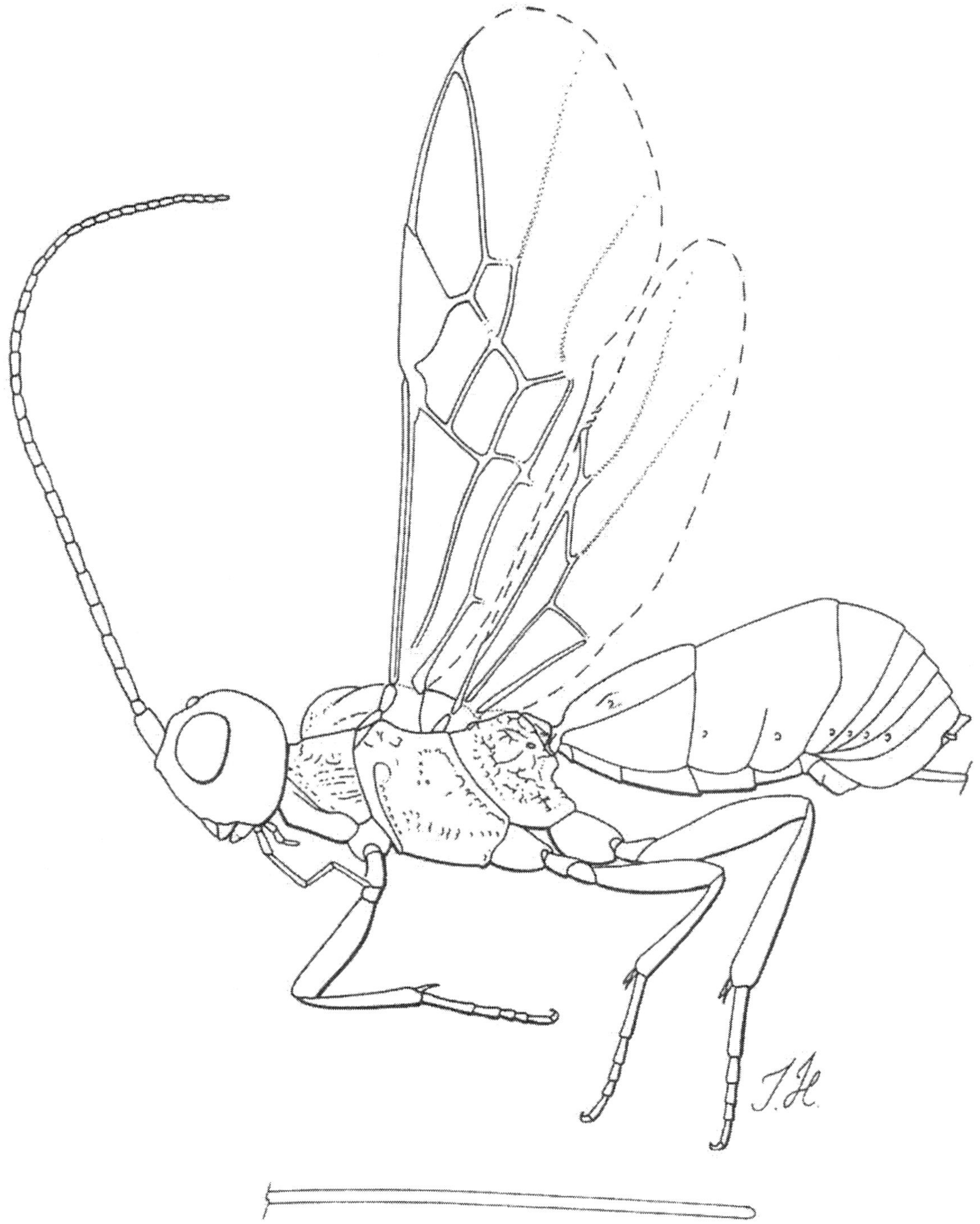
J.H.

SUBFAMILY CENOCOELIINAE

C. VAN ACHTERBERG

Afdeling Entomologie (Hymenoptera), Nationaal Natuurhistorisch Museum, Postbus 9517, 2300 RA Leiden, The Netherlands

INTRODUCTION. The subfamily Cenocoeliinae is a small and rather uniform group, with species 3-11 mm long. Females have a long to medium-sized ovipositor, at least as long as the metasoma. The metasoma is attached high on the propodeum, a unique character within the non-cyclostome Braconidae (but known of some rare Doryctinae). Few genera are known and only two are widespread. The Nearctic and Mexican species were revised by Saffer (1982).

BIOLOGY. The limited information on the biology of Cenocoeliinae is summarized by Shaw and Huddleston (1991). As far as known, all species are solitary koinobiont endoparasitoids of coleopteran larvae with an endophytic way of life. Most of the host records for Nearctic species are from Cerambycidae, but some Neotropical species have been reared from seed-eating Curculionidae (Saffer, 1977, 1982). The known hosts of Cenocoeliinae belong to the coleopteran families Cerambycidae and Curculionidae, and less commonly to the Buprestidae and Scolytidae.

BIOLOGICAL CONTROL POTENTIAL. Potential seems to be limited. In the Neotropics, Cenocoeliinae may be used in biological control programs if seed-eating Curculionidae become a pest.

COMMON GENERA. *Capitonius* Brullé and *Cenocoelius* Haliday are the major New World genera. *Cenocoelius* (= *Promachus* Cresson) is widely distributed in the Holarctic region and is less common in the Neotropical region. Most of the 47 described species of New World Cenocoeliinae belong to the genus *Capitonius*, but only two of these species have been formally transferred from *Cenocoelius* (van Achterberg, 1994). No revision of the Neotropical *Capitonius* is available. The Nearctic *Cenocoelius* and *Capitonius* species are keyed in Saffer (1982) under the genus *Cenocoelius* s. l.

DISTRIBUTION. The Cenocoeliinae are almost worldwide in distribution, but are unknown from the Afrotropical region. Most of the approximately 60 described species occur in the New World, but many species in the Neotropical region are undescribed. Van Achterberg (1994) recognized two tribes: the less typical Ussurohelconini (Oriental, Southeast Palaearctic) and the widespread Cenocoeliini.

DISTINGUISHING FEATURES. Cenocoeliinae have generally a rather stout habitus combined with a comparatively robust head, and a distinctly protruding ovipositor in females. The fore wing venation is characterized by the short second submarginal cell (Figs 1-4). Essential for their recognition is the high insertion of the metasoma (Figs 8, 15, 16) and the presence of the postpectal carina (though the latter is reduced in some Neotropical species). The occipital carina is complete and the wings are subhyaline, banded or completely dark brown.

KEY TO THE NEW WORLD GENERA OF THE SUBFAMILY CENOCOELIINAE

1. Frontal cavity without strong lateral carinae, at most weakly developed anteriorly; frons distinctly convex laterally (Figs 5, 6). Petiole slender, 2.5-4 times longer than its apical width (Fig. 7). Hind coxa of female without depression at inner side. Wings: Fig. 1 .... ***Foenomorpha*** Szépligeti

   Distribution: Suriname and Peru; rare; only one described species. Biology: unknown. Reference: van Achterberg (1994).

- At least half of frontal cavity bordered by a pair of strong lateral carinae; frons laterally only slightly convex (Figs 9, 11, 24). Petiole broader, less than 2.8 times longer than its apical width (Figs 10, 18-20). Hind coxa of female usually with depression at inner side (Figs 12, 14) 2

2(1). Metasomal tergum 3 with acute lateral margin (Fig. 15) and coarsely sculptured (Figs 15, 17). Occipital carina strongly and angularly protruding posteriorly (Fig. 15) ....***Evaniomorpha*** Szépligeti

Distribution: Brazil; rare; known only from type species. Biology: unknown. Reference: van Achterberg (1994).

- Tergum 3 without acute lateral margin (Figs 16, 22) and smooth (Figs 18-20). Occipital carina evenly curved in profile, not conspicuously protruding posteriorly (Fig. 16)...3

3(2). Hind wing M+CU 1.2-2.1 times longer than 1M (Fig. 2); if about twice as long then hind wing cu-a shorter than 1M; hind wing 1M distinctly longer than r-m (Fig. 2), rarely only slightly longer. Depression at inner side of hind coxa of female ending ventrally just basad middle of coxa (as in Fig. 12).......***Capitonius*** Brullé

Distribution: throughout New World; rather commonly collected. Biology: largely unknown, one species has been reared from seed-eating Curculionidae (Saffer, 1977). References: Saffer (1982, Nearctic and Mexican species), van Achterberg (1994).

- Hind wing M +CU 2.2-6.0 times longer than 1M (Figs 3, 4); if about twice as long then hind wing cu-a longer than vein 1M; hind wing 1M usually shorter than r-m or of equal length (Figs 3, 4). Depression at inner side of hind coxa of female ending ventrally at middle of coxa or more distad (Fig. 14)....................4

4(3). Metasomal tergum 3 (measured along midline) about 1.8 times longer than tergum 2 (Fig. 20). Minute posterior scutellar depression present (Fig. 23). Basal half of fore wing much less setose than its apical half, and subbasal cell glabrous. Sternaulus absent medially (Fig. 22). Wings: Fig. 3 .....***Aulacodes*** Cresson

Distribution: U.S.A. (Florida) and Cuba; one described species. Biology: unknown. Reference: van Achterberg (1994).

- Metasomal tergum 3 about equal in length to tergum 2, or shorter (as in Fig. 18). Posterior scutellar depression absent (Fig. 21). Basal half of fore wing similarly setose as its apical half, and subbasal cell densely setose. Sternaulus present medially (Fig. 16). Wings: Fig. 4 ....***Cenocoelius*** Haliday

Distribution: Holarctic, Neotropical. Biology: reared from larvae of wood- and bark- boring Coleoptera (Curculionidae, Cerambycidae, Buprestidae and Scolytidae). References: Saffer (1982, Nearctic and Mexican species), van Achterberg (1994, including *Promachus*).

REFERENCES

Achterberg, C. van 1994. Generic revision of the subfamily Cenocoeliinae Szépligeti (Hymenoptera: Braconidae). Zoologische Verhandelingen Leiden 292: 1-52.

Saffer, B. 1977. A new species of *Cenocoelius* from Costa Rica (Hymenoptera: Braconidae). Proceedings of the Entomological Society of Washington 79: 593-596.

Saffer, B. 1982. A systematic revision of the genus *Cenocoelius* (Hymenoptera, Braconidae) in North America, including Mexico. Polskie Pismo Entomologiczne 52: 73-167.

Shaw, M. R. and Huddleston, T. 1991. Classification and biology of braconid wasps (Hymenoptera: Braconidae). Handbooks for the Identification of British Insects 7(11): 1-126.

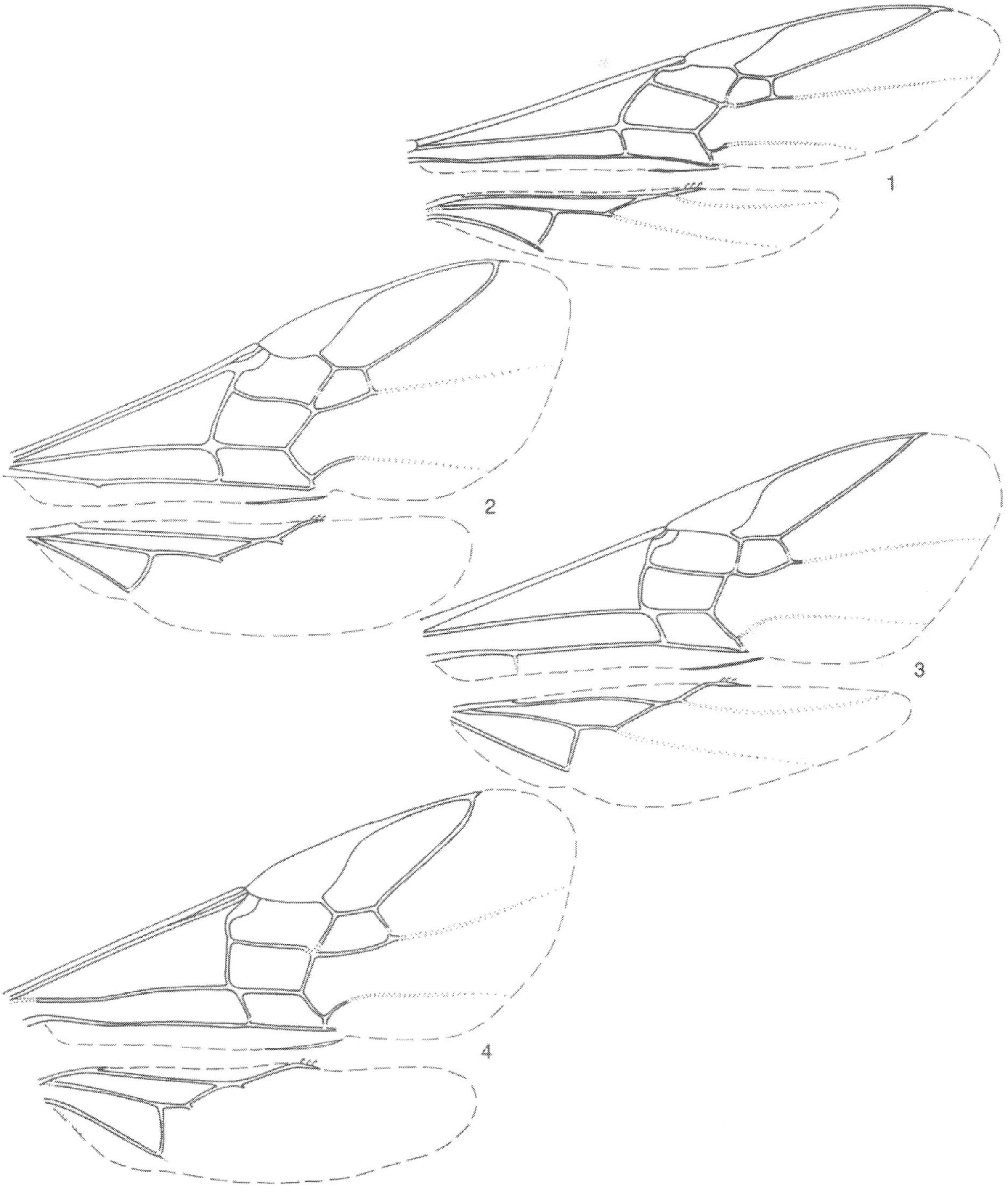

Figs 1-4: Wings of Cenocoeliinae. 1, *Foenomorpha bicolor* Szépligeti: 2, *Capitonius* sp.; 3, *Aulacodes* sp.; 4, *Cenocoelius analis* (Nees).

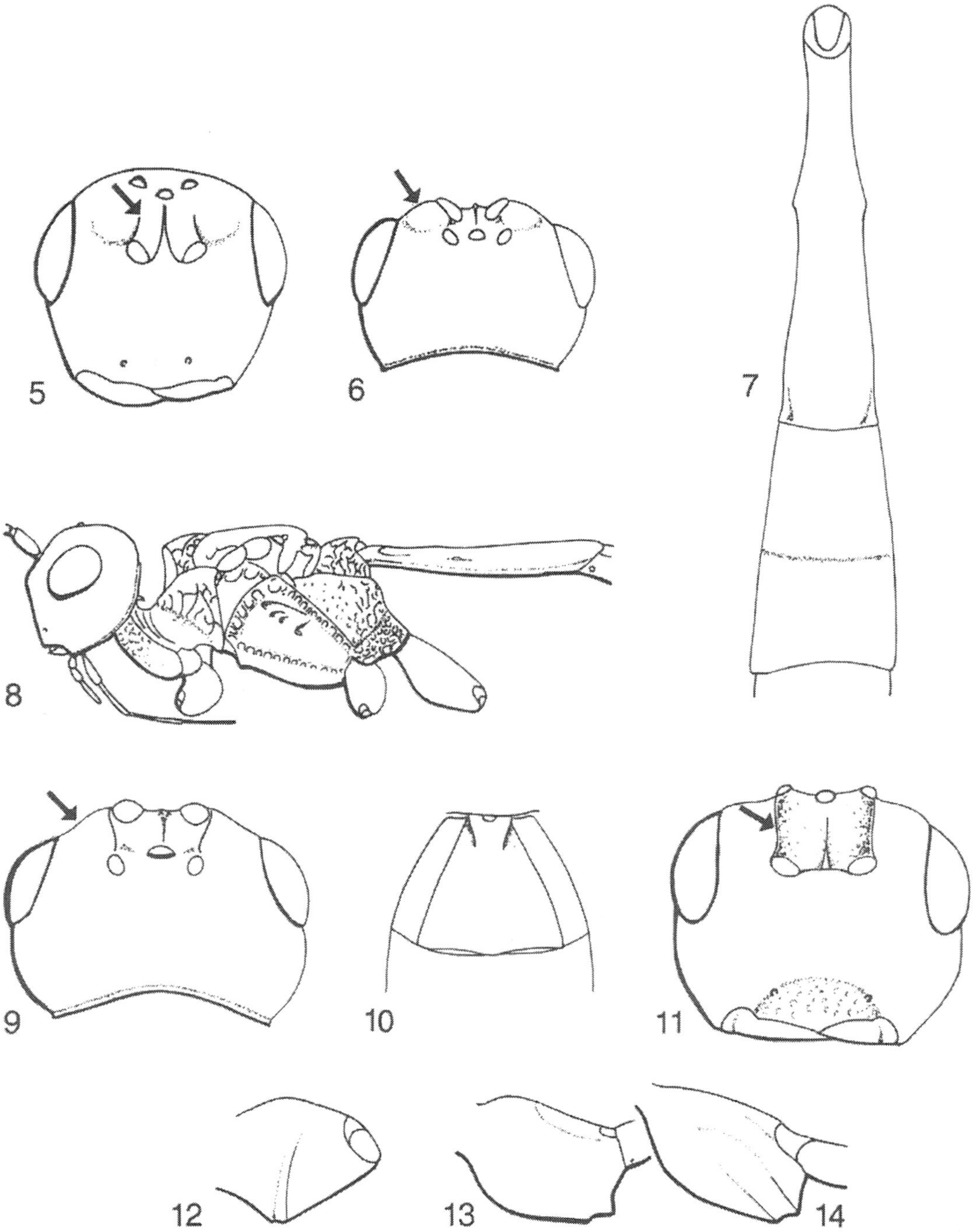

Figs 5-14: Cenocoeliinae. 5, head of *Foenomorpha* sp., frontal view, arrow = poorly developed lateral carina bordering frontal cavity; 6, head of *Foenomorpha* sp., dorsal view, arrow = frons (distinctly convex); 7, petiole of *Foenomorpha* sp., dorsal view; 8, *Foenomorpha* sp., lateral view; 9, head of *Cenocoelius* sp., dorsal view, arrow = frons (slightly convex); 10, petiole of *Cenocoelius* sp.; 11, head of *Cenocoelius* sp., frontal view, arrow = lateral carina bordering frontal cavity; 12-14, hind coxae: 12, *Capitonius* sp.; 13, *Cenocoelius* sp., outer side; 14, *Cenocoelius* sp., inner side. All figures after van Achterberg (1994).

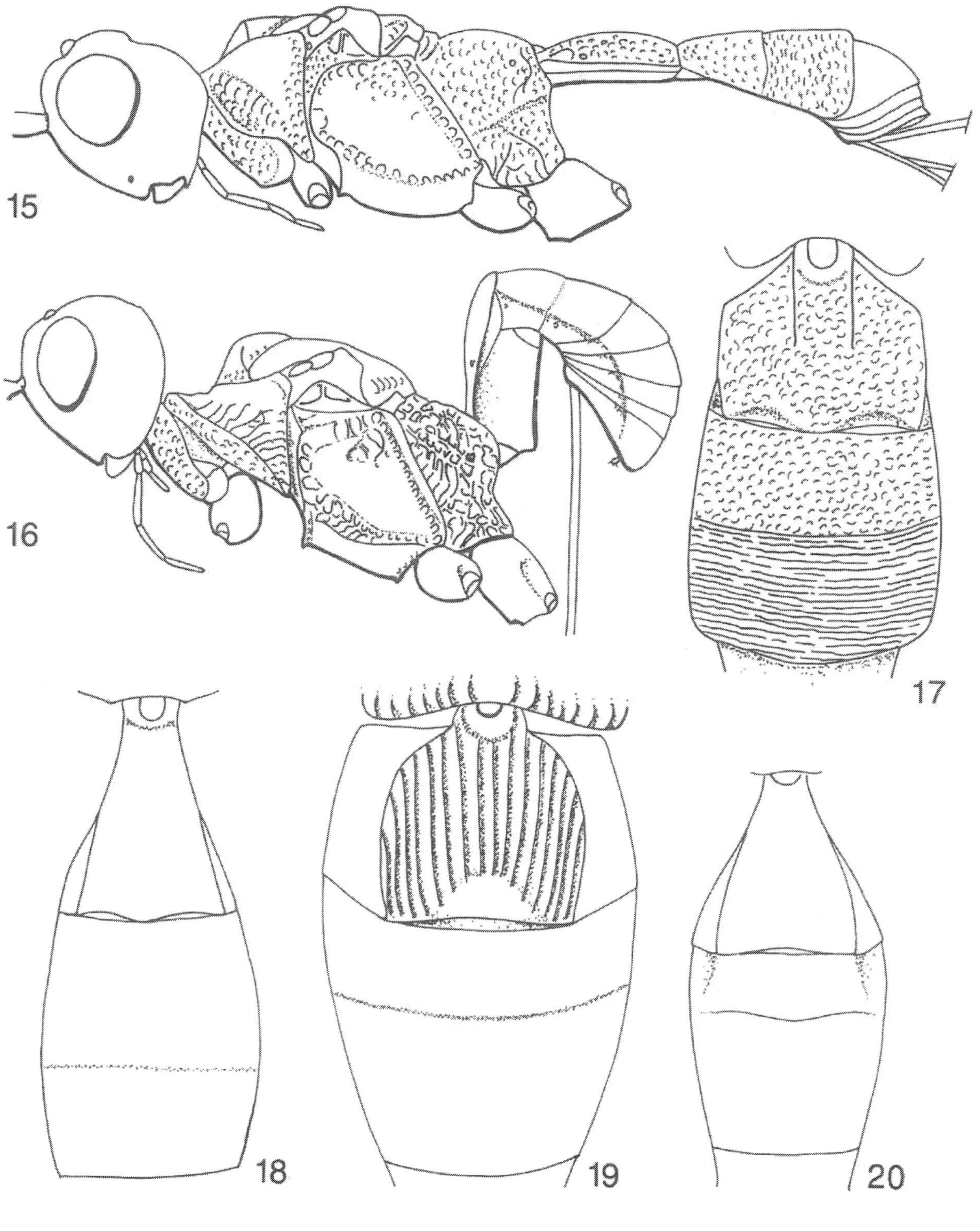

Figs 15-20: Cenocoeliinae. 15, habitus of *Evaniomorpha* sp.; 16, habitus of *Cenocoelius* sp.; 17-20, metasoma, dorsal view: 17, *Evaniomorpha* sp.; 18, *Capitonius* sp.; 19, *Capitonius* sp ; 20, *Aulacodes* sp. All figures after van Achterberg (1994).

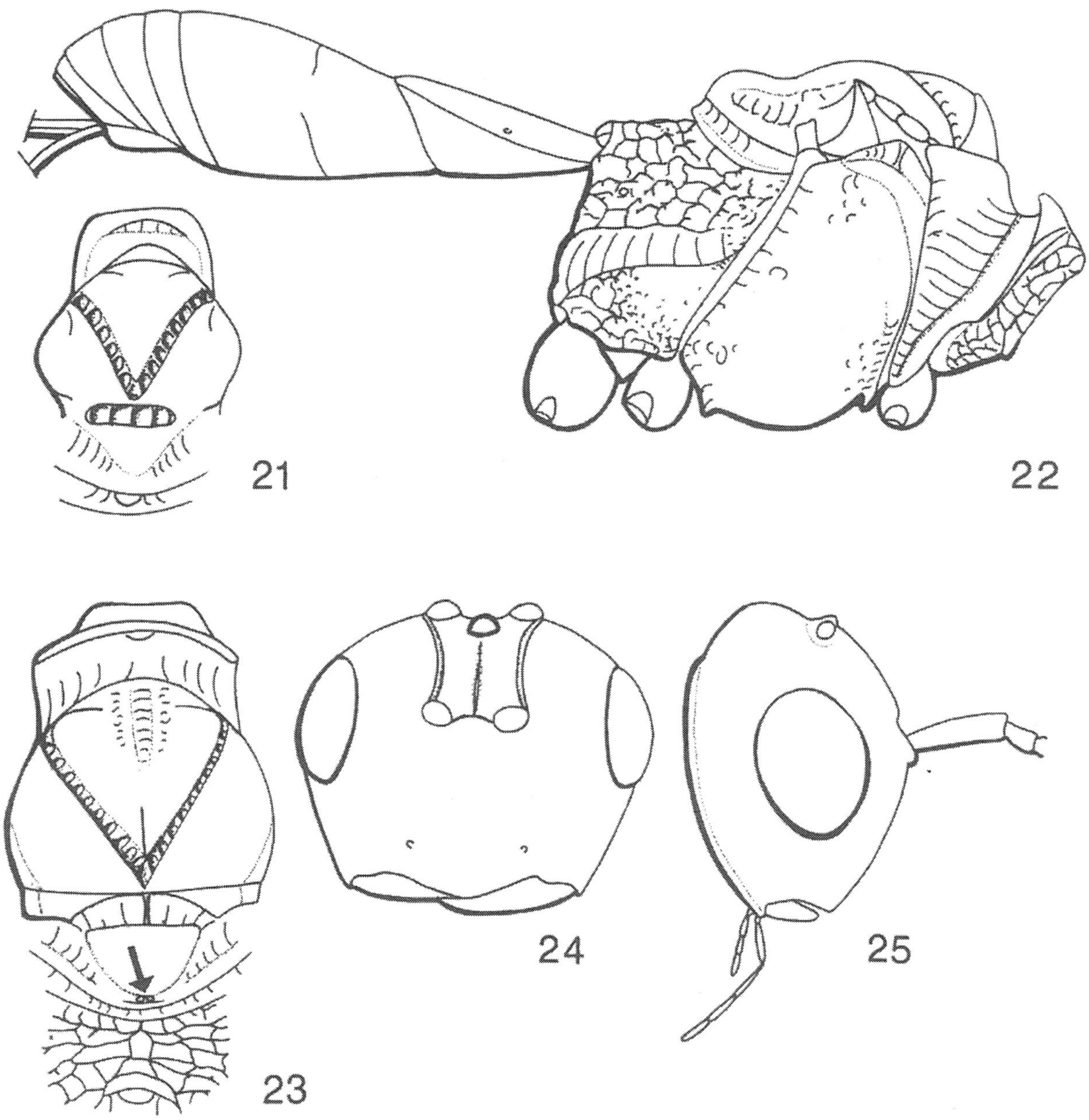

Figs 21-25: Cenocoeliinae, 21, mesosoma of *Cenocoelius* sp., dorsal view; 22, mesosoma and metasoma of *Aulacodes* sp., lateral view; 23, mesosoma of *Aulacodes* sp., dorsal view, arrow = scutellar depression; 24, head of *Aulacodes* sp., frontal view; 25, head of *Aulacodes* sp., lateral view. All figures after van Achterberg (1994).

CENOCOELIINAE UPDATE

C. van Achterberg
Department of Terrestrial Zoology, Naturalis Biodiversity Center, Postbus 9517, 2300 RA Leiden, The Netherlands.
c.vanachterberg@xs4all.nl

The Cenocoeliinae chapter of the manual needs a few small updates. Van der Ent and Shaw (1998) examined the biodiversity of cenoceoliines in Costa Rica along elevation gradients and found 57 morphospecies, mostly occurring below 500 meters. None were found above 1600 meters elevation. Cenocoeliinae are now known from the Afrotropical Region but remain very rarely collected (Pitz and Sharkey 2005). Pitz (2010) lowered the status of *Foenomorpha* Szépligeti to subgenus of *Capitonius* Brullé and revised its 19 species including 16 new species. Pitz and Sharkey (2007) described three aberrant Neotropical species of *Capitonius*.

Annotated list of recent papers on New World Cenocoeliinae

Braet Y, van Achterberg C. 2001. New species of the genera *Foenomorph* Szépligeti (Cenocoeliinae) and *Chelonus* Panzer (Cheloninae) (Hymenopter: Braconidae), from French Guiana, and Brazil. *Zoologische Mededelingen, Leiden* 75(6): 103-118.

Ent L J van der, Shaw SR. 1998. Species richness of Costa Rican Cenocoeliini (Hymenoptera: Braconidae): a latitudinal and altitudinal search for anomalous diversity. *Journal of Hymenoptera Research* 7(1): 15-24.

Ent L J van der, Shaw SR. 1999. A new species of *Capitonius* (Hymenoptera: Braconidae) from Costa Rica with rearing records. *Pan-Pacific Entomologist* 75(2): 112-120.
A new species, *Capitonius tricolorvalvus*, is described and was reared from stem galls of *Cissus verticilliata* and leaf petioles of *Cecropia*.

Pitz KM. 2010. New taxonomic status of *Foenomorpha* Szépligeti as a subgenus of *Capitonius* Brullé (Hymenoptera: Braconidae: Cenocoeliinae) including a revision of species. *Zootaxa* 2607: 1-54.
Foenomorpha Szépligeti is treated as a subgenus of *Capitonius* Brullé and the species are revised and keyed.

Pitz KM, Sharkey MJ. 2005. *Cenocoelius huggerti*, the first record of the subfamily Cenocoeliinae (Hymenoptera: Braconidae) from Africa. *Journal of Hymenoptera Research* 14(1): 92-95.
The distribution of Cenocoeliinae includes the Afrotropical region.

Pitz KM, Sharkey MJ. 2007. Three new species of Cenocoeliinae (Hymenoptera: Braconidae) with novel morphological characteristics and habitat records. *Journal of Hymenoptera Research* 16(1): 167-177.
Includes one species lacking a closed second submarginal cell in the fore wing.

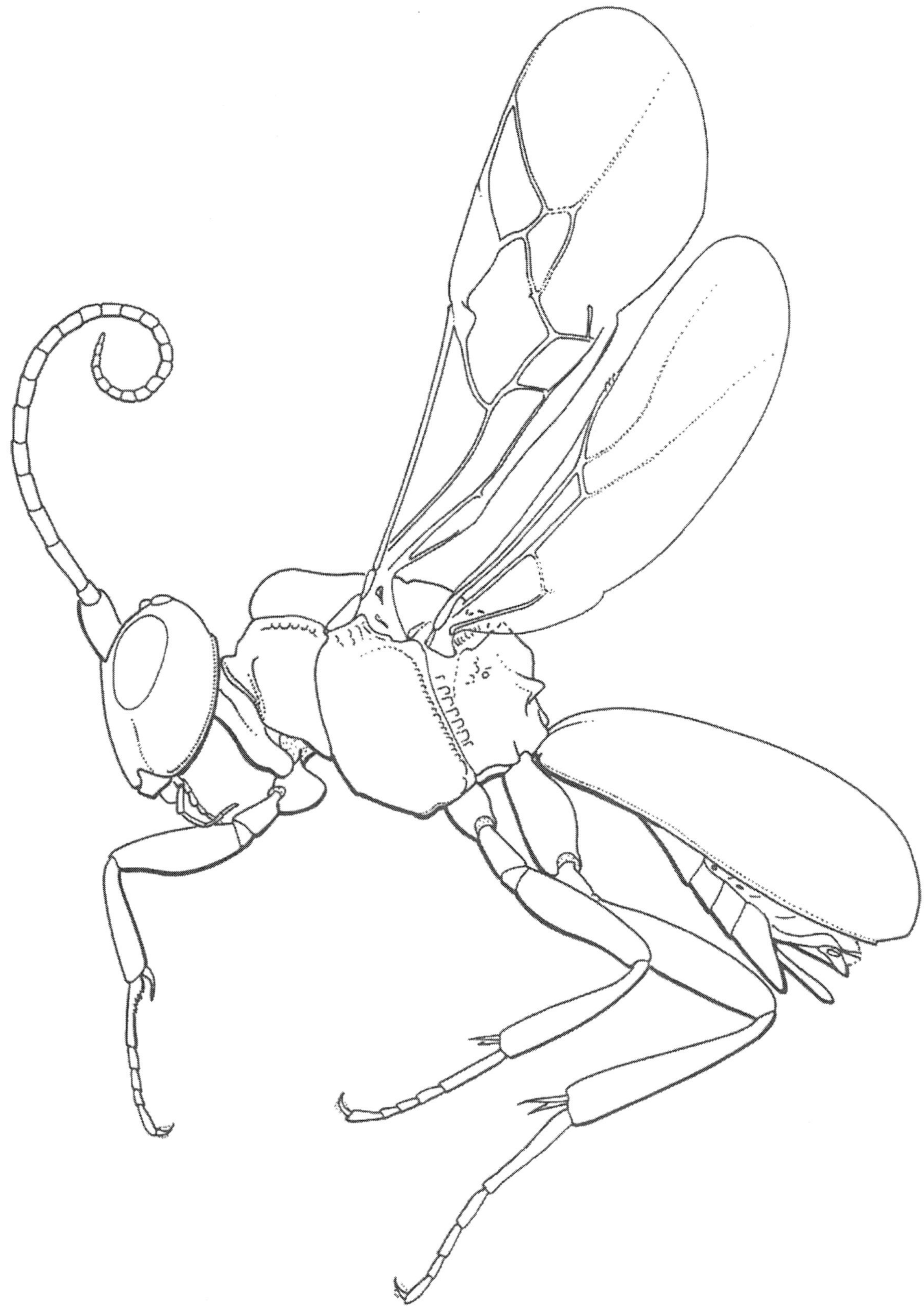

SUBFAMILY CHELONINAE

S. R. SHAW

Department of Ecosystem Science and Management, University of Wyoming, Laramie, Wyoming 82071, U.S.A.

INTRODUCTION. The Cheloninae comprise a large subfamily with more than 500 known species worldwide (Walker and Huddleston, 1987). Adult chelonines have a distinctive carapace that covers the metasoma dorsally, formed by the fusion the first three metasomal terga (Dudarenko, 1974). Although members of several other braconid subfamilies (e.g. Rogadinae, Helconinae) have convergently evolved a metasomal carapace, chelonines are the most commonly encountered braconids that have this characteristic. Most chelonines are black or brown with lighter legs, but yellow, orange, and red body coloration occurs in some tropical species.

BIOLOGY. Chelonines are solitary koinobiont egg-larval endoparasitoids of Lepidoptera that feed largely in concealed situations, especially Tortricoidea and Pyraloidea (Shenefelt, 1973). Many host larvae are borers in stems, buds, or fruit, while others create a cryptic feeding retreat by rolling, folding, or tying leaves with silk. In some cases, host location and oviposition may be simplified by attacking eggs which are numerous and exposed (S. Shaw, 1983). Other species, with long ovipositors, may attack eggs in more concealed situations. Although egg-larval parasitoids occur in other braconid subfamilies (e.g. Alysiinae, Helconinae, Ichneutinae), chelonines are the most common braconids that employ this strategy (Clausen, 1956). Oviposition, though rapid, is relatively slow compared to some endoparasitoids (e.g. microgastrines attacking exposed hosts) because the chorion must be drilled. For example, a species of *Chelonus* near *curvimaculaius* Cameron completes its oviposition sequence in 17 to 21 seconds (Leluk and Jones, 1989). Although the host egg is attacked, the host larva is not killed until after it has matured (Vance, 1932). Chelonine larvae have only three instars (Vance, 1932). The first instar may diapause within the parasitized host until the host larva matures and is ready to pupate (Balduf, 1963). Some species apparently induce a precocious host metamorphosis (Leluk and Jones, 1989). The third (final) instar larva emerges from the host (although its caudal end may remain inside) and completes its feeding from outside. The chelonine larva spins a whitish cocoon and pupates within the cocoon or pupation chamber of the host (Boyce, 1936). More complete reviews of chelonine biology are provided by van Achterberg (1990), M. Shaw and Huddleston (1991), and S. Shaw (1995).

BIOLOGICAL CONTROL POTENTIAL. Chelonines are frequently recovered in biological control programs directed against lepidopterous pests. For example, *Chelonus* (*Microchelonus*) *kellieae* Marsh, a parasitoid of the potato tuberworm, *Phthorimaea operculella* (Zeller), from Costa Rica, has been reared and released in southern California (Marsh, 1979a). *Chelonus insularis* Cresson (= *texanus* Cresson) is a common parasitoid of *Heliothis* and *Spodoptera* that occurs throughout the New World (Marsh, 1978). *Chelonus sonorensis* Cameron has been repeatedly recovered from pyralid stalk-borers in Texas and Mexico (Rodriguez-del-Bosque et al., 1990). *Ascogaster quadridentata* Wesmael has been used as a biological control agent for several tortricoid pests in fruit orchards (S. Shaw, 1983).

COMMON GENERA Populations are locally abundant at times, and a Malaise trap situated in an appropriate habitat may yield large numbers of chelonines. Of the seven genera in the New World, *Chelonus* Panzer, *Ascogaster* Wesmael, and *Phanerotoma* Wesmael tend to be collected most frequently. The adult females spend much time walking on the host food plant in search of host eggs, so large numbers of a given chelonine species can also be collected by sweeping the foliage of host food plants. *Phanerotoma*, whose species are nocturnal, can often be collected at lights. The Nearctic species of *Ascogaster* and *Leptodrepana* Shaw were reviewed by S. Shaw (1983), and the Nearctic species of *Chelonus* (subgenus *Microchelonus* Szépligeti) were studied by McComb (1968). *Chelonus* (subgenus *Chelonus*), although commonly encountered in temperate North America, has not been studied in a comprehensive fashion. Zettel (1989a, b, 1990a, b, 1991, 1992a,b) has recently published a series of papers on *Phanerotoma* and related new genera, including a revision for North America and descriptions of new tribes, genera, and species. Many chelonine species remain undescribed in the Neotropical region. S. Shaw (1983) proposed the genus *Leptodrepana* for a species-group formerly placed in *Ascogaster*. This division works well for the New World species, but proves more difficult with the Old World fauna. Consequently, European authorities have continued to group all these species in a single genus, *Ascogaster* (Huddleston, 1984; van Achterberg, 1990). Nevertheless, this approach creates a problem since S. Shaw (1983) demonstrated that *Ascogaster* would be paraphyletic if *Leptodrepana* were included. Consequently, the continued recognition of *Leptodrepana* as a valid genus in the New World is warranted (Zettel, 1990a). Eventually a phylogenetic analysis of world *Ascogaster* species will be required, and it seems likely that more genera will ultimately be recognized, rather than less. A similar problem exists regarding recognition of *Microchelonus* Szépligeti. *Microchelonus* is retained here as a subgenus of *Chelonus* (following McComb) solely for convenience, to facilitate recognition of certain species.

DISTRIBUTION. Chelonines are cosmopolitan in distribution, and can be found in any vegetated habitats where concealed-feeding Lepidoptera occur. Species of *Leptodrepana* and *Ascogaster* are more common in shrubby areas and forested habitats, while *Chelonus* species are often more heat and drought tolerant, and occur commonly in meadows, prairies, and grasslands. *Phanerotoma* are often more abundant in arid regions, or during dry seasons. *Chelonus* and *Ascogaster* appear more diverse in temperate North America, while in Central and South America *Phanerotoma* and *Leptodrepana* are more frequently encountered.

DISTINGUISHING FEATURES. Chelonines are readily recognized by the metasomal carapace (Figs 7-11), which (except in the comparatively rare *Fischeriella* Zettel) is more complete and encloses more of the ventral portions of the metasoma than most other braconids that have convergently evolved carapace-like morphology. The chelonine carapace involves the complete fusion of the first three terga, and usually conceals the rest of the metasoma when viewed from above. However, in living, critical-point-dried, or chemically-dried specimens the remainder of the metasoma can be seen from the side, or below. In air-dried specimens (including many old museum specimens) the soft parts of the metasoma are contracted into the carapace, and thus concealed. Males of *Chelonus* (subgenus *Microchelonus*) have a peculiar seta-lined cavity or pit at the apex of the carapace (Figs 19, 20), the function of which is unknown. It may play some role in the dispersal of sex pheromones. In other braconids that have carapace-like morphology, the first tergite often is not fused with the second, or more of the metasoma is visible from above, posteriorly. Some of the brachistine helconines (e.g. *Urosigalphus* Ashmead and *Triaspis* Haliday) have a very complete carapace like that of chelonines, but these have more reduced fore wing venation (the r-m cross vein is absent, thus there is no closed second submarginal cell). In contrast, chelonines always have the r-m cross vein present, thus forming a closed second submarginal cell (Figs 1-6).

KEY TO THE NEW WORLD GENERA OF THE SUBFAMILY CHELONINAE

1. Metasomal carapace with 2 distinct and complete transverse grooves (Figs 7-9) 2
- Metasomal carapace without complete transverse grooves, usually forming an evenly sculptured, convex surface (Fig. 10, 11); more rarely, some indication of transverse lines can be seen on the sides, but this is obscured by sculpture on the dorsal surface 7

2(1). Fore wing (RS+M)a widely separated from 1M, arising directly from parastigma, often near middle (Fig. 1) (Tribe Phanerotomini) 3
- Fore wing (RS+M)a joining 1M below the parastigma, or at the base of the parastigma (Fig. 2) (Tribe Pseudophanerotomini) 5

3(2). Metasomal tergum 2 distinctly longer than tergum 3; distal median part of tergum 3 weakly sclerotized; terga 4-8 exposed and desclerotized medially .... ***Fischeriella*** Zettel
Distribution: Ecuador and Peru; rare; one described species. Biology: unknown. Reference: Zettel (1990a).
- Metasomal tergum 2 shorter than tergum 3 (Fig. 8), or about equal (Fig. 7); terga 4-8 usually concealed (Figs 7, 8) or only slightly exposed, and (if visible) well sclerotized 4

4(3). Propodeum coarsely sculptured, with distinct carinae or rugosities (Fig. 12). Margin of clypeus with small teeth (Fig. 13). Wings: Fig. 1 .... ***Phanerotoma*** Wesmael

Distribution: widespread and very common; 50 described species in the New World, with many undescribed Neotropical species remaining. Biology: reared from Pyralidae, Gelechiidae, and Tortricidae. References: Zettel (1992a, revision for North America), Zettel (1989a, 1989b, 1990b, 1991, 1992b, descriptions and keys for some Neotropical species).

\- Propodeum antero-medially smooth and polished, posteriorly and laterally punctate. Margin of clypeus rounded (as in Fig. 14) .... ***Phanerotomoides*** Zettel

Distribution: Brazil; rare; one described species. Biology: unknown. Reference: Zettel (1990a).

5(2). Fore wing m-cu arising directly in line with 2RS. Occipital carina dorsally interrupted.Clypeus without teeth .... ***Leptochelonus*** Zettel

Distribution: Chile; rare; one described species. Biology: unknown. Reference: Zettel (1990a).

\- Fore wing m-cu arising distad 2RS (Fig. 2). Occipital carina complete. Clypeus usually with 2 blunt teeth .6

6(5). Head and mesosoma coarsely sculptured, with coarse pitted depressions. Metasomal carapace strongly arched; apex of female carapace with 2 sharp teeth .... ***Dentigaster*** Zettel

Distribution: Panama, Brazil, and Argentina; uncommon; 3 described species. Reference: Zettel (1990a, description and key to species).

\- Head and mesosoma finely punctate (Fig. 14, 15). Carapace flat and without teeth at apex (Fig. 9). Wings: Fig. 2 .... ***Pseudophanerotoma*** Zettel

Distribution: Nearctic and Neotropical regions; 10 described species. Biology: unknown. Reference: Zettel (1990a, description and keys to subgenera and species).

7(1). Fore wing (RS+M)a absent, thus first submarginal and first discal cells forming a single large confluent cell (Fig. 3, 4). Eyes densely and visibly setose at low magnification (Fig. 16) (genus *Chelonus* Panzer)......8

\- Fore wing (RS+M)a present, thus first submarginal and first discal cells distinct and clearly separated (Fig. 5, 6). Eyes smooth and devoid of setae, or only sparsely setose with minute setae that are clearly visible only at high magnification (Figs 17, 18) 9

8(7). Female with 14 or fewer flagellomeres. Apex of male carapace usually (but not always) with a setose foramen, cavity, or pit (Figs 19, 20). Wings: Fig. 4 .... ***Chelonus*** subgenus ***Microchelonus*** Szépligeti

Distribution: Cosmopolitan; 111 described North American species, numerous undescribed (but only 5 described) Neotropical species. Biology: reared from various microlepidoptera, including Lyonetiidae, Gracillariidae, Momphidae, Coleophoridae, Cosmopterigidae, Gelechiidae, Yponomeutidae, Pyralidae, Tortricidae, and Cochylidae. Reference: McComb (1968, revision for North America).

\- Female with 15 or more flagellomeres. Apex of male carapace without a setose foramen, cavity, or pit (Figs 10, 11). Wings: Fig. 3 .... ***Chelonus*** subgenus ***Chelonus*** Panzer

Distribution: Cosmopolitan; very diverse and abundant group, greatly in need of revision; at least 28 described New World species, numerous undescribed species. Biology: reared from various Lepidoptera, especially Gelechiidae, Pyralidae, and Noctuidae. Reference: Marsh (1979b).

9(7). Ocelli forming an isoceles triangle (Fig. 17). Scutellum sculptured and usually convex. Ventral cavity of female carapace usually distinctly shorter than carapace length, not reaching apex (Fig. 21). Wings: Fig. 5 .... ***Ascogaster*** Wesmael

Distribution: widespread; 11 described Nearctic species, at least 6 undescribed Neotropical species. Biology: mostly reared from Tortricidae; more rarely Geometridae. Reference: S. Shaw (1983, revision for Nearctic Region).

\- Ocelli forming an equilateral triangle (Fig. 18). Scutellum smooth, usually flattened. Ventral cavity of female carapace as long or nearly as long as carapace, about reaching apex which is sometimes flared (Fig. 22). Wings: Fig. 6........ ***Leptodrepana*** Shaw

Distribution: widespread; 4 described Nearctic species, at least 16 undescribed Neotropical species. Biology: mostly unknown; one attacks Plutellidae. Reference: S. Shaw (1983, description and key to species for Nearctic region).

REFERENCES

Achterberg, C. van 1990. Revision of the Western Palearctic Phanerotomini (Hymenoptera: Braconidae). Zoologische Verhandelingen Leiden 255: 3-106.

Balduf, W. V. 1963. A distinct type of host-parasite relation among insects. Annals of the Entomological Society of America 56: 386-391.

Boyce, H. R. 1936. Laboratory breeding of *Ascogaster carpocapsae* Viereck with notes on biology and larval morphology. The Canadian Entomologist 68: 241-246.

Clausen, C. P. 1956. The egg-larval host relationship among the parasitic Hymenoptera. Bolettino del Laboratorio di Zoologia Generale e Agraria Portici 33: 119-133.

Dudarenko, G. P. 1974. Formation of the abdominal carapace in braconids (Hymenoptera: Braconidae) and some aspects of the classification of the family. Entomological Review 53: 80-90.

Leluk, J. and Jones, D. 1989. *Chelonus* sp. near *curvimaculatus* venom proteins: analysis of their potential role and processing during development of host *Trichoplusia ni*. Archives of Insect Biochemistry and Physiology 10: 1-12.

Marsh, P. M. 1978. The braconid parasites (Hymenoptera) of *Heliothis* species (Lepidoptera: Noctuidae). Proceedings of the Entomological Society of Washington SO: 15-36.

Marsh, P. M. 1979a. Descriptions of new Braconidae (Hymenoptera) parasitic on the potato tuberworm and related Lepidoptera from Central and South America. Journal of the Washington Academy of Sciences 69: 12-17.

Marsh, P. M. 1979b. Family Braconidae. Pp. 144-295. In: Krombein, K. V., Hurd, P. D., Smith, D. R., and Burks, B. D. (Eds.). Catalog of Hymenoptera in America North pf Mexico. Smithsonian Institution Press, Washington, D.C.

McComb, C. W. 1968. A revision of the *Chelonus* sub genus *Microchelonus* in North America north of Mexico (Hymenoptera: Braconidae). University of Maryland Agricultural Experiment. Station Bulletin A-149: 1-148.

Rodrigucz-del-Bosque, L. A., Browning, H. A., and Smith, J. W., Jr. 1990. Seasonal parasitism of cornstalk borers (Lepidoptera: Pyralidae) by indigenous and introduced parasites in northeastern Mexico. Environmental Entomology 19: 393-402.

Shaw, M. R. and Huddleston, T. 1991. Classification and biology of braconid wasps (Hymenoptera: Braconidae), Handbooks for the Identification of British Insects 7(11): 1-126.

Shaw, S. R. 1983. A taxonomic study of Nearctic *Ascogaster* and a description of a new genus *Leptodrepana* (Hymenoptera: Braconidae). Entomography 2: 1-54.

Shaw, S. R. 1995. The Braconidae. Pp. 431-464. In: Hanson, P. and Gauld, I. (Eds.). The Hymenoptera of Costa Rica. Oxford University Press, Oxford.

Shenefelt, R. D. 1973. Pars 10. Braconidae 6, Cheloninae. Pp. 813-936. In: Vecht, J. van der and Shenefelt, R. D. (Eds.). Hymenopterorum Catalogus (nova editio) Dr. W. Junk, The Hague.

Vance, A. M. 1932a. The biology and morphology of the braconid *Chelonus annulipes* Wesmael, a parasite. of the European corn borer. United States Department of Agriculture Technical Bulletin 294: 1-48.

Walker, A. K. and Huddleston, T. 1987. New Zealand chelonine braconid wasps (Hymenoptera), Journal of Natural History 21: 339-361.

Zettel, H. 1989a. Beiträge zur Kenntnis Neotropischer Arten der Gattung *Phanerotoma* Wesmael: 1. Die *Ph. bilinea*-gruppe und 2. Die *Ph. attenuata*-gruppe (Hymenoptera: Braconidae, Cheloninae). Linzer Biologische Beiträge 21: 317-338.

Zettel, H. 1989b. Beiträge zur Kenntnis Neotropischer Arten der Gattung *Phanerotoma* Wesmael: 3. Die *Ph. trivittata*-gruppe (Hymenoptera: Braconidae, Cheloninae). Linzer Biologische Beiträge 21: 527-542.

Zettel, H. 1990a. Eine Revision der Gattungen der Cheloninae (Hymenoptera, Braconidae) mit Beschreibungen neuer Gattungen und Arten. Annales Naturhistoriches Museum Wien 91: 147-196.

Zettel, H. 1990b. Beiträge zur Kenntnis Neotropischer Arten der Gattung *Phanerotoma* Wesmael: 4. Die *Ph. fuscovaria*-gruppe (Hymenoptera: Braconidae, Cheloninae). Linzer Biologische Beiträge 22: 3-19.

Zettel, H. 1991. Beiträge zur Kenntnis Neotropischer Arten der Gattung *Phanerotoma* Wesmael: 5. Die *Ph. minuta*-gruppe und 6. Die *Ph. popovi*-gruppe (Hymenoptera: Braconidae, Cheloninae). Linzer Biologische Beiträge 23: 375-385.

Zettel, H. 1992a. Revision der *Phanerotoma*-Arten Nordamerikas (Hymenoptera: Braconidae, Cheloninae). Linzer Biologische Beiträge 24: 275-330. Zettel, H. 1992b. Beiträge zur Kenntnis Neotropischer Arten der Gattung *Phanerotoma* Wesmael: 7. Die *Ph. atriceps*-gruppe (Hymenoptera: Braconidae, Cheloninae). Linzer Biologische Beiträge 24: 663-669.

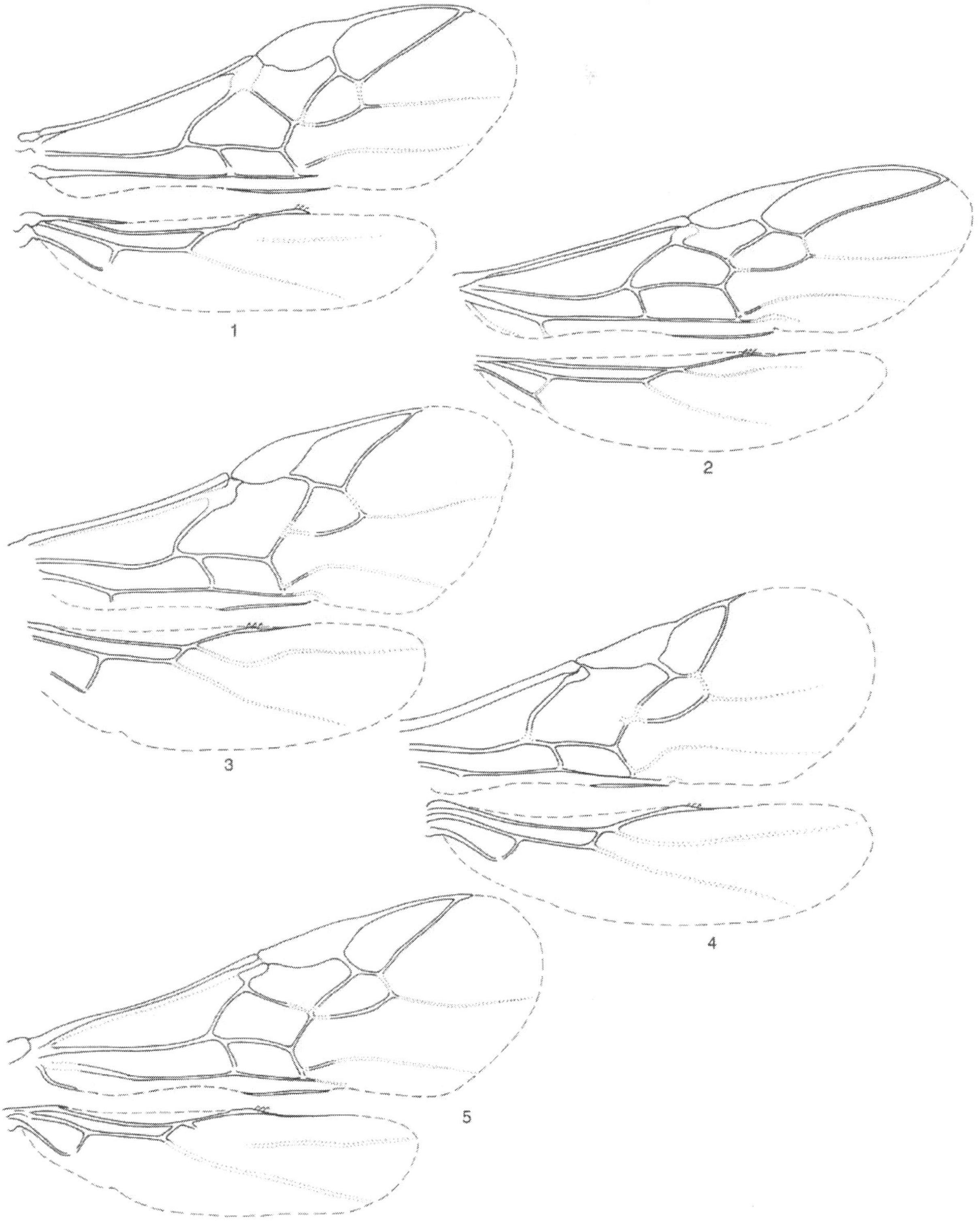

Figs 1-5: Wings of Cheloninae. 1, *Phanerotoma longicauda* Walley; 2, *Pseudophanerotoma pulchra* Zettel; 3, *Chelonus* (*Chelonus*) sp.; 4, *Chelonus* (*Microchelonus*) sp.; 5, *Ascogaster impatientis* Shaw.

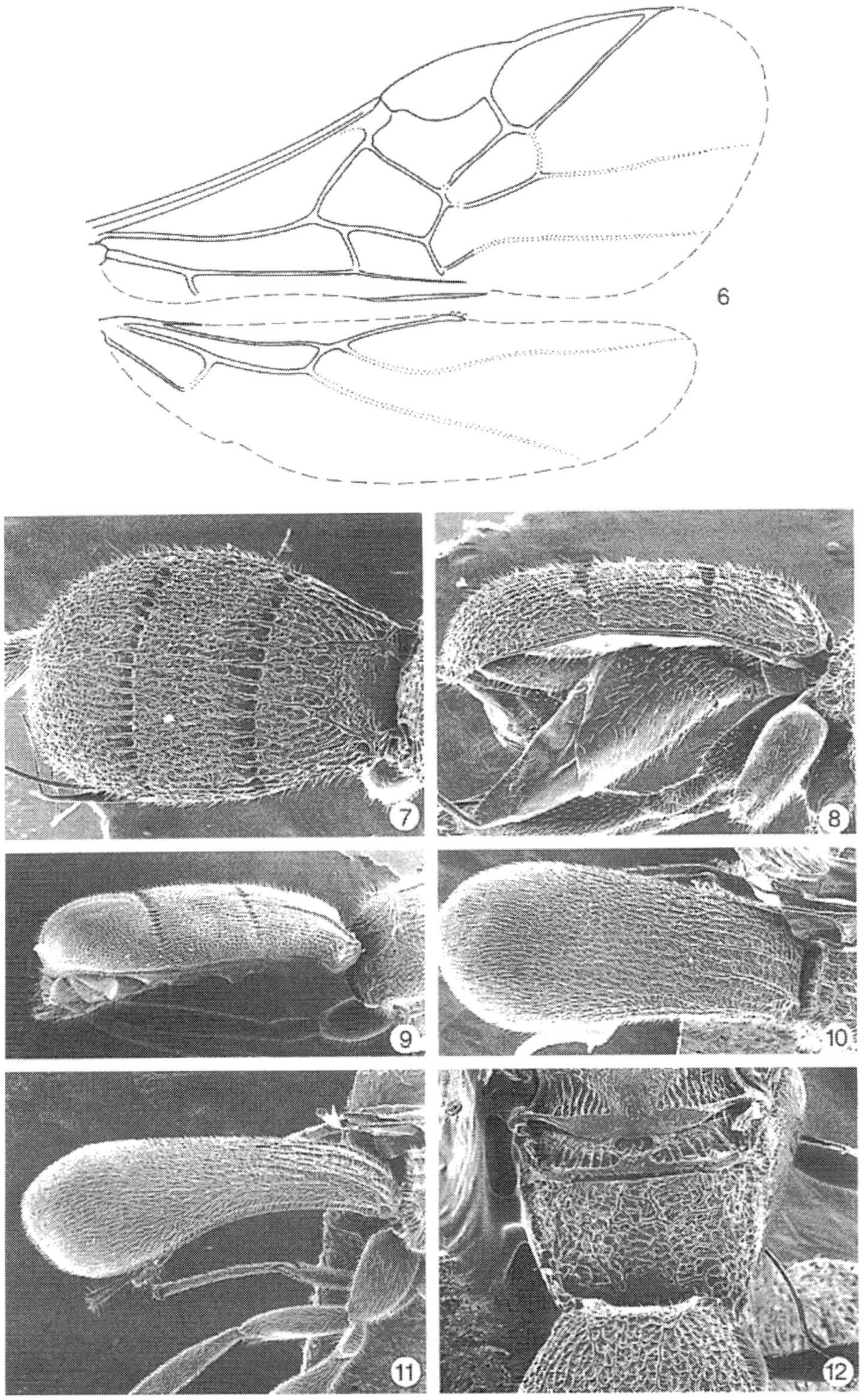

Figs 6-12: Cheloninae. 6, wings of *Leptodrepana* sp.; 7-11, metasoma: 7. *Phanerotoma* sp., dorsal view; 8, Phanerotoma sp., lateral view; 9, *Pseudophanerotoma pulchra* Zettel, dorso-lateral view; 10, *Chelonus (Chelonus)* sp., dorsal view; 11, *Chelonus (Chelonus)* sp., lateral view; 12, propodeum of Phanerotoma sp.

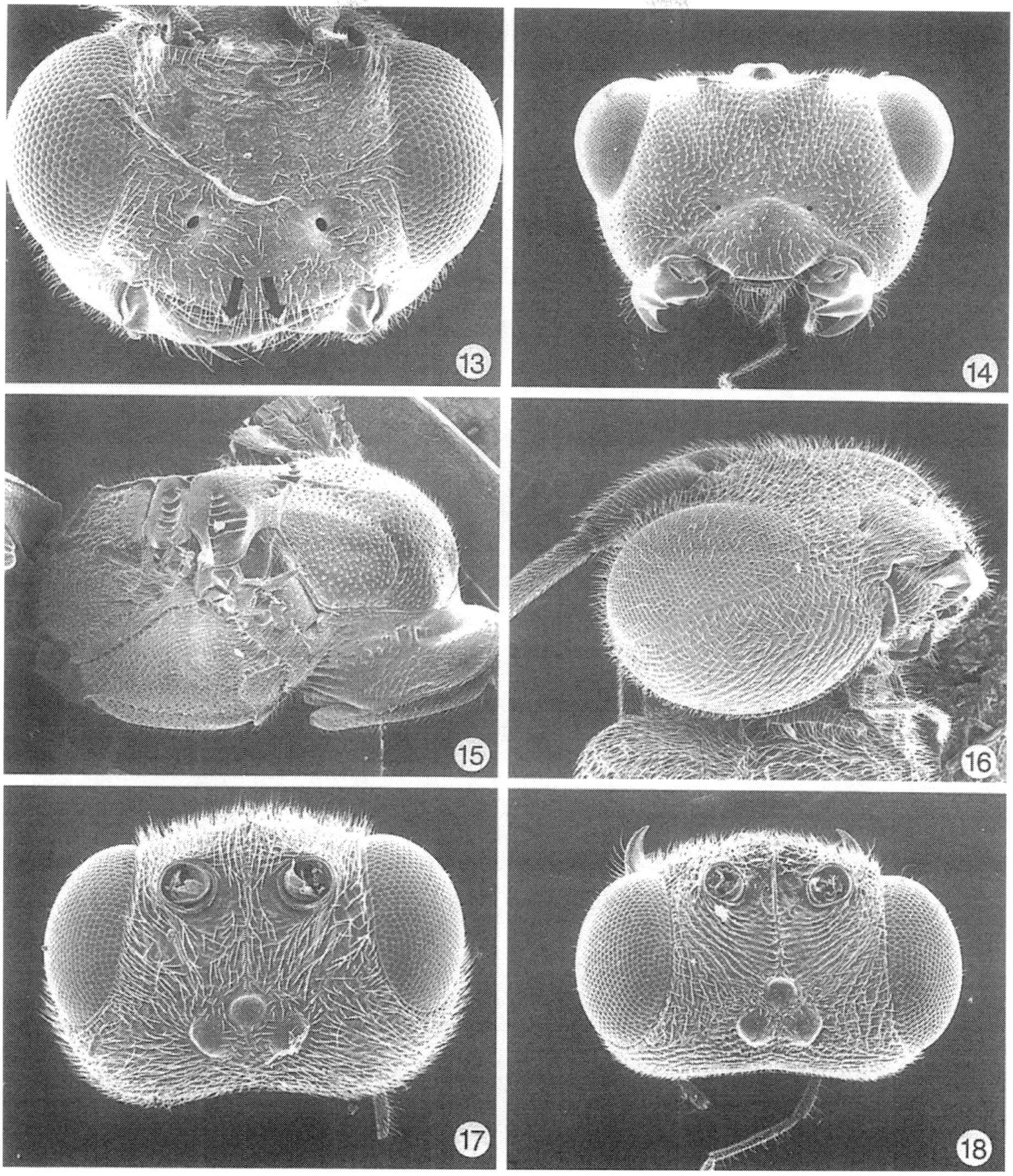

Figs 13-18: Cheloninae. 13, head of *Phanerotoma* sp., frontal view, arrows = small teeth on ventral margin of clypeus; 14, head of *Pseudophanerotoma* pulchra Zettel, frontal view; 15, mesosoma of *Pseudophanerotoma pulchra* Zettel; 16, head of *Chelonus* sp. showing setose eyes; 17, eyes and ocelli of *Ascogaster* sp., dorsal view; 18, eyes and ocelli of *Leptodrepana* sp., dorsal view.

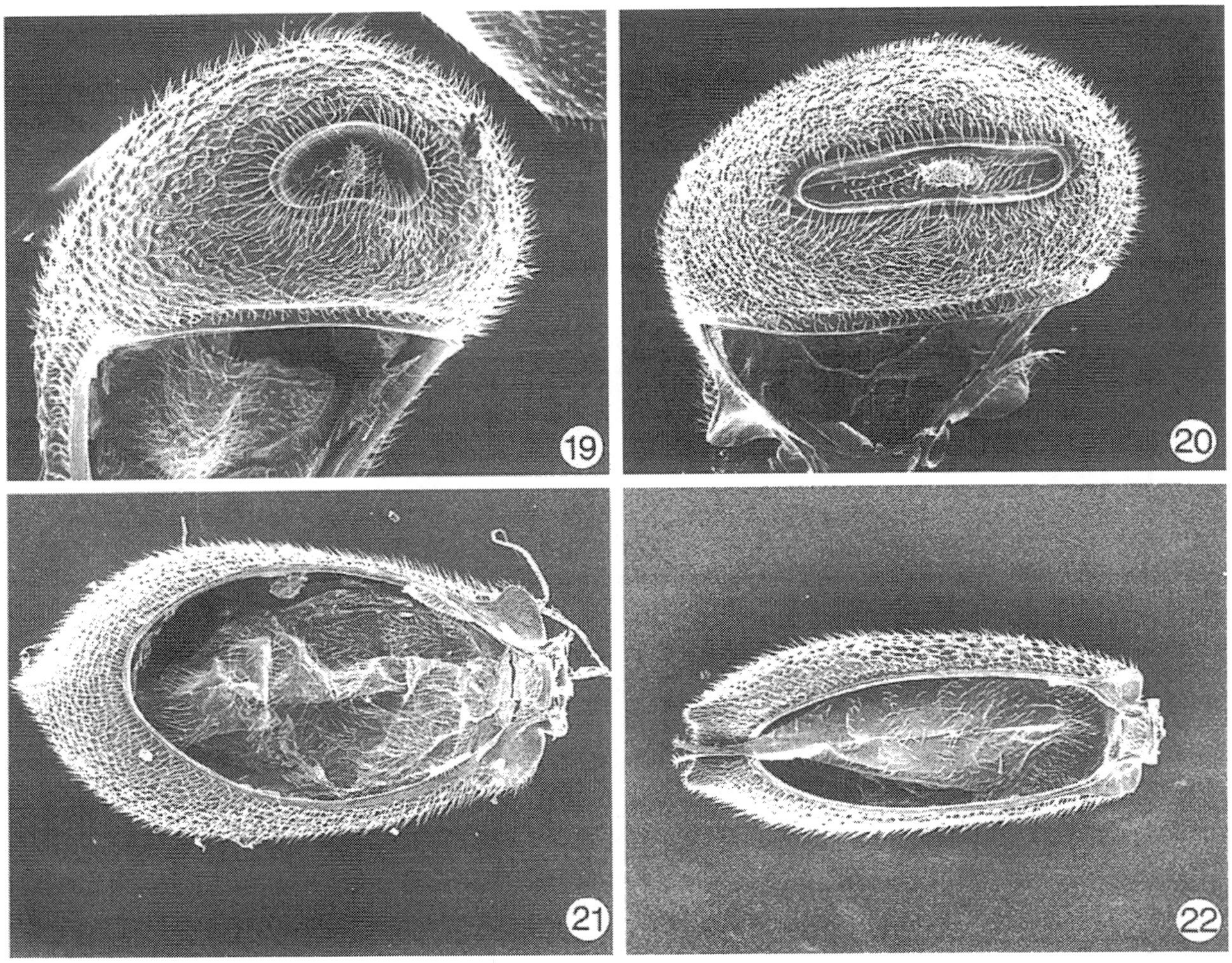

Figs 19-22: Carapaces of Cheloninae. 19, *Chelonus* (*Microchelonus*) sp., posterior view; 20, *Chelonus* (*Microchelonus*) sp., posterior view; 21, *Ascogaster* sp., ventral view; 22, *Leptodrepana* sp., ventral view.

CHELONINAE UPDATE

S.R. Shaw
Department of Ecosystem Science and Management, University of Wyoming, Laramie, Wyoming 82071, U.S.A.
braconid@uwyo.edu

The Cheloninae chapter of the manual continues to work well for the identification of New World genera, although some controversy has centered over the proper definition of the genus *Ascogaster* and whether or not to treat *Leptodrepana* as a separate genus. Kittel *et al.* (2016) concluded that *Ascogaster* is paraphyletic with two clades: "a mostly 'North American' and a mostly 'Australian' clade." They found that the Australian *Ascogaster* lineage is the sister-group to *Leptodrepana, Austroascogaster*, and *Megascogaster* combined, while the North American clade is sister-group to all of them. The potential paraphyly of *Ascogaster* was previously discussed by Shaw (1984), Huddleston (1984), Tang and Marsh (1994) and Shaw (1997). Given their phylogenetic results, Kittel *et al.* (2016) proposed two possible solutions for the classification of chelonine genera: "either the whole clade could be treated as *Ascogaster s.l.* and *Austroascogaster*, *Leptodeprana*, and *Megascogaster* synonymized under it, or the 'North American' clade could be treated as *Ascogaster* s. str. (in which the type species of the genus *Ascogaster instabilis* Wesmael belongs) and a new genus (which is morphologically virtually identical to *Ascogaster*) comprising the 'Australian' species be erected. The genera *Leptodrepana, Megascogaster*, and *Austroascogaster* would, thus, not be affected in the latter scenario. However, before any nomenclatural changes to *Ascogaster* are considered, a more detailed exploration of the characters defining each group would be required, along with a more comprehensive sampling of taxa representing each group." Since the first "solution" would require the extreme measure of sinking three genera, perhaps unnecessarily, and would still leave us with an incomplete understanding of *Ascogaster* phylogeny worldwide, it seems better (for present) to follow the second solution and to continue recognizing *Leptodrepana* as a valid genus, separate from *Ascogaster,* in the New World.

Annotated list of recent papers on New World Cheloninae

Kittel RN, Austin AD. 2014. Synopsis of Australian chelonine wasps (Hymenoptera: Braconidae: Cheloninae) with description of two new genera. *Austral Entomology* 53: 183-202.
Revised the Cheloninae of Australia and described two new genera.

Kittel RN, Austin AD, Klopfstein S. 2015. Molecular and morphological phylogenetics of chelonine parasitoid wasps (Hymenoptera: Braconidae), with a critical assessment of divergence time estimations. *Molecular Phylogenetics and Evolution* 101: 224-241.
A molecular phylogenetic analysis of the Cheloninae was conducted, concluding that the genera comprising the former subfamily Adeliinae (including *Adelius* and *Paradelius*) are nested within the Cheloninae. Therefore, the adeliines are best reclassified as belonging in the subfamily Cheloninae, where they can be treated as tribe Adeliini.

Tang Y, Marsh PM. 1994. A taxonomic study of the genus *Ascogaster* in China (Hymenoptera: Braconidae: Cheloninae). *Journal of Hymenoptera Research* 3: 279-302.
The *Ascogaster* species (including *Leptodrepana*) of China were revised.

Papp J. 2016. First survey of the Neotropical species of *Microchelonus* Szépligeti with descriptions of twenty-five new species (Hymenoptera: Braconidae: Cheloninae). *Acta zoologica Academiae Scientiarum Hungaricae* 62(3): 217-344.

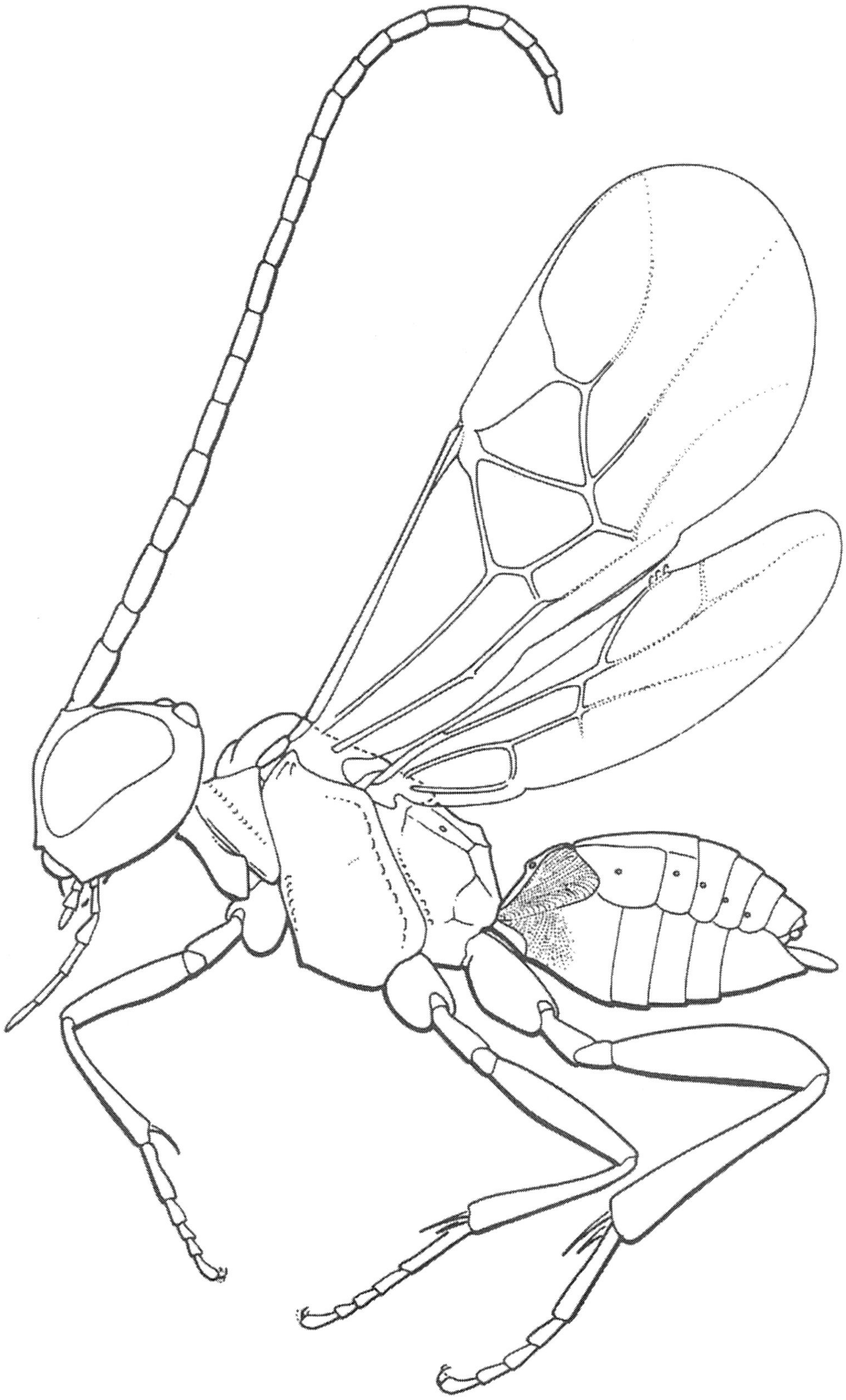

SUBFAMILY DIRRHOPINAE

J.B. WHITFIELD

Department of Entomology, 320 Morrill Hall, University of Illinois, 505 S. Goodwin Ave, Urbana, IL 61801, U.S.A.

INTRODUCTION. Only a single relatively rare genus, *Dirrhope* Förster, is known and so far all specimens have come from Canada, U.S.A., Mexico, Costa Rica and Eurasia. The genus has been placed in the Microgastrinae by Muesebeck (1935), Tobias (1967) and Marsh (1979) and in the Adeliinae by Telenga (1955) and Čapek (1970). Van Achterberg (1984) erected a new subfamily for *Dirrhope* and the genus has generally been treated as belonging to its own distinct subfamily since that time (Quicke and van Achterberg, 1990; Whitfield and Wagner, 1991; Wharton et al., 1992; Whitfield and Mason, 1994-but see Tobias, 1986, who treats it in the Miracinae). Within the Braconidae, the Dirrhopinae is clearly a part of the same general lineage as the Microgastrinae and Adeliinae (Quicke and van Achterberg, 1990; Wharton et al., 1992; Whitfield and Mason, 1994).

BIOLOGY. The only host records are from leaf-mining moth larvae of the family Nepticulidae (van Achterberg, 1984; Whitfield and Wagner, 1991).

BIOLOGICAL CONTROL POTENTIAL. Probably none. Even where their recorded hosts are common, *Dirrhope* species appear to remain rare.

COMMON GENERA. *Dirrhope* is the only described genus and it is rare.

DISTRIBUTION. The subfamily is certainly Holarctic, ranging southwards in the New World to Costa Rica; it remains to be recorded from other regions and may be absent elsewhere.

DISTINGUISHING CHARACTERS. *Dirrhope* species somewhat resemble adeliines or small microgastrines in form but differ from adeliines in not having the first two metasomal tergites fused, and in having the spiracle of the strongly flattened petiole posterior to the midlength of the tergite. Unlike the microgastrines, the first metasomal spiracle is on the median tergite rather than on a lateral tergite, and the first tergite is highly flattened. The fore wing venation is also diagnostic (Fig. 1).

REFERENCES

Achterberg, C. van 1984. Essay on the phylogeny of Braconidae (Hymenoptera: Ichneumonoidea). Entomologisk Tidskrift 105: 41-58.

Čapek, M. 1970. A new classification of the Braconidae (Hymenoptera) based on the cephalic structures of the final instar larvae and biological evidence. The Canadian Entomologist 102: 846-875.

Marsh, P. M. 1979. Family Braconidae. Pp. 144-313. In: Krombein, K.V., Hurd, P. D., Smith, D.R., and Burks, B. D. (Eds.). Catalog of Hymenoptera in America North of Mexico. Smithsonian Institution Press, Washington, D.C.

Muesebeck, C. F. W. 1935. On two little known genera of Braconidae (Hymenoptera). Proceedings of the Entomological Society of Washington 37: 173-177.

Quicke, D. L. J. and Achterberg, C. van 1990. Phylogeny of the subfamilies of the family Braconidae (Hymenoptera: Ichneumonoidea). Zoologische Verhandelingen Leiden 258: 1-95.

Telenga, N. A. 1955. Fauna of the U.S.S.R. Hymenoptera, Vol. 5, no. 4, Braconidae: Microgasterinae and Agathinae. Zoologicheskii Institut Akademii Nauk SSSR. (In Russian.)

Tobias, V. I. 1967. A review of the classification, phylogeny and evolution of the family Braconidae (Hym.). Entomological Review 46: 387-399.

Tobias, V. I. 1986. Keys to the insects of the European part of the USSR. III. Hymenoptera, 4: 1-501. Akademia Nauk, Leningrad. (In Russian.)

Wharton, R. A, Shaw, S. R., Sharkey, M. J., Wahl, D.B., Woolley, J. B., Whitfield, J. B., Marsh, P. M., and Johnson, J. W. 1992. Phylogeny of the subfamilies of the family Braconidae (Hymenoptera: Ichneumonoidea): a reassessment. Cladistics 8: 199-235.

Whitfield, J. B. and Mason, W. R. M. 1994. Mendesellinae, a new subfamily of braconid wasps (Hymenoptera: Braconidae) with a review of relationships within the microgastroid assemblage. Systematic Entomology 19: 61-76.

Whitfield, J. B. and Wagner, D. L. 1991. Annotated key to the genera of Braconidae (Hymenoptera) attacking leaf-mining Lepidoptera in the Holarctic Region. Journal of Natural History 25: 733-754.

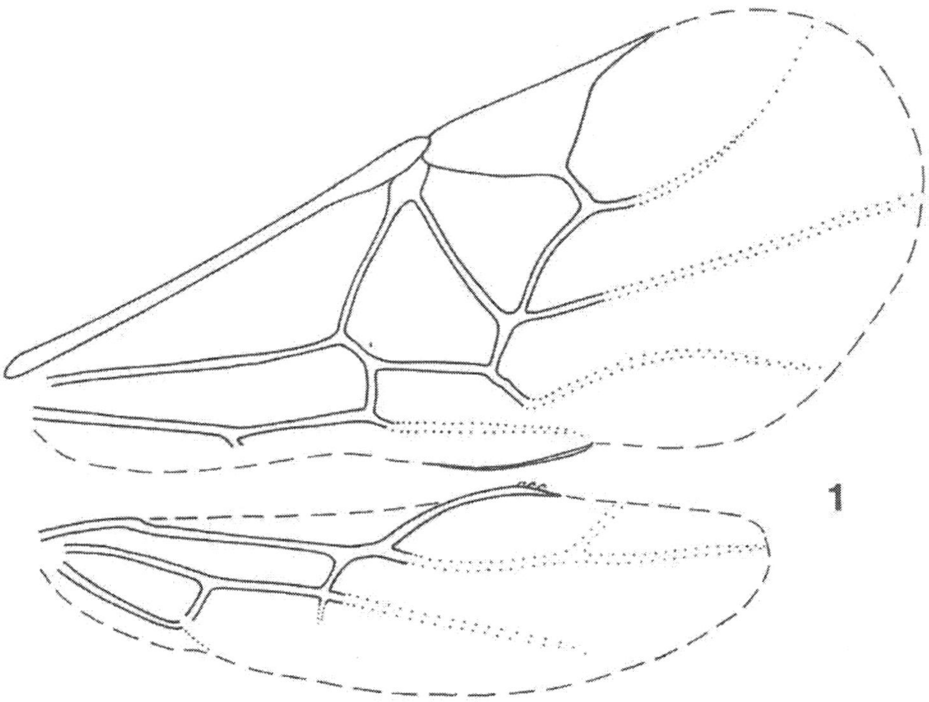

1

Fig. 1: Wings of *Dirrhope* sp.

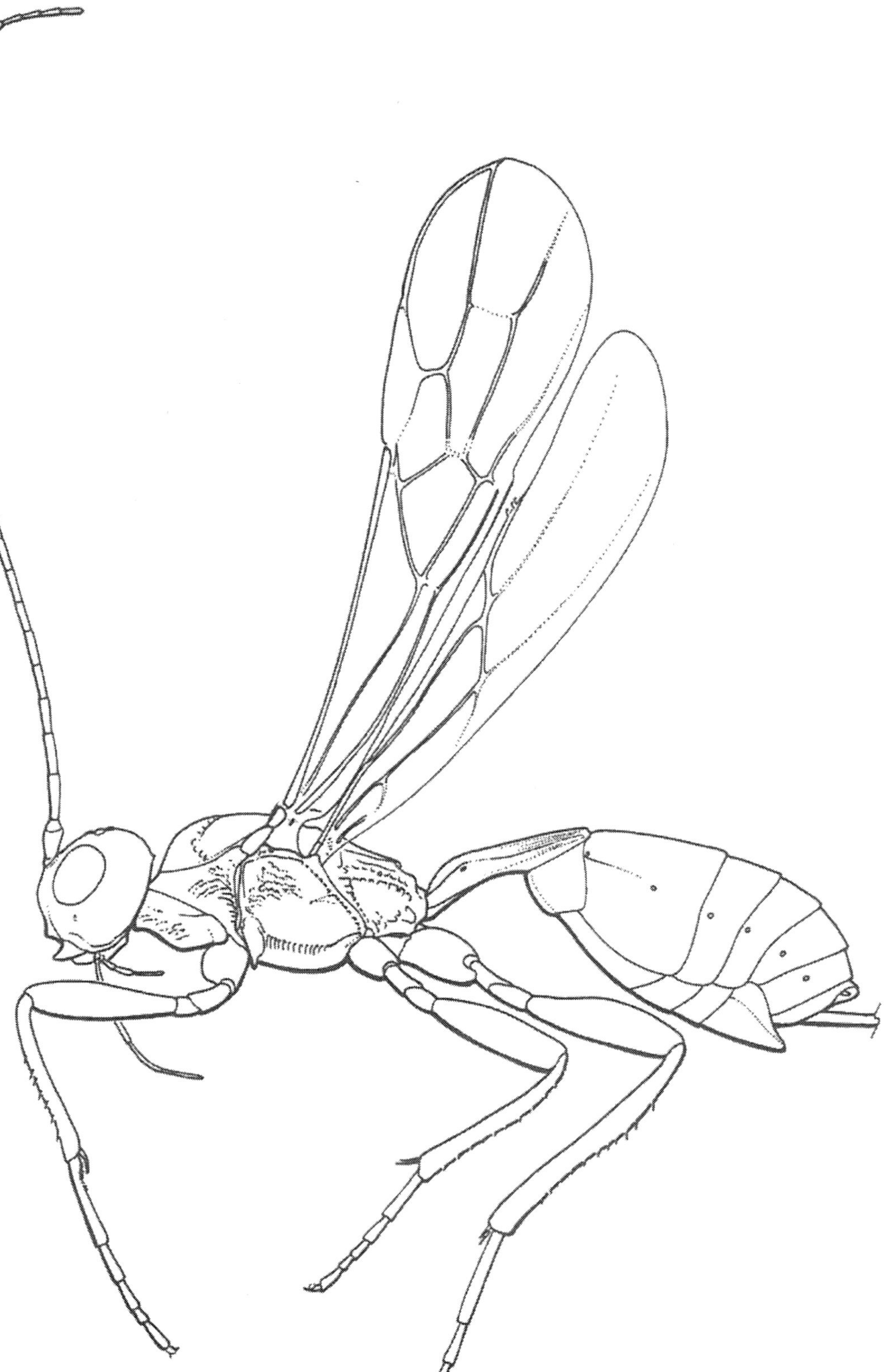

SUBFAMILY DORYCTINAE

P. M. MARSH

Cooperating Scientist, Systematic Entomology Laboratory, U.S. Department of Agriculture, (present address: P.O. Box 384, North Newton, Kansas 67117, U.S.A.)

INTRODUCTION. The Doryctinae is one of the more diverse groups in the Braconidae, with species ranging in size from 1 mm to 25 mm in body length. There are an unusually large number of genera, particularly in the tropics, but many of these genera contain one or only a few described species. In the New World, the most recent studies have been by Marsh (Marsh, 1965, 1966a, b, 1967a, b, 1968, 1969a, b, 1970a, b, 1973, 1976, 1982a, b, 1983a, b, 1984, 1988, 1989, 1991, 1993; Matthews and Marsh, 1969, 1973; Quicke and Marsh, 1992; Ramirez and Marsh, 1996). Although little comprehensive research has been done on tribal classification, Shenefelt and Marsh (1976) catalogued the genera into seven tribes, Fischer (1981a) listed nine tribes and Belokobyl'skij (1992) listed 13 tribes and 20 subtribes. There have been only a few recent attempts to provide a phylogenetic framework for the tribes and subtribes (Belokobyl'skij, 1992, 1995). I have chosen to disregard tribal separations in this discussion, providing a single key to the New World genera. Some species of controversial placement, such as those in the genus *Monitoriella* Hedqvist, have been included in the chapter on the Hormiinae.

BIOLOGY. Very little is known about the habits of this subfamily relative to the number of described species. Most species, where biological records are known, are idiobiont ectoparasitoids of wood-boring beetle larvae but a few attack stem boring lepidopterous and saw fly larvae. The host larva is paralyzed before the egg is deposited and the parasitoid cocoon is usually formed in the host tunnel or mine. One described and several undescribed species of the genus *Allorhogas* Gahan are apparently phytophagous in seeds (Macedo and Monteiro, 1989; Marsh, 1991). Species of *Psenobolus* Reinhard have been reared from figs in Costa Rica where they display extreme sexual dimorphism in the males, similar to that found in chalcid fig wasps. In an unusual host association, one species of the genus *Sericobracon* Shaw has been reared as an endoparasitoid from webspinners of the order Embiidina (Shaw and Edgerly, 1985). *Heterospilus microstigmi* Richards and several related but undescribed species of *Heterospilus* Haliday have been reared from nests of species in the aculeate wasp genus *Microstigmus* Ducke (Matthews, 1968). Another species of *Heterospilus* attacks stem-infesting sawflies and several others attack larvae of stem boring moths. More commonly, however, *Heterospilus* has been reared from wood boring Coleoptera, especially from members of the family Scolytidae.

Although the Doryctinae, and in particular the genus *Heterospilus*, have one of the most diverse host ranges in the Braconidae, most rearings are from larval Coleoptera. All rearing records for *Ecphylus* Förster, for example, are from larvae of Scolytidae, Lyctidae, and Bostrichidae. *Spathius* Nees, and *Notiospathius* Marsh and Matthews are commonly reared from Buprestidae, Cerambycidae, Curculionidae, and Scolytidae. *Doryctes* Haliday is also generally associated with Buprestidae and Cerambycidae, as are a number of other commonly encountered doryctines.

BIOLOGICAL CONTROL POTENTIAL. Although the potential for extensive use in biological control programs exists because their primary hosts are wood and stem boring beetle larvae, the Doryctinae have not been extensively utilized. Several species have been purposely introduced: *Dendrosoter protuberans* (Nees) was introduced from Europe and established on the European elm bark beetle, *Scolytus multistriatus* (Marsham); *Syngaster lepidus* Brullé and *Jarra* spp. (neither included in the generic key below) were recently introduced and possibly established in southern California from Australia against the eucalyptus longhorn borer; *Heterospilus annulatus* Marsh and *H. megalopus* Marsh were studied as parasitoids of the cotton boll weevil in Texas; and

Allorhogas pyralophagus Marsh was introduced but not established against a sugarcane borer in Texas. Several native species of *Spathius* have been studied in bark beetle control programs, and *Heterospilus cephi* Rohwer, which probably arrived in the U.S.A. when its host was introduced, attacks the wheat stem sawfly. Doryctines have only rarely been used in the Neotropical Region for biological control.

COMMON GENERA. The most commonly encountered New World species belong to the genera *Heterospilus*, *Ecphylus*, *Spathius*, and *Notiospathius*. The genus *Heterospilus* is concentrated in the New World with an estimated 200 Nearctic and 300 Neotropical species, 90% of which are undescribed; less than 75 described species occur in the remainder of the World, where relatively few undescribed species are known. No revisions are available and the genus is badly in need of study. *Ecphylus* occurs in the Holarctic, Ethiopian, and Neotropical Regions with 22 described Nearctic species, three Neotropical ones, and eight additional species from the Old World. There are about 25 undescribed species in the Neotropics. The Nearctic species were studied by Marsh (1965). The genus *Spathius* is worldwide in distribution with a concentration of species in the Old World tropics. Most of the Neotropical species differ in wing venation from typical *Spathius* and were placed in the genus *Notiospathius* by Matthews and Marsh (1973). The most recent study was by Matthews (1970) who revised the North American species of *Spathius*.

DISTRIBUTION. The Doryctinae are worldwide in distribution with most species occurring in tropical areas. The number of described species in the New World is about 370 (about 1175 worldwide); the number of genera is about 77 (150 worldwide). As studies are made in both the New and Old World tropics these numbers are expected to increase dramatically. Most of the genera treated here are known only from the New World.

DISTINGUISHING FEATURES. The Doryctinae have a characteristically large head because of the need to chew out of a host tunnel or mine. They are in the cyclostome group of braconids (see section on Braconidae in introductory chapter of manual) with a circular opening present between the clypeus and mandibles when the mandibles are closed (Figs 55, 70). They are distinguished from other cyclostomes by the following combination of characters: the presence of a row of stout spines on the outer edge of the fore tibia that are usually shorter and thicker than normal setae (Figs 65-67); the presence of a flange at the apico-lateral corner of the propleuron just above the fore coxae and extending slightly over the ventro-lateral corner of the pronotum: and the presence (in most species) of a double node near the apex of the dorsal valve of the ovipositor (Quicke et al., 1992).

KEY TO THE NEW WORLD GENERA OF THE SUBFAMILY DORYCTINAE

Wing venation is important in distinguishing genera of the Doryctinae. Most venation features are easily understood as defined in the key, but the genus *Heterospilus* may present problems. This genus is characterized by the absence of a distinct and complete tubular vein 2RS in the fore wing (Fig. 9). In some species, however, this vein may be indicated by a more-or-less complete infuscated line (sometimes referred to as a nebulous vein). Because this is not a distinct tubular vein with obvious connections to other veins, I consider vein 2RS to be absent when it appears to be nebulous. This character appears in couplets 23 and 46.

1. Wings absent or reduced to scale-like pads that are shorter than the mesosoma 2
- Wings present 4

2(1). Femora swollen (Fig. 49). Entirely wingless .... ***Pambolidea*** Ashmead (in part)
Distribution: southern U.S.A. through Mexico; 2 described species. Biology: parasitoids of Scolytidae, Bostrichidae. References: Marsh (1965), Shenefelt and Marsh (1976).
- Femora usually not swollen; if hind femur swollen (males of some species of *Ecphylus*) then wings present as small scale-like pads 3

3(2). Propodeum greatly convex in side view, coriaceous. Scutellum greatly convex .... ***Aptenobracon*** Marsh
Distribution: U.S.A.; one described species. Biology: reared from Curculionidae and Mordellidae in stems. Reference: Marsh (1965).
- Propodeum not convex, sloping from base to apex, rugulose. Scutellum flat....... ***Ecphylus*** Förster (in part)
Distribution: Alaska and Canada to Panama, Caribbean; 25 described New World species plus many undescribed Neotropical species. Biology: parasitoids of Scolytidae, Bostrichidae, Lyctidae. References: Marsh (1965, key to Nearctic species), Shenefelt and Marsh (1976).

4(1). Hind wing with stigma absent (all females will key here, as will many males) 5

- Hind wing with stigma present (Fig. 6) (only males key here, including those of many common species.82
5(4). Fore wing r-m absent (Figs 2, 3) 6
- Fore wing r-m present (Fig. 11), though often weak 23
6(5). Hind wing cu-a absent (Figs 13) 7
- Hind wing cu-a present (Fig. 5) 15
7(6). First subdiscal cell of fore wing open at apex, 2cu-a absent, and/or 2-1A becoming obsolete towards apex (Fig. 2)..8
- First subdiscal cell closed at apex, 2cu-a present and meeting 2-1A..11
8(7). Fore wing with (RS+M)b longer than 2RS (Fig. 1) ....***Percnobracon*** Kieffer

Distribution: Mexico, Peru, Uruguay, Argentina; uncommon; 2 described species. Biology: reared from cecidomyiid galls (needs to be confirmed). References: Kieffer (1910), Muesebeck (1958), Shenefelt and Marsh (1976).

- Fore wing with (RS+M)b shorter than 2RS (Fig. 2) or m-cu directly in line with 2RS and therefore (RS+M)b absent 9
9(8). Antenna with 11 flagellomeres. Wings: Fig. 2 ....***Micrommatus*** Marsh

Distribution: Brazil; one described species. Biology: unknown. Reference: Marsh (1993).

- Antenna with more than 11 flagellomeres..10
10(9). Hind coxa with sharp antero-ventral tubercle at base (as in Figs 46, 63). Mesoscutum coarsely coriaceous (Fig. 51), vertex transversely striate (Fig. 53) ....***Masonius*** Marsh

Distribution: Costa Rica and Panama; one described species. Biology: unknown. Reference: Marsh (1993).

- Hind coxa round at base (as in Fig. 64), at most with slightly rounded protrusion. Mesoscutum and vertex smooth, lacking obvious sculpture ....***Fritziella*** Marsh

Distribution: Brazil; one described species. Biology: unknown. Reference: Marsh (1993).

11(7). Fore wing 1Cu-a absent (Fig. 3)..***Ecphylus*** Förster

New World Distribution: Alaska and Canada to Panama; common, though often overlooked because of small body size; 25 described New World species plus many undescribed Neotropical species. Biology: parasitoids of Scolytidae, Bostrichidae, Lyctidae. References: Marsh (1965, key to Nearctic species), Shenefelt and Marsh (1976).

- Fore wing 1Cu-a present (Fig. 4) 12
12(11). Propodeum at base with laterally flattened lamella which projects dorsally. Wings: Fig. 4 ....***Bohartiellus*** Marsh

Distribution: Costa Rica, Brazil, Bolivia; 2 described, one undescribed species. Biology: unknown. Reference: Marsh (1983b).

- Propodeum without such a projection 13
13(12). Scape with double rim at apex (Fig. 34)..***Achterbergia*** Marsh

Distribution: Costa Rica, Suriname, Ecuador, Peru; one described species. Biology: unknown. Reference: Marsh (1993).

- Scape normal, without double rim 14
14(13). Antenna with 12 flagellomeres..***Ecphyloides*** Marsh

Distribution: Brazil; one described species. Biology: unknown. Reference: Marsh (1993).

- Antenna with more than 12 flagellomeres ....***Aivalykus*** Nixon

Distribution: Indo-Australian, Ethiopian, Nearctic, Neotropical: 4 described species, but only one of these from the New World, several undescribed Neotropical species. Biology: parasitoids of Scolytidae. References: Marsh (1965), Shenefelt and Marsh (1976).

15(6). Fore wing m-cu arising distad 2RS, (RS+M)b absent (Fig. 5) ....***Monolexis*** Förster

Distribution: Cosmopolitan; 8 described species, 2 of these found in the New World. Biology: parasitoids of Lyctidae, Scolytidae. References: Marsh (1965), Shenefelt and Marsh (1976).

- Fore wing m-cu arising basad 2RS (Fig. 6) or directly in line with 2RS, (RS+M)b often present 16
16(15). Hind femur swollen, at most 2.5 times as long as wide (as in Fig. 49) 17
- Hind femur not swollen, at least 3 times as long as wide..19
17(16). Cyclostome opening large and broad, head somewhat prognathous (Figs 37, 38). Wings: Fig. 6 ....***Verae*** Marsh

Distribution: Brazil; one described species. Biology: unknown. Reference: Marsh (1993).

- Cyclostome opening smaller and round, head not prognathous 18
18(17). Stigma of fore wing short and broad, about 1.5 times as long as broad (Fig. 7), vein R1 sometimes absent and vein 3RS sometimes weak apically, not reaching wing apex ....Pambolidea Ashmead

Distribution: southern U.S.A. through Mexico; 2 described species. Biology: parasitoids of Scolytidae, Bostrichidae. References: Marsh (1965), Shenefelt and Marsh (1976).

- Stigma not as broad, about 3 times as long as broad, vein R1 always present (Fig. 8) ....***Hecabolus*** Curtis

Distribution: Palaearctic, Indo-Australian, Nearctic (U.S.A.), Neotropical (Argentina); 5 described species, 2 bf these found in the New World. Biology: parasitoids of Scolytidae, Bostrichidae. Reference: Shenefelt and Marsh (1976).

19(16). Fore wings with infuscate bands (as in Fig. 26) 20

\- Fore wings hyaline, not banded 21

20(19). Notauli absent. Face flat .... ***Percnobraconoides*** Marsh

Distribution: U.S.A. (Arizona), and Mexico; one described species. Biology: possible parasitoid of Bostrichidae. Reference: Marsh (1989).

\- Notauli present. Face swollen (Fig. 52) .... ***Tarasco*** Marsh

Distribution: Mexico; one described species. Biology: unknown. Reference: Marsh (1993).

21(19). Antenna with less than 18 flagellomeres. Hind tibia swollen .... ***Coiba*** Marsh

Distribution: Central America, Caribbean; 3 described species. Biology: unknown. Reference: Marsh (1993).

\- Antenna with more than 18 flagellomeres. Hind tibia not swollen 22

22(21). Groove between metasomal terga 2 and 3 deep, straight, heavily sculptured; tergum 3 with second transverse groove that is also heavily sculptured (Fig. 36) .... ***Caingangia*** Marsh

Distribution: Brazil; one described species. Biology: unknown. Reference: Marsh (1993).

\- Groove between metasomal terga 2 and 3 absent, tergum 2 with semi-circular area at base that is rugose-striate, tergum 3 without transverse groove (Fig. 45) .... ***Leluthia*** Cameron (in part)

Distribution: Nearctic, Neotropical; 3 described species, several undescribed Neotropical species. Biology: parasitoids of Buprestidae. References: Marsh (1967b), Shenefelt and Marsh (1976).

23(5). First and second submarginal cells of fore wing confluent, 2RS absent, rarely (some species of *Heterospilus* that are keyed both ways) represented by an infuscated line (nebulous vein) (Fig. 9) (See note at the beginning of key) 24

\- First and second submarginal cells of fore wing separated, 2RS present as a distinct (tubular) vein 26

24(23). Hind wing cu-a absent or nearly so (Fig. 10). Spines of fore tibia short and blunt .... ***Labania*** Hedqvist

Distribution: Honduras, Costa Rica, Cuba; one described species. Biology: reared from leaf gall on *Ficus colubrinae* in Costa Rica. References: Hedqvist (1963), Wharton (1993, transfer to Doryctinae),

\- Hind wing cu-a present (Figs 9, 11), Spines of fore tibia pointed, may be long or short 25

25(24). Metasomal tergum 2 with 2 posteriorly converging grooves (Fig. 35). Hind coxa round at base, without basal tubercle .... ***Pioscelus*** Muesebeck and Walkley

Distribution: U.S.A. and Canada; 2 described species. Biology: unknown. Reference: Shenefelt and Marsh (1976).

\- Tergum 2 without such longitudinal grooves. (Fig. 50). Hind coxa with distinct basal tubercle (as in Fig. 63). Wings: Fig. 9 .... ***Heterospilus*** Haliday (in part)

Distribution: Cosmopolitan but most species in Western Hemisphere; estimated 200 Nearctic and 300 Neotropical species. Biology: parasitoids mostly of wood boring beetle larvae but some on stem boring sawflies and moths. References: Shenefelt and Marsh (1976), Marsh (1982a, b), Quicke and Marsh (1992).

26(33). Metasoma inserted high on propodeum, distance between insertion of metasoma and hind coxae about equal in length to hind coxa 27

\- Metasoma inserted lower on propodeum close to hind coxae 28

27(26). Hind wing venation normal with veins M+CU, 1M and r-m distinct .... ***Evaniodes*** Szepligeti

Distribution: Brazil and Peru; one described species. Biology: unknown. References: Shenefelt and Marsh (1976),Fischer (1981a).

\- Hind wing with veins 1M and r-m absent, M+CU and cu-a meeting at junction of Sc+R, R1 and RS (Fig. 15) .... ***Pariodes*** Fischer

Distribution: Brazil; one described species. Biology unknown. Reference: Fischer (1981a). Possibly a synonym of *Evaniodes*.

28(26). Hind coxa round at base, without basal tubercle (as in Fig. 64) 29

\- Hind coxa with a more or less distinct antero-ventral basal tubercle (as in Figs 46, 63). This tubercle cannot be seen unless the base of the coxa is clearly visible 44

29(28). Hind wing m-cu strongly curved distally towards apical margin of wing (Figs 11, 12) 30

\- Hind wing m-cu extending straight to posterior margin of wing or weakly curved towards wing base or apex (Fig. 14) 32

30(29). Hind coxa with a long, curved tooth dorsally (Fig. 39). Wings: Fig. 11 .... ***Odontobracon*** Cameron

Distribution: Nearctic and Neotropical; 14 described species. Biology: parasitoids of Cerambycidae. Reference: Marsh (1988).

\- Hind coxa without tooth dorsally 31

31(30). Marginal cell of hind wing divided by weak nebulous cross vein (Fig. 12) .... ***Nervellius*** Roman

Distribution: Neotropical; 4 described species. Biology: unknown. Reference: Marsh (1988).

\- Marginal cell without cross vein (as in Fig. 11) .... ***Holcobracon*** Cameron

Distribution: Indo-Australian, Neotropical (Caribbean); 4 described species. Biology: unknown. Reference: Marsh (1988).

32(29). Fore wing m-cu arising distad 2RS, (RS+M)b absent (Fig. 13) 33

- Fore wing m-cu arising basad 2RS or directly in line with 2RS, (RS+M)b often present (Fig. 14) 35

33(32). Head with rugose or scale-like swellings on each side of frons (Fig. 42). Wings: Fig. 13 .... ***Dendrosoter*** Wesmael

Distribution: Cosmopolitan: 6 of the 25 described species occur in the New World, but thus far recorded only from U.S.A. Biology: parasitoids of wood boring beetles, especially Scolytidae. References: Marsh (1967a), Shenefelt and Marsh (1976).

- Head without such swellings 34

34(33). Petiole at least 1.5 times longer than apical width, apical width 2 times basal width. Hind wing r-m about 0.1-0.2 times length of 1M .... ***Janzenia*** Marsh

Distribution: Mexico and Costa Rica; one described species. Biology: unknown. Reference: Marsh (1993).

- Petiole only slightly longer than apical width, apical width only slightly greater than basal width (Fig. 58). Hind wing r-m 0.3-0.4 times length of 1M .... ***Rhoptrocentrus*** Marshall

Distribution: Holarctic; one described species. Biology: parasitoids of Cerambycidae and Bostrichidae. Reference: Shenefelt and Marsh (1976).

35(32). Occipital carina completely absent (Fig. 41). Pronotum with median tubercle. Hind wing M+CU much longer than 1M (Fig. 16) 36

- Occipital carina present laterally and usually also dorsally (Fig. 40). Pronotum without such a tubercle. Hind wing M+CU at most equal to 1M (Fig. 14), usually much shorter (Figs 17, 19)................................37

36(35). Propleuron with 2 or 4 sharp or blunt tubercles (Fig. 71). Wings: Fig. 16 .... ***Binarea*** Brullé

Distribution: Neotropical; 4 described species, many undescribed. Biology: unknown. Reference: Shenefelt and Marsh (1976).

- Propleuron without such tubercles.. ***Liobracon*** Szépligeti

Distribution: Nearctic, Neotropical; 20 species, several undescribed. Biology: unknown. Reference: Shenefelt and Marsh (1976).

37(35). Hind wing M+CU usually equal to or slightly longer than 1M (Figs 14, 18), rarely slightly shorter 38

- Hind wing M+CU shorter than 1M (Fig. 17), usually much shorter (Fig. 30) 41

38(37). Small species, body length 3 mm or less. Wings hyaline .... ***Psenobolus*** Reinhard

Distribution: Costa Rica and Brazil; 3 described species. Biology: associated with figs, possibly an inquiline. References: Shenefelt and Marsh (1976), Ramirez and Marsh (1996).

- Large species, body length 6-25 mm. Wings either uniformly infuscate or with infuscate bands 39

39(38). Antenna short, 21 flagellomeres or less, all flagellomeres wider than long. Wings: Fig. 14 .... ***Histeromeroides*** Marsh

Distribution: U.S.A. (Florida), Brazil, Ecuador, Peru; one described species. Bi- ology: unknown. Reference: Marsh (1993).

- Antenna longer, much more than 21 flagellomeres, most flagellomeres longer than..............................40

40(39). Hind wing r-m less than half length of M+CU; hind wing cu-a perpendicular to posterior margin of wing or angled slightly toward wing apex (Fig. 18)... ***Monarea*** Szépligeti

Distribution: Neotropical; 3 described species, several undescribed. Biology unknown. Reference: Shenefelt and Marsh (1976).

- Hind wing r-m more than half length of M +CU, often nearly equal; hind wing cu-a angled toward wing base .... ***Megaloproctus*** Schulz

Distribution: Neotropical; 13 described species. Biology: unknown. Reference: Marsh (1983a).

41(37). Scape equal to or longer than first flagellomere 42

- Scape shorter than first flagellomere 43

42(41). Scape with expanded flap on apical anterior edge (Fig. 43). First subdiscal cell of fore wing with small opening at apex, 2cu-a absent (Fig. 17). Metasomal tergum 2 distinctly sculptured at base, but without semicircular or triangular area, remainder of metasoma smooth and shining ....***Rhoptrocentroides*** Marsh

Distribution: Costa Rica, Brazil; one described species. Biology: unknown. References: Marsh (1993). Note: Belokobyl'skij (1995) treated *Rhoptrocentroides* as synonym of *Pseudorhoptrocenirus* Granger from Madagascar, which needs confirmation.

- Scape not expanded on apical rim. First subdiscal cell closed, 2cu-a present and meeting 2-1A. Tergum 2 with semicircular or triangular area at base .... ***Gymnobracon*** Szépligeti

Distribution: Ethiopian, Neotropical; 3 described species, all at least 15 mm long. Biology: unknown. Reference: Shenefelt and Marsh (1976).

43(41). Ovipositor sheaths densely covered with long stout setae (Figs 59, 60)......................... ***Sharkeyella*** Marsh

Distribution: Mexico, Brazil, Bolivia; one described species. Biology: unknown. Reference: Marsh (1993).

- Ovipositor sheaths normal, with sparse covering of short thin setae. Wings: Fig. 19 .... ***Notiospathius*** Matthews and Marsh

Distribution: Neotropical; 14 described species, several undescribed. Biology: unknown. References: Matthews and Marsh (1973), Shenefelt and Marsh (1976).

44(28). Hind wing cu-a absent... ***Mononeuron*** Fischer

Distribution: Brazil; one described species. Biology: unknown, a very small species, reared from plant galls. Reference: Fischer (1981b).

- Hind wing cu-a present 45

45(44). Hind wing M +CU usually equal to or longer than 1M, rarely slightly less but always at least 0.7 times length of 1M (Fig. 20) 46

- Hind wing M+CU shorter than 1M, at most 0.7 times length and usually half length of 1M or less (Fig. 27)65

46(45). First and second submarginal cells of fore wing confluent, fore wing 2RS absent, rarely represented by an infuscate line (nebulous vein) (Fig. 9) (See note at beginning of key)............***Heterospilus*** Haliday (in part)

Distribution: Cosmopolitan but most species in Western Hemisphere; estimated 200 Nearctic and 300 Neotropical species. Biology: parasitoids mostly of wood boring beetle larvae but some on stem boring sawflies and moths. References: Shenefelt and Marsh (1976), Marsh (1982a, b), Quicke and Marsh (1992).

- First and second submarginal cells separated, fore wing 2RS present as a distinct (tubular) vein 47

47(46). First subdiscal cell of fore wing closed at apex, 2cu-a present and distinctly meeting 2-1A (Fig. 20).........48

- First subdiscal cell open at apex, 2cu-a absent, occasionally an infuscate spot or short line present between veins 2-1A and 2CU but no distinct vein present (Fig.22)..53

48(47). Hind wing r-m nearly as long as 1M, and always at least 0.75 times its length (Fig. 20) 49

- Hind wing r-m at most half length of 1M, usually less (Fig. 21)...51

49(48). Mesoscutum not sharply raised anteriorly, nearly on same plane as pronotum (Fig. 46). Sternaulus narrow and usually smooth. Wings: Fig. 20..***Doryctes*** Haliday

Distribution: Cosmopolitan: about 100 described species, many undescribed. Biology: parasitoids of wood boring beetle larvae. References: Marsh (1969a, key to Nearctic species), Shenefelt and Marsh (1976).

- Mesoscutum sharply raised above pronotum anteriorly (Figs 44, 61). Sternaulus 50

50(49). Cyclostome opening wide and oval, nearly as wide as eye height (Fig. 54); mandibles large, clypeus protruding...***Exontsira*** Belokobyl'skij

Distribution: Holarctic; 5 described species. Biology: unknown. Reference: Belokobyl'skij (1982). Note: Belokobyl'skij (1987) synonymized *Exontsira* with *Dolopsidea* Hincks which needs to be confirmed; one species of *Ontsira* is distinct and agrees with *Exontsira* as originally defined by Belokobyl'skij.

- Cyclostome opening small and circular, at most as wide as half height of eye (Fig. 55); mandibles small, clypeus not protruding..***Ontsira*** Cameron

Distribution: Indo-Australian, Holarctic; 22 described species. Biology: parasitoids of various wood boring beetle larvae. References: Marsh (1966a), Shenefelt and Marsh (1976).

51(48). Petiole long and narrow, about 3 times longer than apical width ....***Mixtec*** Marsh

Distribution: Mexico; one described species. Biology: unknown. Reference: Marsh (1993). Note: Belokobyl'skij (1995) treated Mixtec as a synonym of the Eurasian *Hypodoryctes* Kokujev, which needs confirmation.

- Petiole short and broad, apical width equal to or slightly greater than length...52

52(51). Hind wing m-cu curved towards wing apex (Fig. 21) ....***Araucania*** Marsh

Distribution: Ecuador, Peru, Chile, Argentina; 2 described species. Biology: un- known. Reference: Marsh (1993).

- Hind wing m-cu straight and pointed towards wing base (about as in Fig. 23) ....***Guaygata*** Marsh

Distribution: Caribbean; one described species. Biology: unknown. Reference: Marsh (1993).

53(47). Mesopleuron smooth and shining; in most genera, mesosoma generally smooth except for carinae (Fig. 61)..54

- Mesopleuron weakly coriaceous to rugose; entire mesosoma mostly coriaceous or rugose (Fig. 62)......59

54(53). Petiole longer than apical width..55

- Petiole with length equal to or less than apical width...57

55(54). Wings with infuscate bands (as in Fig. 26); second submarginal cell of fore wing long, extending beyond apex of stigma to near wing tip (Fig. 22) ....***Aphelopsia*** Marsh

Distribution: Panama and Ecuador; one described species. Biology: unknown. Reference: Marsh (1993).

- Wings hyaline or entirely infuscate: second submarginal cell short, barely extending beyond apex of stigma (Fig. 23)..56

56(55). Wings entirely infuscate, stigma yellow. Maxillary palp long, extending to middle coxa, longest (apical) segment of palp longer than first flagellomere. Wings: Fig. 23 ....***Waitaca*** Marsh

Distribution: Brazil; one described species. Biology: unknown. Reference: Marsh (1993).

- Wings hyaline, stigma brown. Maxillary palp shorter, not extending beyond middle of mesopleuron, longest (apical) segment shorter than first flagellomere...***Panama*** Marsh

Distribution: Panama; one described species. Biology: unknown. Reference: Marsh (1993).

57(54). Parastigma of fore wing large and swollen, longer than r, parastigma yellow, stigma brown (Fig. 24).........

.... ***Shawius*** Marsh
Distribution: Brazil and Argentina; one described species. Biology: unknown. Reference: Marsh (1993).

- Parastigma small, shorter than r, parastigma same color as stigma 58

58(57). Propodeum smooth, except for carinae, and shining .... ***Heerz*** Marsh
Distribution: Costa Rica and Brazil; 2 described species. Biology: unknown. Reference: Marsh (1993).

- Propodeum rugose, at least laterally (Fig. 62) .... ***Parallorhogas*** Marsh
Distribution: Indo-Australian, Ethiopian, Nearctic, Neotropical; 5 described species, only one of which occurs in the New World (U.S.A. (Texas) and Mexico). Biology: parasitoids of Pyralidae, Bruchidae, Bostrichidae. References: Marsh (1984, 1993).

59(53). Frons distinctly excavated (Fig. 47). Hind wing m-cu usually curved toward wing apex 60

- Frons not excavated. Hind wing m-cu usually straight or angled toward wing base 61

60(59). Petiole short and broad, length less than apical width, without cross carina separating basal and apical areas (Fig. 56). Fore wing r usually longer than 3RSa .... ***Stenocorse*** Marsh
Distribution: Indo-Australian, Nearctic, Neotropical; one described species, widespread in the New World, plus a few undescribed. Biology: parasitoids of Bruchidae. References: Marsh (1968), Shenefelt and Marsh (1976).

- Petiole longer, length at least equal to apical width, usually longer, with cross carina separating basal and apical areas. Fore wing r equal to or shorter than 3RSa (Fig. 25) .... ***Allorhogas*** Gahan
Distribution: Palearctic, Ethiopian, Nearctic, Neotropical; 9 described species plus a few undescribed. Biology: reared from galls and seeds, possibly phytophagous. References: Shenefelt and Marsh (1976), Marsh (1993). Note: *Austrodolops* Blanchard, previously included in Helconinae, is probably a junior synonym of *Allorhogas*, but the holotype needs to be studied before correct placement can be made.

61(59). Wings with infuscate bands (Fig. 26) .... ***Callihormius*** Ashmead
Distribution: Nearctic, Neotropical; 7 described species. Biology: parasitoids of Buprestidae and Cerambycidae. References: Marsh (1966b), Shenefelt and Marsh (1976).

- Wings hyaline or occasionally weakly infuscate throughout 62

62(61). Propleuron with longitudinal carinae on each side, medially concave .... ***Whartonius*** Marsh
Distribution: Mexico, Guatemala, Colombia, Brazil; 2 described species. Biology: parasitoids of Curculionidae. Reference: Marsh (1993).

- Propleuron rounded and without longitudinal carinae 63

63(62). Metasomal tergum 3 with heavily sculptured transverse groove (similar to groove on tergum 3 in Fig. 36), metasomal terga beyond 2 smooth and shining .... ***Sericobracon*** Shaw
Distribution: Trinidad; 2 described species. Biology: parasitoids of Embiidina (webspinners). Reference: Shaw and Edgerly (1985).

- Tergum 3 without transverse groove, all terga sculptured, rugose or coriaceous 64

64(63). Metasomal tergum 2 with median semi-circular rugose area at base, tergum 3 coriaceous (Fig. 45). Wings: Fig. 33 .... ***Leluthia*** Cameron (in part)
Distribution: Nearctic, Neotropical; 3 described species plus many undescribed. Biology: parasitoids of Buprestidae. References: Marsh (1967b), Shenefelt and Marsh (1976).

- Tergum 2 without semi-circular area at base differentiated from remainder of tergum 2 and tergum 3 (Fig. 57) .... ***Glyptocolastes*** Ashmead
Distribution: Nearctic; 2 described species but several undescribed from Neotropical Region. Biology: parasitoids of Bruchidae and Bostrichidae. References: Marsh (1968, 1993).

65(45). Fore wing m-cu arising distad 2RS, (RS+M)b absent (Fig. 27) 66

- Fore wing m-cu arising basad or directly in line with 2RS, (RS+M)b often present (Fig. 28) 69

66(65). Fore and middle femora with blister like swelling anteriorly near base (Fig. 69). Wings: Fig. 27 ....
.... ***Rhaconotus*** Ruthe
Distribution: Cosmopolitan; about 75 described species. Biology: parasitoids of stem boring beetle and lepidopterous larvae. References: Marsh (1976, key to Nearctic species), Shenefelt and Marsh (1976).

- Fore and middle femora without such swellings 67

67(66). Petiole with lateral wing-like projections at base (Fig. 48) .... ***Ptesimogaster*** Marsh
Distribution: Nearctic, Neotropical (Caribbean); 3 described species. Biology: unknown. References: Marsh (1965), Matthews and Marsh (1969), Shenefelt and Marsh (1976).

- Petiole without such projections at base 68

68(67). First subdiscal cell of fore wing dosed at apex, 2cu-a present and distinctly meeting 2-1A (Fig. 29) ....
.... ***Spathius*** Nees
Distribution: Cosmopolitan but rare in Neotropical Region; several hundred described species mostly from Old World tropics. Biology: parasitoids of various wood boring beetle larvae. References: Matthews (1970, key to Nearctic species), Shenefelt and Marsh (1976).

- First subdiscal cell open at apex, 2cu-a absent .... ***Donquickeia*** Marsh
Distribution: Brazil; 2 described species. Biology: unknown. Reference: Marsh (1993). Replacement name for *Quickia* Marsh (Marsh, 1997).

69(65). First subdiscal cell of fore wing open at apex, 2cu-a absent, occasionally an infuscated spot or short line present between 2-1A and 2CUa but no distinct vein present (Fig. 28) 70

\- First subdiscal cell dosed at apex, 2cu-a present and distinctly meeting 2-1A (Fig. 30) 75

70(69). Most of mesosoma smooth and shining, mesonotum occasionally coriaceous, propodeum occasionally rugose 71

\- Most of mesosoma sculptured, rugose or coriaceous, at most mesopleuron smooth below sternaulus..73

71(70). Cyclostome opening wide and oval, nearly as wide as eye height (Fig. 70) ....***Semirhytus*** Szépligeti

Distribution: Neotropical; 2 described species plus several undescribed. Biology: unknown. Reference: Shenefelt and Marsh (1976).

\- Cyclostome opening circular, at most as wide as half height of eye 72

72(71). Scape notched at apex. Clypeus not bordered dorsally by carina. Wings with infuscate bands; hind wing r-m longer than half length of 1M (Fig. 28) ....***Whitfieldiellus*** Marsh

Distribution: Mexico to Panama; one described species. Biology: unknown. Reference: Marsh (1993). Replacement name for *Whitfieldia* Marsh (Marsh, 1997).

\- Scape not notched at apex. Clypeus bordered dorsally by carina. Wings uniformly infuscate; hind wing r-m less than half length of 1M (Fig. 31) ....***Johnsonius*** Marsh

Distribution: Neotropical (Central America, northern South America); 3 described species. Biology: unknown. Reference: Marsh (1993).

73(70). Mesoscutum not sharply raised anteriorly, on nearly same plane as pronotum (Fig. 68); pronotum long and with transverse raised ridge behind head ....***Curtisella*** Spinola

Distribution: Nearctic, Neotropical; 6 described species plus several undescribed. Biology: unknown. Reference: Shenefelt and Marsh (1976).

\- Mesoscutum sharply raised anteriorly, at right angle with pronotum (as in Figs 44, 61); pronotum short and without raised ridge behind head...74

74(73). Metasomal tergum 2 with semi-circular or diamond-shaped rugose area at base (Fig. 45). Wings: Fig. 33.. ....***Leluthia*** Cameron (in part)

Distribution: Nearctic, Neotropical; 3 described species plus many undescribed. Biology: parasitoids of Buprestidae. References: Marsh (1967b), Shenefelt and Marsh (1976).

\- Tergum 2 usually heavily sculptured but without distinct semi-circular or diamond shaped area medially at base ....***Acrophasmus*** Enderlein

Distribution: Nearctic, Neotropical; 10 described species plus many undescribed. Biology: parasitoids of Lyctidae, Bostrichidae. References: Marsh (1968), Shenefelt and Marsh (1976).

75(69). Metasomal tergum 2 with converging grooves at base which enclose triangular or semi-circular area (Fig. 72)..76

\- Tergum 2 without such grooves at base, if triangular or semi-circular area present it is not enclosed by deep grooves (Fig. 73) 79

76(75). Petiole more than twice as long as apical width, usually 3 or more times longer than apical width. Wings: Fig. 30. .... ***Trigonophasmus*** Enderlein

Distribution: Neotropical; 8 very large, described species plus a few undescribed. Biology: unknown. Reference: Shenefelt and Marsh (1976).

\- Petiole at most twice as long as apical width 77

77(76). Petiole with two dorsal carinae extending to apex of tergum...............................***Hybodoryctes*** Szépligeti

Distribution: Indo-Australian, Neotropical; 2 described species. Biology unknown. Reference: Shenefelt and Marsh (1976).

\- Petiole at most with dorsal carinae present at base but not extending beyond basal fourth of tergum ..78

78(77). First flagellomere longer than scape and pedicel combined ....***Pedinotus*** Szépligeti

Distribution: Indo-Australian, Nearctic, Neotropical; 4 described species plus several undescribed. Biology: parasitoids of Scolytidae. Reference: Shenefelt and Marsh (1976).

\- First flagellomere equal to or shorter than combined length of scape and pedicel.....***Osmophila*** Szépligeti

Distribution: Neotropical; 3 described species. Biology: unknown, Reference: Shenefelt and Marsh (1976). *Syngaster*, recently introduced to southern California, keys here though the hind coxa is not strongly tuberculate.

79(75). Frons excavated (as in Fig. 47); distinctly raised ridge present between antennae. Petiole as wide at apex as base ....***Acanthorhogas*** Szépligeti

Distribution: Neotropical; one described species plus a few undescribed. Biology: unknown. Reference: Shenefelt and Marsh (1976).

\- Frons not excavated; without raised ridge between antennae. Petiole wider at apex than at base 80

80(79). Wings uniformly hyaline..***Tripteria*** Enderlein

Distribution: Brazil; one described species. Biology: unknown. Reference: Shenefelt and Marsh (1976).

\- Wings with infuscate bands or otherwise patterned with dark and pale spots 81

81(80). Sculpture at base of metasomal tergum 2 alveolate (Fig. 73). Petiole only slightly longer than apical width. Wings: Fig. 32 ....***Atopodoryctes*** Marsh

Distribution: Ethiopian, Nearctic, Neotropical (Caribbean); one described species. Biology: unknown. Reference: Marsh (1993).

\- Sculpture at base of tergum 2 costate (as in Fig. 50). Petiole nearly twice as long as apical width ....

.... ***Cyphodoryctes*** Marsh

Distribution: Brazil; one described species. Biology: unknown. Reference: Marsh (1993). Replacement name for *Cyrtonion* Marsh (Marsh, 1997).

82(4). Fore wing 2RS absent, or at least not tubular, thus first and second submarginal cells confluent (Fig. 9) .... ***Heterospilus*** Haliday

Distribution: Cosmopolitan but most species in New World; estimated 200 Nearctic and 300 Neotropical species. Biology: parasitoids mostly of wood boring beetle larvae but some on stem boring sawflies and moths. References: Shenefelt and Marsh (1976), Marsh (1982a, b), Quicke and Marsh (1992).

\- Fore wing 2RS present, first and second submarginal cells separated 83

83(82). Fore wing r-m absent (as in Fig. 6) 84

\- Fore wing r-m present 87

84(83). Hind femur swollen, at most 2.5 times as long as apical width (as in Fig. 49) 85

\- Hind femur not swollen, at least 3 times as long as apical width 86

85(84). Cyclostome opening large and broad, head somewhat prognathous (Figs 37, 38). Wings: Fig. 6 .... ***Verae*** Marsh

Distribution: Brazil; one described species. Biology: unknown. Reference: Marsh (1993).

\- Cyclostome opening smaller and round, head not prognathous .... ***Hecabolus*** Curtis

Distribution: Palaearctic, Indo-Australian, Nearctic (U.S.A.), Neotropical (Argentina); 5 described species, 2 of these found in the New World. Biology: parasitoids of Scolytidae, Bostrichidae. Reference: Shenefelt and Marsh (1976).

86(84). Antenna with less than 18 flagellomeres .... ***Coiba*** Marsh

Distribution: Central America, Caribbean; 3 described species. Biology: un- known. Reference: Marsh (1993).

\- Antenna with more than 18 flagellomeres .... ***Caingangia*** Marsh

Distribution: Brazil; one described species. Biology: unknown. Reference: Marsh (1993).

87(83). Head with rugose or scale-like swellings on each side of frons (Fig. 42) .... ***Dendrosoter*** Wesmael

Distribution: Cosmopolitan; 6 of the 25 described species found in the New World, but thus far recorded only from U.S.A. Biology: parasitoids of wood boring beetles, especially Scolytidae. References: Marsh (1967a), Shenefelt and Marsh (1976).

\- Head without such swellings 88

88(87). Petiole more than twice as long as apical width 89

\- Petiole less than twice as long as apical width 90

89(88). Mesoscutum not sharply raised anteriorly, on nearly same plane as pronotum (Fig. 68); pronotum long and with transverse raised ridge behind head .... ***Curtisella*** Spinola

Distribution: Nearctic, Neotropical; 6 described species plus several undescribed. Biology: unknown. Reference: Shenefelt and Marsh (1976).

\- Mesoscutum sharply raised anteriorly, at right angle with pronotum (as in Figs 44, 61); pronotum short and without raised ridge behind head .... ***Acrophasmus*** Enderlein

Distribution: Nearctic, Neotropical; 10 described species plus many undescribed. Biology: parasitoids of Lyctidae, Bostrichidae. References: Marsh (1968), Shenefelt and Marsh (1976).

90(88). First subdiscal cell of fore wing closed, 2cu-a present and meeting 2-1A .... ***Araucania*** Marsh

Distribution: Ecuador, Peru, Chile, Argentina; 2 described species. Biology: unknown. Reference: Marsh (1993).

\- First subdiscal cell open at apex, 2cu-a absent 91

91(90). Fore wings with infuscate bands (Fig. 26) .... ***Callihormius*** Ashmead

Distribution: Nearctic, Neotropical; 7 described species. Biology: parasitoids of Buprestidae and Cerambycidae. References: Marsh (1966b), Shenefelt and Marsh (1976).

\- Fore wings not banded 92

92(91). Propleuron with longitudinal carinae on each side, medially concave .... ***Whartonius*** Marsh

Distribution: Mexico, Guatemala, Colombia, Brazil; 2 described species. Biology: parasitoids of Curculionidae. Reference: Marsh (1993).

\- Propleuron rounded and without longitudinal carinae 93

93(92). Metasomal tergum 2 with median semi-circular rugose area at base delimited by divergent grooves laterally and transverse sinuate groove apically (Fig. 45) .... ***Leluthia*** Cameron

Distribution: Nearctic, Neotropical; 3 described species plus many undescribed. Biology: parasitoids of Buprestidae. References: Marsh (1967b), Shenefelt and Marsh (1976).

\- Tergum 2 without clearly delimited semi-circular area at base, but with divergent grooves present laterally (Fig. 57) .... ***Glyptocolastes*** Ashmead

Distribution: Nearctic; 2 described species but several undescribed from Neotropical Region. Biology: parasitoids of Bruchidae and Bostrichidae. References: Marsh (1968, 1993).

REFERENCES

Belokobyl'skij, S. A. 1982. Braconids of the genera *Exontsira* gen. n., *Ontsira* Cameron and *Hypodoryctes* Kokujev (Hymenoptera, Braconidae) of the Asiatic part of the USSR. Entomologicheskoe Obozrenie 61: 600-613. (In Russian.)

Belokobyl'skij, S. A. 1987. On the status of the genus *Dolopsidea* Hincks (Hymenoptera, Braconidae, Doryctinae). Pp. 73-78. In: Storozheva and Kusakin. (Eds.). New Data on Insect Systematics from the Soviet Far East. Academy of Sciences of the USSR, Vladivostok. (In Russian.)

Belokobyl'skij, S. A. 1992. On the classification and phylogeny of the braconid wasp subfamilies Doryctinae and Exothecinae (Hymenoptera, Braconidae). Part I. On the classification, 1. Entomologicheskoe Obozrenie 71: 900-928. (In Russian.)

Belokobyl'skij. S. A. 1995. Main evolutionary transformations of the morphological structures in the subfamilies Doryctinae and Exothecinae (Hymenoptera, Braconidae). Entomologicheskoe Obozrenie 74: 153-176. (In Russian.)

Fischer, M. 1981a. Versuch einer systematischen Gliederung der Doryctinae, insbesondere der Doryctini, und Redeskription nach Material aus den Naturwissenschaftlichen Museum in Budapest (Hymenoptera, Braconidae). Polskie Pismo Entomologiczne 51: 41-99.

Fischer, M. 1981b. *Mononeuron*, ein neues Doryctinen-genus aus Brasilien (Hymenoptera. Braconidae). Zeitschrift der Arbeitsgemeinschaft Österreichischer Entomologen 33: 47-51.

Hedqvist, K.-J. 1963. Notes on Hormiinae with descriptions of new genera and species (Hym., Ichneumonoidea, Braconidae). Entomologisk Tidskrift 84: 30-61.

Kieffer, J. J. 1910. In: Kieffer, J. J. and Jorgensen, P., Gallen und Gallentiere aus Argentina. Central blatt für Bakteriologie, Parasitenkunde und Infektionskrankheiten. 27: 362-442.

Macedo, M. V. de and Monteiro, R. T. 1989. Seed predation by a braconid wasp, *Allorhogas* sp. (Hymenoptera). Journal of the New York Entomological Society 97: 358-362.

Marsh, P. M. 1965. The Nearctic Doryctinae, I. A review of the subfamily with a taxonomic revision of the tribe Hecabolini (Hymenoptera: Braconidae). Annals of the Entomological Society of America 58: 668-699.

Marsh, P. M. 1966a. The Nearctic Doryctinae, II. The genus *Doryctodes* Hellén (Hymenoptera: Braconidae). Transactions of the American Entomological Society 92: 503-517.

Marsh, P. M. 1966b. The Nearctic Doryctinae, III. The genus *Callihormius* Ashmead (Hymenoptera: Braconidae). Proceedings of the Entomological Society of Washington 68: 240-246.

Marsh, P. M. 1967a. The Nearctic Doryctinae, IV. Notes on the genus *Dendrosoter* Wesmael and description of one new species (Hymenoptera: Braconidae). Proceedings of the Entomological Society of Washington 69: 246-247.

Marsh, P. M. 1967b. The Nearctic Doryctinae, V. The genus *Leluthia* Cameron and comments on the status of the tribe Hecabolini (Hymenoptera: Braconidae). Proceedings of the Entomological Society of Washington 69: 359-364.

Marsh, P. M. 1968. The Nearctic Doryctinae, VI. The genera *Acrophasmus, Glyptocolastes, Doryctinus* and a new genus, *Stenocorse* (Hymenoptera: Braconidae). Proceedings of the Entomological Society of Washington 70: 101-113.

Marsh, P. M. 1969a. The Nearctic Doryctinae, VII. The genus *Doryctes* Haliday (Hymenoptera: Braconidae), Transactions of the American Entomological Society 94: 379-405.

Marsh, P. M. 1969b. Lectotype designations in the genus *Heterospilus* (Hymenoptera: Braconidae). Proceedings of the Entomological Society of Washington 71: 210-214.

Marsh, P. M. 1970a. The Nearctic Doryctinae, VIII. The genera *Liobracon* and *Pedinotus* with notes on the definition of the subfamily (Hymenoptera: Braconidae). Proceedings of the Entomological Society of Washington 72: 313-317.

Marsh, P. M. 1970b. The Nearctic Doryctinae, IX. The genus *Odontobracon* and notes on related genera (Hymenoptera: Braconidae). Pan-Pacific Entomologist 46: 276-283.

Marsh, P. M. 1973. New synonyms and new combinations in North American Doryctinae (Hymenoptera: Braconidae). Journal of the Washington Academy of Sciences 63: 69-72.

Marsh, P. M. 1976. The Nearctic Doryctinae, X. The genus; *Rhaconotus* Ruthe (Hymenoptera: Braconidae). Proceedings of the Entomological Society of Washington 78: 389-403.

Marsh, P. M. 1982a. Descriptions of two new species of *Heterospilus* (Hymenoptera: Braconidae). Boletin de Entomologia Venezolana 2: 57-60.

Marsh, P. M. 1982b. Two new species of *Heterospilus* (Hymenoptera: Braconidae) from Mexico being introduced against the cotton boll weevil, *Anthonomus grandis* (Coleoptera: Curculionidae). Proceedings of the Entomological Society of Washington 84: 849-854.

Marsh, P. M. 1983a. A taxonomic study of the South American genus *Megaloproctus* Schulz (Hymenoptera: Braconidae: Doryctinae). Contributions of the American Entomological Institute 20: 363-380.

Marsh, P. M. 1983b. *Bohartiellus*, a new genus of Doryctinae from South America (Hymenoptera: Braconidae). Pan-Pacific Entomologist 59: 138-151.

Marsh, P. M. 1984. A new species of Braconidae (Hymenoptera) from Mexico introduced into Texas to control a sugar cane borer, *Eoreuma loftini* (Lepidoptera: Pyralidae). (Hymenoptera: Braconidae), Proceedings of the Entomological Society of Washington 86: 861-863.

Marsh, P. M. 1988. Revision of the tribe Odontobraconini in the Western Hemisphere (Hymenoptera: Braconidae-Doryctinae). Systematic Entomology 13: 443-464.

Marsh, P. M. 1989. Notes on Braconidae (Hymenoptera) associated with jojoba (*Simmondsia chinensis*) and descriptions of new species. Pan-Pacific Entomologist 65: 58-67.

Marsh, P. M. 1991. Description of a phytophagous doryctine braconid from Brazil (Hymenoptera: Braconidae). Proceedings of the Entomological Society of Washington 91: 92-95.

Marsh, P. M. 1993. Descriptions of new Western Hemisphere genera of the subfamily Doryctinae (Hymenoptera: Braconidae). Contributions of the American Entomological Institute 28(1): 1-58.

Marsh, P. M. 1997. Replacement names for Western Hemisphere genera of Doryctinae (Hymenoptera: Braconidae), Proceedings of the Entomological Society of Washington 99: 586.

Matthews, R. W. 1968. Nesting biology of the social wasp *Microstigmus comes* (Hymenoptera: Sphecidae, Pemphredoninae). Psyche 75: 23-45,

Matthews, R. W. 1970. A revision of the genus *Spathius* in America north of Mexico (Hymenoptera, Braconidae). Contributions of the American Entomological Institute 4(5): 1-86.

Matthews, R. W. and Marsh, P. M. 1969. On the status of Cresson's *Stenophasmus* species from Cuba (Hymenoptera: Braconidae). Proceedings of the Entomological Society of Washington 71: 94-98.

Matthews, R. W. and Marsh, P. M. 1973. *Notiospathius*, a new Neotropical genus (Hymenoptera: Braconidae). Journal of the Washington Academy of Sciences 63: 73-75.

Muesebeck, C. F. W. 1958. New Neotropical wasps of the family Braconidae (Hymenoptera) in the U. S. National Museum. Proceedings of the United States National Museum 107: 405-461.

Quicke, D. L. J., Fisher, L. C. and Fitton, M. G. 1992. New diagnostic ovipositor characters for doryctine wasps (Hymenoptera, Braconidae). Journal of Natural History 26: 1035-1046.

Quicke, D. L. J. and Marsh, P. M. 1992. Two new species of Neotropical parasitic wasps with highly modified ovipositors (Hymenoptera: Braconidae: Braconinae and Doryctinae). Proceedings of the Entomological Society of Washington 94: 559-567.

Ramirez B., W. and Marsh, P. M. 1996. A review of the genus *Psenobolus* (Hymenoptera: Braconidae) from Costa Rica, an inquiline fig wasp with brachypterous males, with descriptions of two new species. Journal of Hymenoptera Research 5: 64-72.

Shaw, S. R. and Edgerly, J. S. 1985. A new braconid genus (Hymenoptera) parasitizing webspinners (Embiidina) in Trinidad. Psyche 92: 505-511.

Shenefelt, R. D. and Marsh, P. M. 1976. Pars 13. Braconidae 9, Doryctinae. Pp. 1263-1424. In: van der Vecht, J. and Shenefelt, R. D. (Eds.). Hymenopterorum Catalogus (nova edito). Dr. W. Junk, The Hague.

Wharton, R. A. 1993. Review of the Hormiini (Hymenoptera: Braconidae) with a description of new taxa. Journal of Natural History 27: 107-171.

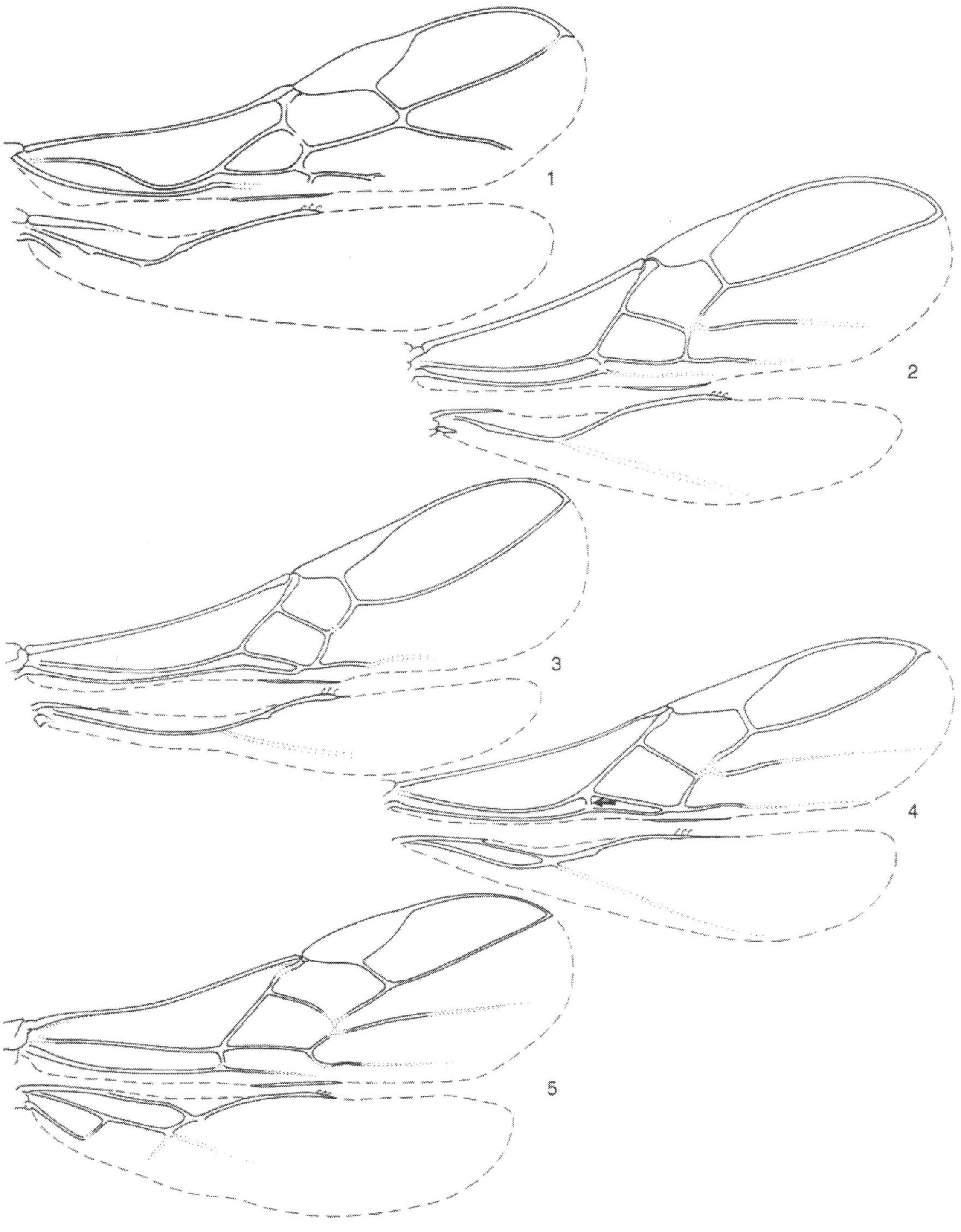

Fig. 1-5: Wings of Doryctinae. 1, *Percnobracon secundus* Muesebeck; 2, *Micrommatus brevicornis* Marsh; 3, *Ecphylus* sp.; 4, *Bohartiellus* sp., arrow = 1cu-a; 5, *Monolexis fuscicornis* Förster.

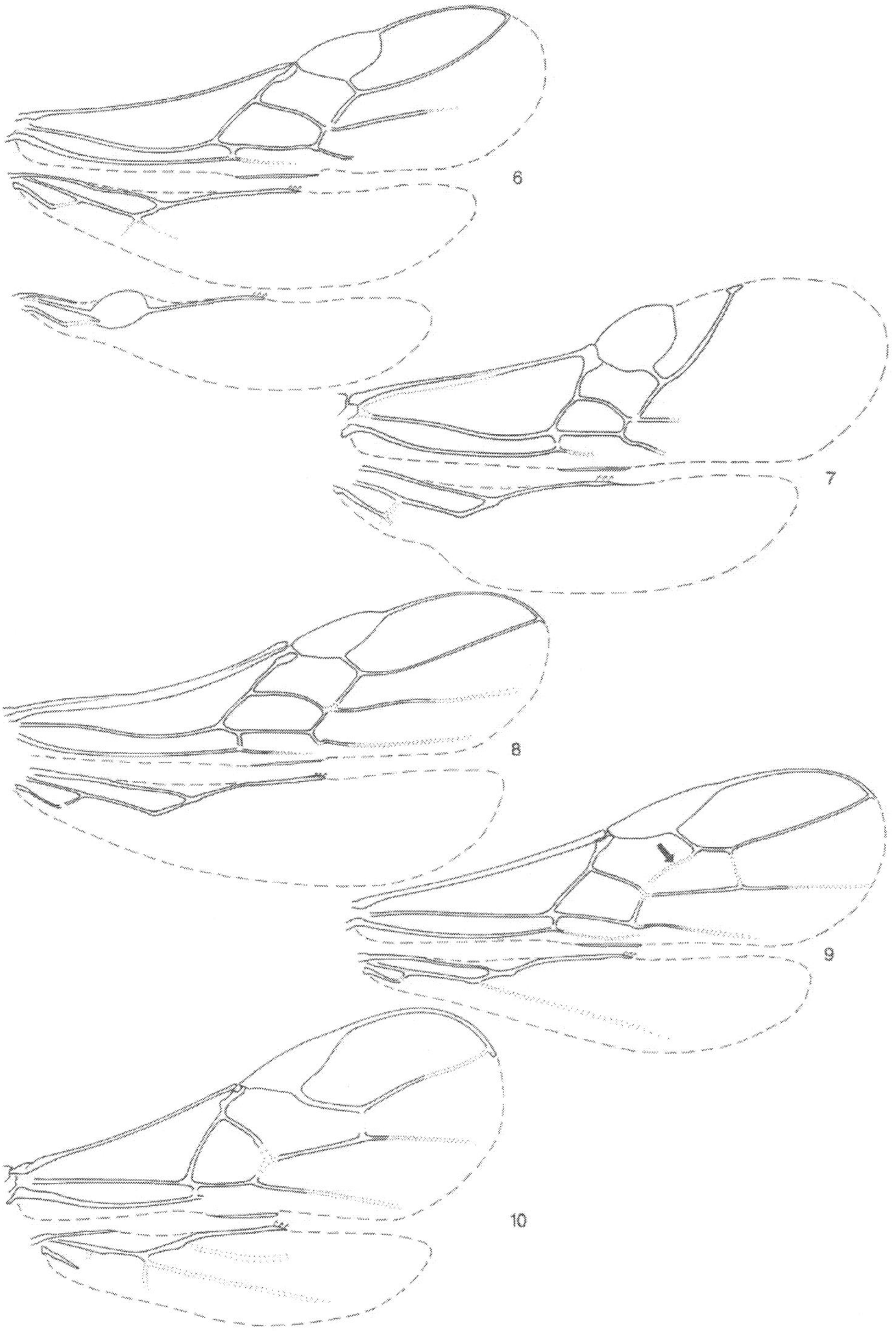

Figs 6-10: Wings of Doryctinae. 6, *Verae pecula* Marsh showing difference in hind wing between male (with stigma) and female; 7, *Pamolidea barberi* Marsh; 8, *Hecabolus*. sp.; 9, *Heterospilus* sp., arrow = weak 2RS; 10, *Labania* sp.

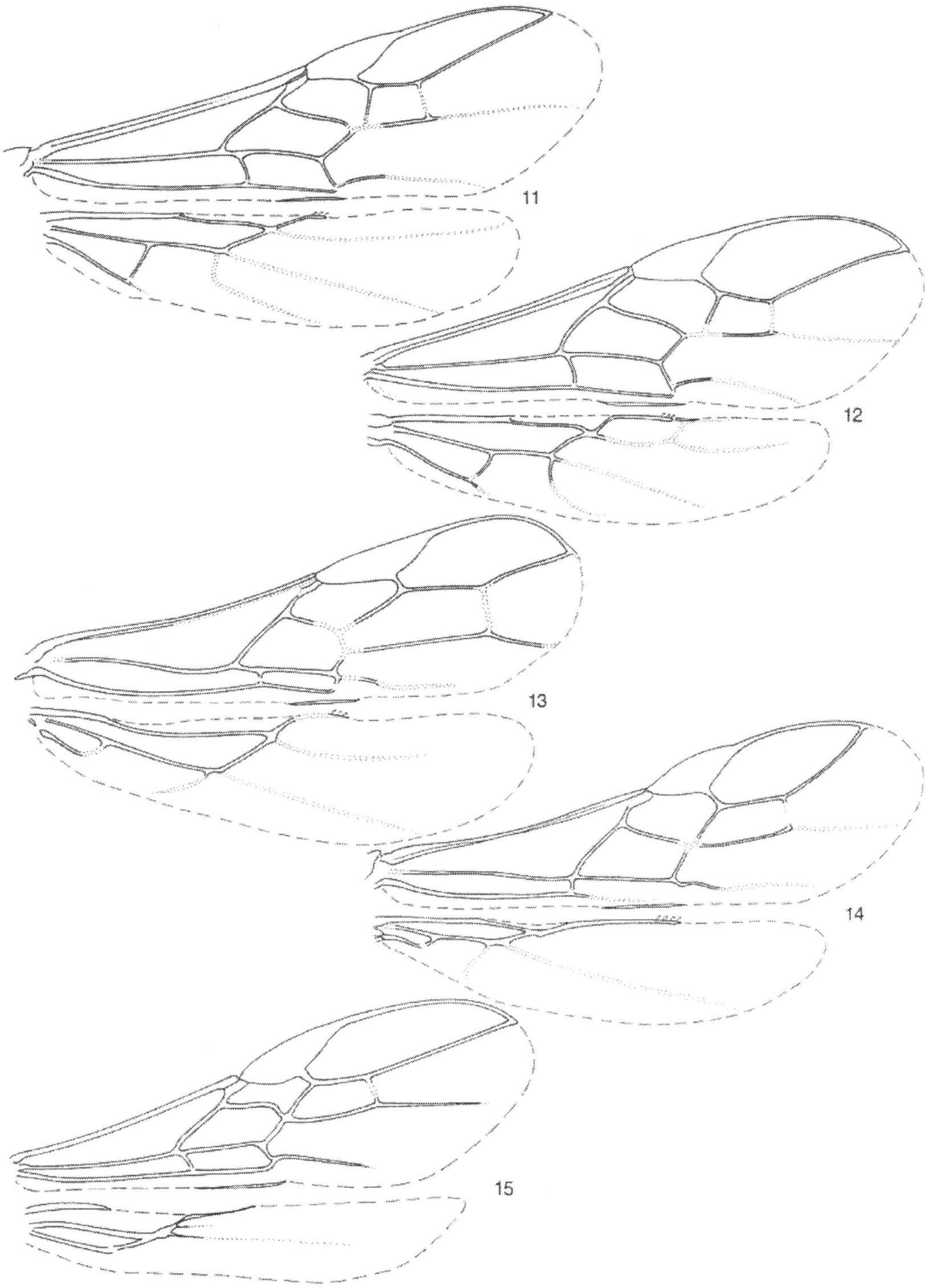

Figs 11-15: Wings of Doryctinae. 11, *Odontobracon* sp.; 12, *Nervellius* sp.; 13, *Dendrosoter* sp.; 14, *Histeromeroides onkoierebrus* Marsh; 15, *Pariodes* sp.

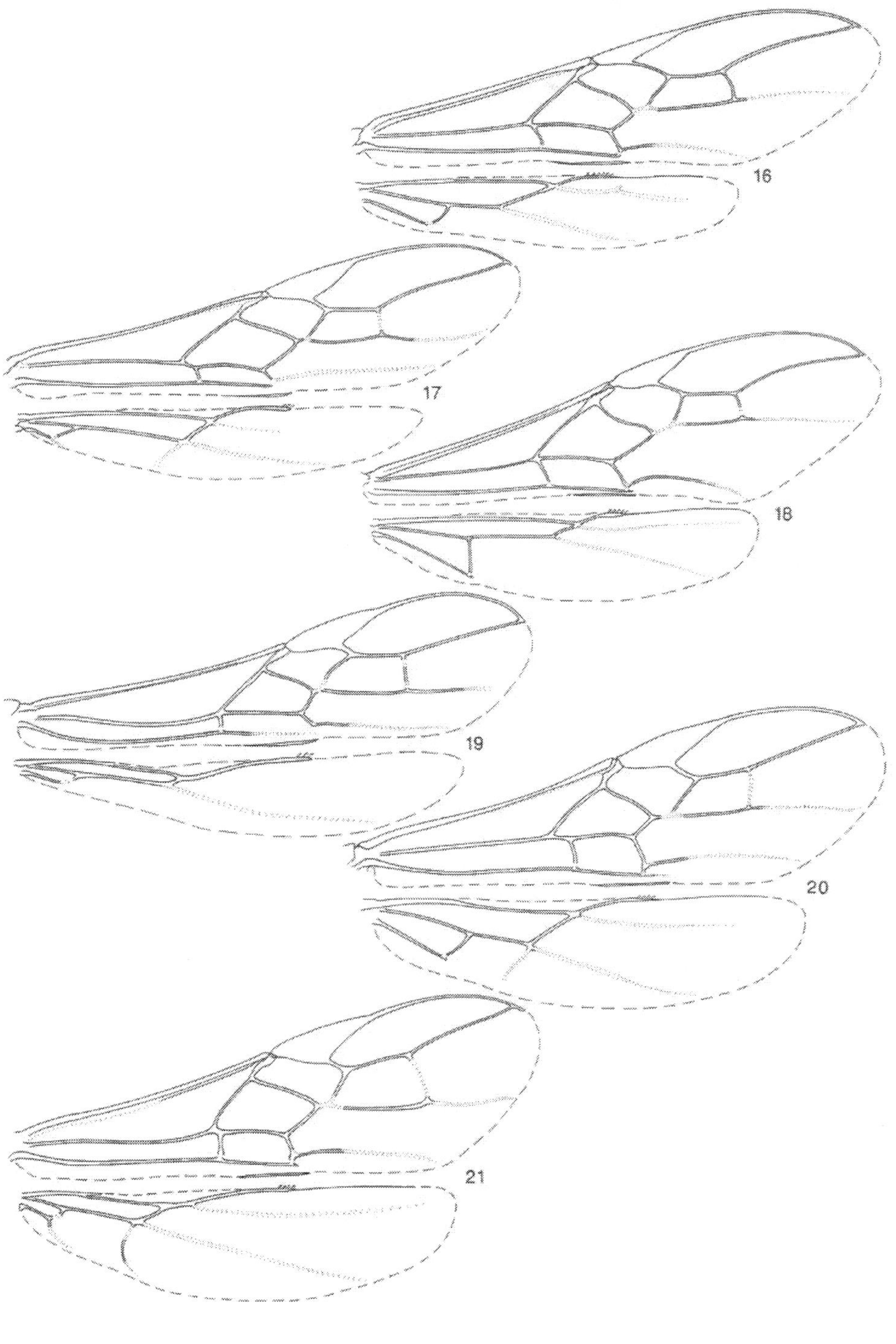

Figs 16-21: Wings of Doryctinae. 16, *Binarea* sp.; 17, *Rhoptroceniroides platyfemur* Marsh; 18, *Monarea* sp.; 19, *Notiospathius* sp.; 20, *Doryctes* sp.; 21, *Araucania maculipennis* Marsh.

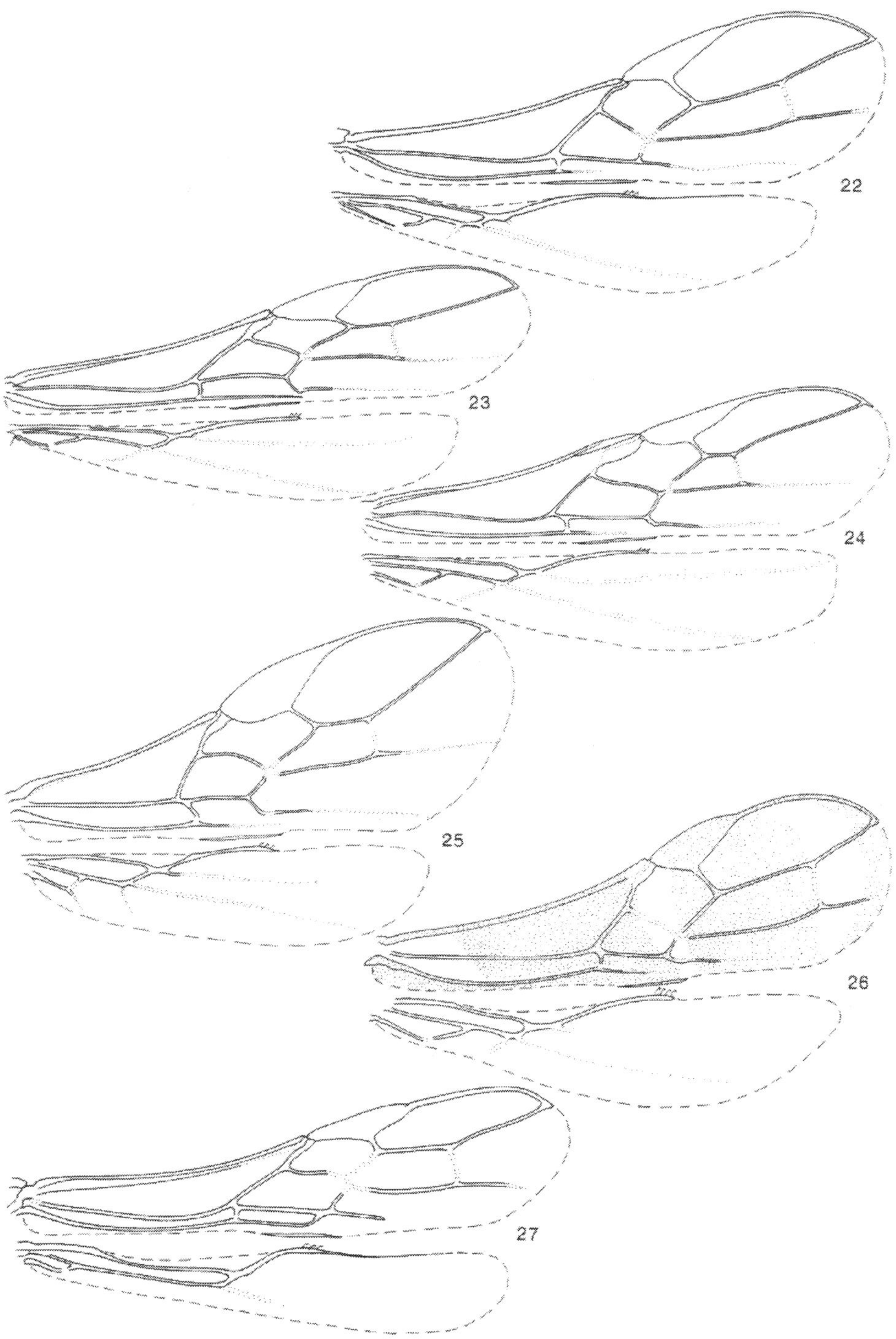

Figs 22-27: Wings of Doryctinae. 22, *Aphelopsia annulicornis* Marsh; 23, *Waitaca flavostigma* Marsh; 24, *Shawius braziliensis* Marsh; 25, *Allorhogas* sp.; 26, *Callihormius* sp.; 27, *Rhaconotus* sp.

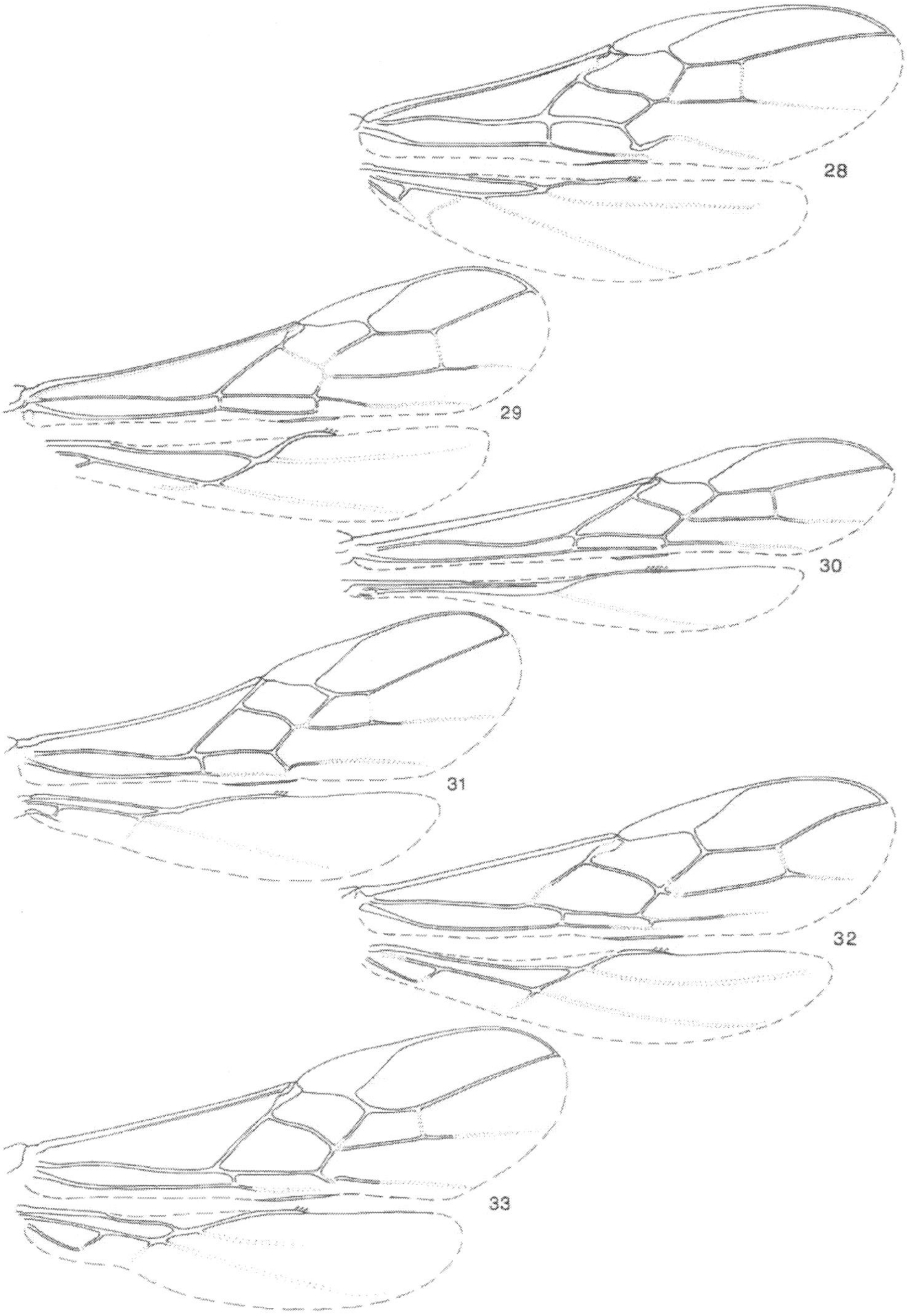

Figs 28-33: Wings of Doryctinae. 28, *Whitfieldiellus variegata* (Marsh); 29, *Spathius* sp.; 30, *Trigonophasmus* sp.; 31, *Johnsonius* sp.; 32, *Atopodoryctes carbonarius* Marsh; 33, *Leluthia* sp.

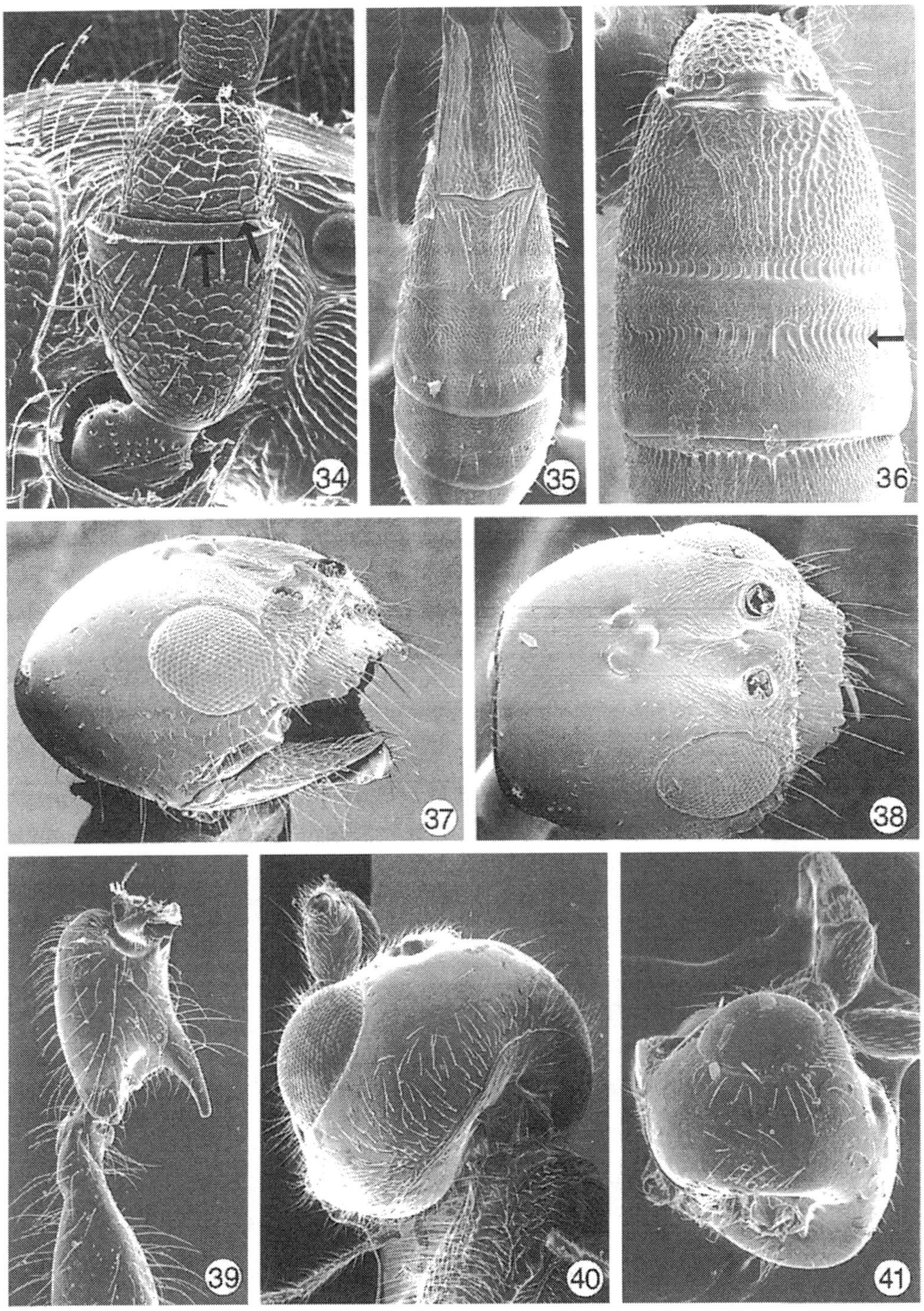

Figs 34-41: Doryctinae. 34, scape of *Achterbergia arawak* Marsh, arrows = double rim at apex; 35, metasoma of *Pioscelus wichitus* (Viereck); 36, metasomal terga 1-3 of *Caingangia flavokolos* Marsh, dorsal view, arrow = tranverse groove on tergum 3; 37, head of *Verae pecula* Marsh, lateral view; 38, head of *Verae pecula* Marsh, dorsal view; 39, hind coxa of *Odontobracon* sp.; 40, head of *Doryctes* sp., postero-lateral view; 41, head of *Liobracon* sp., postero-lateral view. Figures 34 and 36-38 after Marsh (1993).

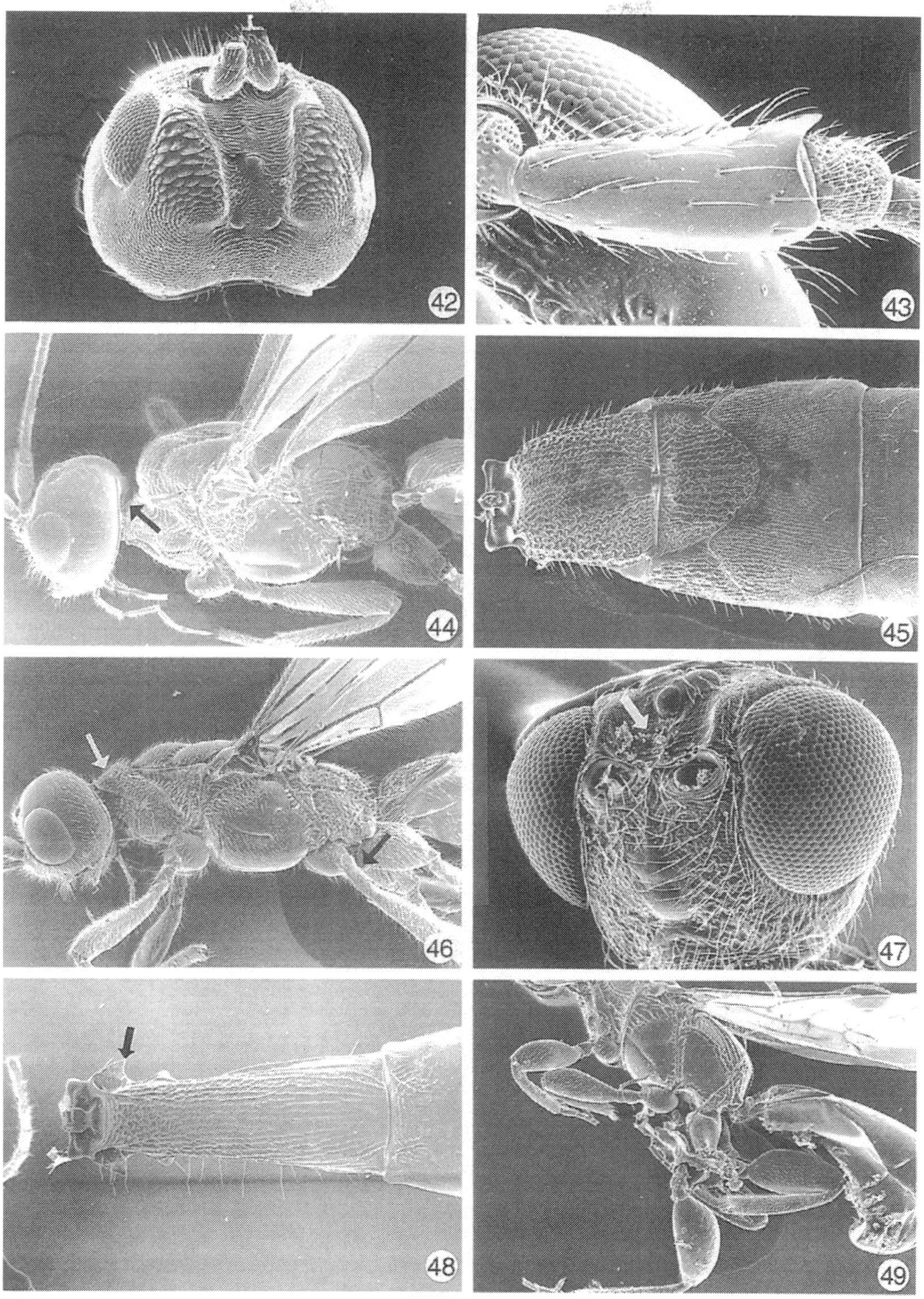

Figs 42-49: Doryctinae. 42, head of *Dendrosoter* sp., dorsal view; 43, scape of *Rhopirocentroides platyfemur* Marsh (after Marsh, 1993); 44, mesosoma of *Ontsira* sp., arrow = pronotum; 45, metasoma of *Leluthia* sp.; 46, mesosoma of *Doryctes* sp., arrows = pronotum and basal tubercle on hind coxa; 47, head of *Allorhogas* sp., arrow = excavated frons (after Marsh, 1993); 48, petiole of *Ptesimogaster parkeri* Marsh, arrow = wing-like projection at base; 49, femora of *Pambolidea barberi* Marsh.

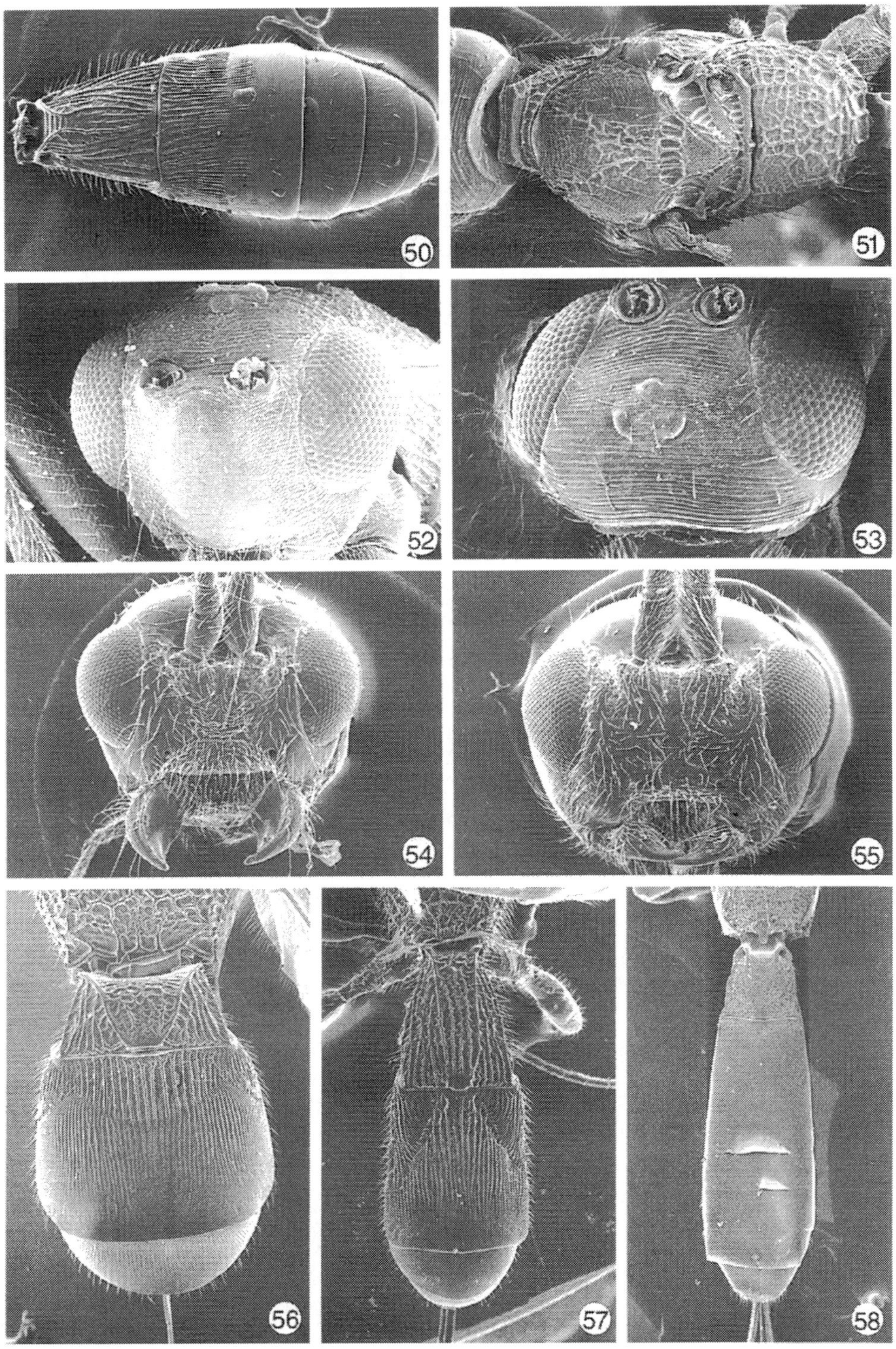

Figs 50-58: Doryctinae. 50, metasoma of *Heterospilus* sp., dorsal view; 51, mesosoma of *Masonius fasciatus* Marsh, dorsal view (after Marsh, 1993); 52, head of *Tarasco spathiformis* Marsh, frontal view (after Marsh, 1993); 53, head of *Masonius fasciaius* Marsh, dorsal view; 54, head of *Exontsira* sp., frontal view; 55, head of *Ontsira* sp., frontal view; 56-58, metasoma: 56, *Stenocorse bruchivorus* (Crawford), dorsal view; 57, *Glyptocolastes caryae* (Ashmead), dorsal view; 58, *Rhoptrocentrus piceus* Marshall, dorsal view.

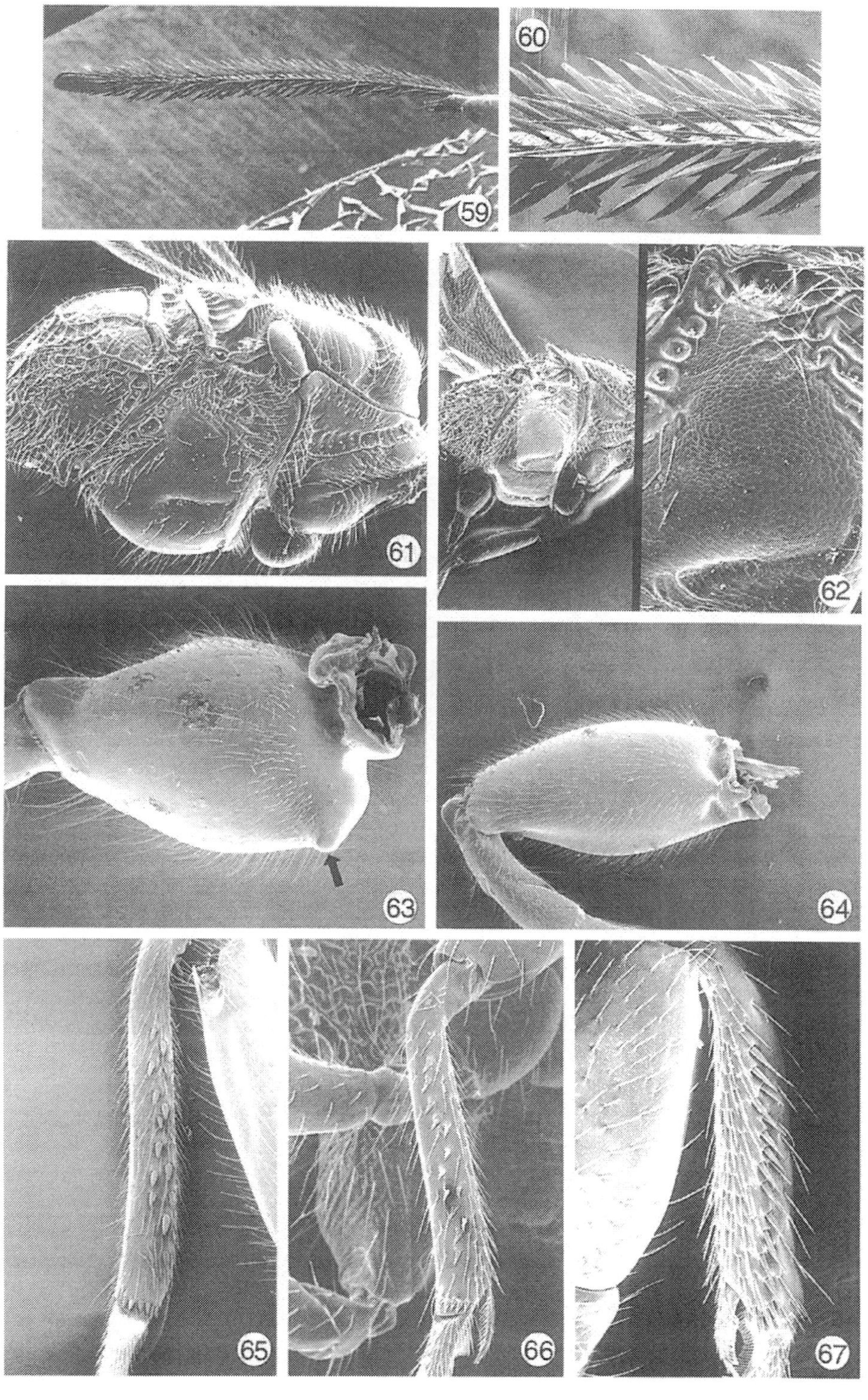

Figs 59-67: Doryctinae. 59, ovipositor of *Sharkeyella pilosus* Marsh; 60, detail of ovipositor of *Sharkeyella pilosus* Marsh; 61, mesosoma of *Parallorhogas* sp., lateral view; 62, mesosoma of *Allorhogas* sp., lateral view, detail showing coriaceous sculpture; 63, hind coxa of *Doryctes* sp., arrow = antero-ventral tubercle; 64, hind coxa of *Megaloproctus* sp.; 65-07, fore tibiae: 65, *Liobracon* sp.; 66, *Heterospilus* sp.; 67, *Rhopirocentroides platyfemur* Marsh. Figures 59-62 after Marsh (1993).

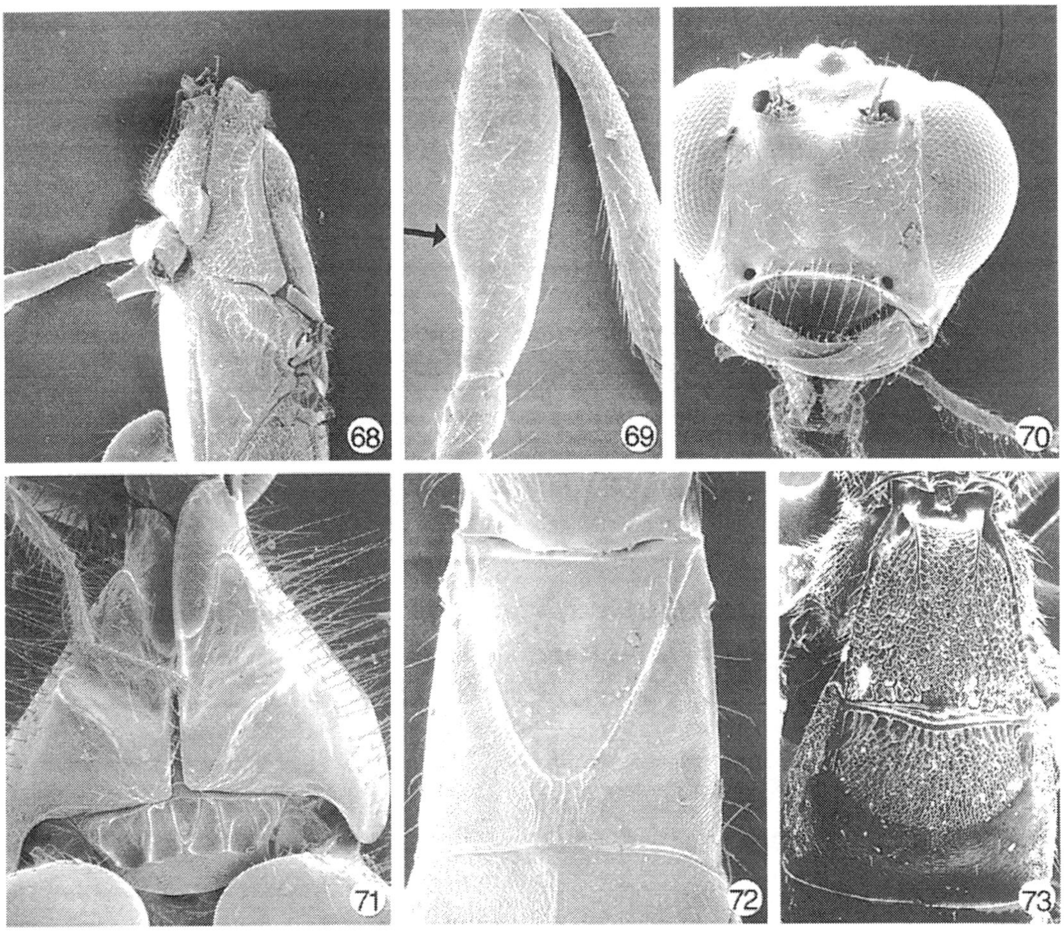

Figs 68-73: Doryctinae. 68, mesosoma of *Curtisella* sp., lateral view; 69, femur of *Rhaconotus* sp., arrow = blister-like swelling; 70, face of *Semirhuius* sp.; 71, propleuron of *Binarea* sp., ventral view; 72, metasomal tergum 2 of *Trigonophasmus* sp.; 73, metasoma of *Atopodoryctes* sp. (after Marsh, 1993).

DORYCTINAE UPDATE

A. Zaldívar-Riverón
Colección Nacional de Insectos, Instituto de Biología, Universidad Nacional Autónoma de México, Ciudad de México, azaldivar@ib.unam.mx

Doryctinae is one of the largest subfamilies of Braconidae, being particularly species-rich in the New World, where the majority of its genera occur. Since the chapter of the manual was published, a total of 34 additional doryctine genera have been described for the New World. In Marsh's (2002) monograph of the Doryctinae of Costa Rica (excluding *Heterospilus*), the author provided a key to doryctine genera from this country, where he included 63 of the currently recognized doryctine genera for the New World.

The most relevant taxonomic changes made at the generic level within the subfamily are detailed below. *Monitoriella* Hedqvist was transferred to the Doryctinae based on molecular and morphological phylogenetic evidence (Zaldívar-Riverón *et al.*, 2006). Ypsistocerini, with its three valid genera (*Ypsistocerus* Cushman, *Termitobracon* Brues and *Embobracon* van Achterberg), is now considered as a tribe of Doryctinae based on two synapomorphies of the venom gland apparatus (Quicke *et al.*, 1992). The recently described, monotypic *Preembobracon zaldivarriveroni* Quicke & Butcher was proposed as belonging to Ypsistocerini (Quicke and Butcher, 2015).

The following synonyms were proposed: *Achterbergia* Marsh= *Aivalykus* Nixon, *Hansonorum* Marsh= *Notiospathius* Matthews & Marsh (De Jesús-Bonilla *et al.*, 2011), *Amazondoryctes* Barbalho & Penteado-Dias= *Pioscelus* Muesebeck and Walkley (Belokobylskij, 2006), *Acrophasmus* Enderlein= *Doryctinus* Roman (Kula and Marsh, 2011), *Ptesimogastroides* Braet & van Achterberg= *Ptesimogaster* Marsh (Zaldívar-Riverón *et al.*, 2012).

A molecular phylogenetic study revealed that *Notiospathius* was polyphyletic (Caccarelli and Zaldívar-Riverón, 2013). This work resulted in the subsequent recognition of two new genera, *Bolivar* (Zaldívar-Riverón *et al.*, 2012) and *Sergey* (Martínez *et al.*, 2016). *Caenophanes* Foerster was recorded for the first time in the New World by Belokobylskij *et al.* (2011), and here I record the occurrence of *Platyspathius* Viereck in the New World (Venezuela) for the first time.

Three comprehensive taxonomic revisions provided keys as follows: the Heterospilini from Costa Rica (Marsh *et al.*, 2013), the gall-associated genera (Zaldívar-Riverón *et al.*, 2014) and the *Pedinotus* genus-group (Samacá-Sáez *et al.*, 2016). Moreover, taxonomic revisions were published for some genera, including *Coiba* (Kula, 2009), the Nearctic species of *Spathius* Nixon (Marsh and Strazanac, 2009) and the Heterospilini from Costa Rica (Marsh *et al.*, 2013). Other regional taxonomic works are the species of *Iare* Barbalho & Penteado-Dias (Martínez *et al.*, 2010) from Mexico, *Allorhogas* Gahan (Martínez and Zaldívar-Riverón, 2012) from Mexico, *Notiospathius* from Mexico and Venezuela (López-Estrada *et al.*, 2012; Resendiz-Flores *et al.*, 2014), *Monitoriella* from Brazil (Shimbori *et al.*, 2011), species of *Hecabolus* Curtis from South America (Sormus De Castro *et al.*, 2013), and the Neotropical species of *Spathius* Nixon (Belokobylskij and Zaldívar-Riverón, 2014) and *Rhaconotus* Ruthe (Belokobylskij and Zaldívar-Riverón, 2015).

Annotated list of recent papers on New World Doryctinae

Barbalho SM, Penteado-Dias AM. 2002. A new genus and species of Doryctinae (Hymenoptera: Braconidae) with fused first and second metasomal tergites. *Entomological News* 113; 179-182.
Description of *Iare*.

Barbalho S M, Penteado-Dias AM, Marsh PM. 1999. Descriptions of new genera from Brazil in the tribes Heterospilini and Spathiini with similar wing venation (Hymenoptera: Braconidae). *Journal of Hymenoptera Research* 8: 139-153.
Description of the genera *Canchim, Jataiella, Heterospathius, Leptodoryctes, Spathospilus* and *Amazondoryctes* (= Pioscelus).

Barbalho SM, Penteado-Dias AM. 2000. *Platydoryctes*, a new Doryctinae genus from Brazil. *Entomological news* 11: 25-31.
Description of *Platydoryctes*.

Belokobylskij SA. 2006. *Neoheterospilus* gen. n., a new genus of the tribe Heterospilini (Hymenoptera: Braconidae, Doryctinae) with highly modified ovipositor and a worldwide distribution. *Insect Systematics and Evolution* 37: 149-178.
Synonymy of *Amazondorcytes* as a junior synonym of *Pioscelus*.

Belokobylskij SA, Samacá-Sáenz A, Zaldívar-Riverón A. 2015. *Mexiare gen. nov.*, a new Doryctinae genus (Hymenoptera: Braconidae) with fused first and second metasomal tergites. *Zootaxa* 3914: 122-130.
Description of *Mexiare*.

Belokobylskij SA, Ceccarelli FS, Zaldívar-Riverón A. 2011. *Kauriphanes* n. gen., a new genus of braconid parasitoid wasp (Hymenoptera: Braconidae: Doryctinae) from New Zealand. *Annales de la Société Entomologique de France* 47: 394-401.
First record of the genus for the New World.

Belokobylskij SA, Zaldívar-Riverón A. 2014. The genus *Spathius* (Hymenoptera, Braconidae, Doryctinae) in Mexico: occurrence of a highly diverse Old World taxon in the Neotropics. *Zookeys* 427: 59-73.
Description of two species of *Spathius* from Mexico

Belokobylskij SA, Zaldívar-Riverón A. 2015. The Neotropical species of the genus *Rhaconotus* Ruthe 1854 (Hymenoptera: Braconidae: Doryctinae). *Annales Zoologici*. 65: 249-268.
Description of new species of *Rhaconotus* from Mexico and Brazil. A key to the Neotropical species of the genus is provided.

Braet Y, Barbalho SM, van Achterberg, C. 2003. Description of four new genera and nine new species of Doryctinae (Hymenoptera: Braconidae) from French Guiana. *Zoologische Mededelingen* 77: 93-125.
Description of *Dapsilitas, Lamquetia, Neostaphius* and *Ondigus*.

Ceccarelli FS, Zaldívar-Riverón A. 2013. Broad Polyphyly and Historical Biogeography of the Neotropical Wasp Genus *Notiospathius* (Braconidae: Doryctinae). *Molecular Phylogenetics and Evolution* 69: 142-152.
Molecular phylogenetic study that revealed *Notiospathius* was polyphyletic.

De Jesús-Bonilla VS, Nunes JF, Penteado-Dias AM, Csösz S, Zaldívar-Riverón A. 2011. A new synonym of the Neotropical parasitoid wasp genus *Notiospathius* (Braconidae: Doryctinae), with redescription of two species and description of five new species from Brazil. *Zookeys* 122: 71-90.
Synonymy of *Hansonorum* with *Notiospathius*. Description and redescription of five and two species, respectively, *Notiospathius* from Brazil.

Gadelha S. de S, Nunes JF, De Oliveira ML. 2016. New genera, species and records of Doryctinae (Hymenoptera: Braconidae) from Brazil. *Zootaxa* 4083: 359-370.
Description of *Tuberatra* and *Parsteres*.

Gadelha S. de S., Nunes JF, Zaldívar-Riverón A, De Oliveira ML. 2016. *Venifurca* n. gen., a new genus of Doryctinae (Hymenoptera: Braconidae) from the Neotropics, and its phylogenetic placement in the subfamily. *Journal of Hymenoptera Research* 51: 91-100.
Description of *Venifurca*.

Gomes SAG, Penteado-Dias AM. 2006. Description of a new genus of Doryctinae wasps (Hymenoptera: Braconidae) from Brazil. *Zoologische Mededelingen* 80: 81-85.
Description of *Lianus*.

Kula RR. 2009. Review of the New World species of *Coiba* Marsh (Hymenoptera: Braconidae: Doryctinae), including descriptions of two new species, new distribution records, and a key to species. *Proceedings of the Entomological Society of Washington* 111: 183-198.
Taxonomic revision of the New World species of *Coiba*.

Kula RR, Marsh PM. 2011. Doryctinae (Hymenoptera: Braconidae) of Konza Prairie Excluding Species of Heterospilus Haliday. *Proceedings of the Entomological Society of Washington* 113: 451-491.
Proposal of *Acrophasmus* as a junior synonym of *Doryctinus, and nomenclatural changes for other genera*.

López-Estrada, Briceño GR, Smith MA, Nunes JF, Penteado-Dias AM, Ceccarelli FS, Clebsch H, Zaldívar-Riverón A. 2012. Seven new species of *Notiospathius* from northwest Venezuela. *Journal of Hymenoptera Research* 29: 37-61.
Description of seven species of *Notiospathius* from Venezuela.

Marsh PM. 2002. The Doryctinae of Costa Rica (excluding the genus *Heterospilus*). *Memoirs of the American Entomological Institute* 70: 1-319.
Taxonomic study of 62 currently recognized doryctine genera from Costa Rica (excluding *Heterospilus*).

Marsh PM, Strazanac JS. 2009. A taxonomic review of the genus *Spathius* Nees (Hymenoptera: Braconidae) in North America and comments on the biological control of the emerald ash borer (Coleoptera: Buprestidae). *Journal of Hymenoptera Research* 18: 80-112.
Taxonomic revision of the Nearctic species of *Spathius*.

Marsh PM, Wild AL, Whitfield JB. 2013. The Doryctinae (Braconidae) of Costa Rica: genera and species of the tribe Heterospilini. *Zookeys* 347: 1-474.
Taxonomic revision of 286 species of Heterospilini from Costa Rica.

Martínez JJ, Ceccarelli SF, Zaldívar-Riverón A. 2010. The genus *Iare* (Hymenoptera: Braconidae: Doryctinae) in Mexico, with the description of two new species. *Zootaxa* 2685: 30-38.
Description of two species of *Iare* from Mexico.

Martínez JJ, Meza-Lázaro RN, Pedraza-Lara C, Zaldívar-Riverón A. 2016. *Sergey* gen. n., a new doryctine genus from temperate forests of Mexico and Cuba (Hymenoptera, Braconidae). *Zookeys* 589: 143-164.

Description of *Sergey*.

Martínez JJ and Zaldívar-Riverón A. 2013. Seven new species of *Allorhogas* (Hymenoptera: Braconidae: Doryctinae) from Mexico. *Revista Mexicana de Biodiversidad* 84: 117-139.
Descripiton of seven species of *Allorhogas* from Mexico.

Nunes JF, Zaldívar-Riverón A, Sormus de Castro C, Marsh PM, Penteado-Dias AM, Briceño R, Martínez JJ. 2012. *Doryctopambolus* Nunes & Zaldívar-Riverón (Braconidae), a new Neotropical doryctine wasp genus with propodeal spines. *Zookeys* 223: 53-67.
Descripiton of *Doryctopambolus*.

Nunes JF, Penteado-Dias AM, Ceccarelli SF, Zaldívar-Riverón A. 2012. Redescription of *Mononeuron duguetiae* Fischer (Braconidae: Doryctinae), a gall associated species on *Duguetia furfurácea* (St. Hill) (Annonaceae). *Journal of Hymenoptera Research* 24: 75-84.
Redescription of the monotypic genus *Mononeuron*.

Quicke DLJ, Butcher BA. 2015. *Preembobracon gen. nov.* (Hymenoptera: Braconidae: Doryctinae: Ypsistocerini: Embobraconina) from Brazil. *Zootaxa* 4000: 275-280.
Description of *Preembobracon*, which here is considered as a junior synonym of *Labania*.

Quicke DLJ, Tunstead J, Falco JV, Marsh PM. 1992. Venom gland and reservoir morphology in the Doryctinae and related braconid wasps (Insecta, Hymenoptera, Braconidae). *Zoologica Scripta*: 21 403-416.
Study of the venom gland apparatus of doryctine genera.

Reséndiz-Flores A, Nunes JF, García-París M, Zaldívar-Riverón A. 2014. Six new species of the parasitoid wasp genus *Notiospathius* (Hymenoptera: Braconidae: Doryctinae) from Mexico. *Revista Mexicana de Biodiversidad* 85: 391-401.
Description of six species of *Notiospathius* from Mexico.

Samacá-Sáez E, Belokobylskij SA, Quicke DLJ, Zaldívar-Riverón A. 2016. Systematics of the Neotropical braconid wasps of the *Pedinotus* genus group (Doryctinae). *Systematic Entomology* 41: 481-491.
Phylogenetic study of the *Pedinotus* genus group and redescription of its genera.

Seltmann K and Sharkey MJ. 2007. A new genus and species of apterous Doryctinae (Hymenoptera: Braconidae) from Costa Rica. *Zootaxa* 10415: 17-24.
Description of *Oroceguera*.

Shimbori EM, Penteado-Dias AM, Nunes JF. 2011. *Monitoriella* Hedqvist (Hymenoptera: Braconidae: Doryctinae) from Brazil, with descriptions of three new species. *Zootaxa* 2921, 28-38.
Description of three species of *Monitoriella*.

Sormus de Castro C, Zaldívar-Riverón A, Briceño-G R, Penteado-Dias A. 2013. The genus *Hecabolus* Curtis 1834 (Braconidae: Doryctinae) in South America, with description of six new species. *Zootaxa* 3664: 377-391.
Description of six species of *Hecabolus* from South America.

Zaldívar-Riverón A, Mori M, Quicke DLJ. 2006. Systematics of the cyclostome subfamilies of braconid parasitic wasps (Hymenoptera: Ichneumonoidea): a simultaneous molecular and morphological Bayesian approach. *Molecular Phylogenetics and Evolution* 38: 130-145.
Phylogenetic study of the cyclostome group of subfamilies based on molecular and morphological evidence.

Zaldívar-Riverón A, Martínez JJ, Ceccarelli SF, De Jesús-Bonilla VS, Rodríguez-Pérez AC, Reséndiz-Flores A, Smith MA. 2012. DNA barcoding a highly diverse group of parasitoid wasps (Braconidae: Doryctinae) from a Mexican nature reserve. *Mitochondrial DNA* 21(S1): 18-23.
DNA barcoding inventory of the Doryctinae from the Chamela region, Jalisco, Mexico. *Ptesimogastroides* was regarded as a junior synonym of *Ptesimogaster.*

Zaldívar-Riverón A, Rodríguez-Jiménez A, Sarmiento CE, Pedraza-Lara C, López-Estrada EK. 2013. Description and phylogenetic relationships of *Bolivar gen. nov.*, a new genus of Neotropical doryctine wasps (Braconidae). *Invertebrate Systematics* 27: 673-688.
Description of *Bolivar.*

Zaldívar-Riverón A, Martínez JJ, Belokobylskij SA, Pedraza-Lara C, Shaw SR, Hanson PE, Varela-Hernández F. 2014. Systematics and evolution of gall formation in the plant-associated wasp genera of the subfamily Doryctinae (Hymenoptera: Braconidae). *Systematic Entomology* 39: 633-659.
Description of *Ficobolus, Pseudopsenobolus* and *Sabinita.*

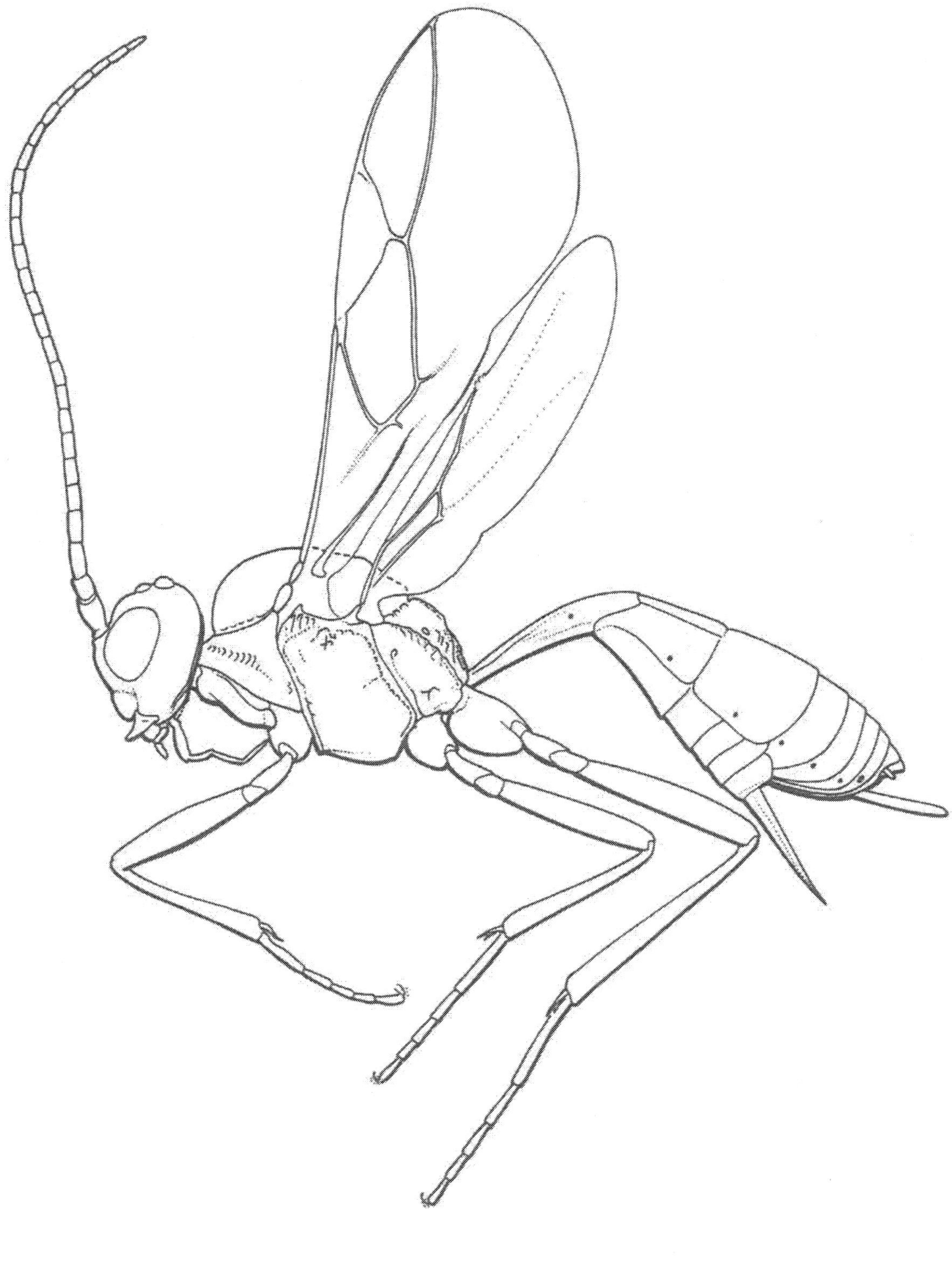

SUBFAMILY EUPHORINAE

S. R. SHAW

Department of Ecosystem Science and Management, University of Wyoming, Laramie, Wyoming 82071, U.S.A

INTRODUCTION. The Euphorinae comprise a large and diverse subfamily represented in the New World by at least 31 genera in nine tribes (S. Shaw, 1985, 1987, 1989). Euphorines are rather unusual hymenopterans in attacking the adult stage of various insects. A sister-group relationship with the Meteorinae is generally acknowledged, and some authorities (e.g. Quicke and van Achterberg, 1990) have included *Meteorus* Haliday in the Euphorinae on morphological grounds. However, S. Shaw (1985) limited the subfamily to include only those lineages that predominantly parasitize the adult stage.

The euphorine genera of the world were reviewed by S. Shaw (1985) and their phylogeny and biology were discussed by S. Shaw (1988). A substantial amount of revisionary work has been done by Loan (1967a, b, 1969, 1970, 1974, 1975, 1980) for various genera in the Nearctic region. Although euphorines are quite diverse in the Neotropics, only a few revisionary studies have been done and most of the Neotropical species are undescribed. Mason (1964) reviewed the species of *Chrysopophthorus* Goidanich, a remarkable group parasitizing adult Chrysopidae. Papp (1965) reviewed the genus *Aridelus* Marshall, which parasitize species of Pentatomidae. Haeselbarth and Loan (1983) described the genus *Townesilitus*, a parasitoid of Chrysomelidae. *Orionis* Shaw and Betelgeuse Shaw are two recently discovered genera that are apparently endemic to Central America (Shaw, 1987, 1989). Van Achterberg (1992) described a new centistine genus, *Centistoides*, from Suriname.

BIOLOGY. Euphorines are solitary or (more rarely) gregarious koinobiont endoparasitoids of adult Coleoptera (e.g. *Perilitus* Nees, *Dinocampus* Förster, *Microctonus* Wesmael, *Centistes* Haliday), adult Hymenoptera (e.g. *Syntretus* Förster), or adult Neuroptera (e.g. *Chrysopopthorus*), and of nymphal and adult Heteroptera and Psocoptera (e.g. *Peristenus* Förster, Nees, *Aridelus*, *Wesmaelia* Förster, *Holdawayella* Loan, *Euphoriella* Ashmead). As a group, the subfamily Euphorinae has a host range that is substantially broader than other braconid subfamilies, presumably because of diversification following the evolution of parasitism of adult insects (Muesebeck, 1936; Tobias, 1966; S. Shaw, 1985, 1988). It is hypothesized that adult-parasitism evolved through parasitism of Chrysomelidae, whose adults and larvae occur on the same plants and have similar feeding habits (Tobias, 1966; S. Shaw, 1988). Subsequent evolution mostly involved diversification among plant-feeding beetles (especially Chrysomelidae, Curculionidae, and Scolytidae), as well as exploitation of some predators (e.g. Carabidae, Coccinellidae). Use of other host orders probably evolved initially through the parasitism of species with similar feeding habits in the same microhabitat. Parasitism of adult wasps by the Syntretini probably involves attacking hosts while they are feeding at flowers (S. Shaw, 1988; Walker et al., 1990). More information on euphorine biology is provided in S. Shaw (1988, 1995a) and M. Shaw and Huddleston (1991).

BIOLOGICAL CONTROL POTENTIAL. Some euphorines have been utilized for biological control, mostly for suppression of pest beetles and bugs. Several species of *Microctonus* have been studied for control of weevils on forage crops in Europe and North America (Loan, 1967a, 1969, 1975). More recently, a South American species, *Microctonus hyperodae* Loan, was imported to New Zealand to aid in the suppression of the Argentine stem weevil (S. Shaw, 1993; McNeill et al., 1993). A Brazilian *Centistes* species is currently being studied as a possible biocontrol agent for the southern corn rootworm (*Diabrotica*) in the United States (S, Shaw, 1995b). Several species of *Peristenus* and *Leiophron* are utilized as biocontrol agents for lygus bugs (Loan and S. Shaw, 1987) and other mirids (Loan, 1980).

Dinocampus coccinellae (Schrank) is a parasitoid of several species of beneficial ladybird beetles, and is thus one of the few braconid species that might sometimes be regarded as a pest. On several occasions it has been reared from ladybird beetles purchased from commercial companies for use in a greenhouse situation, but under normal conditions *Dinocampus* is not abundant enough to justify regarding it as a significant pest.

COMMON GENERA. As compared with other braconid subfamilies, none of the euphorines are generally regarded as "common." Nevertheless, some genera (*Microctonus*, *Peristenus*, *Leiophron*, *Euphoriella*) tend to be collected more often than others. Actually, euphorines are quite diverse, but because of their highly specialized habits and low population levels they tend to be harder to collect than other braconids. More attention needs to be given to rearing euphorines from adult insect hosts, especially in the Neotropics. Some of the main euphorine host groups (Curculionidae, Chrysomelidae) are among the most diverse insects in tropical forest canopies, but this habitat is virtually unexplored for euphorine diversity.

DISTRIBUTION. The Euphorinae are a widespread and cosmopolitan subfamily, in some cases even occurring on remote oceanic islands (e.g. *Dinocampus*). Most of the genera found in the New World range throughout all the temperate and tropical parts of the Nearctic and Neotropical regions. The main distributional limitations are as follows: *Ropalophorus* Curtis, *Holdawayella* Kokujev, *Ecclitura* Kokujev, *Allurus* Förster, *Spathicopis* van Achterberg, *Loxocephalus* Förster, and *Cosmophorus* Ratzeburg are genera (mostly Holarctic) that occur in cool, temperate parts of North America, but apparently do not occur in Central or South America. *Litostolus* van Achterberg and *Marshiella* Shaw are Holarctic genera that range southward to Costa Rica. Records for *Pygostolus* Haliday are scarce, but it is recorded from widely scattered sites in North, Central and South America. The ranges of *Streblocera* Westwood, *Lecythodella* Enderlein, *Centistina* Enderlein, and *Bracteodes* DeSaeger in the New World are mostly limited to the Neotropics, although the latter two genera range into the southern parts of southwestern United States. *Orionis* and *Betelgeuse* are known only from Central America. *Centistoides* is known only from Suriname (van Achterberg, 1992).

DISTINGUISHING FEATURES. The fore wing cross vein 2cu-a is always absent in the Euphorinae, resulting in a characteristic "open" subdiscal cell (Fig. 1). This feature is also true of the Meteorinae (a synapomorphy), and occurs convergently in many other groups, including nearly all Blacinae and a few Helconinae. The fore wing vein 3RS is distinctly curved in the Euphorinae (Fig. 12), a feature that serves well to separate euphorines from meteorines, although a curved 3RS does occur convergently in a few Helconinae and some other groups. Most commonly the fore wing cross vein r-m is absent (thus there is usually no closed second submarginal cell). Within the Euphorinae there are tendencies toward the petiolization of the first metasomal tergum, reduction of wing venation and ovipositor, and shifting of the eyes lower and forward onto the face. Euphorines oviposit by bending the abdomen forwards under the body and poking the ovipositor in front of the face, and female specimens often die in this position, providing a useful diagnostic character for recognizing the subfamily (though aphidiines and meteorines also do this).

KEY TO THE NEW WORLD GENERA OF THE SUBFAMILY EUPHORINAE

1. Fore wing r-m present, thus second submarginal cell present (Figs 1, 2) (Tribe Euphorini, in part) 2

\- Fore wing r-m absent, thus second submarginal cell absent (Fig. 3) 3

2(1). Mesosoma covered with coarse honey-combed, areolate sculpture, which on the mesoscutum completely obscures the notauli (Fig. 25). Petiole somewhat shorter than remainder of metasoma excluding ovipositor (Fig. 26). Malar suture absent between eye and mandible. Shortest distance between eyes of female greater than width of clypeus. Petiole and wing venation (Fig. 1) well sclerotized .... ***Aridelus*** Marshall

Distribution: U.S.A. to Argentina; 4 described New World species, several undescribed Neotropical species. Biology: parasitoids of late instar nymphs and adults of Pentatomidae. Reference: Papp (1965, key).

\- Mesosoma without honey-combed or areolate sculpture except on propodeum (Fig. 27), mesoscutum punctate and with distinct notauli. Petiole as long as or longer than remainder of metasoma. Malar suture present (Fig. 28). Shortest distance between eyes of female equal to or less than width of clypeus. Petiole and wing venation (Fig. 2) desclerotized and often translucent .... ***Chrysopophthorus*** Goidanich

Distribution: U.S.A. to Brazil, Caribbean; 4 described New World species, several undescribed Neotropical species. Biology: parasitoids of adult Chrysopidae. Reference: Mason (1964, key).

3(1). Basal width of petiole equal to or greater than half width of propodeum (Fig. 32). First metasomal spiracle distinctly anterior to middle of segment (Fig. 32). Fore wing 3RS fully developed (Figs 3-6). Ovipositor and sheaths short and extremely broad (Fig. 33) (Tribe Centistini) 4

\- Basal width of petiole less than half width of propodeum, often much narrower; first metasomal segment strongly petiolate (Fig. 29). First metasomal spiracle usually near middle (Fig. 29) or behind middle of segment (Fig. 30); if somewhat anterior to middle (Fig. 31), then fore wing 3RS partly absent (Fig. 16). Ovipositor and sheaths variable in length, but not extremely broad 9

4(3). Fore wing 2M desclerotized and reduced to only a short stub (as in Figs 3, 4), or absent 5

\- Fore wing 2M sclerotized and present as a short but distinct branch (Figs 5, 6) 6

5(4). Hind coxa without an apical tooth (Fig. 34). Tarsal claws simple. Metasomal sternum 4 without a pair of teeth. Fore wing (RS+M)a present (Fig. 3), partly absent (Fig. 4),or completely absent..... ***Centistes*** Haliday

Distribution: Cosmopolitan; 7 described New World species, many undescribed Neotropical species. Biology: parasitoids of adult beetles, especially Curculionidae, Chrysomelidae, and Carabidae. Reference: Shaw (1995b, key to New World species). Includes *Syrrhizus* Förster.

\- Hind coxa with an apical tooth (Fig. 35). Tarsal claws cleft. Metasomal sternum 4 with a pair of teeth (Fig. 35). Fore wing (RS+M)a always present (as in Fig. 3) .... ***Allurus*** Förster

Distribution: eastern Canada; one described species in New World. Biology: parasitoids of adult Curculionidae, Reference: S. Shaw (1985).

6(4). Dorsope absent. Maxillary palp with 3 segments; labial palp with 1 segment. Female hypopygium deeply incised medially .... ***Centistoide***s van Achterberg

Distribution: Suriname; rare; known only from a single specimen. Biology: unknown. Reference: van Achterberg (1992a, description).

\- Dorsope present near base of petiole: Maxillary palp with 5 segments; labial palp with 3 segments. Female hypopygium truncated, not deeply incised medially 7

7(6). Ovipositor, in dorsal view, wider at apex than at base. Malar space about 2 times wider than base of mandible. Fore wing (RS+M)a absent (as in Fig. 24) .... ***Spathicopis*** van Achterberg

Distribution: U.S.A. and Canada; one described species. Biology: unknown. Reference: van Achterberg (1977, description).

\- Ovipositor, in dorsal view, wider at base than at apex. Malar space about as wide as base of mandible. Fore wing (RS+M)a always completely present (Figs 5, 6) 8

8(7). Fore tarsus with combined tarsomeres 3 and 4 longer than tarsomere 5. Apex of petiole as wide as or only slightly wider than base. Apical flagellomere of antenna with sharp spine. Mesosoma often very densely setose ventrally, the setae obscuring the surface. Dorsope of petiole extremely large (3 times wider than spiracle of same segment). Wings: Fig. 5 .... ***Pygostolus*** Haliday

Distribution: Holarctic and Neotropical; 3 described species in New World. Biology: parasitoids of adult Curculionidae. Reference: van Achterberg (1992b, key to Holarctic species).

\- Fore tarsus with combined tarsomeres 3 and 4 shorter than tarsomere 5 (Fig. 36). Petiole nearly twice as wide at apex as at base (Fig. 38). Apical flagellomere without a sharp spine. Mesosoma not densely setose ventrally. Dorsope of petiole not so large (Fig. 38). Wings: Fig. 6............. ***Litostolus*** van Achterberg

Distribution: Canada to Costa Rica; one described Nearctic species, at least 4 undescribed Neotropical species. Biology: unknown. Reference: van Achterberg (1985, description).

9(3). Fore wing (RS+M)a present (Fig. 7), or (less commonly) (RS+M)a absent but m-cu and 2RS also absent (Fig. 12) 10

\- Fore wing (RS+M)a absent, m-cu and 2RS present, thus first submarginal and discal cells are combined to form a single large closed cell (Figs 18-24) 21

10(9). Petiole at least 4 times wider at apex than at base (Fig. 29); sculptured dorsally; not fused ventrally 11

\- Petiole at most 3 times wider at apex than at base, sometimes much narrower (Fig. 31); sculptured or smooth dorsally; sometimes fused ventrally (Tribe Euphorini, in part) 14

11(10). Scape of antenna at most 2 times longer than its greatest width. Labial palp with .... 3 segments. Wings: Fig. 7 (Tribe Perilitini) .... ***Perilitus*** Nees

Distribution: widespread; uncommon; one described Nearctic species. Biology: parasitoids of adult beetles, especially Curculionidae and Chrysomelidae. Reference: S. Shaw (1985).

\- Scape at least 3 times longer than its greatest width (Fig. 37). Labial palp with 2 segments (Fig. 39) (Tribe Dinocampini, in part) 12

12(11). Petiole rugose (Fig. 29). Wings: Fig. 8 .... ***Dinocampus*** Förster

Distribution: Cosmopolitan; one described species in New World. Biology: parasitoids of adult Coccinellidae. Reference: S. Shaw (1985).

- Petiole longitudinally costate or striate (Fig. 40) 13

13(12). Antennal flagellum shorter than mesosoma, with only 8 flagellomeres; weakly clubbed apically (Figs 37, 41); scape about 3 times longer than its greatest width. Fore wing (RS+M)b very short or absent, much shorter than m-cu (Fig. 9) .... ***Ropalophorus*** Curtis

Distribution: Holarctic; one described species in New World. Biology: parasitoids of adult Scolytidae. Reference: Shenefelt (1960).

- Antennal flagellum distinctly longer than mesosoma, with more than 8 flagellomeres; entirely filiform; scape about 5 times longer than its greatest width. Fore wing (RS+M)b long, nearly as long as or longer than m-cu (Fig. 10) .... ***Centistina*** Enderlein

Distribution: Ethiopian. Neotropical, and southwestern Nearctic; several undescribed New World species. Biology: unknown. Reference: S. Shaw (1985).

14(10). Fore wing RS fully developed and distinct throughout its length (Figs Il-14); (RS+M)a either straight or absent 15

- Fore wing RS with at least some portion absent (Figs 15-17); (RS+M)a curved 18

15(14). Petiole with tergum and sternum not fused ventrally (Fig. 43), or fused only basally (Fig. 42); tergum usually sculptured 16

- Petiole with tergum and sternum entirely fused ventrally to form a very long, narrow, unsculptured tube (Fig. 45) 17

16(15). Petiole with tergum and sternum meeting or fused ventrally at the base of the segment (Fig. 42). Occipital carina complete. Fore wing always with first submarginal and discal cells completely closed (Fig. 11) .... ***Peristenus*** Förster

Distribution: mostly Holarctic; 24 described North American species, at least 5 undescribed species from Central America. Biology: parasitoids of late instar nymphs and adults of Miridae. Reference: Loan (1974, revision of Nearctic species).

- Petiole with tergum and sternum entirely separate, not fused ventrally at the base of the segment (Fig. 43). Occipital carina absent or incomplete dorsally (Fig. 44). Fore wing venation variable, but sometimes reduced so that first submarginal and discal cells are open apically (Fig. 12) .... ***Leiophron*** Nees

Distribution: widespread, except Australia; common; 19 described North American species, many undescribed Neotropical species. Biology: parasitoids of late instar nymphs and adults of Miridae. Reference: Loan (1974,revision of Nearctic species).

17(15). Petiole as long as or longer than remainder of metasoma excluding ovipositor (Fig. 45). Frons smooth (Fig. 46) or with a single median carina. Fore wing 3RS short and strongly curved; M+CU absent or very weak basally (Fig. 13) .... ***Wesmaelia*** Förster

Distribution: Holarctic; 2 described species in New World, one undescribed Neotropical species. Biology: parasitoids of late instar nymphs and adults of Nabidae. Reference: S. Shaw (1985), S. Shaw (1997, key to New World species).

- Petiole shorter than remainder of metasoma excluding ovipositor (Fig. 47). Frons rugose (Fig. 48). Fore wing 3RS longer and only weakly curved basally; M +CU entirely present (Fig. 14) .... ***Orionis*** Shaw

Distribution: Central America; 2 described species. Biology: unknown. Reference: S. Shaw (l987, description and key).

18(14). Petiole smooth and polished dorsally (Fig. 49). Propodeum with a distinct posteromedian depression. Eyes apparently bare, without distinct setae. Malar space present; malar suture absent. Wings: Fig. 15 .... ***Holdawayella*** Loan

Distribution: Holarctic and Neotropical: 2 described Nearctic species, and at least one undescribed Neotropical species. Biology: parasitoids of late instar nymphs and adults of Tingidae. Reference: Loan (1967b, description).

- Petiole sculptured dorsally (Fig. 50). Propodeum without a distinct postero-median depression. Eyes minutely, but distinctly setose (Fig. 51, 52). Malar suture present or eye converging on mandible and malar space virtually absent (Figs 51, 52) 19

19(18). Ovipositor short and curved, less than 0.25 times length of petiole, often even shorter and concealed in apex of metasoma (Fig. 53). Fore wing RS entirely absent, or represented only by a short stub off the anterior wing margin (Fig. 16). Petiole with tergum and sternum fused ventrally to form a single tubular segment (Fig. 31) .... ***Euphoriella*** Ashmead

Distribution: widespread; common, but often overlooked because of small size; 10 described Nearctic species: many undescribed Neotropical species. Biology: parasitoids of late instar nymphs and adults of Psocoptera. Reference: Loan and New (1971, key to North American species).

- Ovipositor straight and as long as or longer than petiole, never concealed (Fig. 54). Fore wing vein RS present basally, but absent distally and not reaching wing margin (Figs 17). Petiole with tergum and sternum not fused ventrally 20

20(19). Eyes, in anterior view, slanting towards each other and converging on the lower part of face, eyes at lower end separated by only a short distance that is less than width of clypeus (Fig. 51). Face flat or slightly convex, only sparsely setose .... ***Cryptoxilos*** Viereck

Distribution: widespread; 3 described species in New World; some undescribed Neotropical species. Biology: parasitoids of adult Scolytidae Reference: S. Shaw (1985).

\- Eyes not slanting towards each other nor converging on the lower part of face, widely separated by a distance that is more than the width of the clypeus (Fig. 52). Face below antennae deeply excavated and densely setose (Fig. 52). Wings: Fig. 17 .... ***Plynops*** Shaw

Distribution: southern Florida to South America; 10 species. Biology: unknown. Reference: S. Shaw (1996, description, key to species).

21(9). Fore wing M+CU absent or extensively desclerotized (Figs 18, 19). Hind. wing 1A and cu-a greatly reduced or absent, thus subbasal cell of hind wing open apically or absent (Figs 18, 19). Tarsal claws cleft, thus each claw ending with two sharp tips (Fig. 55). Petiole with tergum and sternum fused ventrally to form a single tubular segment (Fig. 30) (Tribe Syntretini) 22

\- Fore wing M+CU usually present and fully sclerotized (Figs 20-24). Hind wing 1A nearly always present and meeting cu-a, thus closed subbasal cell of hind wing present (Figs 21-24). Tarsal claws simple, with only one sharp tip. Petiole variable, but tergum and sternum commonly not fused ventrally or fused only basally 24

22(21). Mesoscutum punctate, with distinct notauli (Fig. 56). Fronto-clypeal suture absent, not even indicated by a depression .... ***Bracteodes*** de Saeger

Distribution: southern Arizona, Mexico, Costa Rica; rare: one undescribed species in New World. Biology: unknown. Reference: S. Shaw (1985).

\- Mesoscutum smooth and shining, notauli absent. Fronto-clypeal suture present, or atleast indicated by a depression 23

23(22). Propodeum carinate, the postero-median area completely bordered by distinct carinae (Fig. 57). Wings: Fig. 18 .... ***Syntretus*** Förster

Distribution: widespread; Holarctic, Neotropical, and Ethiopian; 3 described and many undescribed species, especially in Neotropics. Biology: parasitoids of adult Ichneumonidae and Apidae. References: Marsh (1979), S. Shaw (1985).

\- Propodeum mostly smooth and polished, sometimes with some carinae at extreme apex but the postero-median area never bordered (Pig. 58). Wings: Fig. 19 .... ***Falcosyntretus*** Tobias

Distribution: Holarctic and Neotropical; only one described New World species, at least 3 undescribcd species. Biology: unknown. References: Tobias (1965), S. Shaw (1985).

24(21). Mandible nearly as wide apically as at base, with a large space between clypeus and mandibles. even when mandibles. are closed (Fig. 59). Antennal bases prominent, anterior margin raised in two rounded projections (Fig. 60). Wings: Fig. 20 (Tribe Cosmophorini) .... ***Cosmophorus*** Ratzeburg

Distribution: Holarctic and Caribbean (Jamaica); 5 described Nearctic species. Biology: parasitoids of adult Scolytidae. Reference: S. Shaw (1985).

\- Mandible usually gradually narrowing from base to apex; without a space between clypeus and mandibles when closed. Base of antenna not prominent, or if somewhat raised then anterior margin even and without projections 25

25(24). Basal 0.25 of petiole not fused ventrally, the tergum and sternum clearly separated (as in Fig. 43) 26

\- Basal 0.25 of petiole fused ventrally 30

26(25). Antennal scape short, at most twice as long as its greatest width (Fig. 65); female flagellum filiform 27

\- Scape at least 4 times longer than its greatest width (Figs 64, 66); female flagellum filiform or variously modified: either serrate, slightly clubbed, with fused flagellomeres or with hook-like projections 28

27(26). Eyes of female not bulging beyond the frons (Fig. 61). Mesoscutum punctate. Hind femora less than 6 times longer than wide. Metasoma beyond petiole not strongly compressed, about as wide as propodeum. Wings: Fig. 21 (Tribe Microctonini, in part) .... ***Microctonus*** Wesmael

Distribution: Cosmopolitan; 26 described and many undescribed species in New World, especially in the Neotropics. Biology: parasitoids of adult beetles, especially Curculionidae, Chrysomelidae, and Carabidae. Reference: Loan (1969, key to Nearctic species).

\- Eyes of female bulging anteriorly beyond the frons (Fig. 62). Mesoscutum finely rugulose (Fig. 63). Hind femora more than 6 times longer than wide. Metasoma beyond petiole strongly compressed, less than half as wide as propodeum (Tribe Loxocephalini) .... ***Loxocephalus*** Förster

Distribution: Holarctic; rare; one described species in New World. Biology: associated with ants, but not reared. Reference; S. Shaw (1985).

28(26). Mesosoma entirely coarsely rugose to rugose-punctate. Basal flagellomeres of female antenna serrate along outer margin; flagellum with 13 flagellomeres; scape about 4 times longer than wide (male unknown) .... ***Betelgeuse*** Shaw

Distribution: Mexico; one described and one undescribed species. Biology: unknown. Reference: S. Shaw (1989, description).

- Mesosomal sculpture varied, but not entirely coarsely rugose to rugose-punctate. Female antenna not serrate, either basal flagellomeres. not modified, or with hook-like projections· along inner margin (Fig. 64); flagellum with 14 or more flagellomeres; scape at least 5 times longer than wide (Tribe Microctonini, in part) 29

29(28). Female antenna raptorial (Fig. 64): scape with a basal horn, prominence, or longitudinal carina; flagellum with opposable basal section of fused flagellomeres. and one or more basal flagellomeres modified with hook-like projections, flagellum filiform apically; males without raptorial modifications, but with elongate scape .... ***Streblocera*** Westwood

Distribution: Pantropical and Palearctic; rare; at least 2 undescribed Neotropical species. Biology: parasitoids of adult beetles, especially Chrysomelidae. Reference: S. Shaw (1985).

- Female antenna not raptorial: scape without a basal horn, prominence, or longitudinal carina: flagellum long, slender, and gradually clubbed apically, without a modified basal section; male unknown .... ***Ecclitura*** Kokujev

Distribution: Holarctic; rare; one undescribed species in eastern United States. Bi- ology: unknown. Reference: S. Shaw (1985).

30(25). Scape long, more than 4 times longer than its greatest width (Fig. 66). Wings: Fig. 22 .... ***Lecythodella*** Enderlein

Distribution: Neotropical; one described and 4 undescribed species. Biology: unknown.

- Antennal scape short, only about 2 times longer than its greatest width (Fig. 65) (Tribe Townesilitini) ...31

31(30). Basal flagellomeres of female cylindrical, not densely setose (Fig. 65); flagellar setae not flattened at tips. Wings: Fig. 23 .... ***Townesilitus*** Haeselbarth and Loan

Distribution: Holarctic and Neotropical; 3 described Nearctic species and at least 4 undescribed Neotropical species. Biology: parasitoids of adult beetles, especially Chrysomelidae. Reference: Haeselbarth and Loan (1983, description).

- Basal flagellomeres of female broad and flattened, densely setose (Fig. 67, 68); flagellar setae flattened at tips. Wings: Fig. 24 .... ***Marshiella*** Shaw

Distribution: Holarctic, U.S.A. and Canada to Costa Rica; 2 described New World species, possibly one undescribed. Biology: reared from Anthicidae. Reference: S. Shaw (1985, description).

REFERENCES

Achterberg, C. van 1977. A new Holarctic genus, *Spathicopis* gen. nov., belonging to the Euphorinae, Centistini (Hym. Braconidae), Entomologische Berichten 37: 27-31.

Achterberg, C. van 1985. The genera and subgenera of Centistini, with description of two new taxa from the Nearctic region (Hymenoptera: Braconidae: Euphorinae). Zoologische Mededelingen Leiden 59: 348'-362.

Achterberg, C. van 1992a. *Centistoides* gen. nov. (Hymenoptera: Braconidae: Euphorinae) from Suriname. Zoologische Mededelingen Leiden 66: 3.;15-348.

Achterberg, C. van 1992b. Revision of the European species of the genus *Pygostolus* Haliday (Hymenoptera: Braconidae: Euphorinae), with a key to the Holarctic species. Zoologische Mededelingen Leiden 66: 349-358.

Haeselbarth, E. and Loan, C. 1983. *Townesiliius*, a new genus for a species group in *Microctonus* (Hymenoptera: Braconidae). Contributions of the American Entomological Institute 20: 384-387.

Loan, C. C. 1967a. Studies on the taxonomy and biology of the Euphorinae (Hymenoptera: Braconidae). II. Host relations of six *Microctonus* species. Annals of the Entomological Society of America 60: 236-240.

Loan, C. C. 1967b. Studies on the taxonomy and biology of the Euphorinae (Hymenoptera: Braconidae). III. A new Canadian genus and species parasitic on Tingidae (Heteroptera). Annals of the Entomological Society of America 60: 240-244.

Loan, C. C. 1969. A summary of the species of *Microcionus* north of Mexico with five new species. Proceedings of the Entomological Society of Washington 71: 404-416.

Loan, C. C. 1970. *Euphoriana* Gahan: a redefinition with new name combinations and a new Canadian species (Hymenoptera: Braconidae: Euphorinae). Proceedings of the Entomological Society of Washington 72: 437-442.

Loan, C. C. 1974. The North American species of *Leiophron* Nees and *Peristienus* Foerster (Hymenoptera: Braconidae: Euphorinae). Transactions of the Royal Entomological Society of London 126: 207-238,

Loan, C. C. 1975. A review of the Haliday species of *Microctonus*, a neotype designation for *M. aeihiops,* and a description of *M. aethiopoidee* n. sp. (Hymenoptera: Braconidae: Euphorinae), Entomophaga 20: 31-41.

Loan, C. C. 1980. Plant bug hosts (Heteroptera: Miridae) of some euphorine parasites (Hymenoptera: Braconidae) near Belleville, Ontario, Canada. Le Naturaliste Canadiene 101: 821-860.

Loan, C. C. and New, T. R 1971. An account of the North American species of *Euphoriella* Ashmead with descriptions of seven new species Hymenoptera: Braconidae, Euphorinae). Proceedings of the Entomological Society of Ontario 102: 92-108.

Loan, C. C. and Shaw, S. R. 1987. Euphorine parasites of *Lygus* and *Adelphocoris* (Hymenoptera: Braconidae and Heteroptera: Miridae). Pp. 69-75 In: Economic importance and biological control of *Lygus* and *Adelphocoris* in North America. United States Department of Agriculture, Agricultural Research Service, ARS-64.

Marsh, P. M. 1979. Family Braconidae. Pp. 144-295. In: Krombein, K. V., Hurd, P. D., Smith, D. R., and Burks, B. D. (Eds.). Catalog of Hymenoptera in America North of Mexico. Smithsonian Institution Press, Washington, D.C.

Mason, W. R. M. 1964. The genus *Chrysopophthorus* Goidanich (Hymenoptera: Braconidae). The Canadian Entomologist 96: 1005-1017.

McNeill, M. R., Phillips, C. B., and Goldson, S. L. 1993. Diagnostic characters and biology of three *Microctonus* spp. (Hymenoptera: Braconidae, Euphorinae) parasitoids of weevils (Coleoptera: Curculionidae) in New Zealand pasture and lucerne. New Zealand Entomologist 16: 39-44.

Muesebeck, C. F. W. 1936. The genera of parasitic wasps of the braconid subfamily Euphorinae, with a review of the Nearctic species (Hymenoptera: Ichneumonoidea). United States Department of Agriculture Miscellaneous Publication 241: 1-38.

Papp, J. 1965. A monograph of the genus *Aridelus* Marsh. (Hymenoptera: Braconidae). Acta Zoologica Hungaricae, 11: 181-201.

Quicke, D. L. J. and Achterberg, C. van 1990. Phylogeny of the subfamilies of the family Braconidae (Hymenoptera: Ichneumonoidea). Zoologische Verhandelingen Leiden 258: 1-95.

Shaw, M. R. and Huddleston, T. 1991. Classification and biology of braconid wasps (Hymenoptera: Braconidae). Handbooks for the Identification of British Insects 7(11): 1-126.

Shaw, S. R. 1985. A phylogenetic study of the subfamilies Meteorinae and Euphorinae (Hymenoptera: Braconidae). Entomography 3: 277-370.

Shaw, S. R. 1987. *Orionis*, a new genus from Central America, with an analysis of its phylogenetic placement in the tribe Euphorini (Hymenoptera: Braconidae). Systematic Entomology 12: 103-109. Shaw, S. R. 1988. Euphorine phylogeny: the evolution of diversity in host-utilization by parasitoid wasps (Hymenoptera: Braconidae). Ecological Entomology 13: 323-335.

Shaw, S. R. 1989. A new Mexican genus and species of Dinocampini with serrate antennae (Hymenoptera: Braconidae: Euphorinae). Psyche 95: 289-297.

Shaw, S. R. 1993. Three new *Microctonus* species indigenous to New Zealand (Hymenoptera: Braconidae), New Zealand Entomologist 16: 29-39.

Shaw, S. R. 1995a. A new species of *Centistes* from Brazil (Hymenoptera: Braconidae: Euphorinae) parasitizing adults of *Diabrotica* (Coleoptera: Chrysomelidae) and key to New World species. Proceedings of the Entomological Society of Washington 97: 153-160.

Shaw, S. R. 1995b. The Braconidae. Pp. 431-464. In: Hanson, P. and Gauld, I. (Eds.). The Hymenoptera of Costa Rica. Oxford University Press, Oxford.

Shaw, S. R. 1996. Plynops, a peculiar new genus and ten. new species in the tribe Euphorini (Hymenoptera: Braconidae). Journal of Hymenoptera Research 5: 166-183.

Shaw, S. R 1997. The Costa Rican species of *Wesmaelia* Foerster with description of a new species (Hymenoptera: Braconidae: Euphorinae). The Pan-Pacific Entomologist 73: 103-109.

Shenefelt, R. D. 1960. *Ropalophorus* Haliday - a genus new to North America (Hymenoptera: Braconidae: Euphorinae). Annals of the Entomological Society of America 53: 542-546.

Tobias, V. I. 1965. Generic groupings and evolution of parasitic Hymenoptera of the subfamily Euphorinae (Hymenoptera: Braconidae). Entomologicheskoe Obozrenie 44: 841-865.

Walker, A. K., Joshi, N. K., and Verma, S. K. 1990. The biosystematics of *Syntretomorpha szaboi* Papp (Hymenoptera: Braconidae: Euphorinae) attacking the Oriental honey bee, *Apis cerana* Fabricius (Hymenoptera: Apidae), with a review of braconid parasitoids attacking bees. Bulletin of Entomological Research 80: 79-83.

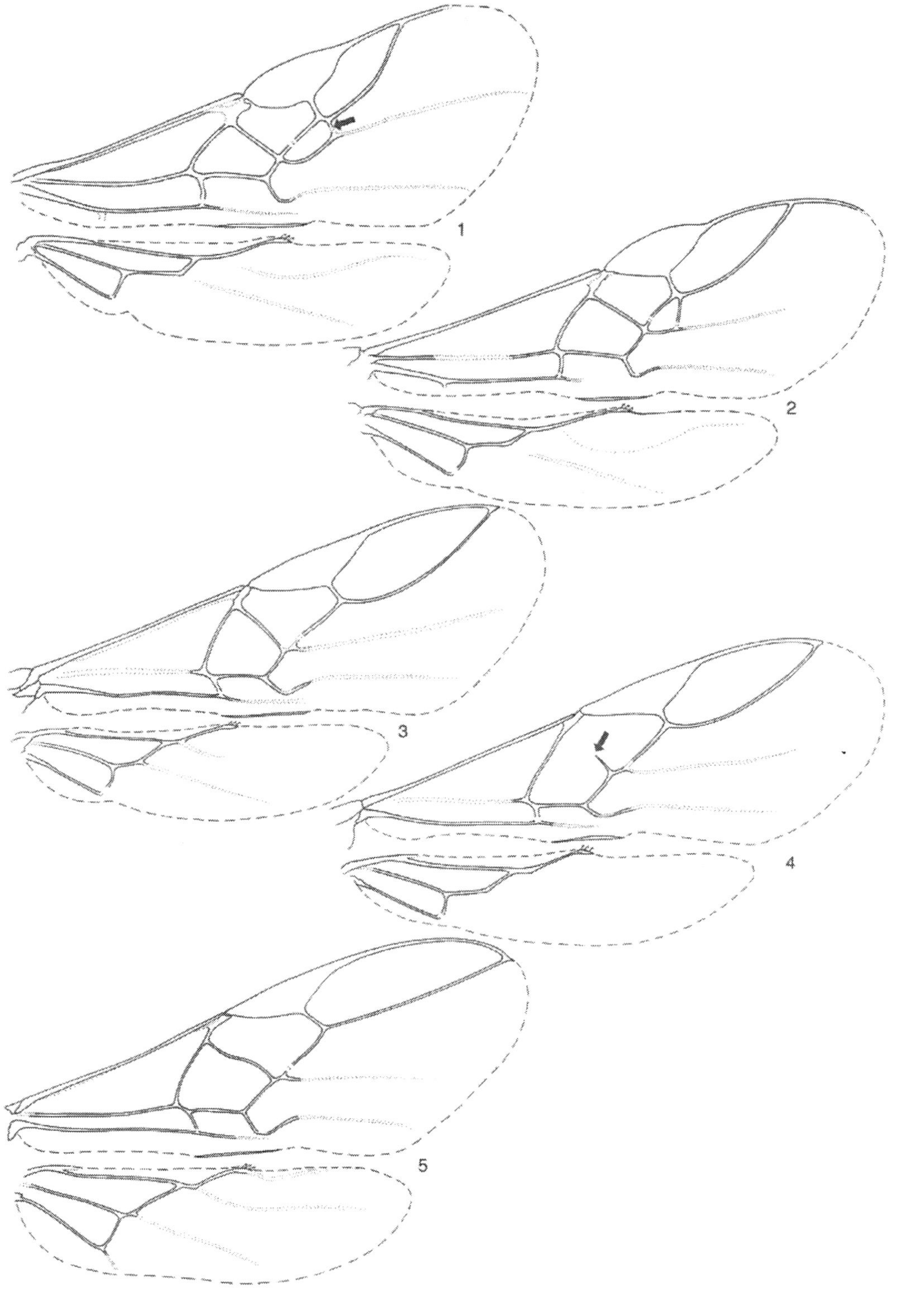

Figs 1-5: Wings of Euphorinae. 1, *Aridelus nigrator* (F.), arrow = r-m; 2, *Chrysopophthorus americanus* Mason; 3, *Centistes gasseni* Shaw; 4, *Centistes* (*Syrrhizus*) sp., arrow = (RS+M)a partially absent; 5, *Pygostolus falcatus* (Nees).

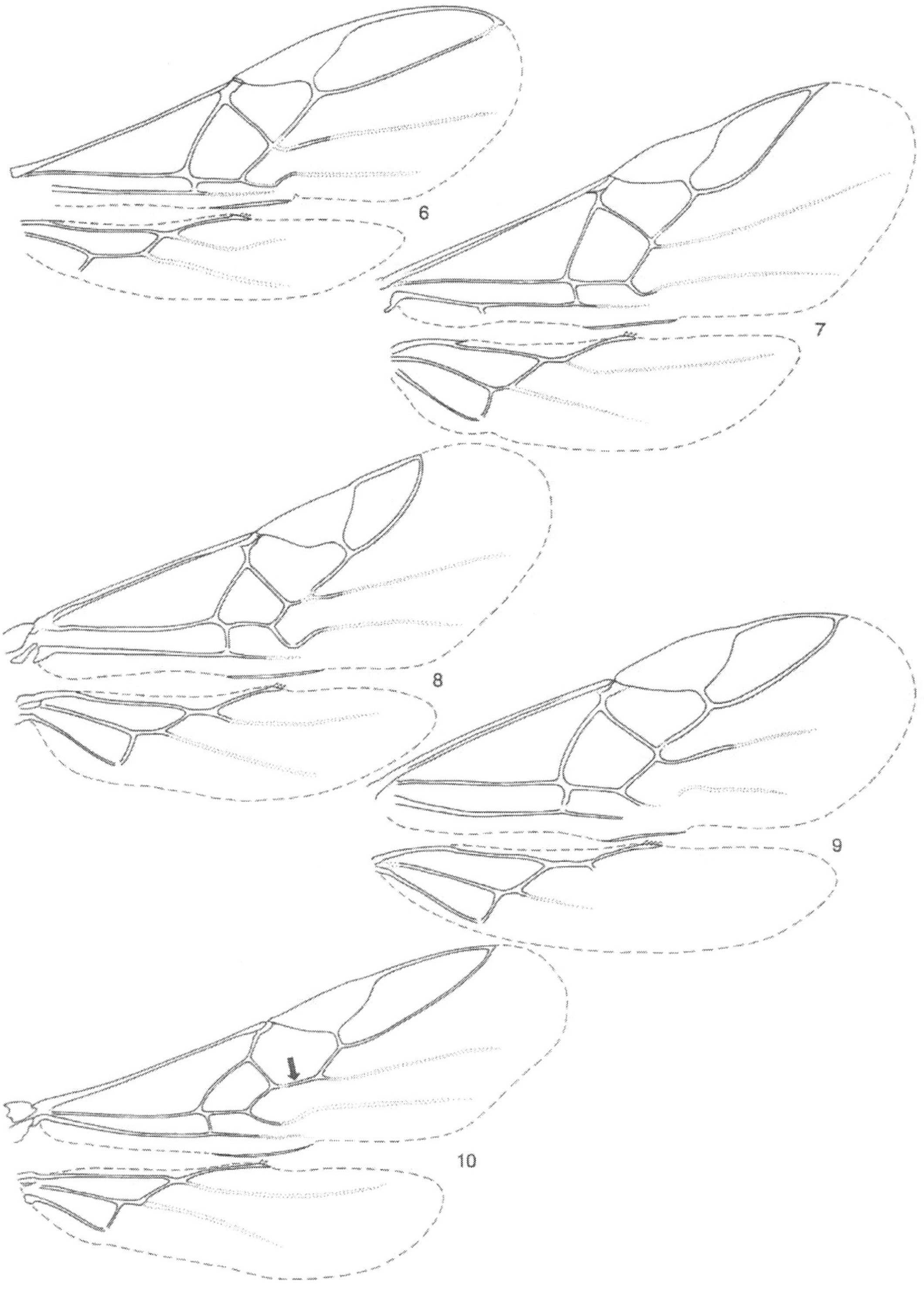

Figs 6-10: Wings of Euphorinae. 6, *Litostolus* sp.; 7, Perilitus sp.; 8, *Dinocampus coccinellae* (Schrank); 9, *Ropalophorus clavicornis* (Wesmael) showing (RS+M)b absent; 10, *Centistina* sp., arrow = (RS+M)b.

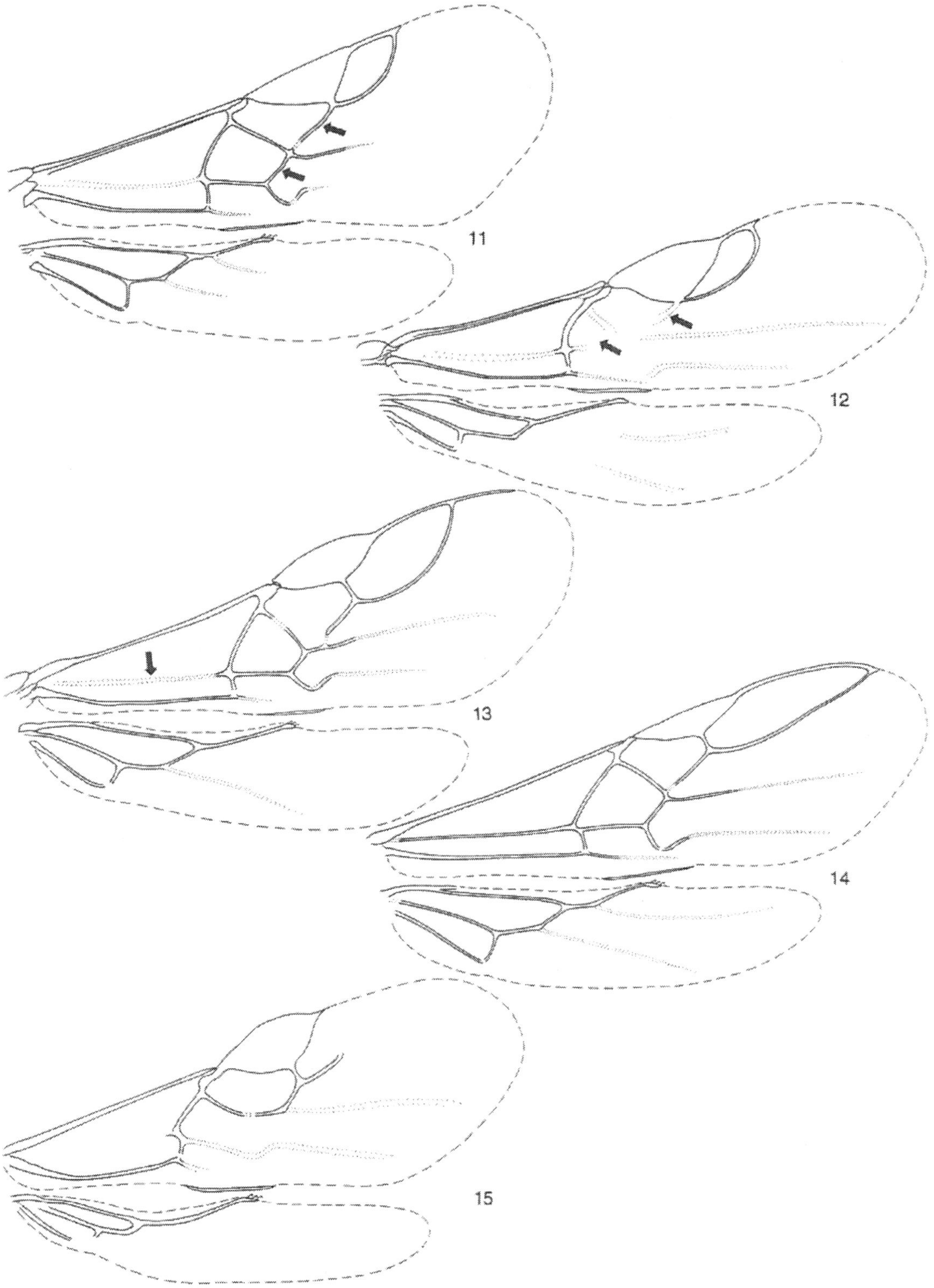

Figs 11-15: Wings of Euphorinae. 11, *Peristenus pallipes* Curtis, arrows = closed 1st submarginal and discal cells; 12, *Leiophron* sp., arrows = open 1st submarginal and discal cells; 13, *Wesmaelia lizanoi* Shaw, arrow = M+CU; 14, *Orionis masoni* Shaw; 15, *Holdawayella* sp.

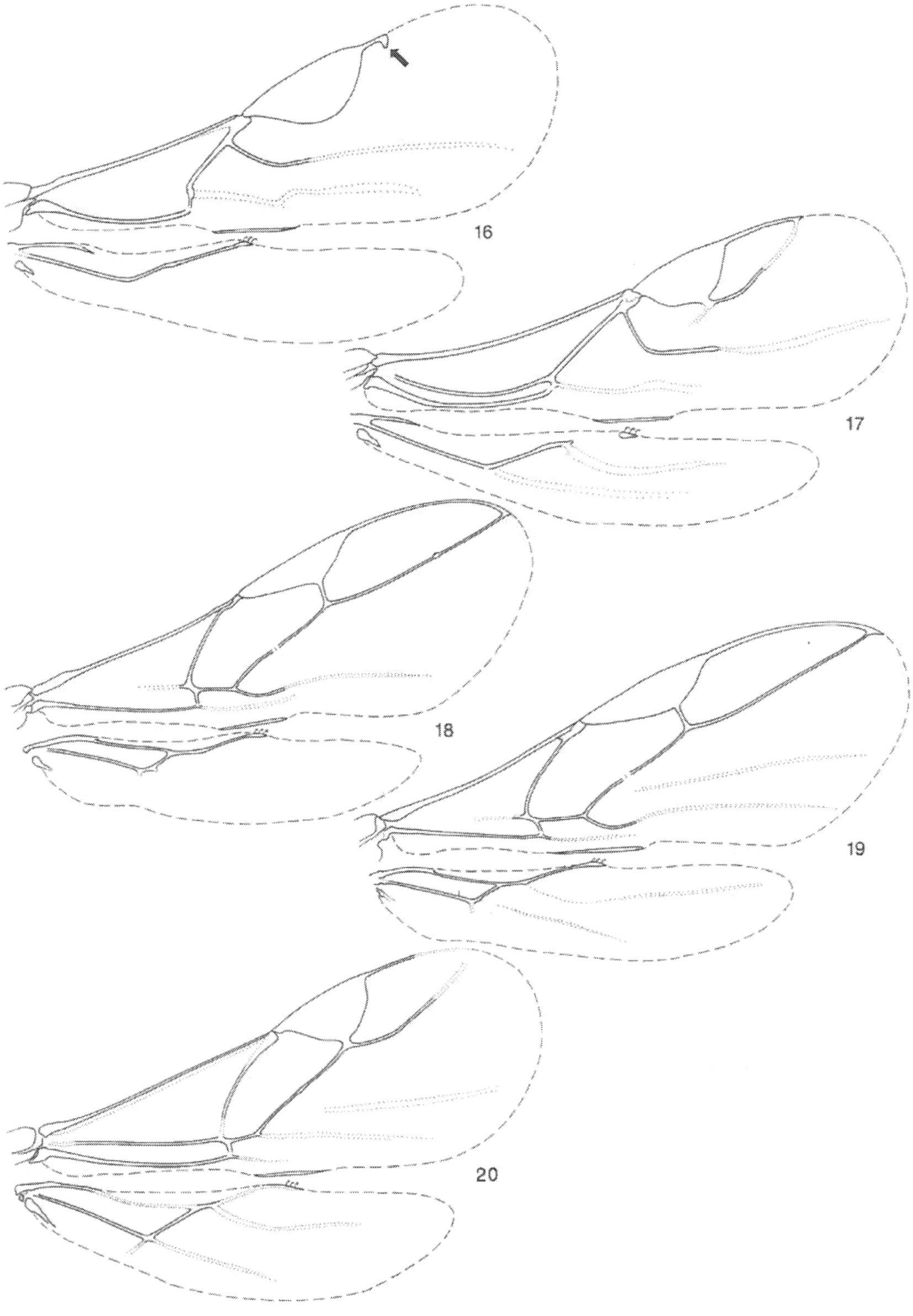

Figs 16-20: Wings of Euphorinae. 16, *Euphoriella* sp., arrow = short stub of RS; 17, *Plynops hansoni* Shaw; 18, *Syntretus* sp.; 19, *Falcosyntretus* sp.; 20, *Cosmophorus* sp.

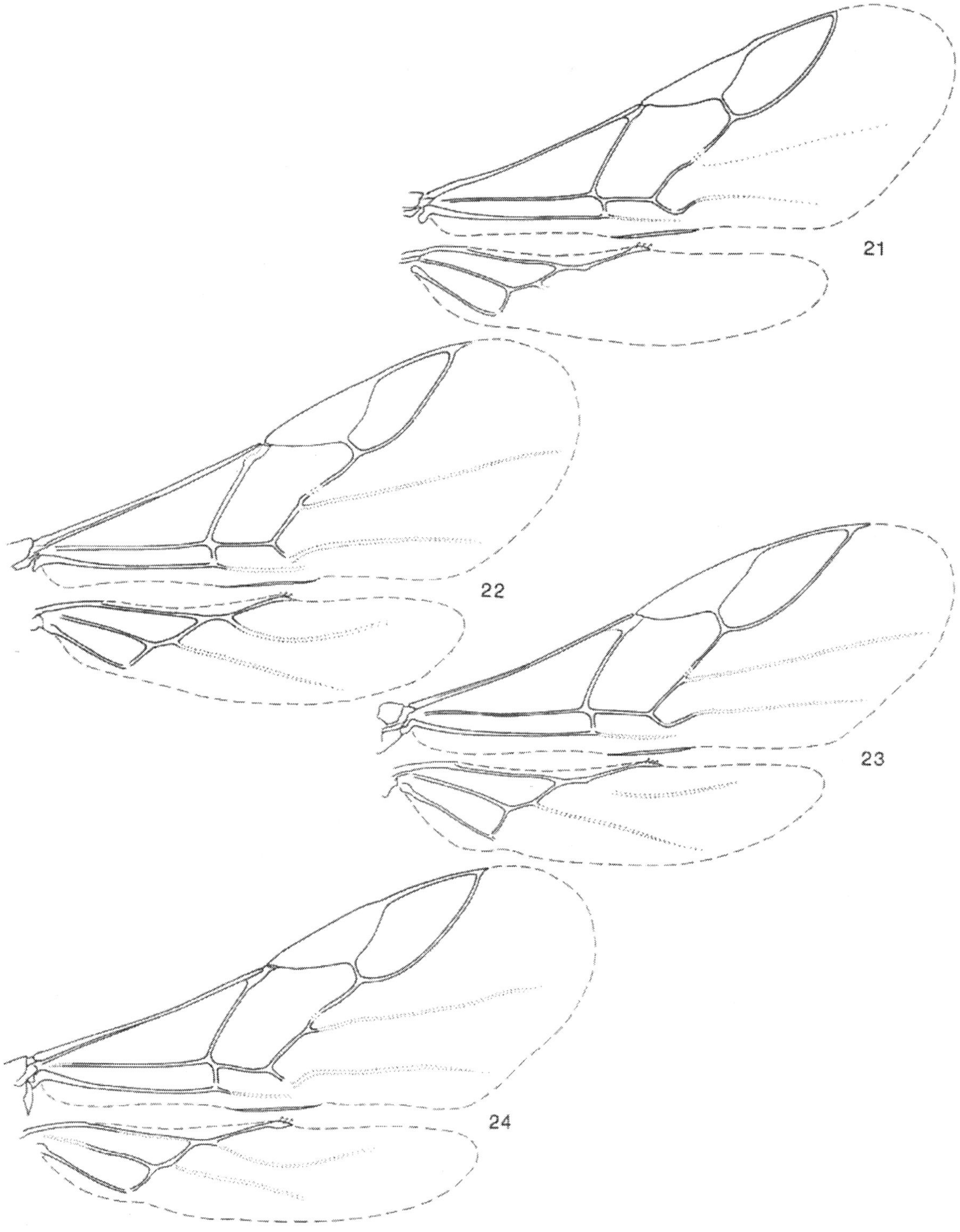

Figs 21-24: Wings of Euphorinae, 21, *Microctonus hyperodae* Loan; 22, *Lecythodella* sp.; 23, *Townesilitus* sp.; 24, *Marshiella* sp.

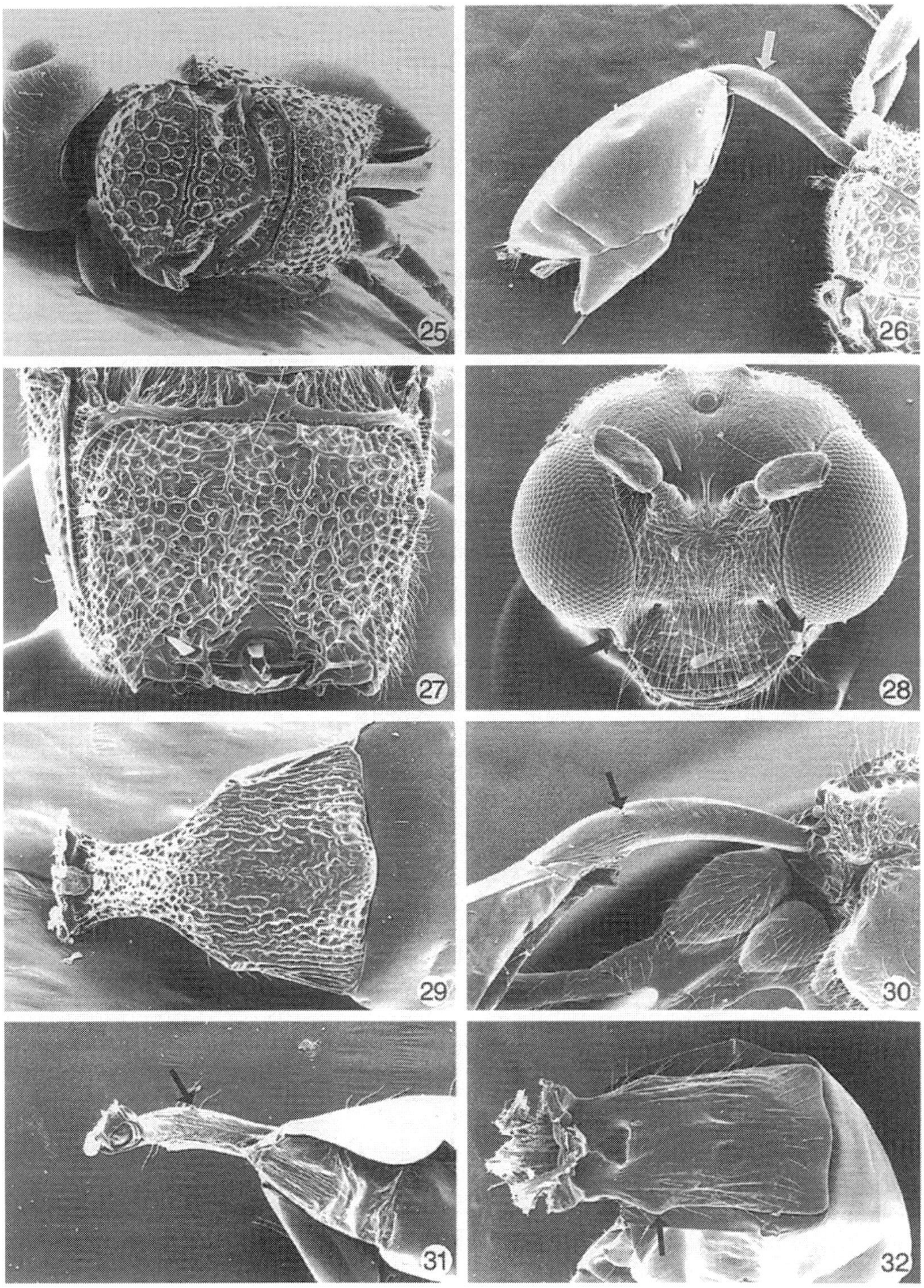

Figs 25-32: Euphorinae. 25, mesosoma of *Aridelus nigrator* (F.), dorsal view; 26, metasoma of *Aridelus nigrator* (F.), lateral view, arrow = petiole; 27, propodeum of *Chrysopophthorus* sp.; 28, head of *Chrysopophthorus* sp., frontal view, arrows = malar suture; 29-'32, petioles: 29, *Dinocampus coccinellae* (Schrank), dorsal view; 30, *Syntretus* sp., lateral view, arrow = spiracle; 31, *Euphoriella* sp., ventral view, arrow = spiracle; 32, *Centistes* sp., dorsal view, arrow = spiracle.

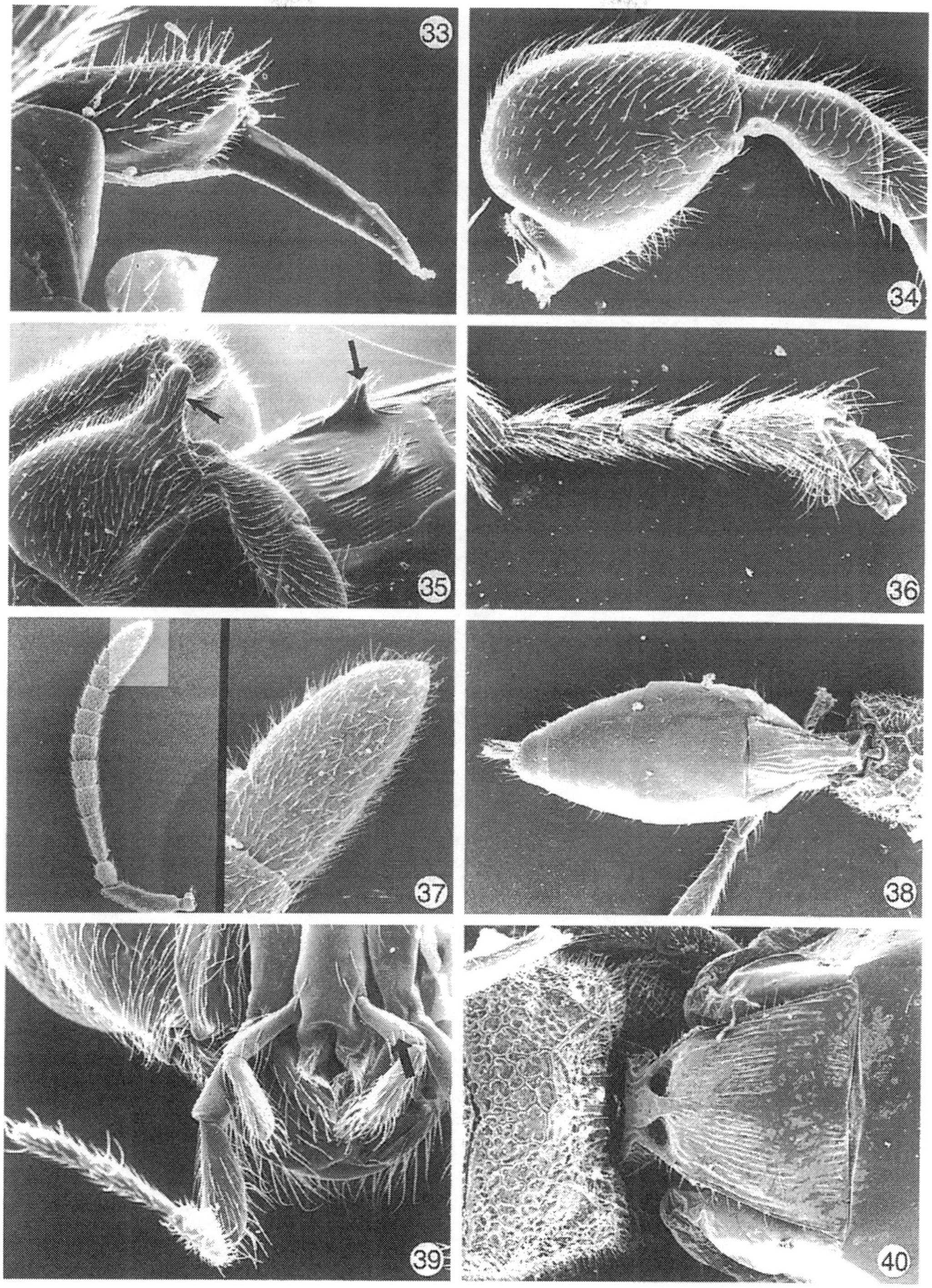

Figs 33-40: Euphorinae. 33, ovipositor and sheath of *Centistes* sp., lateral view; 34, hind coxa of *Centistes* sp.; 35, base of hind leg and metasoma of *Allurus* sp., arrows = apical tooth on coxa and teeth on sternum 4; 36, fore tarsus of *Litostolus* sp.; 37, antenna of *Ropalophorus wisconsinensis* Shenefelt, with close-up of antennal club; 38, metasoma of *Litostolus* sp.,. dorsal view; 39, labial palp of *Dinocampus coccinellae* (Schrank), arrow = basal segment; 40, petiole of *Ropalophorus clavicornis* (Wesmael), dorsal view.

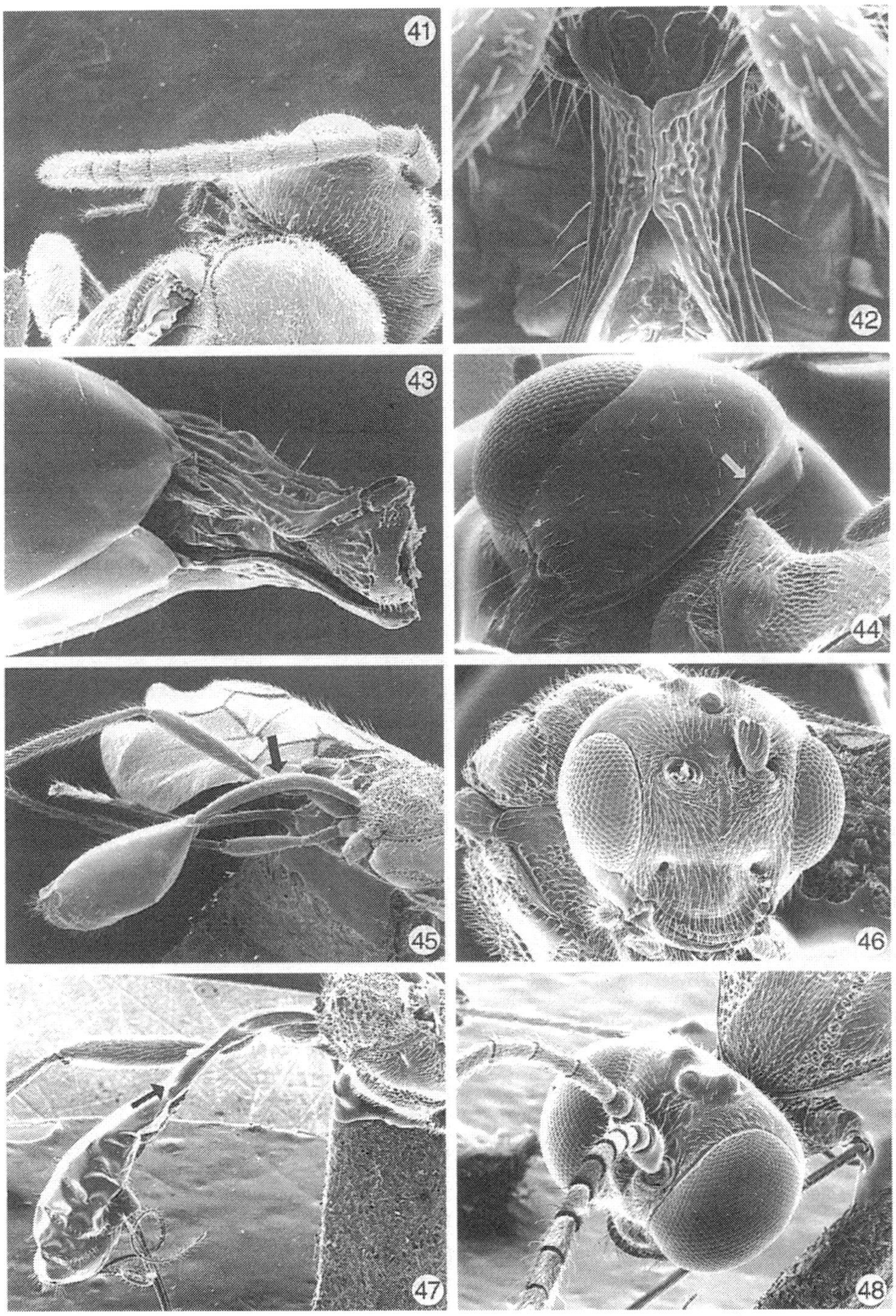

Figs 41-48: Euphorinae. 41, antenna of *Ropalophorus clavicornis* (Wesmael); 42, petiole of *Peristenus* sp., ventral view; 43; petiole of *Leiophron* sp., ventral view; 44, head of *Leiophron*. sp., lateral view, arrow = dorsal end of occipital carina; 45, metasoma of *Wesmaelia lizanoi* Shaw, arrow = petiole; 46, face and frons of *Wesmaelia lizanoi* Shaw; 47, metasoma of *Orionis* sp., lateral view, arrow = posterior end of petiole; 48, head of *Orionis* sp. showing sculpture on frons.

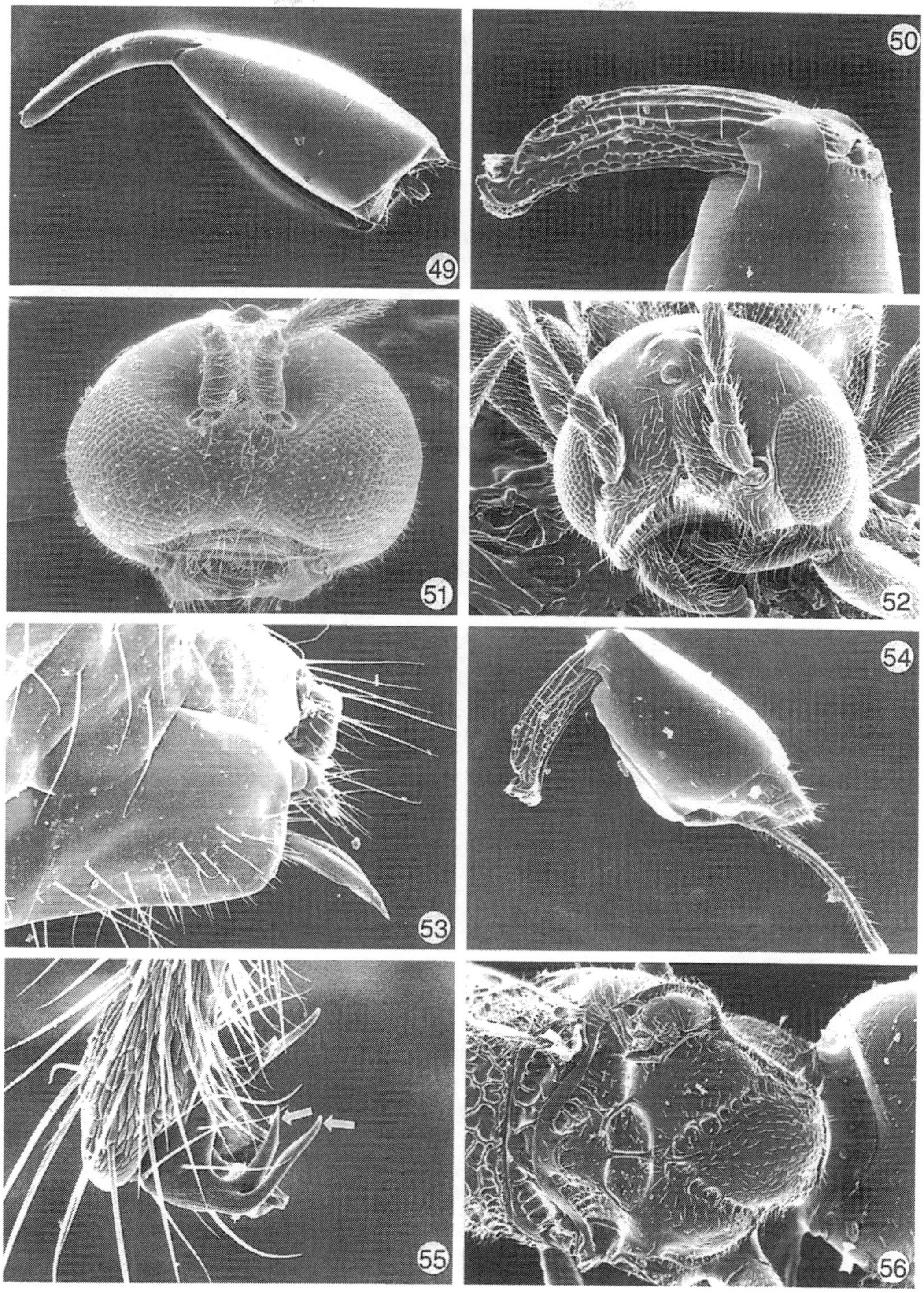

Figs, 49-56: Euphorinae, 49, petiole of *Holdawayella tingiphaga* Loan, lateral view; 50, petiole of *Cryptoxilos convergens* Muesebeck, lateral view; 51, face of *Cryptoxilos convergens* Muesebeck; 52, face of *Plynops hansoni* Shaw; 53, ovipositor of *Euphoriella* sp., lateral view; 54, metasoma of *Cryptoxilos convergens* Muesebeck showing elongate ovipositor; 55, tarsal claws of *Syntretus* sp., arrows = apical points of a single claw; 56, mesonotum of *Bracteodes* sp., dorsal view.

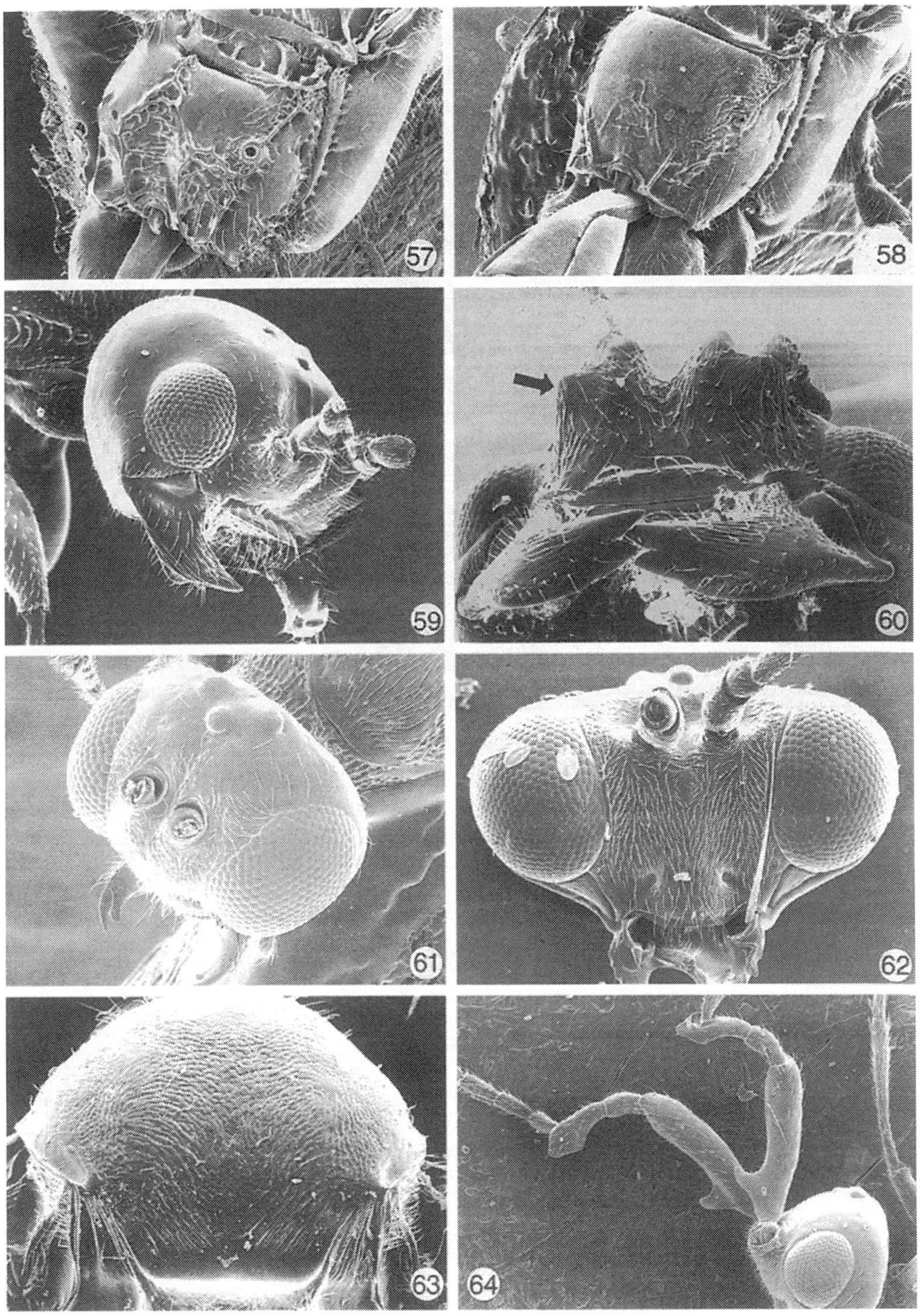

Figs 57-64: Euphorinae. 57, propodeum of *Syntretus* sp.; 58, propodeum of *Falcosyntretus* sp.; 59, head of *Cosmophorus* sp.; 60, face of *Cosmophorus* sp., antero-ventral view, arrow = prominent antennal base; 61, head of *Microctonus* sp., dorsal·view; 62, head of *Loxocephalus boops* (Wesmael), frontal view; 63, mesoscutum of *Loxocephalus boops* (Wesmael); 64, head and base of antenna of *Streblocera* sp.

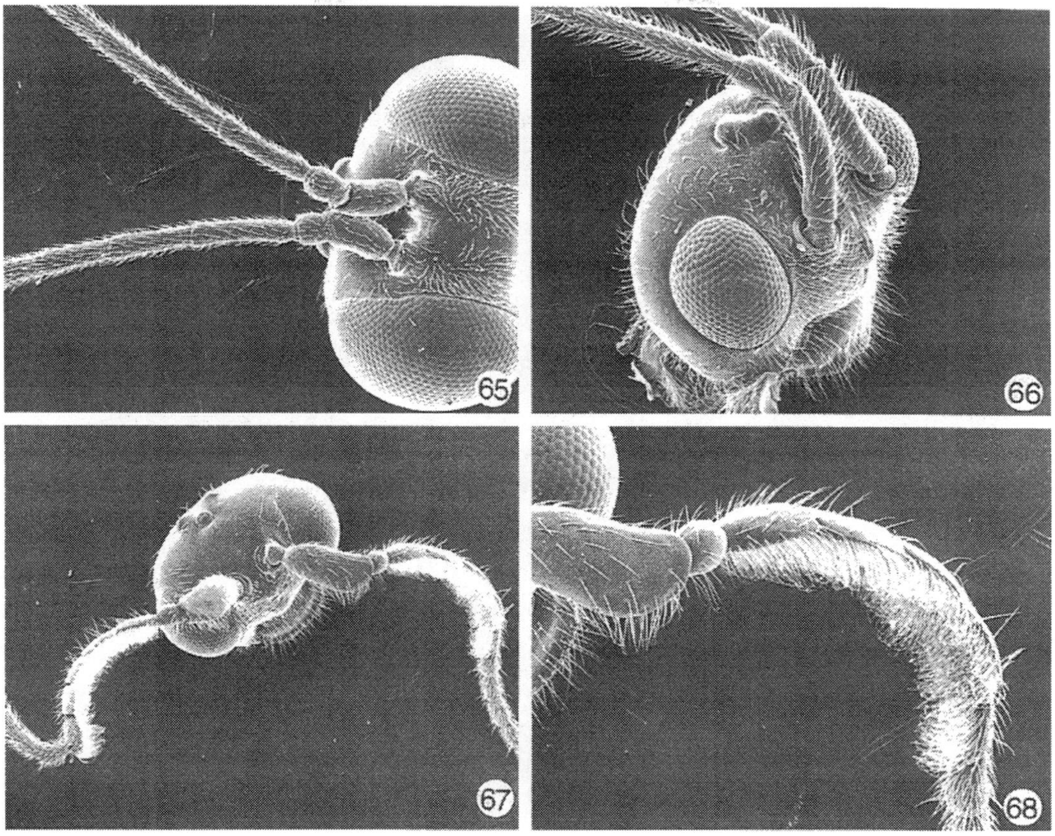

Figs 65-68; Euphorinae, 65, antenna of *Townesilitus* sp.; 66, scape of *Lecythodella* sp.; 67, antenna of *Marshiella* sp.; 68, base of antenna of *Marshiella* sp.

EUPHORINAE UPDATE

S.R. Shaw
Department of Ecosystem Science and Management, University of Wyoming, Laramie, Wyoming 82071, U.S.A.
braconid@uwyo.edu

The Euphorinae chapter of the manual continues to work well for the identification of New World genera, although controversy has continued over the definition of the subfamily and the relationships among genera. Stigenberg *et al.*, examined the phylogenetic relationships within the Euphorinae based on molecular data (2015), and proposed that the subfamily be expanded to include the tribes Meteorini and Planitorini (*Mannokeraia+Planitorus*). They presented a revised higher-level classification of the Euphorinae in which 52 extant genera and 14 tribes are recognized. They reinstated the genus *Microctonus* as belonging to the tribe Perilitini and synonymized *Ussuraridelus* with *Holdawayella*. They proposed the following tribal rearrangements: *Spathicopis* and *Stenothremma* were transferred to Perilitini; *Tuberidelus, Eucosmophorus* and *Plynops* to Cosmophorini; *Ecclitura* to Dinocampini; *Chrysopophthorus, Holdawayella* and *Wesmaelia* to Helorimorphini; and *Proclithroporus* and *Heia* to Townesilitini. The monotypic tribe Cryptoxilonini was synonymized with Cosmophorini. The genera *Pygostolus* and *Litostolus* were placed in a separate tribe, Pygostolini, previously recognized as a subtribe of the Centistini.

Annotated list of recent papers on New World Euphorinae

Achterberg C van. 1994. The Palearctic species of the genus *Chrysopopthorus* Goidanich (Hymenoptera: Braconidae: Euphorinae). *Zoologische Mededelingen, Leiden* 68(25): 301-307.
Revised the Palearctic species of *Chrysopopthorus* and discusses their biology.

Achterberg C van, Quicke DLJ. 2000. The Palaeotropical species of the tribe Cosmophorini Capek (Hymenoptera: Braconidae: Euphorinae) with description of twenty-two new species. *Zoologische Mededelingen, Leiden* 74(19): 283-338.
Reviews the Palaeotropical species of the tribe Cosmophorini and describes twenty-two new species.

Achterberg C van, Haeselbarth E. 2003. Revision of the genus *Syntretus* Foerster (Hymenoptera: Braconidae: Euphorinae) from Europe. *Zoologische Mededelingen, Leiden* 77 (2): 9-78.
Revised the *Syntretus* species of Europe, and treat *Falcosyntretus* as a junior synonym of *Syntretus*.

Achterberg C van, Shaw SR. 2000. Two new species of the genus *Centistina* Enderlein (Hymenoptera: Braconidae: Euphorinae) from Costa Rica. *Zoologische Mededelingen, Leiden* 74(4): 63-73.
Provides descriptions of two new species of *Centistina* from Costa Rica.

Achterberg C van, Shaw SR. 2001. *Tainiterma*, a new genus of the subfamily Euphorinae (Hymenoptera: Braconidae) from Vietnam and China. *Zoologische Mededelingen, Leiden* 75(3): 69-78.
Described *Tainiterma*, a new euphorine genus from Vietnam and China, with a peculiar flattened ovipositor.

Aguirre H, de Almeida LF, Shaw SR. 2016. Revision of the genus *Centistes* (Hymenoptera: Braconidae: Euphorinae: Centistini) of Costa Rica. *Zootaxa* (manuscript accepted for publication)
The *Centistes* species of Costa Rica are revised, including the description of 23 new species, and providing a key to all known Neotropical species.

An T-H, Kim H, Seo H-Y, Kim K-G, Ku D-S. 2015. Notes on the Genus *Cosmophorus* Ratzeburg (Hymenoptera: Braconidae: Euphorinae) from Korea. *Korean Journal of Applied Entomology* 54: 39-41.
Reviewed the species of *Cosmophorus* from Korea.

Belokobylskij SA. 2000a. A new genus and subgenus of the subfamily Euphorinae (Hymenoptera: Braconidae) from East Asia. *Zoologische Mededelingen, Leiden* 73: 255-267.
Describes a new euphorine genus, *Mama*, from East Asia, which has since been discovered in North America.

Belokobylskij SA. 2000b. On the Asian species of the genus *Streblocera* Westwood, 1833 (Hymenoptera: Braconidae, Euphorinae) with a key to the eastern Palaearctic species (continuation). *Entomologische Zeitschrift mit Insektenbörse* 110: 290-297
Provided a key to the eastern Palearctic species of *Streblocera*.

Belokobylskij SA, Ku DS. 1998. Notes on the Korean species of the Genus *Streblocera* (Insecta: Hymenoptera: Braconidae) with description of a new species and a key to Korean species. *Korean Journal of Systematic Zoology* 14: 319-325.
Reviewed the *Streblocera* species of Korea.

Bortoni MA, Shimbori EM, Shaw SR, Da Silva Souza-Gessner C, Penteado-Diaz A. 2016. A review of the genus *Orionis* Shaw (Hymenoptera: Braconidae: Euphorinae) and first records of the genus from South America and Oriental Region. *Zookeys* 4208 (3): 249-260.
Revised the species of *Orionis* and provided the first records of the genus from mainland South America and the Oriental region.

Cabrera Walsh G, Athanas MM, Salles LA, Schroder R.F. 2003. Distribution, host range, and climatic constraints on *Centistes gasseni* (Hymenoptera: Braconidae), a South American parasitoid of cucumber beetles, *Diabrotica* spp. (Coleoptera: Chrysomelidae). *Bulletin of Entomological Research* 93(6): 561-567.
Studied the distribution, host range, and climatic constraints on *Centistes gasseni*, a South American parasitoid of cucumber beetles of biocontrol importance.

Chen X, Achterberg C van. 1997. Revision of the subfamily Euphorinae (excluding the tribe Meteorini Cresson) (Hymenoptera: Braconidae) from China. *Zoologische Verhandelingen* 313: 1-217.
Provided a revision of the subfamily Euphorinae from China.

Chen X, He J, Achterberg C van, Ma Y. 2001. A new species of the genus *Cryptoxilos* Viereck (Hymenoptera: Braconidae: Euphorinae) from China. *Entomologica Fennica* 12: 108-111.
Described a new *Cryptoxilos* species from China.

Chen X, Belokobylskij SA., He J-H, Ma Y. 2000. An Outline of the taxonomic studies of the subfamily Euphorinae (Excluding the genus Meteorus Haliday) (Hymenoptera: Braconidae) from China, with a Checklist of the Chinese species. *Entomotaxonomia* 22(1): 65-70.
Reviewed the Euphorinae of China and provided a checklist of Chinese species.

Day WH, Baird CR, Shaw SR. 1999a. A new native species of *Peristenus* (Hymenoptera: Braconidae) parasitizing *Lygus hesperus* (Hemiptera: Miridae) in Idaho: biology, importance, and description. *Annals of the Entomological Society of America* 92: 370-375.

A new *Peristenus* species parasitizing *Lygus* in Idaho, and its biocontrol significance, was reported.

Day WH, Baird CR, Shaw SR, Tropp JM. 1999b. Discovery of a new species of parasite of *Lygus hesperus* in Idaho, and research for its use for the biological control of *Lygus lineolaris* in alfalfa and other crops in the mid-Atlantic states. *Proceedings of the 33rd Northeast Regional Field Crops Insect Conference* 22.
A new *Peristenus* species parasitizing *Lygus* in Idaho, and its biocontrol significance, was reported.

Pucci TM. 2013. Contributions to the classification of the North American Microtonus (Braconidae: Euphorinae). *Zootaxa* 3725: 1-150.

Foissner W, Achterberg C van. 1997. The valid name for the genus *Loxocephalus* Foerster, 1862 (Insecta, Hymenoptera: Braconidae), preoccupied by *Loxocephalus* Eberhard, 1862 (Protozoa: Ciliophora). *Zoologische Mededelingen, Leiden* 71(4): 31-32.
Showed that the genus name *Loxocephalus* was preoccupied, therefore the correct name for that euphorine genus is *Myiocephalus.*

Fox AS, Shaw SR, Dosdall LM, Lee B. 2004. *Microctonus melanopus* (Ruthe) (Hymenoptera: Braconidae), a parasitoid of adult cabbage seedpod weevil (Coleoptera: Curculionidae): distribution in Southern Alberta and female diagnosis. *Journal of Entomological Science* 39(3): 350-361.
Reported on the discovery of *Microctonus melanopus* (Ruthe) in North America, a known European parasitoid of adult cabbage seedpod weevil, provided diagnostic characters, and discussed the biocontrol significance.

Gloag RS, Shaw SR, Burwell C. 2009. A new species of *Syntretus* (Hymenoptera: Braconidae: Euphorinae) parasitizes the stingless bee *Trigona carbonaria* Smith (Hymenoptera: Apidae: Meliponinae). *Australian Journal of Entomology* 48(1): 8-14.
Described a new species of *Syntretus* that parasitizes adult workers of stingless bees.

Goulet H, Mason PG. 2006. Review of the Nearctic species of *Leiophron* and *Peristenus* (Hymenoptera: Braconidae: Euphorinae) parasitizing *Lygus* (Hemiptera: Miridae: Mirini). *Zootaxa* 1323: 1-118.
Reviewed the Nearctic species of *Leiophron* and *Peristenus* parasitizing *Lygus* bugs.

Lee H-R, An T-H, Byun B-K, Ku DS. 2016. Five newly recorded species of the genus *Streblocera* Westwood (Hymenoptera: Braconidae: Euphorinae) from Korea. *Journal of Asia-Pacific Biodiversity* 9(4): 468-471.
Described five new species of *Streblocera*.

McDonald MJ. 2001. Systematic Studies in Euphorinae (Hymenoptera: Braconidae), with a Revision of the New World Species of *Aridelus*. Ph.D. Thesis, Cornell University, Ithaca, NY.
Revised the New World species of *Aridelus* in a Ph.D. thesis, but this work has not been formally published so the proposed names of his "new species" remain unavailable. Nevertheless, the work remains helpful for correct identification of the already-named species.

Papp J, Chou LY. 1995. The genus *Wesmaelia* Foerster of Taiwan (Hymenoptera: Braconidae: Euphorinae*). Chinese Journal of Entomology* 15(4): 345-354.
Revised the genus *Wesmaelia* for Taiwan and presented a key for the *Wesmaelia* species of the world.

Papp J, Shaw SR. 2000. A study of the genus *Falcosyntretus* Tobias from the new world with five new species and a key to known species (Hymenoptera: Braconidae: Euphorinae). *Proceedings of the Entomological Society of Washington*. 102(3): 634-642.

Poinar G Jr, Shaw SR. 2016. Endoparasitism of a Cretaceous adult weevil by a euphorine wasp (Hymenoptera: Braconidae). *Neures Jahrbuch für Geologie und Paläontologie* 282(1): 109-133.
Reported the discovery of a euphorine wasp cocoon attached to an adult weevil in amber of Cretaceous age.

Robinson WS, Barbosa D, Kulikowski A, Shaw SR. 2015. Cloud forest studies of mating and defensive behavior of the parasitoid wasp, *Napo townsendi* Shaw (Hymenoptera: Braconidae: Euphorinae). *International Journal of Tropical Insect Science* 35(3): 103-116.
The mating and defensive behavior of *Napo townsendi* was explored.

Shaw SR. 1998. In: Wharton RA, Marsh PM, Sharkey MJ. (eds.). *Subfamilia Euphorinae. Manual para los generos de la familia Braconidae (Hymenoptera) de Nuevo Mundo*. Edición por Español. Publicado por the International Society of Hymnopterists. 238-259.
Provided a Spanish-language key to the Euphorinae genera of the New World.

Shaw SR. 2002. Two new species of *Betelgeuse* from Mexico (Hymenoptera: Braconidae: Euphorinae). *Pan-Pacific Entomologist* 78(3): 188-196.
Two new species of *Betelgeuse* from Mexico were described.

Shaw SR. 2004. Essay on the evolution of adult-parasitism in the subfamily Euphorinae (Hymenoptera: Braconidae). *Proceedings of the Russian Entomological Society, St. Petersburg* 75(1): 1-15. (English with Russian abstract).
Discussed the evolution of adult-parasitism in the subfamily Euphorinae, and introduced the term "imagobiosis" for the phenomenon of parasitism of adult insect hosts directly as displayed in many euphorines.

Shaw SR. 2012. A new genus and new species of Dinocampini (Hymenoptera: Braconidae: Euphorinae) from Napo province in Ecuador. *International Journal of Tropical Insect Science* 32: 101-107.
Described a new genus and new species of Dinocampini, *Napo townsendi*, from Napo province in Ecuador.

Shaw SR. 2014. *Planet of the Bugs*. University of Chicago Press, Chicago. 239 pp. ISBN 978-0-226-16361-1 (cloth), ISBN 978-0-226-16375-8 (e-book).
Events leading to the discovery of *Napo townsendi* are discussed in Chapter 10.

Shaw SR, Berry JA. 2005. Two new *Cryptoxilos* species (Hymenoptera: Braconidae: Euphorinae) from New Zealand and Fiji parasitizing adult Scolytidae (Coleoptera). *Invertebrate Systematics* 19: 371-381.
Described two new species of *Cryptoxilos* from New Zealand reared from bark beetles.

Shaw SR, Marsh PM. 2000. Revision of the enigmatic genus *Marshiella* Shaw in the New World with description of three new species (Hymenoptera: Braconidae, Euphorinae). *Journal of Hymenoptera Research* 9(2): 277-287.
Revised the genus *Marshiella*, described three new species, and provided a key to New World species.

Shaw SR, Papp J. 2000. A study of the genus *Falcosyntretus* Tobias from the New World with five new species and a key to the known species (Hymenoptera: Braconidae, Euphorinae). *Proceedings of the Entomological Society of Washington* 102(3): 634-642.
Revised *Falcosyntretus* for the New World, describing five new species, and providing a key to known species.

Shaw SR, Salerno G, Colazza S, Peri E. 2001. First record of *Aridelus rufotestaceus* Tobias (Hymenoptera: Braconidae, Euphorinae) parasitizing *Nezara viridula* nymphs (Heteroptera: Pentatomidae) with observations on its immature stages and development. *Journal of Hymenoptera Research* 10(2): 131-137.

Studied *Aridelus rufotestaceus* Tobias parasitizing *Nezara viridula* nymphs (Pentatomidae) and provided observations on its immature stages and development. The species has been imported and studied in the USA as a biocontrol organism for suppressing Southern Green Stinkbug.

Simbolotti G, Villemant C, Achterberg C van. 2004. Notes on the genus *Mama* Belokobylskij (Hymenoptera: Braconidae: Euphorinae), and on the use of monotype taxa. *Zoologische Mededelingen, Leiden* 78(26): 451-459.

Provided a detailed discussion of diagnostic characters for the genus *Mama*.

Stigenberg J. 2012. *Spathicopis* van Achterberg, 1977 (Braconidae, Euphorinae) a new wasp genus for Sweden, with a spoon shaped ovipositor. *Entomologisk Tidskrift* 133(4): 169-172.

Reported on the discovery of *Spathicopis* in Sweden and provided improved diagnostic characters for the genus, which is known to occur in North America.

Stigenberg J, Boring CA, Ronquist F. 2015. Phylogeny of the parasitic wasp subfamily Euphorinae (Braconidae) and evolution of its host preferences. *Systematic Entomology* 40(3): 570-591.

Examined the phylogenetic relationships within the Euphorinae based on molecular data, and provided a revised tribal classification for the genera.

Williams L, Logarzo GA, Shaw SR, Price LD, Manrique V. 2004. A new species of *Leiophron* (Hymenoptera: Braconidae) from Argentina and Paraguay. *Annals of the Entomological Society of America* 96(6): 834-846.

Described a new species of *Leiophron* from Argentina and Paraguay of biocontrol importance.

Yang Z, Gu Y, Song Y. 2003. A new species in the genus *Ropalophorus* Curtis (Hymenoptera: Braconidae) from China, parasitizing adults of the bark beetle *Ips subelongatus* (Coleoptera: Curculionidae: Scolytinae), with a key to world species of the genus. *Zoologische Mededelingen, Leiden* 77 (36): 631-636.

Described new species of *Ropalophorus* from China, parasitizing adult *Ips* bark beetles, and provided a key to world species of the genus.

Zijp JP, Blommers LHM. 2002. Biology of *Centistes delusorius*, a parasitoid of adult apple blossom weevil. *Agricultural and Forest Entomology* 4: 275-282.

Provided details of the biology of *Centistes delusorius*, a parasitoid of adult apple blossom weevil.

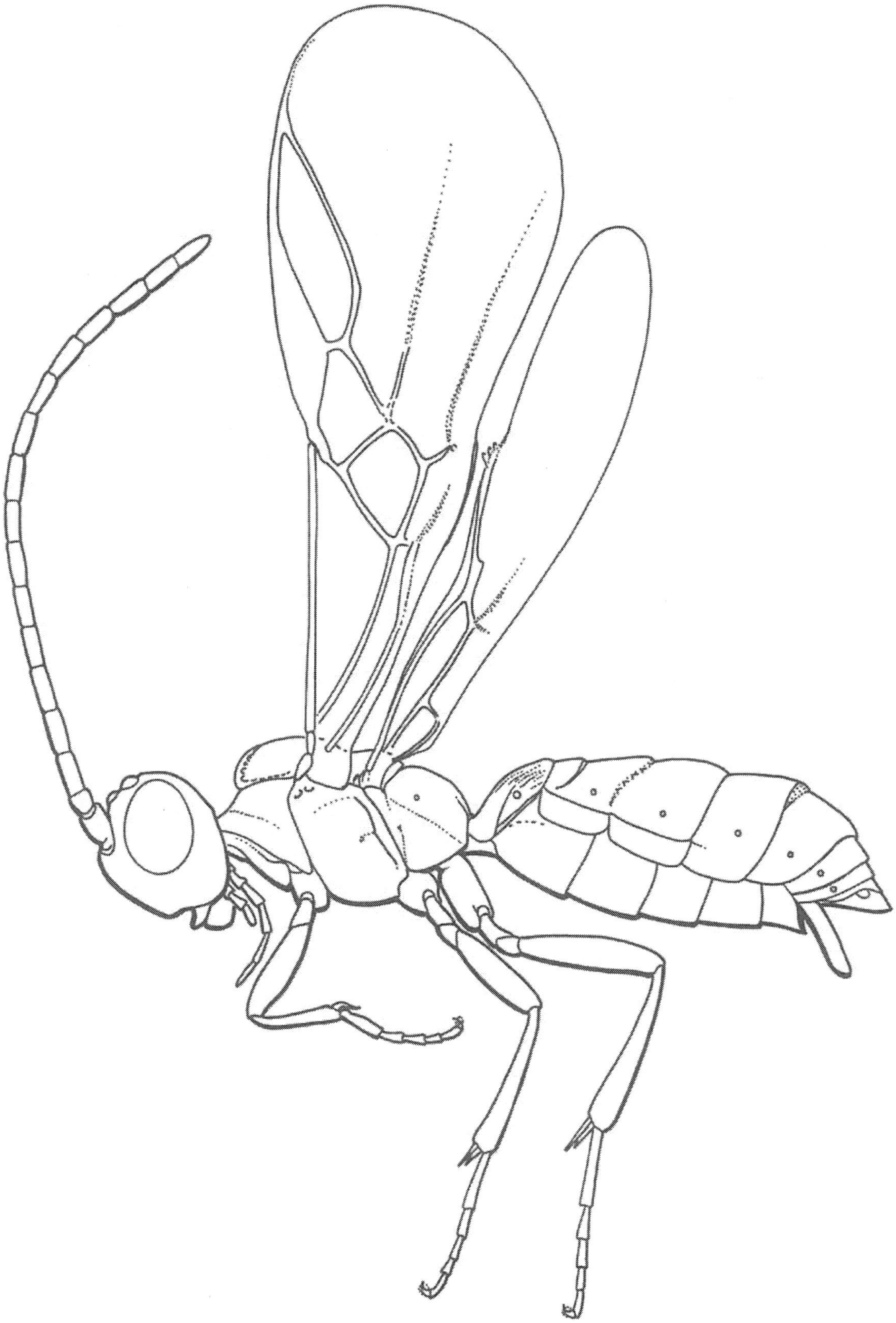

SUBFAMILY GNAMPTODONTINAE

R. A. WHARTON

Department of Entomology, Texas A&M University, College Station, Texas 77843, U.S.A.

INTRODUCTION. The subfamily Gnamptodontinae was established by van Achterberg (1983a) to accommodate a small group of species variously placed with the Opiinae, the Rogadinae (in the broad sense), or the Exothecinae (Marsh, 1979). A relationship to opiines is supported by the presence of Hagen's glands (Buckingham and Sharkey, 1988). The Gnamptodontinae, however, parasitize leaf-mining Lepidoptera rather than the cyclorrhaphous Diptera that serve as hosts for the Opiinae. Fischer (1972), Quicke and van Achterberg (1990), Whitfield and Wagner (1991) and Shaw and Huddleston (1991) provide summaries of available in formation. The latest revisions and keys to species are by Fischer (1977, 1987) and van Achterberg (1983a).

The use of the names *Gnamptodon* Haliday and Gnamptodontinae (rather than the more widely used *Gnaptodon* and Gnaptodontinae) is based on Opinion 1424 (1987) of the Commission on Zoological Nomenclature.

BIOLOGY. Gnamptodontines have been reared exclusively from leaf-mining Lepidoptera of the family Nepticulidae (van Achterberg, 1983a, b; Whitfield and Wagner, 1991; Shaw and Huddleston, 1991). Nearly all of the published host records are for Palaearctic species. Though many of the New World species have been reared, the host data remain largely unpublished (Whitfield and Wagner, 1991). Although apparently koinobiont parasitoids (Shaw and Huddleston, 1991), it is not yet known whether gnamptodontines are endo- or ectoparasitic. Detailed biological studies are lacking.

BIOLOGICAL CONTROL POTENTIAL. There has been no previous usage of gnamptodontines in biological control. Given the host restriction noted above, potential use in biological control would be limited to nepticulid pests (such as those on pecan).

COMMON GENERA. Three genera are presently included in the Gnamptodontinae, but only two of these, *Gnamptodon* and *Pseudognaptodon* Fischer, occur in the New World. Van Achterberg (1983a), following Fischer (1972, 1977, 1981) provisionally included *Liparophleps* Enderlein (= *Plesademon* Fischer) in the Gnamptodontinae. However, I have a series from Costa Rica which confirms the sexually dimorphic nature of the wing vein pattern, and concur with Marsh (1976) that *Liparophleps* is a synonym of *Semirhytus* Szépligeti and belongs in Doryctinae. There are 11 species described from the New World, with eight of these placed in *Gnamptodon* (Fischer, 1977; Marsh, 1979). Most of the New World species are undescribed, and previous characterizations of gnamptodontine genera (van Achterberg, 1983a, Whitfield and Wagner, 1991) need to be altered somewhat to accommodate variation in the New World species. Some New World *Gnamptodon*, for example, have both a postero-median pit on the mesonotum and a sculptured sternaulus. *Pseudognaptodon*, characterized by a wing vein loss relative to *Gnamptodon*, may eventually prove to be simply one or more species groups of *Gnamptodon*.

It is tempting to include *Exodontiella* Wharton in the Gnamptodontinae because of the similarities in body sculpture and venation between *Exodontiella* and the Palaearctic *Gnaptogaster* Tobias. This possibility merits further study, but since so little is known about *Exodontiella*, I prefer to retain it in the Opiinae (Wharton, 1988).

DISTRIBUTION. Worldwide, with *Gnamptodon* especially diverse in Australia and *Pseudognaptodon* containing a large number of undescribed Neotropical species. Gnamptodontines have not been well-studied outside the Palaearctic Region, and Fischer (1977) recorded only 3 species south of the United States.

DISTINGUISHING FEATURES. Gnamptodontines are characterized by the presence of a well-defined line or groove delimiting a basal-median, somewhat rectangular area on the second metasomal tergum (Fig. 3). This basal-median area is described in some publications as a raised swelling or elevation, but in several species it is fairly flat. Additionally, all species are quite small (usually less than 3mm), lack an epicnemial carina, and usually lack an occipital carina.

KEY TO THE NEW WORLD GENERA OF THE SUBFAMILY GNAMPTODONTINAE

1. Fore wing r-m present (Fig. 1) .... ***Gnamptodon*** Haliday
Distribution: Alaska and Canada to Brazil; 8 described New World species (Whit- field and Wagner, 1991). Biology: reared from Nepticulidae. Reference: Fischer (1977).

\- Fore wing r-m absent (Fig. 2) .... ***Pseudognaptodon*** Fischer
Distribution: southeastern United States (North Carolina to Texas) to South America, including Chile; only 3 described species. Biology: reared from Nepticulidae, but no specific host records have been published (Whitfield and Wagner, 1991). Reference: Fischer (1977).

REFERENCES

Achterberg, C. van 1983a. Revisionary notes on the subfamily Gnaptodontinae, with description of eleven new species (Hymenoptera, Braconidae). Tijdschrift voor Entomologie 126: 25-57.

Achterberg, C. van 1983b. The preference of zoophagous Hymenoptera for certain types of plants as shown by the subfamily Gnaptodontinae (Braconidae), Verhandlungen SIEEC X. Budapest: 97-98.

Buckingham, G.R. and Sharkey, M.J. 1988. Abdominal exocrine glands in Braconidae (Hymenoptera). Pp. 199-242. In: Gupta, V. K. (Ed.). Advances in Parasitic Hymenoptera Research. Brill, Leiden.

Fischer, M. 1972. Hymenoptera Braconidae (Opiinae I).Das Tierreich 91: 1-621.

Fischer, M. 1977. Hymenoptera Braconidae (Opiinae II-Amerika). Das Tierreich 96: 1-1001.

Fischer, M. 1981. Eine neue Synonymie bei den Opiinae (Hymenoptera, Braconidae). Zeitschrift der Arbeitsgemeinschaft Österreichischer Entomologen 32:136.

Fischer, M. 1987. Hymenoptera: Opiinae III-äthiopische, orientalische, australische und ozeanische Region. Das Tierreich 104: 1-734.

Marsh, P.M. 1976. Pars 13. Braconidae 9, Doryctinae.P. 1331. In: Vecht, J. van der and Shenefelt, R.D. (Eds.). Hymenopterorum Catalogus (nova editio). Dr. W. Junk, The Hague.

Marsh, P.M. 1979. Family Braconidae. Pp. 144-295. In: Krombein, K.V., Hurd, P.D., Smith, D.R., and Burks, B.D. (Eds.). Catalog of Hymenoptera in America North of Mexico. Smithsonian Institution Press, Washington, D.C.

Opinion 1424. 1987. *Gnamptodon* Haliday, 1833 (Insecta, Hymenoptera): *Bracon pumilio* Nees, 1834 designated as type species. Bulletin of Zoological Nomenclature 44: 55-56.

Quicke, D.L.J. and Achterberg, C. van 1990. Phylogeny of the subfamilies of the family Braconidae (Hymenoptera: Ichneumonoidea). Zoologische Verhandelingen Leiden 258: 1-95.

Shaw, M. R. and Huddleston, T. 1991. Classification and biology of braconid wasps (Hymenoptera: Braconidae). Handbooks for the Identification of British Insects 7(11): 1-126.

Wharton, R.A. 1981. Classification of the braconid subfamily Opiinae (Hymenoptera). The Canadian Entomologist 120: 333-360.

Whitfield, J.B. and Wagner, D. L. 1991. Annotated key to the genera of Braconidae (Hymenoptera) attacking leaf-mining Lepidoptera in the Holarctic Region. Journal of Natural History 25: 733-754.

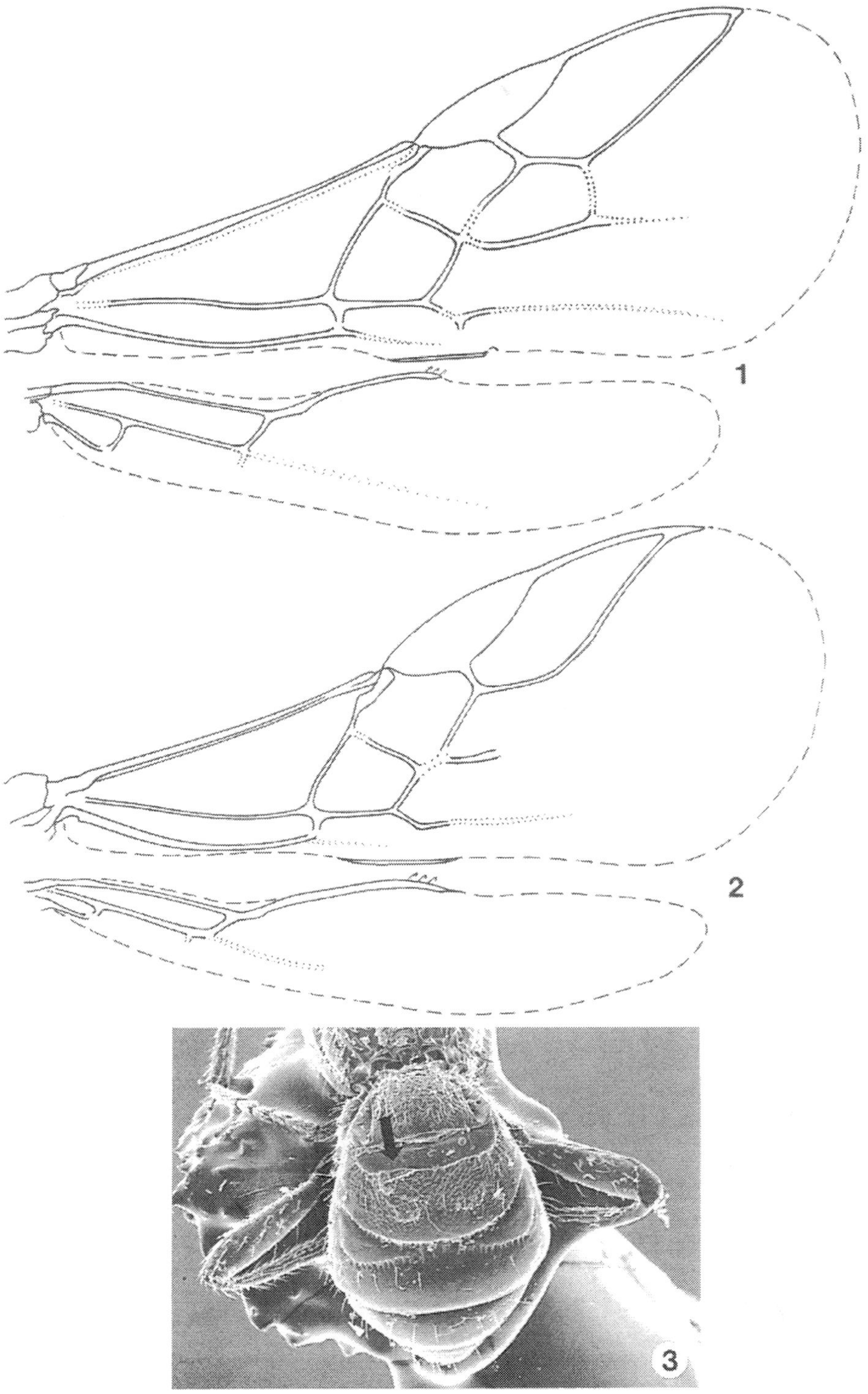

Figs 1-3: Gnamptodontinae. 1, wings of *Gnamptodon* sp.; 2, wings of *Pseudognaptodon* sp.; 3, metasoma of *Gnamptodon* sp., dorsal view, arrow = basal-median rectangular area on metasomal tergum 2.

GNAMPTODONTINAE UPDATE

R. R. Kula

Systematic Entomology Laboratory, Beltsville Agricultural Research Center, Agricultural Research Service, U.S. Department of Agriculture, c/o National Museum of Natural History, Smithsonian Institution, P.O. Box 37012, MRC-168, Washington, DC 20013-7012 U.S.A., Robert.Kula@ars.usda.gov*

Gnamptodontines have been reported as parasitoids primarily of leaf-mining lepidopterans in Nepticulidae (Yu *et al.*, 2012) but also Gracillariidae (Balevski and Tomov, 1997) and Heliozelidae (Williams, 2004). They are apparently koinobiont parasitoids (Shaw and Huddleston, 1991), but it is not yet known whether they are endo- or ectoparasitic. References to information about gnamptodontines published through December 2011 are available in Yu *et al.* (2012).

The presence of Hagen's glands suggested a relationship to opiines (Buckingham and Sharkey, 1988), but gnamptodontines parasitize leaf-mining lepidopterans rather than cyclorrhaphous Diptera as is the case for all species of Opiinae. Zaldívar-Riverón *et al.* (2006) recovered Gnamptodontinae in a clade consisting of Telengaiinae, Exothecinae, Alysiinae, Opiinae, and Braconinae in most analyses, with Gnamptodontinae and Telengaiinae sister taxa. Wharton *et al.* (2006) found Gnamptodontinae sister to either Braconinae or a clade consisting of Alysiinae, Opiinae, and Exothecinae. Further, *Exodontiella* Wharton was recovered with Gnamptodontinae in all analyses and, therefore, Wharton *et al.* (2006) transferred it from Opiinae. Sharanowski *et al.* (2011) consistently recovered a clade composed of Gnamptodontinae, Exothecinae, Alysiinae, and Opiinae. Unlike Zaldívar-Riverón *et al.* (2006), species of *Telengaia* Tobias were not included in either Wharton *et al.* (2006) or Sharanowski *et al.* (2011). Thus, there is a growing body of evidence that gnamptodontines are most closely related to species in the aforementioned subfamilies, but relationships among those subfamilies are unresolved.

Gnamptodontines are characterized by a weakly to strongly defined somewhat rectangular area on the second metasomal tergum. The area is only weakly swollen in species of *Exodontiella* (Wharton *et al.*, 2006). Five genera are presently included in Gnamptodontinae; three of those, *Pseudognaptodon* Fischer, *Gnamptodon* Haliday, and *Exodontiella*, occur in the New World with 28, eight, and two described species, respectively (Yu *et al.*, 2012).

Cirelli *et al.* (2002) described eight species of *Pseudognaptodon* from Brazil and provided a key to New World species. Williams (2004) described 17 species of *Pseudognaptodon* from the New World and divided the genus into two species groups. A key to New World species was provided but did not include the species described in Cirelli *et al.* (2002). Fischer (1977) is the most recent taxonomic treatment of New World species of *Gnamptodon*; the genus is in need of revision. Wharton (1978) proposed *Exodontiella* for two species with exodont mandibles and placed it in Opiinae based on features other than the mandibles. New species of *Exodontiella* have not been described since that time. Wharton (1997) noted similar body sculpture and wing venation between species of *Exodontiella* and the gnamptodontine genus *Gnaptogaster* Tobias, found only in the Palaearctic Region. Ultimately, Wharton *et al.* (2006) transferred *Exodontiella* to Gnamptodontinae based on results from analysis of molecular and morphological data.

Species of Gnamptodontinae are known from all zoogeographic regions except the Oceanic. Species richness is currently highest in the Neotropical and Palaearctic regions (Yu *et al.*, 2012). However, gnamptodontines have not been well-studied outside the Palaearctic Region; thus, global patterns of species richness are unclear at this time.

Recent papers on New World Gnamptodontinae

Balevski N, Tomov R. 1997. Mining moths from family Bucculatricidae, Nepticulidae and Tischeriidae (Lepidoptera) as hosts of parasitoids from family Braconidae (Hymenoptera). *Acta Entomologica* Bulgarica 1-2: 114-119.

Buckingham GR, Sharkey MJ. 1988. Abdominal exocrine glands in Braconidae (Hymenoptera). Pp. 199-242. In: Gupta, V. K. (Ed.). *Advances in Parasitic Hymenoptera Research*. Brill, Leiden.

Cirelli KRN, Braga SMP, Penteado-Dias AM. 2002. New species of *Pseudognaptodon* Fischer (Hymenoptera: Braconidae: Gnamptodontinae) from Brazil. *Zoologische Mededelingen, Leiden* 76: 89-96.

Fischer M. 1977. Hymenoptera Braconidae (Opiinae II-Amerika). *Das Tierreich* 96: 1-1001.

Sharanowski BJ, Dowling APG, Sharkey MJ. 2011. Molecular phylogenetics of Braconidae (Hymenoptera: Ichneumonoidea), based on multiple nuclear genes, and implications for classification. *Systematic Entomology* 36: 549-572.

Shaw MR, Huddleston T. 1991. Classification and biology of braconid wasps (Hymenoptera: Braconidae). *Handbooks for the Identification of British Insects* 7(11): 1-126.

Wharton RA. 1978. Exodontiellini, a new tribe of Opiinae with exodont mandibles (Hymenoptera: Braconidae). *Pan-Pacific Entomologist* 53: 297-303. (Volume dated October 1977; issued 14 February 1978.)

Wharton RA. 1997. Subfamily Gnamptodontinae. Pp. 256-259. In: Wharton, R.A., Marsh, P.M., and Sharkey, M.J. (Eds.). Manual of the New World Genera of the Family Braconidae (Hymenoptera). Special Publication No. 1. International Society of Hymenopterists, Washington, DC.

Wharton RA, Yoder MJ, Gillespie JJ, Patton JC, Honeycutt RL. 2006. Relationships of *Exodontiella*, a non-alysiine, exodont member of the family Braconidae (Insecta, Hymenoptera). *Zoologica Scripta* 35: 323-340.

Williams DJ. 2004. Revision of the genus *Pseudognaptodon* Fischer (Hymenoptera: Braconidae: Gnamptodontinae). *Journal of Hymenoptera Research* 13: 149-205.

Yu DS, van Achterberg K, Horstmann K. 2012. World Ichneumonoidea 2011. Taxonomy, Biology, Morphology and Distribution. USB Flash drive. Taxapad, Ottawa, Canada. www.taxapad.com.

Zaldívar-Riverón A, Mori M, Quicke DLJ. 2006. Systematics of the cyclostome subfamilies of braconid parasitic wasps (Hymenoptera: Ichneumonoidea): a simultaneous molecular and morphological Bayesian approach. *Molecular Phylogenetics and Evolution* 38: 130-145.

*Mention of trade names or commercial products in this publication is solely for the purpose of providing specific information and does not imply recommendation or endorsement by the USDA. USDA is an equal opportunity provider and employer. **The author thanks Robert A. Wharton for permission to use text from the Gnamptodontinae chapter of the Manual.

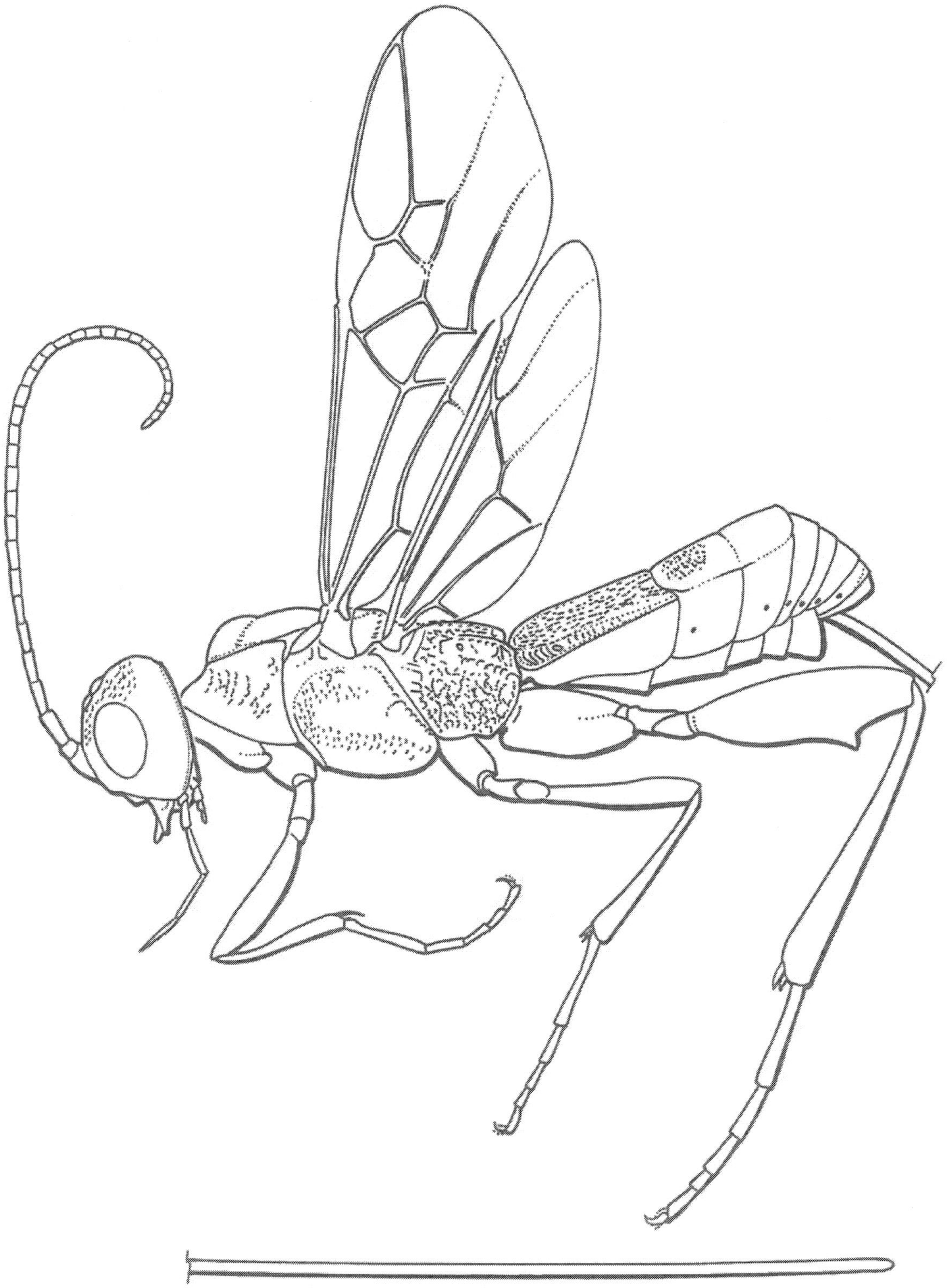

SUBFAMILY HELCONINAE

M. J. SHARKEY

Department of Entomology, University of Kentucky, S-225 Agricultural Science Building North, Lexington, Kentucky 40546, U.S.A.

INTRODUCTION. The subfamily Helconinae, as here defined, contains 14 genera in the New World and about 40 world-wide. The subfamily has been treated in many different manners over the years and in this text I have adopted a wide subfamilial concept. The Helconinae is comprised of four tribes, three of which occur in the New World. The fourth tribe, Brulleiini, has not been found in the New World. Because each of the three tribes occurring in the New World is so distinctive, they are treated separately in the text that follows. Some of these tribes may be monophyletic whereas others, e.g. Diospilini, appear to be less rigorously defined. There are also a number (5-8%) of New World species that do not fit easily into present generic concepts, especially in the Brachistini and Diospilini. Two taxa, Cenocoeliinae and Blacinae, which are sometimes included within the Helconinae, are recognized as separate subfamilies in this volume. They have been treated as subfamilies in recent publications (van Achterberg, 1988, 1994) and inclusion in the Helconinae s. l. would not make the monophyly of the Helconinae any more convincing than it is without them.

KEY TO THE NEW WORLD TRIBES OF THE SUBFAMILY HELCONINAE

1. Second submarginal cell of fore wing open, r-m absent (Figs 9-14) .... **Brachistini**
- Second submarginal cell of fore wing closed, r-m present (Figs 1-8) 2

2(1). Hind wing cross-vein a absent (Figs 5-8) or nearly so (Fig. 4); second submarginal cell of fore wing usually (95%) not narrowed anteriorly, with 3RSa often equal to or longer than 2M (Figs 4-8)............... **Diospilini**
- Hind wing cross-vein a present (Figs 1, 3); second submarginal cell of fore wing narrowed anteriorly, 3RSa shorter than 2M (Figs 1-3) .... **Helconini**

TRIBE HELCONINI

INTRODUCTION. The tribe Helconini is worldwide in distribution and contains about 12 genera, three of which are found in the New World. Shenefelt (1970) cataloged the World species of Helconini. The monotypic genus *Eumacrocentrus* Ashmead is almost certainly a derived member of *Helcon* Nees but formal synonymy requires a detailed cladistic analysis. There are no identification keys to New World species.

BIOLOGY. Members of the Helconini are solitary endoparasitoids of cerambycid beetle larvae and possibly other wood-boring beetles (Tobias, 1967; Shaw and Huddleston, 1991).

BIOLOGICAL CONTROL POTENTIAL. No attempts have been made to use members of the Helconini in classical biological control programs. Because they attack Cerambycidae, helconines may play a role in natural control of potential pests.

COMMON GENERA. Both *Helcon* and *Wroughtonia* Cameron are relatively rarely encountered in temperate North American forests. Because of their large size however, they are usually kept when encountered and thus are present in most collections. In temperate areas of South America *Helcon* is relatively common.

DISTRIBUTION. Cosmopolitan. In the New World, members appear to be restricted to temperate areas. In North America two species are found as far south as Mexico, and in South America some species of *Helcon* are found as far north as the Brazilian state of Paraña.

DISTINGUISHING FEATURES. Members of the Helconini may be recognized by the following combination of characters: hind wing with anal cross vein (a) present (Figs 1-3); fore wing with two anal cross veins (la and 2a) (Fig. l); second submarginal cell of fore wing present and narrowed anteriorly (Figs 1-3). The tribe Helconini contains rather large, robust braconids.

KEY TO THE NEW WORLD GENERA OF THE TRIBE HELCONINI

1. Hind femur with ventral tooth or rugose ridge (Fig. 23). Wings: Fig. 1. .... ***Wroughtonia*** Cameron
   Distribution: Holarctic; 8 described Nearctic species. Biology: endoparasitoids of larval Cerambycidae. References: Marsh (1979), van Achterberg (1987, as *Wroughtonia* and *Helconidea* Viereck).

\- Hind femur lacking ventral tooth or rugose ridge 2

2(1). Fore wing r less than one-third length of 3RSa (Fig. 2)....................................... ***Eumacrocentrus*** Ashmead
   Distribution: eastern U.S.A.; one described species. Biology: unknown, probably an endoparasitoid of larval Cerambycidae. Reference: Marsh (1979).

\- Fore wing r more than half length of 3RSa (Fig. 3) .... ***Helcon*** Nees
   Distribution: Holarctic and Neotropical, widespread in temperate and subtropical New World; 4 described Nearctic species, many undescribed Neotropical species. Biology: endoparasitoids of larval Cerambycidae. Reference: Marsh (1979).

TRIBE DIOSPILINI

INTRODUCTION. The tribe Diospilini is worldwide in distribution and contains about 13 genera, five of which are found in the New World. Shenefelt (1970) cataloged the World species of Diospilini. There are no identification keys to New World species, most of which are undescribed. Papp (1995) presented a key to the Neotropical genera of Diospilini. He includes *Austrodolops* Blanchard and *Aspigonus* Wesmael, which are not included here. Although I have not seen the type of *Austrodolops*, it is clear from the description and illustrations (Blanchard, 1936) that it does not belong in Diospilini. As suggested by Papp, the Doryctinae may be the correct placement for this genus. Papp's inclusion of *Aspigonus* among the Neotropical diospiline genera was based on the description of *A. stramineicolor* by Viereck (1913). Viereck stated that *stramineicolor* was most similar to *Baeacis abietis* (Ratzeburg) but considered *Baeacis* Förster to be a junior synonym of *Aspigonus*. Viereck therefore placed his new species in *Aspigonus*. The species probably belongs to the genus *Baeacis* or *Diospilus* Haliday. Although I have not seen the type species, there is no evidence to suggest that members of the genus *Aspigonus* are found in the New World.

BIOLOGY. The biology of the members of the Diospilini is largely unknown. Where known, species are solitary endoparasitoids of phytophagous beetles such as Curculionidae and Nitidulidae (Shaw and Huddleston, 1991). The biology of one species of *Diospilus* has been described (Osborne, 1960), and although the stage of the host attacked was not noted, egg-larval parasitism does occur in at least one other species (Parrott and Glasgow, 1916).

BIOLOGICAL CONTROL POTENTIAL. No attempts have been made to use members of the Diospilini in classical biological control programs. Because their hosts, especially the Curculionidae, are potential crop pests, some may play a role in natural control.

COMMON GENERA. *Diospilus* has more than 100 species in the New World (mostly undescribed) that are distributed throughout temperate and tropical areas. Other genera are comparatively rare.

DISTRIBUTION. Cosmopolitan.

DISTINGUISHING FEATURES. The Diospilini, as currently defined, does not appear to be monophyletic and there are a number of species that do not fit into current generic concepts. Diagnosis of the group is, therefore, rather difficult. *Diospilus* and *Baeacis* have a peculiar shape to the second submarginal cell in that it is parallel sided or even slightly wider anteriorly than posteriorly (Figs 7, 8). This, in combination with the presence of an occipital carina and non-petiolate first metasomal segment, are sufficient to distinguish them from other braconids. The only known species of *Vadum* Mason is rather slender with two teeth at the apex of the clypeus (Fig. 24). *Taphaeus* Wesmael has the following unique combination of characters; occipital carina present; second submarginal cell of fore wing present and narrowed anteriorly (Fig. 6); IRS of fore wing absent (Fig. 6); 2cu-a of fore wing absent; clypeus lacking teeth; metasoma not petiolate. *Topaldios* Papp is similar to *Taphaeus* except that IRS of the fore wing is present (Fig. 4) and 2cu-a of the fore wing is either present or absent.

KEY TO THE NEW WORLD GENERA OF THE TRIBE DIOSPILINI

1. Fore wing 3RSa shorter than 2M, second submarginal cell thus narrowed anteriorly (Figs 4-6) 2

\- Fore wing 3RSa not shorter than 2M, second submarginal cell therefore not narrowed anteriorly (Figs 7, 8) 4

2(1). Fore wing 1RS present (Fig. 4)..***Topaldios*** Papp

Distribution: Argentina and Chile; 2 described species, several more undescribed. Biology: unknown. Reference: Papp (1995).

\- Fore wing 1RS absent (Figs 5, 6) 3

3(2). Apex of clypeus with two blunt teeth (Fig. 24). Fore wing 2cu-a present (Fig. 5)...................***Vadum*** Mason

Distribution: central and southern U.S.A.; one described species. Biology: unknown. Reference: Mason (1987).

\- Apex of clypeus straight or evenly rounded. Fore wing 2cu-a absent (Fig. 6). ....***Taphaeus*** Wesmael

Distribution: Holarctic; fewer than 10 species, mostly undescribed. Biology: recorded as a parasitoid of Melandryidae in Europe. Reference: Marsh (1979).

4(3). Apex of clypeus with blunt median tooth (Fig. 25). Wings: Fig. 7...***Baeacis*** Förster

Distribution: Cosmopolitan, relatively rare in Holarctic and New World; one un- described Nearctic species reared from *Erinobius bicolor* White. A handful of undescribed Neotropical species will key out here but their exact affinities are un- known. No keys. Biology: recorded as endoparasitoids of Anobiidae and Cerambycidae in Europe. Reference: Marsh (1979).

\- Apex of clypeus without median tooth, clypeus evenly rounded (Fig. 26) or with two teeth (as in Fig. 24). Wings: Fig. 8 ....***Diospilus*** Haliday

Distribution: Cosmopolitan, 5 described Nearctic species; about 20 undescribed Nearctic species are known and many more undescribed Neotropical species. Biology: endoparasitoids of Coleoptera including Anobiidae, Nitidulidae, Curculionidae and perhaps other families. Egg-larval parasitism is known for one European species (Parrott and Glasgow, 1916) but whether this biology is common to all species is unknown. Reference: Mason (1968, re: *Diospilus fomitis* Mason).

TRIBE BRACHISTINI

INTRODUCTION. The tribe Brachistini is cosmopolitan and contains about 11 genera, six of which are found in the New World. Shenefelt (1970) cataloged the World species of Brachistini. Martin (1956) revised and keyed the species of the genera *Aliolus* Say (including the species of *Nealiolus* Mason), *Triaspis* Haliday, *Sehizoprymnus* Förster, and *Urosigalphus* Ashmead in America north of Mexico. Gibson (1972a, b, 1974) described and keyed the New World species of *Urosigalphus*.

BIOLOGY. Members of the Brachistini are known to be egg-larval endoparasitoids of Coleoptera, usually in the families Curculionidae or Bruchidae (Clausen, 1954; Haeselbarth, 1962; Parnell, 1964; Capek, 1971; Alauzet, 1987; Shaw and Huddleston, 1991). There is one report (Charlet, 1994) of *Nealiolus curculionis* (Fitch) depositing eggs into the early first instar larvae of the sunflower stem weevil *Cylindrocopturus adspersus* (LeConte). Fairly complete biological studies have been published on *Eubazus semirugosus* (Nees), a parasite of *Pissodes* spp. (Curculionidae). Several less detailed biologies have been described for species of *Triaspis* (Beirne, 1946; Berry, 1947; Obrtel, 1960). Shaw and Huddleston (1991) summarized the known biological information.

BIOLOGICAL CONTROL POTENTIAL. *Nealiolus curculionis* is an important parasitoid of the plum curculio *Conotrachelus nenuphar* (Hbst.) and *E. semirugosus* is a major parasitoid of *Pissodes* spp. attacking softwood coniferous trees in temperate regions. *Triaspis thoracicus* (Curtis) has been imported to Canada and the United States to combat legume weevils, especially the pea weevil *Bruchus pisorum* (L.). *Triaspis vestiticida* Viereck was introduced to the U.S.A. against the cotton boll weevil *Anthonomus grandis* Boheman, but apparently failed to establish. Many species of Brachistini attack potential beetle pests and are undoubtedly useful in natural control. For example, there are four species of Brachistini that attack sunflower seed weevils in the northern great plains of the U.S.A. (Charlet and Seiler 1994).

COMMON GENERA. *Nealiolus*, *Triaspis* and *Urosigalphus* are the most common genera of Brachistini in the New World. Each has more than 100 species (including undescribed forms).

DISTRIBUTION. Cosmopolitan.

DISTINGUISHING FEATURES. Members of the tribe Brachistini always have the r-m cross vein of the fore wing absent (Figs 9-14) and most have the first three metasomal terga forming a carapace (Figs 15-18). Those that do not form a carapace may be very difficult to distinguish from some *Blacus* Nees (Blacinae), some *Centistini* (Euphorinae), or even some *Orgilus* Haliday (Orgilinae). The following combination of Brachistini characters is usually sufficient: r-m of fore wing absent (Figs 9-14); 2cu-a of fore wing usually (95%) present (Figs 12-14) (not present in *Blacus* (Blacinae) and Centistini); occipital carina complete dorsally; petiole with basal, dorso-lateral pits (dorsope) weak or absent (Figs 21, 22); scutellum without raised lateral margins; subalar depression with transverse carina(e).

KEY TO THE NEW WORLD GENERA OF THE TRIBE BRACHISTINI

1. Metasomal terga 1 and 2 fused (Figs 15-17); metasoma carapace-like (Figs 15-17) 2

\- Metasomal terga 1 and 2 not fused (Fig. 18, 19, 21, 22); metasoma variable, carapace-like (Fig. 18) or not (Fig. 19) 4

2(1). Outer claw of hind leg much larger than inner claw (Fig. 20). Fore wing 2cu-a absent (Fig. 9); fore wing 1CUa subequal to 1CUb (Fig. 9) .... ***Urosigalphus*** Ashmead

Distribution: New World; about 100 species, many undescribed Neotropical species. Biology: emerging from the larval stage (perhaps egg-larval parasitoids); endoparasitoids of Curculionidae and Bruchidae. References: Gibson (1972a, b, 1974, revisions of New World species).

\- Outer claw of hind leg equal in size to inner claw. Fore wing 2cu-a variable, present(Fig. 10) or absent; if fore wing 1CU is intersected by 2cu-a, then 1CUa less than one third length of 1CUb (Fig. 10) 3

3(2). Metasomal carapace with 2 complete transverse grooves, or anterior groove complete and posterior groove present at least laterally (Fig. 16). Wings: Fig. 10 .... ***Triaspis*** Haliday

Distribution: Cosmopolitan; about 100 New World species, many undescribed Neotropical species. Biology: emerging from larval stage (perhaps egg-larva parasitoids); endoparasitoids of Curculionidae and Bruchidae. References: Martin (1956, revision of species north of Mexico).

\- Metasomal carapace without 2 complete transverse grooves, at most with anterior transverse groove complete and posterior groove absent (Fig. 17), usually both grooves completely absent (as in Fig. 15) or both indicated only laterally. Wings: Fig. 11 .... ***Schizoprymnus*** Förster

Distribution: New World, Palaearctic: primarily north temperate; about 10 New World species, 4 described species north of Mexico. Biology: endoparasitoids of Mordellidae and perhaps other Coleoptera: possibly egg-larval parasitoids. Reference: Martin (1956, revision of species north of Mexico).

4(1). Metasomal tergum 3 with lateral tergite separated from median tergite by sharp longitudinal fold on anterior 0.4 or less (Fig. 19) (including *Allodorus* Förster) .... ***Eubazus*** Nees

Distribution: Cosmopolitan; several dozen New World species, somewhat more diverse in the neotropics. Biology: parasites of Curculionidae and Bruchidae and perhaps other families of Coleoptera. No keys to New World species. Some importance in controlling weevil pests of wood (e.g. *Pissodes*). References: Mason (1974, nomenclature and generic systematics). A diverse polyphyletic assemblage.

\- Metasomal tergum 3 with lateral tergite separated from median tergite by sharp longitudinal fold on anterior 0.7 or more (Figs 21, 22) 5

5(4). Lateral tergite of metasomal segment 2 fused with that of segment 3, suture between them absent (Fig. 21) .... ***Nealiolus*** Mason

Distribution: New World, primarily Neotropical, about 80-120 species, mostly undescribed. Biology: reared from Curculionidae. References: Martin (1956, key to 9 species north of Mexico under *Aliolus*), Mason (1974, separation of *Nealiolus* from *Aliolus* and list of included species), Marsh (1979).

\- Lateral tergite of metasomal segment 2 not fused with that of segment 3, suture between them present (Fig. 22 .... ***Aliolus*** Say

Distribution: Holarctic, very rare in Europe, common in America north of Mexico, apparently restricted to temperate regions; perhaps about 30 species, most of which are undescribed. Biology: reared from Cerambycidae, Curculionidae, and Mordellidae. Reference: Martin (1956, key to 15 species north of Mexico).

REFERENCES

Achterberg, C. van 1987. Revision of the European Helconini (Hymenoptera: Braconidae: Helconinae Zoologische Medelingen Leiden 61: 263-285.

Achterberg, C. van 1988. Revision of the subfamily Blacinae Foerster (Hymenoptera, Braconidae). Zoologische Verhandelingen Leiden 249: 1-324.

Achterberg, C. van 1994. Generic revision of the subfamily Cenocoeliinae Szépligeti (Hymenoptera: Braconidae). Zoologische Verhandelingen Leiden 292: 1-52.

Alauzet, C. 1987. Bioecologie de *Eubazus semirugosus*, *Coeloides abdominalis* et *C. sordidator* (Hym: Braconidae) parasites de *Pissodes notatus* (Col: Curculionidae) dans le sud de la France. Entomophaga 32: 39-47.

Beirne, B.P. 1946. Notes on the biology of some hymenopterous parasites of the beech weevil (*Rhynchaenus fagi* L.) (Col.), Proceedings of the Royal Entomological Society of London (A) 21:7-11.

Berry, P.A. 1947. *Anihonomus vestitus* and its natural enemies in Peru, and their importation into the United States. Journal of Economic Entomology 40: 801-804.

Blanchard, E.E. 1936. Descripción de Ichneumonideos Argentines. Anales de la Sociedad ciencias argentina 122: 398-407.

Čapek, M. 1971. The classification of the Braconidae (Hymenoptera) from the viewpoint of their biology. Proceedings of the XIII International Congress of Entomology 1: 120-121.

Charlet, L.D. 1994. Seasonal abundance and impact of the sunflower stem Weevil parasitoid, *Nealiolus curculionis* (Hymenoptera; Braconidae), in the northern Great Plains. Biological Control 4: 26-31.

Charlot, L.D. and Seiler, G.J. 1994. Sunflower seed weevils (Coleoptera: Curculionidae) and their parasitoids from native sunflowers (*Helianthus*) in the northern great plains. Annals of the Entomological Society of America 87: 831-835.

Clausen, C.P. 1954. The egg-larval host relationship among the parasitic Hymenoptera. Bollettino del Laboratorio di Zoologia Generale e Agraria della Facoltá Agraria in Portici 33: 119-133.

Gibson, L.P. 1972a. Revision of the genus *Urosigalphus* of the United States and Canada (Hymenoptera: Braconidae). Miscellaneous Publications of the Entomological Society of America 8: 8:3-134.

Gibson, L.P. 1972b. *Urosigalphus* of Mexico and Central America (Hymenoptera: Braconidae). Miscellaneous Publications of the Entomological Society of America 8:135-157.

Gibson, L.P. 1974. South American *Urosigalphus* (Hymenoptera: Braconidae). Miscellaneous Publications of the Entomological Society of America 9: 201-226

Haeselbarth, E. 1962. Zur Biologie, Entwicklungsgeschichte und Ökologie von *Brachistes atricornis* Ratz. (Hym., Brac.) als eines parasiten von *Pissodes piceae* (Ill.) (Col., Curc.). Zeitschrift für angewandte Entomologie 49: 233-289.

Marsh, P.M. 1979. Family Braconidae. Pp. 14~295. In: Krombein, K.V., Hurd, P.D., Smith, D.R., and Burks, B.D. (Eds.). Catalog of Hymenoptera in America North of Mexico, Smithsonian Institution Press, Washington, D.C.

Martin, J. C. 1956. A taxonomic revision of the triaspidine braconid wasps of Nearctic America (Hymenoptera). Canadian Department of Agriculture Publication 965: 1-157.

Mason, W.R.M. 1968. New Canadian Braconidae (Hymenoptera). The Canadian Entomologist 100: 715-728.

Mason, W.R.M. 1974. A generic synopsis of Brachistini (Hymenoptera: Braconidae) and recognition of the name *Charmon* Haliday. Proceedings of the Entomological Society of Washington 76: 235-246.

Mason, W.R.M. 1987. *Vadum*, a new genus of Nearctic Braconidae (Hymenoptera). Proceedings of the Entomological Society of Washington 89: 325-328.

Obrtel, R. 1960. Premature stages of *Triaspis caudatus* (Nees) (Hym.: Braconidae). Folia Zoologica 9: 35-38.

Osborne, P. 1960. Observations on the natural enemies of *Meligethes aeneus* (F.) and *M. viridescens* (F.) (Coleoptera: Nitidulidae). Parasitology 50: 91-110.

Papp, J. 1995. New diospiline taxa from the Neotropical region (Hymenoptera, Braconidae, Helconinae). Acta Zoologica Academiae Scientiarum Hungaricae 41: 103-114.

Parnell, J.R. 1964. The parasite complex of the two seed beetles *Bruchidius ater* (Marsham) (Coleoptera: Bruchidae) and *Apion fuscirostre* Fabricius (Coleoptera: Bruchidae). Transactions of the Royal Entomological Society of London 116: 73-88.

Parrott, P.J, and Glasgow, H. 1916. The leaf-weevil (*Polydrusus impressifrons* Gyll.). New York State Agricultural Experiment Station Technical Bulletin 56: 1-22.

Shaw, M.R. and Huddleston, T. 1991. Classification and biology of braconid wasps (Hymenoptera: Braconidae). Handbooks for the Identification of British Insects 7(11): 1-126.

Shenefelt, R.D. 1970. Pars 5. Braconidae 2, Helconinae, Calyptinae, Mimagathidinae, Triaspinae. Pp. 177-306. In: Ferrière, Ch. and Vecht, J. van der (Eds.). Hymenopterorum Catalogus (nova editio). Dr. W. Junk, The Hague.

Tobias, V.I. 1967. The Far East as a center of abundance of the tribe Helconini. In: Harmful and useful insects of the Far East. Trudy Zoologicheskogo Instituta Akademii Nauk SSSR 41: 222-238. (In Russian.)

Viereck, H.L. 1913. Descriptions of twenty-three new genera and thirty-one new species of ichneumon-flies. Proceedings of the United States National Museum 46: 359-386.

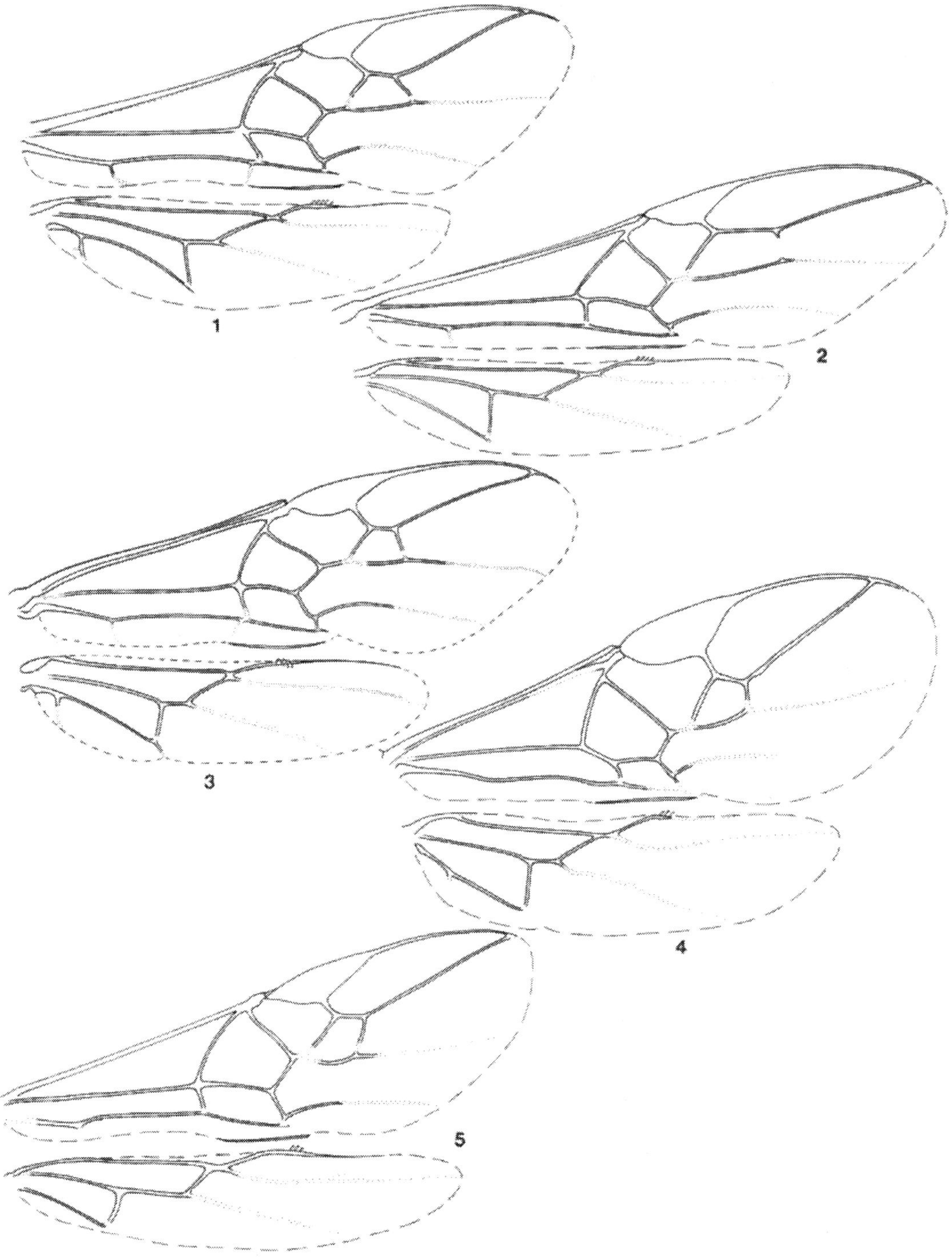

Figs 1-5. Wings of Helconinae. 1, *Wroughtonia* sp.; 2, *Eumacrocenirus* sp.; 3, *Helcon* sp.; 4, *Topaldios* sp.; 5, *Vadum volatum* Mason.

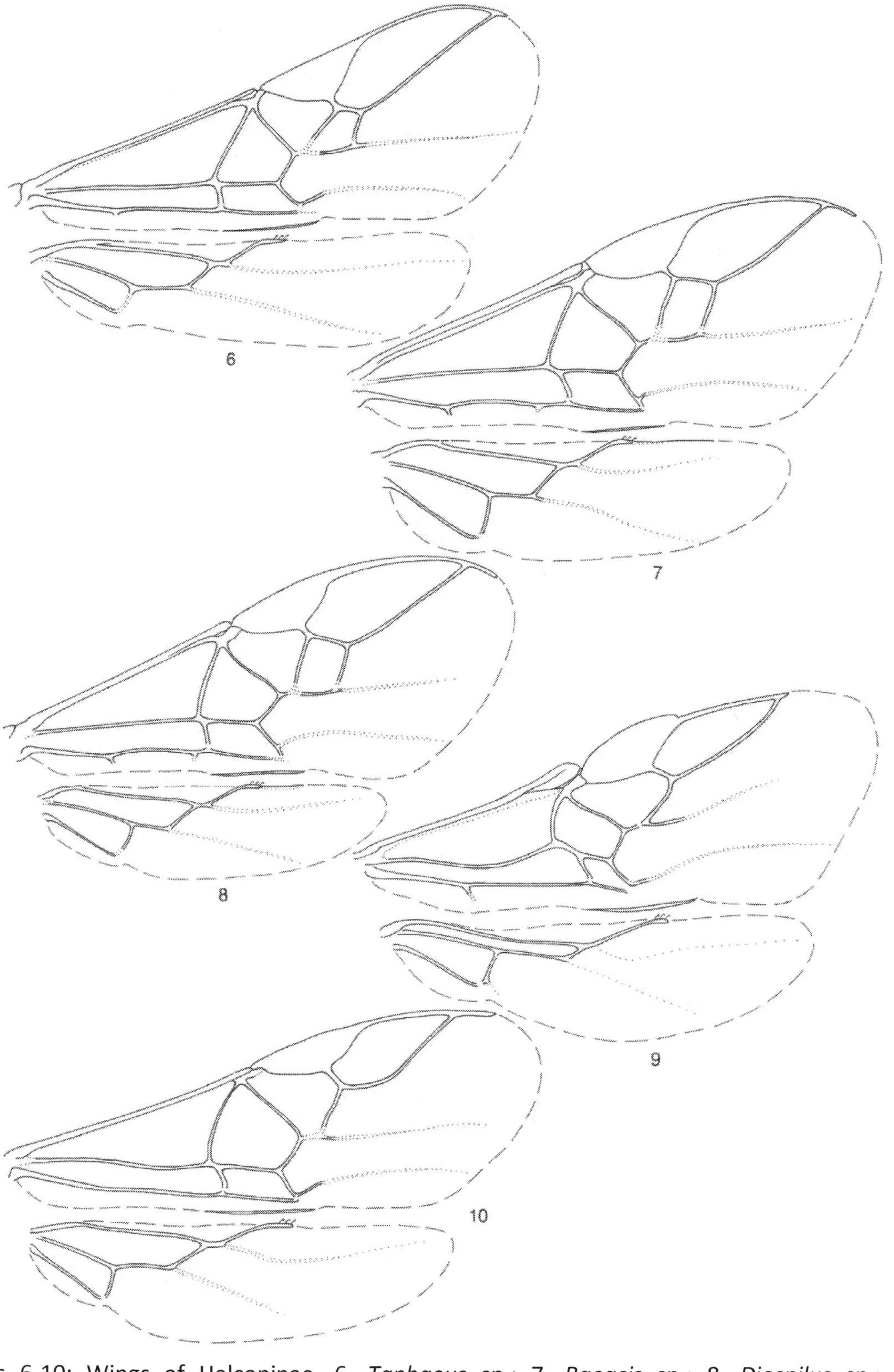

Figs 6-10: Wings of Helconinae. 6, *Taphaeus* sp.; 7, *Baeacis* sp.; 8, *Diospilus* sp.; 9, *Urosigalphus* sp.; 10, *Triaspis* sp.

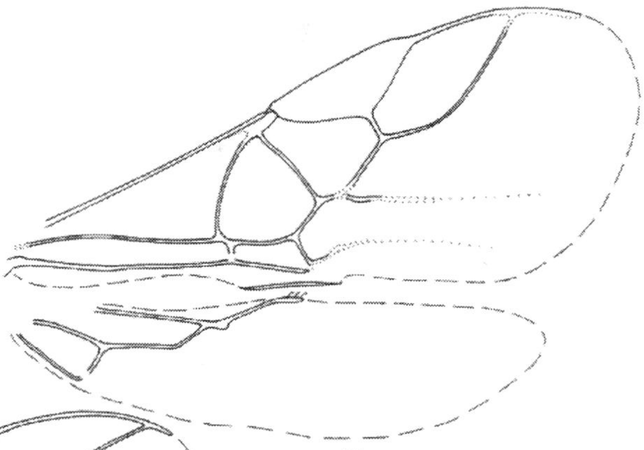

11

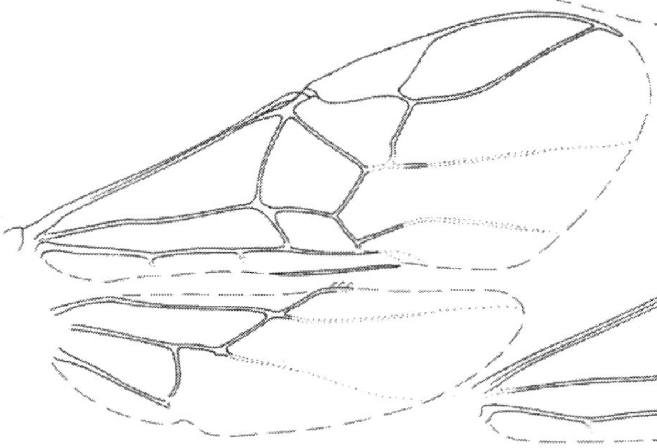

12

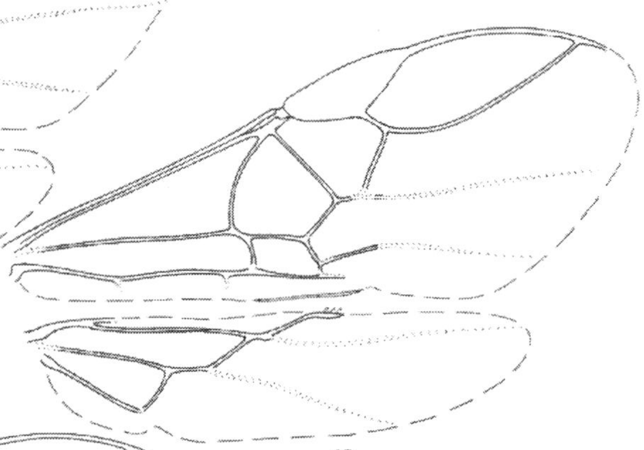

13

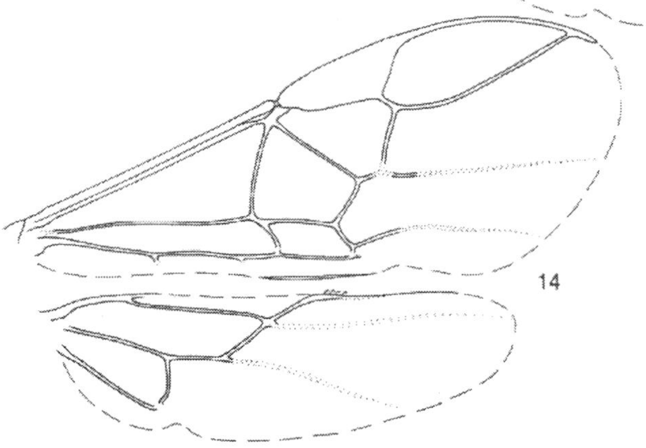

14

Figs 11-14: Wings of Helconinae, 11, *Schizoprymnus* sp.; 12, *Eubazus* sp.; 13, *Nealiolus* sp.; 14, *Aliolus* sp.

Figs 15-22: Helconinae. 15-18, metasomal terga, dorsal view: 15, *Urosigalphus* sp. 16, *Triaspis* sp.; 17, *Schizoprynmus* sp.; 18, *Aliolus* sp.; 19, metasoma of *Eubazus* sp., lateral view, arrow = rounded margin between median and lateral tergites; 20, hind leg of *Urosigalphus* sp. showing enlarged outer tarsal claw; 21-22, metasoma, lateral view; 21, *Nealiolus* sp.; 22, *Aliolus* sp.

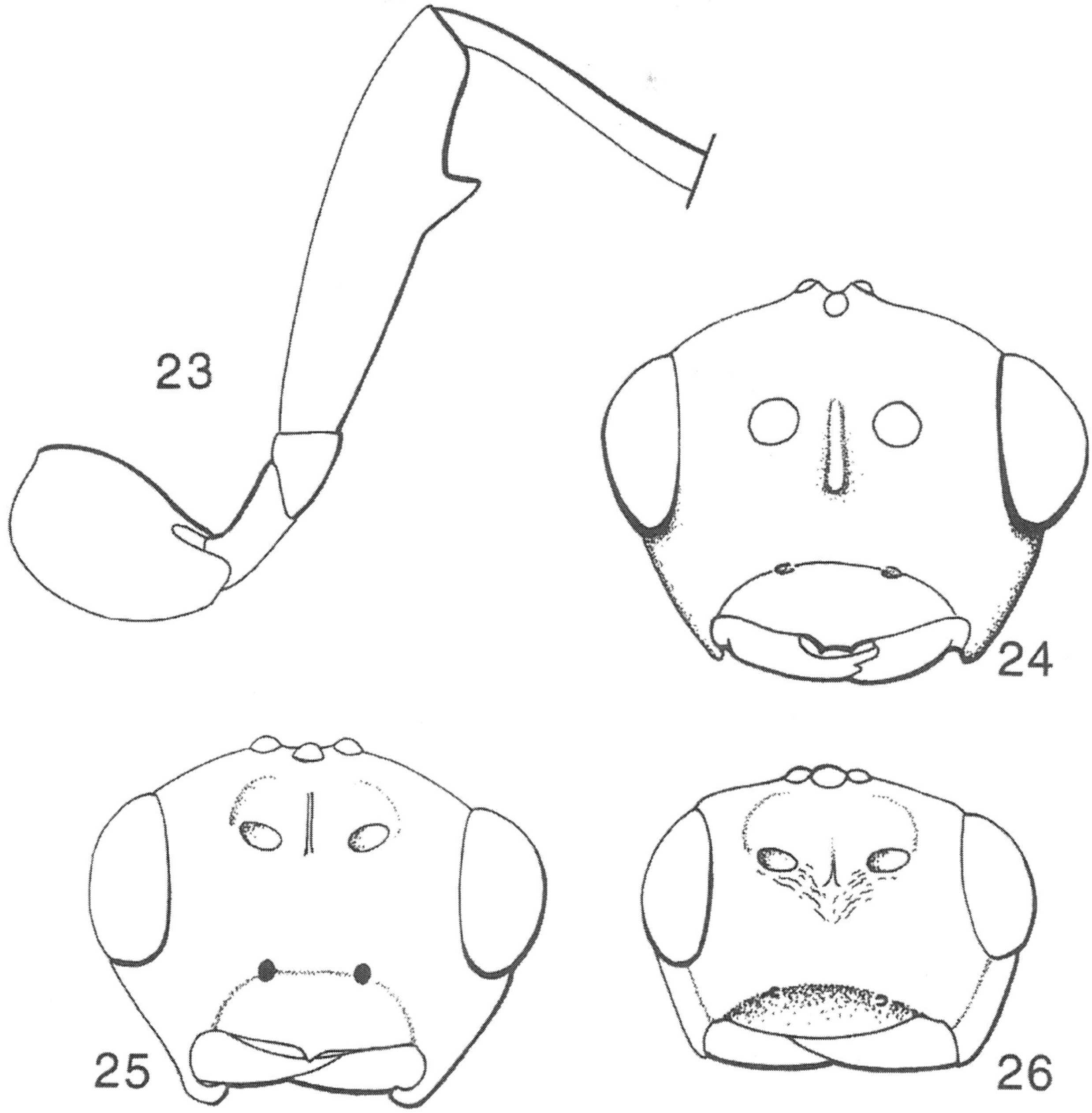

Figs 23-26: Helconinae. 23, hind leg of *Wroughtonia* sp. showing tooth on femur; 24-26, heads frontal views, showing apices of clypei with or without teeth: 24, *Vadum volatum* Mason; 25, *Baeacis* sp.; 26, *Diospilus* sp.

HELCONINAE UPDATE

M. J. Sharkey
Department of Entomology, University of Kentucky, Lexington, Kentucky 40546-0091 U.S.A. msharkey@uky.edu

Van Achterberg (1998) described a new genus of Helconini, *Vervoortihelcon*, from Chile. In the key to genera of Helconini it runs to *Eumacrocentrus*. It is easily distinguished from the latter by its highly sculptured terga.

The image in the Manual of the hind wing of *Eumacrocentrus* (Figure 2 in the Helconinae chapter) is misleading in that crossvein a is not shown. In fact specimens of the genus have this vein and, though weak, it is complete to the hind margin of the wing, as in the Manual's illustration of the wings of *Wroughtonia* (Figure 1 in the Helconinae chapter).

Vadum (Mason) was found to be a primary homonym and the generic name was changed to *Vadumasonium* (Krammerer, 2006).

Annotated list of recent papers on New World Helconinae

Achterberg C van. 1998. *Vervoortihelcon*, a new genus of the subfamily Helconinae Foerster (Hymenoptera: Braconidae) from Chile. *Zoologische Verhandelingen Leiden*. 323: 401-405.

Achterberg C van, Kenis M. 2000. The Holarctic species of the subgenus *Allodorus* Foerster *s.s.* of the genus *Eubazus* Nees (Hymenoptera: Braconidae). *Zoologische Mededelingen, Leiden*. 73(29): 427-455.
A key is provided, and three new species from Canada are described and illustrated.

Blanchard EE. 1936. Descripción de Icneumonoideos Argentinos. *Anales de la Sociedad Cientifica Argentina*. 122: 398-407.
Description of *Austrodolops*.

Kammerer CF. 2006. Notes on some preoccupied names in Arthropoda. *Acta Zootaxonomica Sinica*. 31(2): 269-271.
Vadum changed to *Vadumasonium*.

López Martínez V and Romero Napoles J. 2004. Identification key to the Mexican and Central American species of *Triaspis* Haliday (Hymenoptera: Braconidae), with descriptions of six new species. *Annals of the Entomological Society of America*. 97(1): 15-27.

López Martínez V, Romero Napoles J, Gonzalez Hernandez A, Equihua Martinez A, Ramirez Alarcon S, and Sanchez Garcia JA. 2003. Description of the male of *Triaspis kurtogaster* Martin and comments about the distribution of *T. azteca* Martin and *T. eugenii* Wharton and Lopez-Martinez (Hymenoptera: Braconidae). *Folia Entomologica Mexicana* 42(2): 153-160.

Papp J. 2005. Eleven new *Eubazus* Nees species from the Neotropical region (Hymenoptera: Braconidae: Brachistinae). *Annales Historico-Naturales Musei Nationalis Hungarici* 97: 163-195.

Sharanowski BJ, Dowling APG, Sharkey MJ. 2011. Molecular phylogenetics of Braconidae (Hymenoptera: Ichneumonoidea), based on multiple nuclear genes, and implications for classification. *Systematic Entomology* 36: 549-572.

Wharton RA, López Martínez V. 2000. A new species of *Triaspis* Haliday (Hymenoptera: Braconidae) parasitic on the pepper weevil, *Anthonomus eugenii* Cano (Coleoptera: Curculionidae). *Proceedings of the Entomological Society of Washington* 102(4): 794-801.

SUBFAMILY HISTEROMERINAE

D. L. J. Quicke

Department of Biology, Faculty of Life Sciences, Chulalongkorn University, Bangkok, Thailand.

INTRODUCTION. The Histeromerinae is a very small subfamily of braconids comprising a single genus, *Histeromerus* Wesmael, known from only four described species only one of which occurs in the New World. For a long time *Histeromerus* has been included within the Doryctinae (Shenefelt and Marsh, 1976; Sharkey, 1993) or Braconinae (van Achterberg, 1976). However, recently it has been afforded independent subfamily status (van Achterberg, 1984), and it is currently believed to be one of the most basal groups of Braconidae despite its obvious suite of autapomorphies (Quicke and van Achterberg, 1990; Quicke et al., 1992).

BIOLOGY. Observations on the Eurasian species *H. mystacinus* Wesmael indicate that Histeromerinae are gregarious idiobiont ectoparasitoids of wood-inhabiting beetle larvae, prepupae and pupae (Shaw, 1995). Hosts are paralyzed and the female appears to remain with a single host after oviposition though functional host guarding has not been demonstrated. Females can burrow through solid wood as well as frass-filled tunnels to reach their hosts. Shaw (1995) has further suggested that the ovipositor may not be used for oviposition but rather that the egg may emerge directly from the genital orifice at the ovipositor base. Direct observations of oviposition are needed for confirmation, however. Recorded host taxa include members of the Anobiidae, Buprestidae, Cerambycidae, Cisidae, Elateridae, Lucanidae, Lyctidae and Ptinidae (van Achterberg, 1992). Shaw (1995) cautions against accepting host records not supported by direct observation and identified host remains, and doubts the validity of the smaller of these recorded hosts.

BIOLOGICAL CONTROL POTENTIAL. Probably none, since hosts of at least *H. mystacinus* appear to be associated with dead timber or moribund trees.

COMMON GENERA. *Histeromerus* is often overlooked by hymenopterists though it is frequently encountered by coleopterists collecting from dead wood.

DISTRIBUTION. The only known New World species, *H. canadensis* Ashmead, occurs in Canada, the United States and recently van Achterberg (1992) has discovered it in western Europe (Netherlands). Other species occur in Europe, Taiwan, Papua New Guinea and Australia.

DISTINGUISHING FEATURES. The elongate and distally tapering hind basitarsus (Fig. 4) is totally diagnostic, and within the New World, the shape of the head (Figs 2, 3) is similarly characteristic, though in the Old World, some Bracon species (Braconinae) are convergent in this respect (Sarhan and Quicke, 1991). The pattern of stout setae on the fore tibia is also distinctive (Fig. 5).

REFERENCES

Achterberg, C van 1976. A preliminary key to the subfamilies of Braconidae (Hym.). Tijdschrift voor Entomologie 119: 33-78.

Achterberg, C van 1984. Essay on the phylogeny of Braconidae (Hymenoptera: Ichneumonoidea). Entomologisk Tidskrift 105: 41-58.

Achterberg, C van 1992. Revision of the genus Histeromerus Wesmael (Hymenoptera: Braconidae), Zoologische Mededelingen Leiden 66: 189-196.

Quicke, D. L. J. and Achterberg, C van 1990. Phylogeny of the subfamilies of the family Braconidae (Hymenoptera: Ichneumonoidea). Zoologische Verhandelingen Leiden 258: 1-95.

Quicke, D. L. J., Tunstead, J., Falco, J. V., and Marsh, P. M. 1992. Venom gland and reservoir morphology in the Doryctinae and related braconid wasps (Insecta, Hymenoptera, Braconidae). Zoologica Scripta 21: 403-416.

Sarhan, A. A. and Quicke, D. L. J. 1991. A new species of Bracon from Mongolia (Hym., Braconidae). Entomologist's Monthly Magazine 127: 19-23.

Sharkey, M. J. 1993. Family Braconidae, Pp. 362-395. In: Goulet, H. and Huber, J. T. (Eds.). Hymenoptera of the World: An Identification Guide to Families. Agriculture Canada, Ottawa.

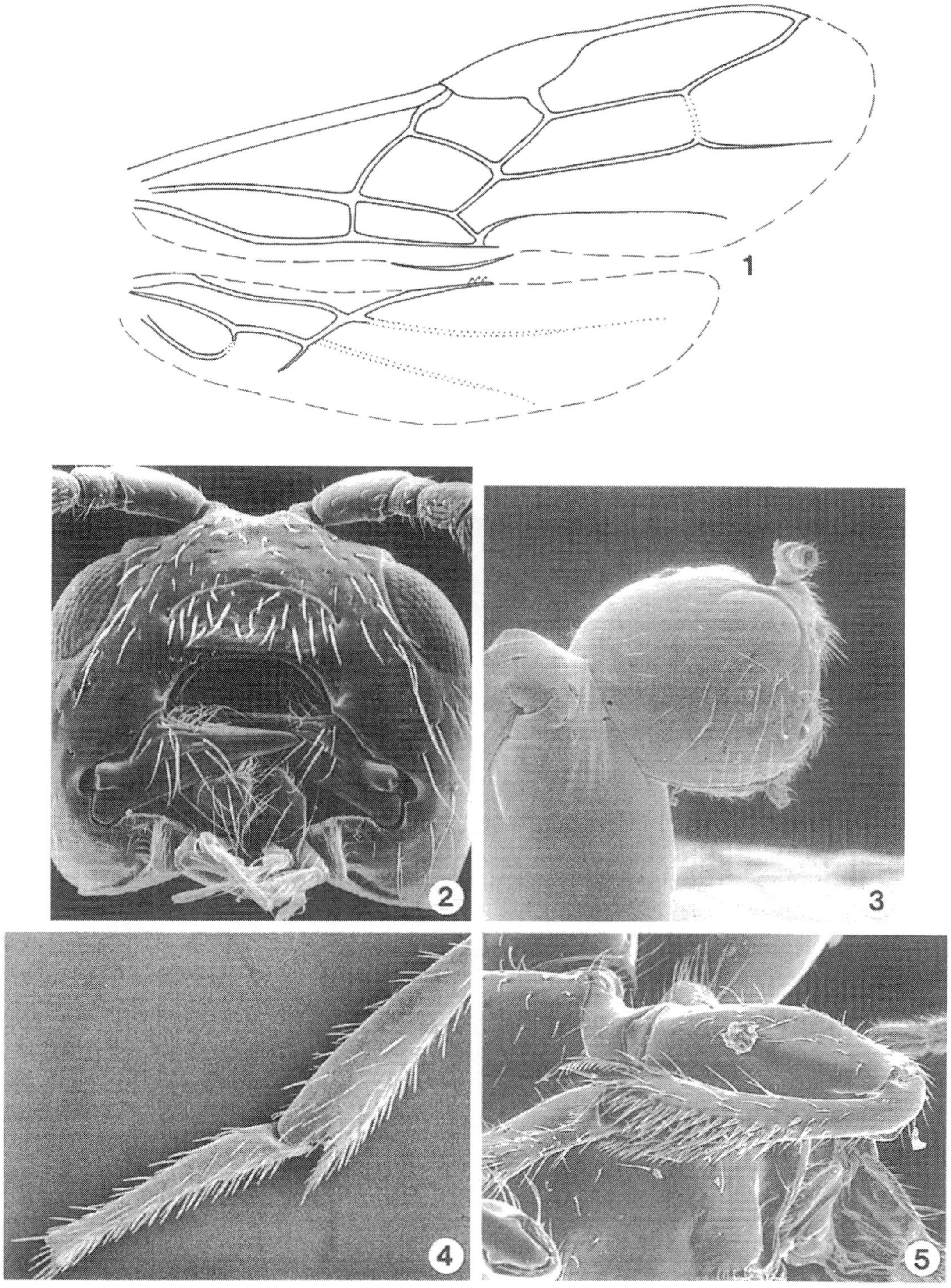

Figs 1-5: Histeromerinae. 1, wings of *Histeromerus canadensis* Ashmead; 2, head of *Histeromeris* sp., frontal view; 3, head of *Histeromeris* sp., lateral view; 4, hind basitarsus of *Histeromeris* sp.; 5, fore tibia of *Histeromeris canadensis* Ashmead.

HISTEROMERINAE UPDATE

D.L.J. Quicke
Department of Biology, Faculty of Life Sciences, Chulalongkorn University, Bangkok, Thailand.
d.quicke@email.com

Based on DNA sequence data, *Histeromerus*, the sole genus of the old Histeromerinae, is now considered to belong to the Rhyssalinae (in the Manual it is treated within the dustbin taxon 'Hormiinae') (Zaldívar-Riverón *et al.*, 2006, Sharanowski *et al.*, 2011, Quicke 2015, Quicke *et al.*, 2016). No new extant species have been described in this group since the 1997 Manual; however, the recent DNA barcoding effort at the University of Guelph clearly shows that two species occur in Canada, one *H. canadensis* the other with CO1 sequence identical to European *H. mystacinus* (BOLD public data 2017). The latter is here recorded from the New World for the first time. The two species may be separated morphologically using van Achterberg's (1992) key.

Recent papers on New World Histeromerinae

Quicke DLJ. 2015. *The Braconid and Ichneumonid Parasitic Wasps: Biology, Systematics, Evolution and Ecology.* Wiley Blackwell, Oxford, 688 pp.

Quicke DLJ, Belokobylskij SA, Smith MA, Rota J, Hrcek J, Butcher BA. 2016. A new genus of rhysipoline wasp (Hymenoptera: Braconidae) with modified wing venation from Africa and Papua New Guinea, parasitoid on Choreutidae (Lepidoptera). *Annales Zoologici* 66: 173-192.

Sharanowski BJ, Dowling AP, Sharkey MJ. 2011. Molecular phylogenetics of Braconidae (Hymenoptera: Ichneumonoidea), based on multiple nuclear genes, and implications for classification. *Systematic Entomology* 36: 549-572.

Zaldívar-Riverón A, Mori M, Quicke DLJ. 2006. Systematics of the cyclostome subfamilies of braconid wasps (Hymenoptera: Ichneumonoidea): a simultaneous molecular and morphological approach. *Molecular Phylogenetics and Evolution* 38: 130-145.

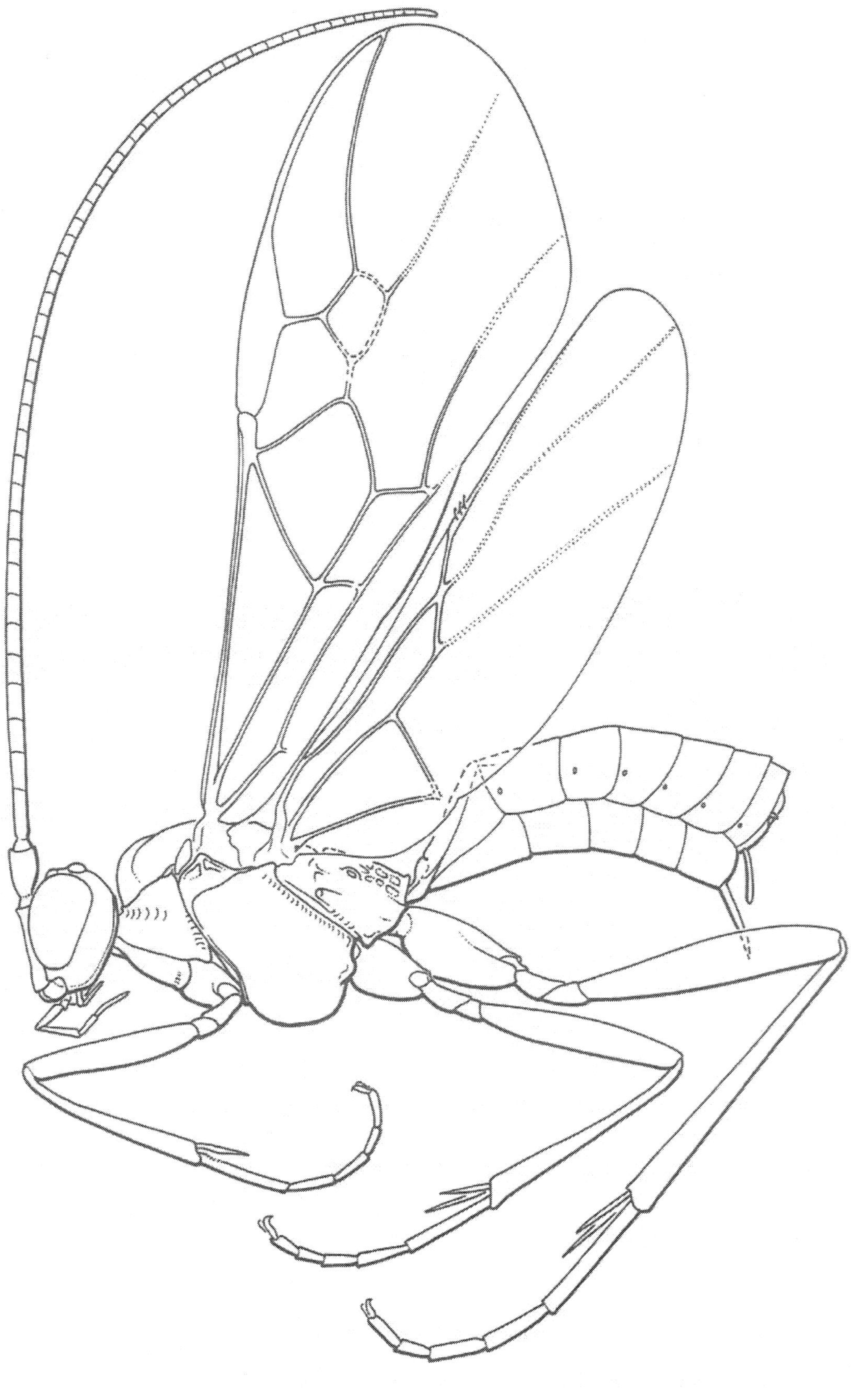

SUBFAMILY HOMOLOBINAE

R. A. WHARTON

Department of Entomology, Texas A&M University, College Station, Texas 77843, U.S.A.

INTRODUCTION. The subfamily Homolobinae was established by van Achterberg (1979) to accommodate species that had been grouped with *Zele* Curtis, and variously placed in the Helconinae, Macrocentrinae, or Zelinae (Meteorinae, or Meteorini of the Euphorinae). Van Achterberg (1979) included four genera and 50 species in the Homolobinae, but *Charmon* Haliday and *Charmontia* van Achterberg (=Charmontini), are morphologically quite distinct (Mason, 197 4; Shaw and Huddleston, 1991). In addition to differences in venation between Charmontini and the remaining Homolobinae (see key below), *Charmon* has ridges along the ovipositor that resemble those of *Macrocentrus* Curtis (Quicke and van Achterberg, 1990; van Achterberg and Quicke, 1992). Charmon, however, lacks teeth on the trochantellus that are characteristic of the Macrocentrinae, and has an antescutal depression similar to *Homolobus* Förster. Thus, rather than create a separate chapter for *Charmon* and *Charmontia*, and treat them as a separate subfamily (as in Quicke and van Achterberg, 1990), I have retained the Charmontini in Homolobinae pending further clarification of its relationships.

Aside from the three papers cited above, little has been published on the Homolobinae subsequent to the excellent revision by van Achterberg (1979). An important exception to this is the series of fine papers by Maetô (1982a, b, c). Eighteen species have been described from the New World.

BIOLOGY. Homolobines are solitary, koinobiont endoparasitoids of Lepidoptera. Detailed studies on the biology of the Homolobinae are lacking. There are, however, numerous rearing records, and these are reviewed by van Achterberg (1979). Both van Achterberg (1979) and Shaw and Huddleston (1991) note that Noctuidae and Geometridae are the most commonly recorded hosts of *Homolobus*. There is only a single host record for *Exasticolus* van Achterberg, and this is from the Lasiocampidae (van Achterberg, 1979). Shaw and Huddleston (1991) suggest that *Charmon* has a broad host range relative to *Homolobus*, and list Tortricidae and Gelechiidae among the more frequently recorded hosts.

The New World species of *Charmon* and *Charmontia* have exceptionally long ovipositors (about 1-2 times as long as the body, depending on the species), with *Charmon* attacking weakly concealed lepidopteran larvae. Hosts for *Charmontia* are unknown. *Exasticolus* and the most commonly encountered New World species of *Homolobus* have very short ovipositors. They attack exposed, nocturnally feeding hosts.

BIOLOGICAL CONTROL POTENTIAL. A small number of *Charmon* originating from Japan and Korea were released in the United States as part of a program directed against the tortricid *Grapholita molesta* (Busck). The introduction was unsuccessful (Clausen, 1978), but van Achterberg (1979) later recorded the same species as widespread in North America. Otherwise, there has been no usage of this group in biological control. Until more information becomes available on host preferences and specificity, there is unlikely to be greater utilization since the known species do not regularly attack insect pests.

COMMON GENERA. *Homolobus* and *Charmon* are the most commonly encountered genera. *Homolobus truncator* (Say), usually recorded in the North American literature as *Zele mellea* (Cresson) prior to van Achterberg's (1979) revision, is particularly widespread and abundant throughout North and Central America.

DISTRIBUTION. *Exasticolus* is restricted to the New World, with both Nearctic and Neotropical species. *Charmontia* is known only from Chile. The two species of *Charmon* recorded from the New World (van Achterberg, 1979) are both Holarctic and *Homolobus* is worldwide in distribution. At least two of the *Homolobus* species recorded from the New World are also present in both the Palaearctic and Oriental Regions.

DISTINGUISHING FEATURES. *Charmon* and *Charmontia* are readily recognized by the presence of only two submarginal cells (fore wing cross vein r-m absent) in combination with the presence of an anal cross vein (a) in the hind wing (Fig. 1) (Mason, 1974). *Homolobus* and *Exasticolus* resemble some of the large macrocentrines (such as *Austrozele*) in overall size and shape, and in the presence of long tibial spurs (Fig. 6), but have a well-developed occipital carina (absent in macrocentrines), and an antescutal depression mid-dorsally behind the pronotal collar (Figs 9, 10). The occipital carina and antescutal depression, in combination with the presence of r-m in the fore wing and a square to rectangular second submarginal cell, can be used to separate *Homolobus* and *Exasticolus* from all other New World, non-cyclostome Braconidae.

KEY TO THE NEW WORLD GENERA OF THE SUBFAMILY HOMOLOBINAE

1. Fore wing r-m absent (Figs 1, 3) (Charmontini) 2
- Fore wing r-m present (Fig. 2) (Homolobini) 3

2(l). Fore wing 1M and (RS+M)a meeting at 1RS, the latter distinct (Fig. 3). Third segment of labial palp equal in length to second .... ***Charmontia*** van Achterberg

Distribution: Chile, known only from a single specimen. Biology: unknown. Reference: van Achterberg (1979).

- 1RS absent, 1M and (RS+M)a separated at parastigma (Fig. I). Third segment of labial palp much shorter than second segment (Fig. 5), sometimes absent .... ***Charmon*** Haliday

Distribution: across northern Canada to northern Mexico (Durango and Baja California); 2 described species in New World. Biology: reared in North America from Tortricidae (at least 13 species), Gelechiidae (5 species), Oecophoridae (one species) and Pyralidae (one species) (Allen, 1962 as *Eubadizon*, Marsh, 1979). References: Mason (1974), van Achterberg (1979, key to species).

3(l). Inner side of hind tibia with a well-developed apical comb (Fig. 7). Fore wing (RS+M)a evenly curved (Fig. 4 .... ***Exasticolus*** van Achterberg

Distribution: Canada (Ontario) to Peru and Argentina, not recorded from Chile or Caribbean; 3 described species. Biology: one specimen reared from Lasiocampidae in Costa Rica. Reference: van Achterberg (1979, key to species).

- Inner side of hind tibia without a well-developed apical comb (Fig. 8). Fore wing vein (RS+M)a straight (Fig. 2), without distally-directed curvature .... ***Homolobus*** Förster

Distribution: Alaska and Canada to Argentina, Chile, including Caribbean; common; 12 described species in New World. Biology: reared from Noctuidae and Geometridae. References: van Achterberg (1979, key to species), Shaw and Huddleston (1991).

REFERENCES

Achterberg, C. van 1979. A revision of the subfamily Zelinae auct. (Hymenoptera, Braconidae). Tijdschrift voor Entomologie 122: 241-479.

Achterberg, C. van and Quicke, D.L.J. 1992. Phylogeny of the subfamilies of the family Braconidae: A reassessment assessed. Cladistics 8: 237-264.

Allen, H.W. 1962. Parasites of the Oriental fruit moth in the eastern United States. United States Department of Agriculture Technical Bulletin 1265: 1-139.

Maetô, K. 1982a. The genus *Homolobus* Foerster of Japan (Hymenoptera, Braconidae). I. Subgenus *Homolobus*. Kontyu 50: 314-323.

Maetô, K. 1982b. The genus *Homolobus* Foerster of Japan (Hymenoptera, Braconidae). II. Subgenera *Chartolobus*, *Apatia*, *Phylacter* and *Oulophus*. Kontyu 50: 425-433.

Maetô, K. 1982c. A new species of the genus *Homolobus* (Hymenoptera, Braconidae) from New Guinea. Kontyu 50: 527-530.

Marsh, P.M. 1979. Family Braconidae. Pp. 144-295. In: Krombein, K. V., Hurd, P. 0., Smith, D. R., and Burks, B.D. (Eds.). Catalog of Hymenoptera in America North of Mexico, Smithsonian Institution Press, Washington, D.C.

Mason, P.M. 1979. A generic synopsis of Brachistini (Hymenoptera: Braconidae) and recognition of the name *Charmon* Haliday. Proceedings of the Entomological Society of Washington 76: 235-246.

Quicke, D.L.J. and Achterberg, C. van 1990. Phylogeny of the subfamilies of the family Braconidae (Hymenoptera: Ichneumonoidea). Zoologische Verhandelingen Leiden 258: 1-95.

Shaw, M.R. and Huddleston, T. 1991. Classification and biology of braconid wasps (Hymenoptera: Braconidae). Handbooks for the Identification of British Insects 7(11): 1-126.

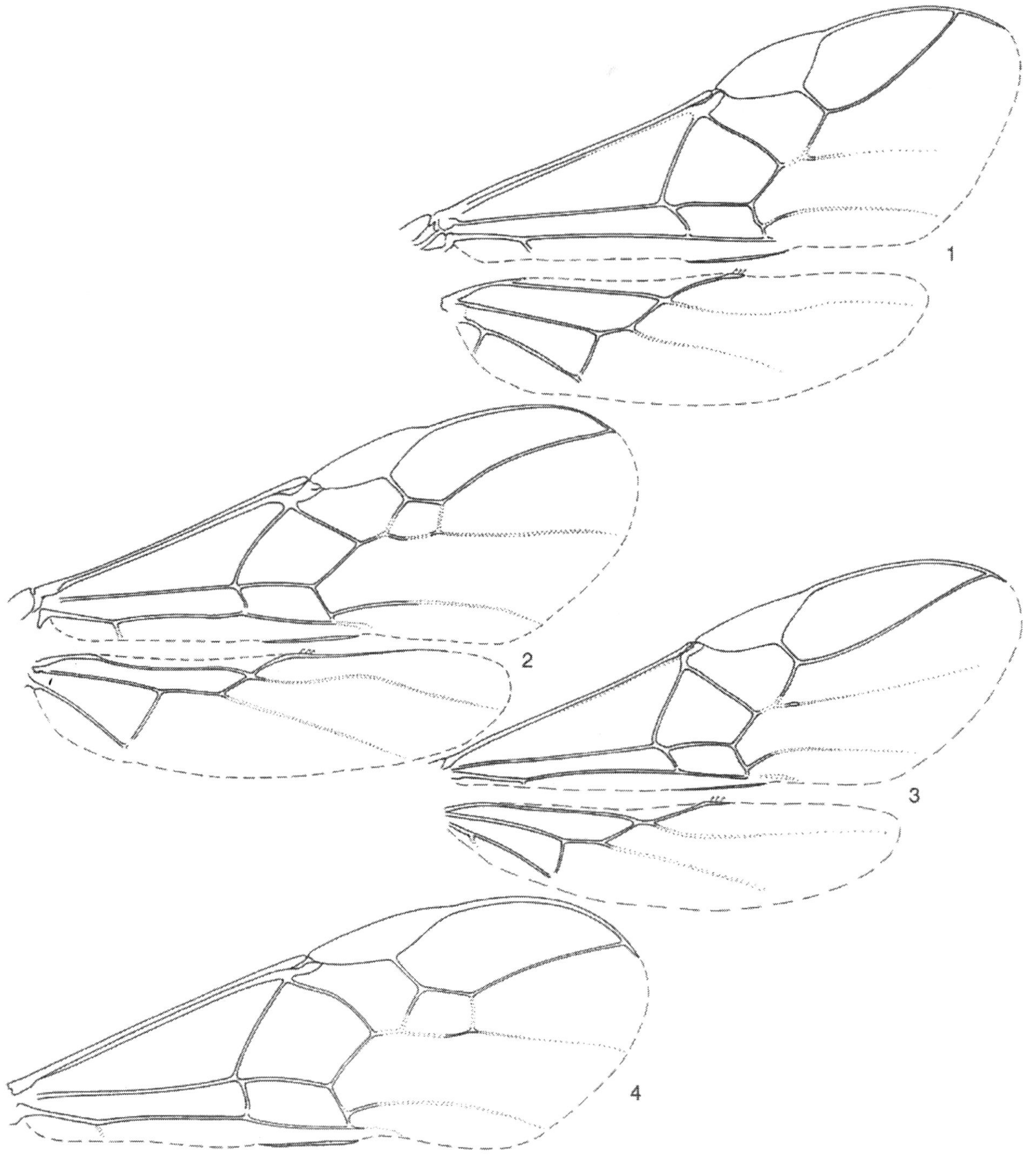

Figs 1-4: Wings of Homolobinae. 1, *Charmon extensor* (L.); 2, *Homolobus truncator* (Say); 3, *Charmontia inopina* van Achterberg; 4, *Exasticolus* sp.

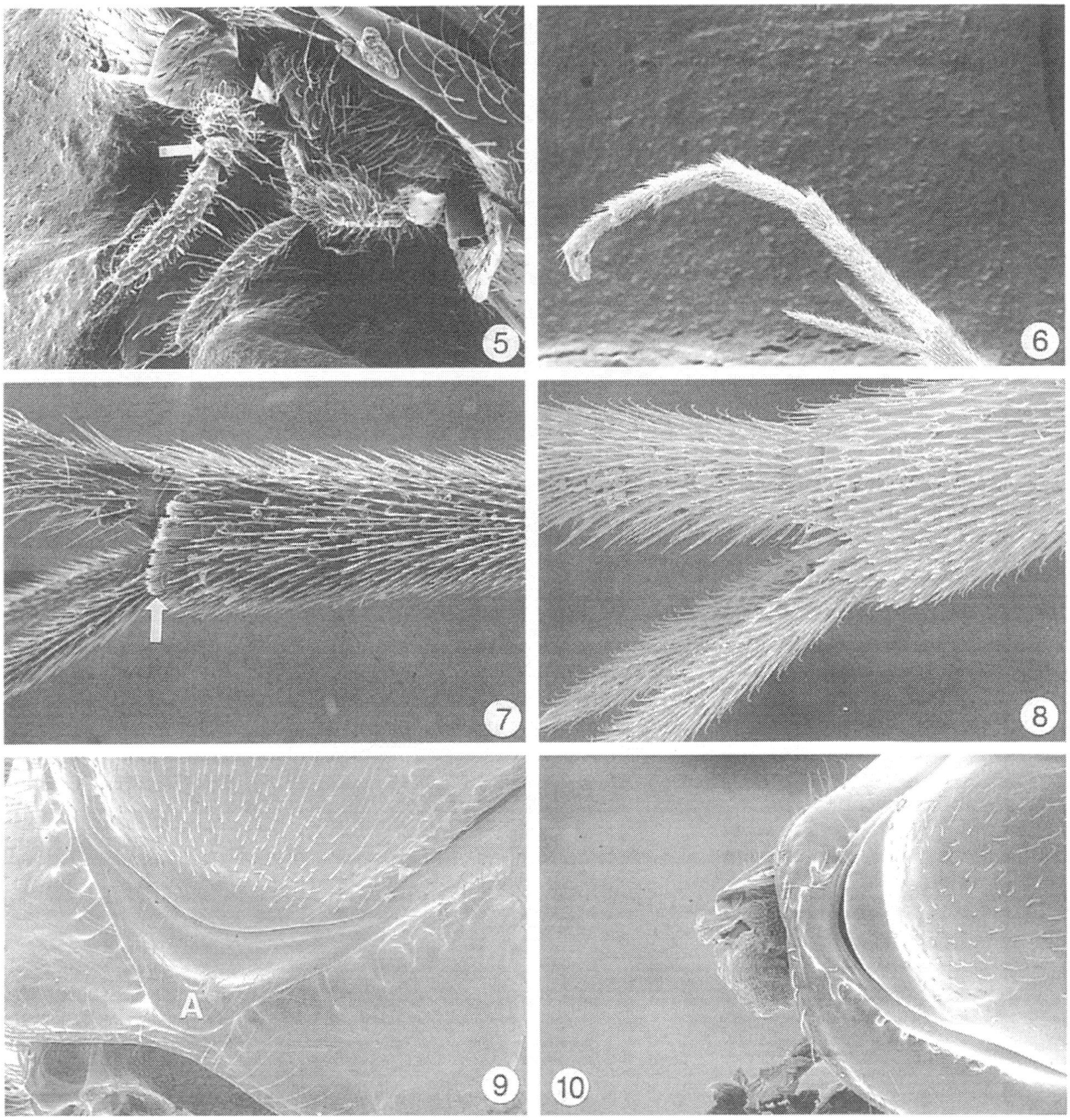

Figs 5-10: Homolobinae. 5, mouthparts of *Charmon* sp., arrow = third segment of labial palp; 6, tibial spurs and hind tarsus of *Homolobus truncator* (Say); 7, hind tibia and basitarsus of *Exasticolus* sp., arrow = apical comb of hind tibia; 8, hind tibia and basitarsus of *Homolobus* sp.; 9, pronotum and anterior portion of mesoscutum of *Homolobus truncator* (Say), dorsal view, A = antescutal depression; 10, antescutal depression of *Charmon* sp.

HOMOLOBINAE UPDATE

C. van Achterberg
Department of Terrestrial Zoology, Naturalis Biodiversity Center, Postbus 9517, 2300 Leiden, The Netherlands
c.vanachterberg@xs4all.nl

The Homolobinae chapter of the manual is partly outdated. According to the analysis by Sharanowski *et al.* (2011) the Charmontini (as Charmontinae) are sister Amicrocentrinae. In 2001 Braet & van Achterberg published a key to the species of *Exasticolus* van Achterberg and included one new species. Van Achterberg and Shaw (2009) published a key to the Neotropical species of the subgenus *Oulophus* van Achterberg and included a new species of *Homolobus* Foerster. Finally, López-Martinez *et al.* (2011) published an updated key to the species of *Exasticolus* van Achterberg, included a new species, and treated *E. rosenbergi* (Viereck) as a valid species.

Annotated list of recent papers on New World Homolobinae

Achterberg C van, Shaw SR. 2009. A new species of the genus *Homolobus* Foerster from Ecuador (Hymenoptera: Braconidae: Homolobinae). *Zoologische Mededelingen, Leiden* 83(4): 805-810.

Homolobus fuscinervis van Achterberg & Shaw is described and included in a key to the species of the subgenus *Oulophus* van Achterberg.

Braet Y, Achterberg C van. 2001. Notes on the genera *Exasticolus* van Achterberg (Homolobinae) and *Orgilus* Haliday (Orgilinae) (Hymenoptera: Braconidae), with the description of three new species from French Guiana. *Zoologische Mededelingen, Leiden* 75(5): 89-102.

A key to species of the genus *Exasticolus* van Achterberg is given, and one new species is described.

López-Martinez V, Delfin-Gonzalez H, Achterberg, K van, Alia-Tejaca I. 2011. A new species of the genus *Exasticolus* van Achterberg (Hymenoptera: Braconidae: Homolobinae) from Mexico. *Studies on Neotropical Fauna and Environment* 46(1): 59-62.

Exasticolus xmatkuilensis López-Martínez, Delfin-González & van Achterberg is described and included in a key to species of the genus *Exasticolus* van Achterberg.

Penteado-Dias AM, Figueiredo MLC, Dias MM, Osório TC, Cruz I. 2006. First host records for *Exasticolus fuscicornis* (Cameron, 1887) (Hymenoptera: Braconidae: Homolobinae). *Zoologische Mededelingen, Leiden* 80-1(8): 109-111.

Exasticolus fuscicornis (Enderlein) was reared from a *Leuciris* sp. (Geometridae) and *Spodoptera frugiperda* (Smith) (Noctuidae) in Brazil.

Sharanowski BJ, Dowling APG, Sharkey MJ. 2011. Molecular phylogenetics of Braconidae (Hymenoptera: Ichneumonoidea), based on multiple nuclear genes, and implications for classification. *Systematic Entomology* 36: 549-572.

Phylogeny and relationships of the Charmontinae and Homolobinae.

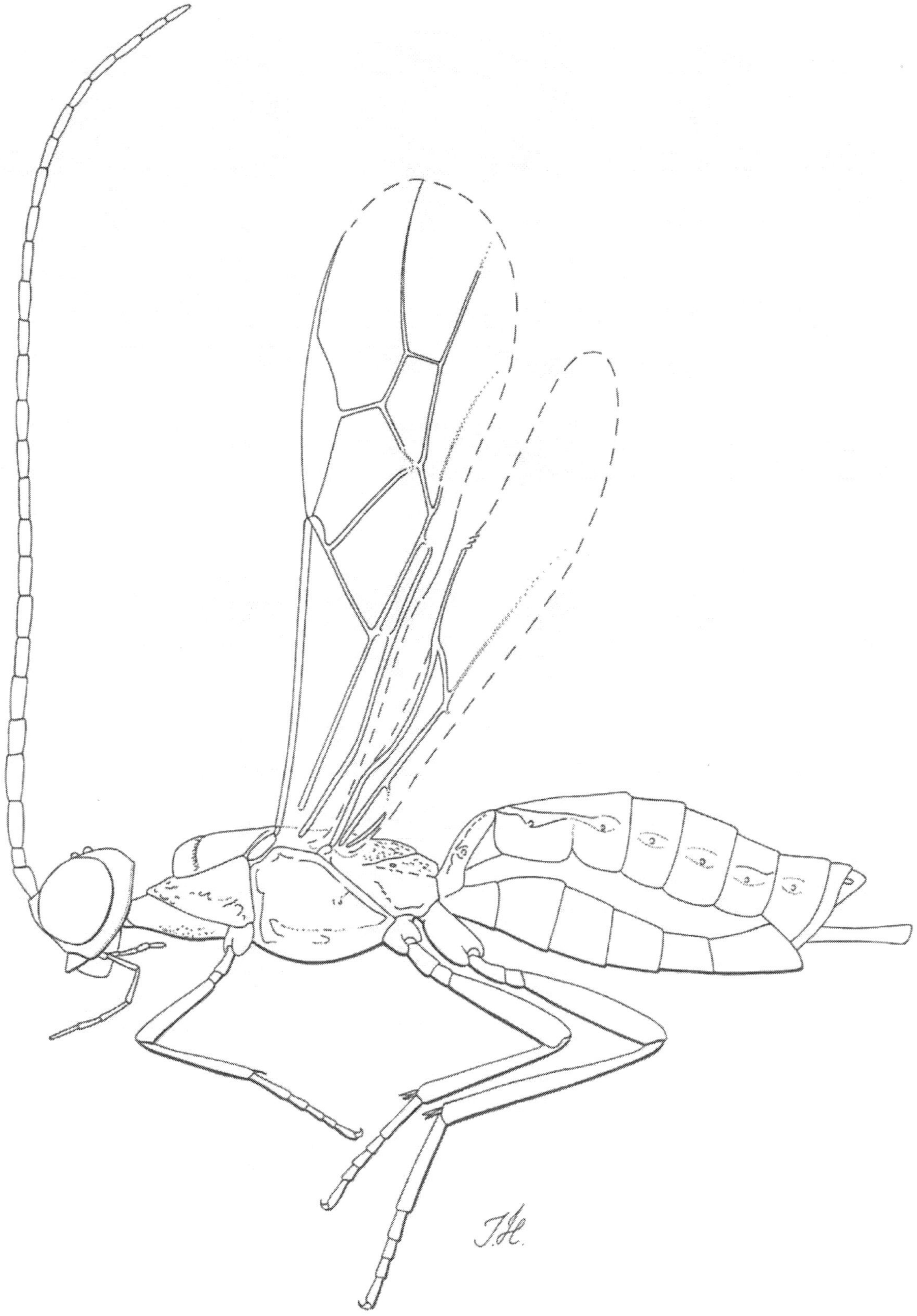
J.H.

SUBFAMILY HORMIINAE

J. B. WHITFIELD

Department of Entomology, 320 Morrill Hall, University of Illinois, 505 S. Goodwin Ave, Urbana, IL 61801 U.S.A.

R. A. WHARTON

Department of Entomology, Texas A&M University, College Station, Texas 77843, U.S.A.

INTRODUCTION. The hormiines are a morphologically and biologically diverse assemblage of external parasitoids of larval Lepidoptera, Coleoptera, and less commonly Hymenoptera and Diptera. The tribes included here under Hormiinae are not necessarily closely related to one another; some, in fact, are more closely related to members of other subfamilies (especially Doryctinae, Rogadinae, Opiinae, and Alysiinae) than to one another. Despite some recent attention to the study of relationships among these groups (van Achterberg, 1983, 1984, 1991, 1995; Belokobyl'skij, 1984, 1987a, 1992, 1993a, b; Belokobyl'skij and Tobias, 1986; Whitfield and van Achterberg, 1987; Wharton, 1988, 1993; Whitfield, 1988, 1992; Quicke and van Achterberg, 1990; Wharton et al., 1992; Quicke, 1993), there is as yet no consensus on how best to classify these diverse groups into subfamilies. Here we treat them all together, solely for convenience, as the subfamily Hormiinae. The genera in the key have been variously classified as Hormiinae (especially Hedqvist, 1963; Wharton, 1993), Rogadinae (e.g. Wharton, 1988), Doryctinae (especially Belokobyl'skij and Tobias, 1986), Exothecinae (e.g. Shenefelt, 1975) and even, rarely, Braconinae or Opiinae. Van Achterberg (1995) uses additional subfamily names. The tribal names Avgini, Monitoriellini, Rhyssalini, Rhysipolini, Pambolini, Acrisidini, Hormiini, Hydrangeocolini, Lysitermini and Exothecini have been used to subdivide this assemblage even further (Belokobyl'skij, 1992, 1993a). The Central and South American faunas of this complex have received very little attention and several potential new genera remain undescribed. *Hormiopius* Blanchard may also belong in the Hormiinae (Wharton, 1993), but is not included in the key below because type material could not be located.

BIOLOGY. All of the genera treated here are either known to be or suspected to be ectoparasitoids, usually attacking concealed hosts. An important exception is *Monitoriella* Hedqvist, which is phytophagous (Infante et al., 1995). Many or most species usually placed in the Rhyssalini, Hormiini, Pambolini and probably also Hydrangeocolini are gregarious. Most of the Hormiinae are also idiobionts (Askew and Shaw, 1986) and develop rapidly on late-instar host larvae. Exceptional are *Rhysipolis* Förster and possibly also several related genera, in which the hosts regain activity after parasitization and are often able to crawl away and spin a cocoon before succumbing to the parasitoid larvae (Shaw, 1983; Whitfield, 1992). As far as is known, the vast majority of the species parasitize hosts that are concealed in some way (e.g. leaf-miners, leaf-rollers, leaf-tiers, stem-borers, seed-borers, gall-formers, or borers in fungi).

BIOLOGICAL CONTROL POTENTIAL. Few species in this subfamily have been used in classical biological control, and for the New World, usage has been confined to *Colastes* Haliday and *Shawiana* van Achterberg on leaf-mining Agromyzidae and Tenthredinidae. Nevertheless, others may prove useful in the regulation of stem and twig borers, leaf-miners, and leaf-rolling and leaf-tying lepidopteran larvae. *Allobracon* Gahan, *Chremylus* Haliday, *Hormius* Nees, *Oncophanes* Förster and *Rhysipolis* all have at least one species that has been reared from an insect pest (see footnotes in key below), but none of these has been evaluated.

COMMON GENERA. The most commonly encountered genera are *Hormius*, *Parahormius* Nixon, *Pambolus* Haliday, *Oncophanes*, *Rhysipolis*, *Colastes* and (in the Neotropics) *Allobracon* and *Hydrangeocola Brèthes*, They are rarely among the dominant components of the fauna at any locality.

DISTRIBUTION. *Allobracon*, *Lysitermus* Förster, *Aspilodemon* Fischer and *Hydrangeocola* are predominantly if not entirely tropical in the New World; most other genera are more widespread but some genera that attack Symphyta (e.g. *Shawiana*, *Xenarcha* Förster) tend to be restricted to north temperate areas. *Acrisis* Förster, *Proacrisis* Tobias, *Colastes*, *Shawiana*, *Xenarcha*, *Dolopsidea* Hincks, *Neurocrassus* Snoflák, *Oncophanes*, *Pseudobathystomus* Belokobyl'skij and *Rhyssalus* Haliday are Holarctic. *Allobracon*, *Aspilodemon*, *Catharoctonus* Viereck and a new genus from Chile are all endemic to the New World. Wharton (1993) has suggested that *Hydrangeocola*, another New World endemic, has its closest relationship with the Australian *Opiopterus* Szépligeti.

DISTINGUISHING FEATURES. The Hormiinae belong to the cyclostome group of subfamilies (see discussion of braconid classification in the introductory chapter of the manual), with most species having a clearly defined cyclostome configuration to the clypeus and labrum (Fig. 22). Hormiines are mainly recognized by the absence of characters associated with the other cyclostomes such as the Rogadinae *sensu stricto* (medial longitudinal carina on the petiole, endoparasitism leading to mummification of host larvae), the Doryctinae *sensu stricto* (a row of short, stout setae on the prothoracic tibia), or with the Braconinae (reduction in number of maxillary palp segments to five, loss of propodeal carination, flattening and reduction of the dorsal carinae of the petiole, loss of the occipital carina). Most of these distinguishing features have exceptions that challenge the definition of each of the subfamilies involved: a new subfamily arrangement is badly needed.

KEY TO THE NEW WORLD GENERA OF THE SUBFAMILY HORMIINAE

1. Petiole partially to almost entirely desclerotized, depigmented and somewhat membranous posteriorly (Fig. 41). Wings: Fig. 3 .... ***Allobracon*** Gahan

Distribution: Texas and Florida to Peru and Brazil, including Caribbean; 11 described and many undescribed species. Biology: reared from leaf-mining Lepidoptera and Coleoptera (Buprestidae and Chrysomelidae) (Wharton, 1993). References: Clark (1965, as *Leurinion*, Wharton (1993).

\- Petiole uniformly pigmented and heavily sclerotized throughout 2

2(1). Brachypterous or apterous; wings, when present, short and without recognizable venation 3

\- Macropterous, some veins may be lost (especially fore wing r-m and hind wing RS), but venation otherwise normal 5

3(2). Epicnemial carina absent (Fig. 23) .... ***Parahormius*** Nixon (in part)

Distribution: Canada to Mexico; 5 described species in New World and many undescribed, though only a few, rare, undescribed species key here. Biology (host records only): reared only from lepidopteran larvae in the families Coleophoridae, Cosmopterygidae, Gelechiidae, and Lyonetiidae (Nixon, 1940; Hedqvist, 1963; Whitfield and Wagner, 1991). Reference: Wharton (1993, key to New World species).

\- Epicnemial carinae present (Fig. 24) 4

4(3). Propodeum with a pair of lateral spines (Fig. 25). Metasomal terga uniformly sclerotized .... ***Pambolus*** Haliday (in part)

Distribution: Canada to Brazil; particularly diverse in Neotropics, but only 2 described species in New World; brachypterous species relatively rare. Biology: reared from Buprestidae and Chrysomelidae in Palaearctic region, but records need verification. References: Belokobyl'skij (1986, 1987a, 1993a), who also recognizes *Dimeris* Ruthe and *Phaenodus* Förster rather than treating these as synonyms of *Pambolus*.

\- Propodeum lacking lateral spines. At least metasomal 2+3 desclerotized medially (Fig. 40) .... ***Hormius*** Nees (in part)

Distribution: Cosmopolitan; large genus, but only a few undescribed species key here. Biology: gregarious ectoparasitoids reared from Gelechiidae, Tortricidae, and possibly Coleophoridae in New World, and several other lepidopteran families elsewhere (Voukassovitch, 1927; Basinger, 1938; Shenefelt, 1975); little biological information available for New World species. Reference: Wharton (1993).

5(2). Fore wing with either 2RS or r-m (Fig. 1) absent 6

\- Both 2RS and r-m present, though r-m usually unpigmented 9

6(5). Fore wing 2RS absent; r-m present. Metasoma in the form of a 3-segmented carapace (Fig. 32) .... ***Lysitermus*** Förster (in part)

Distribution: U.S.A. (Texas) to Costa Rica, Caribbean; only 2 described species in New World. Biology: essentially unknown, but the type species was reared from bagworms (Psychidae) in Europe. References: van Achterberg (1991, Palaearctic species); Wharton (1993).

\- Fore wing 2RS present; r-m absent (Fig. 1). Metasoma carapace-like or not 7

7(6). Epicnemial carina absent (Fig. 23) .... ***Parahormius*** Nixon (in part)

Distribution: Canada to Mexico; only a few undescribed species key here. Biology (host records only): reared only from lepidopteran larvae in the families Coleophoridae, Cosmopterygidae, Gelechiidae, and Lyonetiidae (Nixon, 1940; Hedqvist, 1963; Whitfield and Wagner, 1991). Reference: Wharton (1993, key to New World species).

- Epicnemial carina present (Fig. 24) (some Betylobraconinae may also key here)..8

8(7). First subdiscal cell of fore wing closed, 2-1A present and well-defined. Metasomal terga 2+3 (and often 4-6) desclerotized medially (Fig. 40). Hind tibia of male not unusually thickened ...***Hormius*** Nees (in part)

Distribution: Cosmopolitan; only one described and a few undescribed species key here. Biology: gregarious ectoparasitoids reared from Gelechiidae, Tortricidae, and possibly Coleophoridae in New World, and several other lepidopteran families elsewhere (Voukassovitch, 1927; Basinger, 1938; Shenefelt, 1975); little biological information available for New World species. References: Belokobyl'skij and Tobias (1986, key to Palaearctic species), Wharton (1993).

- First subdiscal cell open, 2-1A absent or nearly so (Fig. 1). Terga 2+3 normally sclerotized or even somewhat carapace-like. Hind tibia of male unusually thickened ....***Acrisis*** Förster

Distribution: Canada to Costa Rica; only 2 described species in New World. Biology: essentially unknown (Shaw and Huddleston, 1991). References: Muesebeck (1935), Tobias (1983), Belokobyl'skij (1993a).

9(5). Epicnemial carina absent (Fig. 23) 10

- Epicnemial carina present (Fig. 24, also Fig. 31 but weak) 16

10(9). Occipital carina completely absent (Fig. 37) ....**undescribed genus #1**

We have seen several species from the Neotropical Region that appear to represent undescribed genera, but further study is needed before they can be adequately described. The most commonly encountered of the species we have seen key here.

11(10). Fore wing m-cu arising distad 2RS (Figs. 2, 3, 7-10, 14, 15, 17) 12

- Fore wing m-cu arising basad 2RS, or occasionally directly in line with2RS (Exotheccini) (Figs. 5, 6, 13, 16, 18-20)..13

12(11). Metasomal terga 2+3 sculptured and fully sclerotized dorsally (Fig. 34). Wings: Fig. 10 ....***Monitoriella*** Hedqvist (in part)

Distribution: Mexico to Peru, Trinidad; uncommon; 5 described species. Biology: phytophagous, forming galls on *Philodendron* and perhaps other plants (Wharton, 1993; Infante et al., 1995). References: Hedqvist (1963, description, key), Wharton (1993), Infante et al. (1995). Probably related to Doryctinae.

- Metasomal terga 2+3 not sculptured, mostly desclerotized medially (Fig. 40). Wings: Fig. 14 ....***Parahormius*** Nixon (in part)

Distribution: Canada to Costa Rica; moderately common; 5 described New World species, most New World species undescribed. Biology (host records only): reared only from lepidopteran larvae in the families Coleophoridae, Cosmopterygidae, Gelechiidae, and Lyonetiidae (Nixon, 1940; Hedqvist, 1963; Whitfield and Wagner, 1991). Reference: Wharton (1993, key to New World species).

13(11). First, second, and usually all of third metasomal terga uniformly and finely granular (Fig. 29). Mesoscutum dull, not polished and shining. Fore wing 2cu-a often absent.........***Xenosternum*** Muesebeck

Distribution: central U.S.A. to central Mexico; only one described species. Biology: essentially unknown; reared from Gelechiidae, Cosmopterygidae, and possibly Gracillariidae (Copenhafer and Parker, 1938; Pinto et al., 1987). References: Muesebeck (1935), van Achterberg (1983), Belokobyl'skij (1993a, tribal placement and relationship to *Avga*).

- Metasomal terga smooth or striate, rarely somewhat rugose. Mesoscutum largely polished. Fore wing 2cu-a present...14

14(13). Pronotum usually sculptured dorsally and usually with transverse, often crenulate sulcus (Fig. 26), never largely unsculptured with large round median pit. Fore wing r arising from approximately basal third of stigma (Fig. 5) ....***Colastes*** Haliday

Distribution: Alaska and Canada to Mexico, Brazil; number of described species in New World uncertain because the species have not been studied following separation of *Colastes* and *Shawiana* by van Achterberg (1983). Biology: detailed biology available for at least one Holarctic species (Shaw, 1983); solitary, idiobiont ectoparasitoids on a wide variety of concealed hosts in at least 4 orders: Lepidoptera, Coleoptera, Hymenoptera, and Diptera, with most records from leaf-miners. References: Mason (1968), van Achterberg (1983, redefinition, revision of Palaearctic species), Belokobyl'skij and Tobias (1986).

- Pronotum dorsally largely unsculptured, with large, round median pit (Fig. 27). Origin of radius variable... 15

15(14). Fore wing r arising basad middle of stigma (Fig. 18), often from proximal third of stigma (as in Fig. 5)........ ....***Shawiana*** van Achterberg

Distribution: Alaska and Canada to Mexico; at least 2 described species in New World. Biology: solitary, primarily idiobiont ectoparasitoids on concealed tenthredinid sawflies, with one Palaearctic species reared from leaf-mining Coleoptera (Dowden, 1941; Shaw, 1983, as *Phanomeris*). References: Muesebeck (1932), van Achterberg (1983, description, revision of Palaearctic species), Shaw and Huddleston (1991), Belokobyl'skij (1993a, as subgenus of *Colastes*).

- Fore wing r arising from middle of stigma (Fig. 19) ....***Xenarcha*** Förster

Distribution: Alaska and Canada to Mexico; only one described species in New World. Biology: solitary, more or less idiobiont ectoparasitoids in which there is some host growth and activity after egg deposition but not subsequent to egg hatch; reared in Palaearctic region mostly from leaf-mining Tenthredinidae (Shaw, 1983). Reference: van Achterberg (1983).

16(9). First three metasomal terga carapace-like, the carapace heavily sculptured (Fig. 32), bearing T2 and T3 spiracles along dorso-lateral margin (4th and following terga hidden beneath carapace in normal resting position but may be distended in dead specimens, depending on manner of preparation). Fore wing 1-1A curved, its middle portion not distinctly separated from posterior margin of wing; fore wing m-cu arising distad 2RS (Fig. 9) .... ***Lysitermus*** Förster (in part)

Distribution: U.S.A. (Texas) to Costa Rica, Caribbean; relatively rare; only 2 described species in New World. Biology: essentially unknown, but the type species was reared from bagworms (Psychidae) in Europe. References: van Achterberg (1991, Palaearctic species), Wharton (1993).

\- Metasoma rarely carapace-like; if carapace present, then not as above (species of *Oncophanes* and *Lysitermoides* with a metasomal carapace have T2 and T3 spiracles in membranous lateral tergites). Wing venation variable, often not as above 17

17(16). Fore wing stigma linear, either indistinguishable from R1 or slightly thickened, with r arising basad midlength when stigma is perceptible; fore wing m-cu arising distad 2RS; second submarginal cell very long, 3RSa usually at least twice as long as 2RS (Figs 2, 8) (Hydrangeocolini) 18

\- Stigma broad, never linear; position of m-cu variable, often arising basad 2RS; size of second submarginal cell variable, often short 19

18(17). Fore wing stigma linear, indistinguishable from R1; fore wing 2a always absent (Fig. 2) .... ***Aspilodemon*** Fischer

Distribution: Mexico to Brazil, Chile, though not recorded from Caribbean; about 7 described species. Biology: unknown. References: Fischer (1966, description and key, as Opiinae); Wharton (1988).

\- Stigma distinct, though very narrow; 2a usually present (Fig. 8) .... ***Hydrangeocola*** *Brèthes*

Distribution: Chile; about 13 described species. Biology: essentially unknown, though the type species was reared from a gall on Hydrangea (Brèthes, 1927). References: Fischer (1968, as *Kephalosema*, Wharton (1988, 1993).

19(17). Antenna with only 10 flagellomeres, distinctly shorter than body. Petiole very broad, only 0.5-0.7 times as long as apical width (Fig. 30). Wings: Fig. 7 .... ***Chremylus*** Haliday

Distribution: eastern U.S.A. and Chile; one species, probably more widely distributed than published records suggest because of its host association with stored product pests. Biology: gregarious parasitoids of case-bearing tineid larvae, but also recorded from Coleoptera (Mason, 1948). References: Belokobyl'skij (1987, l993a), Wharton (1993).

\- Antenna with 11 or more flagellomeres, often as long as or longer than body. Petiole more slender, at least as long as apical width (e.g. Fig. 28) 20

20(19). Fore wing m-cu clearly arising basad 2RS (Figs 4, 13) 21

\- Fore wing m-cu arising distad 2RS (Fig. 17), occasionally arising directly in line with 2RS (*Oncophanes* and *Rhyssalus* will key through both halves of this couplet) 28

21(20). Propodeum with wide, transverse, crenulate groove near base (Fig. 33). Ovipositor and sheaths extremely short, barely exerted. Wings: Fig. 4 .... ***Cantharoctonus*** Viereck

Distribution: Canada to Peru, though not recorded from Caribbean; 3 described species. Biology: reared from *Bucculatrix* (Mason, 1968). References: Shenefelt (1975), Whitfield and van Achterberg (1987).

\- Propodeum normal, with very narrow, barely visible basal transverse groove (as in Figs 28, 31). Ovipositor and sheaths variable but often at least half as long as hind tibia 22

22(21). Occipital and hypostomal carinae not meeting; either occipital carina absent ventrally near base of mandible (Fig. 36) or the two carinae reaching mandible and narrowly separated throughout (Fig. 39) (most Pambolini and Rhysipolini) 23

\- Occipital and hypostomal carinae meeting well above base of mandible (Fig. 38) (some Rhyssalini) 26

23(22). Propodeum with a pair of postero-lateral spines (Fig. 25), sometimes (in small specimens) reduced to small bumps at junction of dorsal and posterior angles of propodeum. Occipital carina incomplete ventrally (Fig. 36). Wings: Fig. 13 .... ***Pambolus*** Haliday (in part)

Distribution: Canada to Brazil; particularly diverse in Neotropics, but only 2 described species in New World. Biology: reared from Buprestidae and Chrysomelidae in Palaearctic region, but records need verification. References: Belokobyl'skij (1986, 1987a, 1993a), who also recognizes *Dimeris* Ruthe and *Phaenodus* Förster rather than treating these as synonyms of *Pambolus*.

\- Propodeum lacking distinct spines. Occipital carina complete, extending to base of mandible and narrowly separated from hypostomal carina (Fig. 39) 24

24(23). Fore wing 2a present, although not always strongly pigmented throughout (Fig. 21) .... **undescribed genus #2**

Distribution: Chile and adjacent Andean regions; several species have been collected, but have not yet received taxonomic attention. Biology: unknown.

\- Fore wing 2a absent 25

25(24). Fore wing 1M and (RS+M)a strongly swollen into a stigma-like structure (may be only in males) (Fig. 11) .. .. ***Neurocrassus*** *Snoflák*

Distribution: a single, unidentified species has been recorded from central U.S.A. (Whitfield, 1988). Biology: unknown for New World; the single described species was reared from *Stenostola* in Europe (Starý, 1957). References: *Snoflák* (1945); Whitfield (1988).

- Fore wing 1M and (RS+M)a normal, not swollen (Fig. 16) .... ***Rhysipolis*** Förster

Distribution: Alaska and Canada to Mexico; only 3 described species in New World, 2 that are Holarctic and one apparently endemic. Biology: solitary koinobiont ectoparasitoids of late instar leaf-miners and leaf-tiers that later change leaves and/or otherwise feed and/or pupate elsewhere (Shaw, 1983); New World records confined to Gracillariidae. References: Shenefelt (1975), Belokobyl'skij and Tobias (1986), Belokobyl'skij (1993a), Nobles and Whitfield (in review, revision of Nearctic species).

26(22). At least metasomal tergum 2 mostly sculptured (as in Figs 29, 30). Wings: Fig. 12 32

- Metasomal tergum 2 mostly smooth, unsculptured (as in Fig. 35) 27

27(26). Median mesonotal lobe strongly raised above lateral lobes anteriorly; metasomal terga 2+3 somewhat enlarged, well sclerotized and highly polished. Wings: Fig. 6 .... ***Dolopsidea*** Hincks

Distribution: central U.S.A.; very rare; known only from one undetermined species in New World. Biology: essentially unknown, rearing of the Palaearctic type species from an anobiid needs confirmation. References: Shenefelt (1975, as *Rhyssalus*), Belokobyl'skij (1987b, 1992), Whitfield (1988).

- Median mesonotal lobe slightly if at all raised above lateral lobes; metasomal terga 2+3 weakly sclerotized, never highly polished. Wings: Fig. 17 .... ***Rhyssalus*** Haliday (in part)

Distribution: U.S.A. and Canada; the few species occurring in the New World are in need of revision. Biology: essentially unknown, though specimens have been reared from beetle-infested wood (Shaw and Huddleston, 1991). References: Shenefelt (1975), Whitfield (1988), Quicke and van Achterberg (1990), Belokobyl'skij (1992).

28(20). Fore wing 2CUb arising above middle of first subdiscal cell, often directly in line with 1CU (Fig. 10). Metasoma sculptured, at least tergum 2 striate and/or rugose (Fig. 34). Large species, usually greater than 5 mm in length (Monitoriellini) .... ***Monitoriella*** Hedqvist (in part)

Distribution: Mexico to Peru, Trinidad; uncommon; 5 described species. Biology: phytophagous, forming galls on *Philodendron* and perhaps other plants (Wharton, 1993; Infante et al. 1995). References: Hedqvist (1963, description, key), Wharton (1993), Belokobyl'skij (1993a), Infante et al. (1995). Probably related to Doryctinae.

- If tergum 2 with sculpture, then either 2CUb arising at or below middle of first subdiscal cell or the species are very small (at most 2 mm) and first subdiscal cell is open 29

29(28). Metasoma dorsally largely desclerotized beyond petiole; at least terga 2+3 completely desclerotized medially (Fig. 40), never with rugose, granular, or striate sculpture. Fore wing 1cu-a essentially interstitial with 1M; fore wing vein 2CUa absent or nearly so, 2CU usually arising from m-cu or directly in line with 1CU (as in Fig. 14) .... ***Hormius*** Nees (in part)

Distribution: Cosmopolitan; common; about 18 described and many undescribed species in New World. Biology: gregarious ectoparasitoids reared from Gelechiidae, Tortricidae, and possibly Coleophoridae in New World, and several other lepidopteran families elsewhere (Voukassovitch, 1927; Basinger, 1938; Shenefelt, 1975); little biological information available for New World species. References: Belokobyl'skij and Tobias (1986, key to Palaearctic species), Belokobyl'skij (1989, Australian species), Wharton (1993).

- Metasoma dorsally with terga 2 and/or 3 often sculptured, nearly always normally sclerotized. Fore wing 1cu-a nearly always arising distinctly distad 1M; when first subdiscal cell present, 2CUb nearly always arising from or below the middle of its distal border (2CUa therefore present)....................................... 30

30(29). Fore wing 3RSb incomplete, not extending to wing margin as a tubular vein (Fig. 15). Small species less than 2 mm in length .... ***Proacrisis*** Tobias

Distribution: Canada to Mexico, though no records have been previously published for this genus in the New World. Biology: unknown. References: Tobias (1983), Belokobyl'skij (1993a).

- Fore wing 3RSb extending to wing margin as a tubular vein. Body length variable, but often greater than 2 mm in length (some Rhyssalini) 31

31(30). At least metasomal tergum 2 with sculpture (as in Figs 29, 30) 32

- Metasomal tergum 2 smooth, unsculptured (as in Fig. 33) 33

32(26,31). Three basal segments of metasoma carapace-like, longitudinally striate, with interconnecting sculpture between the striae; fourth metasomal tergum largely or entirely concealed beneath third......... .... ***Lysitermoides*** van Acherberg

Distribution: eastern North America (southern Ontario to Florida); 2 described species. Biology: unknown. Reference: van Achterberg (1995).

- Three basal segments of metasoma not as above: fourth segment exposed about as as much as third; sculpture otherwise, often coriaceous.. ***Oncophanes*** Förster (in part)

Distribution: Alaska and Canada to Mexico; about 8 described species in New World. Biology: gregarious ectoparasitoids of Tortricidae and other leaf-rolling or leaf-tying microlepidopteran larvae (Shaw, 1983; Knight and Croft, 1983; Shaw and Huddleston, 1991). References: Muesebeck (1935, revision), Belokobyl'skij (1993a).

33(32). Mesonotal midpit distinct, broad, and unusually shallow; notauli narrow, deep, sculptured throughout, extending to midpit region. Male of described species with thickened hind tibia. Wings: Fig. 17 ....

.... ***Rhyssalus*** Haliday (in part)

Distribution: U.S.A. and Canada; the few species occurring in-the New World are in need of revision. Biology: essentially unknown, though specimens have been reared from beetle-infested wood (Shaw and Huddleston, 1991). References: Shenefelt (1975), Whitfield (1988), Quicke and van Achterberg (1990), Belokobyl'skij (1992).

\- Mesonotal midpit absent (as in Fig. 26), posterior region of mesoscutum striate; notauli weak, usually represented posteriorly only by broad, shallow, unsculptured depressions. Male tibia not thickened .... ***Pseudobathystomus*** Belokobyl'skij

Distribution: U.S.A. and Canada; rare; known only from one species in New World. Biology: unknown. Reference: Belokobyl'skij (1993a).

REFERENCES

Achterberg, C. van 1983. Revisionary notes on the Palaearctic genera and species of the tribe Exothecini Foerster (Hymenoptera, Braconidae). Zoologische Mededelingen Leiden 57: 339-355.

Achterberg, C. van 1984. Essay on the phylogeny of the Braconidae (Hymenoptera: Ichneumonoidea). Entomologisk Tidskrift 105: 41-58.

Achterberg, C. van 1991. Revision of the genera of the Afrotropical and W. Palaearctical Rogadinae Foerster (Hymenoptera: Braconidae). Zoologische Verhandelingen Leiden 273: 1-102.

Achterberg, C. van 1995. Generic revision of the subfamily Betylobraconinae (Hymenoptera: Braconidae) and other groups with modified fore tarsus. Zoologische Verhandelingen Leiden 298: 1-242.

Askew, R.R. and Shaw, M. R. 1986. Parasitoid communities: their size, structure and development. Pp. 225-264. In: Waage, J. and Greathead, D. (Eds.). Insect Parasitoids, 13th Symposium of The Royal Entomological Society of London. Academic Press, London.

Basinger, A. J. 1938. The orange tortrix, *Argyrotaenia citrana*. Hilgardia 11: 635-669.

Belokobyl'skij, S. A. 1984. On the division of the tribe Exothecini s. I. (Hymenoptera, Braconidae) in two with the description of a new genus and subgenus. Zoologicheskii Zhurnal 63: 1019-1025. (In Russian.)

Belokobyl'skij, S. A. 1986. A review of the Palaearctic species of the genera *Pambolus* Hal. and *Dimeris* Ruthe (Hymenoptera, Braconidae). Proceedings of the Zoological Institute, Leningrad 159: 18-37. (In Russian.)

Belokobyl'skij, S. A. 1987a. Structure of the male genitalia in the braconid subfam. Doryctinae (Hymenoptera, Braconidae)- its evolution and significance in classification of the group. Morphological Foundations for Insect Phylogeny 69: 209-219. (In Russian.)

Belokobyl'skij, S. A. 1987b. On the status of the genus *Dolopsidea* Hincks (Hymenoptera, Braconidae, Doryctinae). Pp. 73-78. In: Storozheva and Kusakin (Eds.). New Data on Insect Systematics from the Soviet Far East. USSR Academy of Sciences, Vladivostok. (In Russian.)

Belokobyl'skij, S. A. 1992. On the classification and phylogeny of the braconid wasps of the subfamilies Doryctinae and Exothecinae (Hymenoptera, Braconidae). Part I. On the classification, 1. Entomologicheskoe Obozrenie 71: 900-928. (In Russian.)

Belokobyl'skij, S. A. 1993a. On the classification and phylogeny of the braconid wasp subfamilies Doryctinae and Exothecinae (Hymenoptera: Braconidae). Part I. On the classification, 2. Entomologicheskoe Obozrenie 72: 143-164. (In Russian.)

Belokobyl'skij, S. A. 1993b. On the classification and phylogeny of the braconid wasp subfamilies Doryctinae and Exothecinae (Hymenoptera: Braconidae). Part II. On the phylogeny. Entomologicheskoe Obozrenie 72: 891-914. (In Russian.)

Belokobyl'skij, S. A. and Tobias, V. I. 1986. Subfamily Doryctinae. Pp. 21-72. In: Tobias V. I. (Ed.). Identification of Insects of European U.S.S.R. Vol. III, part V. Hymenoptera: Braconidae. Akademia Nauk, Leningrad. (In Russian.)

Brèthes, J. 1927. Nouveaux Hymenopteres parasites du Chile. Revista Chilena de Historia Natural 31: 194-200.

Clark, J. A. J. 1965. Six new species of the S. American genus *Leurinion* Muesebeck (Hym., Braconidae, Hormiini). Annals and Magazine of Natural History 13: 645-660.

Copenhafer, L. M. and Parker, R. L. 1938. *Ornix pruniuorelia* Chambers (Lepidoptera, Tineidae), a pest of the apple tree in the lower Missouri River Valley region. Journal of the Kansas Entomological Society 11: 37-53.

Dowden, P. B. 1941. Parasites of the birch leaf-mining sawfly (*Phyllotoma nemorata*). United States Department of Agriculture Technical Bulletin 757: I-55.

Fischer, M. 1966. *Aspilodemon*, ein neues Opiinen Genus aus Brasilien (Hymenoptera, Braconidae). Entomophaga 11: 161-176.

Fischer, M. 1968. *Kephalosema*, ein neues Horrniinen-Genus aus Chile (Hymenoptera, Braconidae). Polskie Pismo Entotnologiczne 38: 791-805.

Hedqvist, K.J. 1963. Notes on Hormiinae with description of new genera and species (Hym., Ichneumonoidea, Braconidae). Entomologisk Tidskrift 84: 30-61.

Infante, F., Hanson, P., and Wharton, R. 1995. Phytophagy in the genus *Monitoriella* (Hymenoptera: Braconidae) with description of new species. Annals of the Entomological Society of America 88: 406-415.

Marsh, P. M. 1979. Family Braconidae. Pp. 144-295. In: Krombein, K. V., Hurd, P. D., Smith, D.R., and Burks, B. D. (Eds.). Catalog of Hymenoptera in America North of Mexico. Smithsonian Institution Press, Washington, D.C.

Mason, H. C. 1948. *Chremylus rubiginosus* (Nees), a braconid parasite of the casemaking clothes moth. Annals of the Entomological Society of America 41: 28-40.

Mason, W.R. M. 1968. New Canadian Braconidae (Hymenoptera). The Canadian Entomologist 100: 715-728.

Muesebeck, C. F. W. 1932. Two new species of *Phanomeris* Foerster (Hymenoptera: Braconidae) parasitic on leaf-mining sawflies. Proceedings of the Entomological Society of Washington 34: 81-83.

Muesebeck, C. F. W. 1935. On the genus *Oncophanes* Foerster, with descriptions of two new related genera (Hymenoptera: Braconidae). Annals of the Entomological Society of America 28: 241-250.

Nixon, G. E. J. 1940. New genera and species of Hormiinae, with a note on *Hormiopterus* Giraud (Hym., Braconidae). Annals and Magazine of Natural History 11: 473~493.

Nobles, L. S. and Whitfield, J. B. A revision of the Nearctic species of *Rhysipolis* Foerster (Hymenoptera: Braconidae), Submitted to Journal of Natural History.

Pinto, J. D., Frommer, s I., Manweiler, S. A. 1987. The insects of jojoba, *Simmondsia chinensis*, in natural stands and plantations in southwestern North America. The Southwestern Entomologist 12: 287-298.

Quicke, D. L. J. 1993. The polyphyletic origin of endoparasitism in the cyclostome lineages of Braconidae (Hymenoptera): a reassessment. Zoologische Mededelingen Leiden 67: 159-177.

Quicke, D. L. J. and Achterberg, C. van 1990. Phylogeny of the subfamilies of the family Braconidae. Zoologische Verhandelingen Leiden 258: 3-95.

Shaw, M. R. 1983. On evolution of endoparasitism: the biology of some genera of Rogadinae (Braconidae). Contributions of the American Entomological Institute 20: 307-328.

Shaw, M. R. and Huddleston, T. 1991. Classification and biology of braconid wasps (Hymenoptera: Braconidae). Handbooks for the Identification of British Insects 7(11): 1-126.

Shenefelt, R. D. 1975. Pars 12. Braconidae 8, Exothecinae Rogadinae. Pp. 1115-1262. In: Vecht, J. van der and Shenefelt, R. D. (Eds.). Hymenopterorum Catalogus (nova edito). Dr. W. Junk, The Hague.

Snoflák, J. 1945. *Neurocrassus* gen. n. *tesari* sp. n. Folia Entomologica 8: 25-27.

Starý, P. 1957. Notes on the Braconidae (Hym.) of Czechoslovakia IV. (Part I). Acta Societas Entomologicae èechoslovenlee 54: 277-292.

Tobias, V. I. 1983. Towards an understanding of the genus *Acrisis* Foerster, 1862, of the tribe Acrisidini Hellen, 1957 (Hymenoptera, Braconidae, Doryctinae), Trudy Vsesoyuznogo Entomologicheskogo Obshchestva 65: 155-168. (In Russian.)

Voukassovitch, P. 1927. Observations sur les parasites et hyperparasites du papillon *Psammotis hyalinalis* HB. Compres Rendus des Séances de la Socièté de Biologie 93: 235-237.

Watanabe, C. 1968. A revision of the genus *Acanthormius* Ashmead, with descriptions of six new species (Hymenoptera, Braconidae), Insecta Matsumurana 30: 57-66.

Wharton, R. A. 1988. Classification of the braconid subfamily Opiinae (Hymenoptera). The Canadian Entomologist 120: 333-360.

Wharton, R. A. 1993. Review of the Hormiini (Hymenoptera: Braconidae) with a description of new taxa. Journal of Natural History 27: 107-171.

Wharton, R. A., Shaw, S. R., Sharkey, M., Wahl, D.B., Woolley, J. B., Whitfield, J.B., Marsh, P. M. and Johnson, J. W. 1992. Phylogeny of the subfamilies of the family Braconidae (Hymenoptera: Ichneumonoidea): a reassessment. Cladistics 8: 199-235.

Whitfield, J. B. 1988. Taxonomic notes on Rhyssalini and Rhysipolini (Hymenoptera: Braconidae) with first Nearctic records of three genera. Proceedings of the Entomological Society of Washington 90: 471-473.

Whitfield, J. B. 1992. The polyphyletic origin of endoparasitism in the cyclostome lineages of Braconidae (Hymenoptera). Systematic Entomology·17: 273-286.

Whitfield, J. B. and Achterberg, C. van 1987. Clarification of the taxonomic status of the genera *Cantharoctonus* Viereck, *Noserus* Foerster and *Pseudagava* Tobias (Hymenoptera: Braconidae). Systematic Entomology 12: 509-518.

Whitfield, J. B. and Wagner, D, L 1991. Annotated key to the genera of Braconidae (Hymenoptera) attacking leaf-mining Lepidoptera in the Holarctic region. Journal of Natural History 25: 733-754.

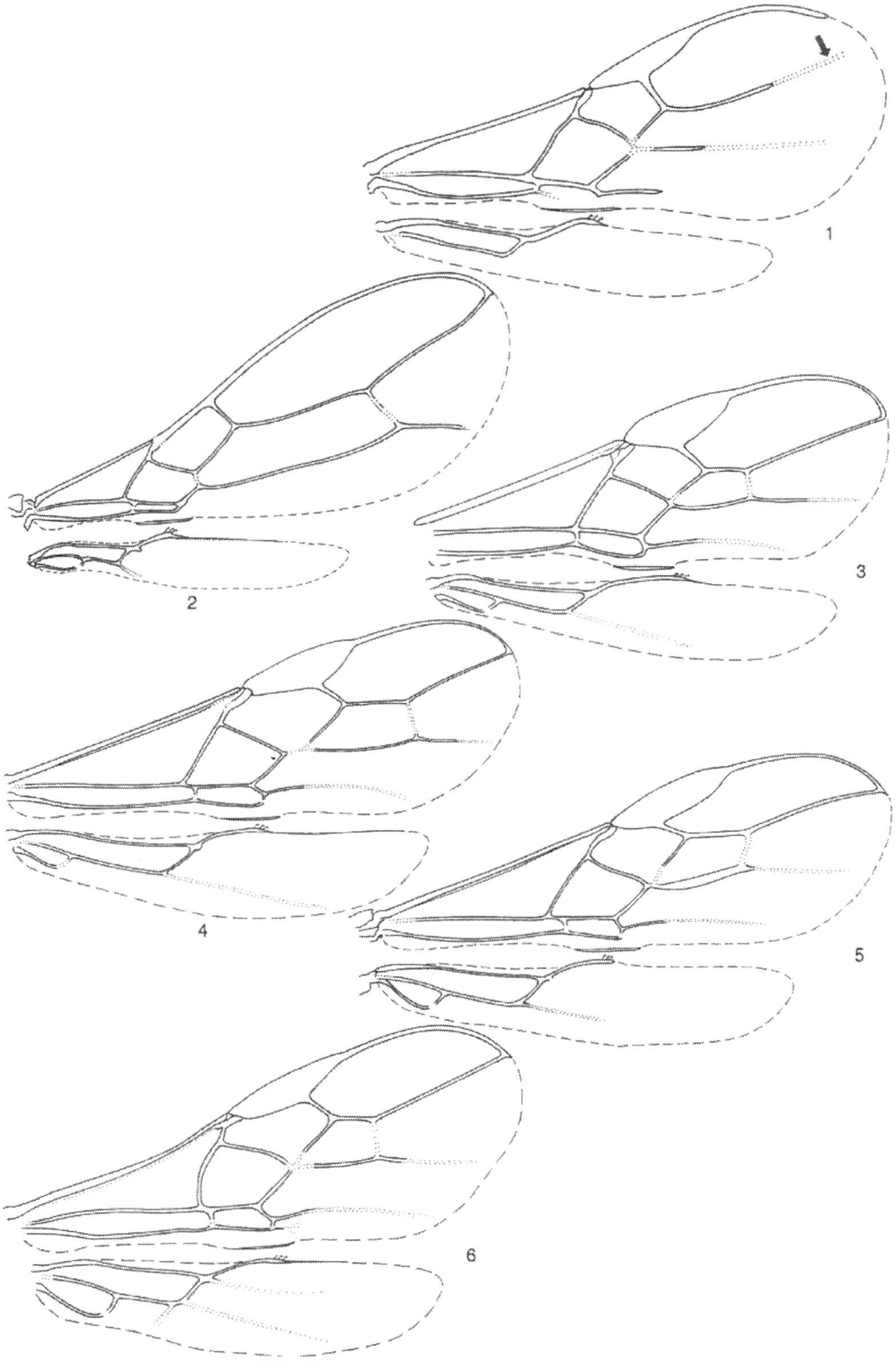

Figs 1-6: Wings of Hormiinae. 1, *Acrisis* sp., arrow = 3RSb not reaching wing margin; 2, *Aspilodemon* sp.; 3, *Allobracon* sp.; 4, *Cantharoctonus* sp.; 5, *Colastes* sp.; 6, *Dolopsidea* sp.

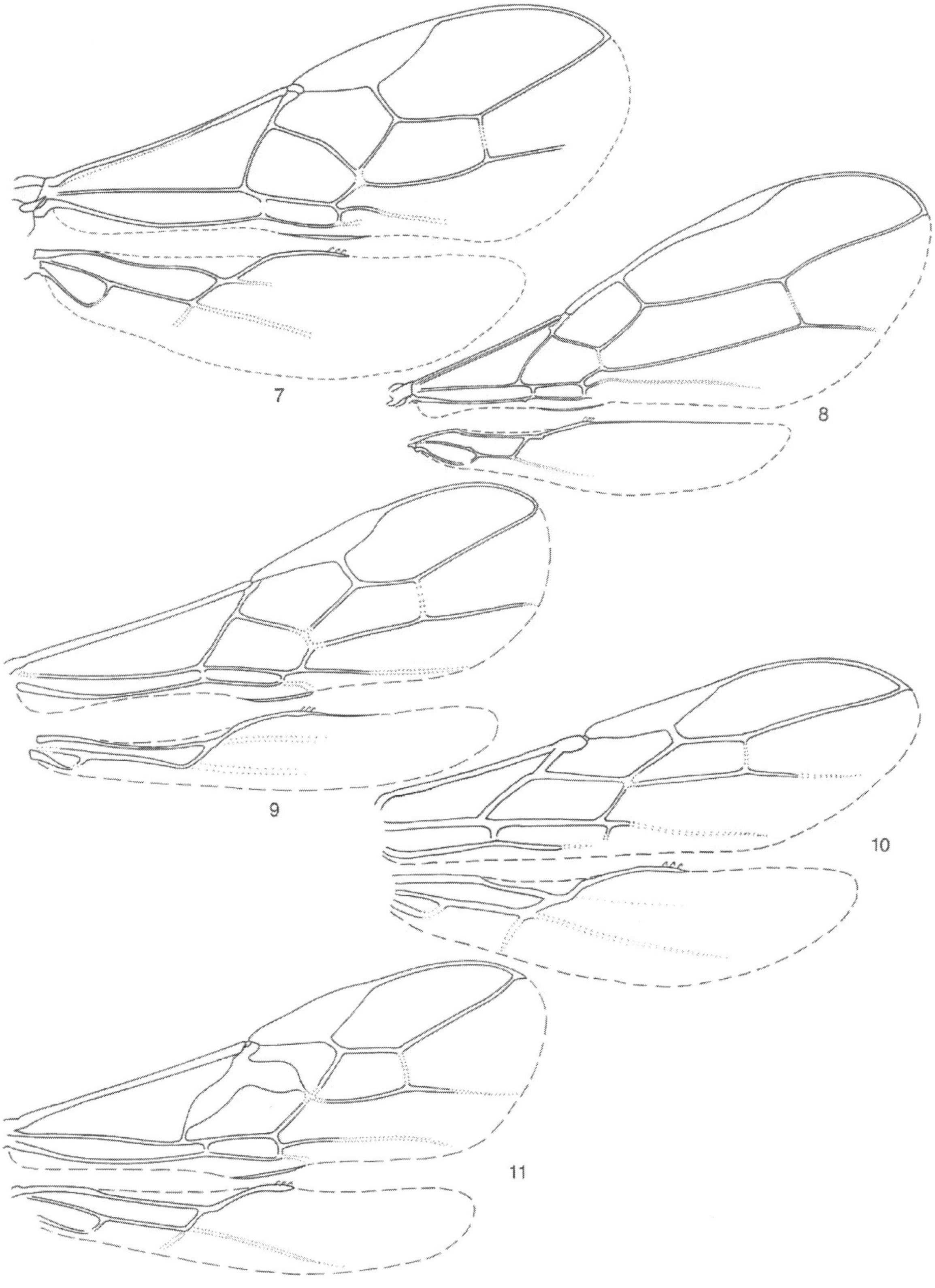

Figs 7-11: Wings of Hormiinae. 7, *Chremylus elaphus* Haliday; 8, *Hydrangeocola* sp.; 9, *Lysitermus* sp.; 10, *Monitoriella* sp.; 11, *Neurocrassus* sp.

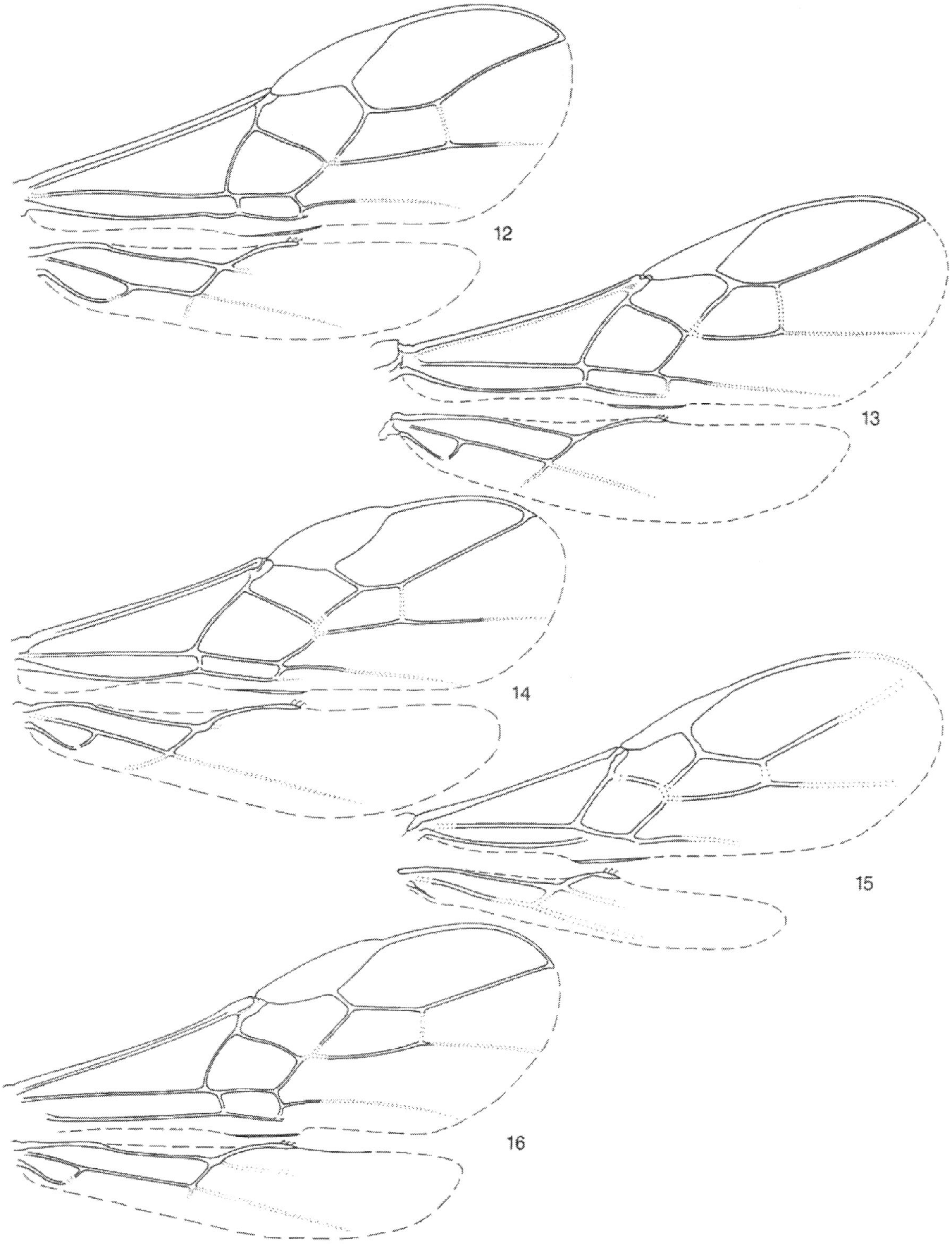

Figs 12-16: Wings of Hormiinae. 12, *Oncophanes* sp.; 13, *Pambolus* sp.; 14, *Parahormius* sp.; 15, *Proacrisis* sp.; 16, *Rhysipolis* sp.

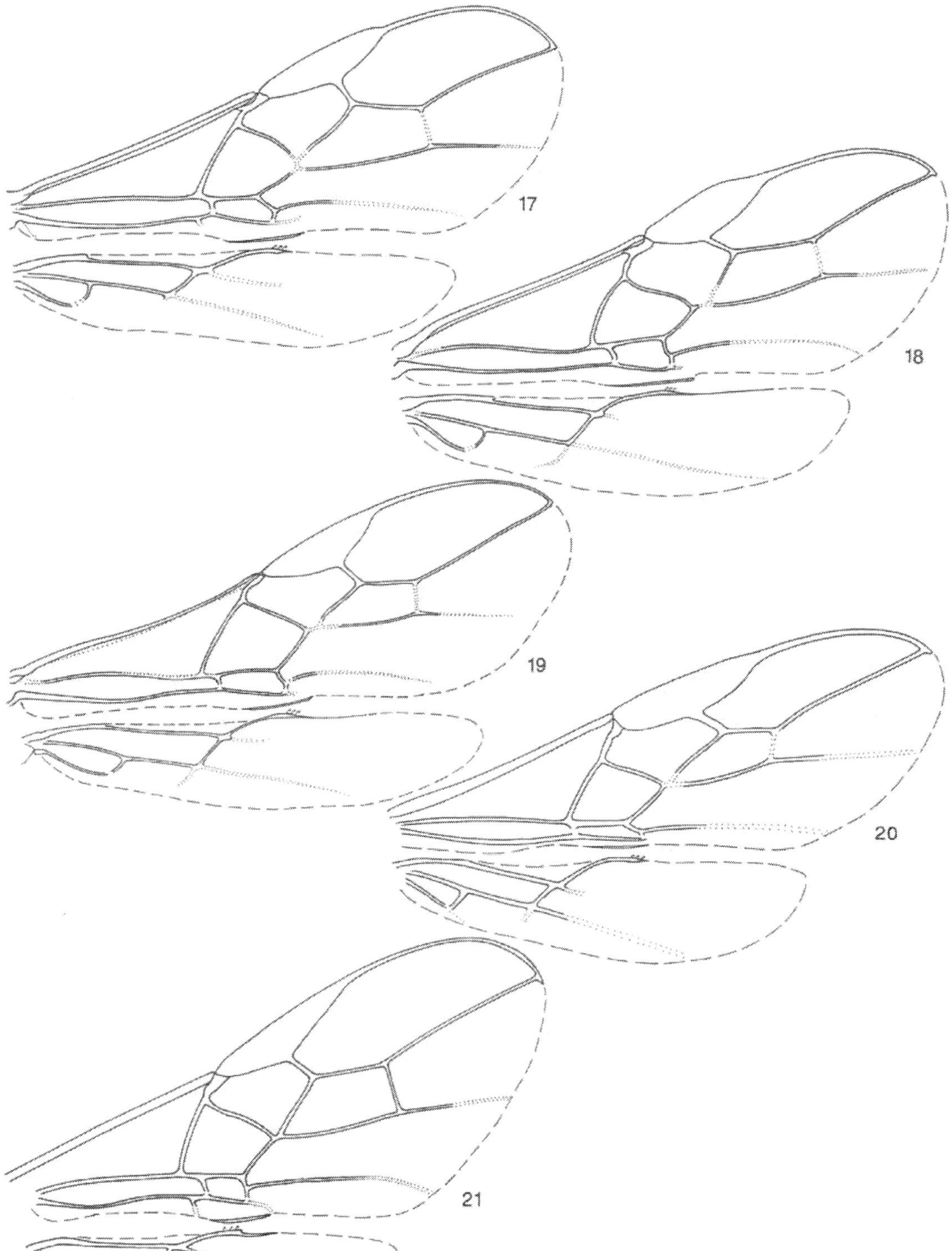

Figs 17-21: Wings of Hormiinae. 17, *Rhyssalus* sp.; 18, *Shawiana* sp.; 19, *Xenarcha* sp.; 20, *Xenosternum* sp.; 1, undescribed genus #2.

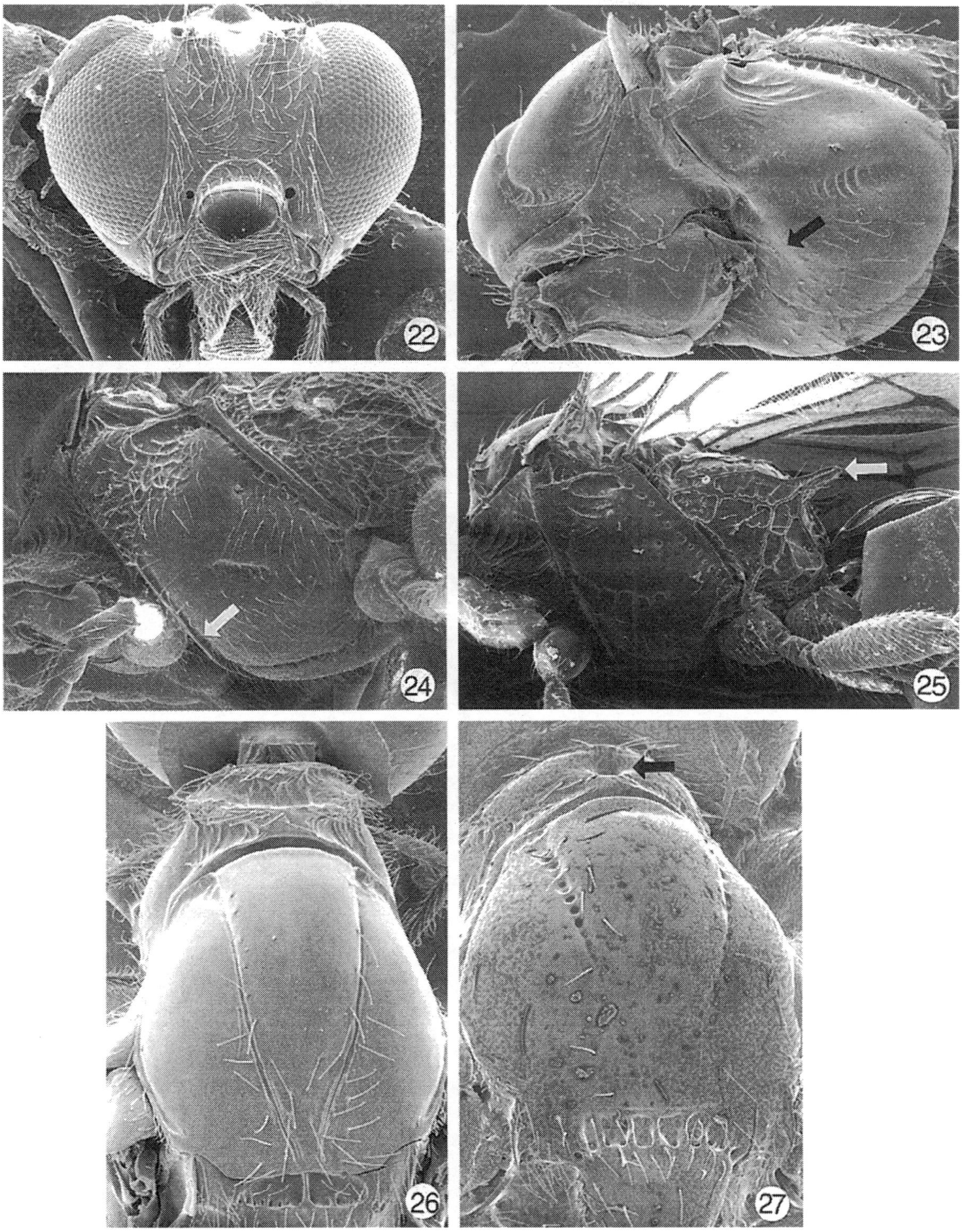

Figs 22-27: Hormiinae. 22, head of *Rhysipolis* sp., frontal view; 23-25, mesosoma: 23, *Parahormius* sp., lateral view, arrow = epicnemial carina absent; 24, *Clinocentrus* sp., lateral view, arrow = epicnemial carina; 25, *Pambolus* sp., lateral view, arrow = propodeal spine; 26, pronotum and mesoscutum of *Colastes* sp., dorsal view; 27, pronotum and mesoscutum of *Xenarcha* sp., dorsal view, arrow = median pit on pronotum.

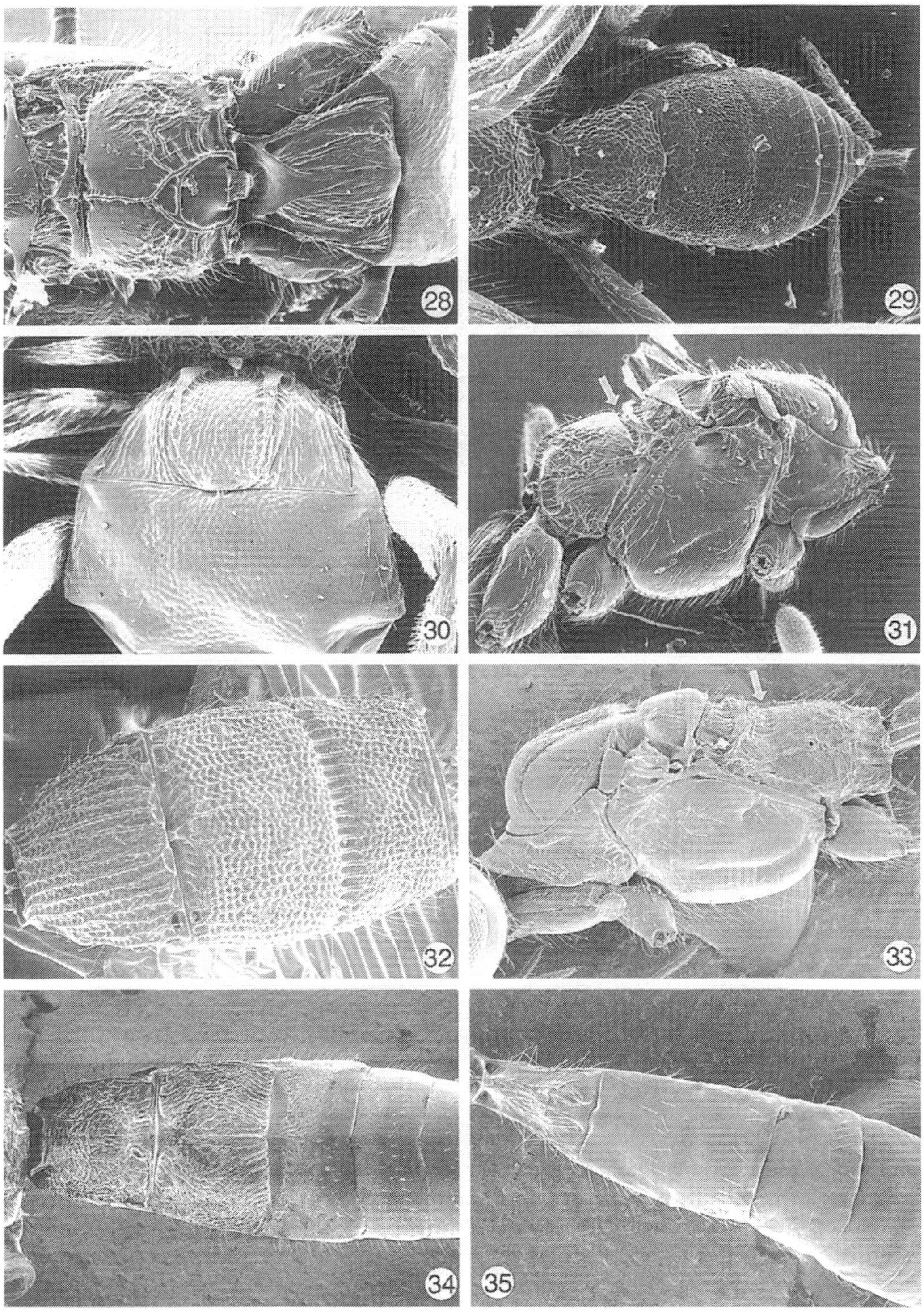

Figs 28-35: Hormiinae. 28, propodeum and petiole of *Rhysipolis* sp., dorsal view; 29, metasoma of *Xenosternum* sp.; 30, petiole and metasomal tergum 2 of *Chremylus elaphus* Haliday, dorsal view; 31, mesosoma of *Rhysipolis* sp., lateral view, arrow = basal transverse groove of propodeum; 32, metasoma of *Lysitermus* sp., dorsal view; 33, mesosoma of *Cantharoctonus* sp., lateral view, arrow = broad, basal transverse groove of propodeum: 34, metasoma of *Monitoriella elongata* Hedqvist, dorsal view (after Infante et al., 1995); 35, metasoma of *Cantharoctonus* sp., dorsal view.

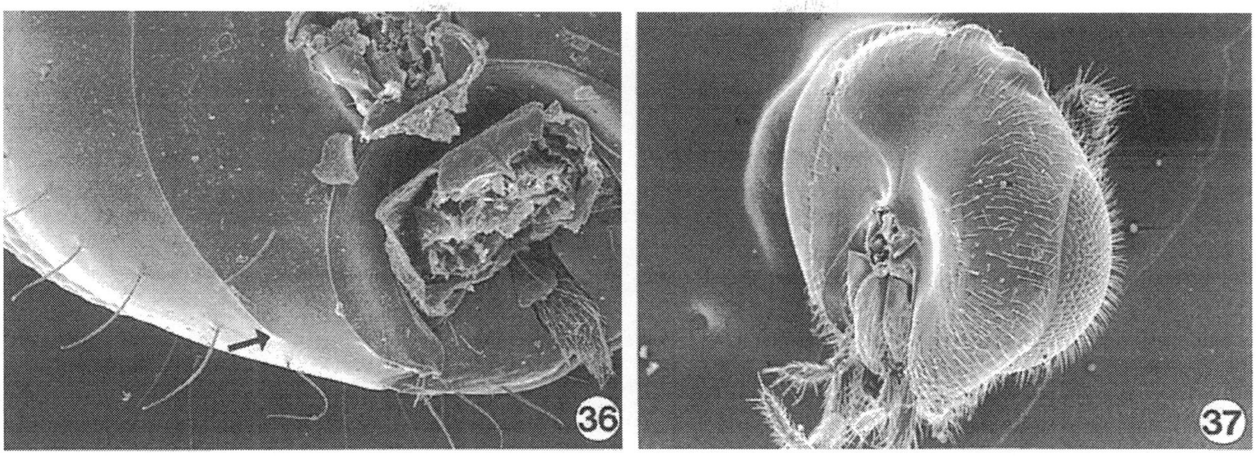

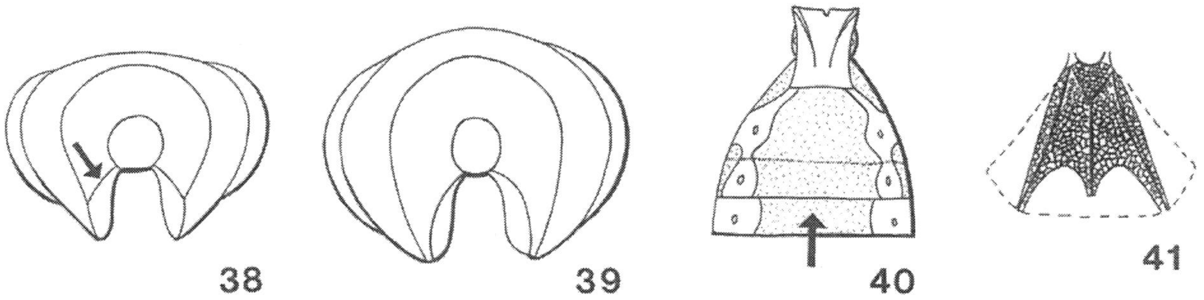

Figs 36-41: Hormiinae. 36-39, heads: 36, *Parahormius* sp., posterior view, arrow = ventral end of occipital carina: 37, Cotesia sp., posterior view showing absence of occipital carina; 38, *Rhysipolis* sp., posterior view, arrow = hypostomal carina; 39, *Oncophanes* sp., posterior view; 40, metasoma of *Hormius* sp., dorsal view, arrow = desclerotized median portion of terga 2-4; 41, petiole of *Allobracon* sp., dorsal view (after Wharton, 1993).

HORMIINAE UPDATE

D.L.J. Quicke
Department of Biology, Faculty of Life Sciences, Chulalongkorn University, Bangkok, Thailand.
d.quicke@email.com

The genera treated under the Hormiinae in the 1997 Manual comprise a polyphyletic assemblage of basically small-bodied cyclostome braconids (Zaldívar-Riverón *et al.*, 2006) that do not belong to the large subfamilies Aphidiinae, Alysiinae, Braconinae, Doryctinae, Opiinae or Rogadinae, or the distinctive Gnamptodontinae. The key to genera in the 1997 Manual includes two then undescribed genera. One of these, 'undescribed genus #2', was subsequently described as *Andesipolis* (Whitfield *et al.*, 2004), the other has not been described. The systematic placement of *Andesipolis* has been controversial (Townsend and Shaw, 2009; Sharanowski *et al.*, 2011, Quicke, 2015). Morphologically and biologically it resembles *Rhysipolis*, but molecular data clearly place it far more basally in the cyclostomes, and it is currently regarded as a member of the Mesostoinae (Quicke 2015; Shimbori *et al.*, 2017). Scatolini *et al.* (2002) described a new genus, *Pseudorhysipolis*, from Brazil.

The key to the genera works reasonably well but the genera are now split between the subfamilies Doryctinae, Exothecinae, Hormiinae, Lysiterminae, Pambolinae, Mesostoinae, Rhysipolinae and Rhyssalinae, as follows.

Acrisis, *Dolopsidea*, *Histeromerus* (treated as a monotypic subfamily in the 1997 Manual), *Lysitermoides*, *Oncophanes*, *Proacrisis*, *Pseudobathystomus* and *Rhyssalus* belong to the Rhyssalinae (Belokobylskij and Koponen, 2004).

Allobracon is not assignable to any currently recognized subfamily - it appears, based on DNA sequence data, to be an isolated genus related to the Braconinae, Exothecinae, Gnamptodontinae, Opiinae and Alysiinae clade.

Andesipolis, *Aspilodemon*, *Avga* and *Hydrangeocola* belong to the Mesostoinae.

Cantharoctonus, *Rhysipolis* and *Neurocrassus* (probably) all belong to the Rhysipolinae.

Chremylus is currently classified in the Pambolinae, but molecular data suggest a different affiliation (Quicke *et al.*, 2016).

Colastes, *Shawiana*, *Xenarcha* and *Xenosternum* belong to the Exothecinae.

Hormius and *Parahormius* remain in the Hormiinae.

Lysitermus belongs to the Lysiterminae.

Monitoriella is a derived phytophagous member of the Doryctinae (Zaldívar-Riverón *et al.*, 2007).

Pambolus, and probably also *Pseudorhysipolis*, belong to the Pambolinae.

Annotated list of recent papers on New World Exothecinae, Hormiinae, Pambolinae, Lysiterminae, Mesostoinae, Rhysipolinae Rhyssalinae (including Histeromerini), and related enigmatic genera

Belokobylskij SA, Koponen M. 2004. A new subgenus and species of the genus *Pseudobathystomus* Belokobylskij (Hymenoptera: Braconidae: Rhyssalinae) from the Canary Islands. Entomologica Fennica 15: 225-230.
Key to species of *Pseudobathystomus* but does not refer to New World material.

Belshaw R, Grafen A, Quicke DLJ. 2003. Inferring life history from ovipositor morphology in parasitoid wasps using phylogenetic regression and discriminant analysis. *Zoological Journal of the Linnaean Society* 139: 213-228.
Morphology-based predictions that *Allobracon* and *Aspilodemon* will be ectoparasitoids.

Braet Y, Achterberg C van. 2003. New species of *Pambolus* Haliday and *Phaenocarpa* Foerster (Hymenoptera: Braconidae: Pambolinae, Alysiinae) from French Guyana, Suriname and Panama. *Zoologische Mededelingen, Leiden* 77: 153-179.
Descriptions of seven new Neotropical *Pambolus* species and a key to all Neotropical species.

Figueroa JI, Martínez AM, Mejía A, Flores A, Sánchez JA, López V. 2010. *Aspilodemon* Fischer (Hym.: Braconidae, Hormiinae), nuevo registro para México. pp. 929-933. In: Cruz M., S.G., J. Tello E, A. Mendoza E. and A. Morales M. (Eds.). Entomología Mexicana Vol. IX. Sociedad Mexicana de Entomología.
Northern-most distribution record of *Aspilodemon*.

Martínez JJ, Fadia S, Ceccarelli FS, Zaldívar-Riverón A. 2012. Two new species of *Pambolus* (Hymenoptera, Braconidae) from Jamaica. *Journal of Hymenoptera Research* 24(1): 85-93.

Oda RAM, de Macêdo MV, Quicke DLJ. 2001. First biological data for *Aspilodemon* Fischer (Hymenoptera: Braconidae: Hydrangeocolinae): parasitoids of cecidomyiid fly galls on Asteraceae in Brazil. *Journal of Hymenoptera Research* 10: 126-130.
First published host record for *Aspilodemon*.

Papp J. 2000. *Pambolus oblongispina* sp. n. from Honduras (Hymenoptera: Braconidae: Pambolinae). *Genus* 11(1): 87-93.
Includes an illustrated key to the three Neotropical species of *Pambolus* known at that time and a redescription of *Pambolus longicornis* (Enderlein, 1920).

Penteado-Dias AM, Achterberg C van. 2004. A new species of *Allobracon* Gahan (Braconidae: Hormiinae) from Brazil, and the first record of green Hormiinae *Zoologische Mededelingen, Leiden* 78: 241-247.
Includes color photographs of an *Allobracon* that has green pigmentation.

Quicke DLJ. 2015. *The Braconid and Ichneumonid Parasitic Wasps: Biology, Systematics, Evolution and Ecology*. Wiley Blackwell, Oxford, 688 pp.
Critically acclaimed book on all aspects of Ichneumonoidea (Annotation by M. Sharkey).

Quicke DLJ, Belokobylskij SA, Smith MA, Rota J, Hrcek J, Butcher BA. 2016. A new genus of rhysipoline wasp (Hymenoptera: Braconidae) with modified wing venation from Africa and Papua New Guinea, parasitoid on Choreutidae (Lepidoptera). *Annales Zoologici* 66: 173-192.

Scatolini D, Penteado-Dias AM, Achterberg C van. 2002. *Pseudorhysipolis gen. nov.* (Hymenoptera: Braconidae: Rhysipolinae), with nine new species from Brazil, Suriname and Panama. *Zoologische Mededelingen, Leiden* 76: 109-131.
Original description of *Pseudorhysipolis.*

Sharanowski BJ, Dowling APG, Sharkey MJ. 2011. Molecular phylogenetics of Braconidae (Hymenoptera: Ichneumonoidea), based on multiple nuclear genes, and implications for classification. *Systematic Entomology* 36: 549-572.
Molecular phylogenetic study including several taxa treated in this taxa and notably *Andesipolis*.

Shimbori EM, Penteado-Dias AM. 2007. New species of *Allobracon* Gahan, 1915 (Hymenoptera: Braconidae: Hormiinae), from Brazilian Atlantic forest. *Studies on Neotropical Fauna and Environment* 42(3).
Descriptions of twelve new species of *Allobracon*.

Shimbori EM, Penteado-Dias AM, Nunes J. 2011. *Monitoriella* Hedqvist (Hymenoptera: Braconidae: Doryctinae) from Brazil, with descriptions of three new species. *Zootaxa* 2921: 28-38.

Shimbori EM, Yamada MV, Penteado-Dias AM. 2012. Two new species of *Cantharoctonus* Viereck (Hymenoptera: Braconidae: Rhysipolinae) from Brazil. *Zoologische Mededelingen, Leiden* 86: 497-504.

Shimbori EM, Souza-Gessner CDSS, Penteado-Dias AM, Shaw SR. 2017. A revision of the genus *Andesipolis* (Hymenoptera: Braconidae: Mesostoinae) and redefinition of the subfamily Mesostoinae. *Zootaxa* 4216: 101-152.
Key to all species of *Andesipolis* and descriptions of 22 new species.

Spencer L, Whitfield JB. 1999. Revision of the Nearctic species of *Rhysipolis* Förster (Hymenoptera: Braconidae). *Transactions of the American Entomological Society* 125: 295-324.
Key to all nine species of *Rhysipolis* from the Nearctic with descriptions of seven new species and new host records (all Gracillariidae).

Townsend AC, Shaw SR. 2009. A new species of *Andesipolis* Whitfield & Choi from the eastern Andes of Ecuador with notes on biology and classification (Hymenoptera: Braconidae: Rhysipolinae). *Journal of Insect Science*, 9(36).

Whitfield JB, Choi W-Y, Suh K-I. 2004. *Andesipolis*, a puzzling new genus of cyclostome Braconidae (Hymenoptera) from the Chilean Andes, with descriptions of three new species. *Zootaxa* 438: 1-15.

Zaldívar-Riverón A, Mori M, Quicke DLJ. 2006. Systematics of the cyclostome subfamilies of braconid wasps (Hymenoptera: Ichneumonoidea): a simultaneous molecular and morphological approach. *Molecular Phylogenetics and Evolution* 38: 130-145.

Zaldívar-Riverón A, Belokobylskij SA, León-Regagnon V, Martinez JJ, Briceno R, Quicke, DLJ. 2007. A single origin of gall association in a group of parasitic wasps with disparate morphologies. *Molecular Phylogenetics and Evolution* 44: 981-999.

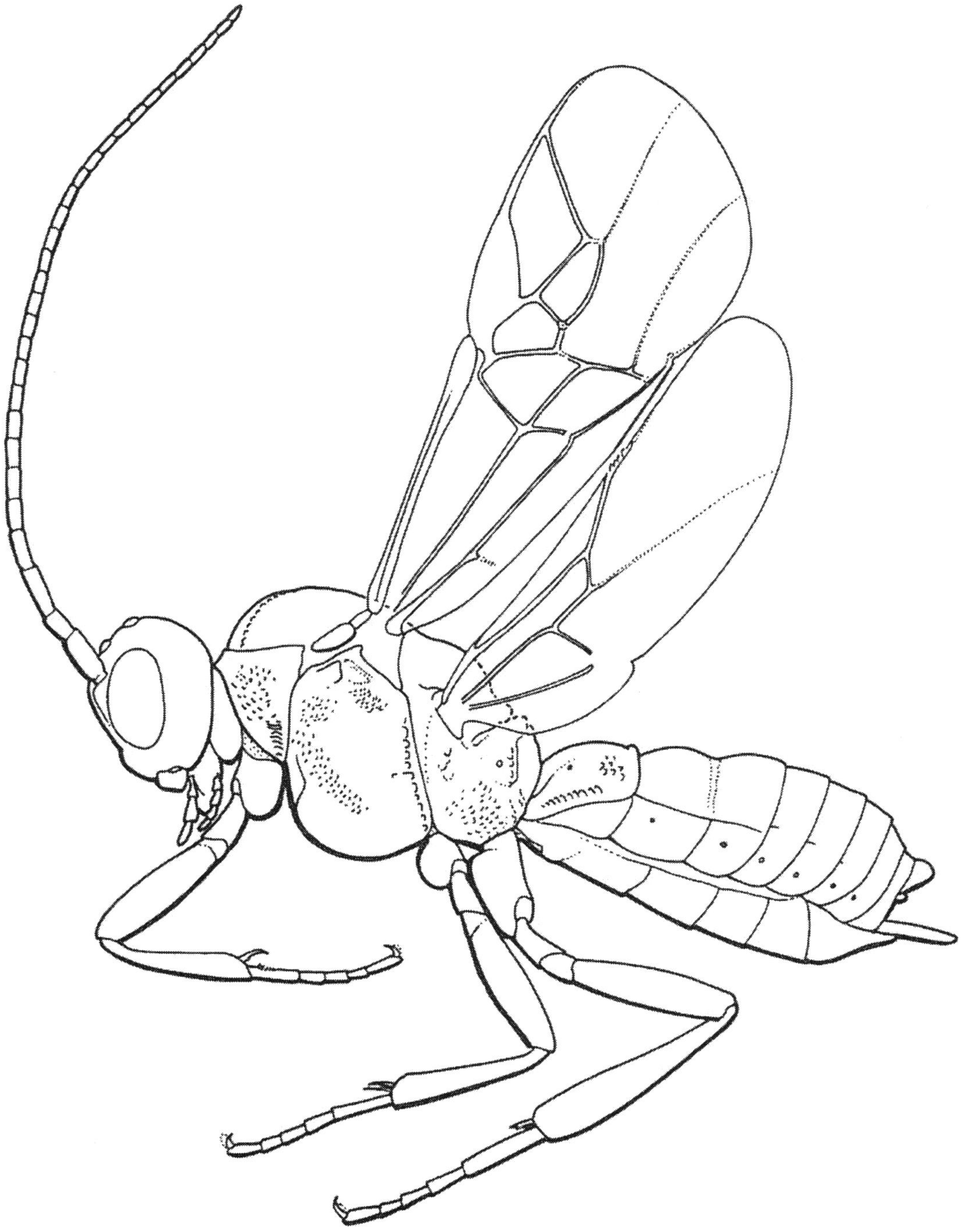

SUBFAMILY ICHNEUTINAE

M. J. SHARKEY

Department of Entomology, University of Kentucky, S-225 Agricultural Science Building North, Lexington, Kentucky 40546, U.S.A.

INTRODUCTION. The subfamily Ichneutinae contains nine genera, all of which are found in the New World. Mason (1969) proposed a number of new genera and described new species. Shenefelt (1973) cataloged the world species of Ichneutinae. Sharkey and Wharton (1994) subsequently revised and keyed the genera, described new species, and conducted a cladistic analysis.

BIOLOGY. Most members of the Ichneutinae are koinobiont endoparasites of sawflies of the families Argidae and Tenthredinidae, however one lineage, comprised of the genera *Oligoneurus* Szépligeti, *Paroligoneurus* Muesebeck, and *Lispixys* Mason, has switched to leaf-mining lepidopterous hosts. The few species with known biologies suggest that all members of the family may be egg-larval parasitoids, attacking the egg but emerging from the last larval instar. The biology of the sawfly parasitoids is reviewed by Shaw and Huddleston (1991).

BIOLOGICAL CONTROL POTENTIAL. At least some members of the Ichneutinae may be important in the natural control of pest species of sawflies and Lepidoptera. To my knowledge none has been used in a classical biological control program.

COMMON GENERA. Most genera are relatively rare; however, *Proterops* Wesmael and *Oligoneurus* contain dozens of species in the New World. Both of these genera, as well as *Ichneutes* Nees, are in great need of revision.

DISTRIBUTION. Cosmopolitan. *Oligoneurus, Paroligoneurus, Proterops*, and *Ichneutes* are also found outside the New World; the remaining genera are New World endemics. Contrary to popular belief, ichneutines are most diverse in humid tropical regions. Members of *Ichneutes* are exceptional, in that they are restricted to northern temperate and boreal regions with some of the species being Holarctic.

DISTINGUISHING FEATURES. Ichneutines are characterized by the possession of one or more small spines at the apex of the fore tibia (Fig. 12). Additionally, setae of the ovipositor sheath are restricted to the apex (Figs 20, 21); the ovipositor is always shorter than the body length, often barely exerted (Fig. 20, 21); and the oral cavity is not cyclostome (Fig. 15). One small species group of *Paroligoneurus* has exodont mandibles (Fig. 16).

KEY TO THE NEW WORLD GENERA OF THE SUBFAMILY ICHNEUTINAE

1. Venation reduced, fore wing RS not reaching wing margin. (Figs 1-3) 2

\- Venation complete, or nearly so, fore wing RS reaching wing margin (Figs 4-10) 4

2(1). Hind wing 1-1A complete, meeting or nearly meeting cu-a (Fig. 1) .... ***Oligoneurus*** Szépligeti

Distribution: Cosmopolitan; 2 described New World species, many undescribed. Biology: reared from *Bucculatrix* and possibly other leaf-mining Lepidoptera. References: Mason (1969), Whitfield and Wagner (1991), Sharkey and Wharton (1994).

\- Hind wing 1-1A incomplete, clearly separated from cu-a (Figs 2, 3) 3

3(2). Ventral side of female flagellomeres in apical 0.75 of antenna with 2 widely spaced longitudinal placodes and with fields of modified sensillae between them (Fig. 11). Petiole often raised medially, always without narrow median longitudinal depression (as in Fig. 17). Wings: Fig. 2... ***Paroligoneurus*** Muesebeck

Distribution: Canada to Ecuador; 11 described New World species, many undescribed. Biology: reared from Nepticulidae and possibly other leaf-mining Lepidoptera. References: Mason (1957, as *Muesebeckia* Mason; 1991, as *Anaprixia* Ma- son; 1969), Whitfield and Wagner (1991), Sharkey and Wharton (1994).

\- Female flagellomeres with evenly spaced longitudinal placodes and without distinct fields of modified sensillae. Petiole with narrow median longitudinal depression (Fig. 18). Wings: Fig. 3........ ***Lispixys*** Mason

Distribution: southern U.S.A. to Peru; 2 described species. Biology: unknown, but hosts are probably larvae of leaf-mining Lepidoptera. References: Mason (1969), Sharkey and Wharton (1994).

4(1). Pigmented portion of hind wing RS absent or represented by short stub (shorter than r-m) on R (Fig. 4) .... ***Ichneutes*** Nees

Distribution: U.S.A. and Canada; 7 described species in New World. Biology: egg- larval parasitoids of tenthredinid sawflies. References: Shenefelt (1973, catalog), Sharkey and Wharton (1994).

\- Pigmented portion of hind wing RS ending near wing margin (Figs 5-10) 5

5(4). Occipital carina present. Wings: Fig. 5 .... ***Helconichia*** Sharkey and Wharton

Distribution: Brazil, Ecuador, northern Argentina; 3 described species. Biology: unknown. Reference: Sharkey and Wharton (1994).

\- Occipital carina absent 6

6(5). Epicnemial carina present (as in Fig. 13). Wings: Fig. 6 .... ***Muesonia*** Sharkey and Wharton

Distribution: Mexico to Venezuela and Ecuador; 3 described species. Biology: unknown. Reference: Sharkey and Wharton (1994).

\- Epicnemial carina absent 7

7(6). Notaulus smooth. Hind tarsal claw with rounded basal tooth. Wings: Fig. 7 .... ***Proterops*** Wesmael

Distribution: Canada to Argentina; 3 described New World species, several undescribed. Biology: reared from Argidae. References: Shenefelt (1973, catalog, mostly as *Ichneutidea* Ashmead), Sharkey and Wharton (1994).

\- Notaulus sculptured, at least some crenulae anteriorly. Hind tarsal claw simple, lacking basal tooth 8

8(7). Longitudinal carina on propleuron present (Fig. 14). Petiole without pair of longitudinal carinae (Fig. 17). Female without patch of setae at apex of hypopygium (as in Fig. 20). Wings: Figs 8, 9 .... ***Masonbeckia*** Sharkey and Wharton

Distribution: Mexico to northern Argentina, not recorded from Chile or Caribbean; 6 described species. Biology: one specimen reared from an undetermined sawfly (probably Argidae). Reference: Sharkey and Wharton (1994).

\- Longitudinal carina of propleuron absent (as in Fig. 13). Petiole with pair of dorsal longitudinal carinae (Fig. 19). Female with patch of setae at apex of hypopygium (Fig. 21). Wings; Fig. 10 .... ***Hebichneutes*** Sharkey and Wharton

Distribution: Mexico to Brazil; 4 described species. Biology: unknown. Reference: Sharkey and Wharton (1994).

REFERENCES

Mason, W. R. M. 1957. A new genus and species of Microgastrinae (Hymenoptera: Braconidae). The Canadian Entomologist 89: 355-357.

Mason, W. R. M. 1969. Muesebeckiini, a new tribe of Braconidae (Hymenoptera). Proceedings of the Entomological Society of Washington 71: 263-278.

Mason, W. R. M. 1991. A new genus of exodont Ichneutinae (Hymenoptera: Braconidae). Insecta Mundi 5: 95-99.

Sharkey, M. J. and Wharton, R. A. 1994. A revision of the genera of the world Ichneutinae (Hymenoptera: Braconidae). Journal of Natural History 28: 873-912.

Shaw, M. R. and Huddleston, T. 1991. Classification and biology of braconid wasps (Hymenoptera: Braconidae). Handbooks for the Identification of British Insects. 7(11): l-126.

Shenefelt, R. D. 1973. Pars 9. Braconidae 5, Microgastrinae and Ichneutinae. Pp. 669-812. In: Vecht, J. van der and Shenefelt, R. D. (Eds.). Hymenopterum Catalogus (nova editio). Dr. W. Junk, The Hague.

Whitfield, J. H. and Wagner, D. L. 1991. Annotated key to the genera of Braconidae (Hymenoptera) attacking leafmining Lepidoptera in the Holarctic region. Journal of Natural History 25: 733-754.

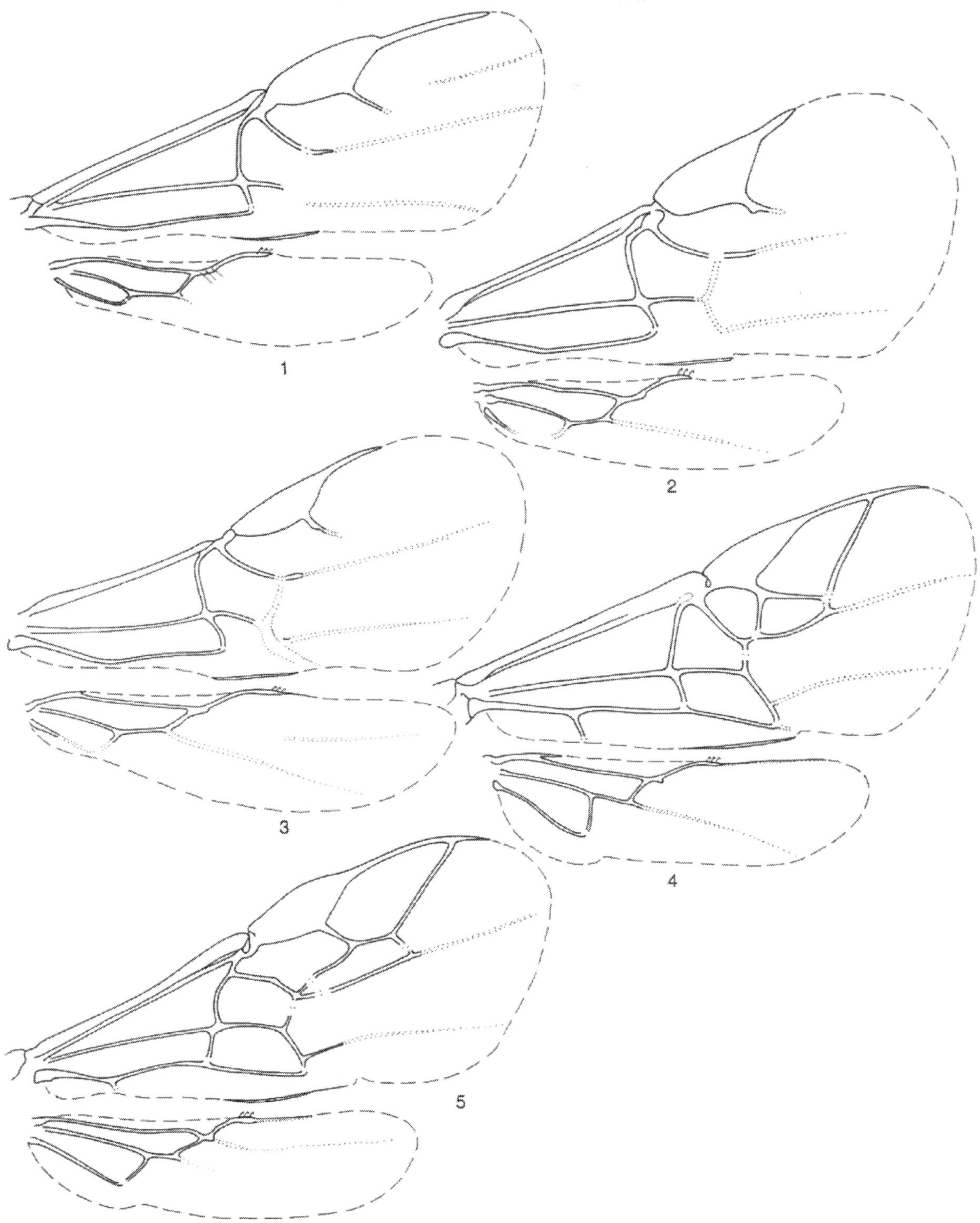

Figs 1-5: Wings of Ichneutinae. 1, *Oligoneurus* sp:; 2, *Paroligoneurus eximius* (Mason); 3, *Lispixys avispa* Mason; 4, *Ichneutes* sp.; 5, *Helconichia areolata* Sharkey and Wharton.

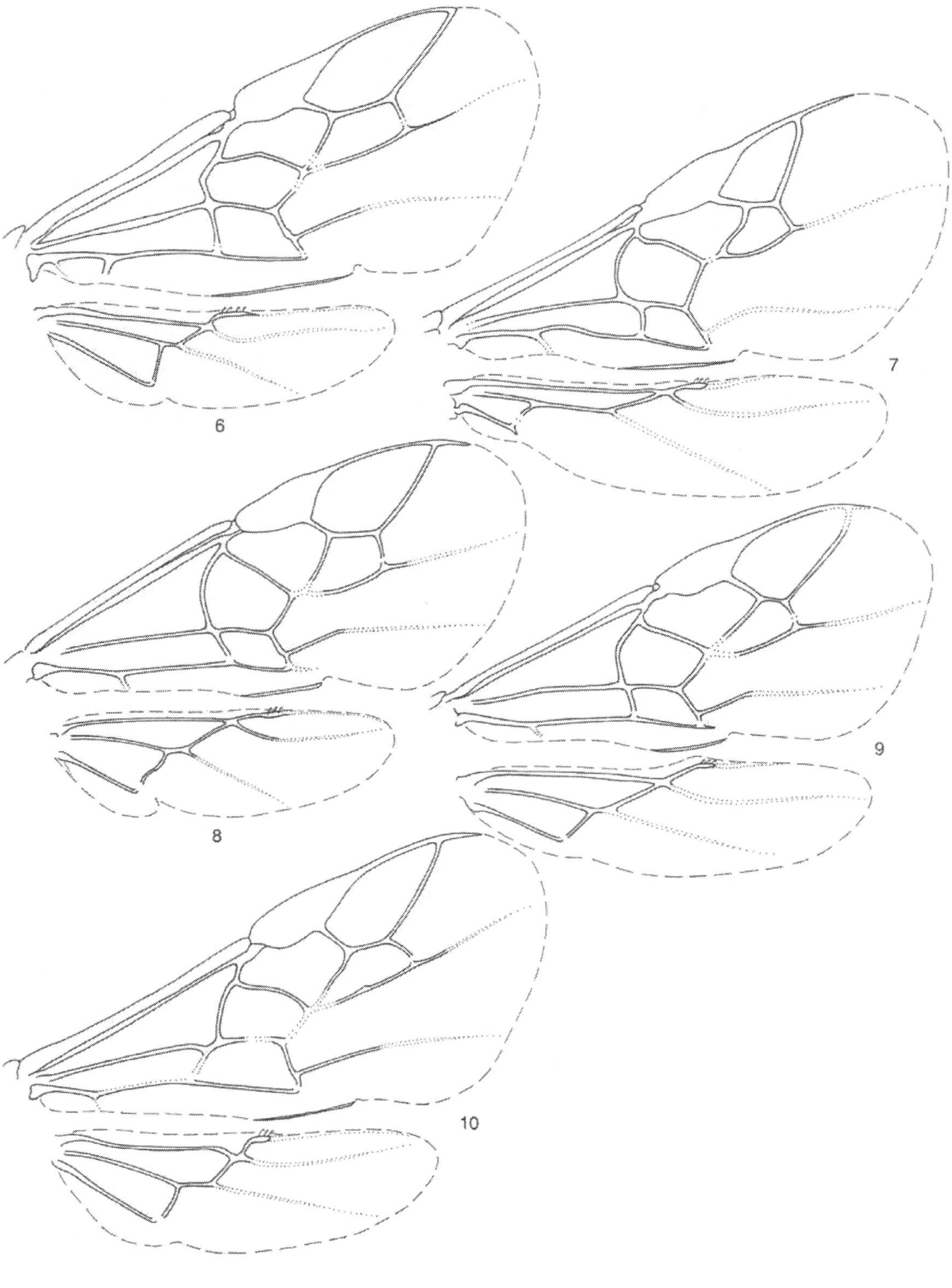

Figs 6-10: Wings of Ichneutinae. 6, *Muesonia stamina*. Sharkey and Wharton; 7, *Proterops proteroptoides* (Viereck); 8, *Masonbeckia compressicornis* Sharkey and Wharton; 9, *Masonbeckia flavofemorata* Sharkey and Wharton; 10, *Hebichneutes tricolor* Sharkey and Wharton.

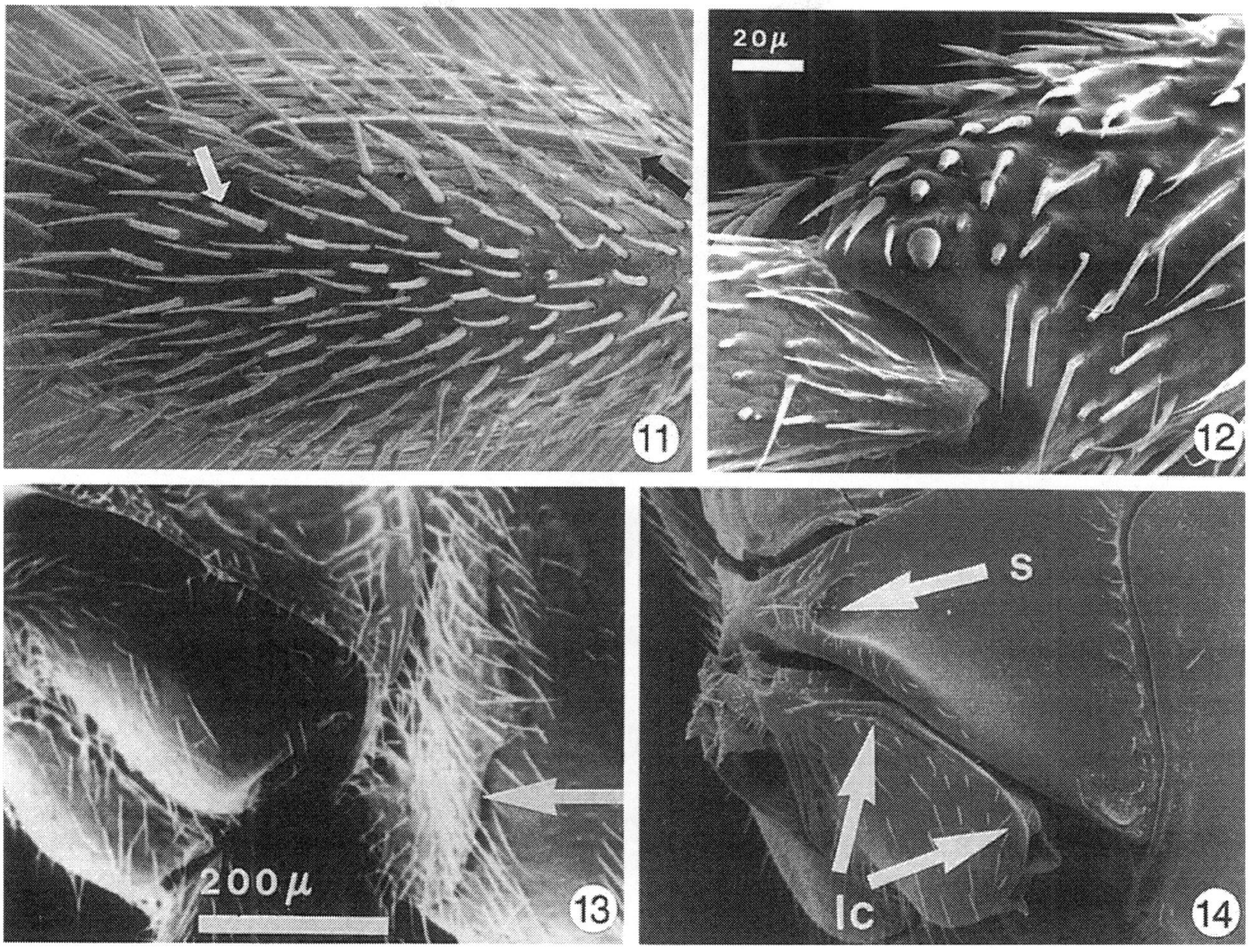

Figs 11-14: Ichneutinae. 11, female flagellomeres of *Paroligoneurus eximius* (Mason), ventral view, arrows = longitudinal placode and modified sensilla; 12, apex of fore tibia of *Ichneutes* sp.; 13, propleuron and anteroventral portion of mesopleuron of *Ichneutes* sp., arrow = epicnemial carina of mesopleuron, 14, propleuron of *Masonbeckia compressicornis* Sharkey and Wharton, lc = longitudinal carina of propleuron, s = subpronope. All figures after Sharkey and Wharton (1994).

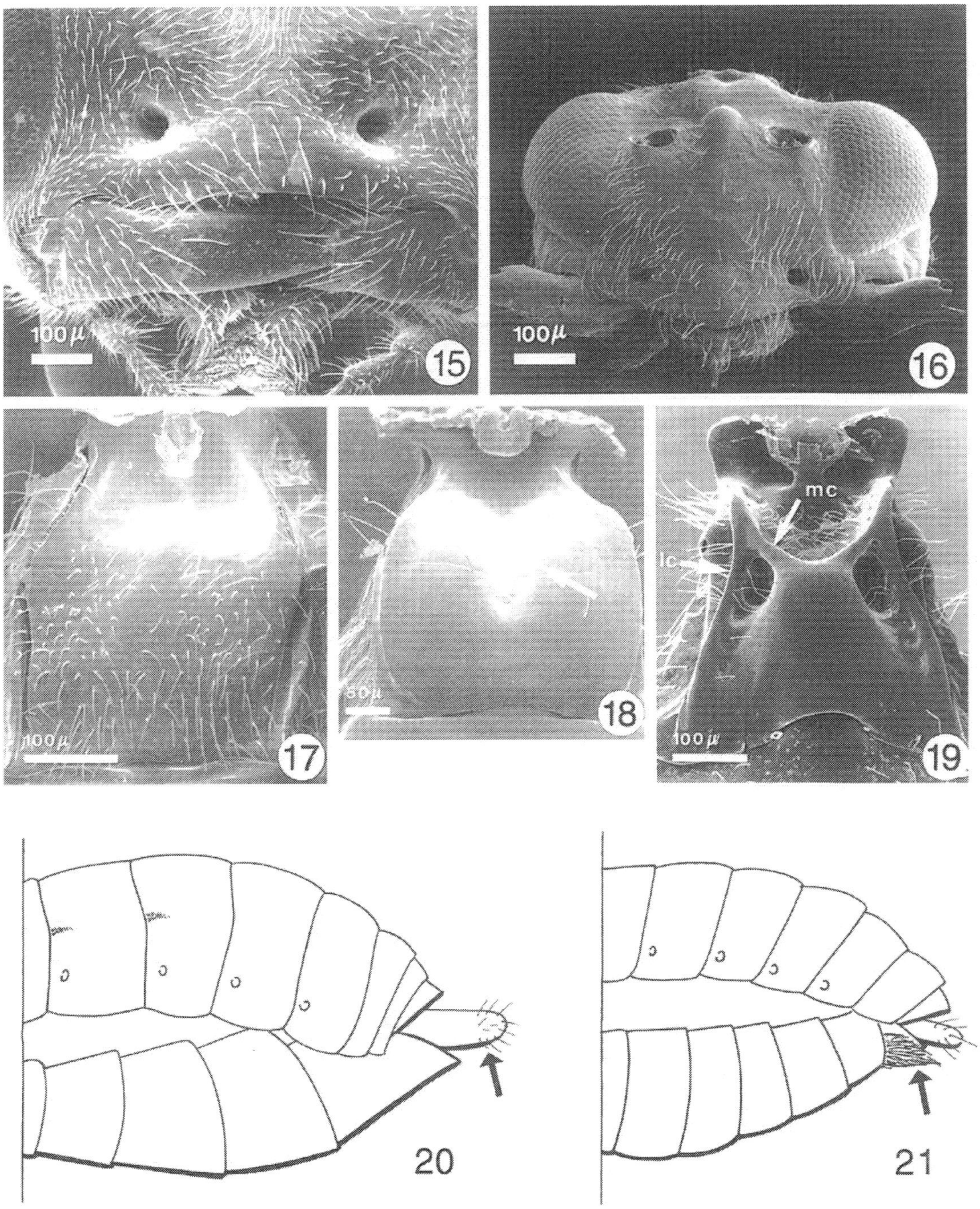

Figs 15-21: Ichneutinae. 15, head of *Proterops proteroptoides* (Viereck), frontal view; 16, head of *Paroligoneurus* sp. showing exodont mandibles; 17-19, petioles: 17, *Masonbeckia compressicornis* Sharkey and Wharton; 18, *Lispixys avispa* Mason, arrow = median longitudinal depression; 19, *Hebichneutes* sp. le = dorsal longitudinal carina, mc = median carina, 20, metasoma of *Proterops* sp., arrow = ovipositor sheath; 21, metasoma of *Hebichneutes* sp., arrow = apical setae. Figs 15-19 after Sharkey and Wharton (1994).

ICHNEUTINAE UPDATE

M.J. Sharkey

Department of Entomology, University of Kentucky, Lexington, Kentucky 40546-0091 U.S.A. msharkey@uky.edu

Sharanowski and Sharkey (2007) described three new species of *Helconichia* and updated the key to all known species. Kula (2016) described one new species of *Paroligoneurus* Muesebeck. No new extant genera were described since the Manual's publication but Fisher *et al.* (2015) described a new genus and species, *Colemanus keeleyorum,* from a fossil found in western North America.

Recent papers on New World Ichneutinae

Fisher JR, Tucker EM, Sharkey MJ. 2015. *Colemanus keeleyorum* (Braconidae, Ichneutinae s. l.): a new genus and species of Eocene wasp from the Green River Formation of western North America. *Journal of Hymenoptera Research* 44: 57-67.

Kula RR. 2016. Cardiochilinae and Ichneutinae (Hymenoptera: Braconidae) of Konza Prairie. *Proceedings of the Entomological Society of Washington* 118: 273-288.

Sharanowski BJ, Sharkey MJ. 2007. Description of three new species of *Helconichia* Sharkey & Wharton (Hymenoptera: Braconidae: Ichneutinae) with a revised key to all species. *Zootaxa* 1502: 45-53.

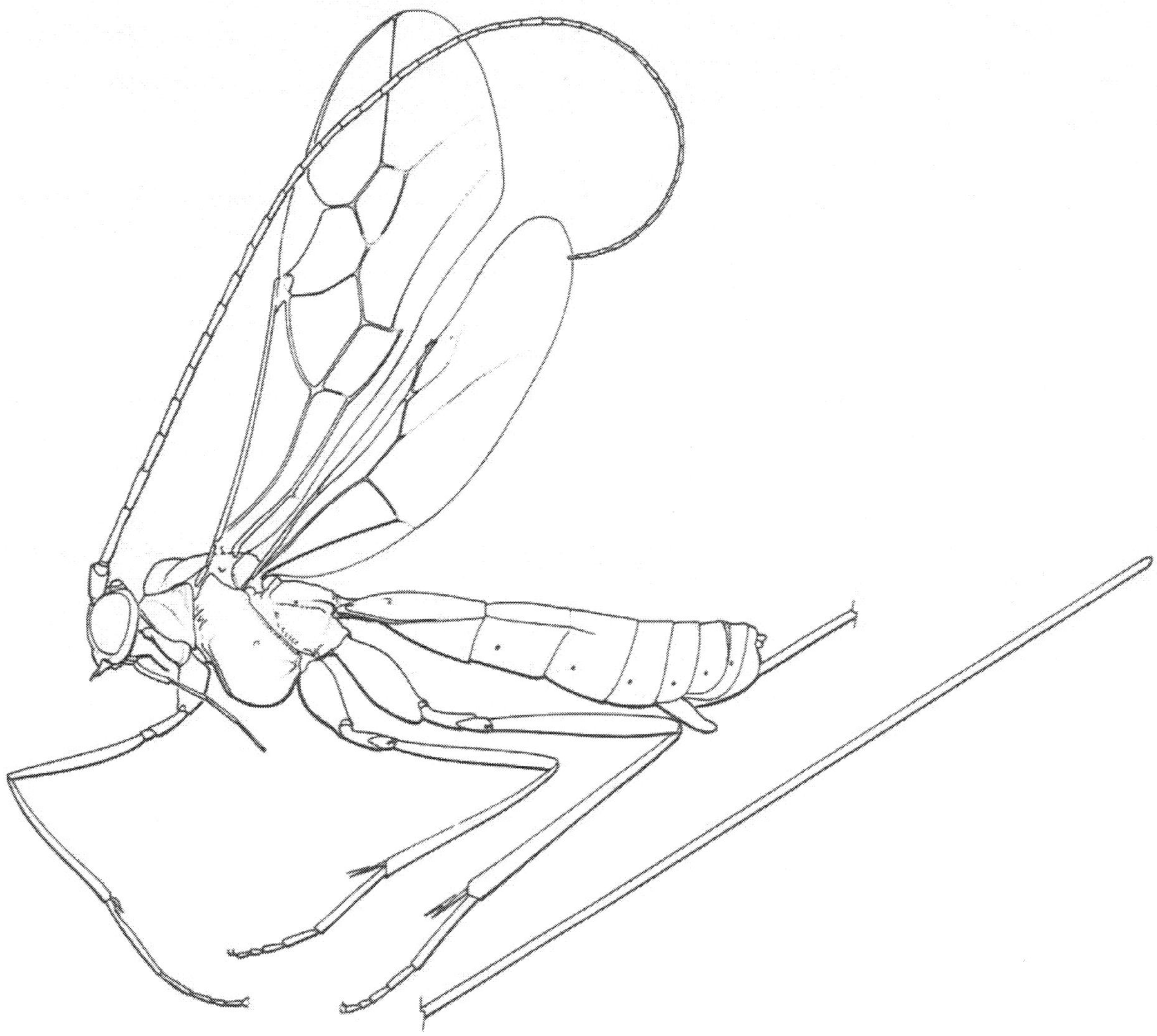

SUBFAMILY MACROCENTRINAE

R. A. WHARTON

Department of Entomology, Texas A&M University, College Station, Texas 77843, U.S.A.

INTRODUCTION. The subfamily Macrocentrinae contains only four genera in the New World. Most of the described species belong to the genus Macrocentrus Curtis and Muesebeck (1932) has revised the North American Species. The latest generic synopsis is by van Achterberg (1993), although Muesebeck (1932) also provides some important details regarding generic-level characteristics. Shaw and Huddleston (1991) provide a useful summary of the literature, with special reference to the biology.

BIOLOGY. The Macrocentrinae include both solitary and gregarious koinobiont endoparasitoids of lepidopterous larvae (Clausen, 1940; Shaw and Huddleston, 1991). Where known, the gregarious species are polyembryonic. Detailed biological information is available for two species, *M. ancylivorus* Rohwer (Daniel, 1932; Finney et al., 1947) and *M. grandii* Goidanich (Parker, 1931; Ding et al., 1989). *Macrocentrus grandii* is referred to as *M. cingulatum* Brischke (or with authorship incorrectly attributed to Reinhard) in some recent literature (see van Achterberg and Haeselbarth, 1983), and as *M. gifuensis* Ashmead (a misidentification) in some of the more important earlier publications. Although only a single individual of *M. ancylivorus* emerges from its host, initial development is polyembryonic (Daniel, 1932). Details of the development of other macrocentrines are largely unknown. It is possible, however, that those species that emerge as solitary parasitoids (at least in *Macrocentrus* and *Hymenochaonia* Dalla Torre) are initially polyembryonic in their development. New World macrocentrines have been reared from 10 families of Lepidoptera.

Many of the macrocentrines are pale colored and crepuscular to nocturnal. The species of *Macrocentrus* and *Hymenochaonia* have long ovipositors, often laying their eggs in early instars of hosts concealed in leaf whorls, stems, tubers, or rolled leaves. *Austrozele* Roman and *Dolichozele* Viereck, on the other hand, contain species with short ovipositors, and the few available records indicate that they attack more exposed hosts such as arctiids and geometrids.

BIOLOGICAL CONTROL POTENTIAL. Both *M. ancylivorus* and *M. grandii* have been used extensively in biological control programs directed against pyralids, gelechiids, and tortricids. One of the first mass rearing programs utilizing braconids for augmentative releases was established for one of these species (Finney et al., 1947). A few other macrocentrines have also been used in biological control (e.g. Smith et al., 1993), and some, including *Hymenochuonia delicata* (Cresson), a parasitoid of tortricids (Allen, 1962), have been widely distributed as a result.

COMMON GENERA. *Macrocentrus* and *Hymenochaonia* are the major New World genera. There are about 100 described species of *Macrocentrus* worldwide, and about one-third of these occur in the New World. Most of the *Hymenochaonia* are tropical, and undescribed species are common in collections. The number of described species in *Dolichozele* and *Austrozele* is uncertain since several of the Neotropical species with short ovipositors (especially those described by Enderlein (1920)) need to be examined for correct generic placement following van Achterberg's (1993) revision of macrocentrine genera. At least one species of *Austrozele* is commonly encountered in southeastern U.S.A., but *Dolichozele* appears to be relatively rare.

DISTRIBUTION. Macrocentrines are found throughout the world. The four genera treated here are apparently widespread in the New World, occurring from at least southern United States through Brazil (though the range of *Austrozele* needs verification). *Macrocentrus* is the dominant genus in the north temperate zone. *Austrozele* and *Hymenochaonia* range into southern Canada, and both *Macrocentrus* and *Hymenochaonia* have been recorded from Argentina (De Santis, 1967). The Chilean fauna is essentially unstudied. *Dolichozele* and *Hymenochaonia* are endemic to the New World.

DISTINGUISHING FEATURES. Macrocentrines can be recognized by the absence of an occipital carina and the presence of small teeth on the trochantellus, With the exception of *M. incompletus* Muesebeck, all described species have three submarginal cells (Figs 1-3).

KEY TO THE NEW WORLD GENERA OF THE SUBFAMILY MACROCENTRINAE

1. Petiole with laterope absent or nearly so (Fig. 4), rarely shallowly impressed; petiole flat or convex basal-medially (Fig. 5)..2

\- Petiole with laterope deep and large (Figs 6, 7); petiole nearly always slightly concave basal-medially (Fig. 7)..3

2(1). Ovipositor long, ovipositor sheath at least as long as metasoma (Fig. 8). Length of inner spur of hind tibia 0.4-0.6 times length of hind basitarsus (about as in Fig. 10). Wings: Fig. 2......***Hymenochaonia*** Dalla. Torre

Distribution: southern Canada to Argentina; only 5 described species. Biology: Allen (1962); reared primarily from Pyralidae and Tortricidae (Marsh, 1979). Reference: van Achterberg (1993).

\- Ovipositor short, about equal to apical height of metasoma (Fig. 9). Length of inner spur of hind tibia 0.6-0.8 times length of hind basitarsus (Fig. 11), except in *D. gravitarsis*............................***Dolichozele*** Viereck

Distribution: southern and western U.S.A. to Brazil, Caribbean. Biology: largely unknown, one species reared from Arctiidae (Marsh, 1979). Reference: van Achterberg (1993).

3(1). Ovipositor long, ovipositor sheath nearly always at least as long as metasoma (Fig. 8). Length of inner spur of hind tibia 0.- 0.5 times length of hind basitarsus (Fig. 10). Wings: Fig. 3 ....***Macrocentrus*** Curtis

Distribution: throughout New World, but not yet recorded from Chile; commonly collected; 40 described species in New World. Biology: nearly all of the detailed work has been on polyembryonic species (overview in Shaw and Huddleston, 1991); Tortricidae and Pyralidae are the most frequently recorded hosts. Reference: Muesebeck (1932, key to Nearctic species).

\- Ovipositor short, about equal to apical height of metasoma (Fig. 9). Length of inner spur of hind tibia 0.5-0.8 times length of hind basitarsus (Fig. 11). Wings: Fig. 1 ....***Austrozele*** Roman

Distribution: southern Canada to Brazil, Caribbean. Neotropical records based on transfer of *Paniscozele* Enderlein to *Austrozele* (van Achterberg, 1993), but at least some of these Neotropical species may actually belong in *Dolichozele*. Biology: New World species have been reared primarily from Geometridae. Reference: Mason (1977).

REFERENCES

Achterberg, C. van 1993. Revision of the subfamily Macrocentrinae Foerster (Hymenoptera: Braconidae) from the Palaearctic region. Zoologische Verhandelingen Leiden 286: 1-110.

Achterberg, C. van and Haeselbarth, E. 1983. Revisionary notes on the European species of *Macrocentrus* Curtis *sensu stricto* (Hymenoptera: Braconidae). Entomofauna 4: 37-59.

Allen, H. W. 1962. Parasites of the Oriential fruit moth in the Eastern United States. United States Department of Agriculture Technical Bulletin 1265: 19, 25-26.

Clausen, C. P. 1940. Entomophagous Insects. McGraw-Hill, New York and London. 688 pp.

Daniel, D. M. 1932. *Macrocentrus ancylivorus* Rohwer, a polyembryonic braconid parasite of the Oriental fruit moth. New York State Agricultural Experiment Station Technical Bulletin 187: 1-101.

De Santis, L. 1967. Catalogo de las Himenopteros Argentinos de la Serie Parasitics. incluyendo Bethyloidea. Provincia de Buenos Aires Gobernacion, Comision de Investigacion Cientifica, La Plata. 337 pp.

Ding, D., Swedenborg, P. D., and Jones, R. L. 1989. Chemical stimuli in host-seeking behavior of *Macrocentrus grandii* (Hymenoptera: Braconidae). Annals of the Entomological Society of America 82: 232-236.

Enderlein, G. 1920. Zur Kenntnis aussereuropäischer Braconiden. Archiv für Naturgeschichte A(84): 51-224.

Finney, G. L., Flanders, S. E., and Smith, H. S. 1947. Mass culture of *Macrocentrus ancylivorus* and its host, the potato tuber moth. Hilgardia 17: 437-483.

Marsh, P. M. 1979. Family Braconidae. Pp. 144-295. In: Krombein, K. V., Hurd, P. D., Smith, D. R., and Burks, B. D. (Eds.). Catalog of Hymenoptera in America North of Mexico. Smithsonian Institution Press, Washington, D.C.

Mason, W. R. M. 1977. The identity of *Macrocentrus uniformis* Provancher (nec Cresson), description of a sibling species and a possible grooming organ (Hym.: Braconidae). Naturaliste Canadienne 103: 513-515.

Muesebeck, C. F. W. 1932. Revision of the Nearctic ichneumon-flies belonging to the genus *Macrocentrus*. Proceedings of the United States National Museum 80(2923):1-55.

Parker, H. L. 1931. *Macrocentrus gifuensis* Ashmead, a polyembryonic braconid parasite in the European corn borer. United States Department of Agriculture Technical Bulletin 230: 1-62.

Shaw, M. R. and Huddleston, T. 1991. Classification and biology of braconid wasps (Hymenoptera: Braconidae). Handbooks for the Identification of British Insects 7(11): 1-126.

Smith, J. W., Jr., Wiedenmann, R. N., and Overholt, W. A. 1993. Parasites of Lepidopteran Stemborers of Tropical Gramineous Plants, ICIPE Science Press, Nairobi. 89 pp.

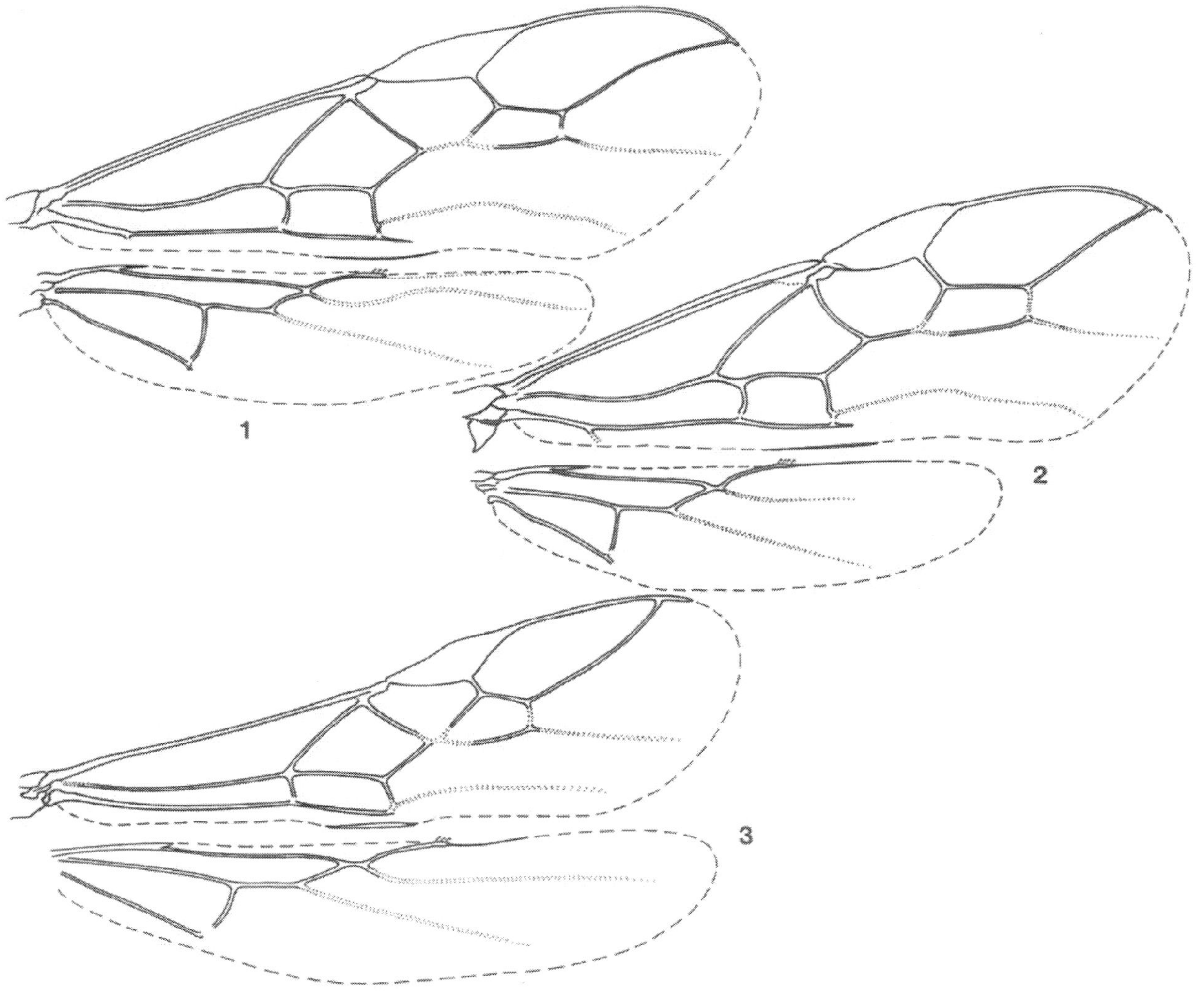

Figs 1-3: Wings of Macrocentrinae. 1, *Austrozele* sp.; 2, *Hymenochaonia delicata* (Cresson); 3, *Macrocentrus* sp.

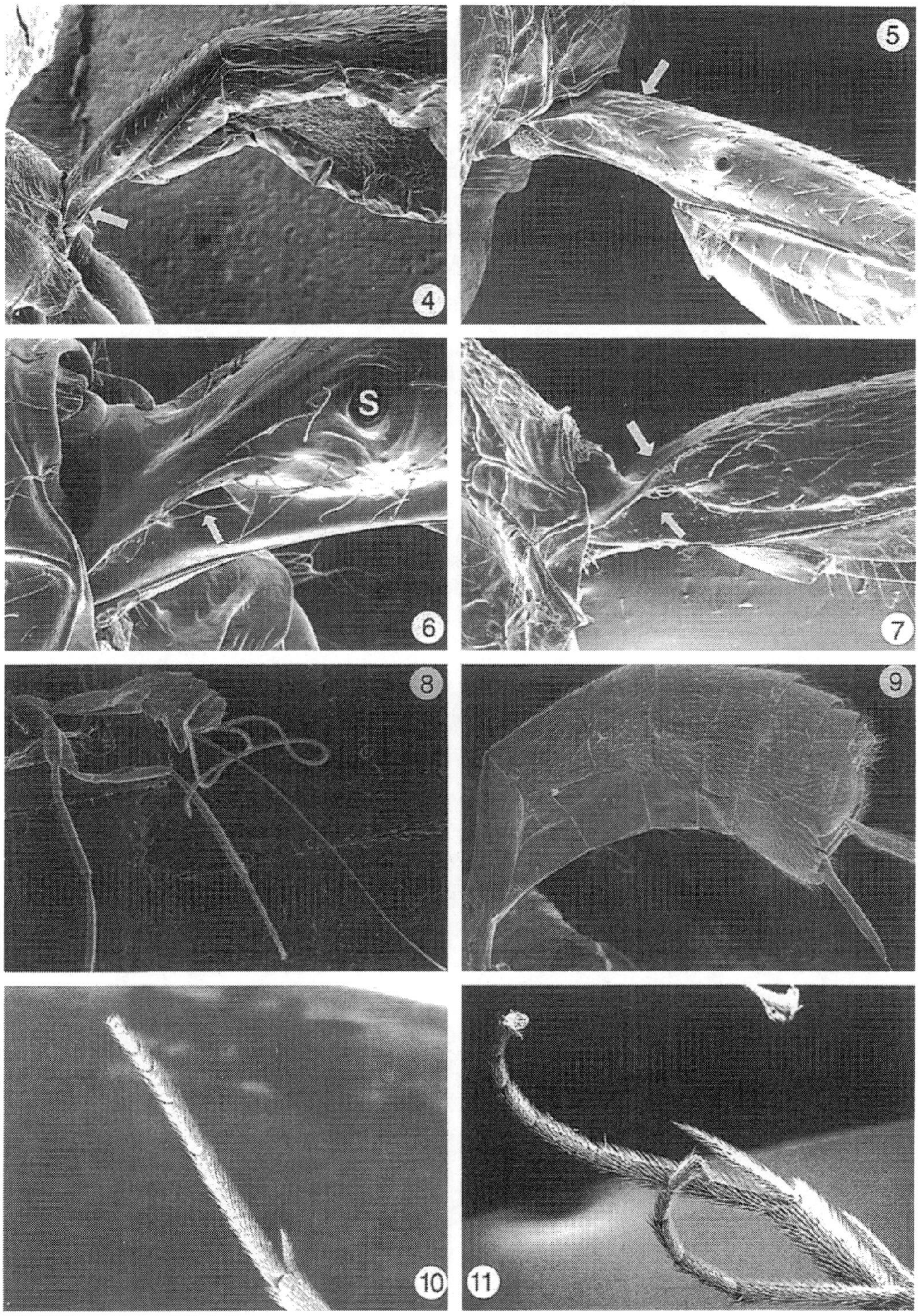

Figs 4-11: Macrocentrinae. 4, petiole of *Hymenochaonia* sp., arrow = laterope (nearly absent); 5, petiole of *Hymenochaonia*.sp., arrow = weakly convex dorsal surface; 6, base of petiole of *Austrozele* sp., arrow = laterope, s = spiracle; 7, base of petiole of *Macrocentrus* sp., arrows = concave dorsal surface and laterope 8, metasoma and ovipositor of *Hymenochaonia* sp., lateral view; 9, metasoma and ovipositor of *Austrozele* sp.; 10, hind tibial spurs and tarsi of *Macrocentrus* sp.; 11, hind tibial spurs and tarsi of *Austrozele* sp.

MACROCENTRINAE UPDATE

C. van Achterberg
Department of Terrestrial Zoology, Naturalis Biodiversity Center, Postbus 9517, 2300 Leiden, The Netherlands
c.vanachterberg@xs4all.nl

Ahlstrom (2005) published a review of the Macrocentrinae north of Mexico.

Annotated list of recent papers on New World Macrocentrinae

Ahlstrom KR. 2005. Revision of the subfamily Macrocentrinae (Hymenoptera: Braconidae) in America north of Mexico. Thomas Say Publications in Entomology: Monographs. Entomological Society of America, Lanham. i-v, 1-274.

The Nearctic species of Macrocentrinae were revised.

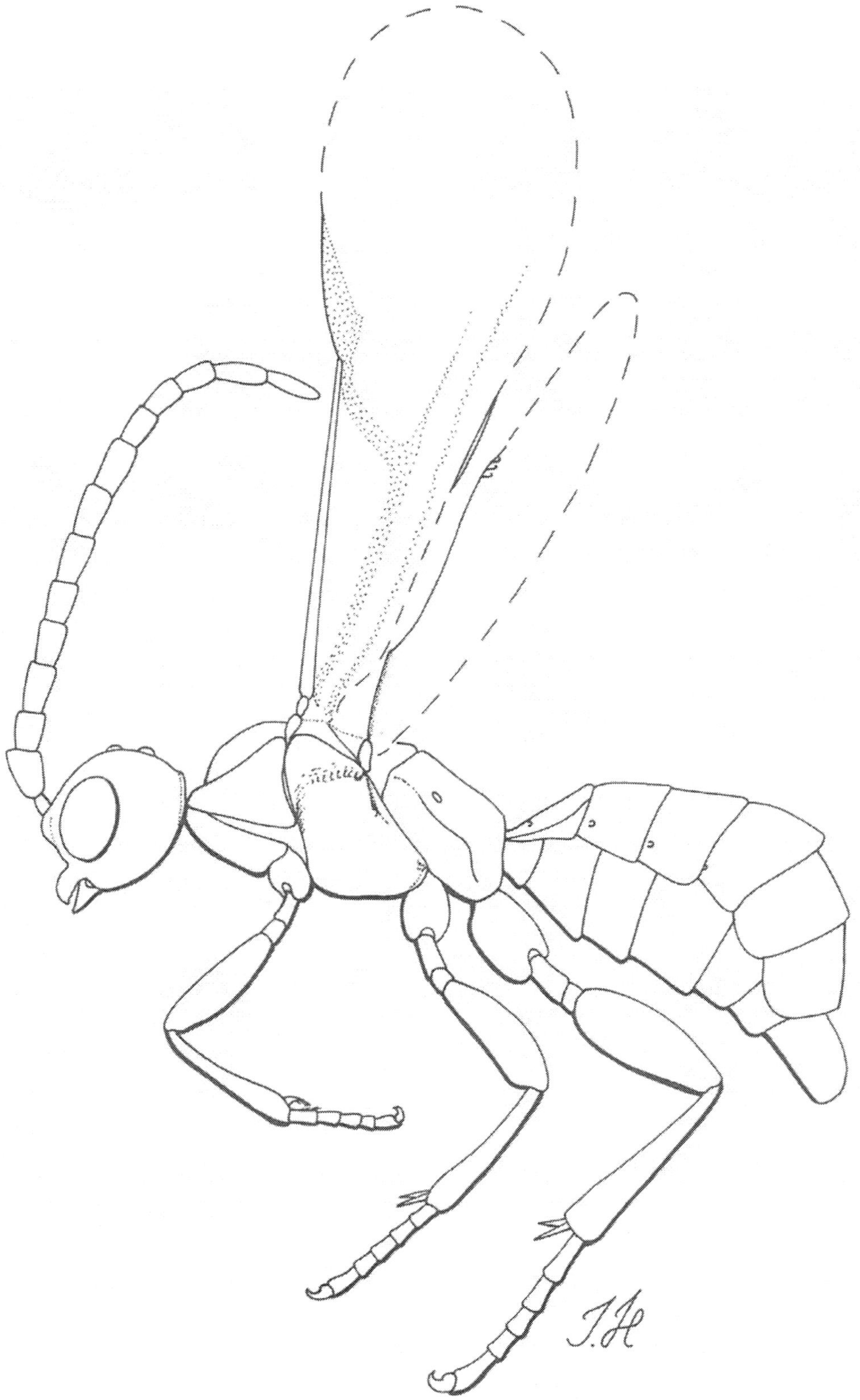
J.H

SUBFAMILY MASONINAE

C. VAN ACHTERBERG

Afdeling Entomologie (Hymenoptera), Nationaal Natuurhistorisch Museum, Postbus 9517, 2300 RA Leiden, The Netherlands

INTRODUCTION. The subfamily Masoninae is a small group with species 1-2 mm long. Females have a short, protruding ovipositor that is much shorter than the metasoma. Two tribes are recognized by van Achterberg (1995): Mannokeraiini (Australian) and Masonini (widespread). Two genera are known and one occurs in the New World. The species are described by van Achterberg (1995).

BIOLOGY. Unknown.

BIOLOGICAL CONTROL POTENTIAL. Unknown.

COMMON GENERA. *Masona* van Achterberg is the only genus present in the New World. It has been collected primarily in flight interception traps (including the wingless females).

DISTRIBUTION. Two of the five described species occur in southeastern U.S.A. The other three species are known only from Australia. There is also a report of an undescribed species from Tanzania (van Achterberg, 1995).

DISTINGUISHING FEATURES. Masoninae have wingless females with an ant-like habitus (Fig. 6), head enlarged behind eyes (Figs 3, 6), antenna distinctly shorter than body and with 14-19 segments (Fig. 5), and head in New World species distinctly prognathous (Fig. 6). The winged males are mainly characterized by the short face and head (Figs 2, 4) and the wings almost devoid of venation (Fig. 1). Essential for their recognition relative to other New World Braconidae is the upcurved ovipositor (Fig. 6), the shortened fore tarsi of females (Fig. 7) and labial palp with one segment in both sexes (Fig. 6). The occipital carina is complete dorsally (Fig. 4).

REFERENCE

Achterberg, C. van 1995. Generic revision of the subfamily Betylobraconinae (Hymenoptera: Braconidae) and other groups with modified fore tarsus. Zoologische Verhandelingen Leiden 298: 1-242.

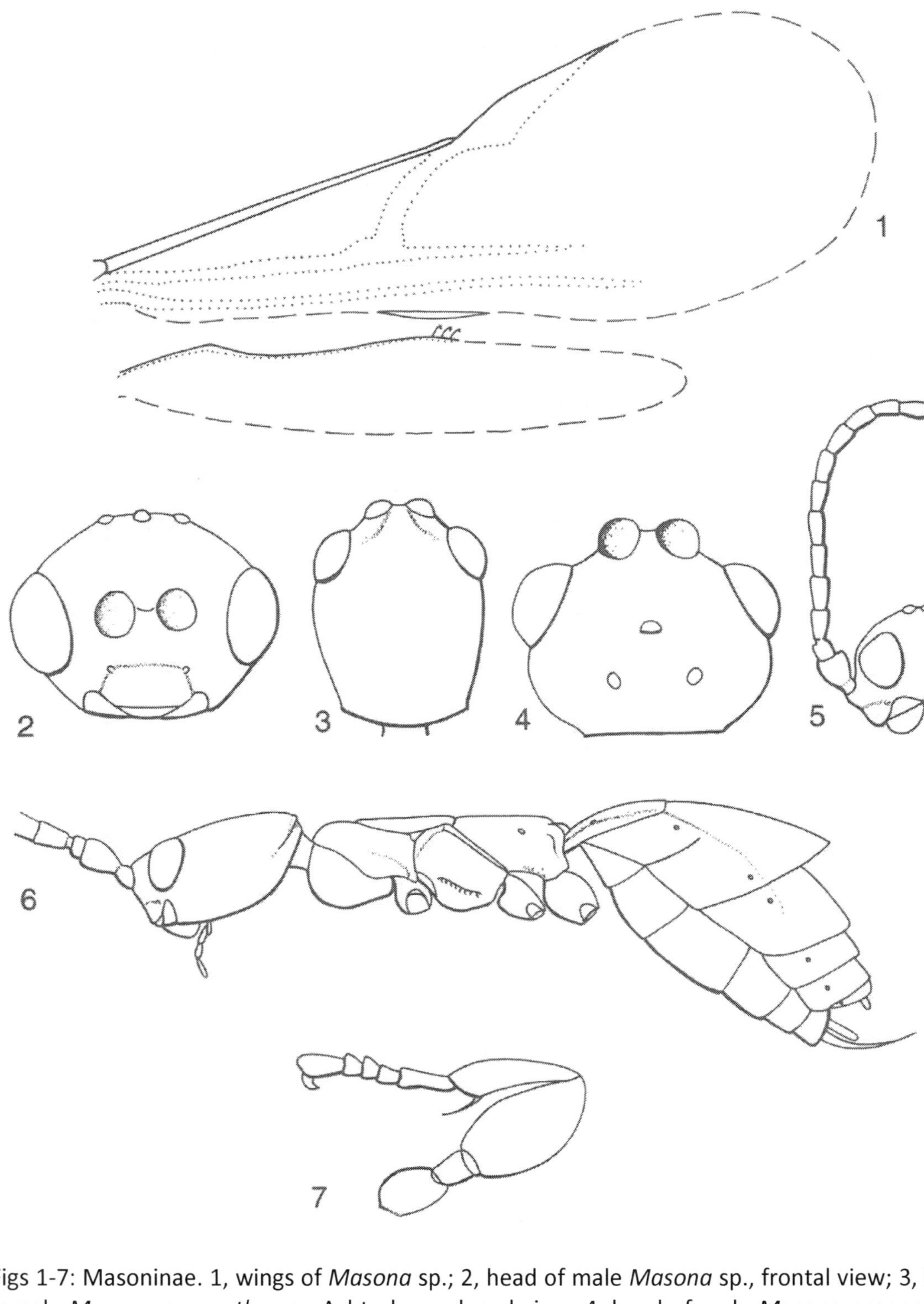

Figs 1-7: Masoninae. 1, wings of *Masona* sp.; 2, head of male *Masona* sp., frontal view; 3, head of female *Masona prognatha* van Achterberg, dorsal view; 4, head of male *Masona prognatha* van Achterberg, dorsal view; 5, head of *Masona prognatha* van Achterberg, lateral view; 6, habitus of female *Masona prognatha* van Achterberg; 7, fore leg of female *Masonia prognatha* van Achterberg. All figures after van Achterberg (1995).

MASONINAE UPDATE

D.L.J. Quicke
Department of Biology, Faculty of Life Sciences, Chulalongkorn University, Bangkok, Thailand.
d.quicke@email.com

The Masoninae was originally described comprising two tribes: Mannokeraiini and Masonini (van Achterberg, 1995). The former is entirely Australian and has since been shown to belong to the Euphorinae (Belshaw and Quicke, 2002; Sharanowski *et al.*, 2011; Stigenberg *et al.*, 2015).

No new extant species have been described in the Masoninae since the 1997 Manual. However, two amber fossils from the New World have been described from Miocene Dominican amber, one in the extant genus *Masona* (van Achterberg, 2001) and one in a new, probably extinct, genus (Engel and Bennett, 2008).

Recent papers on New World Masoninae

Achterberg C van. 2001. The first known fossil Masoninae (Hymenoptera: Braconidae) from Miocene Dominican amber. *Zoologische Mededelingen, Leiden* 75**:** 393-396.

Belshaw R, Quicke DLJ. 2002. Robustness of ancestral state estimates: evolution of life history strategy in ichneumonoid parasitoids. *Systematic Biology* 51: 450-477.

Engel MS, Bennett DJ. 2008. *Anoblepsis*, a new, bizarre braconid wasp genus in Dominican amber (Hymenoptera: Braconidae). *Journal of the Kansas Entomological* Society 81**:** 368-372.

Sharanowski BJ, Dowling APG, Sharkey MJ. 2011. Molecular phylogenetics of Braconidae (Hymenoptera: Ichneumonoidea), based on multiple nuclear genes, and implications for classification. *Systematic Entomology* 36, 549-72.

Stigenberg J, Boring CA, Ronquist F. 2015. Phylogeny of the parasitic wasp subfamily Euphorinae (Braconidae) and evolution of its host preferences. *Systematic Entomology* 40: 570-591.

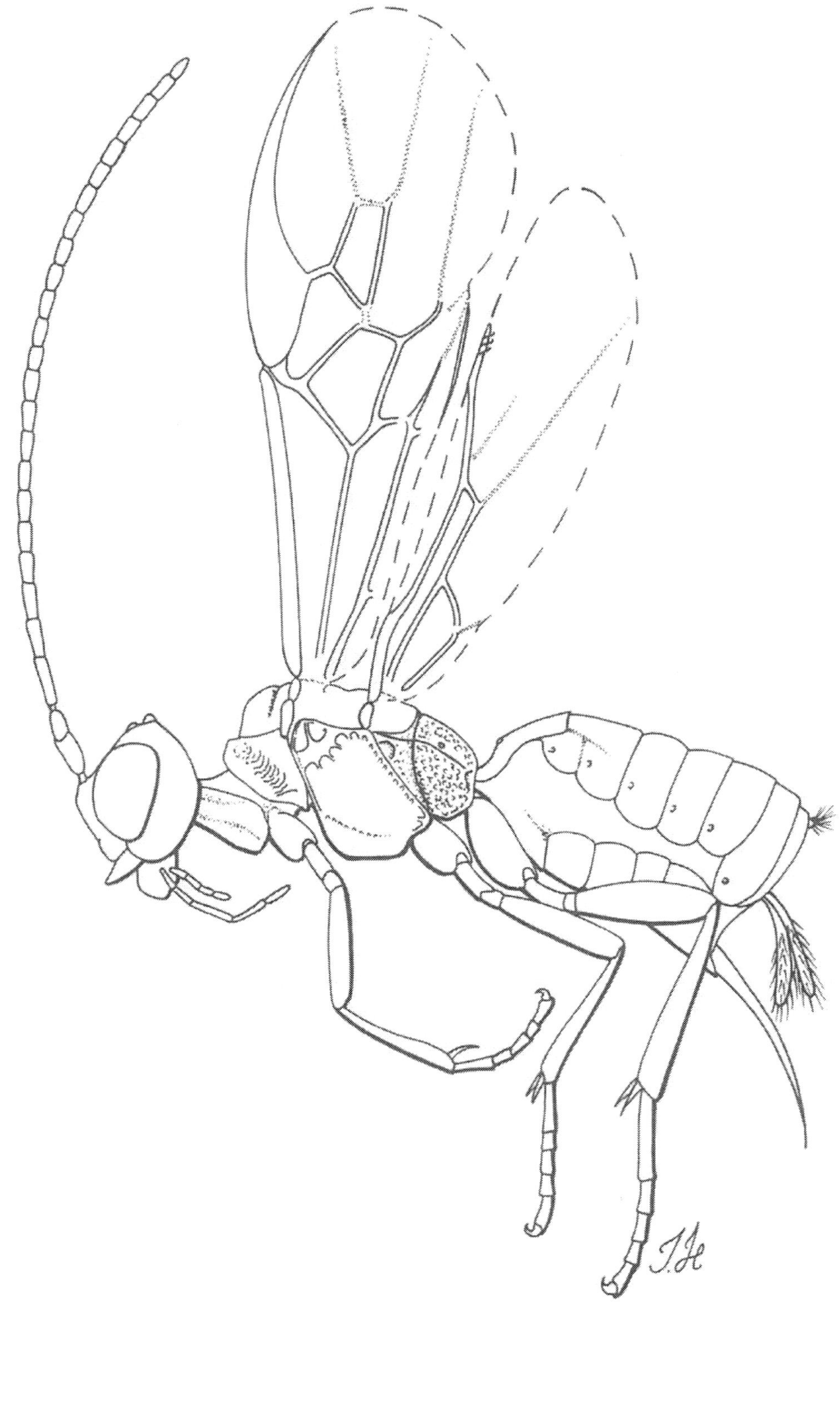
J.H.

SUBFAMILY MENDESELLINAE

J. B. WHITFIELD

Department of Entomology, 320 Morrill Hall, University of Illinois, 505 S. Goodwin Ave, Urbana, IL 61801 U.S.A.

INTRODUCTION. The subfamily Mendesellinae was recently described (Whitfield and Mason, 1994) and has thus not been included in any previous key to braconid genera (e.g. Marsh et al., 1987). It is represented by two genera and nine described species, although single specimens apparently representing additional undescribed species have been seen in collections.

BIOLOGY. Only one species of the subfamily has been reared from its host *Epsilogaster bicolor* Whitfield and Mason is a parasitoid of a momphid stem-galler on *Cephalanthus occidenialis* L. (buttonbush).

BIOLOGICAL CONTROL POTENTIAL. So little is known of the biology of the species that it would be difficult to predict whether mendesellines might be useful in biological control.

COMMON GENERA. Two genera are known, *Epsilogaster* Whitfield and Mason and *Mendesella* Whitfield and Mason. The species are generally rare in collections, although *E. bicolor* appears to be relatively common at some localities.

DISTRIBUTION. The subfamily Mendesellinae is entirely restricted to the New World. *Epsilogaster* is found in South, Central and North America (as far north as southeastern U.S.A.), while *Mendesella* has so far only been found in northern and central rain forest regions of South America.

DISTINGUISHING FEATURES. Mendesellines resemble Miracinae in having the lateral tergites of the first metasomal tergum, and often the central portions of the second metasomal tergum, desclerotized. The fore wing possesses a distinct and usually at least partly tubular r-m cross vein which apically closes off a moderately large cell (Figs 1and2), however, and the spiracles of the petiole are on the lateral edges of the median tergite (Figs 3 and 4).

KEY TO THE NEW WORLD GENERA OF THE SUBFAMILY MENDESELLINAE

1. Median tergite of petiole broader posteriorly than anteriorly (Fig. 3). Wings (Fig. 1) at least partially infuscate. Notauli usually meeting posteriorly in small depression. Ovipositor apically sinuate (Fig. 5). Hypopygium apically acute (Fig. 5).. ***Mendesella*** Whitfield and Mason

\- Median tergite broader anteriorly than posteriorly (Fig. 4). Wings (Fig. 2) hyaline. Notauli usually not quite meeting posteriorly. Ovipositor weakly decurved but not sinuate (Fig. 6). Hypopygium apically with small truncation or emargination (Fig. 6).. ***Epsilogaster*** Whitfield and Mason

REFERENCES

Marsh, P. M., Shaw, S. R., and Wharton, R. A. 1987. An identification manual to the North American genera of the family Braconidae (Hymenoptera). Memoirs of the Entomological Society of Washington 13: 1-98.

Whitfield, J. B. and Mason, W. R. M. 1994. Mendesellinae, a new subfamily of braconid wasps (Hymenoptera: Braconidae) with a review of relationships within the microgastroid assemblage. Systematic Entomology 19: 61-76.

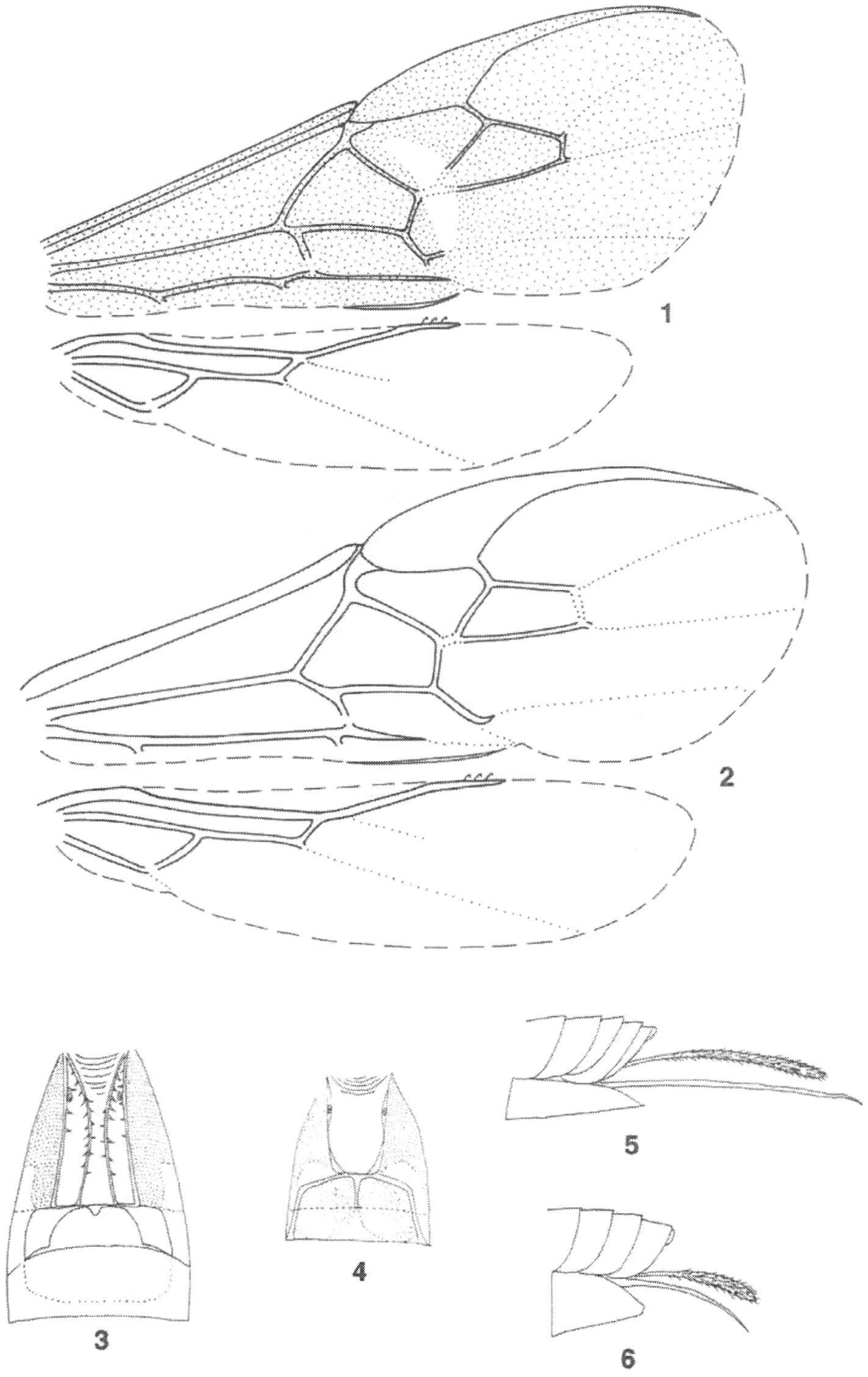

Figs 1-6: Mendesellinae. 1, wings of *Mendesella magna* Whitfield and Mason; 2, wings of *Epsilogaster panama*. Whitfield and Mason; 3, anterior metasomal tergites of *Mendesella amazonica* Whitfield and Mason; 4, anterior metasomal tergites of *Epsilogaster bicolor* Whitfield and Mason; 5, apex of metasoma and ovipositor *Mendesella braziliense* Whitfield and Mason, lateral view; 6, apex or metasoma and ovipositor of *Epsilogaster tico* Whitfield and Mason, lateral view. Figs 3-6 after Whitfield and Mason (1994).

MENDESELLINAE UPDATE

Jose Fernández-Triana
Canadian National Collection of Insects, 960 Carling Avenue, Ottawa, Canada, jose.fernandez@agr.gc.ca

There has not been much change to our knowledge of New World Mendesellinae during the past 20 years. At the generic level there is no change, with only two genera known: *Epsilogaster* Whitfield & Mason and *Mendesella* Whitfield & Mason (Whitfield 1997). Since the publication of the Manual in 1997, six new species have been described (Valerio and Whitfield, 2000; Yamada and Penteado-Dias, 2002; Figueroa *et al.*; 2004; Pitz, 2004), with the known diversity for New World Mendesellinae now standing at 15 species (Yu *et al.*, 2016).

Recent papers on New World Mendesellinae

Figueroa JI, Valerio AA, Lopez V, Whitfield JB, Sharkey MJ. 2004. Two new species of *Epsilogaster* Whitfield & Mason (Hymenoptera: Braconidae) from Mexico and Costa Rica. *Pan-Pacific Entomologist* 79(3-4): 198-206.

Pitz KM. 2004. *Epsilogaster fullertoni*, a new species of *Epsilogaster* Whitfield and Mason (Hymenoptera: Braconidae: Mendesellinae) from Florida, USA. *Zootaxa* 716: 1-7.

Valerio AA, Whitfield JB. 2000. Taxonomic notes on Costa Rican Mendesellinae (Ichneumonoidea: Braconidae), with description of a new Central American species of *Mendesella*. *Journal of Hymenoptera Research* 9(2): 271-276.

Valerio, A.A. and Whitfield, J.B.. 2002. *Epsilogaster faviolae*, a new species of Mendesellinae from Colombia (Hymenoptera: Braconidae) *Zootaxa* 41: 1-7.

Whitfield JB. 1997. Subfamily Mendesellinae. In: Wharton RA, Marsh PM, Sharkey MJ "Manual of the New World genera of the family Braconidae (Hymenoptera)." International Society of Hymenopterists. Special Publication No. 1. 439 pp., pp. 320-323.

Yamada MV, Penteado-Dias AM. 2002. *Mendesella jaraguaiensis*, a new species of Mendeselline wasp (Hymenoptera: Braconidae) from Brazil. *Entomological News* 113(4): 236-238.

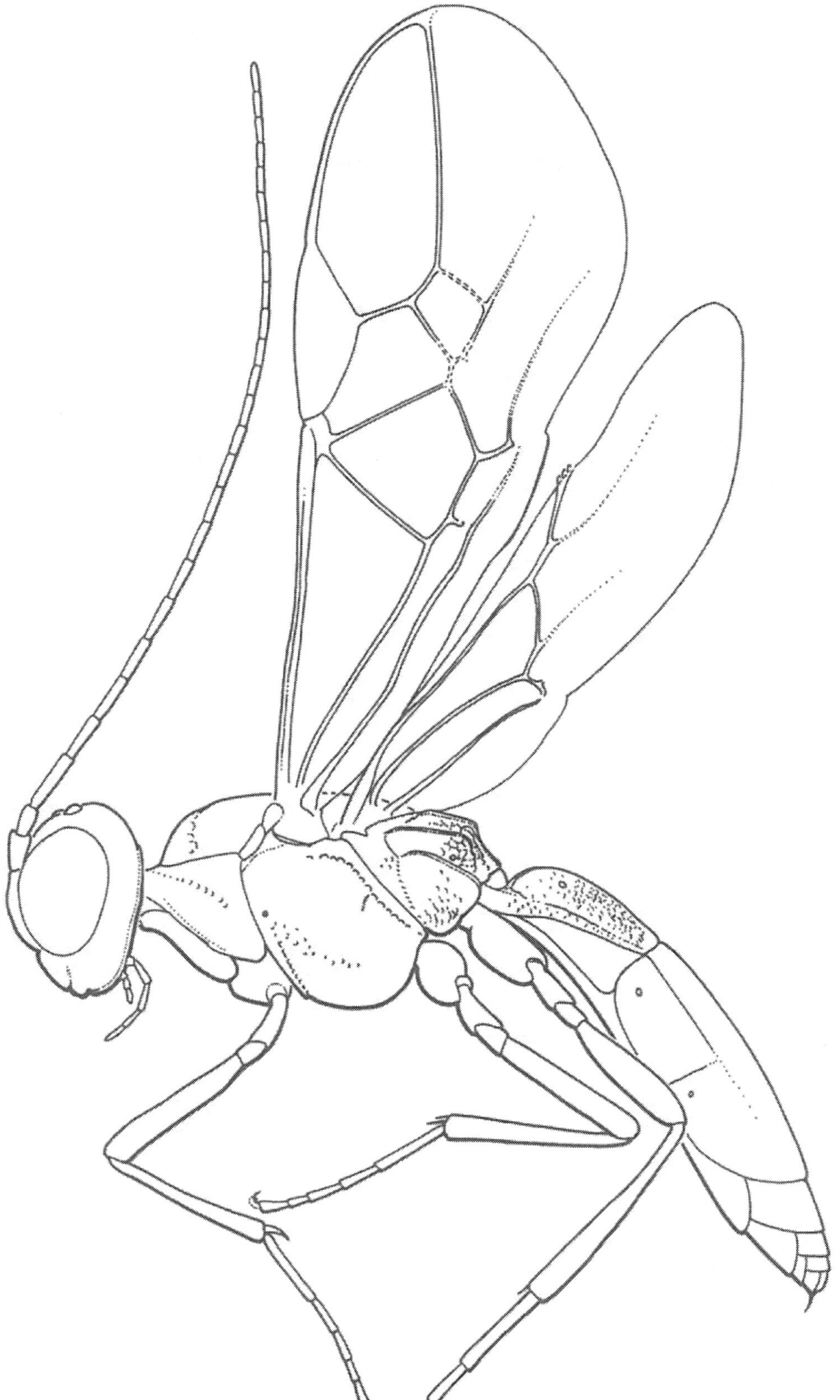

SUBFAMILY METEORIDEINAE

M. J. SHARKEY

Department of Entomology, University of Kentucky, S-225 Agricultural Science Building North, Lexington, Kentucky 40546, U.S.A.

INTRODUCTION. The subfamily Meteorideinae contains two genera, *Meteoridea* Ashmead and *Pronkia* van Achterberg. Only *Meteoridea* is found in the New World. Van Achterberg (1990) proposed the genus *Pronkia*, and presented a key to separate members of *Pronkia* from those of *Meteoridea*. Shenefelt (1970) cataloged the world species of *Meteoridea*. Since then, Austin and Wharton (1992) have described a species of *Meteoridea* from Australia, and Belokobyl'skij (1993) has described a species from the Russian Far East.

BIOLOGY. Members are known to be larval-pupal endoparasitoids of lepidopterous larva. Eggs are laid into the host larva and the adult parasitoid emerges from the pupal stage of the host. The largest collections of *Meteoridea* (Canadian National Collection) have been taken from light traps, suggesting that at least some species are nocturnal.

BIOLOGICAL CONTROL POTENTIAL. Some members of the Meteorideinae may be important in natural control since they have been reared from pest species of Lepidoptera. In general, however, they are seldom collected and poorly studied. None has been used in a classical biological control program.

DISTRIBUTION. Cosmopolitan. No species has yet been described from the Neotropical region, but the Canadian National Collection has a number of undescribed species from the Neotropics.

DISTINGUISHING FEATURES. Meteorideines can be recognized by the following combination of characters: hind wing with vein CUb present although often very small (see habitus): fore wing with second submarginal cell quadrate; terminal segments of metasoma laterally compressed; ovipositor very small, barely (if at all) exerted from the apex of the metasoma.

REFERENCES

Achterberg, C. van 1990. *Pronkia*, a new genus of the Meteorideinae (Hymenoptera: Braconidae) from New Zealand. Zoologische Mededelingen Leiden 64: 169-175.

Austin, A. D. and Wharton, R. A. 1992. New records of subfamilies, tribes and genera of Braconidae (Insecta: Hymenoptera) from Australia, with description of seven new species. Transactions of the Royal Society of Southern Australia 116: 41-63.

Belokobyl'skij, S. A. 1993. Contribution to the taxonomy of Braconidae (Hymenoptera) of the Russian Far East. Russian Entomological Journal 2: 87-103.

Shenefelt, R. D. 1970. Pars 5. Braconidae 2, Helconinae, Calyptinae, Mimagathidinae, Triaspinae. Pp. 177-305. In: Ferrière, Ch. and Vecht, J. van der (Eds.). Hymenopterorum Catalogus (nova editio). Dr. W. Junk, The Hague.

METERORIDEINAE UPDATE

M.J. Sharkey

Department of Entomology, University of Kentucky, Lexington, Kentucky 40546-0091 U.S.A. msharkey@uky.edu

Annotated list of recent papers on New World Meteorideinae.

Penteado-Dias AM. 1996. New Neotropical species of the genus *Meteoridea* Ashmead (Hymenoptera: Braconidae: Meteorideinae). *Zoologische Mededelingen, Leiden* 70(12): 197-205.

Presented a key to all species of *Meteoridea* including two new Neotropical species.

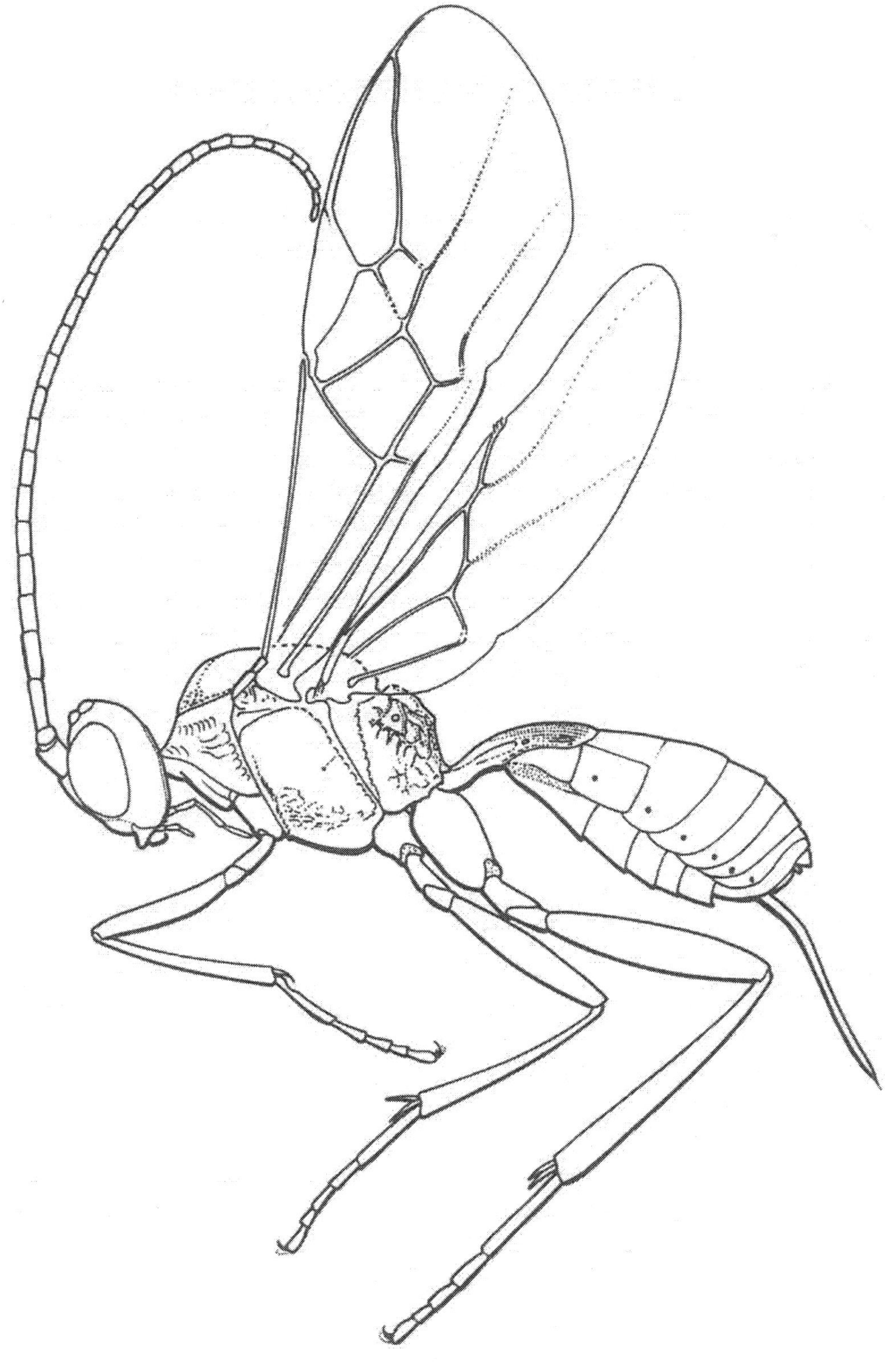

SUBFAMILY METEORINAE

S. R. SHAW

Department of Plant, Soil, and Insect Sciences, University of Wyoming, Laramie, Wyoming 82071-3354, U.S.A.

INTRODUCTION. The Meteorinae (sensu S. Shaw, 1985, 1994; Maetô, 1990; M. Shaw and Huddleston, 1991), sistergroup of the Euphorinae (*sensu stricto*), is a moderately large subfamily comprising at least 174 species worldwide (S. Shaw, 1988). Meteorines are sometimes classified as a tribe within the Euphorinae (e.g. Marsh, 1979; Gauld and Bolton, 1988; Quicke and van Achterberg, 1990).

BIOLOGY. The Meteorinae are solitary or gregarious koinobiont endoparasitoids of larval Coleoptera or Lepidoptera, and many species of *Meteorus* Haliday have broad host ranges (van Achterberg, 1979; West and Miller, 1989; Maet6, 1990). The vast majority of meteorines are solitary parasitoids attacking exophytic (exposed feeding) lepidopteran larvae, and many are nocturnally active. Others utilize hosts that are only weakly concealed (e.g. in leaf rolls or under webbing). The most frequently utilized hosts are Noctuidae, Geometridae, and Tortricidae, but many other lepidopterans including Hepialidae, Zygaenidae, Tineidae, Gelechiidae, Pyralidae, Papilionidae, Lycaenidae, Nymphalidae, Lasiocampidae, Thaumetopoeidae, and Arctiidae are also parasitized (Muesebeck, 1923; Huddleston, 1980; S. Shaw, unpublished data). The solitary parasitoids of arboreal Lepidoptera typically emerge from the host larva and pupate away from the host remains in a pendant cocoon that is often suspended by a long slender thread (Clausen, 1940; Gauld and Bolton, 1988), and it is from this characteristic cocoon that the genus *Meteorus* gained its name. The gregarious *Meteorus* species usually form their cocoons together in a loose, irregular heap (Huddleston, 1980), although the terminal threads may still be present, but intertwined (S. Shaw, 1985). A few highly specialized gregarious species from Africa and Sri Lanka form a very organized spherical cocoon mass, with all the exit holes facing outwards (Nixon, 1943; Huddleston, 1983). Those species attacking Coleoptera larvae typically utilize hosts concealed in wood, bark or fungus, especially Cerambycidae, Tenebrionidae, Scolytidae, Biphyllidae, Melandryidae, and Ciidae (Huddleston, 1980). As far as known, the *Meteorus* species attacking coleopterous larvae form stalkless cocoons within the beetle gallery (DeLeon, 1933; Mason, 1973), and these are regarded as relatively primitive (Mason, 1973; Maetô, 1990).

BIOLOGICAL CONTROL POTENTIAL. Many of the hosts of meteorines are considered pest insects, consequently several *Meteorus* species have been studied for purposes of biological control (Huddleston, 1980, 1983; Muesebeck, 1923). *Meteorus versicolor* (Wesmael) was introduced into the United States in the early 1900s to control the brown tail moth in New England (Muesebeck, 1923). Several other species appear to be important for natural control in forest ecosystems. *Meteorus trachynotus* Viereck has been studied as a parasitoid of the spruce budworm (McGugan, 1955) and *Meteorus hypophloei* Cushman as a parasitoid of bark beetles (DeLeon, 1933). Repeated importation of *Meteorus pulchricornis* (Wesmael) against gypsy moth was summarized by Coulson et al. (1986). Agricultural applications have been fewer, but *Meteorus rubens* (Nees) is an important, widespread parasitoid of various cutworms, and *Meteorus tersus* Muesebeck is a parasitoid of the avocado looper in California (Muesebeck, 1956). *Meteorus papiliovorus* Zitani attacks papilionids on Citrus in Costa Rica (Zitani et al., 1997).

COMMON GENERA. Species of the genus *Meteorus* are moderately common in many habitats. *Meteorus* adults can be collected by sweeping foliage, rearing host larvae, or from Malaise traps. They are also commonly attracted to lights at night, where they can be collected with an aspirator. The pendant cocoons of many species are rather conspicuous, and can be collected on the foliage and branches of the host food plant. The North American species of *Meteorus* were revised by Muesebeck (1923). However, that work should be used in combination with Huddleston (1980), which clarifies the status of some Holarctic species. Most of the Neotropical species are undescribed, and there are no revisionary works for *Meteorus* of that region. Costa Rican *Meteorus* are currently being revised by N. Zitani.

The placement of the less common genus *Zele* Curtis (=Zemiotes Förster) has been controversial, but following the phylogentic studies of Maetô (1990) it appears best to treat *Zele* as a specialized lineage within the Meteorinae. However, this poses a new problem since the continued recognition of Zele as a valid genus renders *Meteorus* paraphyletic. Maetô concluded that "after worldwide investigations... *Meteorus* should be rearranged into several monophyletic genera." In view of the wide range of morphological variation, behaviors, and host associations within *Meteorus* it seems inevitable that the large genus must eventually be subdivided into multiple genera. However, as Maetô observed, to be done properly this will necessarily involve phylogenetic studies of the species-groups on a worldwide basis. Thus, although it is tempting to put aside the problem by placing all members of the subfamily in a single genus *Meteorus*, doing so would not negate the need for comprehensive phylogenetic studies in this group.

DISTRIBUTION. The Meteorinae are cosmopolitan in distribution, and can be found in most habitats where plant-feeding Lepidoptera or wood-boring Coleoptera larvae feed.

DISTINGUISHING FEATURES. As in the Euphorinae, meteorines lack the fore wing cross vein 2cu-a, resulting in an "open" first subdiscal cell (Figs 1, 2). The fore wing vein 3RSb is straight in the Meteorinae and approaches the apex of the wing (thus a long, pointed marginal cell is present), unlike the Euphorinae which usually have vein 3RS distinctly curved, often meeting the wing margin well before the apex. The fore wing cross vein r-m is always present in meteorines, forming a characteristic rhombic or quadrate second submarginal cell. Euphorines almost always lack an r-m cross vein and second submarginal cell, but if present the second submarginal cell is of a much different shape (see key to the Euphorinae, couplet 1). The petiolization of the first metasomal tergum in the Meteorinae is distinctive (Fig. 3), and serves well to separate meteorines from diospiline Helconinae that have similar wing venation. In Meteorinae, the median longitudinal carina of the propodeum is divided at the extreme apex of the segment, a possible defining synapomorphy of the subfamily (Maetô, 1990).

KEY TO THE NEW WORLD GENERA OF THE SUBFAMILY METEORINAE

1. Hind wing vein RS bending towards the wing margin, thus marginal cell narrowing apically (Fig. 1). Metasomal terga with setae arranged in a single subapical row per tergum (Fig. 3). ....***Meteorus*** Haliday
   Distribution: Cosmopolitan; 47 described and numerous undescribed New World species. Biology: parasitoids of larval Coleoptera and Lepidoptera. References: Muesebeck (1923), Huddleston (1980).

\- Hind wing vein RS bending away from the wing margin, thus marginal cell broadening apically (Fig. 2). Metasomal terga (except tergum 2) with setae scattered over terga (Fig.4). ....***Zele*** Curtis
   Distribution: Holarctic south to Venezuela; 7 described and at least one undescribed New World species. Biology: parasitoids of lepidopteran larvae. References: Muesebeck (1923), van Achterberg (1979, 1984).

REFERENCES

Achterberg, C. van 1979. A revision of the subfamily Zelinae auct. (Hymenoptera: Braconidae). Tijdschrift voor Entomologie 122: 241-479.

Achterberg, C. van 1984. Addition to the revision of the genus *Zele* (Hymenoptera: Braconidae). Entomologische Berichten 44: 110-112.

Clausen, C. P. 1940. Entomophagous Insects. McGraw-Hill, New York. 688 pp.

Coulson, J., Fuester, R. W., Schaefer, P. W., Ertle, L. R., Kelleher, J. S., and Rhoades, L. D. 1986. Exploration for and importation of natural enemies of the gypsy moth, *Lymantria dispar* (L.) (Lepidoptera: Lymantriidae), in North America: an update. Proceedings of the Entomological Society of Washington 88: 461-475.

DeLeon, D. 1933. Notes on the biology of *Meteorus hypophloei* Cushman. Bulletin of the Brooklyn Entomological Society 28: 32-37.

Gauld, I. D. and Bolton, B. (Eds.) 1988. The Hymenoptera. British Museum (Natural History), London, Oxford University Press, Oxford. 332 pp.

Huddleston, T. 1980. A revision of the western Palearctic species of the genus *Meteorus* (Hymenoptera: Braconidae). Bulletin of the British Museum of Natural History (Entomology) 41: 1-58.

Huddleston, T. 1983. *Meteorus* (Hymenoptera: Braconidae) of Australia and New Guinea. Systematic Entomology 8: 393-420.

Maetô, K. 1990. Phylogenetic relationships and host associations of the subfamily Meteorinae Cresson (Hymenoptera: Braconidae). Japanese Journal of Entomology 58: 383-396.

Marsh, P. M. 1979. Family Braconidae. Pp. 144-295. In: Krombein, K. V., Hurd, P. D., Smith, D. R., and Burks, B. D. (Eds.). Catalog of Hymenoptera in America North of Mexico. Smithsonian Institution Press, Washington, D.C.

Mason, W.R. M. 1973. Recognition of *Zemiotes* (Hymenoptera: Braconidae). Proceedings of the Entomological Society of Washington 75: 213-215.

McGugan, B. M. 1955. Certain host-parasite relationships involving the spruce budworm. The Canadian Entomologist 87: 178-187.

Muesebeck, C. F. W. 1923. A revision of the North American species of ichneumon-flies belonging to the genus Meteorus Haliday. Proceedings of the United States National Museum 63: 1-44.

Muesebeck, C. F. W. 1956. Two new braconid parasites of the avocado looper (Hymenoptera: Braconidae). The Pan-Pacific Entomologist 32: 25-28. Nixon, G. E. J. 1943. A synopsis of the African species of *Meteorus* (Hymenoptera: Braconidae). Bulletin of Entomological Research 34: 53-64.

Quicke, D. L. J. and Achterberg, C. van 1990. Phylogeny of the subfamilies of the family Braconidae. Zoologische Verhandelingen Leiden 258: 1-95.

Shaw, M. R. and Huddleston, T. 1991. Classification and biology of braconid wasps (Hymenoptera: Braconidae). Handbooks for the Identification of British Insects 7(11): 1-126.

Shaw, S. R. 1985. A phylogenetic study of the subfamilies Meteorinae and Euphorinae (Hymenoptera: Braconidae): Entomography 3: 277-370.

Shaw, S. R. 1988. Euphorine phylogeny: the evolution of diversity in host-utilization by parasitoid wasps (Hymenoptera: Braconidae). Ecological Entomology 13: 323-335.

Shaw, S. R. 1995. The Braconidae. Pp. 431-464 In: Hanson, P. and Gauld, I. (Eds.). The Hymenoptera of Costa Rica. Oxford University Press, Oxford.

West, K. J. and Miller, J. C. 1989. Patterns of hostexploitation by *Meteorus communis* (Hymenoptera: Braconidae). Environmental Entomology 18: 537-540.

Zitani, N. M., Shaw, S. R., and Janzen, D. H. 1997. Description and biology of a new species of *Meteorus* Haliday (Hymenoptera: Braconidae, Meteorinae) from Costa Rica, parasitizing larvae of *Papilio* and *Parides* (Lepidoptera: Papilionidae). Journal of Hymenoptera Research 6(1): 178-185.

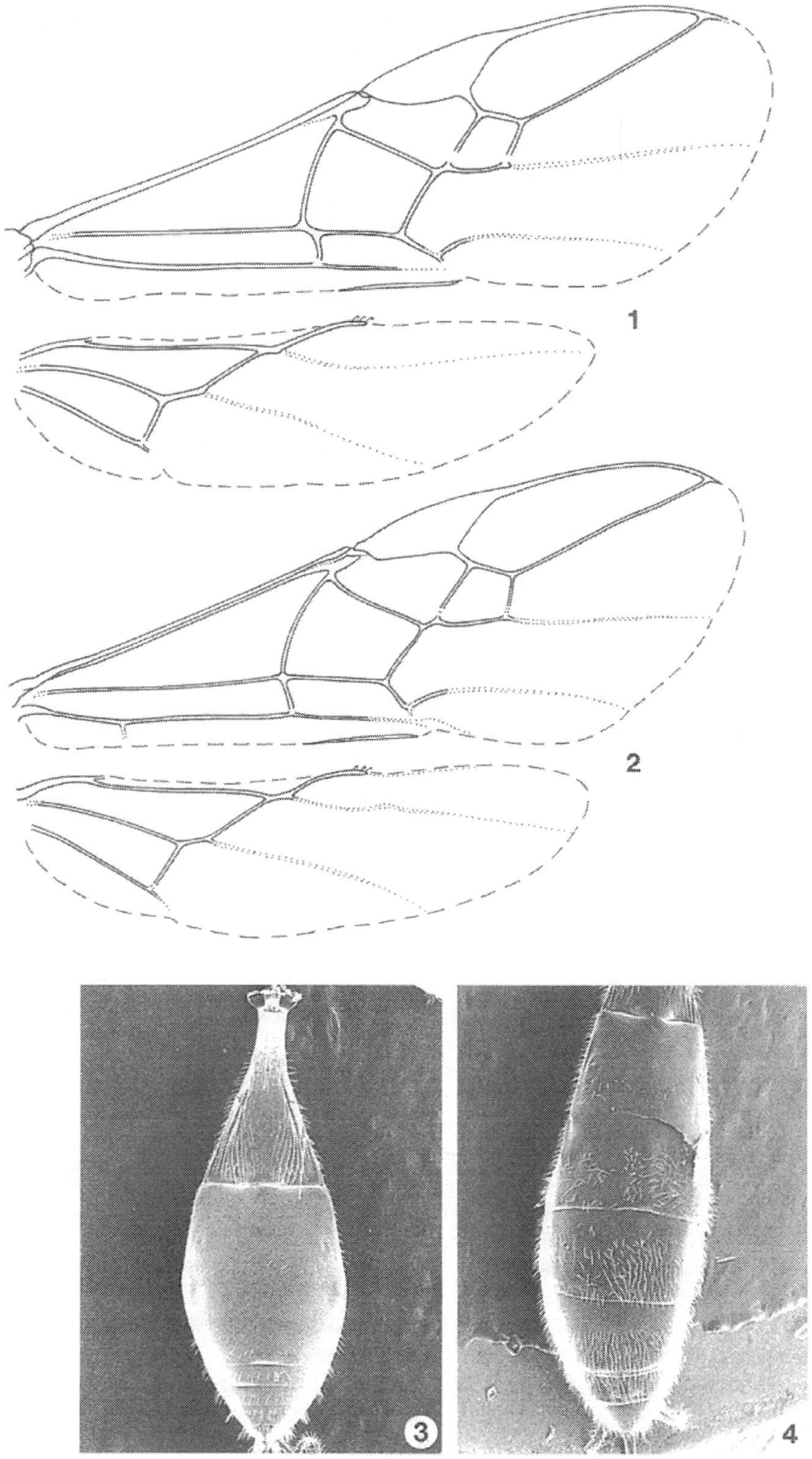

Figs 1-4: Meteorinae. 1, wings of *Meteorus* sp.; 2, wings of *Zele reticulatus* (Muesebeck): 3, metasoma of *Meteorus* sp., dorsal view; 4, metasoma of *Zele* sp., dorsal view.

METEORINAE UPDATE

S.R. Shaw
Department of Ecosystem Science and Management, University of Wyoming, Laramie, Wyoming 82071, U.S.A.
braconid@uwyo.edu

The Meteorinae chapter of the manual continues to work well for the identification of the two New World genera. However, a phylogenetic analysis by Stigenberg *et al.* (2015) concluded that "genus" *Zele* is nested within *Meteorus*; therefore *Meteorus* (as currently defined) is a paraphyletic genus. Their conclusion corroborates earlier analyses done by Stigenberg and Ronquist (2011) and Maetô (1990). However, Stigenberg *et al.* (2015) decided not to change the taxonomic status of either genus, until more comprehensive evidence can be studied. The situation is complicated by the fact that *Zele* is the older name, but has comparatively few species and is less well-known. Synonymizing the two genera is not a desirable outcome, since most of the literature is under the name *Meteorus*, and making it a junior synonym of *Zele* would result in much confusion over many well-known species of biological control significance. Eventually, the "genus" *Meteorus* will probably need to be divided into several smaller monophyletic genera (as has already been done with *Apanteles*), but doing so will indeed require much more study of meteorines on a worldwide basis. Given these complications, Aguirre *et al.* (2015) continued to recognize *Meteorus* and *Zele* as defined by Shaw (1997) and Stigenberg *et al.* (2015).

The North American species of *Meteorus* have not been revised since the work of Muesebeck (1923), but the recent revision of European species by Stigenberg and Ronquist (2011) provides keys, descriptions, and illustrations that can be effectively used to correctly identify several widespread Holarctic species that occur in North America.

Revisions, keys to species, new host records, and descriptions of new species of Neotropical *Meteorus* have been published recently for countries including Colombia (Aguirre *et al.*, 2011), Costa Rica (Zitani *et al.*, 1998; Shaw and Nishida 2005; Barrantes *et al.*, 2011) and Ecuador (Shaw and Jones 2009; Aguirre *et al.*, 2010; Jones and Shaw 2012; Aguirre and Shaw 2014a, 2014b; Aguirre *et al.*, 2015). The most recent estimation of *Meteorus* diversity worldwide is 332 described species, but a large number of new species remain to be discovered in the tropics (Aguirre *et al.*, 2015).

Annotated list of recent papers on Meteorinae

Aguirre H, Shaw SR, Jones GZ. 2010. A new *Meteorus* Haliday species from Colombia and Ecuador (Hymenoptera: Braconidae). *Zootaxa* 2453: 55-61.
Meteorus gigas is described from Colombia and Ecuador.

Aguirre H, Sarmiento CE, Shaw SR. 2011. Taxonomic revision and morphometric analysis of *Meteorus* Haliday, 1835 (Hymenoptera: Braconidae: Meteorinae) from Colombia. *Zootaxa* 2938: 1-68.
Nineteen new *Meteorus* species are described and 16 other species are recorded for the first time from Colombia.

Aguirre H, Shaw SR. 2014a. *Meteorus* Haliday (Hymenoptera: Braconidae) parasitoids of Pyralidae: description and biology of two new species and first record of *Meteorus desmiae* Zitani, 1998 from Ecuador. *Journal of Natural History* 48: 2375-2388.
The biology of three Neotropical *Meteorus* associated with Pyralidae is reviewed.

Aguirre H, Shaw SR. 2014b. Neotropical species of *Meteorus* Haliday (Hymenoptera: Braconidae: Meteorinae) parasitizing Arctiinae (Lepidoptera: Noctuoidea: Erebidae). *Zootaxa* 3779(3): 353-367.
A comprehensive key for the Neotropical *Meteorus* attacking Arctiinae is provided.

Aguirre H, Shaw SR, Berry JA, de Sassi C. 2014. Description and natural history of the first micropterous *Meteorus* species: *M. orocrambivorus* sp. n. (Hymenoptera, Braconidae, Euphorinae), endemic to New Zealand. *Journal of Hymenoptera Research* 38: 45-57.
Describes the first known micropterous species of *Meteorus* and its biology.

Aguirre H, de Almeida LF, Shaw SR, Sarmiento CE. 2015. An illustrated key to Neotropical species of the genus *Meteorus* Haliday (Hymenoptera: Braconidae: Euphorinae). *ZooKeys* 489: 33-94.
An illustrated key to 75 *Meteorus* species from the Neotropical region, including descriptions of 11 new species from Bolivia, Costa Rica, and Ecuador, is presented.

Barrantes G, Triana E, Shaw SR, Jones GZ. 2011. Characteristics of the cocoon and natural history of the gregarious *Meteorus restionis* sp. n. (Hymenoptera, Braconidae, Meteorinae) from Costa Rica. *Journal of Hymenoptera Research* 20: 9-21.
A new species of a gregarious *Meteorus* is described, which constructs an unusual cocoon mass. Cocoons of *M. restionis* are attached perpendicularly along a long, rope-like, suspended cable.

Jones GZ, Shaw SR. 2012. Ten new species of *Meteorus* (Braconidae: Hymenoptera) from Ecuador reared at the Yanayacu Biological Center for Creative Studies. *Zootaxa* 3547: 1-23.
Ten new species of *Meteorus* were described, along with detailed biological information. Four species were reared from Arctiinae caterpillars, two from Nymphalidae, and one each from Apatelodidae, Limacodidae, Megalopygidae, and Noctuidae.

Shaw SR, Jones GZ. 2009. A new species of solitary *Meteorus* (Hymenoptera: Braconidae) reared from caterpillars of toxic butterflies (Lepidoptera: Nymphalidae) in Ecuador. *Journal of Insect Science* 9(34): 1-8.
A new species of *Meteorus* was reared from chemically-defended butterfly caterpillars in Ecuador.

Shaw SR, Nishida K. 2005. A new species of gregarious *Meteorus* (Hymenoptera: Braconidae) reared from caterpillars of *Venadicodia caneti* (Lepidoptera: Limacodidae) in Costa Rica. *Zootaxa* 1028: 49-60.

Stigenberg J, Ronquist F. 2011. Revision of the Western Palearctic Meteorini (Hymenoptera, Braconidae), with a molecular characterization of hidden Fennoscandian species diversity. *Zootaxa* 3084: 1-95.
Treats the meteorines as a tribe within the subfamily Euphorinae. Provides identification keys, descriptions, and biological data for the western Palearctic species, some of which have circumpolar distributions and occur in boreal North America.

Stigenberg J, Boring CA, Ronquist F. 2015. Phylogeny of the parasitic wasp subfamily Euphorinae (Braconidae) and evolution of its host preferences. *Systematic Entomology* 40(3): 570-591.
Treats the meteorines as a tribe within the subfamily Euphorinae.

Zitani NM, Shaw SR, Janzen DH. 1997. Description and biology of new species of *Meteorus* Haliday (Hymenoptera: Braconidae, Meteorinae) from Costa Rica, parasitizing larvae of *Papilio* and *Parides* (Lepidoptera: Papilionidae). *Journal of Hymenoptera Research* 6: 178-185.
Provides a key to 19 *Meteorus* species from Costa Rica, descriptions of 14 new species, and the first host records of the genus from Hesperiidae and Megalopygidae.

Zitani NM, Shaw SR, Janzen DH. 1998. Systematics of Costa Rica *Meteorus* (Hymenoptera: Braconidae: Meteorinae) species lacking a dorsope. *Journal of Hymenoptera Research* 7(2): 182-208.
Describes the new species *Meteorus papiliovorus* and provided the first host record of a New World *Meteorus* reared from Papilionidae.

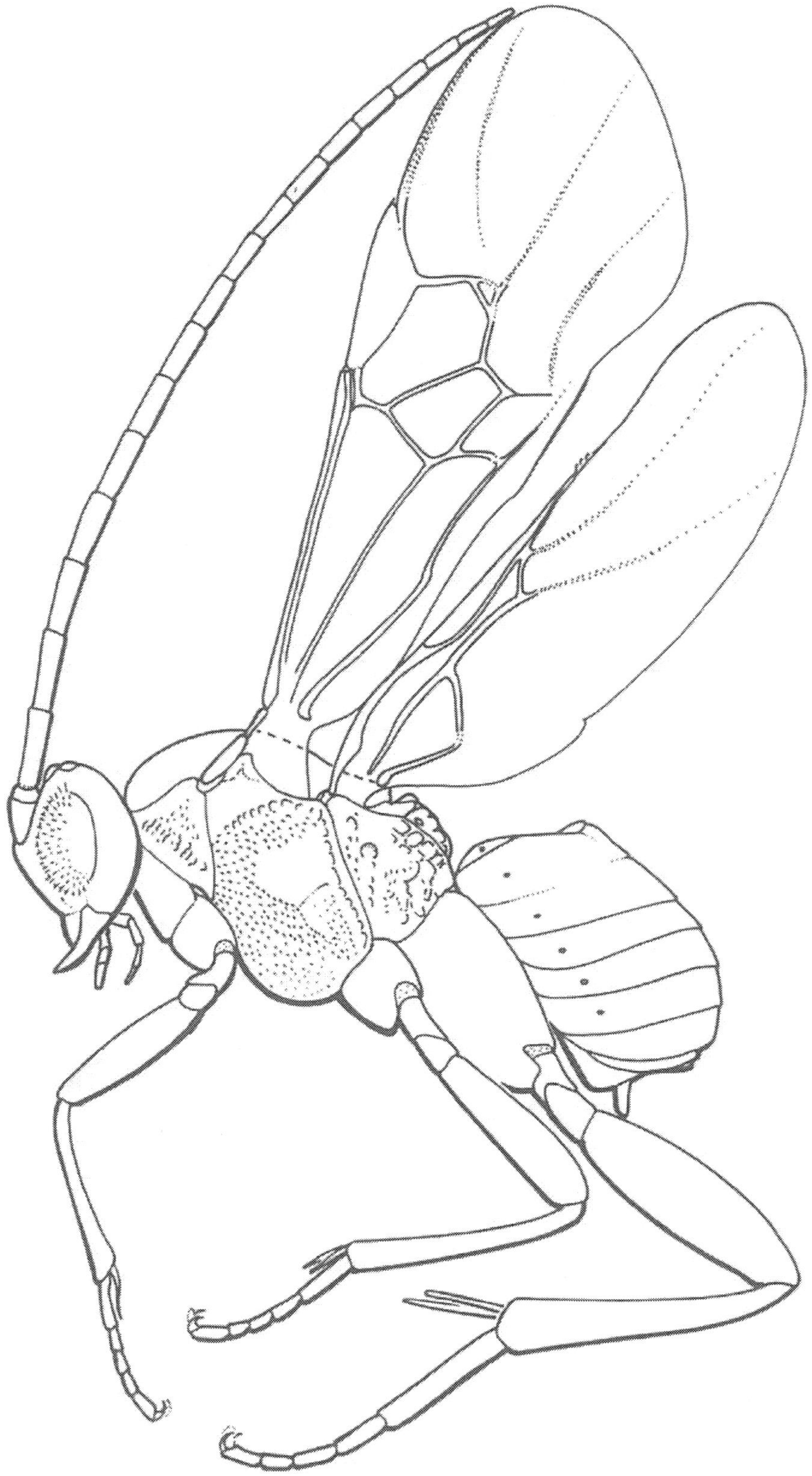

SUBFAMILY MICROGASTRINAE

J. B. WHITFIELD

Department of Entomology, 320 Morrill Hall, University of Illinois, 505 S. Goodwin Ave, Urbana, IL 61801 U.S.A.

INTRODUCTION. Wasps of the subfamily Microgastrinae are the most important single group of parasitoids of Lepidoptera in the world, both in economic terms and in numbers of species. Mason (1981) estimated that the actual world total is likely between 5,000 and 10,000 species; at present over 1,400 species have been described.

The classification of Microgastrinae has become more complex over the last three decades. Nixon (1965) erected two new genera for Neotropical species, and resurrected other generic names from obscurity, ncreasing the number of genera recognized from the New World from approximately eight to 13. With Mason's (1981) landmark but controversial reclassification, the New World species were redistributed into 41 genera, 18 of them newly created. Mason's (1981) classification resulted in the elevation to generic rank of many of Nixon's (1965) species groups of *Apanteles* Förster. Mason's revision thus resulted in a much narrower concept for *Apanteles*, though this genus still contains an exceptionally large number of described species from the New World.

Since the last available catalogs of World species (Shenefelt, 1972, 1973) and North American species (Marsh, 1979), not only has Mason's (1981) reclassification appeared, but several additional new works, describing new genera, and introducing changes in the definitions of genera, have been published (Papp, 1979, 1986; Williams, 1985, 1988; Austin and Dangerfield, 1992; Whitfield, 1995a, b). Some confusion in the application of the generic names *Microgaster* Latreille and *Microplitis* Förster has also arisen (van Achterberg, 1982; Papp, 1984; Tobias, 1986; Mason, 1986; Whitfield, 1987) but this has now been resolved by the International Commission of Zoological Nomenclature (ICZN, 1988), preserving the traditional usage of these names. There is no single reference from which the current generic placements of all New World microgastrine species can be obtained. Although a new checklist of the North American species of Microgastrinae has been produced (Whitfield, 1995b), a comparable work for the Neotropical species is not yet available. Thus, the precise number of described New World species is presently unknown for most of the genera treated here.

In addition to the publication by Mason (1981), five new microgastrine genera have been recently proposed, *Glabromicroplitis* Papp (1979, 1986), *Lathrapanteles* Williams (1985), *Austrocotesia* Austin and Dangerfield (1992), *Napamus* Papp (1993), and *Xanthapanteles* Whitfield (1995a). *Austrocotesia* was originally reported only from the Australasian region, but specimens have since been seen from South America. This genus is therefore included in the key below, as are *Wilkinsonellus* Mason and *Xenogaster* Mason, which were also not previously recorded in the New World. *Glabromicroplitis* has not been incorporated into the key, since this name has not been widely accepted. *Glabromicroplitis* was proposed for a morphologically derived species group within *Microplitis*, and its application to the world fauna is somewhat problematical. Austin and Dangerfield (1993) thus synonymized it with *Microplitis. Napamus* has not been found in the New World as yet, and *Xanthapanteles* has so far been collected only in Argentina.

Several new changes in the diagnostic features used to recognize the genera have been incorporated into the key, based on a more extensive survey of the Neotropical species of Microgastrinae than was available to Mason prior to his 1981 reclassification. When there is some doubt about the distinctness of some genera, this has been noted in the key in the hope that this will encourage others to look more closely at the systematics of these groups. In the key, no attempt has been made to use Mason's (1981) tribal classification. Not only does this simplify using the key, but the tribal classification has come, to a large extent justifiably, under some recent attack (Walker et al., 1990; Austin, 1990).

For the most part, North American specimens can be reliably determined to genus using the keys of either Mason (1981) or Marsh et al. (1987). The outdated revisions of the North American species of Microgastrinae by Muesebeck (1920, 1922) are still useful for identification of many of the species, especially in the eastern United States, but new keys are badly needed. In the few cases where new revisions are available, they are listed under the appropriate generic name in the key below.

No significant species-level revisions of the Mexican, Central American or South American faunas of Microgastrinae have ever been published. However, a number of the South American species have been described in a wide variety of publications, especially by Ashmead, Blanchard, Brethes, Cameron; and Muesebeck. Shenefelt (1972, 1973) provides outdated but useful lists of the World species.

BIOLOGY. Microgastrine species attack virtually the entire taxonomic and biological spectrum of Lepidoptera, with the possible exception of Hepialidae and a few other primitive lineages of Lepidoptera. All species are koinobiont, endoparasitoids of larvae, and exit the host to pupate. A few species spin their cocoons within the cocoon or shelter of the prepupal host (*Deuterixys* Mason, some *Pholetesor* Mason). Probably the majority of the species are solitary, but a very large number of species are gregarious, and from larger hosts such as Sphingidae and Saturniidae, hundreds of larvae may emerge to spin their cocoons either singly or in masses, and the structure of these cocoon masses can be diagnostic. An excellent review of the biology of Microgastrinae is provided by Shaw and Huddleston (1991). Some additional information on tropical genera is provided in Mason (1981).

BIOLOGICAL CONTROL POTENTIAL. More than 100 species in this group have been used or investigated worldwide in the biological control of pest Lepidoptera, and this total is likely to rise. It would be impossible to review the applied literature on microgastrines here. A wide taxonomic variety of lepidopteran hosts is represented among the pests regulated by these wasps, including noctuids, gelechiids, and pyralids in field crops, tortricids and lymantriids in forests, gracillariids in orchards, and tineids in stored products. Examples relevant to the New World include *Cotesia marginiventris* (Cresson) on beet armyworm and green cloverworm (Ruberson and Whitfield, 1996), *Pseudapanteles dignus* (Muesebeck) on tomato pinworm (Cardona and Oatman, 1971*), Cotesia flavipes* Cameron on sugarcane borer (Fuchs et al., 1979), *Apanteles fumiferanae* Viereck on spruce budworm (Lewis, 1960), *Cotesia melanoscela* (Ratzeberg) on gypsy moth (Clausen, 1978), *Pholetesor pedias* Nixon on apple leafminers (Laing and Heraty, 1981), and *Apanteles carpatus* (Say) on the webbing clothes moth (Fallis, 1942).

COMMON GENERA. In the New World, the dominant genera are *Apanteles* Förster, *Cotesia* Cameron, *Diolcogaster* Ashmead, *Dolichogenidea* Viereck, *Glyptapanteles* Ashmead and *Microplitis*. Each of these genera is represented by at least 50-100 species, and *Apanteles* possibly by as many as 500 New World species. In Central and South America *Hypomicrogaster* Ashmead, *Pseudapanteles* Ashmead and *Rhygoplitis* Mason may also be locally abundant and diverse, while *Promicrogaster* Brues and Richardson is usually not so abundant but is species-rich in the Neotropics. In North America *Microgaster* can also be locally abundant.

DISTRIBUTION. Microgastrines are found on all continents and from tropical to arctic climates. Many of the genera have more restricted distributions, or are more diverse in certain regions. More specific information is included under each genus in the key.

DISTINGUISHING FEATURES. All Mictogastrinae have exactly 18 antennal segments (though these may appear subdivided when placode sensilla are evenly arranged in two ranks per flagellomere). They also have the spiracles of the first metasomal segment (= petiole) situated on the lateral tergites, rather than on the median tergite, and tend to have apically reduced wing venation. These features should serve to distinguish microgastrines from all other braconids.

KEY TO THE NEW WORLD GENERA OF THE SUBFAMILY MICROGASTRINAE

It is essential to prepare specimens so that the propodeum, petiole, and the vannal lobe of the hind wing are clearly visible. Unassociated males will be difficult to identify in the latter part of this key, especially past couplet 51.

1. Metasomal terga 1-3 (rarely also 4) fused to form a carapace-like structure covering remainder of metasoma (Figs 49, 86; or resembling Fig. 72 but with terga 1 and 2 fused and tergum 3 also sculptured)2

\- Metasomal terga 1-3 variable, sometimes enlarged and heavily sculptured (Figs 44, 47, 48) but still with tergum 1 normally articulating with tergum 2 4

2(1). Second submarginal cell of fore wing closed distally by r-m, often forming small areolet (Fig. 3). Epicnemial carina always absent ....***Diolcogaster*** Ashmead (in part)

Distribution: Cosmopolitan, more diverse in tropics. Biology: solitary or gregarious, mostly on various macrolepidoptera. References: Muesebeck (1922, as *Protomicroplitis*, in part), Nixon (1965, as *Protomicroplitis*, in part), Mason (1981).

\- Second submarginal cell open distally; r-m absent or fused with other veins (Figs 4, 16). Epicnemial carina present or absent 3

3(2). Mesopleuron with complete epicnemial carina. Metasoma completely carapace-like and strongly convex dorsally (Fig. 49). Scutellum projecting postero-medially as a small tooth or flange. Body usually black. Wings: Fig. 4 ....***Fornicia*** Brullé

Distribution: Pantropical. Biology: largely, or perhaps exclusively, parasitoids of Limacodidae. References: Papp (1980), Mason (1981), Austin (1987).

\- Epicnemial carina absent. Metasoma finely sculptured (Fig. 86) and somewhat flattened dorsally. Scutellum not projecting as tooth or flange, body usually yellow ....***Xanthapanteles*** Whitfield (in part)

Distribution: Argentina; rare; one described species. Biology: unknown. Reference: Whitfield (1995a).

4(1). Second submarginal cell of fore wing closed distally by r-m, forming tiny to moderate sized cell or areolet. (a few genera are variable or difficult to distinguish with this character and they are keyed through both halves of the couplet) (Figs 1, 3, 5-10, 12, 13, 15, 17) 5

\- Second submarginal cell open distally, either through loss of r-m or fusion of this vein with other veins (Figs 2, 4, 11, 14, 16) 33

5(4). Propodeum with a well-developed areola defined by lateral carinae, at least posteriorly (may be variable in shape or definition anteriorly) (Figs 30-33, 35, 36, 48) 6

\- Propodeum without a distinct areola; often with either a median longitudinal carina (Figs 38-44), or with poorly defined pattern of carinae (Figs 34, 37, 46) or, in some species, essentially unsculptured (Fig. 45) 9

6(5). Propodeum with a median longitudinal carina bisecting the areola, the areola small and sometimes more weakly defined (Fig. 36). Metasomal tergum 2 much broader than long and usually relatively smooth (Fig. 53). Second submarginal cell of fore wing extremely small (Fig. 5) ....***Hypomicrogaster*** Ashmead (in part)

Distribution: New World, especially Neotropics. Biology: solitary on microlepidoptera, including some leaf-miners. References: Muesebeck (1958, as *Microgaster*, in part), Nixon (1965).

\- Propodeum without median longitudinal carina (Figs 30-32, 35, 48), or with a short median carina extending only over anterior half of propodeum or less (Fig. 33) 7

7(6). Propodeum with median longitudinal carina extending over anterior third of propodeum, anterior to areola; areola usually pentagonal. Median tergite of petiole relatively slender, not covering full width of dorsal surface. Hypopygium relatively short, evenly sclerotized to median fold (Figs 83, 90). Extremely rare in New World ....***Austrocotesia*** Austin and Dangerfield

Distribution: primarily Australasian but at least one undescribed species in South and Central America. Biology: unknown. Reference: Austin and Dangerfield (1992).

\- Propodeum with areola covering most of central length of propodeum, often with one or two very short longitudinal carinae leading from areola to anterior margin; areola usually more or less oval (Figs 30-33). Median tergite of petiole relatively broad, covering most or all of dorsal width of first tergum. Hypopygium variable, but often long and desclerotized medially into a series of expandable pleats (Fig. 96) 8

8(7). Petiole relatively flat, smooth and broader posteriorly, sometimes weakly punctate (Fig. 88). Hind wing cu-a sinuate (as in Fig. 17) .... ***Dasylagon*** Muesebeck

Distribution: South American, especially Brazil. Biology: reared from Sesiidae and Thyrididae, may attack other families as well. References: Muesebeck (1958), Nixon (1965).

\- Petiole somewhat barrel-shaped in outline to slightly narrower posteriorly, longitudinally aciculo-rugose to aciculo-punctate (Fig. 52). Hind wing cu-a straight or nearly .... ***Papanteles*** Mason

Distribution: Neotropical. Biology: unknown. References: Nixon (1965, as *virbius*-group of *Hypomicrogaster*), Mason (1981).

9(5). Petiole with strong, usually narrow, antero-median longitudinal groove over most of length (Figs 59, 61, 63-65, 68, 72, 79) 10

\- Petiole with broad antero-median depression but without strong antero-median longitudinal groove (there may be a weak postero-median depression or weak anteromedian groove that is shorter than half length of petiole) (Figs 51-54, 56-58, 60, 62, 66, 67, 69, 71, 73-77) 16

10(9). Mesopleuron with distinct (sometimes complete) epicnemial carina (Fig. 25); mesosoma very coarsely sculptured. Metasomal terga 2 and 3 smooth and without well- defined median raised and sculptured area. Wings: Fig. 15 .... ***Snellenius*** Westwood

Distribution: Pantropical. Biology: usually gregarious on Noctuidae and Sphingidae. References: Nixon (1965), Mason (1981), Austin and Dangerfield (1993).

\- Mesopleuron without distinct epicnemial carina; mesosoma variable but rarely extremely coarsely sculptured except on propodeum. Metasomal terga 2 and 3 variable but usually with well-defined median area that is raised and often also sculptured 11

11(10). Median tergite of petiole at least 1.5 times longer than maximum width and usually either much more slender than this, or narrowing posteriorly (Figs 79, 91, 95). Hind coxa variable 12

Petiole broad, somewhat rectangular in outline, often but not always strongly sculptured. Hind coxa greatly enlarged 14

12(11). Antennal placodes arranged in two ranks, antennal segments thus appearing to be divided into two halves (Fig. 20). Wings: Fig. 3 .... ***Diolcogaster*** Ashmead (in part)

Distribution: Cosmopolitan, more diverse in tropics. Biology: solitary or gregarious mostly on various macrolepidoptera. References: Muesebeck (1922, as *Protomicroplitis*, in part), Nixon (1965, as *Protomicroplitis*, in part), Mason (1981).

\- Antennal placodes arranged irregularly, rarely in roughly 3 ranks, thus antennal segments not appearing to be divided into two halves (Fig. 21) 13

13(12). Vannal lobe of hind wing reduced (Fig. 6), not clearly set off from rest of margin of wing by a notch (as in Fig. 12). Propodeum smooth or nearly so, with strong median carina. Large and often strikingly colored .... ***Larissimus*** Nixon

Distribution: Neotropical. Biology: *L. cassander* Nixon has recently been reared from *Bertholdia* (Arctiidae) in Brazil (Penteado-Diaz, 1997). References: Nixon (1965), Mason (1981).

\- Vannal lobe strongly convex and separated from rest of margin of wing by a notch (Fig. 93). Propodeum strongly rugose with a median carina. Petiole extremely narrow, with sides broadly but weakly emarginate (Fig. 79) .... ***Protomicroplitis*** Ashmead

Distribution: New World. Biology: reared from Noctuidae. References: Muesebeck (1922, as *Protomicroplitis*, in part), Mason (1981, restriction of generic definition).

14(11). Metasomal tergum 2 usually with raised median area; when this is well-defined, the median area is much longer than wide (Fig. 72). Wings: Fig. 3 .... ***Diolcogaster*** Ashmead (in part)

Distribution: Cosmopolitan, more diverse in tropics. Biology: solitary or gregarious mostly on various macrolepidoptera. References: Muesebeck (1922, as *Protomicroplitis*, in part), Nixon (1965, as *Protomicroplitis*, in part), Mason (1981).

\- Metasomal tergum 2 subquadrate and often sculptured, but without clearly defined median raised area 15

15(14). Hind wing r-m (typically spectral and therefore difficult to see) delimiting a cell that is much taller than wide (i.e. longer measured on axis from anterior to posterior edge of wing than on axis from basal to distal ends of wing) (Fig. 17). Ovipositor sheaths 0.3-1.0 times as long as hind tibia and setose over most of length. Terga: Fig. 61 .... ***Xanthomicrogaster*** Cameron

Distribution: Neotropical. Biology: unknown. References: Nixon (1965), Mason (1981).

\- Hind wing r-m present but not delimiting a cell shaped as above. Ovipositor sheaths short, polished, with setae concentrated apically (Fig. 83). Terga: Fig. 81 .... ***Exix*** Mason

Distribution: New World. Biology: poorly known but some reared from Noctuidae. Reference: Mason (1981).

16(9). Propodeum with clearly defined transverse carina in addition to median longitudinal carina (Figs 39, 41, 42) 17
- Propodeum without clearly defined transverse carina, sometimes also without distinct median longitudinal carina 19
17(16). Mesonotum with well-defined, although sometimes unsculptured, notauli (Fig. 23). Vannal lobe of hind wing strongly reduced (Fig. 12) .... ***Prasmodon*** Nixon (in part)
Distribution: South America. Biology: unknown. References: Nixon (1965), Mason (1981).
- Mesonotum without well-defined notauli. Vannal lobe well-developed, although sometimes slightly flattened subapically 18
18(17). Vannal lobe of hind wing slightly flattened distally and with sparse fringe of setae (Fig. 92); second submarginal cell of fore wing very small (as in Fig. 5). Body length usually less than 2 mm .... ***Clarkinella*** Mason
Distribution: New World, especially Neotropical. Biology: unknown. Reference: Mason (1981).
- Vannal lobe convex distally and uniformly fringed with setae (Fig. 93); second submarginal cell moderate to large (Fig. 9). Body length usually greater than 2 mm .... ***Paroplitis*** Mason (in part)
Distribution: Holarctic, Oriental. Biology: unknown, probably attacks microlepidoptera, Reference: Mason (1981).
19(16). Head antero-posteriorly thin, with long and bilobed glossa (Fig. 18). Ovipositor sheaths long to very long. Hypopygium long and medially desclerotized into expandable pleats (Fig. 96) 20
- Head not usually antero-posteriorly thin and glossa not deeply bilobed (Fig. 19). Ovipositor sheaths variable. Hypopygium variable but often evenly sclerotized to median fold 21
20(19). Metasomal tergum 2 broader than long (Fig. 56). Ovipositor sheaths apically sinuate. Body usually mostly black. Wings: Fig. 13 .... ***Promicrogaster*** Brues and Richardson (in part)
Distribution: New World, especially Neotropical. Biology: poorly known but apparently attack microlepidoptera concealed inside plant and fungal tissues. References: Nixon (1965, as *Promicrogaster*, in part), Mason (1981, restriction of generic concept).
- Metasomal tergite 2 longer than broad (Fig. 91). Ovipositor sheaths not apically sinuate. Most species largely or entirely yellow .... ***Sendaphne*** Nixon (in part)
Distribution: Neotropical. Biology: unknown. References: Nixon (1965), Mason (1981).
21(19). Mesoscutum with well-defined notauli, at least over anterior half (Figs 22, 23) (*Microplitis* will key both ways) 22
- Mesoscutum without clearly defined notauli (Fig. 24) 25
22(21). Metasomal tergum 2 with median tergite strongly sculptured and broad, rectangular to subrectangular and covering most or all of dorsal width of tergum (as in Fig. 59); with no narrow, median raised region on this tergite. Flagellomeres with placodes arranged irregularly (Fig. 21), thus flagellomeres not appearing subdivided. Mesosoma: shaped as in Fig. 29 .... ***Hygroplitis*** Thomson (in part)
Distribution: Holarctic. Biology: mostly reared from stem-boring Pyraloidea. References: Nixon (1968, as *Microgaster*, in part), Mason (1981).
- Metasomal tergum 2 unsculptured (or only very slightly sculptured), and longer than wide in some species; occasionally with narrow, median raised region on this tergite. Flagellomeres usually with 2 ranks of placodes, thus often appearing subdivided (Fig. 20) 23
23(22). Vannal lobe of hind wing strongly reduced (Fig. 12) .... ***Prasmodon*** Nixon (in part)
Distribution: South America. Biology: unknown. References: Nixon (1965), Mason (1981).
- Vannal lobe not reduced 24
24(23). Propodeum completely smooth and polished. Ovipositor sheaths as long as or longer than hind tibia. Second submarginal cell of fore wing large (Fig. 10); hind wing 2r-m absent, even as a spectral vein .... ***Pelicope*** Mason
Distribution:western North America. Biology: reared from *Mesepiola specca* Davis in seedpods of *Nolina parryi* Watson (Agavaceae), Reference:Mason (1981).
- Propodeum variously sculptured, usually strongly rugose, and often showing traces of a median carina. Ovipositor sheaths short and with setae concentrated apically. Second submarginal cell small (Fig. 8); hind wing 2r-m typically present but not pigmented .... ***Microplitis*** Förster (in part)
Distribution: Cosmopolitan; common. Biology: solitary or gregarious, almost always on macrolepidoptera (especially Noctuidae and Sphingidae). References: Muesebeck (1922), Nixon (1970), Papp (1984), Austin and Dangerfield (1993).
25(21). Hind coxa small, usually not extending beyond petiole. Hind tibial spurs half as long as hind basitarsus or shorter 26

- Hind coxa longer, usually extending nearly to end of metasomal tergum 2. Hind tibial spurs usually longer (if not, then ovipositor is longer than hind tibia and propodeum is completely unsculptured)......28

26(25). Metasomal tergum 2 with well-defined median raised area (Fig. 80). Legs and antennae of females (not males) short and stout. Female flagellomeres with only a single rank of placodes ....**_Venanus_** Mason (in part)

Distribution: New World, mainly Andean. Biology: solitary on microlepidoptera, especially leaf-miners and needle-miners. Reference: Mason (1981).

- Metasomal tergum 2 without well-defined median area bounded laterally by grooves (Figs 69, 78). Legs and antennae of both sexes of normal shape and length in most species. Female flagellomeres usually with placodes in 2 ranks...27

27(26). Ovipositor sheaths about half as long as hind tibia and setose over most of length. Legs of both sexes short and stout. Female flagellomeres with only a single rank of placodes. Relatively rare. Wings: Fig. 9.... ....**_Paroplitis_** Mason (in part)

Distribution: Holarctic, Oriental. Biology: unknown, probably attacks microlepidoptera. Reference: Mason (1981).

- Ovipositor sheaths short, with setae concentrated distally. Legs variable but typically not unusually stout. Flagellomeres with 2 ranks of placodes in both sexes (Fig. 20). Extremely common. Wings: Fig. 8 ....**_Microplitis_** Förster (in part)

Distribution: Cosmopolitan; common. Biology: solitary or gregarious, almost always on macrolepidoptera (especially Noctuidae and Sphingidae). References: Muesebeck (1922), Nixon (1970), Papp (1984), Austin and Dangerfield (1993).

28(25). Metasomal tergum 2 with median tergite strongly sculptured and broad, rectangular to subrectangular and covering most or all of dorsal width of tergum; no narrow median raised area on this tergite (Figs 54, 59, 60) 29

- Metasomal tergum 2 with median tergite either narrower than tergum, not rectangular or subrectangular or tergum 2 with a raised median area...31

29(28). Flagellomeres with placodes arranged irregularly, not in two ranks (Fig. 21). Dorsal surface of mesonotum, scutellum, propodeum, and anterior metasomal terga forming a nearly straight line in profile (Fig. 29) ....**_Hygroplitis_** Thomson (in part)

Distribution: Holarctic. Biology: mostly reared from stem-boring Pyraloidea. References: Nixon (1968, as *Microgaster*, in part), Mason (1981).

- Flagellomeres with placodes in two ranks on most flagellomeres (Fig. 20). Propodeum more steeply sloped, so that mesosoma appears more normally "humpbacked" than above..30

30(29). Metasomal tergum 2 rugulose to aciculo-rugose, longer than tergum 3, which is often also sculptured, at least anteriorly (Fig. 60). Wings: Fig. 7 ....**_Microgaster_** Latreille

Distribution: mostly Holarctic but also Southeast Asia; rare or absent elsewhere. Biology: usually solitary on microlepidoptera and very rarely on macrolepidoptera. References: Muesebeck (1922), Nixon (1968), Mason (1981), Austin and Dangerfield (1992).

- Metasomal tergum 2 rugulose to aciculo-rugose but shorter than tergum 3, which is never sculptured (Fig. 54). Wings: Fig. 1 ....**_Choeras_** Mason (in part)

Distribution: Cosmopolitan; common in temperate zones of Southern Hemisphere. Biology: solitary on microlepidoptera. References: Nixon (1965, as part of several species groups of *Apanteles, Hypomicrogaster*, and *Promicrogaster*), Mason (1981), Williams (1988), Austin and Dangerfield (1992).

31(28). Metasomal tergum 2 usually strongly transverse and without strongly raised median area (Fig. 54). Ovipositor sheaths 0.4-0.8 times as long as hind tibia, setose over most of length. Hypopygium long and medially desclerotized in expandable pleats (Fig. 96). Wings: Fig. 1 ....**_Choeras_** Mason (in part)

See footnote under couplet 30.

- Metasomal tergum 2 medially with a slender, raised, usually longitudinal area that may be differently sculptured than lateral areas (Figs 73, 80). Ovipositor sheaths short, barely exerted, with a few very short setae concentrated apically. Hypopygium short and evenly sclerotized to median fold (Fig. 85) 32

32(31). Body small, usually less than 2.5 mm. Mesosoma dorso-ventrally slightly flattened. Legs and antennae of females (not males) short and stout, with only a single rank of placodes on the female flagellomeres. Metapleuron smooth ....**_Venanus_** Mason (in part)

Distribution: New World, mainly Andean. Biology: solitary on microlepidoptera, especially leaf-miners and needle-miners. Reference: Mason (1981).

- Body larger, usually greater than 3.0 mm. Mesosoma of normal proportions. Legs and antennae of both sexes of normal length and with two ranks of placodes on most flagellomeres (Fig. 20). Metapleuron rugose to punctate ....**_Rasivalva_** Mason

Distribution: Cosmopolitan except perhaps Australia. Biology: solitary on macrolepidoptera, especially Geometridae, Noctuidae, and Arctiidae, Reference: Mason (1981).

33(4). Propodeum with a well-developed areola defined by lateral carinae, at least posteriorly (may be variable in shape or definition anteriorly) (Figs 30-33, 35, 36, 48) 34

- Propodeum without a distinct areola; often with a median longitudinal carina (Figs 38-44) or with poorly defined pattern of carinae (Figs 34, 46) or, in some species, essentially unsculptured (Fig. 45)................44

34(33). Propodeum with a median longitudinal carina bisecting the areola, the areola being small and sometimes more weakly defined (Fig. 36). Metasomal tergum 2 much broader than long and usually relatively smooth (Fig. 53). Wings: Fig. 5 ....***Hypomicrogaster*** Ashmead (in part)

Distribution: New World, especially Neotropics. Solitary on microlepidoptera, including some leaf-miners. References: Muesebeck (1958, as *Microgaster*, in part), Nixon (1965).

- Propodeum without median longitudinal carina, or with a median carina extending only over anterior 0.25 to 0.5 of length, anterior to the areola 35

35(34). Petiole relatively broad, weakly sculptured, nearly always with a postero-median groove that is bounded anteriorly and antero-laterally by a pair of diverging ridges (Fig. 57). Gena with whitish to pale yellow-brown spot. Hypopygium evenly sclerotized to median fold (Fig. 90)...........................***Alphomelon*** Mason

Distribution: New World, mostly Neotropical. Biology: poorly known, 2 species have been reared from Hesperiidae. References: Muesebeck (1920, as *Apanteles*, in part), Mason (1981).

- Petiole variable in shape and sculpture but never with the above pattern of ridges and groove. Gena usually same color as rest of head in New World species. Hypopygium variable but often desclerotized into expandable pleats (Fig. 96).. 36

36(35). Metasomal terga 1 and 2 (and to a lesser extent, 3) broad, covering entire width of tergum and with a pebble-grained, coriaceous surface (Fig. 86) ....***Xanthapanteles*** Whitfield (in part)

Distribution: Argentina; rare; one described species. Biology: unknown. Reference: Whitfield (1995a).

- Metasomal terga 1-3 variable in shape but never with pebble-grained, coriaceous surface: usually smooth, punctate, aciculo-rugose or rugulose 37

37(36). Posterior courses of notauli (usually not clearly impressed) with somewhat denser, more longitudinally elongate, and confluent punctation (Fig. 24). Vannal lobe of hind wing not uniformly fringed with setae, slightly to strongly flattened or slightly concave in outline subapically and either setae absent or more sparsely setose over this flattened or concave portion (Figs 92, 94). Hypopygium nearly always medially desclerotized into expandable pleats (Fig. 96)..***Apanteles*** Förster (in part)

Distribution: Cosmopolitan; common and extremely diverse (probably nearly 1000 species in New World). Biology: almost always solitary on microlepidoptera of all kinds, very rarely gregarious on macrolepidoptera, References: Muesebeck (1920, 1958), Nixon (1965), Mason (1981), many other works for Old World species. *Cecidobracon* Kieffer and Jorgensen may belong here, but its identity is uncertain since the type could not be located.

- Mesoscutum with even, distinct punctation even along posterior courses of notauli, or punctation indistinct posteriorly. Vannal lobe of hind wing evenly convex to slightly flattened subapically and evenly fringed with setae (Fig. 93). Hypopygium variable 38

38(37). Metasomal terga 1 and 2 with median tergites broad, rugose to aciculo-rugose, covering entire dorsal width of these terga; tergum 3 often at least partially sculptured anteriorly (Figs 47, 48).......................39

- Metasomal terga 1 and 2, or at least tergum 1 posteriorly, with median tergites slightly to much narrower than dorsal surface of tergum; sculpturing variable but tergum 3 usually not sculptured 40

39(38). Hypopygium sharply folded medially but evenly sclerotized (Figs 89, 90), sometimes slightly more translucent and flexible postero-medially. Body length usually less than 2.2 mm. Wings: Fig. 11 ....***Pholetesor*** Mason (in part)

Distribution: Mostly Holarctic but a few species in mountains of Neotropics and a few introduced elsewhere for biological control. Biology: solitary, almost always on lepidopteran leaf-miners but a few species on needle-miners and leaf skeletonizers. References: Muesebeck (1920, as Apanteles, in part), Mason (1981), Whitfield (in review).

- Hypopygium long -and medially desclerotized into expandable pleats (Fig. 96). Body length variable but usually greater than 2.3 mm. Propodeum: Fig. 35 ....***Exoryza*** Mason (in part)

Distribution: Holarctic and Oriental. Biology: reared from Pyraloidea in grass stems, including rice. Reference: Mason (1981).

40(38). Hypopygium sharply folded medially and evenly sclerotized, though sometimes slightly more translucent and flexible postero-medially, never with expandable pleats (Figs 89, 90). Body length often less than 2.2 mm 41

- Hypopygium long and medially desclerotized into expandable pleats (Fig. 96). Body length variable but usually greater than 2.3 mm 42

41(40). Ovipositor sheaths short, barely exerted, with setae only over apical third (Fig. 83). Petiole usually with slight median depression subapically...***Parapanteles*** Ashmead

Distribution: New World (especially Neotropical) and Australasia. Biology: gregarious on Noctuidae and Notodontidae, for the few species known. Reference: Mason (1981).

- Ovipositor sheaths usually at least as long as hind basitarsus, and setose over rnost of length (Figs 89, 90). Petiole usually strongly rugose and without any defined sub- apical depression. Wings: Fig. 11.......... ....***Pholetesor*** Mason (in part)

Distribution: mostly Holarctic but a few species in mountains of Neotropics and a few introduced elsewhere for biological control. Biology: solitary, almost always on lepidopteran leaf-miners but a few species on needle-miners and leaf skeletonizers. References: Muesebeck (1920, as *Apanteles*, in part), Mason (J981), Whitfield (in review).

42(40). Vannal lobe of hind wing slightly to strongly flattened to concave subapically, with fringe of setae sparser or absent over the flattened or concave part (Figs 92, 94). Petiole usually slightly to strongly narrowing posteriorly (Figs 50, 55, 56) ....***Apanteles*** Förster (in part)

Distribution: Cosmopolitan, common and extremely diverse (probably nearly 1000 species in New World). Biology: almost always solitary on microlepidoptera of all kinds, very rarely gregarious on macrolepidoptera. References: Muesebeck (1920, 1958), Nixon (1965), Mason (1981), many other works for Old World species. *Cecidobracon* Kieffer and Jorgensen may belong here, but its identity is uncertain since the type could not be located.

- Vannal lobe of hind wing evenly convex and uniformly fringed with setae (Fig. 93). Petiole usually barrel-shaped or nearly parallel-sided and broader posteriorly to subequal in width anteriorly and posteriorly (males may be somewhat narrowed posteriorly)..43

43(42). Metasomal tergum 2 more or less rectangular and subequal in length to tergum 3, and with a straight posterior margin separating tergum 2 from tergum 3; tergum 3 usually also sculptured, at least anteriorly (Fig. 47). Propodeum usually strongly rugose with areola represented by a median groove which may have indistinct borders...***Exoryza*** Mason (in part)

Distribution: Holarctic and Oriental. Biology: reared from Pyraloidea in grass stems, including rice. Reference: Mason (1981).

- Metasomal tergum 2 usually not rectangular or subrectangular, usually slightly longer medially than laterally and shorter than tergum 3; tergum 3 usually unsculptured (Fig. 51). Propodeum variously sculptured but areola usually defined postero-laterally by distinct carinae (Fig. 32) but areola sometimes reduced (Fig. 45 ....***Dolichogenidea*** Viereck (in part)

Distribution: Cosmopolitan; large genus. Biology: almost always solitary on microlepidoptera, very rarely gregarious on macrolepidoptera. References: Muesebeck (1920, as *Apanteles*, in part), Nixon (1965, as *Apanteles*, in part), Mason (1981), many other sources for Old World species.

44(33). Petiole with strong, usually narrow, antero-median longitudinal groove (Figs 44, 59, 61, 65, 68, 72, 79).45

- Petiole usually with broad antero-median depression but without strong antero-median longitudinal groove (there may be a weak postero-median depression)...50

45(44). Metasomal tergum 3 slightly broader than tergum 2 (and often broader anteriorly than posteriorly) so that lateral margin between terga 2 and 3 appears slightly indented in dorsal view. Ovipositor sheaths very short and weakly to strongly decurved. Propodeum mostly weakly sculptured to polished except for strong median longitudinal carina (Fig. 44). Wings: Fig. 2... ....***Deuterixys*** Mason (in part)

Distribution: Cosmopolitan, except perhaps Africa. Biology: solitary on Bucculatrigidae and perhaps also Lyonetiidae. Reference: Whitfield (1985).

- Metasomal tergum 3 either same width as tergum 2 (at least anteriorly), slightly narrower, or not clearly delimited laterally. Ovipositor sheaths variable, often short but rarely strongly decurved if short. Propodeum variable in background sculpturing..46

46(45). Propodeum with distinct transverse carina in about anterior third in addition to median longitudinal carina (Fig. 42). Ovipositor sheaths long, usually about 0.6-0.7 times length of hind tibia. Note: a few intermediates between this genus and *Pseudapanteles* are known (compare to couplet 44)...................... ....***Xenogaster*** Mason

Distribution: South and Central America and South Africa. Biology: unknown. References: Nixon (1965, as *insolens*-group of *Apanteles*), Mason (1981).

- Propodeum without distinct transverse carina (at least not well-defined relative to background sculpturing). Ovipositor sheaths variable 47

47(46). Glossa strongly bilobed apically and usually somewhat elongate (Fig. 18). Ovipositor sheaths usually at least two-thirds as long as hind tibia. Petiole nearly parallel-sided to strongly narrowing posteriorly. Metasomal tergum 2 slightly shorter than tergum 3 and with median tergite usually subtriangular to strongly transverse (Fig. 68) .... ***Pseudapanteles*** Ashmead

Distribution: New World, especially Neotropics. Biology: solitary on leaf-mining and stem-mining microlepidoptera, especially Gelechiidae. References: Muesebeck (1920, as Apanteles, in part), Nixon (1965, as *sesiae*- and *annulicornis*- groups of *Apanteles*), Mason (1981).

\- Glossa not or only weakly bilobed apically (Fig. 19). Ovipositor sheaths, petiole, and metasomal tergum 2 not usually matching above description (except some *Iconella* see couplet 49)......................................48

48(47). Metasomal terga 1 and 2 strongly sculptured (Figs 58, 59). Hypopygium long, and medially desclerotized into expandable pleats (Fig. 96). Ovipositor sheaths usually 0.65-1.0 times as long as hind tibia. Background sculpturing of propodeum (excluding median longitudinal carina) rugose. Wings: Fig. 14........ .... ***Rhygoplitis*** Mason (in part)

Distribution: New World, especially southeastern United States, Mexico, Caribbean, Central and northern South America. Biology: solitary on Pyraloidea and microlepidoptera, References: Muesebeck (1920, as *Apanteles*, in part), Nixon (1968, as *Microgaster*, in part), Mason (1981).

\- Metasomal terga 1 and 2 usually polished or with little sculpturing. Hypopygium variable. Ovipositor sheaths variable but often not as long as above. Background sculpturing of propodeum variable but sometimes not sculptured significantly 49

49(48). Petiole slightly to strongly narrowing posteriorly. Metasomal tergum 2 shorter than tergum 3, usually strongly transverse to subtriangular and broader posteriorly (Fig. 62). Hind wing cu-a sinuate (as in Fig. 17). Ovipositor sheaths at least 0.66 times as long as hind tibia. Hypopygium medially desclerotized into expandable pleats (Fig. 96). Body usually mostly black .... ***Iconella*** Mason (in part)

Distribution: more or less cosmopolitan but rare in Neotropics. Biology: solitary on microlepidoptera, especially concealed feeders. References: Muesebeck (1920, as Apanteles, in part), Nixon (1965, as *sundanus*-group and part of *merula*-group of *Apanteles*), Mason (1981).

\- Petiole narrowly elongate, with sides broadly but weakly emarginate (Fig. 79). Metasomal tergum 2 as long as tergum 3 and usually with poorly delimited, slightly raised median longitudinal region. Hind wing cu-a straight to slightly curved back towards wing base posteriorly, not sinuate (Fig. 16). Ovipositor sheaths shorter than above. Hypopygium evenly sclerotized to median fold (Fig. 83, 84). Body usually mostly yellow .... ***Wilkinsonellus*** Mason

Distribution: Pantropical, Biology: unknown. References: Nixon (1965, as *daira*- and *henicopus*-groups of *Apanteles*), Mason (1981).

50(44). Glossa long and bilobed apically (Fig. 18). Ovipositor long (often as long as or longer than hind tibia) 51

\- Glossa not strongly bilobed apically (Fig. 19); or if rarely so then ovipositor very short...........................52

51(50). Metasomal tergum 2 longer than broad (Fig. 91). Apex of ovipositor not sinuate. Body usually xanthic...... .... ***Sendaphne*** Nixon (in part)

Distribution: Neotropical. Biology: unknown. References: Nixon (1965), Mason (1981).

\- Metasomal tergum 2 shorter than broad (Figs 52, 56). Apex of ovipositor sinuate in profile. Body usually mostly blackish. Wings: Fig. 13 .... ***Promicrogaster*** Brues and Richardson (in part)

Distribution: New World, especially Neotropics. Not common; many undescribed species. Biology: poorly known, probably mostly solitary on microlepidoptera; at least one species reared from bracket fungi. References: Nixon (1965; as *Promicrogaster*, in part), Mason (1981).

52(50). Hypopygium moderately long to long and usually medially desclerotized into expandable pleats (Fig. 96) (a few groups are somewhat difficult to classify using this character and they are keyed both ways). Ovipositor sheaths at least half as long as hind tibia, often longer 53

\- Hypopygium usually relatively short, occasionally long but evenly sclerotized to median fold, although sometimes slightly more translucent and weakened apico-medially or with narrow split antero-medially (Figs 83-85, 89, 90). Ovipositor sheaths variable but often very short and usually attached near base of second valvifer 61

53(52). Mesoscutum with notauli impressed, the notauli usually broad and punctate to rugulose (Figs 22, 23). Propodeum, petiole and metasomal tergum 2 strongly sculptured (Figs 58, 59); propodeum also with median longitudinal carina (as in Fig. 38). Wings: Fig. 14 .... ***Rhygoplitis*** Mason (in part)

Distribution: New World, especially southeastern United States, Mexico, Caribbean, Central and northern South America. Biology: solitary on Pyraloidea and microlepidoptera. References: Muesebeck (1920, as *Apanteles*, in part), Nixon (1968, as *Microgaster*, in part), Mason (1981).

\- Mesoscutum either without discernible notauli or only with slightly different sculpturing marking their posterior ends. Propodeum and metasomal terga variable but not usually as above, typically either without median carina on propodeum or without strong sculpturing on tergites or propodeum 54

54(53). Propodeum with median longitudinal carina (Figs 37, 38, 40, 43, 46) (a few exceptions key both ways) .55

- Propodeum without median longitudinal carina, often with weak pair of diverging or subparallel short carinae present posteriorly as stubs marking the postero-lateral margins of an indistinct areola or median depression (Figs 34, 45) 56

55(54). Hind wing cu-a usually sinuate posteriorly (as in Fig. 17), rarely straight; vannal lobe of hind wing subapically flattened to concave and usually more sparsely setose over flattened or concave portion (Figs 92, 94). Scutellum laterally with triangular polished areas ("lunules") (Fig. 40) ....
....***Iconella*** Mason (in part)

Distribution: more or Jess cosmopolitan but rare in Neotropics. Biology: solitary on microlepidoptera, especially concealed feeders. References: Muesebeck (1920, as *Apanteles*, in part), Nixon (1965, as sundan.us-group and part of the *merula*-group of *Apanteles*), Mason (1981).

- Hind wing cu-a outwardly convex, angling back towards wing base posteriorly; vannal lobe of hind wing evenly convex to slightly flattened and typically evenly fringed with setae (Fig. 93). Scutellum laterally with semicircular polished areas ("lunules") (Fig. 38)...***Choeras*** Mason (in part)

Distribution: Cosmopolitan; common in temperate zones of Southern Hemisphere. Biology: solitary on microlepidoptera. References: Nixon (1965, as part of several species groups of Apanteles, *Hypomicrogaster*, and *Promicrogaster*, Mason (1981), Williams (1988), Austin and Dangerfield (1992).

56(54). Medio-apical borders of metasomal terga 4-6 (sometimes others) desclerotized, often appearing "puckered" in dried specimens. Fore wing r slanting strongly outwards towards wing tip (as in Fig. 14); R1 short, usually not extending more than two-thirds distance from distal end of stigma to end of 3RS fold along wing margin. Metasomal terga usually more densely setose than in most microgastrines (Fig. 55) ....***Illidops*** Mason

Distribution: mostly Holarctic but possibly also from arid regions of Neotropics. Biology: reared from Psychidae, Pyralidae, Scythrididae. References: Mason (1981), Wharton (1983).

- Medio-apical borders of metasomal terga 4-6 not desclerotized. Fore wing r variable but rarely slanting so strongly outwards; R1 variable but usually extending at least two-thirds distance to 3RS fold. Metasomal terga not unusually setose...57

57(56). Propodeum posteriorly with either a single median stub of a carina, a median longitudinal band of rugosity or no discernible short carinae 58

- Propodeum with weak pair of diverging or subparallel short carinae present posteriorly as stubs marking the postero-lateral margins of an indistinct areola or median depression; if carinae not clear, then at least with median depression (Figs 34, 35; extremely reduced form: Fig. 45) 59

58(57). Hind wing cu-a usually sinuate posteriorly (as in Fig. 17), rarely straight; vannal lobe subapically flattened to concave and usually sparsely setose over flattened or concave portion (Figs 92, 94). Scutellum laterally with triangular polished areas ("lunules") (Fig. 40)...................................***Iconella*** Mason (in part)

Distribution: more or less cosmopolitan but rare in Neotropics. Biology: solitary on microlepidoptera, especially concealed feeders. References: Muesebeck (1920, as *Apanteles*, in part), Nixon (1965, as *sundanus*-group and part of *merula*-group of *Apanteles*), Mason (1981).

- Hind wing cu-a outwardly convex, angling back towards wing base posteriorly; vannal lobe evenly convex to slightly flattened and typically evenly fringed with setae (Fig. 93). Scutellum laterally with semicircular polished areas ("lunules") (Fig. 38)..***Choeras*** Mason (in part)

Distribution: Cosmopolitan; common in temperate zones of Southern Hemisphere. Biology: solitary on microlepidoptera. References: Nixon (1965, as part of several species groups of *Apanteles*, *Hypomicrogaster*, and *Promicrogaster*, Mason (1981).

59(57). Vannal lobe of hind wing slightly to strongly flattened to concave subapically, fringe of setae reduced or absent over the flattened or concave part (Figs 92, 94). Petiole usually slightly to strongly narrowing posteriorly (Figs 50, 52). Posterior courses of notauli indistinct but often marked by slightly more longitudinally confluent punctation (Fig. 24). Extremely common......................***Apanteles*** Förster (in part)

Distribution: Cosmopolitan; common and extremely diverse (probably nearly 1000 species in New World). Biology: almost always solitary on microlepidoptera of all kinds, very rarely gregarious on macrolepidoptera. References: Muesebeck (1920, 1958), Nixon (1965), Mason (1981), many other works for Old World species. *Cecidobracon* Kieffer and Jorgensen may belong here, but its identity is uncertain since the type could not be located.

- Vannal lobe of hind wing evenly convex and uniformly fringed with setae (Fig. 93). Petiole usually barrel-shaped or subparallel-sided and either broader posteriorly or subequal in width anteriorly and posteriorly (males may be somewhat narrowed posteriorly). Punctations of posterior portion of mesoscutum, if present, usually uniformly distinct and round 60

60(59). Metasomal tergum 2 more or less rectangular and subequal in length with tergum 3, and with a straight posterior margin separating tergum 2 from tergum 3; tergum 3 usually also sculptured, at least anteriorly (Fig. 47). Propodeum usually strongly rugose with areola represented by a median groove which may have indistinct borders (Fig. 35). Rare .... ***Exoryza*** Mason (in part)

Distribution: Holarctic and Oriental. Biology: reared from Pyraloidea in grass stems, including rice. Reference: Mason (1981).

- Metasomal tergum 2 usually not rectangular or subrectangular, usually slightly longer medially than laterally and shorter than tergum 3; tergum 3 usually unsculptured (Fig. 51). Propodeum variously sculptured (Figs 32, 45) but areola usually defined postero-laterally by distinct carinae. Common............. .... ***Dolichogenidea*** Viereck (in part)

Distribution: Cosmopolitan, large genus. Biology: almost always solitary on microlepidoptera, very rarely gregarious on macrolepidoptera. References: Muesebeck (1920, as *Apanteles*, in part), Nixon (1965, as *Apanteles*, in part), Mason (1981), many other sources for Old World species.

61(52). Metasomal tergum 2 usually slightly broader anteriorly than posteriorly and always narrower than tergite 3 so that lateral margin between tergum 2 and tergum 3 appears slightly indented in dorsal view (Fig. 44). Ovipositor sheaths very short and weakly to strongly decurved. Propodeum mostly weakly sculptured to polished except for strong median longitudinal carina (Fig. 44). Wings: Fig. 2 .... ***Deuterixys*** Mason (in part)

Distribution: Cosmopolitan, except perhaps Africa. Biology: solitary on Bucculatrigidae and perhaps also Lyonetiidae. Reference: Whitfield (1985).

- Metasomal tergum 2 weakly to strongly broadening posteriorly to meet tergite 3 laterally. Ovipositor sheaths variable but rarely both strongly decurved and short. Propodeum rarely with combination of polished background and strong median carina 62

62(61). Ovipositor sheaths long (at least two-thirds as long as hind tibia), setose (sometimes sparsely so) over most of length. Male genital capsule often slightly to greatly enlarged (males for which this is not true can be difficult to distinguish from those of *Glyptapanteles* without associated females). Mostly in western mountains and northern localities of Nearctic region; relatively rare in Neotropics, and there only at high elevations 63

- Ovipositor sheaths short, weakly exserted from hypopygium in most specimens and usually less than half as long as hind tibia (Figs 83-85, 89), attached below mid-height of second valvifers and with setae concentrated apically. Male genitalia of normal size. Several large and commonly encountered genera key here 64

63(62). Ovipositor sheaths arising below mid-height of second valvifers (this character may be difficult to assess in air-dried specimens without relaxing the specimen or dissecting it) ....***Lathrapanteles*** Williams

Distribution: New World, especially Nearctic. Biology: gregarious on stem-borers, especially Noctuidae, in *Typha* and perhaps other rushes and reeds. Reference: Williams (1985).

- Ovipositor sheaths arising at or above mid-height of second valvifers (often in upper third) and often strongly sickle-shaped .... ***Sathon*** Mason

Distribution: Holarctic, Neotropical, Australasian (the Australian species only provisionally placed in *Sathon*). Biology: solitary or gregarious on various micro- and macrolepidoptera. References: Williams (1988); for unusual Australasian species associated with this genus see Austin and Dangerfield (1992).

64(62). Petiole narrower posteriorly than anteriorly, usually evenly narrowing ·posteriorly. Median part of tergum 2, when fully defined, subtriangular and wider posteriorly, usually unsculptured (Figs 74, 75, 77) (a few intermediate species are keyed both ways) 65

- Petiole broader posteriorly, barrel-shaped, or parallel-sided. Tergum 2 usually not as above, but if trapezoidal or even subtriangular, then petiole is mostly to entirely sculptured...................................... 68

65(64). Vannal lobe of hind wing subapically flattened or even concave, setae usually sparse or absent over flattened part of margin (Figs 92, 94). Apical setae of ovipositor sheath sparse and reduced in size, less than half as long as apical width of sheaths (Fig. 85) 66 Vannal lobe of hind wing evenly convex, or if slightly flattened, then almost always with even fringe of setae (Fig. 93). Apical setae of ovipositor sheath sparse and reduced in size, less than half as long as apical width of sheaths (Fig. 85) 66

- Vannal lobe of hind wing evenly convex, or if slightly flattened, then almost always with even fringe of setae (Fig. 93). Apical setae of ovipositor sheats longer, at least some of them about as long as apical width of sheaths and often relatively numerous (in *Pholetesor*, sheaths are setose over most of length) (Figs 83, 84, 89, 90) 67

66(65). Metasomal tergum 2 with median area defined by grooves diverging at an angle less than 90-100°; apical margin of tergum 2 often poorly defined (Fig. 77). Antennae of female short and thick, without false divisions (i.e. without 2 ranks of placodes) .... ***Venanides*** Mason

Distribution: Pantropical. Biology: gregarious, especially on Pyraloidea and Tortricoidea. References: Nixon (1965, as *congoensis*-group of Apanteles), Mason (1981).

- Metasomal tergum 2 with median area defined by grooves diverging at an angle greater than 110-120° (Fig. 76); apical margin of tergum 2 often concave and median region sometimes strongly raised. Antennae of female with 2 ranks of placodes over .... ***Distatrix*** Mason

Distribution: Pantropical, a few ranging into temperate zones. Biology: attack various macrolepidoptera; gregarious ones often make cocoons that mimic flower heads; solitary ones often make single stalked cocoons. References: Nixon (1965, as *formosus*-group of *Apanteles*), Mason (1981), Whitfield and Scaccia (1996).

67(65). Ovipositor sheaths usually clearly arising at about mid-height of second valvifers, slender proximally, with long distal portion that is setose over most of length (Figs 89, 90). Anterior margin of metanotum with pointed, sublateral setiferous projections (Fig. 27, 28, 34) .... ***Pholetesor*** Mason (in part)

Distribution: mostly Holarctic but a few species in mountains of Neotropics and a few introduced elsewhere for biological control. Biology: solitary, almost always on lepidopteran leaf-miners but a few species on needle-miners and leaf skeletonizers. References: Muesebeck (1920, as *Apanteles*, in part), Mason (1981), Whitfield (in review).

- Ovipositor sheaths arising near base (ventral end) of second valvifers, relatively broad proximally and with setae concentrated apically (Figs 83, 84). Anterior margin of metanotum with poorly defined (at most weakly protruding) sublateral setiferous projections (Fig. 26) .... ***Glyptapanteles*** Ashmead

Distribution: Cosmopolitan; common and diverse. Biology: solitary or gregarious on macrolepidoptera, including Pyraloidea. References: Muesebeck (1920, as Apan- teles, in part), Nixon (1965, as *vitripennis-*, *octonarius-*, *pallipes-*, *siderion-*, *demeter-*, *fraternus-*, and *triangulator*-groups of *Apanteles*), Mason (1981), Papp (1983, as the above Nixon groups plus *thompsoni-* and *liparidis*-groups).

68(64). Ovipositor sheaths usually clearly arising at about mid-height of second valvifers, slender proximally, with long distal portion that is setose over most of length (Figs 89, 90). Propodeum sometimes weakly sculptured throughout, or alternatively medially more strongly sculptured (Fig. 34), but never with indication of median longitudinal carina .... ***Pholetesor*** Mason (in part)

Distribution: mostly Holarctic but a few species in mountains of Neotropics and a few introduced elsewhere for biological control. Biology: solitary, almost always on lepidopteran leaf-miners but a few species on needle-miners and leaf skeletonizers. References: Mason (1981), Muesebeck (1920, as *Apanteles*, in part), Mason (1981), Whitfield (in review).

- Ovipositor sheaths arising near base (ventral end) of second valvifers, relatively broad proximally and with setae concentrated apically (Figs 83, 84). Propodeum usually strongly rugose, sometimes with discernible median longitudinal carina (Fig. 46), occasionally weakly sculptured and without median carina (Fig: 45)... 69

69(68). Metasomal tergum 2 with median tergite narrower anteriorly than posterior end of petiole, and usually trapezoidal to subtriangular in shape (Fig. 71). Propodeum usually not strongly sculptured but if so, without median longitudinal carina. Apico-ventral margin of apical segment of fore tarsus usually excavated, with conspicuous curved seta over this excavation .... ***Protapanteles*** Ashmead

Distribution: Holarctic; rare elsewhere except at high elevations. Biology: attack macrolepidoptera, but often solitary. References: Muesebeck (1920, as *Apanteles*, in part), Nixon (1965, as *popularis*-group of *Apanteles*), Mason (1981).

- Metasomal tergum 2 usually broader anteriorly than posterior end of petiole, often subquadrate in shape (Fig. 70). Propodeum usually strongly rugose and often with visible median longitudinal carina. Apico-ventral margin of apical segment of fore tarsus entire .... ***Cotesia*** Cameron

Distribution: Cosmopolitan; common and diverse, especially in temperate zones. Biology: solitary or gregarious on a wide variety of macrolepidoptera. References: Muesebeck (1920, as *Apanteles*, in part), Nixon (1965, as *glomeratus*-group of *Apanteles*), Mason (1981), Papp (1986, 1987, as *glomeratus*-group of *Apanteles*).

REFERENCES

Achterberg, C. van 1982. Notes on some type species described by Fabricius of the subfamilies Braconinae, Rogadinae, Microgastrinae and Agathidinae (Hymenoptera: Braconidae). Entomologische Berichten 42: 133-139.

Austin, A. D. 1987. A review of the Braconidae (Hymenoptera) that parasitize Limacodidae in Southeast Asia, particularly those associated with coconut and oil palm. Pp. 139-184. In: Cock, M. J. W., Godfray, H. C. J., and Holloway, J. D. (Eds.). Slug and Nettle Caterpillars. CAB International, London.

Austin, A. D. 1990. Revision of the enigmatic Australasian. genus *Miropotes* Nixon (Hymenoptera: Braconidae: Microgastrinae), with comments on the phylogenetic importance of the female ovipositor system. Systematic Entomology 15: 43-68.

Austin, A. D. and Dangerfield, P. C. 1992. Synopsis of Australasian Microgastrinae (Hymenoptera: Braconidae), with a key to genera and description of new taxa. Invertebrate Taxonomy 6: 1-76.

Austin, A. D. and Dangerfield, P. C. 1993. Systematics of Australian and New Guinean *Microplitis* Foerster and Snellenius Westwood (Hymenoptera: Braconidae: Microgastrinae), with a review of their biology and host relationships. Invertebrate Taxonomy 7: 1097-1166.

Cardona, C. and Oatman, E. R. 1971. Biology of *Apanteles dignus* (Hymenoptera: Braconidae), a primary parasite of the tomato pinworm. Annals of the Entomological Society of America 64: 996-1007.

Clausen, C. P. 1978. Introduced parasites and predators of arthropod pests and weeds: a world review. Agriculture Handbook No. 480. United States Government Press, Washington, D. C. 545 pp.

Fallis, A. M. 1942. The life cycle of *Apanteles carpaius* (Say) (Hymenoptera: Braconidae), a parasite of the webbing clothes moth *Tineola bissseliella* Hum. Canadian Journal of Research, Section D (Zoological Science) 20: 13-19.

Fuchs, T. W., Huffman; F. R., and Smith, J. W., Jr. 1979. Introduction and establishment of *Apanteles flavipes* (Hym.: Braconidae) on *Diairaea saccharalis* (Lep.: Pyralidae) in Texas. Entomophaga 24: 109-114.

International Commission on Zoological Nomenclature. 1988. Opinion 1510. *Microgaster* Latreille, 1804 (Insecta, Hymenoptera): *Microgaster australis* Thomson, 1895 designated as the type species. Bulletin of Zoological Nomenclature 45: 239-240.

Laing, J. E. and Heraty, J.M. 1981. Establishment in Canada of the parasite *Apanteles pedias* Nixon on the spotted tentiform leafminer, *Phyllonorycter blancardella* (Fabr.). Environmental Entomology 10: 933-935.

Lewis, F. B. 1960. Factors affecting assessment of parasitization by *Apanteles fumiferanae* Vier. and *Glypta fumiferanae* (Vier.) on spruce budworm larvae. Canadian Entomologist 92: 881-891.

Marsh, P. M. 1979. Family Braconidae. Pp. 144-295. In: Krombein, K. V., Hurd, P. D., Smith, D. R., and Burks, B. D. (Eds.). Catalog of Hymenoptera in America North of Mexico. Smithsonian Institution Press, Washington, D.C.

Marsh, P. M., Shaw, S. R., and Wharton, R. A. 1987. An identifitation manual for the North American genera of the family Braconidae (Hymenoptera). Memoirs of the Entomological Society of Washington 13: 1-98.

Mason, W. R. M. 1981. The polyphyletic nature of *Apanteles* Foerster (Hymenoptera: Braconidae): a phylogeny and reclassification of Microgastrinae). Memoirs of the Entomological Society of Canada 115: 1-147.

Mason, W.R. M. 1986. *Microgaster* Latreille, 1804 (Insecta, Hymenoptera): proposed designation of *Microgaster australis* Thomson, 1804 as type species. Z.N. (S.) 1397. Bulletin of Zoological Nomenclature 43: 173-174.

Muesebeck, C. F. W. 1920. A revision of the North American species of the ichneumon-flies belonging to the genus *Apanteles*. Proceedings of the United States National Museum 58: 483-576.

Muesebeck, C. F. W. 1922. A revision of the North American ichneumon-flies belonging to the subfamilies Neoneurinae and Microgasterinae. Proceedings of the United States National Museum. 61: 1-76.

Muesebeck, C. F. W. 1958. New neotropical wasps of the family Braconidae (Hymenoptera) in the U.S. National Museum. Proceedings of the United States National Museum 107: 405-461.

Nixon, G. E. J. 1965. A reclassification of the tribe Microgasterini (Hymenoptera: Braconidae). Bulletin of the British Museum (Natural History) Entomology Supplement 2: 1-284.

Nixon, G. E. J. 1968. A revision of the genus *Microgaster* Latreille (Hymenoptera: Braconidae). Bulletin of the British Museum (Natural History) Entomology 22: 33-72.

Papp, J. 1979. Braconidae (Hymenoptera) from Tunisia, 1. Polia Entomologica Hungarica 32: 175-187.

Papp, J. 1980. On the genus *Fornicia* Brullé with two new Oriental species (Hymenoptera, Braconidae: Microgasterinae). Folia Entomologia Hungarica 33: 305-311.

Papp, J. 1983. A survey of the European species of *Apanteles* Foerster (Hymenoptera, Braconidae: Microgastrinae), VIL The *carbonarius-, circumscriptus-, fraternus-, pallipes-, parasitellae-, vitripennis-, liparidis-, octonarius- and thompsoni*-group. Annales historico-naturales Musei Nationalis Hungarici 75: 247-283.

Papp, J. 1984. Palaearctic species of *Microgaster* Latreille (= *Microplitis* Foerster) with description of seven new species (Hymenoptera, Braconidae, Microgastrinae). Entomologische Abhandlungen und Berichten des Staatlichen Museums für Tierkunde Dresden 47: 95-140.

Papp, J. 1986a. First survey of the *Glabromicroplitis* Papp species of the Holarctic Region, with taxonomical remarks of three *Microgaster* Latreille species (Hymenoptera, Braconidae: Microgastrinae). Annales historico-naturales Musei Nationalis Hungarici 78: 249-253.

Papp, J. 1986b. A survey of the European species of *Apanteles* Foerst. (Hymenoptera, Braconidae: Microgastrinae), IX. The *glomeratus*- group 1. Annales historico-naturales Musei Nationalis Hungarici 78: 225-247.

Papp, J. 1987. A survey of the European species of *Apanteles* Foerst. (Hymenoptera, Braconidae: Microgastrinae), VJI. The *glomeratus*- group 2 and the *cultellatus*-group. Annales historico-naturales Musei Nationalis Hungarici 79: 207-258.

Papp, J. 1993. New braconid wasps (Hymenoptera, Braconidae) in the Hungarian Natural History Museum 4. Annales historico-naturales Musei Nationalis Hungarici 85: 155-180.

Penteado-Diaz, A. M. 1997. First host record for the genus *Larissimus* (Hymenoptera: Braconidae). Entomological News 108: 66.

Ruberson, J. R. and Whitfield, J. B. 1996. Facultative egg-larval parasitism of the beet armyworm, *Spodoptera exigua* (Lepidoptera: Noctuidae) by *Cotesia marginiventris* (Hymenoptera: Braconidae). Florida Entomologist 79: 296-302.

Shaw, M. R. and Huddleston, T. 1991. Classification and biology of braconid wasps (Hymenoptera: Braconidae). Handbooks for the Identification of British Insects 7(11): 1-126.

Shenefelt, R. D. 1972. Pars S. Braconidae 4, Microgasterinae *Apanteles*. Pp. 429-668. In: Vecht, J. van der and Shenefelt, R D. (Eds.), Hymenopterorum Catalogus (nova editio). Dr. W. Junk, The Hague.

Shenefelt, R D. 1973. Pars 9. Braconidae 5, Microgasterinae and Ichneutinae. Pp. 669-805. In: Vecht, J. van der and Shenefelt, R. D. (Eds.). Hymenopterorum Catalogus (nova editio). Dr. W. Junk, The Hague.

Tobias, V. I. 1986. Subfamily Microgastrinae. Pp. 344-459. In: Tobias VJ. (Ed.). Identification of the Insects of European USSR. Volume III, Part IV, Hymenoptera, Braconidae. Akademia Nauk, Leningrad. (In Russian.)

Walker, A. K., Kitching, I. J. and Austin, A. D. 1990. A reassessment of the phylogenetic relationships within the Microgastrinae (Hymenoptera: Braconidae). Cladistics 6: 291-306.

Wharton, R. A. 1983. New species of Illidops and *Bracon* (Hymenoptera: Braconidae) of potential use in biological control. The Canadian Entomologist 115: 667-672.

Whitfield, J. B. 1985. The Nearctic species of *Deuterixys* Mason (Hymenoptera: Braconidae), Part-Pacific Entomologist 61: 60-67.

Whitfield, J. B. 1987. Comments on the proposed designation of *Microgaster australis* Thomson, 1895 as type species of *Microgaster* Latreille, 1804 (Insecta, Hymenoptera) (Case 2397). Bulletin of Zoological Nomenclature 44: 47.

Whitfield, J.B. 1995a. *Xanthapanteles*, a new genus of Microgastrinae (Hymenoptera: Braconidae) from South America. Proceedings of the Entomological Society of Washington 97: 879-883.

Whitfield, J. B. 1995b. Annotated checklist of the Microgastrinae of North America North of Mexico (Hymenoptera: Braconidae). Journal of the Kansas Entomological Society 68: 245-262.

Whitfield, J.B. (in review). A revision of the Nearctic species of the genus *Pholetesor* Mason (Hymenoptera: Braconidae).

Whitfield, J. B. and Scaccia, B. 1996. A new species of *Distatrix* (Hymenoptera: Braconidae) from California, with biological notes. Proceedings of the Entomological Society of Washington 98: 308-313.

Williams, D. J.M. 1985. The New World genus *Lathrapanteles* n. gen.: phylogeny and placement in the Microgastrinae (Hymenoptera: Braconidae: Cotesiini). Canadian Journal of Zoology 63: 1962-1981.

Williams, D. J. M. 1988. Classification, phylogeny and zoogeographic studies of species of Sathon Mason (Hymenoptera: Braconidae). Quaestiones Entomologicae 24: 529-639.

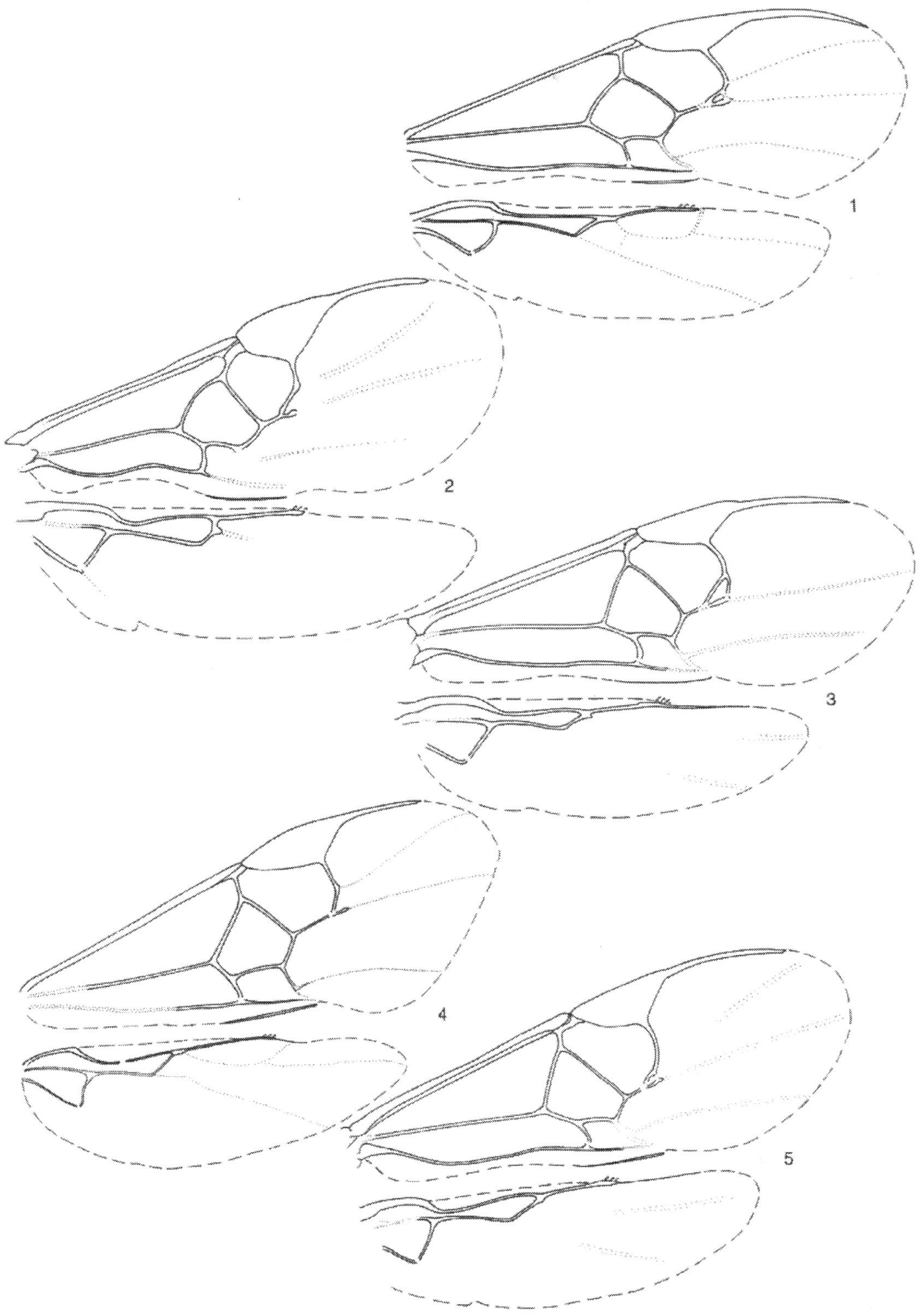

Figs 1-5: Wings of Microgastrinae. 1, *Choeras* sp.; 2, *Deuterixys* sp.; 3, *Diolcogaster* sp.; 4, *Fornicia* sp.; 5, *Hypomicrogaster* sp.

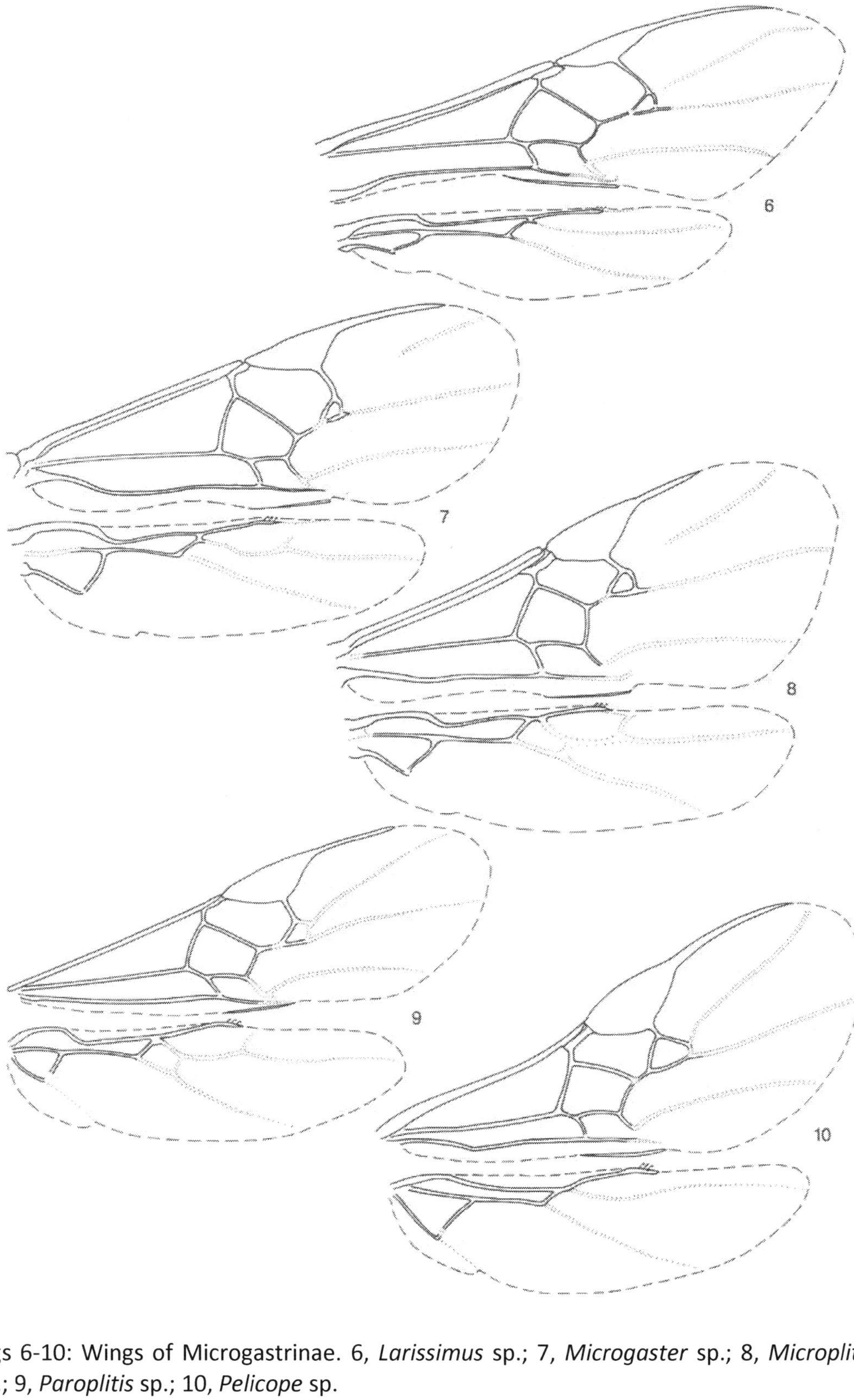

Figs 6-10: Wings of Microgastrinae. 6, *Larissimus* sp.; 7, *Microgaster* sp.; 8, *Microplitis* sp.; 9, *Paroplitis* sp.; 10, *Pelicope* sp.

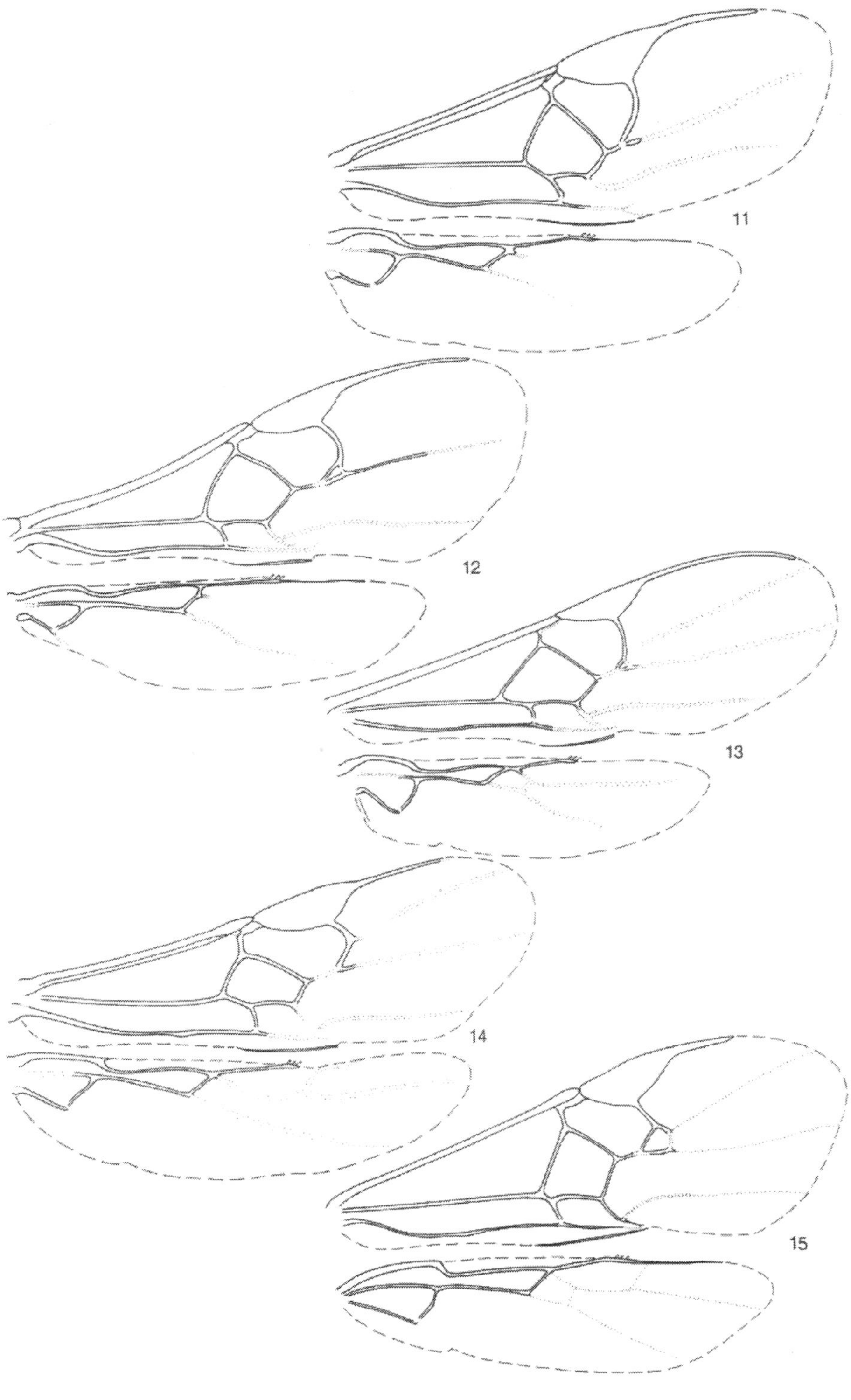

Figs 11-15: Wings of Microgastrinae. 11, *Pholetesor* sp.; 12, *Prasmodon* sp.; 13, *Promicrogaster* sp.; 14, *Rhygoplitis* sp.; 15, *Snellenius* sp.

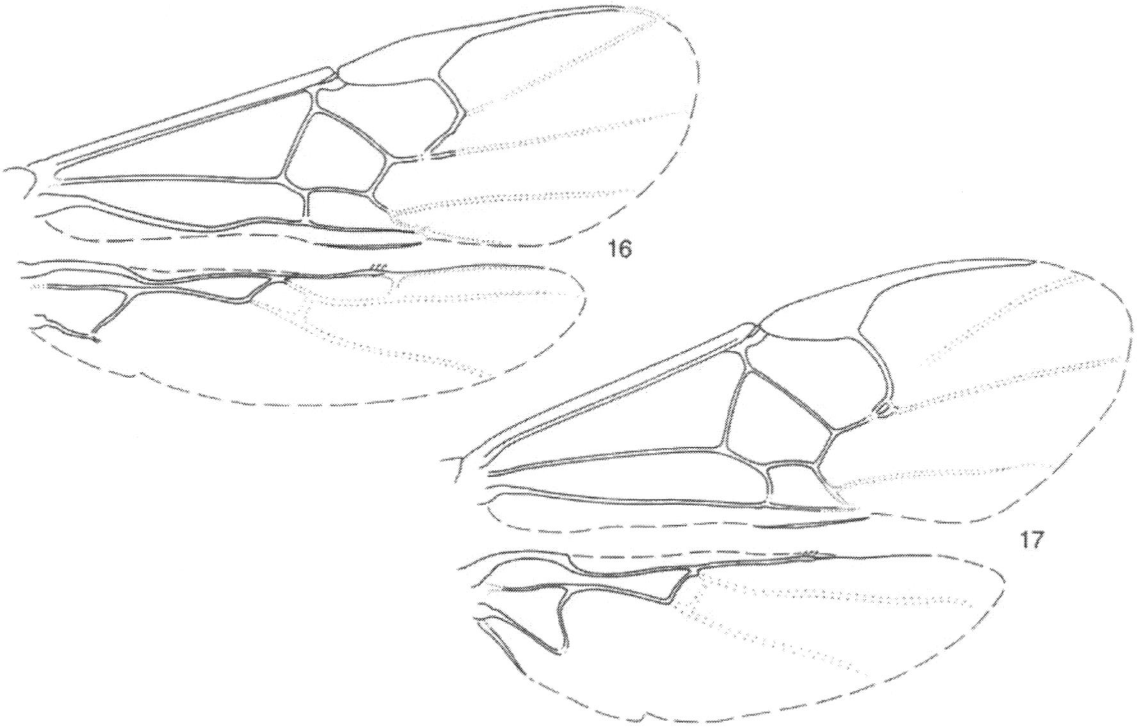

Figs 16-17: Wings of Microgastrinae. 16, *Wilkinsonellus* sp.; 17, *Xanthomicrogaster* sp.

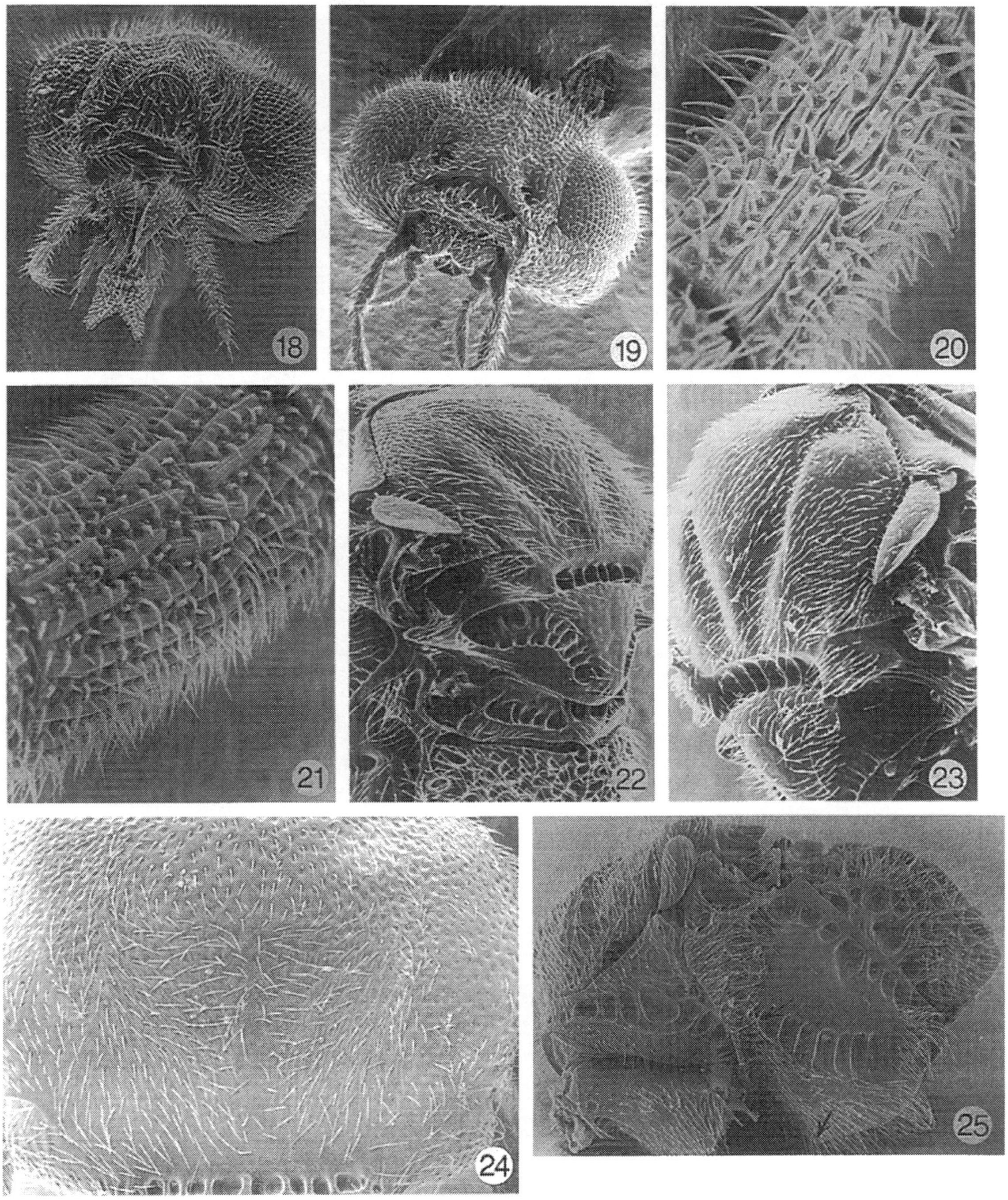

Figs 18-25: Microgastrinae. 18, head of *Pseudapanteles* sp., frontal view, showing elongate glossa; 19, head of *Apanteles* sp., frontal view, showing short glossa; 20, flagellomere of *Venanus pinicola* Mason; 21, flagellomere of *Protomicroplitis calliptera* (Say); 22-24, mesoscutum: 22, *Rhygoplitis aciculatus* (Ashmead), dorso-lateral view; 23, *Prasmodon* sp., dorso-lateral view; 24, *Apanteles* sp., dorsal view; 25, mesosoma of *Snellenius atratus* Shenefelt, lateral view, arrows = epicnemial carina. Figures 18, 20-23, and 25 after Mason (1981).

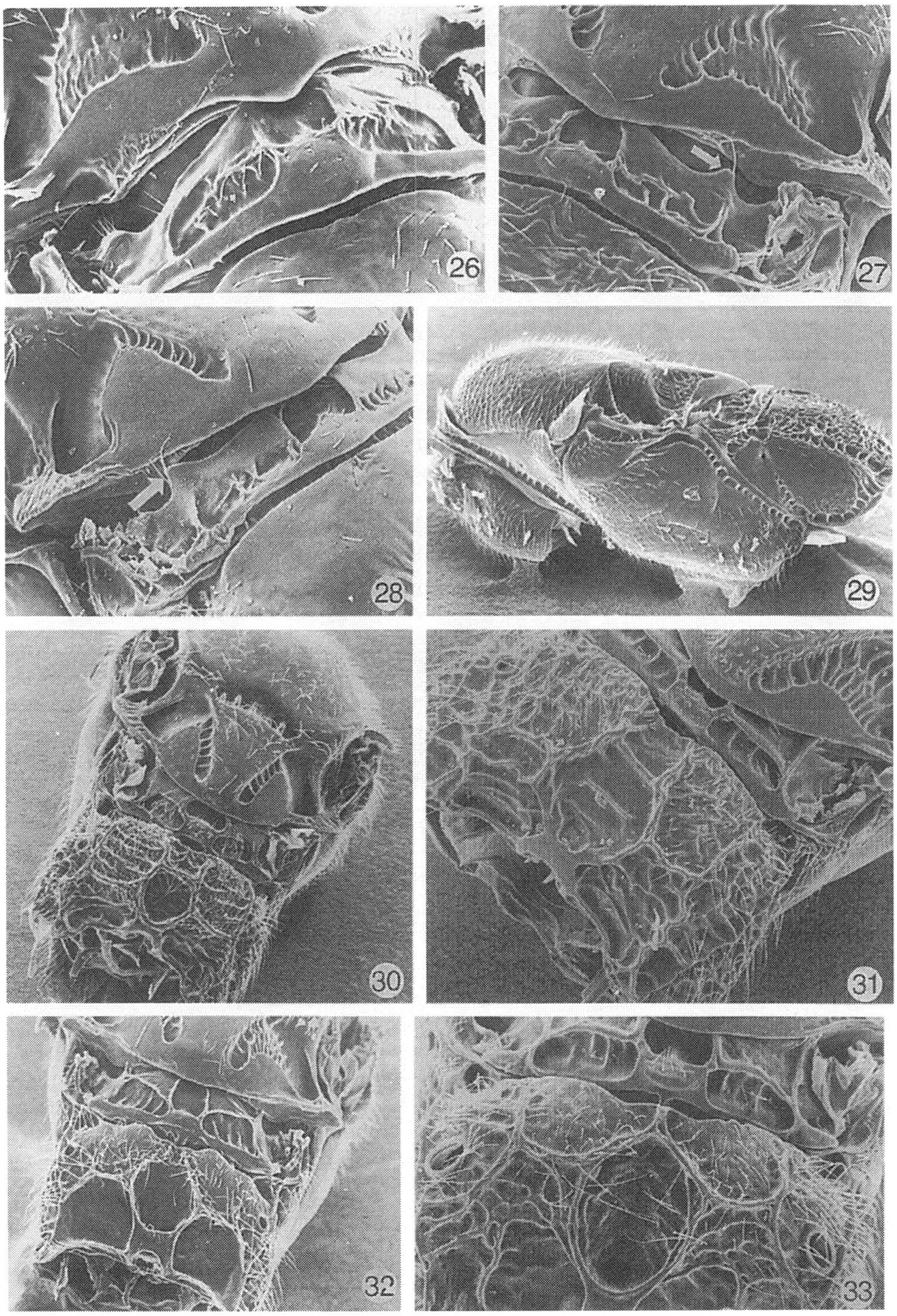

Figs 26-33: Microgastrinae. 26-28, metanotum: 26, *Distatrix* sp., dorsal view; 27, *Pholetesor ornigis* (Weed), dorsal view, arrow = sublateral setiferous projection; 28, *Promicrogaster* sp., dorsal view, arrow = sublateral setiferous projection; 29, mesosoma of *Venanus pinicola* Mason, lateral view; 30, propodeum, metanotum and scutellum of *Alphomelon nigriceps* (Ashmead); 31-33, propodeum: 31, *Apanteles* sp.; 32, *Dolichogenidea lacteicolor* (Viereck); 33, *Papanteles peckorum* Mason. All figures after Mason (1981).

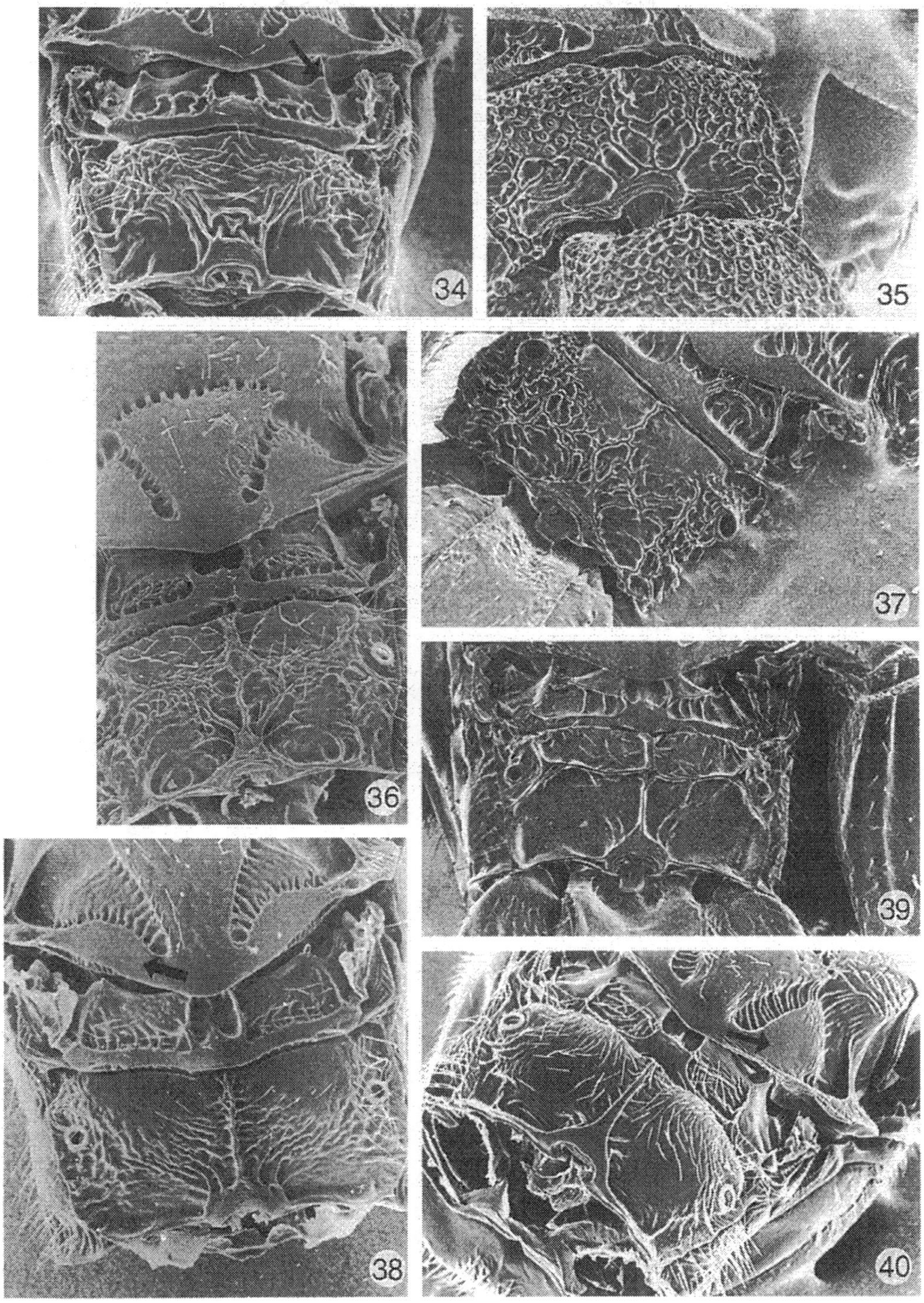

Figs 34-40: Microgastrinae. 34-37, propodeum: 34, *Pholetesor ornigis* (Weed), arrow = sublateral setiferous projection of metanotum; 35, *Exoryza minnesota* Mason; 36, *Hypomicrogaster zonaria* (Say); 37, *Paroplitis beringianus* Mason; 38, mesosoma of *Choeras consimilis* (Viereck), dorsal view, arrow = semicircular polished areas; 39, propodeum of *Clarkinella edithae* Mason; 40, mesosoma of *Iconella etiellae* (Viereck), dorsal view, arrow = triangular polished areas. All figures after Mason (1981).

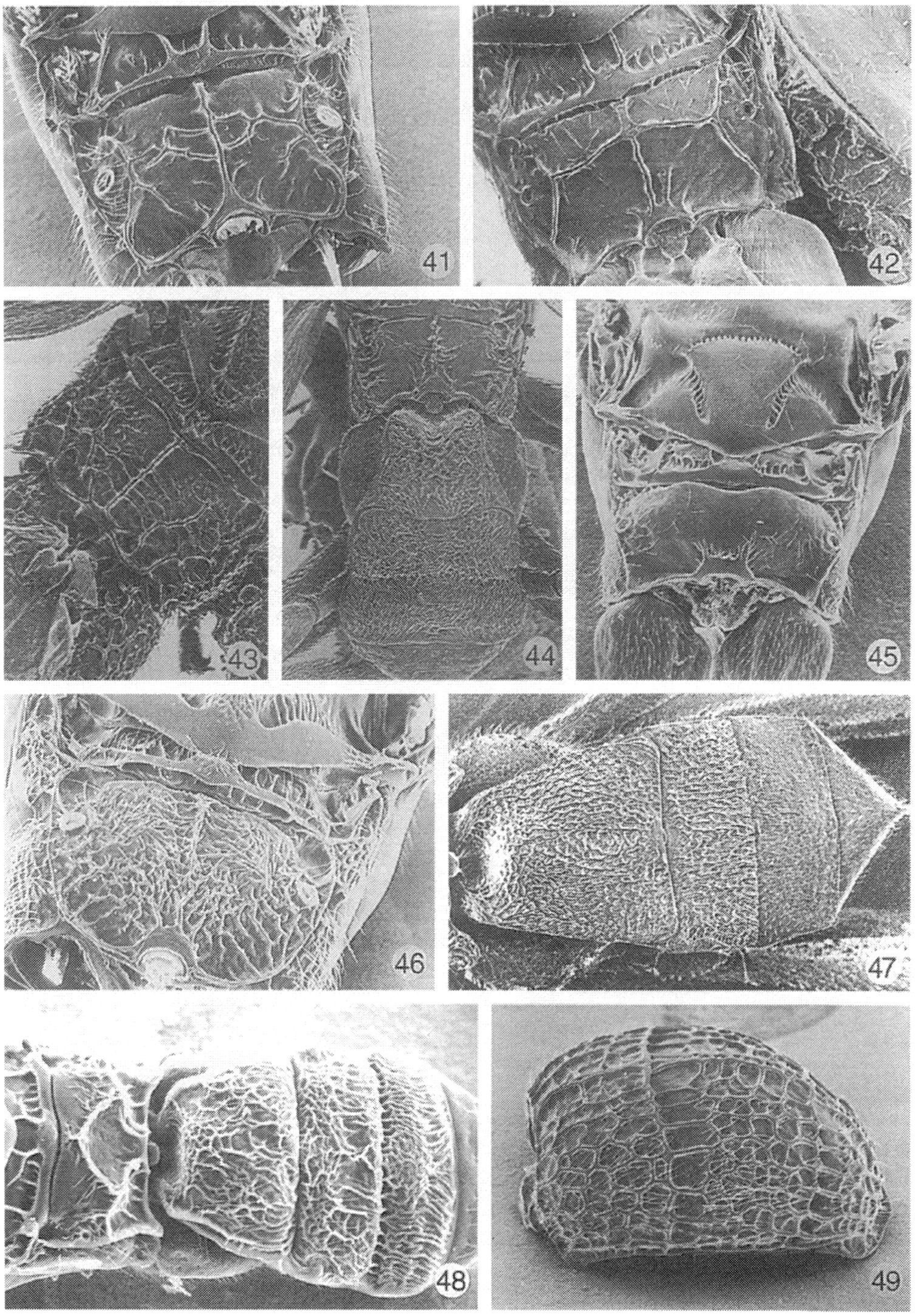

Figs 41-49: Microgastrinae. 41-43, propodeum: 41, *Prasmodon* sp.; 42, *Xenogaster insolens* (Wilkinson); 43, *Wilkinsonellus henicopus* (de Saeger); 44, propodeum and metasoma of *Deuterixys carbonarius* (Wesmael), dorsal view; 45, propodeum of Dolichogenidea sp.; 46, propodeum of *Rasivalva stigmatica* (Muesebeck); 47, anterior metasomal tergites of *Exoryza* sp., dorsal view; 48, propodeum and anterior metasomal tergites of *Pholetesor bucculatricis* (Muesebeck), dorsal view; 49, metasoma of *Fornicia jarmilae* Mason, lateral view. All figures after Mason (1981).

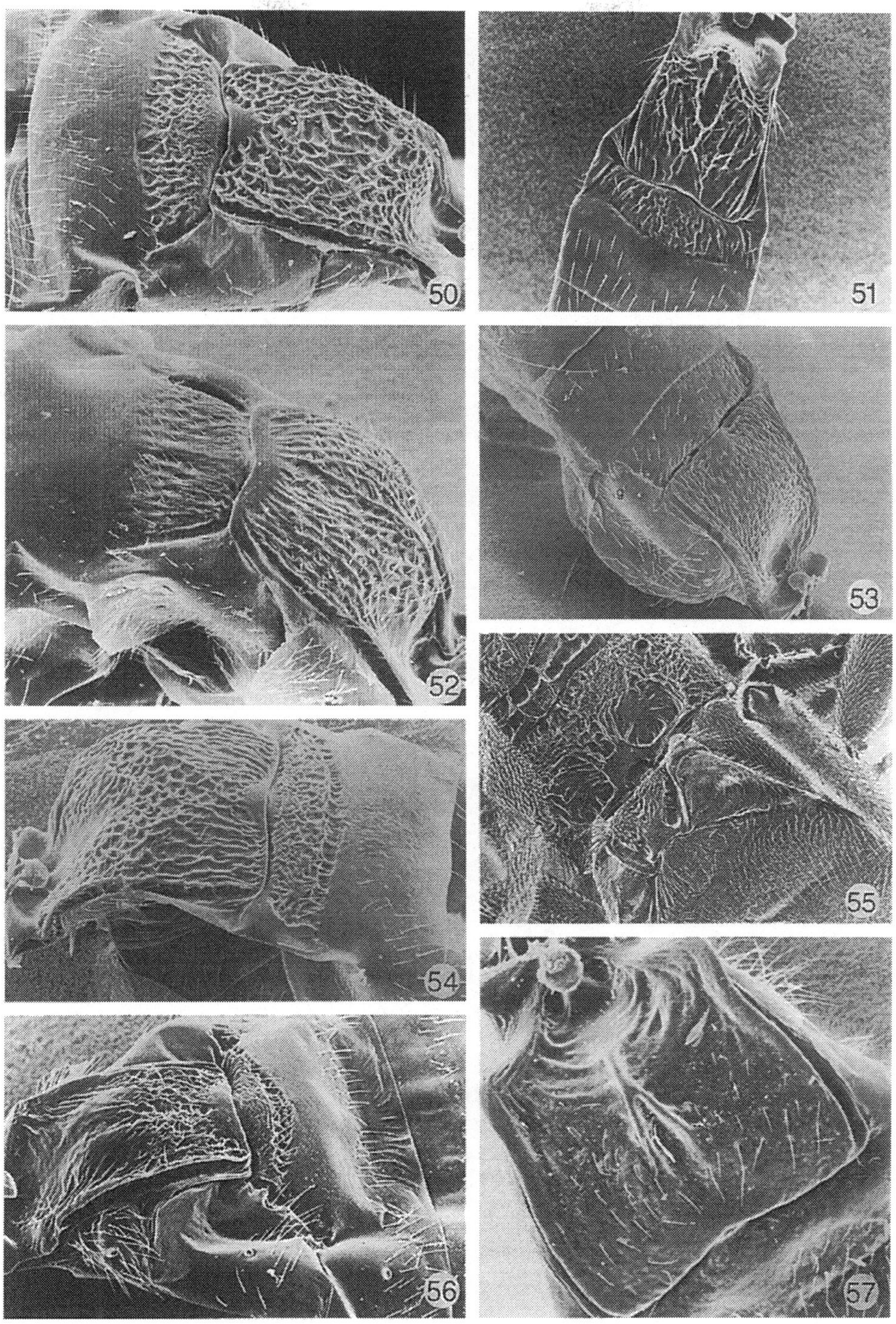

Figs 50-57: Microgastrinae. 50-54, anterior metasomal tergites: 50, *Apanteles* sp., dorsal view; 51, *Dolichogenidea lacteicolor* (Viereck), dorsal view; 52, *Papanteles peckorum* Mason, dorsal view; 53, *Hypomicrogaster zonaria* (Say), dorsal view; 54, *Choeras consimilis* (Viereck), dorso-lateral view; 55, propodeum and anterior metasomal tergites of *Illidops* sp., postero-dorsal view; 56, anterior metasomal tergites of *Promicrogaster* sp., dorso-lateral view; 57, petiole of *Alphomelon nigriceps* (Ashmead), dorsal view. All figures after Mason (1981).

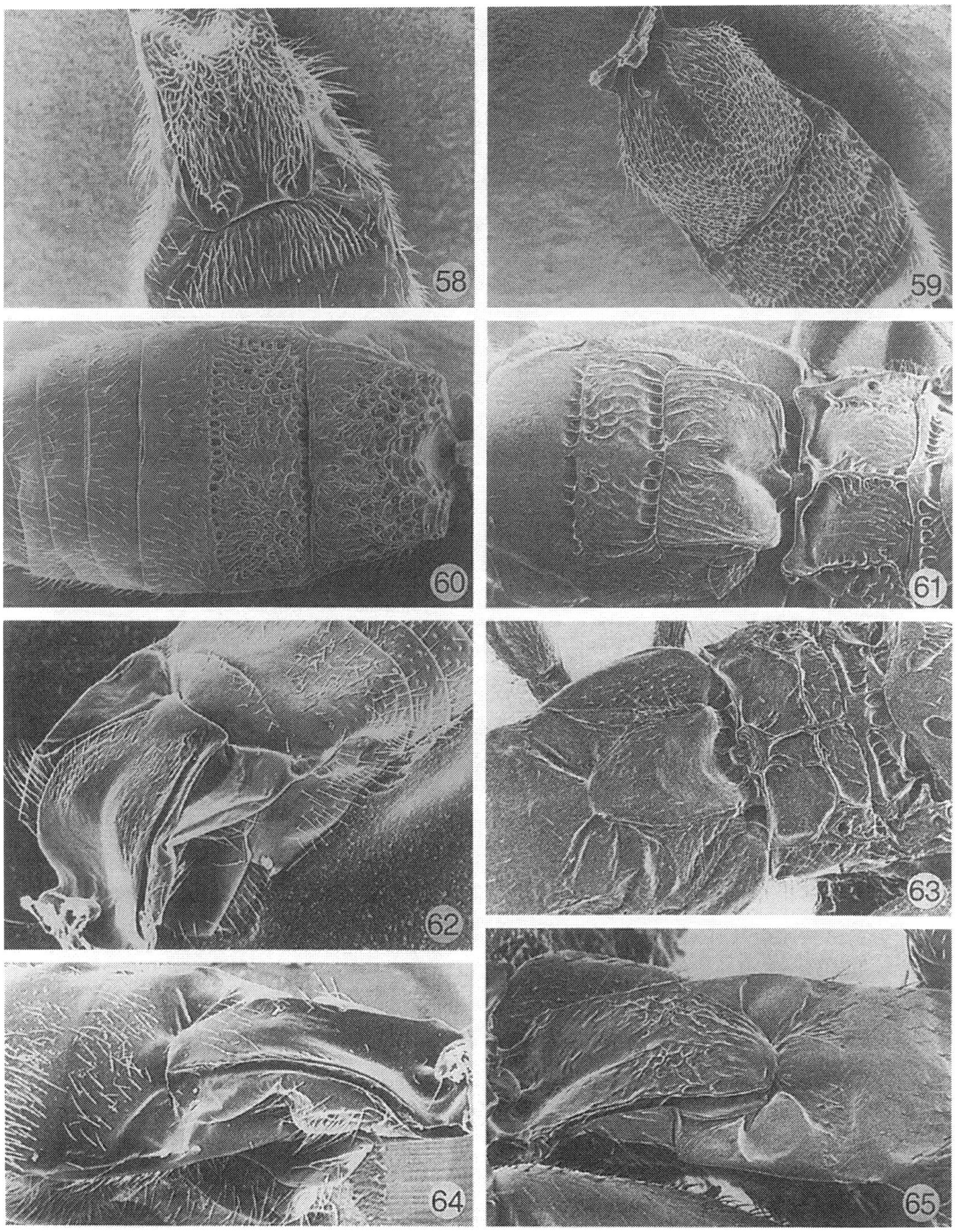

Figs 58-65: Anterior metasomal tergites of Microgastrinae. 58; *Rhygoplitis aciculatus* (Ashmead), dorsal view; 59, *Rhygoplitis terminalis* (Gahan), dorsal view; 60, *Microgaster canadiensis* Muesebeck, dorsal view; 61, *Xanthomicrogaster forticeps* (Cameron), dorsal view; 62, *Iconella* sp., dorsal view; 63, *Clarkinella* sp., dorsal view; 64, *Prasmodon* sp., dorsal view; 65, *Xenogaster insolens* (Wilkinson), dorsal view. All figures after Mason (1981).

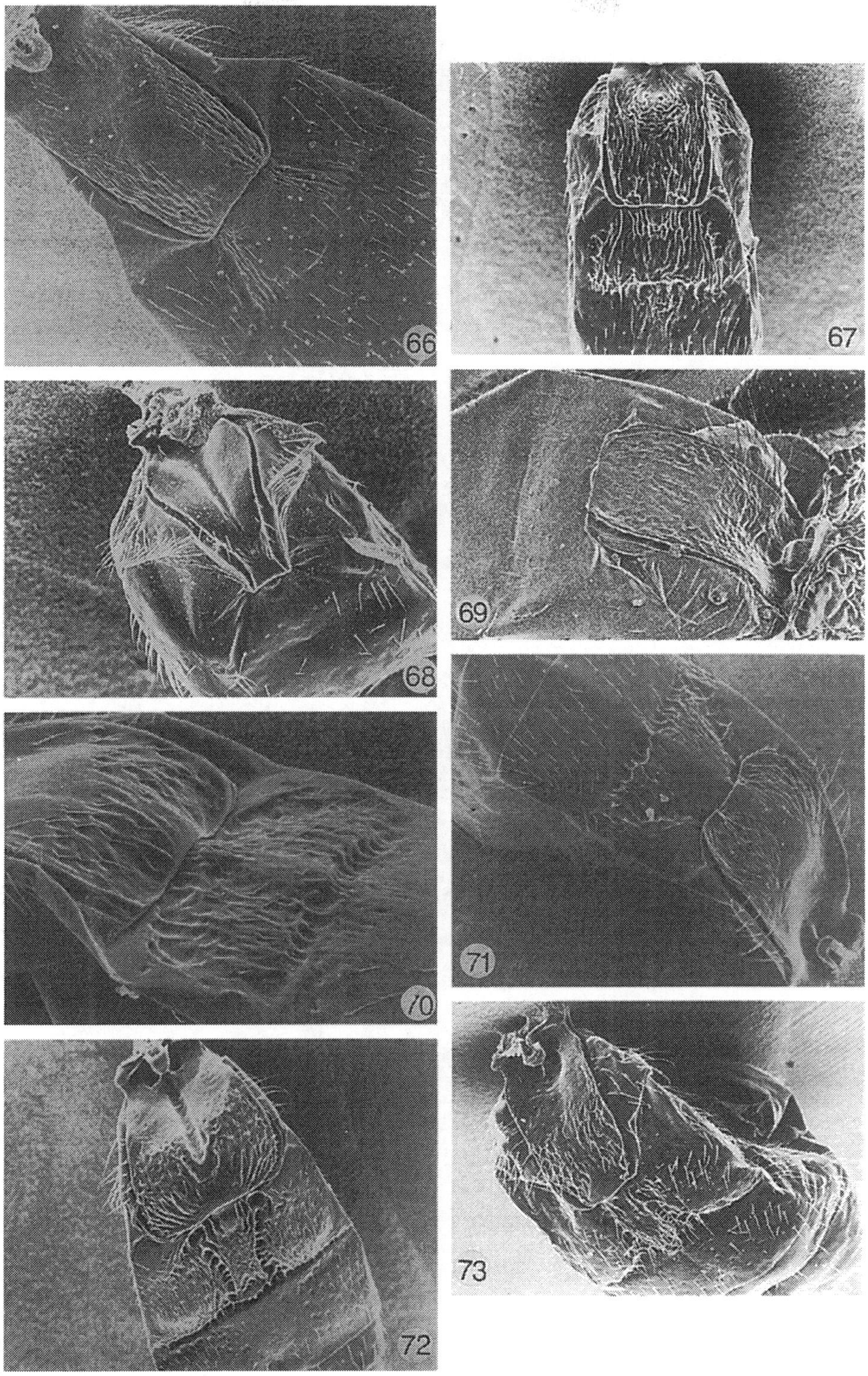

Figs 66-73: Anterior metasomal tergites of Microgastrinae. 66, *Sathon* sp., dorsal view; 67, *Pholetesor ornigis* (Weed), dorsal view; 68, *Pseudapanteles* sp., dorsal view; 69, *Paroplitis beringianus* Mason, dorsal view; 70, *Cotesia* sp., dorso-lateral view; 71, *Protapanteles paleacritae* (Riley), dorsal view; 72, *Diolcogaster facetosa* (Weed), dorsal view; 73, *Rasivalva stigmatica* (Muesebeck), dorsal view. All figures after Mason (1981).

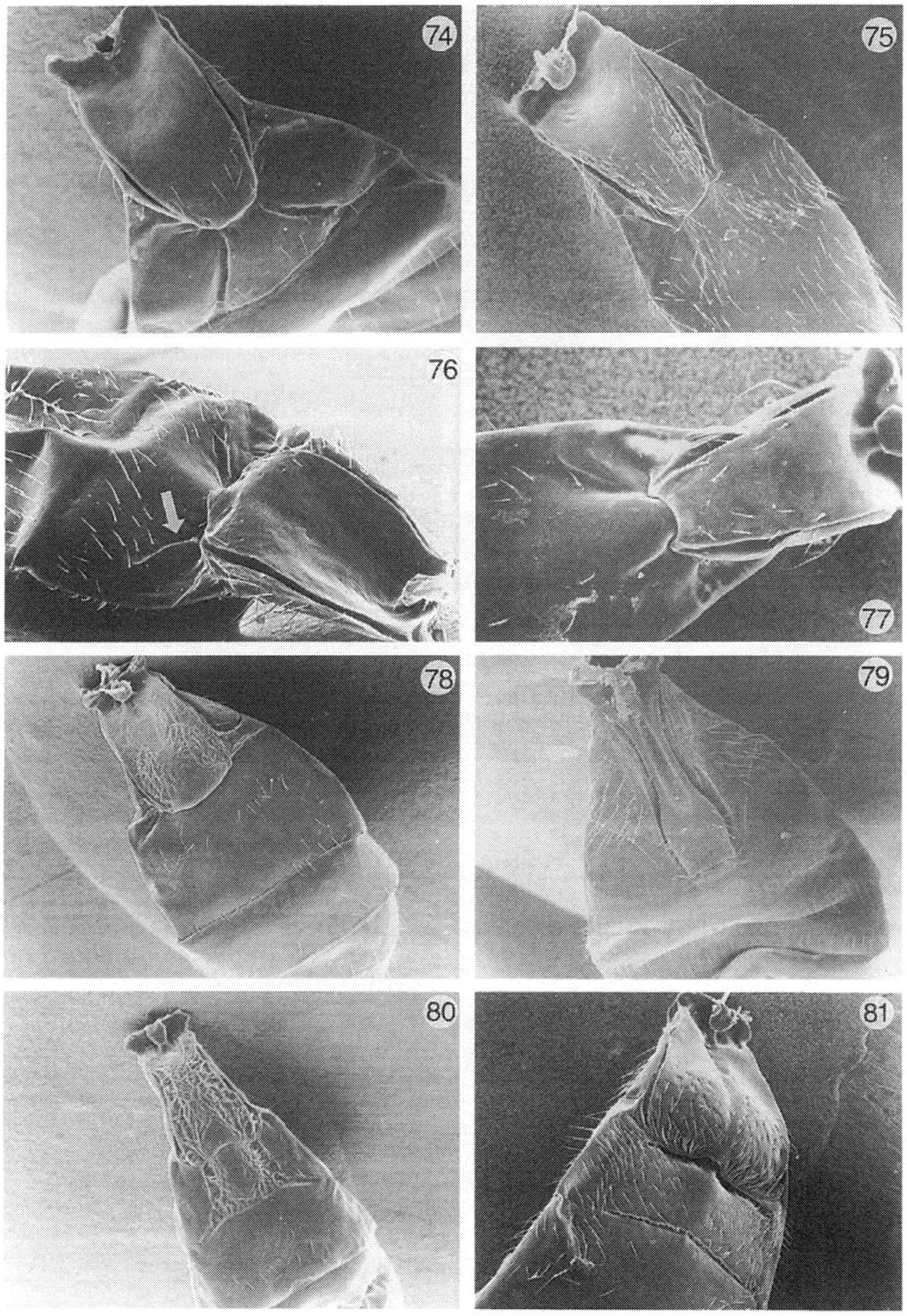

Figs 74-81: Anterior metasomal tergites of Microgastrinae. 74, *Glyptapanteles militaris* (Walsh), dorsal view; 75, *Glyptapanteles* sp. (*vitripennis* group), dorsal view; 76, *Distatrix papilionis* (Viereck), dorsal view, arrows = grooves on tergum 2; 77, *Venanides xeste* Mason, dorsal view; 78, Microplitis sp., dorsal view; 79, *Protomicroplitis calliptera* (Say), dorsal view; 80, *Venanus pinicola* Mason, dorsal view; 81, *Exix mexicana* Mason, dorsal view. All figures after Mason (1981).

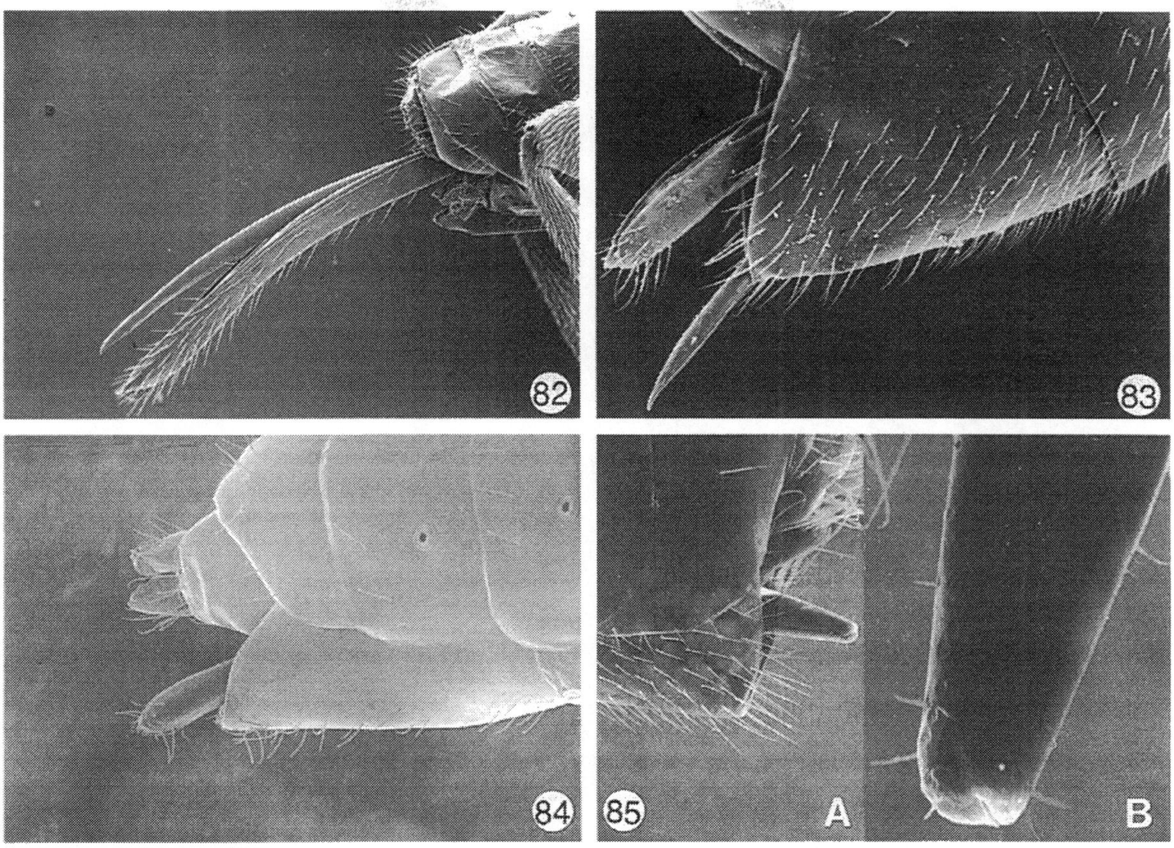

Figs 82-85: Ovipositor and sheath of Microgastrinae, 82, *Apanteles* sp., lateral view; 83, *Glyptapanteles* sp., lateral view; 84, *Cotesia* sp., lateral view; 85, *Distatrix* sp., lateral view, A = full length, B = close up to show apical setae.

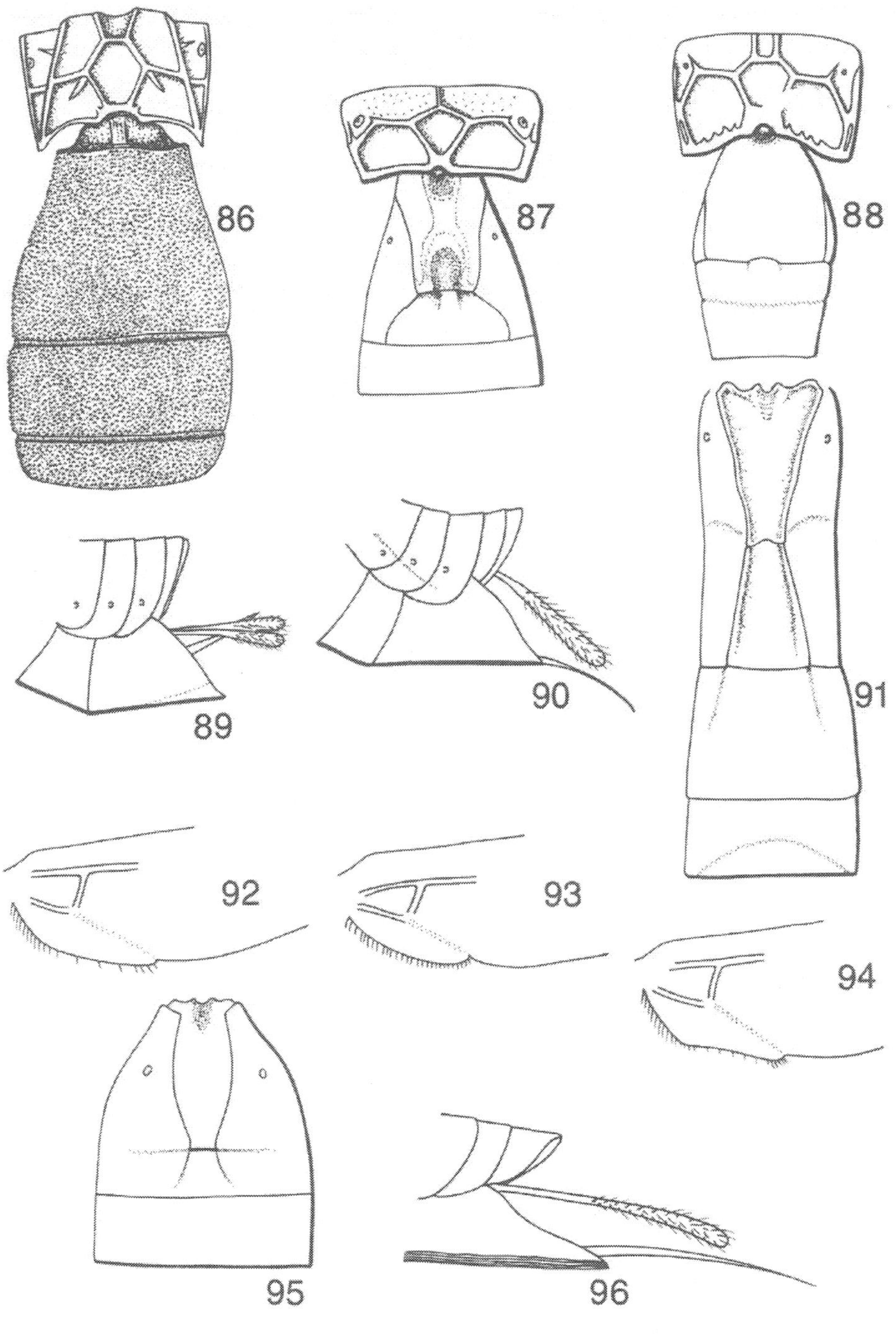

Figs 86-96: Microgastrinae. 86, propodeum and metasomal terga 1-3 of *Xanthapanteles cameronae* Whitfield, dorsal view; 87, propodeum and anterior metasomal tergites of *Parapanteles* sp., dorsal view; 88, propodeum and anterior metasomal tergites of Dasylagon sp., dorsal view; 89, apex of metasoma, female of *Pholetesor bedelliae* (Viereck), lateral view; 90, apex of metasoma, female of *Pholetesor ornigis* (Weed), lateral view; 91, anterior metasomal tergites of *Sendaphne* sp., dorsal view; 92-94, vannal lobe of hind wings: 92, *Apanteles* sp. (in part); 93, *Dolichogenidea* sp.; 94, *Apanteles* sp. (in part); 95, anterior metasomal tergites of *Pelicope yuccamica* Mason, dorsal view; 96, apex of metasoma of *Apanteles* sp., lateral view.

MICROGASTRINAE UPDATE

Jose Fernández-Triana
Canadian National Collection of Insects, 960 Carling Avenue, Ottawa, Canada, jose.fernandez@agr.gc.ca

The Microgastrinae chapter of the manual is outdated. The biggest changes since the publication of the Manual in 1997 are the descriptions of three new genera: *Cuneogaster* Choi and Whitfield, *Mariapanteles* Whitfield and Fernández-Triana, and *Keylimepie* Fernández-Triana (Choi and Whitfield 2006, Whitfield *et al.*, 2012, Fernández-Triana and Boudreault 2016), as well as the finding in the region for the first time species of two other genera: *Austrocotesia* and *Wilkinsonellus* (Valerio and Whitfield 2005, Arias-Penna *et al.*, 2013). Within the past 15 years a significant number of taxonomic revisions of New World fauna have been published resulting in almost 400 new species being described in the genera *Alphomelon* Mason, *Apanteles* Förster, *Austrocotesia* Austin & Dangerfield, *Deuterixys* Mason, *Distatrix* Mason, *Hypomicrogaster* Ashmead, *Microplitis* Förster, *Parapanteles* Ashmead, *Prasmodon* Nixon, *Promicrogaster* Brues & Richardson, *Protomicroplitis* Ashmead, *Pseudapanteles* Ashmead, *Sendaphne* Nixon, *Snellenius* Westwood, *Venanus* Mason and *Wilkinsonellus* Austin & Dangerfield (Scatolini and Penteado-Dias 1999; Whitfield *et al.*, 2001, 2002; Deans *et al.*, 2003; Janzen *et al.*, 2003; Whitfield and Oltra-Moscardo 2004; Valerio *et al.*, 2004, 2005; Valerio and Whitfield 2005; Whitfield and Oltra 2005; Whitfield 2006; Grinter *et al.*, 2009; Valerio *et al.*, 2009, 2015; Fernandez-Triana 2010, 2015, 2016, b; Whitfield *et al.*, 2011; Arias-Penna *et al.*, 2013; Fernández-Triana *et al.*, 2013, 2014a-e, 2015, Fernández-Triana *et al.*, 2015; Sánchez-García *et al.*, 2015, Fernández-Triana and Boudreault 2016). These revisions have almost doubled the number of known species for the region, with close to 900 described species of Microgastrinae now recorded for the New World (Yu *et al.*, 2016).

Recent papers on New World Microgastrinae

Arias D, Whitfield J, Janzen D. 2013. Three new species in the genus *Wilkinsonellus* (Braconidae, Microgastrinae) from the Neotropics, and the first host record for the genus. *ZooKeys* 302: 79-95.

Choi WY, Whitfield JB. 2006. *Cuneogaster*, a new genus of the subfamily Microgastrinae (Hymenoptera: Braconidae) from the Neotropical region. *Proceedings of the Entomological Society of Washington* 108(1): 119-124.

Deans AR, Whitfield JB, Janzen DH. 2003. Taxonomy and natural history of the microgastrine genus *Alphomelon* Mason (Hymenoptera: Braconidae). *Journal of Hymenoptera Research* 12(1): 1-41.

Fernández-Triana J. 2010. Eight new species and an annotated checklist of Microgastrinae (Hymenoptera: Braconidae) from Canada and Alaska. *Zookeys* 63: 1-53.

Fernández-Triana J. 2015. A revision of the genus *Protomicroplitis* Ashmead (Hymenoptera, Braconidae, Microgastrinae), with the description of a new species. *Zootaxa* 4039(4): 529-542.

Fernández-Triana J, Cardinal S, Whitfield J, Smith M, Janzen D. 2013. A review of the New World species of the parasitoid wasp *Iconella* (Hymenoptera, Braconidae, Microgastrinae). *ZooKeys* 321: 65-87.

Fernández-Triana J, Whitfield J, Smith M, Hallwachs W, Janzen D. 2014a. First record of the genus *Venanus* (Hymenoptera: Braconidae: Microgastrinae) in Mesoamerica, with the description of two new species from Costa Rica. *Biodiversity Data Journal* 2: e4167.

Fernández-Triana JL, Whitfield JB, Smith MA, Braet Y, Hallwachs W, Janzen DH. 2014b. Review of the Neotropical genus *Prasmodon* (Hymenoptera, Braconidae, Microgastrinae), with emphasis on species from Area de Conservación Guanacaste, northwestern Costa Rica. *Journal of Hymenoptera Research* 37: 1-52.

Fernández-Triana J, Janzen D, Hallwachs W, Whitfield J, Smith M, Kula R. 2014c. Revision of the genus *Pseudapanteles* (Hymenoptera, Braconidae, Microgastrinae), with emphasis on the species in Area de Conservación Guanacaste, northwestern Costa Rica. *ZooKeys* 446: 1-82.

Fernández-Triana JL, Whitfield JB, Smith MA, Hallwachs W, Janzen DH. 2014d. Revision of the Neotropical genus *Sendaphne* Nixon (Hymenoptera, Braconidae, Microgastrinae). *Journal of Hymenoptera Research* 41: 1-29.

Fernández-Triana J, Whitfield J, Rodriguez J, Smith M, Janzen D, Hajibabaei M, Burns J, Solis A, Brown J, Cardinal S, Goulet H, Hebert P. 2014e. Review of *Apanteles sensu stricto* (Hymenoptera, Braconidae, Microgastrinae) from Area de Conservación Guanacaste, northwestern Costa Rica, with keys to all described species from Mesoamerica. *ZooKeys* 383: 1-565.

Fernández-Triana JL, Whitfield JB, Smith MA, Kula RR, Hallwachs W, Janzen DH. 2015. Revision of the genera *Microplitis* and *Snellenius* (Hymenoptera, Braconidae, Microgastrinae) from Area de Conservación Guanacaste, Costa Rica, with a key to all species previously described from Mesoamerica. *Deutsche Entomologische Zeitschrift* 62(2): 137-201.

Fernández-Triana J, Boudreault C. 2016. *Keylimepie peckorum* gen. n. and sp. n., (Hymenoptera, Braconidae) from southern Florida, U.S., the first known brachypterous member of the subfamily Microgastrinae. *ZooKeys* 584: 95-107.

Fernández-Triana J, Boudreault C, Dapkey T, Smith MA, Rodriguez J, Hallwachs W, Janzen DH. 2016a. Revision of the genus *Promicrogaster* (Hymenoptera, Braconidae, Microgastrinae) from Area de Conservación Guanacaste, Costa Rica, with a key to all species previously described from Mesoamerica. *Journal of Hymenoptera Research* 50: 25-79.

Fernández-Triana JL, Whitfield JB, Smith MA, Dapkey T, Hallwachs W, Janzen DH. 2016b. Review of the world species of *Exoryza* (Hymenoptera, Braconidae, Microgastrinae), with description of five new species. *Deutsche Entomologische Zeitschrift* 63(2): 195-210.

Grinter CC, Whitfield, JB, Connahs, H, Dyer LA, Hallwachs W, Janzen DH. 2009. A key to New World *Distatrix* Mason (Hymenoptera: Braconidae), with descriptions of six new reared Neotropical species. *Journal of Insect Science* 9 (25): 1-15.

Janzen DH, Walker AK, Whitfield JB, Delvare G, Gauld ID. 2003. Host specificity and hyperparasitoids of three new Costa Rican species of *Microplitis* Foerster (Hymenoptera: Braconidae: Microgastrinae), parasitoids of sphingid caterpillars. *Journal of Hymenoptera Research* 12 (1): 42-76.

Sánchez-García JA, Figueroa JI, Whitfield JB. 2015. A new species of *Apanteles* Foerster (Hymenoptera: Braconidae) parasitic on two blackberry leafrollers (Lepidoptera: Tortricidae) in Mexico. *Journal of the Kansas Entomological Society* 88(1): 10-15.

Scatolini D, Penteado-Dias AM. 1999. A new species of *Sendaphne* (Hymenoptera, Braconidae) from Brazil. *Entomological News* 110(1): 53-55.

Valerio AA, Deans AR, Whitfield JB. 2004. Review of the microgastrine braconid wasp genus *Exoryza*, with description of a new species, *E. monocavus*, from Central America. *Zootaxa* 526: 1-11.

Valerio AA, Whitfield JB. 2005. Two new species of the genus *Austrocotesia* Austin & Dangerfield (Hymenoptera: Braconidae) from the Andean region of South America. *Zootaxa* 888: 1-11.

Valerio AA, Rodriguez JJ, Whitfield JB, Janzen DH. 2005. *Prasmodon zlotnicki*, a new Neotropical species of the genus *Prasmodon* Nixon (Braconidae: Microgastrinae) from Costa Rica, with the first host records for the genus. *Zootaxa* 1016: 29-38.

Valerio AA, Whitfield JB, Janzen DH. 2009. Review of world *Parapanteles* Ashmead (Hymenoptera: Braconidae: Microgastrinae), with descriptions of fourteen new Neotropical species and the first description of the final instar larvae. *Zootaxa* 2084: 1-49.

Valerio AA, Whitfield JB. 2015. Taxonomic review of the genus *Hypomicrogaster* Ashmead (Hymenoptera: Braconidae: Microgastrinae), with descriptions of 40 new species. *Zootaxa* 3979(1): 001-098.

Whitfield JB. 2006. Revision of the Nearctic species of the genus *Pholetesor* Mason (Hymenoptera: Braconidae). *Zootaxa* 1144: 1-94.

Whitfield JB, Oltra-Moscardo MT. 2004. The Neotropical species of *Deuterixys* Mason (Hymenoptera: Braconidae). *Journal of Hymenoptera Research* 13(1): 134-148.

Whitfield JB, Cameron SA, Ramirez SR, Roesch K, Messinger S, Taylor M, Cole D. 2001. Review of the *Apanteles* species (Hymenoptera: Braconidae) attacking Lepidoptera in *Bombus* (*Fervidobombus*) (Hymenoptera: Apidae) colonies in the New World, with description of a new species from South America. *Annals of the Entomological Society of America* 94: 851-857.

Whitfield JB, Benzing A, Ponce F. 2002. Review of the *Glyptapanteles* species (Hymenoptera: Braconidae, Microgastrinae) attacking noctuids in field crops in the Neotropical Region, with descriptions of two new species from the Ecuadorian Andes. *Journal of Hymenoptera Research* 11(1): 152-165.

Whitfield JB, Rasmussen C, Arias Penna DC. 2011. Review of the New World microgastrine braconid wasp genus *Venanus* Mason, with descriptions of three new Neotropical species. *Annals of the Entomological Society of America* 104: 1119-1127.

Whitfield J, Fernández-Triana J, Janzen D, Smith M, Cardinal S. 2012. *Mariapanteles* (Hymenoptera, Braconidae), a new genus of Neotropical microgastrine parasitoid wasp discovered through biodiversity inventory. *ZooKeys* 208: 61-80.

Yu DSK, van Achterberg C, Horstmann K. 2016. Taxapad 2016, Ichneumonoidea 2015. Database on flash-drive. Nepean, Ontario, Canada. http://www.taxapad.com.

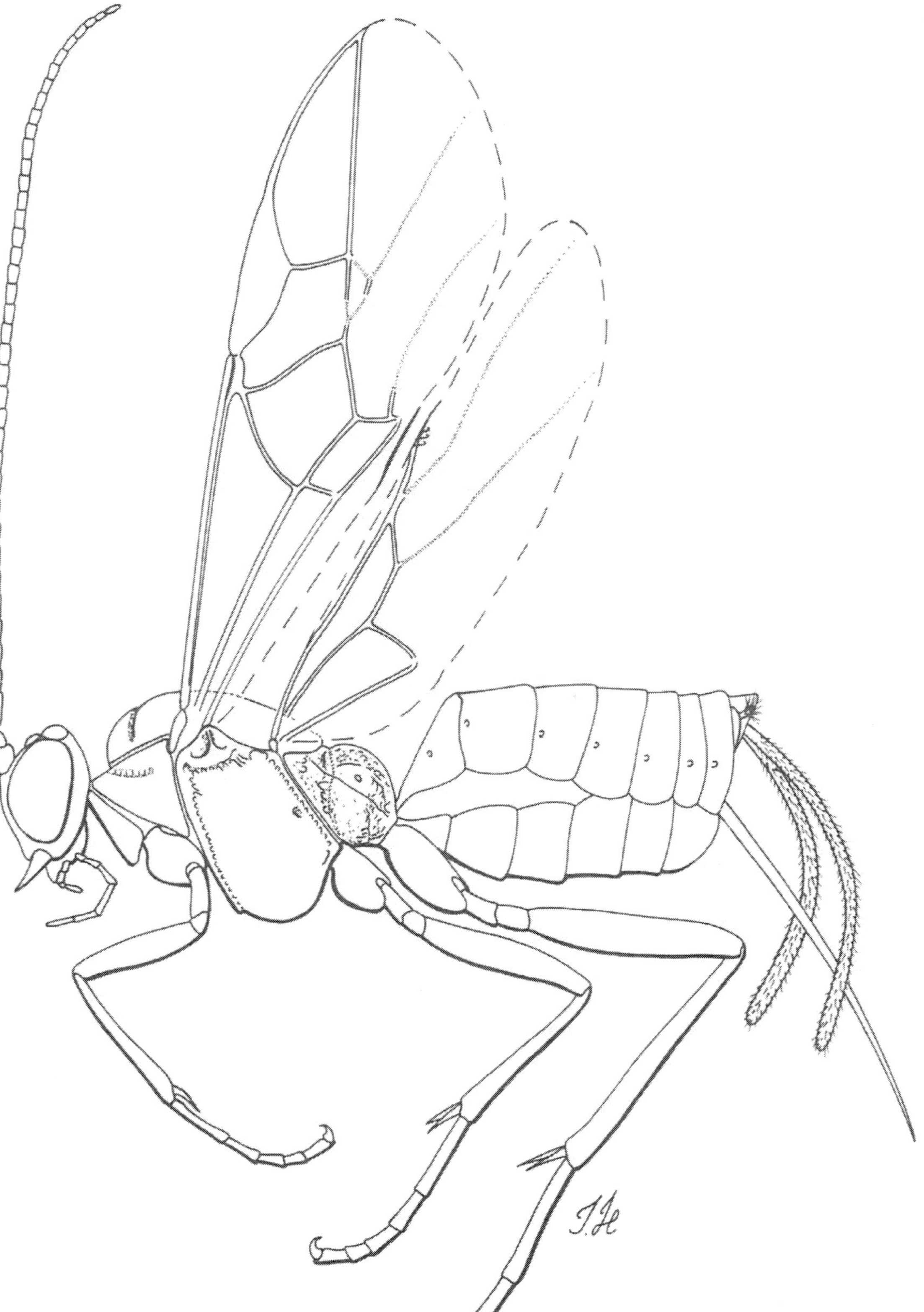

SUBFAMILY MICROTYPINAE

C. VAN ACHTERBERG

Afdeling Entomologie (Hymenoptera), Nationaal Natuurhistorisch Museum, Postbus 9517,
2300 RA Leiden, The Netherlands

INTRODUCTION. The subfamily Microtypinae is a small and rather uniform group, with species 2-6 mm long. Females have a long ovipositor, longer than the metasoma. Few genera are known and only two occur in the New World. The Nearctic species were revised by Čapek and van Achterberg (1992).

BIOLOGY. The few data known on the biology of Microtypinae are summarized by Čapek and van Achterberg (1992). As far as is known, all species are solitary koinobiont endoparasitoids of microlepidopteran larvae with an endophytic way of life. The known hosts of Microtypinae belong to the lepidopteran families Pyralidae, Tortricidae, Gelechiidae, and Yponomeutidae.

BIOLOGICAL CONTROL POTENTIAL. Though they have not been used in biocontrol, some species of *Microtypus* Ratzeburg attack microlepidopteran pests of trees such as *Dioryctria auranticella* (Grote) and *Acrobasis* spp.

COMMON GENERA. *Microtypus* is widely distributed in the Holarctic region.

DISTRIBUTION. The Microtypinae are almost worldwide in distribution, but are unknown from the Australian region. Most of the nine described species occur in the Holarctic region.

DISTINGUISHING FEATURES. Microtypinae are rather slender with a medium sized head, long antenna, and distinctly protruding ovipositor in females. Essential for their recognition is the presence of the small, triangular second submarginal cell of the fore wing (Figs 1, 2) combined with the absence or weak development of the dorsal carinae of the petiole (Figs 9, 11). The occipital carina is either complete (Fig. 10) or reduced mid-dorsally (Fig. 12).

KEY TO THE NEW WORLD GENERA OF THE SUBFAMILY MICROTYPINAE

1. Notauli largely absent (Fig. 4). Occipital carina absent mid-dorsally (Fig. 12). Petiole robust, without constriction behind spiracles (Fig. 11). Wing and body: Figs 2, 4, 6, 8, 11, 12, 14 .... ***Neomicrotypus*** van Achterberg
   Distribution: Chile; known only from a single specimen. Biology: unknown. Reference: van Achterberg (1992).

- Notauli complete (Fig. 3). Occipital carina present mid-dorsally (Fig. 10). Petiole more slender, with constriction behind its spiracles (Fig. 9). Wings and body: Figs 1, 3, 5, 7, 9, 10, 13....***Microtypus*** Ratzeburg
   Distribution: Canada and Alaska to southern Texas; 2 described species in New World. Biology: endoparasitoids of concealed microlepidopteran larvae (Pyralidae, Tortricidae, Gelechiidae, Yponomeutidae). References: Čapek and van Achterberg (1992).

REFERENCES

Achterberg, C. van 1992. Revision of the genera of the subfamily Microtypinae (Hymenoptera: Braconidae). Zoologische Mededelingen Leiden 66: 369-380.

Čapek, M. and Achterberg, C. van 1992. A revision of the genus *Microtypus* Ratzeburg (Hymenoptera: Braconidae). Zoologische Mededelingen Leiden 66: 323-338.

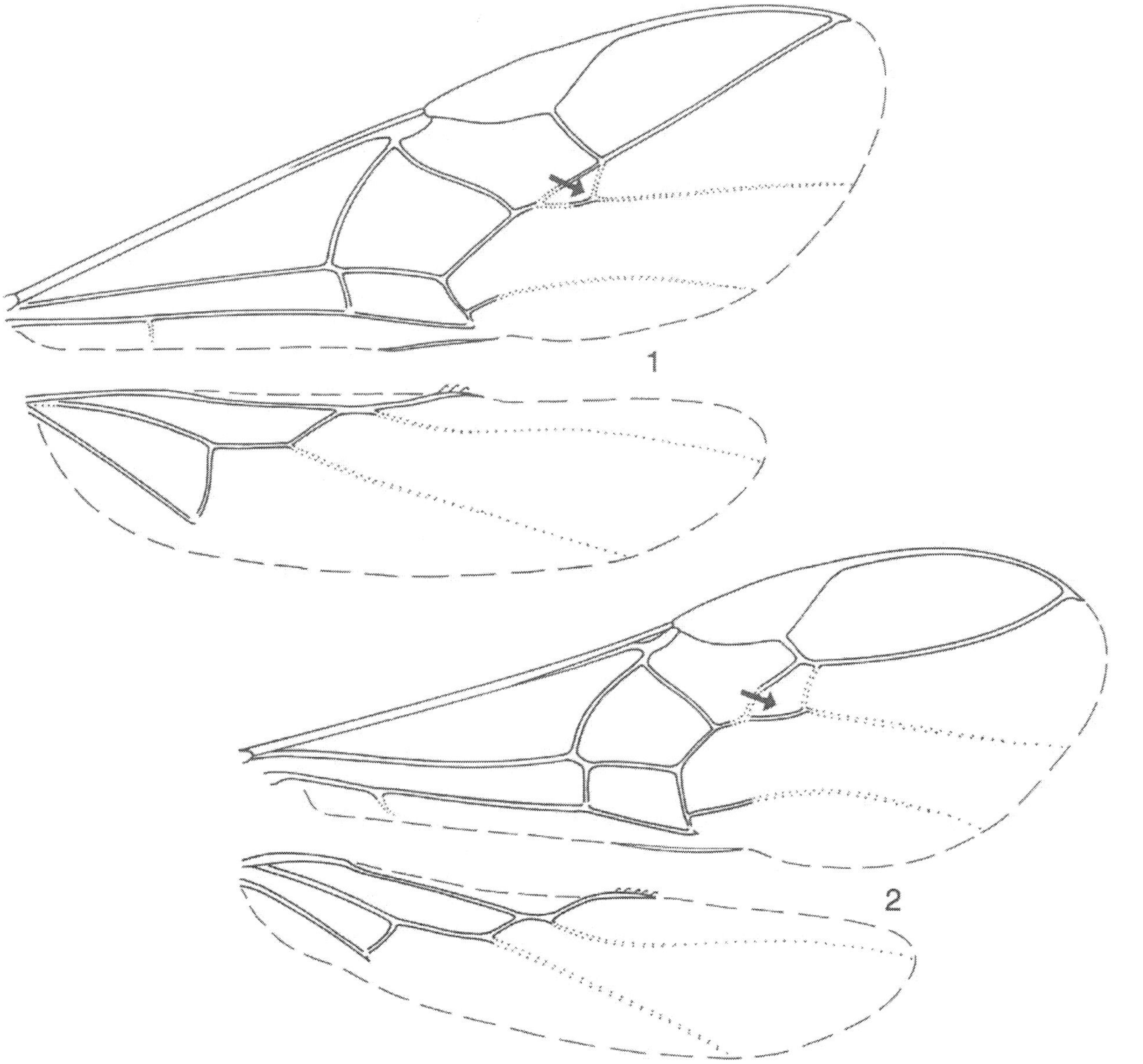

Figs 1-2: Wings of Microtypinae, arrows = second submarginal cell. 1, *Microtypus* sp.; 2, *Neomicrotypus penai.*

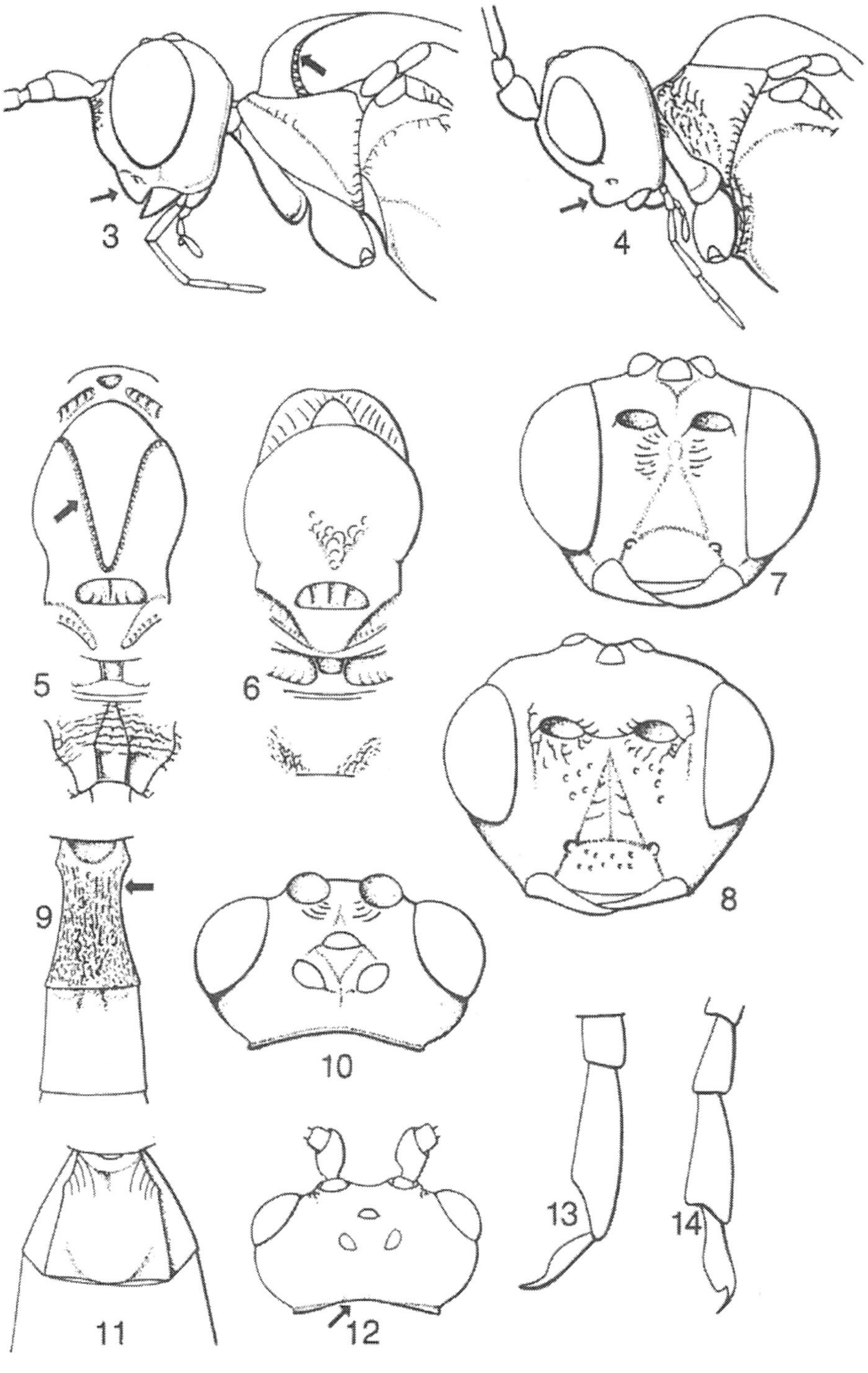

Figs 3-14: Microtypinae. 3-4, heads and anterior portions of mesosoma, arrows = clypeus and notauli: 3, *Microtypus* sp.; 4, *Neomicrotypus penai* van Achterberg; 5-6, mesosoma, dorsal view: 5, *Microtypus* sp., arrow = notaulus; 6, *N. penai*; 7-8, heads, frontal view: 7, *Microtypus* sp.; 8, *N. penai*; 9, petiole of *Microtypus* sp., arrow = constriction behind spiracle; 10, head of *Microtypus* sp., dorsal view; 11, petiole of *N. penai*; 12, head of *N. penai*, dorsal view, arrow = occipital carina absent mid-dorsally; 13-14, tarsal claws: 13, *Microtypus* sp.; 14, *N. penai*. All figures after van Achterberg (1992).

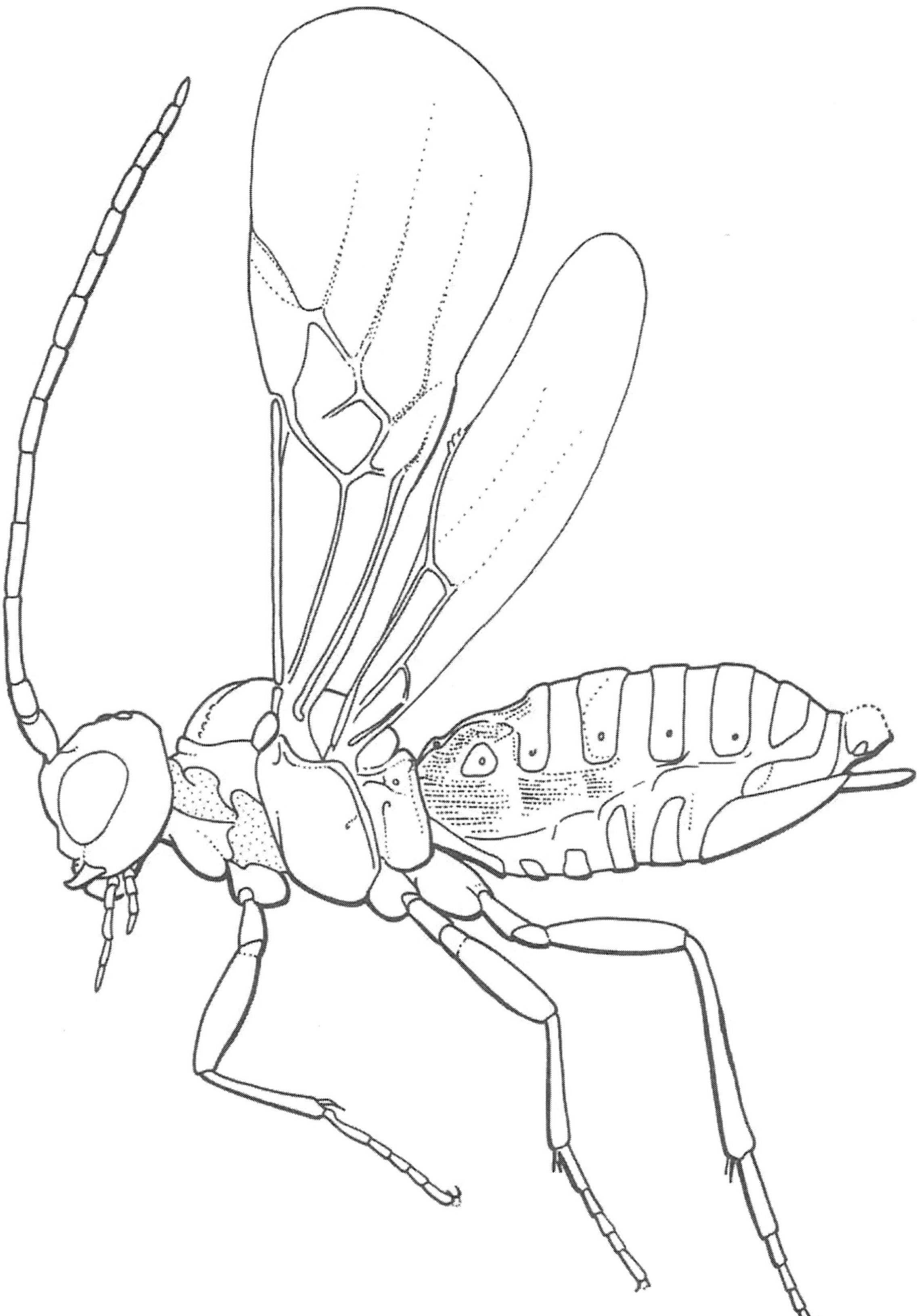

SUBFAMILY MIRACINAE

J. B. WHITFIELD

Department of Entomology, 320 Morrill Hall, University of Illinois, 505 S. Goodwin Ave, Urbana, IL 61801 U.S.A.

INTRODUCTION. The subfamily Miracinae has been considered part of the Microgastrinae (Muesebeck, 1922; Nixon, 1965; Marsh, 1979) but more recently has been accepted as a distinct subfamily (Mason, 1981, 1983; Quicke and van Achterberg, 1990; Wharton et al., 1992; Whitfield and Mason, 1994). Only one genus, *Mirax* Haliday, is generally recognized, although a second generic name, *Centistidea* Rohwer, is available and occasionally used. Probably about six genera occur worldwide, at least two or three of which are found in the New World, but the group has never been seriously studied on a world basis.

BIOLOGY. Many of the species have been reared, always from leaf-miners, usually from Nepticulidae or Heliozelidae but also from some Gracillariidae and Tischeriidae (Whitfield and Wagner, 1991). The larvae are endoparasitoids: the adults emerge from the host cocoon.

BIOLOGICAL CONTROL POTENTIAL. *Mirax* has been occasionally reared in large numbers from leaf-miners on tropical trees and shrubs, including coffee (Muesebeck, 1937). There has been little attempt to use these species in biological control, however.

COMMON GENERA. One genus, *Mirax*, is usually recognized. The species are widely distributed and locally abundant, although not very well represented in collections.

DISTRIBUTION. The subfamily, and the genus *Mirax*, appear to be distributed worldwide. The species appear to be especially well represented in tropical regions.

DISTINGUISHING FEATURES. The narrow first metasomal tergite along with the striate, membranous lateral tergites (Fig. 2) will separate miracines from all other subfamilies except Mendesellinae. From that subfamily miracines differ in having vein r-m of the fore wing absent (Fig. 1), and the Y-shaped structure of the second metasomal tergum (best seen with a light microscope) is also unique.

REFERENCES

Marsh, P. M. 1979. Family Braconidae. Pp. 144-295. In: Krombein, K. V., Hurd, P. D., Smith, D.R., and Burks, B. D. (Eds.). Catalog of Hymenoptera in America North of Mexico. Smithsonian Institution Press, Washington, D.C.

Mason, W. R. M. 1981. The polyphyletic nature of *Apanteles* Foerster (Hymenoptera: Braconidae): a phylogeny and reclassification of Microgastrinae. Memoirs of the Entomological Society of Canada 115: 1-147.

Mason, W.R. M. 1983. A new South African subfamily related to Cardiochilinae (Hymenoptera: Braconidae). Contributions to the American Entomological Institute 20: 49-62.

Muesebeck, C. F. W. 1922. A revision of the North American ichneumon-flies, belonging to the subfamilies Neoneurinae and Microgasterinae. Proceedings of the United States National Museum 61: 1-76.

Muesebeck, C. F. W. 1937. A new West Indian species of *Mirax* Haliday parasitic on the coffee leafminer (Hymenoptera: Braconidae). Proceedings of the Entomological Society of Washington 39: 139-141.

Nixon, G. E. J. 1965. A reclassification of the tribe Microgasterini. Bulletin of the British Museum (Natural History) Entomology Supplement 2: 1-284.

Quicke, D. L. J. and Achterberg, C. van 1990. Phylogeny of the subfamilies of the family Braconidae (Hymenoptera: Ichneumonoidea). Zoologische Verhandelingen Leiden 258: 1-95.

Wharton, R. A., Shaw, S. R., Sharkey, M. J., Wahl, D.B., Woolley, J. B., Whitfield, J. B., Marsh, P. M., and Johnson J. W. 1992. Phylogeny of the subfamilies of the family Braconidae (Hymenoptera: Braconidae): a reassessment. Cladistics 8: 199-235.

Whitfield, J. B. and Mason, W. R. M. 1994. Mendesellinae, a new subfamily of braconid wasps (Hymenoptera: Braconidae) with a review of relationships within the microgastroid assemblage. Systematic Entomology 19: 61-76.

Whitfield, J. B. and Wagner, D. L. 1991. Annotated key to the genera of Braconidae (Hymenoptera) attacking leafmining Lepidoptera in the Holarctic Region. Journal of Natural History 25: 733-754.

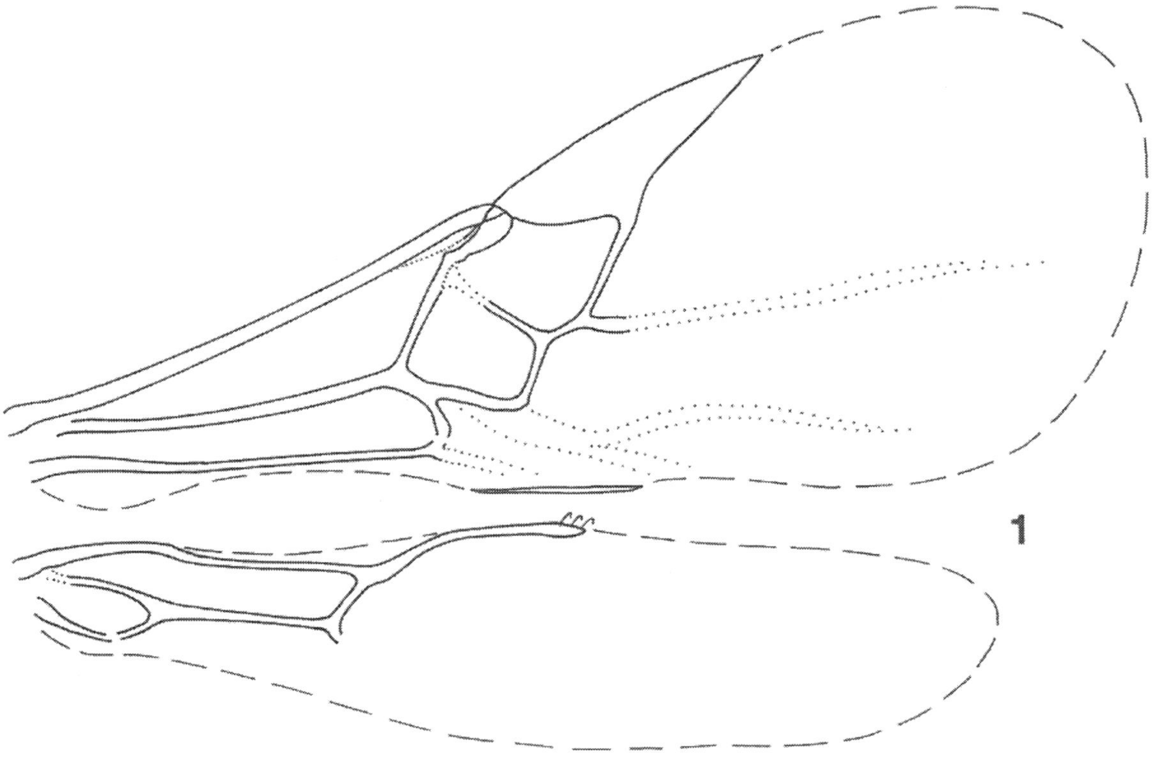

1

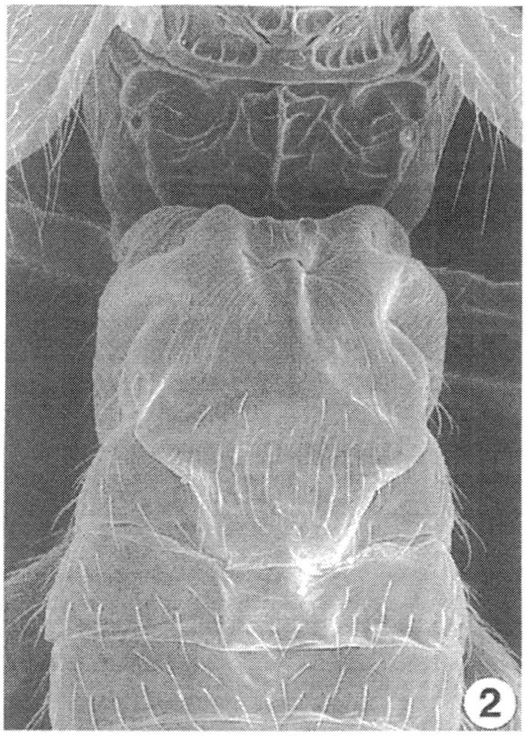

Figs 1-2: Miracinae. 1, wings of *Mirax* sp.; 2, metasoma of *Mirax* sp., dorsal view.

MIRACINAE UPDATE

Jose Fernández-Triana
Canadian National Collection of Insects, 960 Carling Avenue, Ottawa, Canada, jose.fernandez@agr.gc.ca

There has been little change in our knowledge of New World Miracinae over the past 20 years. At the generic level *Mirax* Haliday is usually the only genus recognized (e.g., Whitfield, 1997; Papp, 2013), although some authors continue to recognize a second genus, *Centistidea* Rohwer, 1914, (e.g. Penteado-Dias, 1999; van Achterberg and Mehrnejad, 2002). The biggest changes since the publication of the Manual in 1997 are the description of 13 new species for the region (Penteado-Dias, 1999; Papp 2013; Cauich-Kumul *et al.*, 2014). Those revisions have doubled the known diversity, with 25 species of Miracinae now recorded for the New World (Yu *et al.*, 2016).

Recent papers on New World Miracinae

Achterberg C van, Mehrnejad MR. 2002. The braconid parasitoids (Hymenoptera: Braconidae) of *Kermania pistaciella* Amsel (Lepidoptera: Tineidae: Hieroxestinae) in Iran. *Zoologische Mededelingen, Leiden* 76(2): 27-39.

Cauich-Kumul R, Lopez-Martinez V, De Jesus Garcia-Ramirez M, Delfin-Gonzalez H, Burgos-Solorio A. 2014. Two new species of braconid wasps (Hymenoptera: Braconidae: Miracinae: *Mirax* and Rogadinae: *Choreborogas*) from Mexico. *Florida Entomologist* 97(3): 902-910.

Papp, J. 2013. Eleven new *Mirax* Haliday, 1833 species from Colombia and Honduras and key to the sixteen neotropical *Mirax* species (Hymenoptera: Braconidae: Miracinae). *Acta Zoologica Academiae Scientiarum Hungaricae* 59(2): 97-129.

Penteado-Dias AM. 1999. New species of parasitoids on *Perileucoptera coffeella* (Guerin-Meneville) (Lepidoptera, Lyonetiidae) from Brazil. *Zoologische Mededelingen, Leiden* 73(1-11): 189-197.

Whitfield JB. 1997. Subfamily Miracinae. In: Wharton RA, Marsh PM, Sharkey MJ "Manual of the New World genera of the family Braconidae (Hymenoptera)." International Society of Hymenopterists. Special Publication No. 1. 439 pp., pp. 370-373.

Yu DSK, van Achterberg C, Horstmann K. 2016. Taxapad 2016, Ichneumonoidea 2015. Database on flash-drive. Nepean, Ontario, Canada. http://www.taxapad.com.

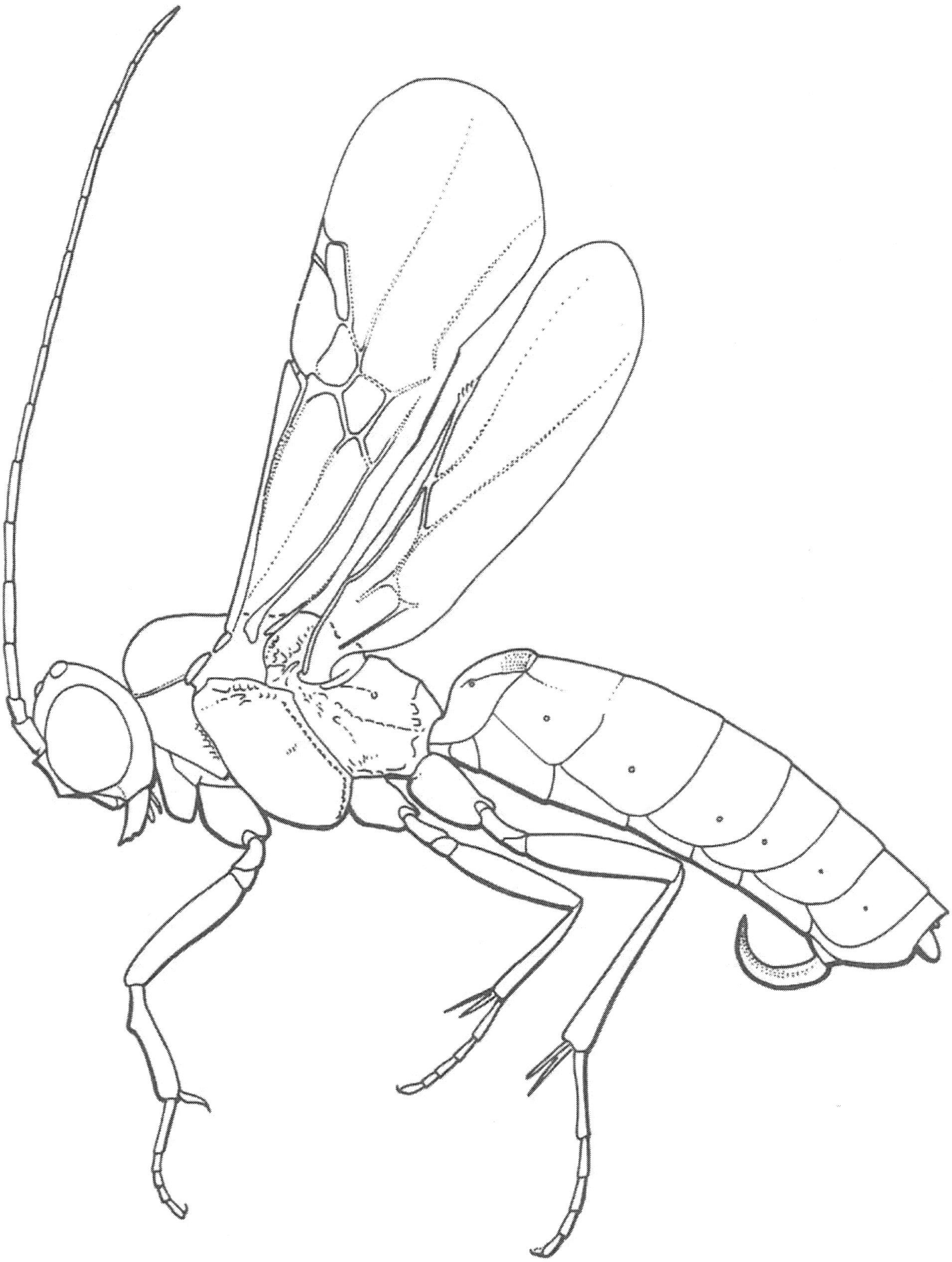

SUBFAMILY NEONEURINAE

S. R. SHAW

Department of Plant, Soil, and Insect Sciences, University of Wyoming, Laramie, Wyoming 82071-3354, U.S.A.

INTRODUCTION. The subfamily Neoneurinae comprises a small group of relatively rare braconids that are associated with ants of the *Formica fusca* L. and *F. rufa* L. species groups (Shenefelt, 1969). They are assumed to be parasitoids of adult *Formica* ants, developing in the abdomen of their host (Marsh et al., 1987; Quicke and van Achterberg, 1990; M. Shaw and Huddleston, 1991). New World neoneurines have been fairly well studied taxonomically. *Neoneurus* Haliday was recently revised by S. Shaw (1992). *Elasmosoma* Ruthe was studied by Muesebeck (1922) and subsequently revised by Huddleston (1976), who treated five species in North America. There are apparently some additional undescribed species in this group.

BIOLOGY. There is little detailed information on the biology and behavior of neoneurines, but they are thought to be endoparasitoids (Huddleston, 1976; Marsh, 1979; M. Shaw and Huddleston, 1991). Although oviposition by neoneurines into the abdomens of adult worker ants has been observed by several authors (Olivier, 1893; Pierre, 1893; Wasmann, 1897; Donisthorpe, 1927; Kariya, 1932; S. Shaw, 1992, 1993), the details of larval development are not known. The process of egg deposition is extremely rapid, taking only a fraction of a second. Wasmann (1897) reported rearing an adult of *Elasmosoma* from a cocoon found attached to the abdomen of a dead worker ant (Formica sp.), and Donisthorpe (1927) apparently reared *Neoneurus aucius* (Thomson) from an observation nest. Neoneurines fly close to the ground and stay very near nests and trails of *Formica* ants. Usually they are found hovering over, or perching near, ant nests and trails, where they can be easily collected with an aspirator. *Neoneurus* females are distinctive in having a highly modified fore leg involving the compression of the fore femur, shortening of the fore tibia and development of a tibial carina, associated sharp tubercles and spines, enlargement of the tibial spur, shortening of the fore tarsus, and enlargement of the fore pulvillus.

BIOLOGICAL CONTROL POTENTIAL. Neoneurines have not been studied for biological control applications. Under natural conditions they are never abundant enough to result in significant suppression of their host ants.

COMMON GENERA. Only two genera of neoneurines (*Neoneurus* and *Elasmosoma*) occur in the New World, and neither is generally thought of as common. However, their rarity in collections is apparently due to their unusual and highly specialized habits.

DISTRIBUTION. The Neoneurinae occur throughout the Holarctic region. In the New World they occur across North America, with records as far south as Portal, Arizona and the Big Bend region of Texas. None have yet been recorded from the Neotropics, although there are unconfirmed reports of neoneurines from Central America.

DISTINGUISHING FEATURES. The fore wing venation of the Neoneurinae is distinctive: the vein C +SC+ R is quite thick and gradually swollen apically near the stigma, and the veins IM and IRS form a sharp angle at the junction of (RS+M)a (Figs 1, 2). Neoneurines also have the maxillary palp reduced to only two segments.

KEY TO THE NEW WORLD GENERA OF THE SUBFAMILY NEONEURINAE

1. Fore wing 3RS complete to wing margin, thus forming a small marginal cell close to stigma (Fig. 1). Antenna with 14 flagellomeres, antenna longer than head and mesosoma combined. Petiole longer than wide (Fig. 3) .... ***Neoneurus*** Haliday
Distribution: Holarctic, 7 described North American species. Biology: associated with formicine ants. Reference: S. Shaw (1992).

\- Fore wing 3RS desclerotized and absent, or indicated apically only by spectral venation, thus marginal cell incomplete and open apically (Fig. 2). Antenna of female with 11 flagellomeres, males with 12-13 flagellomeres, antenna shorter than head and mesosoma combined. Petiole shorter than wide, or about as long as wide (Fig. 4) .... ***Elasmosoma*** Ruthe
Distribution: Holarctic: 5 described North American species. Biology: associated with formicine ants. Reference: Huddleston (1976).

REFERENCES

Donisthorpe, H. St. J. K. 1927. The guests of British ants, their habits and life histories. G. Routledge and Sons Ltd., London. 244 pp.

Huddleston, T. 1976. A revision of *Elasmosoma* Ruthe (Hymenoptera: Braconidae) with two new species from Mongolia. Annales Historico-Naturales Musei Nationalis Hungarici 68: 215-225.

Kariya, M. 1932. Observations on a braconid species parasitic on ants, Konchu Sekai 36: 298-301 (In Japanese.)

Marsh, P. M. 1979. Family Braconidae. Pp. 144-295. In: Krombein, K. V., Hurd, P. D., Smith, D. R., and Burks, B. D. (Eds.). Catalog of Hymenoptera in America North of Mexico. Smithsonian Institution Press, Washington, D.C.

Marsh, P. M., Shaw, S. R., and Wharton, R. 1987. An identification manual for the North American genera of the family Braconidae (Hymenoptera). Memoirs of the Entomological Society of Washington 13: 1-98.

Muesebeck, C. F. W. 1922. A revision of the North American ichneumon-flies belonging to the subfamilies Neoneurinae and Microgasterinae. Proceedings of the United States National Museum 61: 1-76.

Olivier, E. 1893. [no title.] Annales Societe Entomologique France 1893: 71.

Pierre, A. 1893. Un parasite ·des fourmis *Elasmosoma berolinense* Ruthe. Revue Scientifique Bourbon Centre France 6: 112-114.

Quicke, D. L. J. and Achterberg, C. van 1990. Phylogeny of the subfamilies of the family Braconidae (Hymenoptera: Ichneumonoidea). Zoologische Verhandelingen Leiden 258: 1-95.

Shaw, M. R. and Huddleston, T. 1991. Classification and biology of braconid wasps (Hymenoptera: Braconidae). Handbooks for the Identification of British Insects 7(11): 1-126.

Shaw, S. R. 1992. Seven new North American species of *Neoneurus* (Hymenoptera: Braconidae). Proceedings of the Entomological Society of Washington 94: 26-47.

Shaw, S. R. 1993. Observations on the ovipositional behavior of *Neoneurus mantis*, an ant-associated parasitoid from Wyoming. Journal of Insect Behavior 6: 649-658.

Shenefelt, R. D. 1969. Pars 4. Braconidae 1, Hybrizoninae, Euphorinae, Cosmophorinae, Neoneurinae, Macrocentrinae. Pp. 1-176. In: Ferrière, Ch. and Vecht, J, van der (Eds.), Hymenopterorum Catalogus (nova editio), Dr. W. Junk, The Hague.

Wasmann, E. 1897. In: Marshall, T. A. Les Braconides. Pp. 188-189 In: Andre, E. (Ed.). Species des Hymenopteres d'Europe et d'Algerie, Volume V. Dubosclard, Paris.

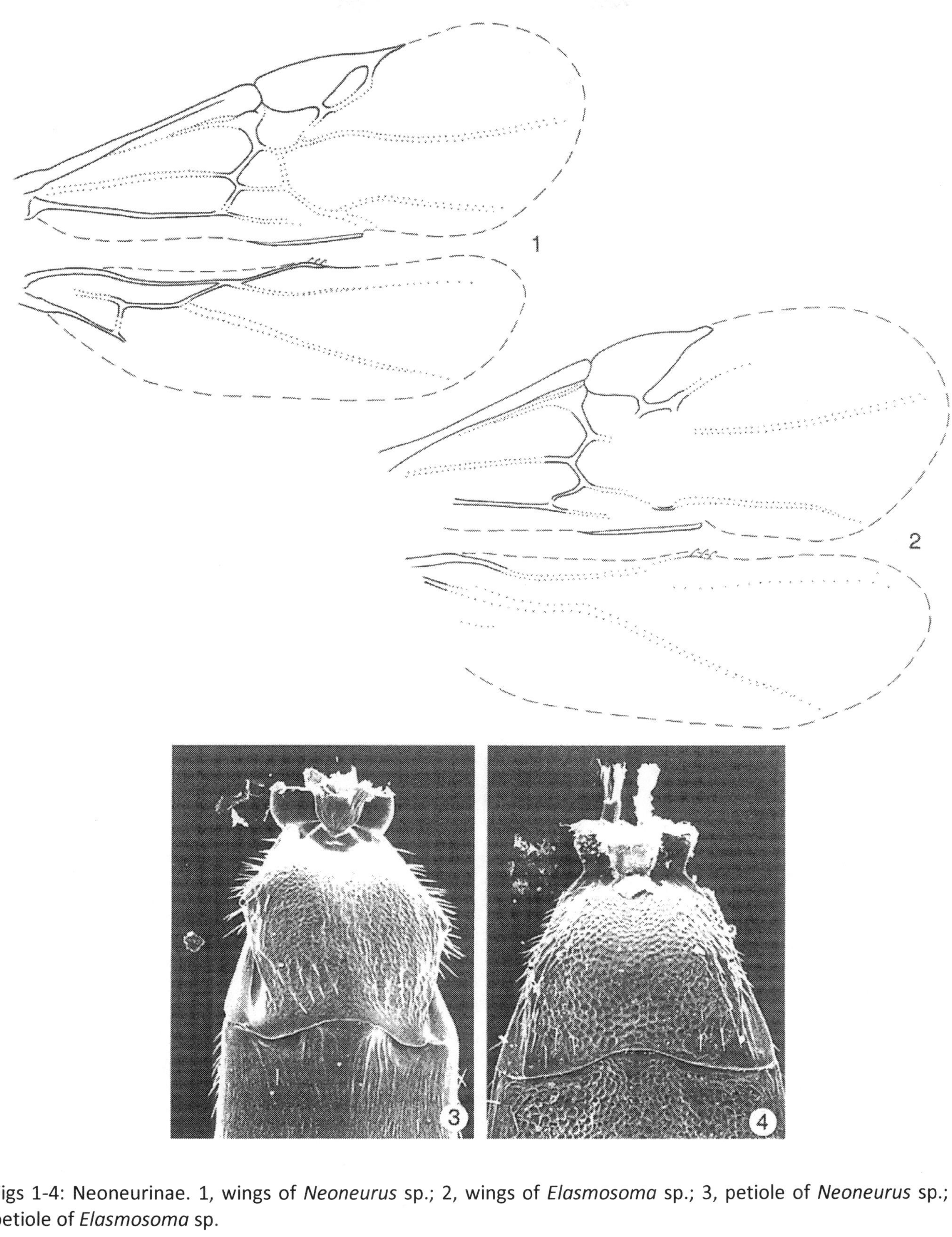

Figs 1-4: Neoneurinae. 1, wings of *Neoneurus* sp.; 2, wings of *Elasmosoma* sp.; 3, petiole of *Neoneurus* sp.; 4, petiole of *Elasmosoma* sp.

NEONEURINAE UPDATE

S.R. Shaw
Department of Ecosystem Science and Management, University of Wyoming, Laramie, Wyoming 82071, U.S.A.
braconid@uwyo.edu

Molecular phylogenetic analyses by Sharanowski *et al.,* (2011) and Stigenberg *et al.* (2015) concluded that neoneurines are nested inside the Euphorinae, corroborating the conclusion that the neoneurines should be classified as a tribe in the Euphorinae. Even before these publications Belokobylskij (2000) placed the neoneurines in the subfamily Euphorinae. Gómez Durán and van Achterberg (2011) have also adopted this placement. The Neoneurinae chapter of the manual continues to work well for the identification of the two known New World genera.

Annotated list of recent papers on New World Neoneurini

Belokobylskij SA (Ed). 2000. Key to the Insects of Russian Far East, Vol. 4. Dal'nauka, Vladivostok.
Includes six genera in a study of the Neoneurini of Eastern Russia (*Elasmosoma*, *Euneoneurus*, *Kollasmosoma*, *Neoneurus*, *Parelasmosoma* and *Sinoneoneurus*). Currently only two of those (*Elasmosoma* and *Neoneurus*) are known from North America, but it remains possible that some of them simply haven't been discovered yet.

Gómez Durán J-M, Achterberg C van. 2011. Oviposition behaviour of four ant parasitoids (Hymenoptera, Braconidae, Euphorinae, Neoneurini and Ichneumonidae, Hybrizontinae), with the description of three new European species. *ZooKeys* 125: 59-106.
The oviposition behavior of three neoneurine species attacking ants, representing three genera, was observed and filmed. The slow-motion movies are available on YouTube (search for *Elasmosoma, Kollasmosoma* and *Neoneurus*). They described three new neoneurine species from Europe and provided keys to the Palearctic species of the genera *Neoneurus* and *Kollasmosoma*.

Poinar G Jr. 2004. Behavior and development of *Elasmosoma* species (Braconidae: Hymenoptera), an endoparasitoid of *Formica* ants (Formicidae: Hymenoptera). *Parasitology* 128: 1-11.
The egg-laying behavior, embryonic development, larval development, and cocoon-forming behavior of an undescribed species of *Elasmosoma* was described.

Poinar G Jr, Miller JC. 2002. First fossil record of endoparasitism of adult ants (Formicidae: Hymenoptera) by Braconidae (Hymenoptera). *Annals of the Entomological Society of America* 9(1): 41-43.
They described a neoneurine wasp larva emerging from an adult ant (*Lasius* sp.) found as an inclusion in Eocene Baltic amber. This demonstrates that ant parasitism by neoneurines was established at least 40 million years ago. They speculated that the larva could be that of an extinct neoneurine, *Elasmosomites primordialis* Brues, which was already known from Baltic amber.

Sharanowski BJ, Dowling APG, Sharkey, MJ. 2011. Molecular phylogenetics of Braconidae (Hymenoptera: Ichneumonoidea), based on multiple nuclear genes, and implications for classification. *Systematic Entomology* 36: 549-572.
Treated the neoneurines as the tribe Neoneurini in the subfamily Euphorinae.

Shaw SR. 2007. A new species of *Elasmosoma* Ruthe (Hymenoptera: Braconidae: Neoneurinae) from the northwestern United States associated with western thatching ants, *Formica obscuripes* Forel, and *Formica obscuriventris clivea* Creighton (Hymenoptera: Formicidae). *Proceedings of the Entomological Society of Washington* 109(1): 1-8.

The new species observed by Poinar (2004) was subsequently described and named *Elasmosoma michaeli* by Shaw (2007) based on material from Oregon, Washington, and Wyoming. Shaw (2007) also provides an identification key to the described North American species of *Elasmosoma*.

Stigenberg J, Boring CA, Ronquist F. 2015. Phylogeny of the parasitic wasp subfamily Euphorinae (Braconidae) and evolution of its host preferences. *Systematic Entomology* 40(3): 570-591.

Conducted a molecular phylogenetic analysis concluding that *Elasmosoma* and *Neoneurus* are nested inside the subfamily Euphorinae.

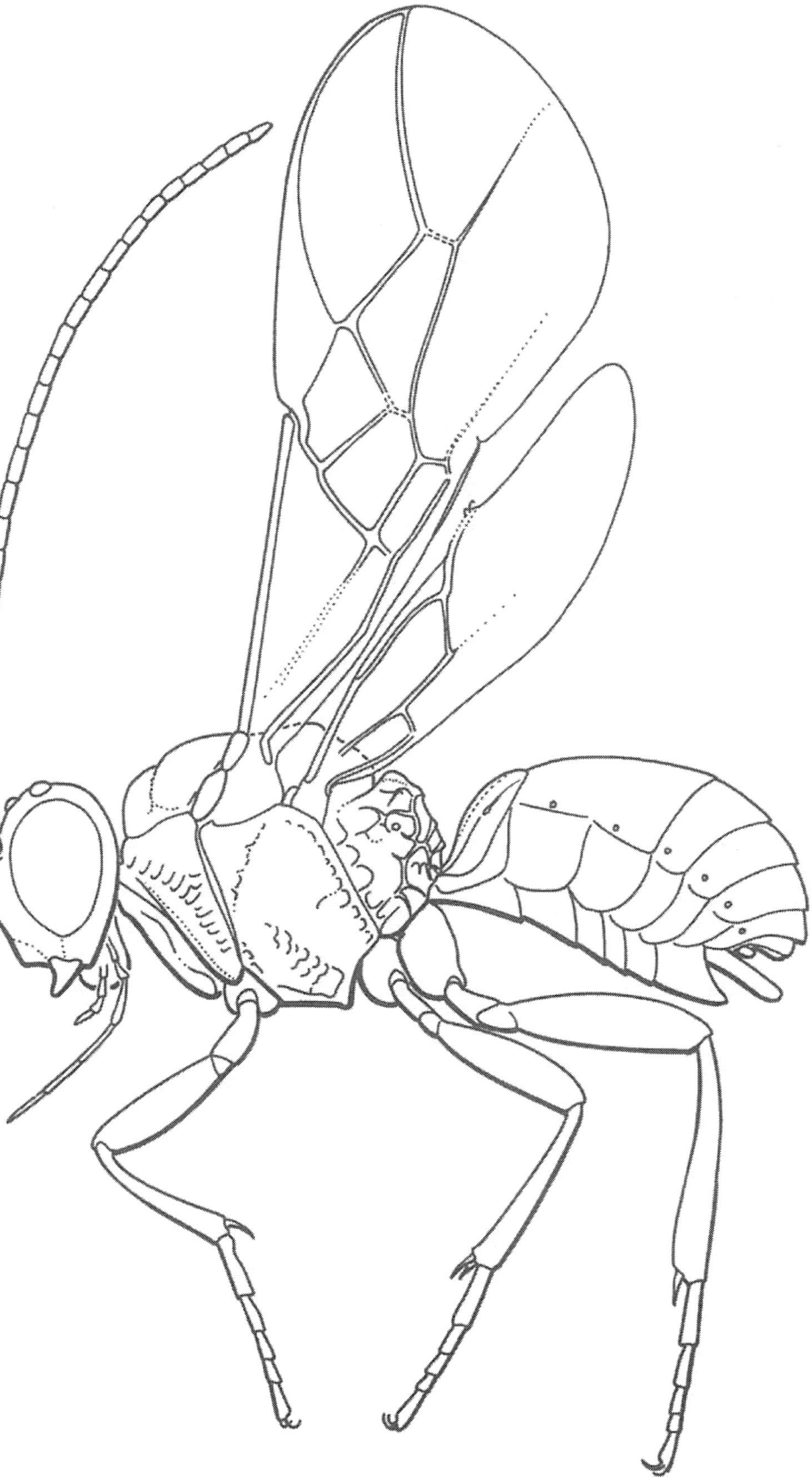

SUBFAMILY OPIINAE

R. A. WHARTON

Department of Entomology, Texas A&M University, College Station, Texas 77843, U.S.A.

INTRODUCTION. The Opiinae is one of the larger subfamilies of the Braconidae, containing over 1300 described species worldwide. Fischer, in a long series of papers beginning in 1956, revised the Opiinae on a world basis, considerably expanding our knowledge of the diversity of this economically important group of dipterous parasitoids. References to Fischer's publications can be found in his particularly useful regional summaries (Fischer, 1972, 1977, 1987), but Fischer's (1971) catalog of the World Opiinae has been rendered largely obsolete because of these same monographs. Several opiine species have been described from the New World since Fischer's (1977) monograph (Wharton and Marsh, 1978; Fischer, 1978, 1979a, b, 1980a, b, 1983a, b; De Santis, 1982). Additional publications focusing on other aspects of opiine systematics (Fischer, 1980b, 1984, 1986; De Santis, 1980; Wharton and Gilstrap, 1983; Wharton, 1983, 1984, 1987, 1988; Tobias and Jakimavicius, 1986; Buckingham and Sharkey, 1988; van Achterberg and Maetô, 1990) have also substantially improved our understanding of the New World fauna.

During the first half of the twentieth century, much of the research on the Opiinae was based on the conservative classification of Gahan (1915), in which most species were placed, ungrouped, in the genus *Opius* Wesmael. This classification could not adequately accommodate the huge number of species added by Fischer, and was gradually modified, culminating in the classification presented in Fischer (1972). Fischer (1972) still placed the vast majority of the opiine species in *Opius*, but segregated these into 36 subgenera, based in part on the earlier studies of Förster (1862) and Thomson (1895). New subgenera have since been added to this total (e.g. Fischer, 1984, 1986). Fischer (1972) also included 23 genera in the Opiinae, and placed these in four tribes. His tribal and generic concepts have been critically examined by Tobias (1967, 1977), Wharton and Marsh (1978), van Achterberg (1983), Wharton (1987, 1988) and van Achterberg and Maetô (1990), resulting in several changes. A great deal of work still remains, however, before a satisfactory higher classification can be achieved. Consequently, generic names may remain unstable for some time.

The relationship between opiines and alysiines is discussed in the chapter on the Alysiinae. The two groups form a monophyletic unit, linked to the cyclostome branch of the Braconidae by the morphology of their venom gland, and defined by their biology as koinobiont endoparasitoids of cyclorrhaphous Diptera (Wharton, 1988).

BIOLOGY. Opiines oviposit in either the egg or larval stage of their host and emerge from the host puparium (Pemberton and Willard, 1918a; van den Bosch and Haramoto, 1951; Prokopy and Webster, 1978). Detailed biologies are available for a few of the species attacking leaf-mining agromyzids and fruit-infesting tephritids, but most of our information on the Opiinae is limited to host records. Confirmed host records are available for about 275 species (Fischer, 1971, 1972, 1977, 1987). Even in the Nearctic region, which is fairly well studied, hosts are known for only 29% of the described species. All of these species are solitary endoparasitoids, with the vast majority of the rearings from either Agromyzidae or Tephritidae. Although at least 13 other families of Diptera have been recorded as hosts of the Opiinae (Fischer, 1971), many of these, particularly from the earlier literature, need confirmation. Some are clearly erroneous, as are all records from non-dipterous hosts. With the exception of *Opius*, there is some degree of host specificity at the generic level for most of the opiines. The species of *Ademon* Haliday all attack Ephydridae, and the species of *Biosteres* Förster s. s. (as defined by Wharton, 1988) have been reared almost exclusively from Anthomyiidae and plant-infesting Scathophagidae. Few exceptions are known (Fischer, 1969c, 1971). Where known, all species of *Doryctobracon* Enderlein and *Psyttalia* Walker are parasitoids of Tephritidae, as are most of those in *Diachasmimorpha* Viereck and *Utetes* Förster. Those species that attack fruit-infesting tephritids, even though they may have a fairly broad host range, have not been reared from tephritids infesting flower heads even though they will, on rare occasion, attack a gall-forming tephritid. Fischer (1969a, b, c) provides some interesting information on host ranges of several of the more commonly reared opiine species.

Research on the biology of tephritid parasitoids has covered a wide array of topics, from abdominal glands and their secretions (Hagen, 1953; Teles da Silva and Palma, 1985; Williams et al., 1988; Buckingham and Sharkey, 1988), acoustic signaling (Sivinski and Webb, 1989), and symbiotic viruses (Lawrence and Akin, 1990) to basic searching behavior (Glas and Vet, 1983). Good reviews of opiine developmental biology are provided by Fischer (1972) and Shaw and Huddleston (1991). Publications by Lawrence (1982, 1986a, b, 1988, 1990) and Lawrence et al. (1990) on host regulation and developmental physiology of the widely introduced *Diachasmimorpha longicaudaia*. (Ashmead) were a logical extension of the work exemplified by Greany et al. (1976, 1977), Lawrence et al. (1976) and Ashley et al. (1977), and nicely complement the meticulous studies of earlier workers such as Pemberton and Willard (1918a, b), Lathrop and Newton (1933), van den Bosch and Haramoto (1951), Newell et al. (1952), Nishida (1955), Balduf (1959), Stavraki-Paulopoulou (1966), and Kaya and Nishida (1968) on other parasitoids of tephritids. Recently, Wong et al. (1990) and Ramadan et al. (1989a, 1991, 1992, 1994) and Ramadan and Beardsley (1992) have examined many facets of the basic biology of the opiines introduced to Hawaii for tephritid biological control.

Studies on the opiine parasitoids of agromyzids have lagged somewhat behind those on tephritid parasitoids. Cameron's (1941) classic work on *Opius ilicie* Nixon was one of the first to thoroughly examine one of the agromyzid parasitoids. Minkenberg and van Lenteren (1986) summarized much of the subsequent behavioral work on the European species *Opius* pallipes Wesmael, and Petitt et al. (1992) and Petitt and Wietlisbach (1993) have extended this line of research to the morphologically similar New World species *O. dissitus* Muesebeck. Petitt's (1988) work on *O. dissitus* has made this one of the best-known opiine parasitoids of Agromyzidae in the New World. Another New World species commonly found on agromyzid pests, *O. dimidiaius* (Ashmead), has also been studied in some detail (Nelson and Roitberg, 1993). As with tephritid parasitoids, the number of studies on the biology of non-economically important hosts (e.g. Potter, 1985) is minimal. Similarly, few details have been published on those opiines attacking stem and seed-feeding agromyzids. De Stephani Perez (1902), Bengtsson (1926), Bremer and Kaufmann (1928), Burghele (1959), Wharton et al. (1990), and Grijpma and van Achterberg (1991) provide details on opiines attacking hosts other than Agromyzidae and Tephritidae.

BIOLOGICAL CONTROL POTENTIAL. Opiine braconids have been used primarily for the biological control of fruit-infesting tephritids and leaf-mining agromyzids. Although opiines have dominated the literature on the biological control of tephritids (Silvestri, 1914; Willard and Mason, 1937; Clausen et al., 1965; Bennett et al., 1977; Wharton, 1989; Knipling, 1992; Waterhouse, 1993; Baranowski et al., 1993), comparatively few species have been used against agromyzid pests (McLeod, 1954; Greathead, 1975; Minkenberg and van Lenteren, 1986; van der Linden, 1986; Neuenschwander et al., 1987; Petitt, 1984, 1988, 1993).The moderately successful programs against *Ceratitis capitata* (Wiedemann) (Medfly) and *Bactrocera dorsalis* (Hendel) (Oriental fruit fly) in Hawaii (Clausen et al., 1965; Gilstrap and Hart, 1987; Wharton, 1989) have resulted in a long-standing emphasis on opiines for the biological control of tephritids. Unfortunately, importation programs against tephritid pests in other parts of the world have seldom, if ever, achieved the degree of control experienced in Hawaii, even though species such as *D. longicaudata* have been established in several countries. The introduction of multivoltine, tropical tephritid parasitoids against univoltine tephritid pests has been particularly unsuccessful. Augmentation programs, however, have been more successful. Inundative releases of *Psyttalia concolor* (Szépligeti), for example, have been used for many years in southern Europe for managing populations of the olive fly, *Dacus oleae* (Gmelin) (e.g. Kapatos et al., 1977). Wong et al. (1991) and Knipling (1992) have been particularly active in promoting the use, for augmentation, of several of the opiine species established in Hawaii. The apparent success of some of these efforts (Wong et al., 1992; Sivinski et al., 1996) has prompted increased attention to details of mass rearing (Carey et al., 1988; Ramadan et al., 1989b; Sivinski and Smittle, 1990; Wong and Ramadan, 1992; Sivinski et al., 1994) as well as renewed interest in looking for alternative species to use in biocontrol. Regional surveys of Neotropical tephritids and their natural enemies (e.g. Aluja et al., 1990) are now fairly common. The references contained in the papers cited in this chapter provide a thorough introduction for those interested in the voluminous literature on the relatively few species of Opiinae used in tephritid biocontrol.

Much of the effort on the biological control of agromyzid pests has been directed towards some of the more successful eulophid parasitoids. Nevertheless, there is potential for the use of opiines, perhaps in combination with other natural enemies, since they are frequently reared from leaf-mining pests (Harding, 1965; Lema and Poe, 1979; Hendrikse and Zucchi, 1979; Hendrikse, 1980; Hendrikse et al., 1980; Kumar, 1984; Woets and van der Linden, 1982; Murphy, 1984) and are easier and less expensive to rear than most eulophids. Petitt (1993), in particular, has made excellent use of augmentative releases of *O. dissitus* for controlling species of *Liriomyza* Mik in Florida.

Fischer (1971) gives a good summary of the value of opiines in the natural control of leaf-mining Anthomyiidae.

COMMON GENERA. Recent advances in our understanding of the relationships within the Opiinae have led to significant changes in the generic names applied to most of the parasitoids of fruit-infesting tephritids. As noted by Shaw and Huddleston (1991), this leaves us with the problem of relating the biological literature to meaningful generic names. The two most widely used generic names in the Opiinae, *Opius* and *Biosteres*, have been the focus of much of this, recent work. The vast majority of the opiine species are still currently placed in *Opius*. *Opius* has long been recognized as a heterogeneous assemblage, but continued segregation of well-defined, monophyletic groups such as *Utetes* (Wharton, 1988)and *Euopius* Fischer s. s. (Fischer, 1972; Wharton, 1988) will help reduce *Opius* to a meaningful unit. The generic name *Biosteres* is perhaps most frequently encountered in publications on the biological control of fruit-infesting Tephritidae. Wharton (1988), however, restricted *Biosteres* to a relatively small group of species that attack leaf-mining Anthomyiidae. In the New World, nearly all of the tephritid parasitoids formerly placed in *Biosteres* are now in *Doryctobracon*, *Fopius* Wharton, or *Diachasmimorpha*. Wharton (1987) originally described *Fopius* as a subgenus of *Rhynchosteres* Fischer, and the latter may, with further study, prove to be the most appropriate generic name for these Old World species that have been introduced to the New World.

DISTRIBUTION. *Opius*, as currently defined, is widely distributed on all continents. *Utetes* and *Ademon* are also widespread, and *Utetes* has distinct tropical and north-temperate components in the New World. *Biosteres* s. s. is predominantly Holarctic, with some New World species occurring at higher elevations at least as far south as southern Mexico. *Doryctobracon* and *Exodontiella* Wharton are New World endemics, the former distributed throughout Latin America with its main tephritid hosts in the genus *Anastrepha* Schiner. *Fopius* and *Psyttalia* are Old World endemics with a few species introduced to the New World for biological control of tephritids. Although Wharton (1987, 1988) restricted *Diachasmimorpha* to Indo-Pacific species with an apically sinuate ovipositor, van Achterberg and Maetô (1990) have applied this name to Wharton's (1988) New World *mexicanus* Cameron species group which lacks the sinuate ovipositor, but otherwise closely resembles the Indo-Pacific species.

The Chilean opiine fauna is very poorly known. Based on what is known for other braconid subfamilies, however, there are likely to be some endemic elements, as well as some faunal elements common to other parts of South America that are missing in Chile.

DISTINGUISHING FEATURES. Although Fischer (1987) retained *Gnamptodon* Haliday in the Opiinae, most other workers have followed van Achterberg (1983) in placing this genus in a separate subfamily. Following exclusion of *Gnamptodon*, parasitoids of leaf-mining Nepticulidae (Lepidoptera), opiines can be defined as endoparasitoids of cyclorrhaphous Diptera, emerging from the host puparium. This biological feature serves to separate them from all but the Alysiinae. Alysiines have outwardly directed, non-overlapping (= exodont) mandibles. Opiines, with the exception of the very rarely encountered *Exodontiella* (see Wharton, 1988), do not.

Other features useful for identifying members of the Opiinae, such as tergal gland morphology, basal displacement of the radial cross vein (r) in the fore wing, loss of the epicnemial carina and often at least part of the occipital carina, are homoplasies: either not restricted to the Opiinae, or not true for all members of the subfamily. The Opiinae is thus one of the more difficult subfamilies to define as a monophyletic group on the basis of either external or internal morphology.

In many opiines there is a gap between the clypeus and mandibles when the mandibles are closed. This gives the appearance of a cyclostome opening similar to that found in the Braconinae, Doryctinae, Hormiinae, and Rogadinae. In the Opiinae, however, the labrum is flat or nearly so, and the clypeus is generally of a different form than that of more typical cyclostomes. This feature is discussed in more detail by Wharton (1988).

KEY TO NEW WORLD GENERA OF OPIINAE

Most species from general collections will key to *Opius* in couplet 14. Unless you are working specifically with parasitoids of Tephritidae, you can ignore *Psyttalia*. The tibial carina in couplet 3 will be difficult to see if the specimen is not positioned correctly and properly curated, but is essential for correct identification of *Utetes*.

1. Mandibles exodont: outwardly directed and not overlapping when closed (as in Fig. 10).. ***Exodontiella*** Wharton

Distribution: south-central Canada and western U.S.A.; known only from 3 species (one undescribed) and 5 specimens. Biology: unknown. References: Wharton (1977, 1988), Quicke and van Achterberg (1990).

\- Mandibles variable, but at least apical teeth always overlapping when closed (Figs 11, 22-25)..................2

2(1). Epicnemial carina present ventrally (Fig. 12). Fore wing 3RSb incomplete, not reaching wing margin (Fig. 1).. ***Ademon*** Haliday

Distribution: U.S.A. and Canada; 3 described species in New World. Biology: parasitoids of *Hydrellia* Robineau-Desvoidy (Ephydridae) in aquatic plants (De Stephani Perez, 1902; Burghele, 1959). Reference: Marsh (1979).

\- Epicnemial carina absent (Fig. 13). Fore wing 3RSb nearly always complete to wing margin as a tubular vein (Fig. 2)..3

3(2). Base of hind tibia with a sharp carina dorso-posteriorly (Fig. 26). Fore wing 2RS shorter than 3RSa (Fig. 2). Clypeal margin truncate to concave, thus forming a distinct opening between clypeus and mandibles when mandibles closed (Fig. 11). Hind wing m-cu absent (Fig. 2) .... ***Utetes*** Förster

Distribution: Canada to Argentina, including Caribbean; at least 35 described New World species. Biology: parasitoids of fruit-infesting Tephritidae (Balduf, 1959; Wharton and Marsh, 1978; Maier, 1981; Juliano and Borowicz, 1987). References: Fischer (1972, 1977, 1983c), Wharton (1988, elevation to generic rank and inclusion of *Bracanastrepha* Brèthes as a subgenus of *Utetes*).

\- Base of hind tibia without such a carina (Fig. 27). Relative lengths of fore wing 2RS and 3RSa variable, 2RS sometimes shorter than 3RSa (Figs 3, 4). Clypeus and hind wing m-cu variable: often as above, but frequently without distinct opening between clypeus and mandibles when mandibles closed (Figs 22-24); hind wing m-cu sometimes present (Figs 3, 4)..4

4(3). Occipital carina absent both dorsally and laterally (Figs 15, 17)..5

\- Occipital carina present, at least laterally (usually absent dorsally) (Figs 14, 16)..7

5(4). Fore wing 2RS distinctly shorter than 3RSa, with 3RSa greater than or equal to 1.2 times longer than 2RS; second submarginal cell relatively large (Figs 2, 5) .... ***Opius*** Wesmael (in part)

Distribution: southern U.S.A. to Argentina, including Caribbean; about45 described New World species key here. Biology: *O. paroulus* Wesmael and related species (as defined by Wharton, 1987, 1988) attack Agromyzidae; the few records for the remaining species are almost exclusively from Tephritidae. Except for the true *Bracanastrepha* (which key to couplet 3 above), all species in Fischer's (1972) Desmiostomatini key here or in couplet 6. References: Fischer (1977), Wharton (1983, 1988, discussion of Desmiostomatini and recognition of the *Opius bellus* Gahan species group).

\- Fore wing 2RS approximately equal to or longer than 3RSa, with 3RSa never more than 1.1 times longer than 2RS; second submarginal cell small (Figs 3, 4)..6

6(5). Fore wing m-cu arising distad 2RS, (RS+M)b absent (Fig. 3).................... ***Diachasmimorpha*** Viereck (in part)

Distribution: southern U.S.A. through Mexico; 3 described species: *D. tryoni* (Cameron) introduced to several tropical and subtropical localities for tephritid biocontrol, with establishment uncertain; plus 2 endemic species attacking Tephritidae in southwestern U.S.A. and Mexico. References: Pemberton and Willard (1918a, b), Ramadan et al. (1989a, b).

\- Fore wing m-cu arising basad or directly in line with 2RS, (RS+M)b often present (Fig. 4).. ***Doryctobracon*** Enderlein

Distribution: Mexico to Argentina, including Florida and Caribbean; endemic to the New World, with 14 described species. Biology: parasitoids of Tephritidae, primarily the genus *Anastrepha* (Keilin and Picado, 1913; Baker et al., 1944; Hernandez- Ortiz et al., 1994 and references therein). References: Fischer (1977), Wharton and Marsh (1978).

7(4). Fore wing r arising from extreme base of stigma; stigma long, narrow, weakly expanded apically (Fig. 6).. ***Eurytenes*** Marshall

Distribution: Canada to Mexico; uncommon; only one described species in New World. Biology: parasitoids of Agromyzidae breeding in grasses (Poaceae) (Fischer, 1969c). References: Fischer (1972), Wharton (1988).

\- Fore wing r arising distinctly distad extreme base of stigma; shape of stigma variable (Figs 5, 7, 8) 8

8(7). Clypeus large, without opening between clypeus and mandibles when mandibles closed (Figs 22, 23) and mandibles with large basal tooth (Figs 22, 23) 9

\- Clypeus variable, often short, with consequent opening between clypeus and mandibles when mandibles closed (as in Fig. 11). If opening absent or nearly so, mandible without basal tooth (Fig. 24, 25). Species in which the occurrence of an opening is difficult to determine or the basal mandibular tooth is poorly defined can be taken through both halves of the couplet 10

9(8). Fore wing 2RS equal to or longer than 3RSa (Fig. 7) ....***Biosteres*** Förster

Distribution: Canada and Alaska to Mexico; about 20 described species in New World. Biology: parasitoids of leaf-mining Anthomyiidae (Bengsston, 1926; Bremer and Kaufmann, 1928). References: Fischer (1977), Marsh (1979), Wharton (1987, 1988, restricted definition).

\- Fore wing 2RS shorter than 3RSa (Fig. 5) ....***Opius*** Wesmael (in part)

Distribution: Canada and Alaska to Argentina; about 35 described New World species key here. Biology: mostly parasitoids of Agromyzidae (Minkenberg and van Lenteren, 1986; Petitt and Wietlisbach, 1993). Reference: Fischer (1977).

10(8). Second submarginal cell of fore wing short, with 2RS equal to or longer than 3RSa (Figs 3, 8); stigma short, broad, discrete (Figs 3, 8). Mesoscutum with deep midpit (Fig. 18) or with completely crenulate notauli meeting posteriorly in a shallow depression (Fig. 28) 11

\- Second submarginal cell long, with 3RSa longer than 2RS (Figs 5, 9). Stigma in most species long and narrow (as in Figs 1, 6). Mesonotum frequently without midpit (Fig. 19); notauli, if present, almost never crenulate posteriorly 14

11(10). Broad opening present between clypeus and mandibles when mandibles closed (as in Fig. 11), clypeus short ....***Diachasma*** Förster

Distribution: U.S.A. and Canada; 3 0£ the 5 described New World species currently placed in *Diachasma* appear to be more closely related biologically and morphologically to New World *Diachasmimorpha* than to the type species of *Diachasma*. Biology: New World species are parasitoids of Tephritidae (Porter, 1928; Boush and Baerwald, 1967; Cameron and Morrison, 1977; Glas and Vet, 1983). References: Muesebeck (1956), Fischer (1977), Wharton and Marsh (1978).

\- Opening between clypeus and mandibles absent or nearly so when mandibles closed (Figs 24, 25), clypeus not short 12

12(11). Notauli crenulate, meeting posteriorly in shallow midpit (Fig. 28). Vertex, especially around ocellar triangle, distinctly sculptured (Fig. 29). Wings: Fig. 8 ....***Fopius*** Wharton

Distribution: endemic to Afrotropics and Indo-Pacific, but introduced and established at least in Costa Rica (Wharton et al., 1981). Biology: parasitoids of Tephritidae (van den Bosch and Haramoto, 1951, as *Opius*; Ramadan et al., 1992, as *Biosteres*). References: Fullaway (1951), Wharton (1987), van Achterberg and Maetô (1990).

\- Notauli unsculptured (Fig. 18). Vertex at most shallowly punctate (Fig. 14, 15) 13

13(12). Ovipositor sinuate at apex (Fig. 20). Notauli usually deeply impressed to midpit (Fig.18) ....***Diachasmimorpha*** Viereck s. s.

Distribution: endemic to Indo-Pacific, but one species, *D. longicaudata*, widely introduced in New World; established in Florida, Mexico to Costa Rica, and probably various parts of South America and Caribbean. Biology: parasitoids of fruit-infesting Tephritidae (Greany et al., 1976, 1977; Lawrence, 1982, 1986a, b, 1988, 1990; Williams et al., 1988; Lawrence et al., 1990; Ramadan et al., 1991; Sivinski et al., 1996). References: Fullaway (1951), Wharton and Gilstrap (1983), Wharton (1987, 1988), van Achterberg and Maetô (1990).

\- Ovipositor not sinuate apically (as in Fig. 21). Notauli usually very shallowly impressed or absent in front of midpit (as in Fig. 19) ....***Diachasmimorpha*** Viereck s. l.

Distribution: Canada to Mexico; about 5 described New World species. Biology: parasitoids of Tephritidae (Lathrop and Newton, 1933, as *Opius*; Wharton and Marsh, 1978, as *Biosteres*). References: Wharton (1988, as *mexicanus* species group); van Achterberg and Maetô (1990).

14(10). Having all of the following characters. Metasomal tergum 2 shorter than tergum 3 (Fig. 31), though this character is hard to see. Fore wing m-cu arising basad 2RS, (Rs+M)b present (Fig. 9) and at least partly thickened; fore wing 1-1A widely separated from wing margin (Fig. 9). Clypeus short, widely separated from mandibles (as in Fig. 11) ....***Psyttalia*** Walker

Distribution: Old World endemic; introduced several times in New World (from France and Hawaii) with at least one recovery in Costa Rica; establishment uncertain, thus rarely encountered. Biology: parasitoids of Tephritidae (Newell et al., 1952; Fischer, 1972; Kapatos et al., 1977). References: Fischer (1971, 1987), Wharton (1988).

\- Differing in one or more of the above characters, Metasomal tergum 2 nearly always equal to or longer than tergum 3 (Fig. 30). Fore wing m-cu usually arising distad 2RS (Fig. 5), seldom (20% of specimens) arising basad or directly in line with 2RS, thus (RS+M)b usually absent; fore wing 1-1A usually (90% of specimens) bent towards and nearly touching wing margin (as in Fig. 6). Clypeus variable, often (> 50%) extending to mandibles in species with long ovipositors (as in Fig. 24) .... ***Opius*** Wesmael

Distribution: Canada and Alaska to Argentina, including Caribbean, over 400 described New World species. Biology: parasitoids of Agromyzidae, Anthomyiidae, Chloropidae, Dropsophilidae, Ephydridae, Lonchaeidae, Scathophagidae, and Tephritidae (Fischer, 1971). References: Fischer (1977), Marsh (1979). Includes *Euopius* Fischer (Wharton, 1988).

REFERENCES

Achterberg, C. van 1983. Revisionary notes on the subfamily Gnaptodontinae, with description of eleven new species (Hymenoptera, Braconidae). Tijdschrift voor Entomologie 126: 25-57.

Achterberg, C. van and Maetô, K. 1990. Two new and aberrant species of Braconidae (Hymenoptera) from Japan. Zoologische Mededelingen Leiden 64: 59-70.

Aluja, M., Guillen, J., Liedo, P., Cabrera, M., Rios, E., de la Rosa, G., Celedonio, H., and Mota, D. 1990. Fruit infesting tephritids (Dipt: Tephritidae) and associated parasitoids in Chiapas, Mexico. Entomophaga 35: 39-48.

Ashley, T. R., Greany, P. D., and Chambers, D. L. (1976) 1977. Adult emergence in *Biosteres* (*Opius*) *longicaudatus* and *Anastrepha suspensa* in relation to the temperature and moisture concentration of the pupation medium. Florida Entomologist 59: 391-396.

Baker, A. C., Stone, W. E., Plummer, C. C., and McPhail, M. 1944. A review of studies on the Mexican fruit fly and related Mexican species. United States Department of Agriculture Miscellaneous Publication 531: 1-155.

Balduf, W. V. 1959. Obligatory and facultative insects in rose hips; their recognition and bionomics. Ilinois Biological Monographs 26: 32-40. Baranowski, R., Glenn, H., and Sivinski, J. 1993. Biological control of the Caribbean fruit fly (Diptera: Tephritidae). Florida Entomologist 76: 245-251.

Bengtsson, S. 1926. Braconologische Notizen, II. Zur Kenntnis der Schmarotzer der Rübenfliege (Pegomyia hyoscyami Panz.). Entomologisk Tidskrift 47: 51-64.

Bennett, F. D., Yaseen, M., Beg, M. N., and Sommeijer, M. J. 1977. *Anastrepha* spp.-investigations on their natural enemies and establishment of *Biosteres longicaudatus* in Trinidad, West Indies. Commonwealth Institute of Biological Control Technical Bulletin 18: 1-12.

Bosch, R. van den and Haramoto, F. H. 1951. *Opius oophilus* Fullaway, an egg-larval parasite of the Oriental fruit fly discovered in Hawaii. Proceedings of the Hawaiian Entomological Society 14: 251-255.

Boush, G. M. and Baerwald, R. J. 1967. Courtship behavior and evidence for a sex pheromone in the apple maggot parasite, *Opius alloeus*. Annals of the Entomological Society of America 60: 865-866.

Bremer, H. and Kaufmann, 0. 1928. Die natürlichen Feinde der Rübenfliege. Arbeiten biologische Reichsanstalt für Land- und Forstwirtschaft 16: 520-555.

Buckingham, G. R. and Sharkey, M. R. 1988. Abdominal exocrine glands in Braconidae (Hymenoptera). Pp. 199-242. In: Gupta, V. K. (Ed.). Advances in Parasitic Hymenoptera Research. Brill, Leiden.

Burghele, A. 1959. Contributii la studiul unor Himenoptere care paraziteaza stadii acvatice de insecte. Analele dell'Universita C. I. Parhon, Serie St. Naturii 22: 143-169.

Cameron, E. 1941. The biology and post-embryonic development of *Opius ilicis* n. sp., a parasite of the holly leaf-miner (*Phytomyza ilicis* Curt.). Parasitology 33: 151-169.

Cameron, P. J. and Morrison, F. 0. 1977. Analysis of mortality in the apple maggot, *Rhagoletis pomonella* (Diptera: Tephritidae), in Quebec. Canadian Entomologist 109: 769-788.

Carey, J. R., Wong, T. T. Y., and Ramadan.M. M. 1988. Demographic framework for parasitoid mass rearing: Case study of *Biosteres tryoni*, a larval parasitoid of tephritid fruit flies. Theoretical Population Biology 34: 279-296.

Clausen, C. P., Clancy, D. W., and Chock, Q. C. 1965. Biological control of the Oriental fruit fly (*Dacus dorsalis* Hendel) and other fruit flies in Hawaii. United States Department of Agriculture Technical Bulletin 1322: 1-102.

De Santis, L. 1980. Catálogo de los himenópteros brasileños de la serie parasitica incluyendo Bethyloidea. Publicación de la Universidad Federal de Parana. 395 pp.

De Santis, L. 1982. Tres notas sabre dipteros agromicidos de la Republica Argentina y sus Himenopteros parastoides (Insecta). III. Dos nuevos Himenopteros parasitoides de Agromycidos argentinos (Braconidae y Pteromalidae*)*. Revista de la Sociedad Entomólogica Argentina 41: 326-330.

De Stephani Perez, T. 1902. Osservazioni biologiche sopra un Braconide acquatico, *Giardinaia urinator*, e descrizione di due altri Imenotteri nuovi. Zoologische Jahrbücher Abteilung für Systematik Ökologie und Geographie der Tiere 15: 625-634.

Fischer, M. 1969a. Notes on reared Opiinae (Hymenoptera, Braconidae). Proceedings of the Royal Irish Academy 67: 217-228.

Fischer, M. 1969b. Die von Dr. H. Buhr gezüchteten Opiinae. Zeitschrift für angewandte Zoologie 56: 65-88.

Fischer, M. 1969c. Über die von Dr. J. T. Nowakowski aus Agromyzidae und Ephydridae gezüchteten Opiinae (Hymenoptera, Braconidae). Polskie Pismo Entomologiczne 39: 369-380.

Fischer, M. 1971. Hym. Braconidae. World Opiinae. Index of Entomophagous Insects. Le Francois, Paris. 189 pp.

Fischer, M. 1972. Hymenoptera: Braconidae (Opiinae I). Das Tierreich 91: 1-620.

Fischer, M. 1977. Hymenoptera: Braconidae (Opiinae II-Amerika). Das Tierreich 96: 1-1001.

Fischer, M. 1978. Übersicht über die Neotropischen Arten der Untergattung *Opius* Wesmael s. str. mit Beschreibung von acht neuen Spezies und weitere Angaben über amerikanische Opiinen (Hymenoptera, Braconidae). Bollettino dell' Istituto di Entomologia Agraria 10: 137-168.

Fischer, M. 1979a. Neotropische Opiinae: Neue Arten der Gattungen *Euopius* Fischer und *Opius* Wesmael (Hymenoptera, Braconidae), Annalen des Naturhistorischen Museum Wien 82: 479-516.

Fischer, M. 1979b. Zur Kenntnis der Artenvielfalt bei den Opiinen-Wespen in der neotropischen Region (Hymenoptera, Braconidae, Opiinae). Polskie Pismo Entomologiczne 49: 227-297.

Fischer, M. 1980a. Neue Forschungsergebnisse in der Taxonomie neotropischer Opiinae (Hymenoptera, Braconidae). Polskie Pismo Entomologiczne 50: 215-269.

Fischer, M. 1980b. Zur Synonymie von Opiinae, eine Neubeschreibung, Redeskriptionen und Ergänzungen von Beschreibungen (Hymenoptera, Braconidae). Instituto di Entomologia Agraria dell'Universita degli Studi ed Osservatorio Regionale per le Malattie delle Piante Palermo 10: 227-250.

Fischer, M. 1983a. Neue Opiinae der Gattungen *Biosteres* Förster, *Aspilodemon* Fischer und *Opius* Wesmael aus der neotropischen Region (Hymenoptera, Braconidae). Sitzungsberichte der Österreichischen Akademie der Wissenschaften, Mathematisch-Naturwissenschaft Klasse I 192: 59-93.

Fischer, M. 1983b. Neubeschreibungen von neotropischen Opiinae aus den Gattungen *Desmiostoma*, Bracanasirepha und *Opius* (Hymenoptera, Braconidae). Entomologische Abhandlungen und Berichte aus dem Staatlichen Museum für Tierkunde in Dresden 47: 65-94.

Fischer, M. 1983c. Illustrierte Redeskriptionen von Opiinen aus der USSR und neuer Bestimmungsschlüssel zu den palaarktischen Arten des subgenus *Utetes* Förster (Hymenoptera, Braconidae). Annalen des Naturhistorischen Museum in Wien 85: 13-37.

Fischer, M. 1984. Deskriptionen und Redescriptionen von Opiinae (Hymenoptera, Braconidae) aus der palaarktischen Region. Annalen des Naturhistorischen Museum Wien 86: 85-131.

Fischer, M. 1986. Neue Bestimmungsschlüssel für palaarrktische Opiinae, neue Subgenera, Redeskriptionen und eine neue Art (Hymenoptera, Braconidae). Annalen des Naturhistorischen Museum Wein 88/89: 607-662.

Fischer, M. 1987. Hymenoptera: Opiinae III-äthiopische, orientalische, australische und ozeanische Region. Das Tierreich 104: 1-734.

Förster, A. 1862. Synopsis der Familien und Gattungen der Braconen. Naturhistorischer Verein der Rheinlande Verhandelingen 19: 225-288.

Fullaway, D. T. 1951. Review of the Indo-Australasian parasites of the fruit flies (Tephritidae). Proceedings of the Hawaiian Entomological Society 14: 243-250.

Gahan, A B. 1915. A revision of the North American ichneumon-flies of the subfamily Opiinae. Proceedings of the United States National Museum 49: 63-95.

Gilstrap, F. E. and Hart, W. G. 1987. Biological control of the Mediterranean fruit fly in the United States and Central America. United States Department of Agriculture ARS-56: 1-64.

Glas, P. C. G. and Vet, L. E. M. 1983. Host-habitat location and host location by *Diachasma alloeum* Muesebeck (Hym.; Braconidae), a parasitoid of *Rhagoletis pomonella* Walsh. (Dipt., Tephritidae). Netherlands Journal of Zoology 33: 41-54.

Greany, P. D., Ashley, T. R., Baranowski, R. M., and Chambers, D. L. 1976. Rearing and life history studies on *Biosteres* (*Opius*) *longicaudatus* (Hym.: Braconidae), Entomophaga 21: 207-215.

Greany, P. D., Tumlinson, J. H., Chambers, D. L., and Boush, G. M. 1977. Chemically-mediated host finding by *Biosteres* (*Opius*) *longicaudatus*, a parasitoid of tephritid fruit fly larvae. Journal of Chemical Ecology 3: 189-195.

Greathead, D. J. 1975. Biological control of the beanfly, *Ophiomyia phaseoli* [Dipt.: Agromyzidae], by *Opius* spp. [Hym.: Braconidae] in the Hawaiian Islands. Entomophaga 20: 313-316.

Grijpma, P. and Achterberg, C. van 1991. Host and host location of *Atoreuteus striatus* Foerster (Hymenoptera: Braconidae: Opiinae). Entomologische Berichten 51: 79-81.

Hagen, K. S. 1953. A premating period in certain species of the genus *Opius* (Hymenoptera: Braconidae). Proceedings of the Hawaiian Entomological Society 15: 115-116.

Harding, J. A. 1965. Parasitism of the leaf miner *Liriomvza munda* in the winter garden area of Texas. Journal of Economic Entomology 58: 442-443.

Hendrikse, A. 1980. A method for mass rearing two braconid parasites (*Dacnusa sibirica* and *Opius pallipes*) of the tomato leafminer (*Liriomvza bryoniae*). Mededelingen van de Faculteit Landbouwwetenschappen Rijksuniversiteit Gent 45: 563-571.

Hendrikse, A. and Zucchi, R. 1979. The importance of observing parasite behavior for the development of biological control of the tomato leafrniner. Mededelingen van de Faculteit Landbouwwetenschappen Rijksuniversiteit Gent. 44: 107-116.

Hendrikse, A., Zucchi, R., Lenteren, J. C. van, and Woets, J. 1980. *Dacnusa sibirica* Telenga and *Opius pallipes* Wesmael (Hym., Braconidae) in the control of the tomato leafminer *Liriomyza bryoniae* Kalt. BulJetin IOBC/WPRS 3: 83-98.

Hernandez-Ortiz, V., Perez-Alonso, R., and Wharton, R. A. 1994. Native parasitoids associated with the genus *Anasirepha* (Dipt.: Tephritidae) in Los Tuxtlas, Veracruz, Mexico. Entomophaga 39: 171-178.

Juliano, S. A. and Borowicz, V. A. 1987. Parasitism of a frugivorous fly, *Rhagoletis corniuora*, by the wasp *Opius richmondi*: relationships to fruit and host density. Canadian Journal of Zoology 65: 1326-1330.

Kapatos, E., Fletcher, B. S., Pappas, S., and Laudeho, Y. 19?'7. The release of *Opius concolor* and *O. concolor* var. *siculus* (Hym.: Braconidae) against the spring generation of *Dacus oleae* (Diptera: Trypetidae) on Corfu. Entomophaga 22: 265-270.

Kaya, H. K. and Nishida, T. 1968. Superparasitism by *Opius oophilus*, a parasite of the Oriental fruit fly. Hawaii Agricultural Experiment Station Technical Bulletin 73: 1-31.

Knipling, E. F. 1992. Principles of insect parasitism analyzed from new perspectives: practical implications for regulating insect populations by biological means. United States Department of Agriculture Handbook 693: 1-337.

Kumar, A. 1984. Incidence of pupal parasitism on *Chromatomyia horticola* (Gour.) (Diptera: Agromyzidae) in some parts of Uttar Pradesh. Annals of Entomology 2: 37-44.

Lathrop, F. H. and Newton, R. C. 1933. The biology of *Opius melleus* Gahan, a parasite of the blueberry maggot. Journal of Agricultural Research 46: 143-160.

Lawrence, P. 0. 1982. *Biosteres longicaudatus*: developmental dependence on host (*Anastrepha suspensa*) physiology. Experimental Parasitology 53: 396-405.

Lawrence, P. 0. 1986a. Host-parasite hormonal interactions: an overview. Journal of Insect Physiology 32: 295-298.

Lawrence, P. 0. 1986b. The role of 20-hydroxyecdysone in the moulting of *Biosteres longicaudatus*, a parasite of the Caribbean fruit fly, *Anastrepha suspensa*. Journal of Insect Physiology 32: 329-337.

Lawrence, P. 0. 1988. Superparasitism of the Caribbean fruit fly, *Anastrepha suspensa* (Diptera: Tephritidae), by *Biosteres longicaudatus* (Hymenoptera: Braconidae): implications for host regulation. Annals of the Entomological Society of America 81: 233-239.

Lawrence, P. 0. 1990. Serosal cells of *Biosteres longicaudatus* (Hymenoptera: Braconidae): ultrastructure and release of polypeptides. Archives of Insect Biochemistry and Physiology 13: 199-216.

Lawrence, P. 0. and Akin, D. 1990. Virus-like particles from the poison glands of the parasitic wasp, *Biosteres longicaudatus* (Hymenoptera: Braconidae). Canadian Journal of Zoology 68: 539-546.

Lawrence, P. 0., Baker, F. C., Tsai, L. W., Miller, C. A., Schooley, D. A., and Geddes, L. G: 1990. JH III levels in larvae and pharate pupae of *Anastrepha suspensa* (Diptera: Tephritidae) and in larvae of the parasitic wasp *Biosteres longicaudatus* (Hymenoptera: Braconidae). Archives of Insect Biochemistry and Physiology 13: 53-62.

Lawrence, P. 0., Baranowski, R. M., and Greany, P. D. 1976. Effect of host age on development of *Biosteres* (=*Opius*) *longicaudatus*, a parasitoid of the Caribbean fruit fly, *Anastrepha suspensa*. Florida Entomologist 59: 33-39.

Lema, K. M. and Poe, S. L. 1979. Age specific mortality of *Liriomyza sativae* due to *Chrysonotomyia formosa* and parasitization by *Opius dimidiaius* and *Chrysonotomyia formosa*. Environmental Entomology 8: 935-937.

Linden, A. van der 1986. Addition of the exotic leafminer parasites *Chrysocharis parksi* and *Opius dimidiatus* to the native Dutch parasite complex on tomato. Mededelingen van de Faculteit Landbouwwetenschappen Rijksuniversiteit Gent 51: 1009-1016.

Maier, C. T. 1981. Parasitoids emerging from puparia of *Rhagoletis pomonella* (Diptera: Tephritidae) infesting hawthorn and apple in Connecticut. The Canadian Entomologist 113: 867-870.

Marsh, P. M. 1979. Family Braconidae. Pp. 144-295. In: Krombein, K. V., Hurd, P. D., Smith, D. R., and Burks, B. D. (Eds.). Catalog of Hymenoptera in America North of Mexico. Smithsonian Institution Press, Washington, D.C.

McLeod, J. H. 1954 (1953). Statuses of some introduced parasites and their hosts in British Columbia. Proceedings of the Entomological Society of British Columbia 50: 19-27.

Minkenberg, 0. P. J.M. and Lenteren, J.C. van 1986. The leafminers *Liriomyza bryoniae* and *L. trifolii* (Diptera, Agromyzidae), their parasites and host plants: a review. Agricultural University Wageningen Papers 86-2: 1-50.

Muesebeck, C. F. W. 1956. On *Opius ferrugineus* Gahan and two closely similar new species. Entomological News 67: 99-102.

Murphy, S. T. 1984. Final report on surveys in the Neotropical/Nearctic Region for natural enemies of *Liriomyza trifolii* and *L. sativae* (Diptera, Agromyzidae). In: FAO/CIBC contract#: TCP/SEN/2202-1 AGOA. Commonwealth Institute of Biological Control West Indian Station, Curepe, Trinidad. 28 pp.

Nelson, J. M. and Roitberg, B. D. 1993. Factors governing host discrimination by *Opius dimidiatus* (Ashmead) (Hymenoptera: Braconidae). Journal of Insect Behavior 6: 13-24.

Neuenschwander, P., Murphy, S. P., and Coly, E. V. 1987. Introduction of exotic parasitic wasps for the control of *Liriomuza trifolii* (Dipt., Agromyzidae) in Senegal. Tropical Pest Management 33: 290-297.

Newell, I. M., Mitchell, W. C., and Rathburn, F. L. 1952. Infestation norms for *Dacus cucurbitae* in *Momordica balsamina*, and seasonal differences in activity of the parasite, *Opius fletcheri*. Proceedings of the Hawaiian Entomological Society 14: 497-508.

Nishida, T. 1955. Natural enemies of the melon fly, *Dacus cucurbitae* Coq. in Hawaii. Annals of the Entomological Society of America 48: 171-178.

Pemberton, C. E. and Willard, H. F. 1918a. Interrelations of fruit-fly parasites in Hawaii. Journal of Agricultural Research 12: 285-303.

Pemberton, C. E. and Willard, H.F. 1918b. A contribution to the biology of fruit-fly parasites in Hawaii. Journal of Agricultural Research 15: 419-466.

Petitt, F. L. 1984. Oviposition behavior of *Opius dissitus* and its use for management of *Liriomyza sativae* on greenhouse tomatoes. Pp. 81-86. In: Poe, S. L. (Ed.). Proceedings of the Fourth Annual Industry Conference on Leafminer. Society of American Florists, Alexandria, Virginia.

Petitt, F. L. 1988. Temperature-dependent development and influence of larval instars of *Liriomyza sativae* Blanchard on parasitization by *Opius dissitus* Muesebeck. Ph. D. dissertation, University of Florida, Gainesville.

Petitt, F. L. 1993. Biological control in the integrated pest management programs at The Land, Epcot Center. Bulletin IOBC/WPRS 16: 129-132.

Petitt, F. L., Tutlings, T. C. J., and Wolf, S. P. 1992. Adult experience modifies attraction of the leafminer parasitoid *Opius dissitus* (Hymenoptera: Braconidae) to volatile semiochemicals. Journal of Insect Behavior 5: 623-634.

Petitt, F. L. and Wietlisbach, D. 0. 1993. Effects of host instar and size on parasitization efficiency and life history parameters of *Opius dissitus*. Entomologia Experimentalis et Applicata 66: 22.7-236.

Porter, B. A. 1928. The apple maggot. United States Department of Agriculture Technical Bulletin 66: 1-48.

Potter, D. A. 1985. Population regulation of the native holly leafminer, *Phytomyza ilicola* Loew (Diptera: Agromyzidae) on American holly. Oecologia 66: 499-505.

Prokopy, R. J. and Webster, R. P. 1978. Oviposition deterring pheromone of *Rhagoletis pomonella*; a kairomone for its parasitoid *Opius lectus*. Journal of Chemical Ecology 4: 481-494.

Quicke, D. L. J. and Achterberg, C. van 1990. Phylogeny of the subfamilies of the family Braconidae (Hymenoptera: Ichneumonidae), Zoologische Verhandelingen Leiden 258: 1-95.

Ramadan, M. M. and Beardsley, J. W. 1992. Supernumerary molts in first instar *Biosteres tryoni* (Cameron) (Hymenoptera: Braconidae: Opiinae). Proceedings of the Hawaiian Entomological Society 31: 235-237.

Ramadan, M. M., Wong, T. T. Y., and Beardsley, J. W. 1989a. Survivorship, potential, and realized fecundity of *Biosteres tryoni* [Hymenoptera: Braconidae], a larval parasitoid of *Ceratitis capitata* [Diptera: Tephritidae]. Entomophaga 34: 291-297.

Ramadan, M. M., Wong, T. T. Y., and Beardsley, J.W., Jr. 1989b. Insectary production of *Biosteres* tryoni (Cameron) (Hymenoptera: Braconidae), a larval parasitoid of *Ceratitis capitata* (Wiedemann) (Diptera: Tephritidae). Proceedings of the Hawaiian Entomological Society 29: 41-48.

Ramadan, M. M., Wong, T. T. Y., and Beardsley, J. W. 1992. Reproductive behavior of *Biosteres arisanus* (Sonan) (Hymenoptera: Braconidae), an egg-larval parasitoid of the Oriental fruit fly. Biological Control. 2: 28-34.

Ramadan, M. M., Wong, T. T. Y., and Mclnnis, D. 0.1994. Reproductive biology of *Biosteres arisanus* (Sonan) (Hymenoptera: Braconidae: Opiinae), an egg larval parasitoid of *Bactrocera dorsalis* (Hendel). Biological Control 4: 93-100.

Ramadan, M. M., Wong, T. T. Y., and Wong, M.A. 1991. Influence of parasitoid size and age on male mating success of Opiinae (Hymenoptera: Braconidae), larval parasitoids of fruit flies (Diptera: Tephritidae). Biological Control 1: 248-255.

Shaw, M. R. and Huddleston, T. 1991. Classification and biology of braconid wasps. Handbooks for the Identification of British Insects 7(11): 1-126.

Silvestri, F. 1914. Report of an expedition to Africa in search of the natural enemies of fruit flies (Trypaneidae) with descriptions, observations and biological notes. Bulletin of the Division of Entomology, Board of Agriculture and Forestry, Territory of Hawaii 3: 1-176.

Sivinski, J., Calkins, C. 0., and Baranowski, R. 1994. A container for eclosion and release of adult insects prior to mass release. Florida Entomologist 77: 513-515.

Sivinski, J., Calkins, C. 0., Baranowski, R., Harris, D., Brambila, J., Diaz, J., Burns, R., Holler, T., and Dodson, G. 1996. Suppression of a Caribbean fruit fly (*Anastrepha suspensa* (Loew) Diptera: Tephritidae) population through augmented releases of the parasitoid *Diachasmimorpha longicaudata* (Ashmead) (Hymenoptera: Braconidae). Biological Control 6: 177-185.

Sivinski, J. and Smittle, B. 1990. Effects of gamma radiation on the development of the Caribbean fruit fly (*Anastrepha suspensa*) and the subsequent development of its parasite *Diachasmimorpha longicaudata*. Entomologia Experimentalis et Applicata 55: 295-297.

Sivinski, J. and Webb, J. C. 1989. Acoustic signals produced during courtship in *Diachasmimorpha* (=*Biosteres*) *longicaudata* (Hymenoptera: Braconidae) and other Braconidae. Annals of the Entomological Society of America 82: 116-120.

Stavraki-Paulopoulou, H. 1966. Contribution a l'étude de la capacité reproductrice et de la fécondité réelle d'*Opius concolor* Szépl. (Hymenoptera-Braconidae). Annales des épiphyties 17: 391-435.

Teles da Silva, M. and Palma, M. d. F. M. 1985. Pygidial gland in males of *Doryctobracon areolatus* (Szépligeti) (Hymenoptera, Braconidae). Revista Brasiliera de Entomologia 29: 571-574.

Thomson, C. G. 1895. Bidrag till Braconidernas kannedom. Opuscula Entomologica 20: 2141-2339. Tobias, V. I. 1967. A review of the classification, phylogeny and evolution of the family Braconidae (Hymenoptera). Entomological Review 46: 38.7-399.

Tobias, V. I. 197.7. The genus *Opius* Wesm. (Hymnoptera, Braconidae) as parasites of fruit flies (Diptera, Tephritidae). Entomological Review 56: 132-139.

Tobias, V. I. and Jakimavicius, A. 1986. Subfamily Opiinae. Pp. 7-100. In: Tobias, V. I. (Ed.). Identification of Insects of European USSR. Volume Ill, Part V. Academia Nauk, Leningrad. (In Russian.)

Waterhouse, D. F. 1993. Biological Control: Pacific Prospects-Supplement 2. Australian Centre for International Agricultural Research, Canberra. 138 pp.

Wharton, R. A. 1977. Exodontiellini, a new tribe of Opiinae with exodont mandibles (Hymenoptera: Braconidae). Pan-Pacific Entomologist 53: 297-303.

Wharton, R. A. 1983. Variation in *Opius hirtus* Fischer and discussion of *Desmiostoma* Foerster (Hymenoptera: Braconidae). Proceedings of the Entomological Society of Washington 85: 327-330.

Wharton, R. A. 1984. The status of certain Braconidae (Hymenoptera) cultured for biological control programs, and description of a new species of *Macrocentrus*. Proceedings of the Entomological Society of Washington 86: 902-912.

Wharton, R. A. 1987. Changes in nomenclature and classification of some opiine Braconidae (Hymenoptera). Proceedings of the Entomological Society of Washington 89: 61-73.

Wharton, R. A. 1988. Classification of the braconid subfamily Opiinae (Hymenoptera). The Canadian Entomologist 120: 333-360.

Wharton, R. A. 1989. Classical Biological Control of Fruit-infesting Tephritidae. Pp. 303-313. In: Robinson, A. S. and Hooper, G. (Eds.). Fruit Flies, Their Biology, Natural Enemies and Control. Elsevier Science Publishers B.V., Amsterdam.

Wharton, R. A., Austin, A. D., and Thomson, J. A. 1990. Two new species of *Opius* Wesmael (Hymenoptera: Braconidae) parasitic on Drosophilidae (Diptera) in Australia. Journal of the Australian Entomological Society 29: 11-18.

Wharton, R. A. and Gilstrap, F. E. 1983. Key to and status of opiine braconid (Hymenoptera) parasitoids used in biological control of *Ceratitis* and *Dacus s. l.* (Diptera: Tephritidae). Annals of the Entomological Society of America 76: 721-742.

Wharton, R. A., Gilstrap, F. E., Rhode, R. H., Fischel, M., M., and Hart, W. C. 1981. Hymenopterous egg-pupal and larval-pupal parasitoids of *Ceratitis capitata* and *Anastrepha* species in Costa Rica. Entomophaga 26: 285-290.

Wharton, R. A. and Marsh, P. M. 1978. New World Opiinae (Hymenoptera: Braconidae) parasitic on Tephritidae (Diptera). Journal of the Washington Academy of Sciences 68: 147-167.

Willard, H. F. and Mason, A. C. 1937. Parasitization of the Mediterranean fruit fly in Hawaii, 1914-33. United States Department of Agriculture Circular 439: 1-18.

Williams, H., Wong, M.A., Wharton, R. A., and Vinson, S. B. 1988. Hagen's gland morphology and chemical content analysis for three species of parasitic wasps (Hymenoptera: Braconidae). Journal of Chemical Ecology 14: 1727-1736.

Woets, J. and Linden, A. van der 1982. On the occurrence of *Opius pallipes* Wesmael and *Dacnusa sibirica* Telenga (Braconidae) in cases of natural control of the tomato leafminer *Liriomyza bryoniae* Kalt. (Agromyzidae) in some large greenhouses in the Netherlands. Mededelingen van de Faculteit Landbouwwetenschappen Rijksuniversiteit Gent 47: 533-540.

Wolcott, G. N. 1948. The insects of Puerto Rico: Hymenoptera. The Journal of Agriculture of the University of Puerto Rico 32: 749-873.

Wong, T. T. Y. and Ramadan, M. M. 1992. Mass rearing biology of larval parasitoids (Hymenoptera: Braconidae: Opiinae) of tephritid flies (Diptera: Tephitidae) in Hawaii. Pp. 405-426. In: Anderson, T. E. and Leppla, N. C. (Eds.). Advances in Insect Rearing for Research and Pest Management. Westview Press, Boulder, Colorado.

Wong, T. T. Y., Ramadan, M. M., Herr, J.C., and Mclnnis, D. 0. 1992. Suppression of a Mediterranean fruit fly (Diptera: Tephritidae) population with concurrent parasitoid and sterile fly releases in Kula, Maui, Hawaii. Journal of Economic Entomology 85: 1671-1681.

Wong, T. T. Y., Ramadan, M. M., Mclnnis, D. 0., and Mochizuki, N. 1990. Influence of cohort age and host age on oviposition activity and offspring sex ratio of *Biosteres tryoni* (Hymenoptera: Braconidae), a larval parasitoid of *Ceratitis capitata* (Diptera: Tephriditidae), Journal of Economic Entomology 83: 779-783.

Wong, T. T. Y., Ramadan, M. M., Mclnnis, D. 0., Mochizuki, N., Nishimoto, J. I., and Herr, J. C. 1991. Augmentative releases of *Diachasmimorpha tryoni* (Hymenoptera: Braconidae) to suppress a Mediterranean fruit fly (Diptera: Tephritidae) population in Kula, Maui, Hawaii. Biological Control 1: 2-7.

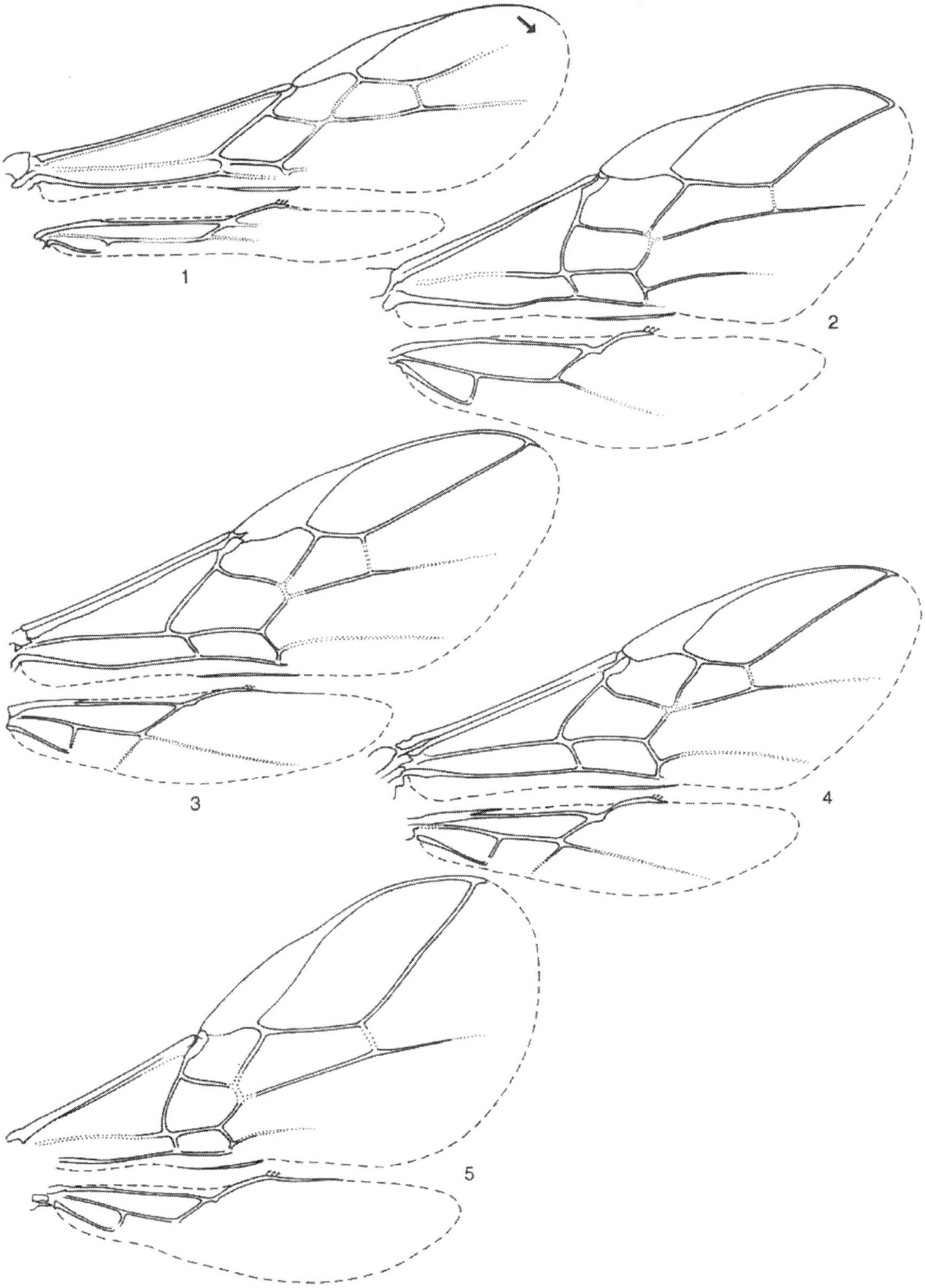

Figs 1-5: Wings of Opiinae. 1, *Ademon* sp., arrow = 3RSb not reaching wing margin; 2, *Utetes richmondi* Gahan; 3, *Diachasmimorpha longicaudata* (Ashmead); 4, *Doryctobracon areolatus* (Szépligeti): 5, *Opius* sp.

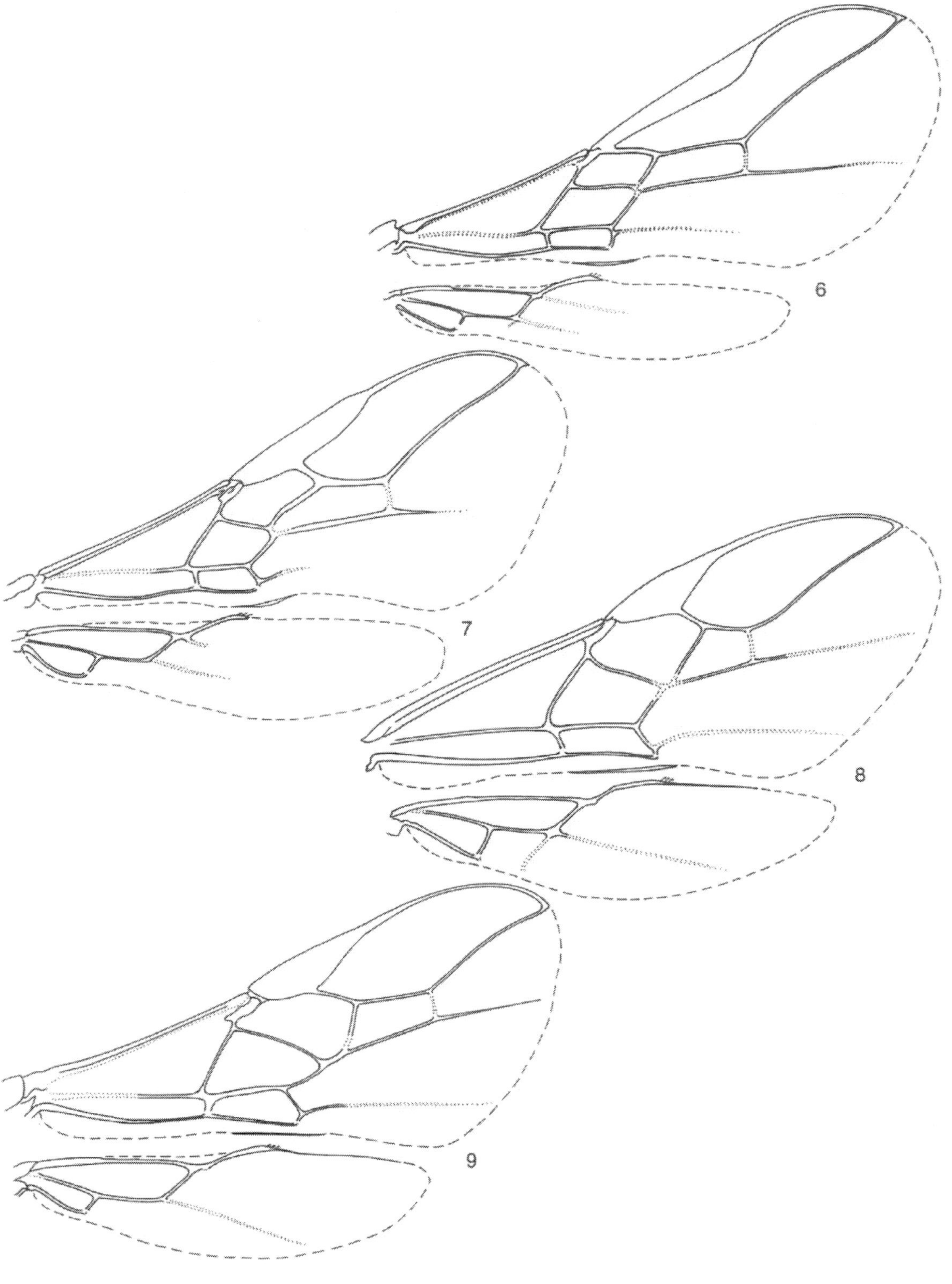

Figs 6-9: Wings of Opiinae. 6, *Eurytenes* sp.; 7, *Biosteres* sp.; 8, *Fopius arisanus* (Sonan); 9, *Psyttalia incise* (Silvestri).

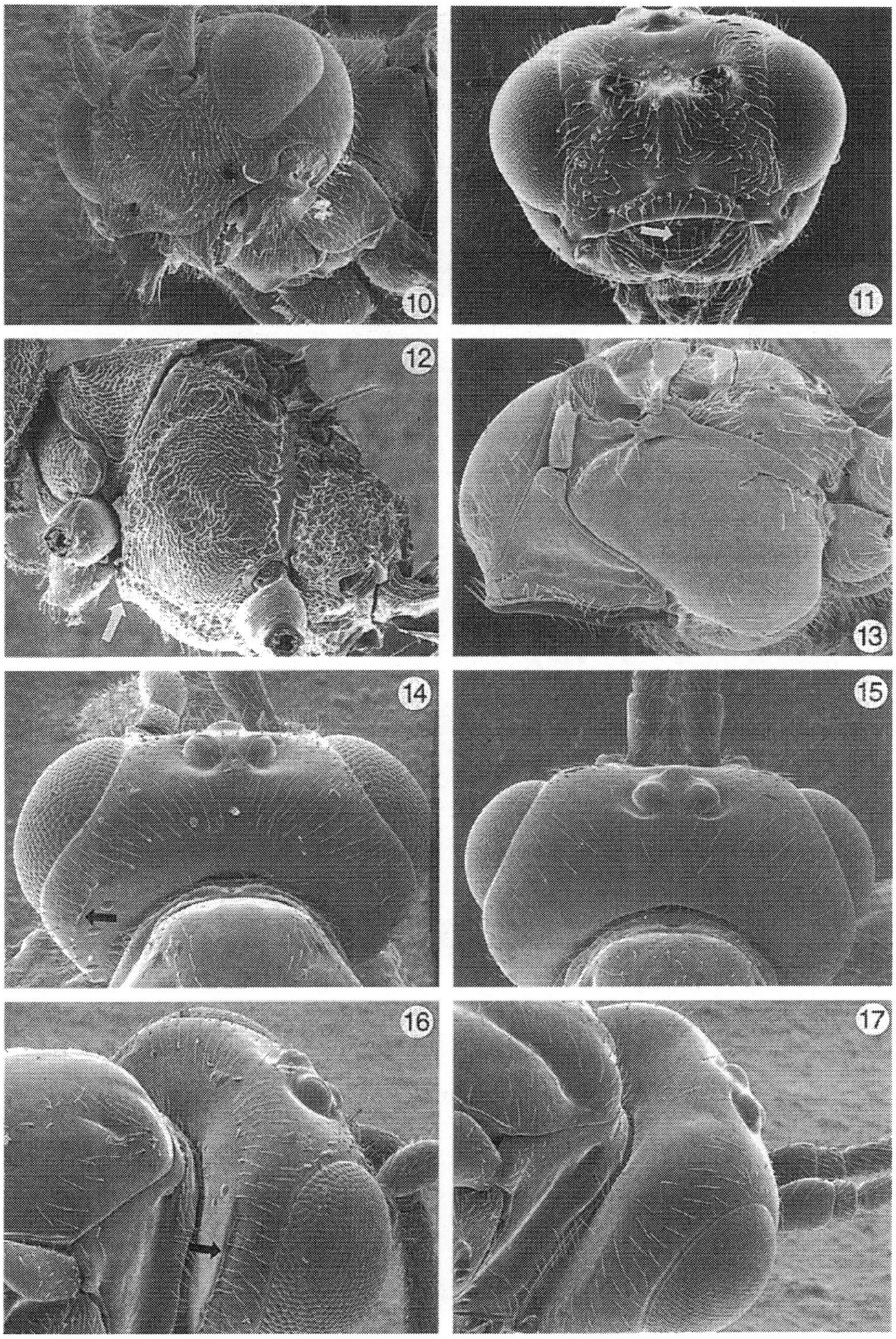

Figs 10-17: Opiinae. 10, head of *Paroligoneurus* sp. (Ichneutinae) showing exodont mandibles similar to those of *Exodontiella*; 11, face of *Utetes* sp., arrow = opening between clypeus and mandibles; 12, mesosoma of *Ademon* sp., lateral view, arrow = epicnemial carina; 13, mesosoma of *Opius dissitus* Muesebeck, lateral view showing absence of epicnemial carina (after Wharton 1984); 14-17, heads: 14, *Psyttalia* sp., dorsal view, arrow = occipital carina: 15, *Diachasmimorpha tryoni* (Cameron), dorsal view showing absence of occipital carina; 16, *Psyttalia* sp., lateral view, arrow = occipital carina: 17, *Diachasmimorpha tryoni* (Cameron), lateral view showing absence of occipital carina.

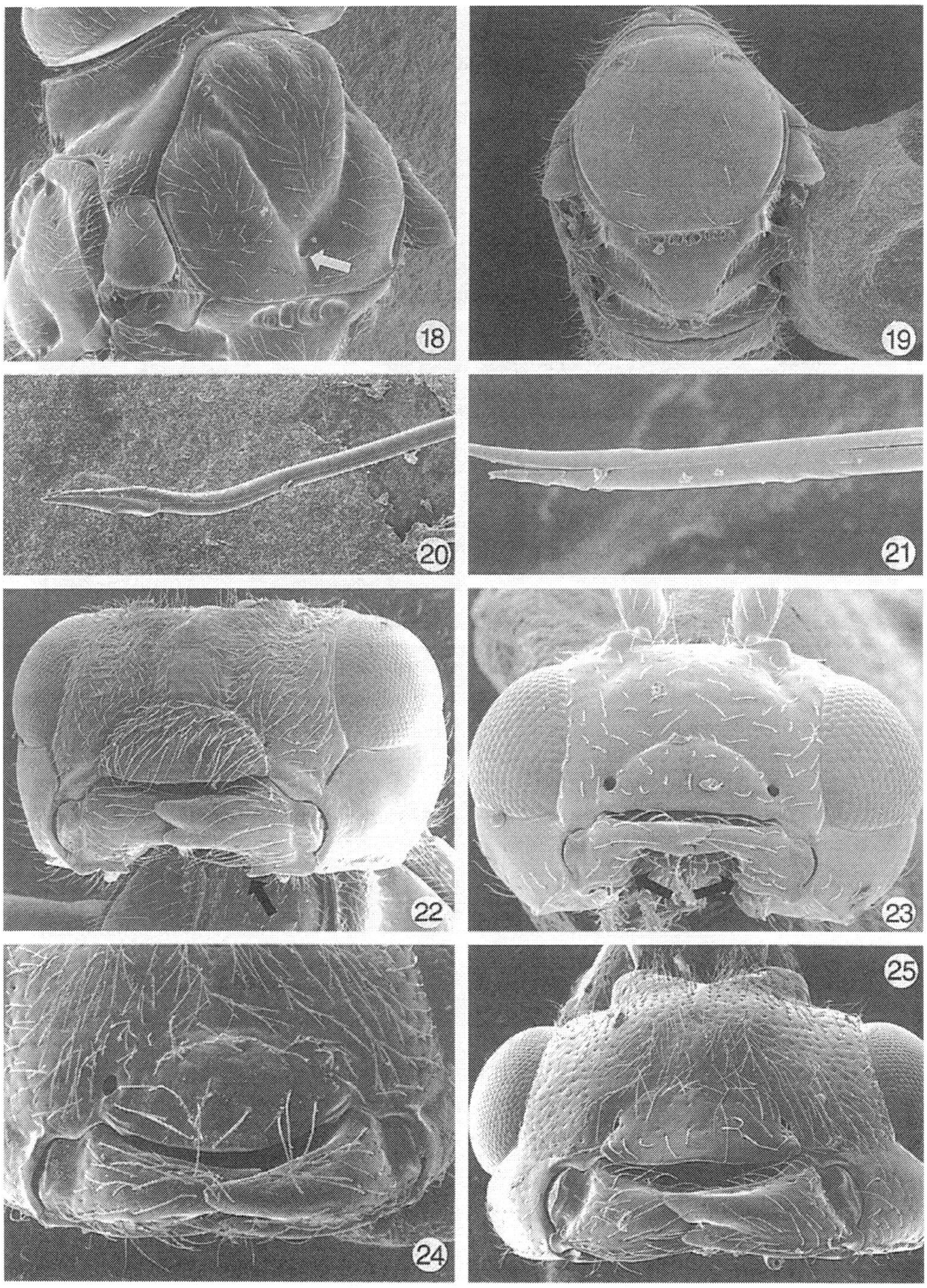

Figs 18-25: Opiinae. 18, mesoscutum of *Diachasmimorpha s.s.*, arrow = deep midpit; 19, mesoscutum of *Opius dissitus* Muesebeck showing absence of notauli and midpit (after Wharton 1984); 20, ovipositor of *Diachasmimorpha tryoni* (Cameron) showing sinuate condition; 21, ovipositor of *Doryctobracon* sp.; 22-25, head, frontal view, showing clypeus and mandibles: 22, *Biosteres* sp., arrow = basal tooth of mandible; 23, *Opius dissitus* Muesebeck, arrows = basal tooth of mandible (after Wharton 1984); 24, *Fopius arisanus* (Sonan); 25, *Diachasmimorpha longicaudata* (Ashmead).

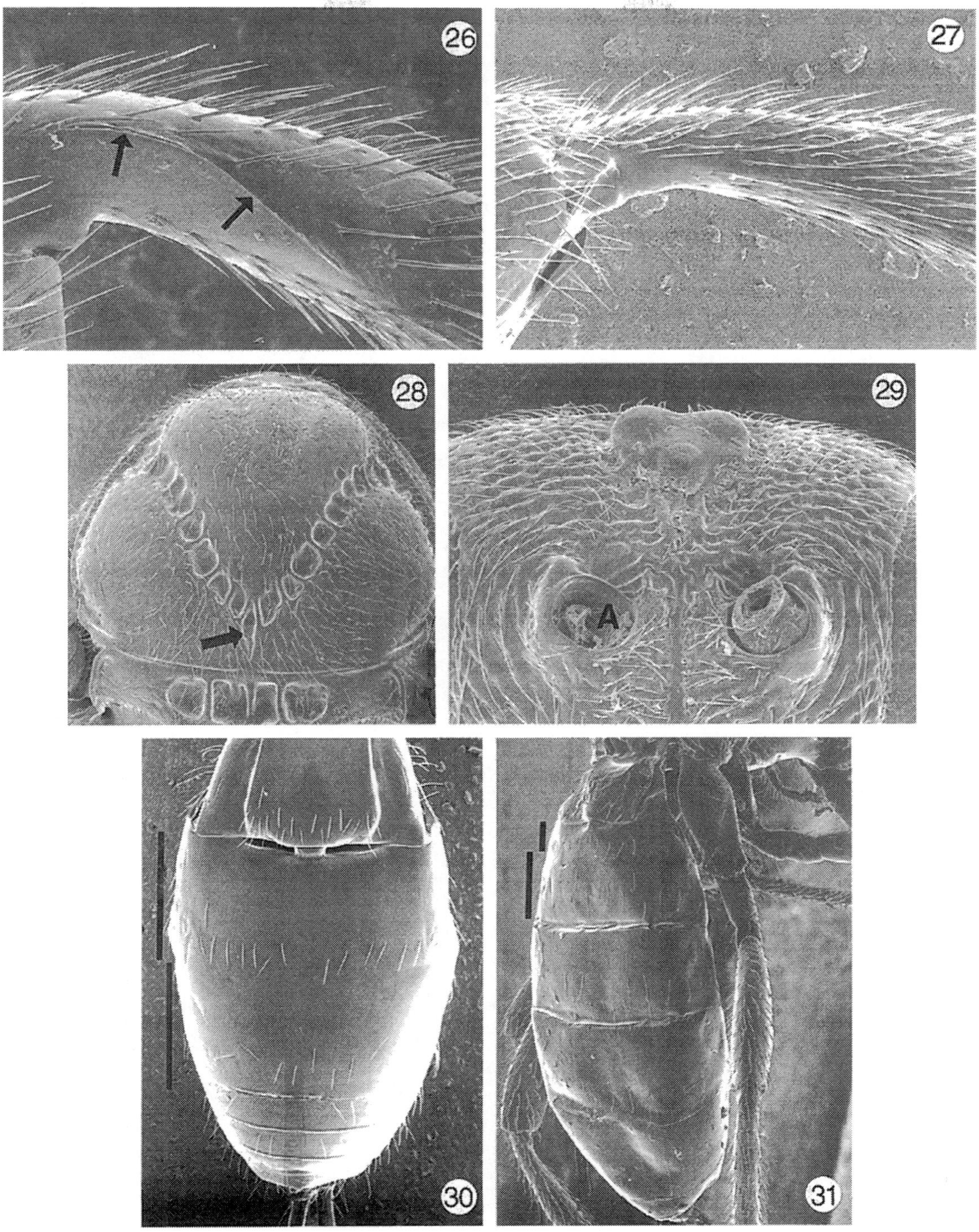

Figs 26-31: Opiinae. 26, hind tibia of *Utetes* sp., posterior view, arrows = sharp carina at base; 27, hind tibia of *Opius* sp., posterior view showing absence of carina; 28, mesoscutum of *Fopius* sp. showing crenulate notauli, arrow = midpit (after Wharton and Gilstrap 1983); 29, top of head of *Fopius arisanus* (Sonan), anterodorsal view showing heavily sculptured frons and vertex, A = antennal socket; 30, metasoma of *Opius* sp., dorsal view, bars showing relative lengths of metasomal terga 2 and 3; 31, metasoma of *Psyttalia* sp., dorsal view, bars showing relative lengths of metasomal terga 2 and 3.

OPIINAE UPDATE

R.R. Kula
Systematic Entomology Laboratory, Beltsville Agricultural Research Center, Agricultural Research Service, U.S. Department of Agriculture, c/o National Museum of Natural History, Smithsonian Institution, P.O. Box 37012, MRC-168, Washington, DC 20013-7012 U.S.A. Robert.Kula@ars.usda.gov*

Opiinae currently contains currently 714 species in the New World, with 259 and 467 in the Nearctic and Neotropical regions, respectively (Yu *et al.*, 2012). Many species remain undescribed (R. Kula, pers. obs.) despite extensive publications on the subfamily (e.g., Fischer, 1977).

Taxonomic publications on New World opiines from 1997 to present treat Neotropical species primarily but also Nearctic species. Wharton (1997a) provided diagnostic data and a key to opiine parasitoids of fruit-infesting tephritid flies worldwide; *Bellopius* Wharton was described as a new subgenus of *Opius* Wesmael, and the taxonomic status of several New World species was revised. Van Achterberg and Salvo (1997) described a new genus and six new species from Argentina, revised the taxonomic status of one genus and several species, and provided a key to opiine species reared from agromyzid flies in Argentina. Fischer (1999) proposed the genus *Neotropopius* Fischer, included a key to genera in the *Opius* genus group, and provided keys for identifying subgenera of *Opius*. Fischer (2001) described five new species from the Neotropical Region and provided keys to species groups and their constituent species. Kula (2003) described two new species from the Nearctic Region and proposed a new species synonymy. Fischer (2004, 2005, 2014) described two, three, and one new species, respectively, from the Neotropical Region. Ovruski *et al.* (2005) provide taxonomic updates for opiines known from Argentina, including several taxon names in van Achterberg and Salvo (1997). Walker and Wharton (2011) reviewed the New World species of *Eurytenes* Förster, including a key to species, as well as descriptions of four new species. Wharton *et al.* (2012) described four new species from Mexico, revised the taxonomic status of three species, and provided a key to the New World species of *Diachasmimorpha* Viereck. Figueroa *et al.* (2013) described a new species of *Eurytenes* from Mexico and updated the key from Walker and Wharton (2011). Wharton *et al.* (2013) described a new species from Trinidad, defined a new species group within *Opius*, considered species from different *Opius* subgenera members of the new species group, and provided a key for identifying the included species. Wharton and Norrbom (2013) described 14 new species and provided a key to opiine parasitoids of flower-, stem-, and leaf-infesting tephritids in the New World.

Species of Opiinae and Alysiinae have been considered closely related, but phylogenetic relationships among them are unresolved (see update for Alysiinae). Opiinae classification in the latter half of the 20th century consisted of most species in *Opius sensu lato*, which was further divided into subgenera (see Wharton, 1988 and Wharton, 1997b for more detailed discussions). Van Achterberg and Salvo (1997) revised the status of *Phaedrotoma* Förster from subgenus of *Opius s.l.* to genus, thereby splitting *Opius s.l.* into two genera. Kula (2003) recommended continued use of the subgeneric classification pending development of species groups within *Opius s.l.* and subsequent assessment of those groups for monophyly via cladistic analyses.

Most species of Opiinae attack agromyzid or tephritid flies; opiines have been used to control pests in those families, particularly fruit-infesting tephritids (Yu *et al.*, 2012). Publications on host use since Wharton (1997b) reflect the dominance of agromyzid and tephritid parasitism (van Achterberg and Salvo, 1997; Fischer, 2005; Wharton *et al.*, 2012; Wharton and Norrbom, 2013). With respect to opiines that parasitize fruit-infesting tephritids, Murillo *et al.* (2016) described and compared anatomical features of first instar larvae for five species that parasitize New World species of *Anastrepha* Schiner. Competition among those species was discussed given the larval morphological features.

Recent papers on New World Opiinae

Achterberg C van, Salvo A. 1997. Reared Opiinae (Hymenoptera: Braconidae) from Argentina. *Zoologische Mededelingen, Leiden* 71: 189-214.

Figueroa JI, Sánchez-García JA, Martínez AM, Pineda S. 2013. New species of *Eurytenes* Foerster (Hymenoptera, Braconidae, Opiinae) from Mexico. *Journal of the Kansas Entomological Society* 86: 117-121.

Fischer M. 1977. Hymenoptera: Braconidae (Opiinae II-Amerika). *Das Tierreich* 96: 1-1001.

Fischer M. 1999. Zur Evolution und zum System der *Opius*-verwandten Gattungen der Unterfamilie Opiinae mit einer erweiterten Aufteilung dieses Gattungs-Komplexes (Hymenoptera, Braconidae, Opiinae). *Linzer Biologische Beiträge* 31: 277-336.

Fischer M. 2001. Sieben neue Opiinae (Insecta: Hymenoptera: Braconidae) in der Sammlung des Naturhistorischen Museums Wien. *Annalen des Naturhistorischen Museums in Wien. Serie B für Botanik und Zoologie* 103: 303-325.

Fischer M. 2004. Einige neue Brackwespen (Insecta: Hymenoptera: Braconidae) und weitere Formen der Kiefer- und Madenwespen (Alysiinae, Opiinae). *Annalen des Naturhistorischen Museums in Wien. Serie B für Botanik und Zoologie* 105: 277-318.

Fischer M. 2005. Some new Opiinae (Insecta: Hymenoptera: Braconidae) in the Natural History Museum Vienna. *Annalen des Naturhistorischen Museums in Wien. Serie B für Botanik und Zoologie* 106: 107-133.

Fischer M. 2014. Winzige neue Arten der Kiefer- und Madenwespen aus Ecuador (Insecta: Hymenoptera: Braconidae: Alysiinae, Exothecinae, Opiinae). *Annalen des Naturhistorischen Museums in Wien. Serie B für Botanik und Zoologie* 116: 21-39.

Kula, RR. 2003. Morphological variation in *Opius* Wesmael (Hymenoptera: Braconidae) with an emphasis on Nearctic species in the subgenus *Gastrosema* Fischer. *Journal of Hymenoptera Research* 12: 278-302.

Murillo, FD, Liedo, P, Nieto-Lopez MG, Cabrera-Mireles H, Barrera JF, Montoya P. 2016. First instar larvae morphology of Opiinae (Hymenoptera: Braconidae) parasitoids of *Anastrepha* (Diptera: Tephritidae) fruit flies. Implications for interspecific competition. *Arthropod Structure & Development* 45: 294-300.

Ovruski SM, Wharton RA, Schliserman P, Aluja M. 2005. Abundance of *Anastrepha fraterculus* (Diptera: Tephritidae) and its associated native parasitoids (Hymenoptera) in "feral" guavas growing in the endangered northernmost Yungas forests of Argentina with an update on the taxonomic status of opiine parasitoids previously reported in this country. *Environmental Entomology* 34: 807-818.

Walker AK, Wharton RA. 2011. A review of New World *Eurytenes s. str.* (Hymenoptera, Braconidae, Opiinae). *Journal of Hymenoptera Research* 20: 23-46.

Wharton R, Daniels S, Shirley X, Restuccia D. 2013. An opiine Braconidae (Hymenoptera) reared from Richardiidae (Diptera) and recognition of a new species group of *Opius s.l. ZooKeys*: 289: 65-101.

Wharton R, Norrbom AL. 2012. New species and host records of New World, mostly Neotropical, opiine Braconidae (Hymenoptera) reared from flower-infesting, stem-galling, and stem-mining Tephritidae (Diptera). *ZooKeys* 349: 11-72.

Wharton R, Ward L, Miko I. 2012. New Neotropical species of Opiinae (Hymenoptera, Braconidae) reared from fruit-infesting and leaf-mining Tephritidae (Diptera) with comments on the *Diachasmimorpha mexicana* species group and the genera *Lorenzopius* and *Tubiformopius. ZooKeys* 243: 27-82.

Wharton RA. 1988. Classification of the braconid subfamily Opiinae (Hymenoptera) *Canadian Entomologist* 120: 333-360.

Wharton RA. 1997a. Generic relationships of opiine Braconidae (Hymenoptera) parasitic on fruit-infesting Tephritidae (Diptera). *Contributions of the American Entomological Institute* 30: 1-53.

Wharton RA. 1997b. Subfamily Opiinae. Pp. 378-395 In: Wharton, R.A., Marsh, P.M., and Sharkey, M.J. (Eds.). Manual of the New World Genera of the Family Braconidae (Hymenoptera). Special Publication No. 1. International Society of Hymenopterists, Washington, DC.

Yu DS, van Achterberg K, Horstmann K. 2012. World Ichneumonoidea 2011. Taxonomy, Biology, Morphology and Distribution. USB Flash drive. Taxapad, Ottawa, Canada. www.taxapad.com.

*Mention of trade names or commercial products in this publication is solely for the purpose of providing specific information and does not imply recommendation or endorsement by the USDA. USDA is an equal opportunity provider and employer.

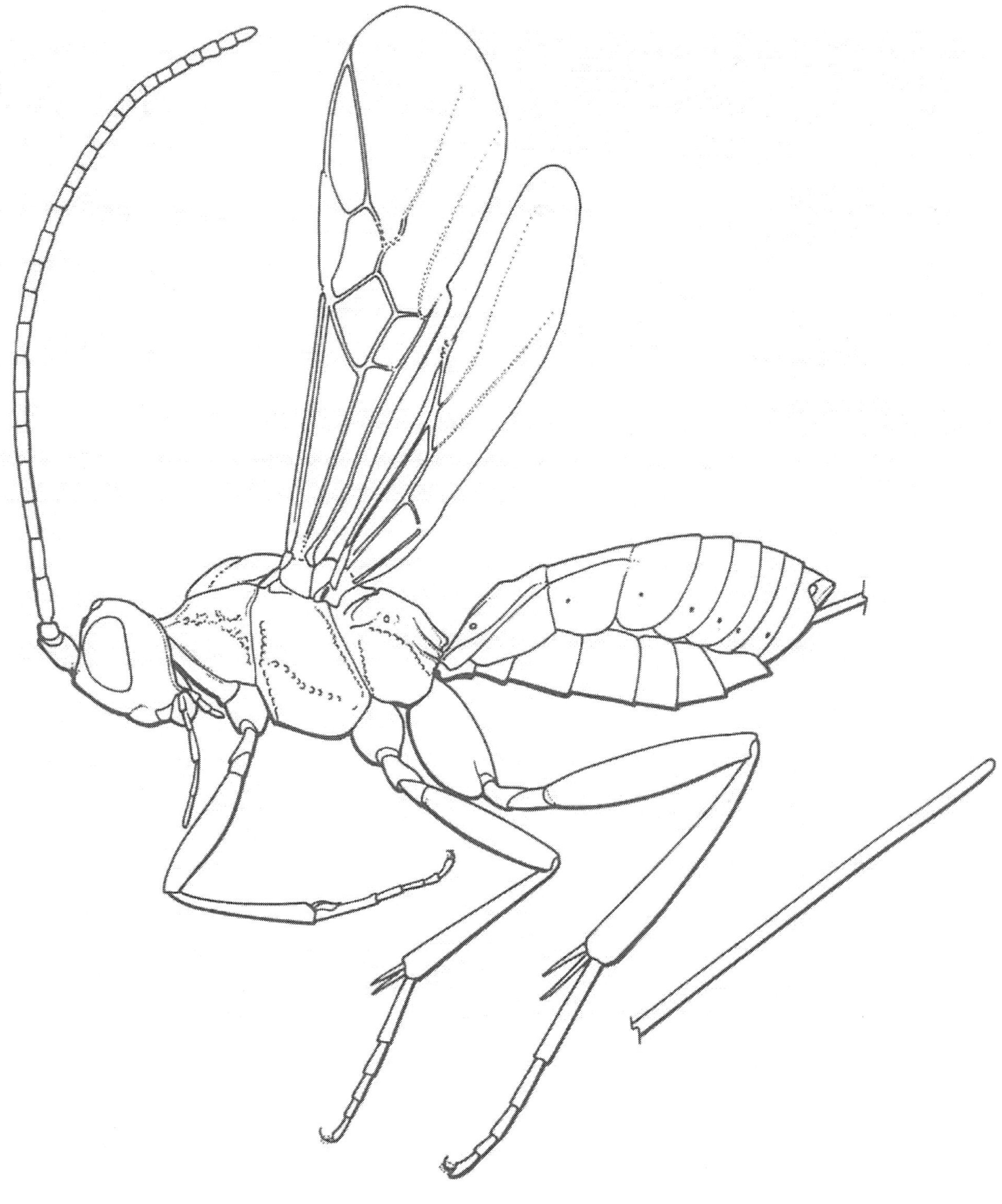

SUBFAMILY ORGILINAE

C. VAN ACHTERBERG

Afdeling Entomologie (Hymenoptera), Nationaal Natuurhistorisch Museum, Postbus 9517,
2300 RA Leiden, The Netherlands

INTRODUCTION. The subfamily Orgilinae is a small and uniform group, with species usually 4-7 mm long. Most females have a long to medium-sized ovipositor. There are 10 genera in the Orgilinae, five of which are recorded from the New World. Many species, especially from the Neotropical area, are undescribed. The Nearctic species of the genus *Orgilus* Haliday were revised by Muesebeck (1970), and some New World species were also described in the subfamily revision by van Achterberg (1987). Van Achterberg (1987) recognized three tribes: the poorly known Neotropical Antestrigini and the widespread tribes Mimagathidini and Orgilini.

BIOLOGY. The few data known on the biology of Orgilinae are aptly summarized by Shaw and Huddleston (1991). As far as known, all species are solitary koinobiont endoparasites of lepidopteran larvae. The largest genus, *Orgilus*, contains species attacking microlepidopteran larvae with a partly endophytic way of life, especially those with leaf-mining or tunneling early instars, whose activities are detectable by extruded frass. The known hosts of Orgilinae belong to Coleophoridae, Gelechiidae, Tortricidae, Pyralidae and Oecophoridae.

BIOLOGICAL CONTROL POTENTIAL. Several species of *Orgilus* have been used for biological or integrated control of lepidopteran pests. For example, *Orgilus longiceps* Muesebeck was imported from Japan and probably unsuccessfully established in the U.S.A. against *Grapholitha molesta* (Busck) (Muesebeck, 1970); *O. gossypii* Muesebeck was released against *Pectinophora gossypiella* (Saunders); and *Orgilus obscurator* Nees against *Rhyacionia* spp. Another orgiline, *Stantonia carpocapsae* Muesebeck, was released against *Carpocapsa pomonellae* (Linnaeus).

COMMON GENERA. *Orgilus* and *Stantonia* Ashmead are the major New World genera. *Orgilus* is mainly concentrated in the Holarctic region and encompasses most species of the Orgilinae. *Stantonia* has a pantropical distribution. No revisions of the New World *Stantonia* and the Neotropical *Orgilus* are available.

DISTRIBUTION. The Orgilinae are worldwide in distribution with most of the approximately 250 described species in the Holarctic region. Three of the genera treated here, *Antestrix* van Achterberg, *Bentonia* van Achterberg and *Podorgilus* van Achterberg, are apparently endemic to the New World.

DISTINGUISHING FEATURES. The Orgilinae have a slender body, moderately long antenna, and a distinctly protruding ovipositor. The venation (Figs 1, 3) is diagnostic, with second submarginal cell of fore wing absent or (rarely) small, vein 3RS of fore wing straight and cross vein 2cu-a of fore wing present. The occipital carina is usually present laterally, and there are usually tiny spines present near the apex of the hind tibia (Figs 12, 14).

KEY TO THE NEW WORLD GENERA OF THE SUBFAMILY ORGILINAE

1. Fore wing 1RS present (Fig. 1). Occipital flange strongly protruding posteriorly (Fig. 10). Scutellum with distinct posterior scutellar depression (Fig. 5, arrow). Tribe Antestrigini............... ***Antesirix*** van Achterberg
   Distribution: Chile and Argentina; rare; 2 described species. Biology: unknown. Reference: van Achterberg (1987, key to species).

\- Fore wing 1RS absent (Figs 2-4). Occipital flange not or only slightly protruding (Fig. 11). Scutellum without distinct posterior scutellar depression (Figs 6, 7)..2

2(1). Hind wing M +CU much shorter than 1M (Fig. 2); posterior margin of hind wing concave subbasally (Fig. 2); fore wing r-m present (Fig. 2), resulting in a triangular second submarginal cell. Tribe Mimagathidini... 3

- Hind wing M+CU about as long as 1M or distinctly longer (Fig. 3); posterior margin of hind wing convex or nearly straight subbasally (Figs 3, 4); fore wing r-m absent, thus triangular second submarginal cell absent (Figs 3, 4). Tribe Orgilini 4

3(2). Scutellum with a deep median longitudinal depression (Figs 6, 7). Propodeum with high lamella posteriorly (Fig. 8)..***Bentonia*** van Achterberg

Distribution: Mexico, Costa Rica, Brazil; 2 described species. Biology: one specimen of the type species was reared from *Anticarsia gemmatalis* Huebner. Reference: van Achterberg (1992, key to species).

- Scutellum evenly convex (Fig. 13). Propodeum without high posterior lamella (Fig. 9). Wings: Fig. 2 ....***Stantonia*** Ashmead

Distribution: U.S.A. to Argentina, mainly Neotropical; 5 described New World species. Biology: largely unknown, some New World species have been reared from Tortricidae and Pyralidae (Muesebeck, 1938). Reference: van Achterberg (1987, general diagnosis; no keys available for New World species).

4(2). Hind wing cu-a vertical or weakly angled towards base (Fig. 3). Hind basitarsus slender (Figs 16). Fore wing r shorter than 2RS (Fig. 3) ....***Orgilus*** Nees

Distribution: Cosmopolitan; commonly collected; about 120. described species in New World. Biology: overview by Shaw and Huddleston (1991). References: Muesebeck (1970, key to Nearctic species), van Achterberg (1987, subgeneric grouping, key to New World species of some subgenera).

- Hind wing cu-a strongly angled towards base (Fig. 4). Hind basitarsus strongly widened (Fig. 15). Fore wing r distinctly longer than 2RS (Fig. 4)..***Podorgilus*** van Achterberg

Distribution: Brazil and Bolivia; rare; one described species. Biology: unknown. Reference: van Achterberg (1994).

REFERENCES

Achterberg, C. van 1987. Revisionary notes on the subfamily Orgilinae (Hymenoptera: Braconidae). Zoologische Verhandelingen Leiden 242: 1-111.

Achterberg, C. van 1992. *Bentonia* gen. nov. (Hymenoptera: Braconidae: Orgilinae) from Brazil. Zoologische Mededelingen Leiden 66: 339-344.

Achterberg, C. van 1994. Two new genera of the tribe Orgilini Ashmead (Hymenoptera: Braconidae: Orgilinae). Zoologische Mededelingen Leiden 68: 173-190.

Muesebeck, C.F.W. 1938. Two reared North American species of the genus *Stantonia* Ashmead (Hymenoptera: Braconidae). Proceedings of the Entomological Society of Washington 40: 89-91.

Muesebeck, C.F.W. 1970. The Nearctic species of *Orgilus* Haliday (Hymenoptera: Braconidae). Smithsonian Contributions to Zoology 30: 1-104.

Shaw, M. R. and Huddleston, T. 1991. Classification and biology of braconid wasps (Hymenoptera: Braconidae). Handbooks for the Identification of British Insects 7(11): 1-126.

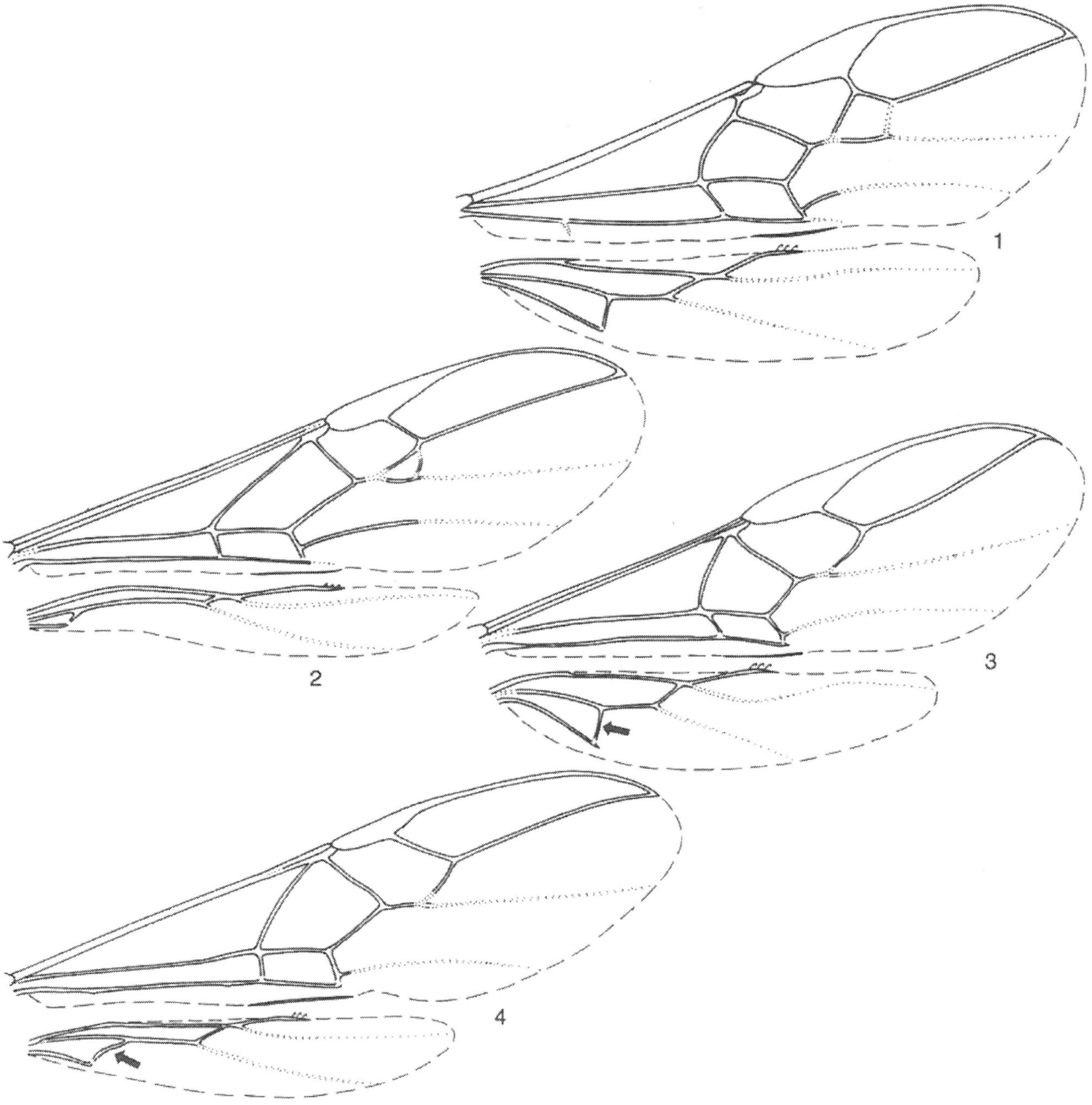

Figs 1-4: Wings of Orgilinae. 1, *Antestrix* sp.; 2, *Stantonia* sp.; 3, *Orgilus* sp., arrow hind wing cu-a; 4, *Podorgilus* sp., arrow = hind wing cu-a.

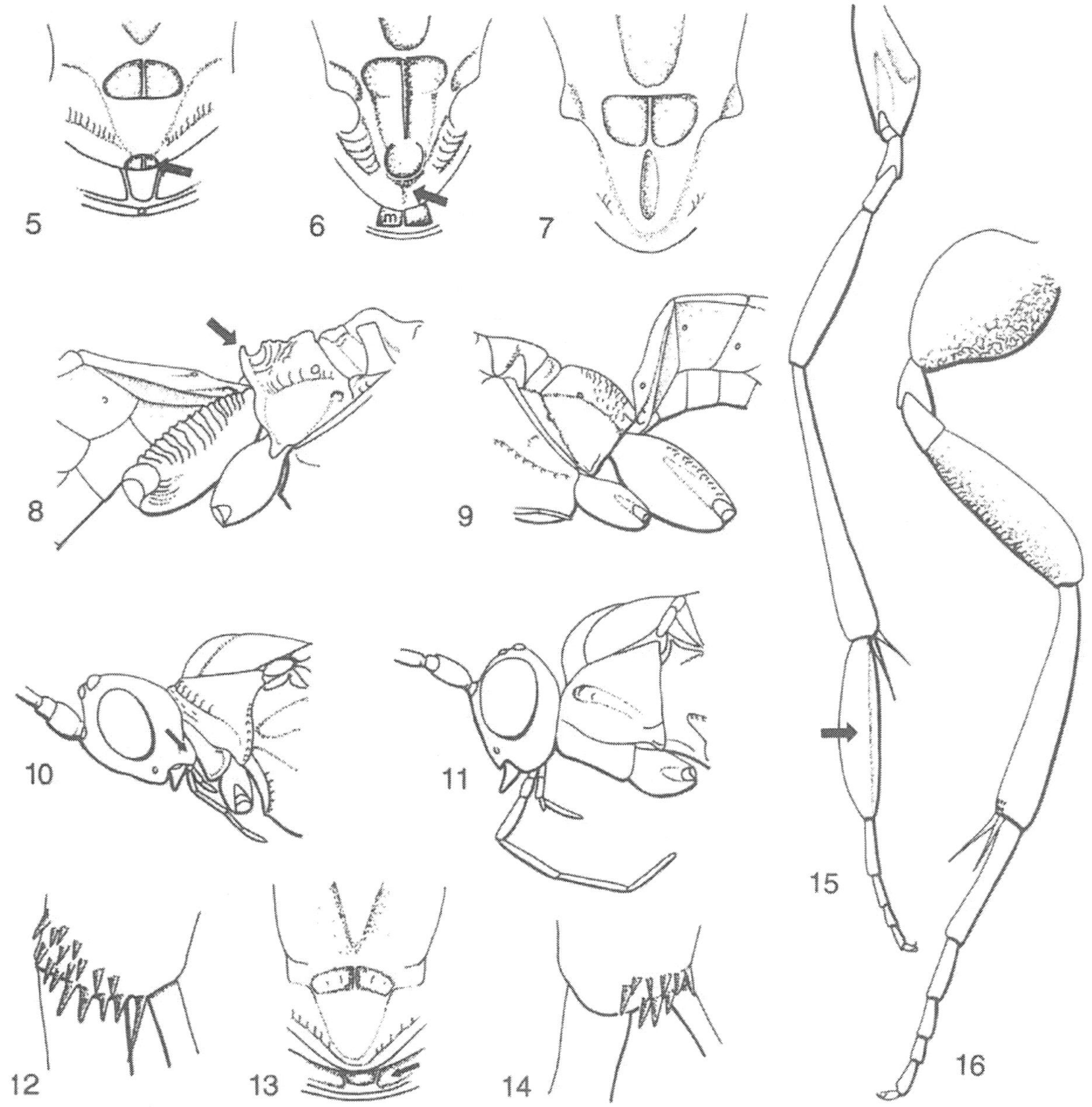

Figs 5-16: Orgilinae. 5-7, scutellar regions: 5, *Antestrix* sp., dorsal view, arrow = posterior scutellar depression; 6, *Bentonia* sp., dorsal view, arrow = posterior scutellar depression absent, m = median depressions on metanotum: 7, *Bentonia* sp., dorsal view; 8, propodeum and base of metasoma of *Bentonia* sp., lateral view, arrow = propodeal lamella; 9, propodeum and base of metasoma of *Stantonia* sp., lateral view; 10, head and anterior part of mesosoma of *Antestrix* sp., lateral view, arrow = occipital flange; 11,. head and anterior part of mesosoma of *Orgilus* sp., lateral view; 12, apex of hind tibia of Orgilinae showing tiny spines; 13, scutellar region of *Stantonia* sp., dorsal view, arrow = metasoma; 14, apex of hind tibia of Orgilinae showing tiny spines; 15, hind leg of *Podorgilus* sp., arrow = enlarged basitarsus: 16, hind leg of *Orgilus* sp. All figures after van Achterberg (1992, 1994).

ORGILINAE UPDATE

C. van Achterberg
Department of Terrestrial Zoology, Naturalis Biodiversity Center, Postbus 9517, 2300 Leiden, The Netherlands.
c.vanachterberg@xs4all.nl

The Orgilinae chapter of the manual is outdated. There are 13 genera in the Orgilinae. Braet and van Achterberg (2003) added a new genus (*Doryctorgilus* Braet and van Achterberg) and two new species of *Podorgilus* van Achterberg. Braet and Quicke (2004) included in their revision of the genus *Stantonia* Ashmead 28 New World species and synonymized *Bentonia* van Achterberg with *Stantonia* Ashmead.

Annotated list of recent papers on New World Orgilinae

Boughton AJ, Kula RR, Gates M, Zhang Y, Nunez M, O'Connor J, Whitfield JB, Center TD. 2012. Parasitoids attacking larvae of a recently introduced weed biological control agent, *Neomusotima conspurcatalis* (Lepidoptera: Crambidae): key to species, natural history, and integrative taxonomy. *Annals of the Entomological Society of America* 105: 753-767.
Braet Y, Achterberg C van. 2001. Notes on the genera *Exasticolus* van Achterberg (Homolobinae) and *Orgilus* Haliday (Orgilinae) (Hymenoptera: Braconidae), with the description of three new species from French Guiana. *Zoologische Mededelingen, Leiden*. 75(5): 89-102.
Braet Y, Achterberg C van. 2003. *Doryctorgilus gen. nov.* and other new taxa, with a study of the internal microsculpture of the ovipositor in the subfamily Orgilinae Ashmead (Hymenoptera: Braconidae). *Zoologische Mededelingen, Leiden* 77(6): 127-152.
The elongate hind trochantellus, the row of spines on fore and middle tibiae and the absence of the metapleural flange separates *Doryctorgilus* Braet and van Achterberg from all other genera of Orgilinae. In addition, two new species of *Podorgilus* van Achterberg are described and a key to species is included.

Braet Y, Quicke DLJ 2004. A phylogenetic analysis of the Mimagathidini with revisionary notes on the genus *Stantonia* Ashmead, 1904 (Hymenoptera: Braconidae: Orgilinae). *Journal of Natural History* 38(12): 1489-1589.
New World species of *Stantonia* are keyed and the genus *Bentonia* is included in *Stantonia*.

Kula RR, Boughton AJ, Pemberton RW. 2010. *Stantonia pallida* (Ashmead) (Hymenoptera: Braconidae) reared from *Neomusotima conspurcatalis* Warren (Lepidoptera: Crambidae), a classical biological control agent of *Lygodium microphyllum* (Cav.) R. Br. (Polypodiales: Lygodiaceae). *Proceedings of the Entomological Society of Washington* 112: 61-68.

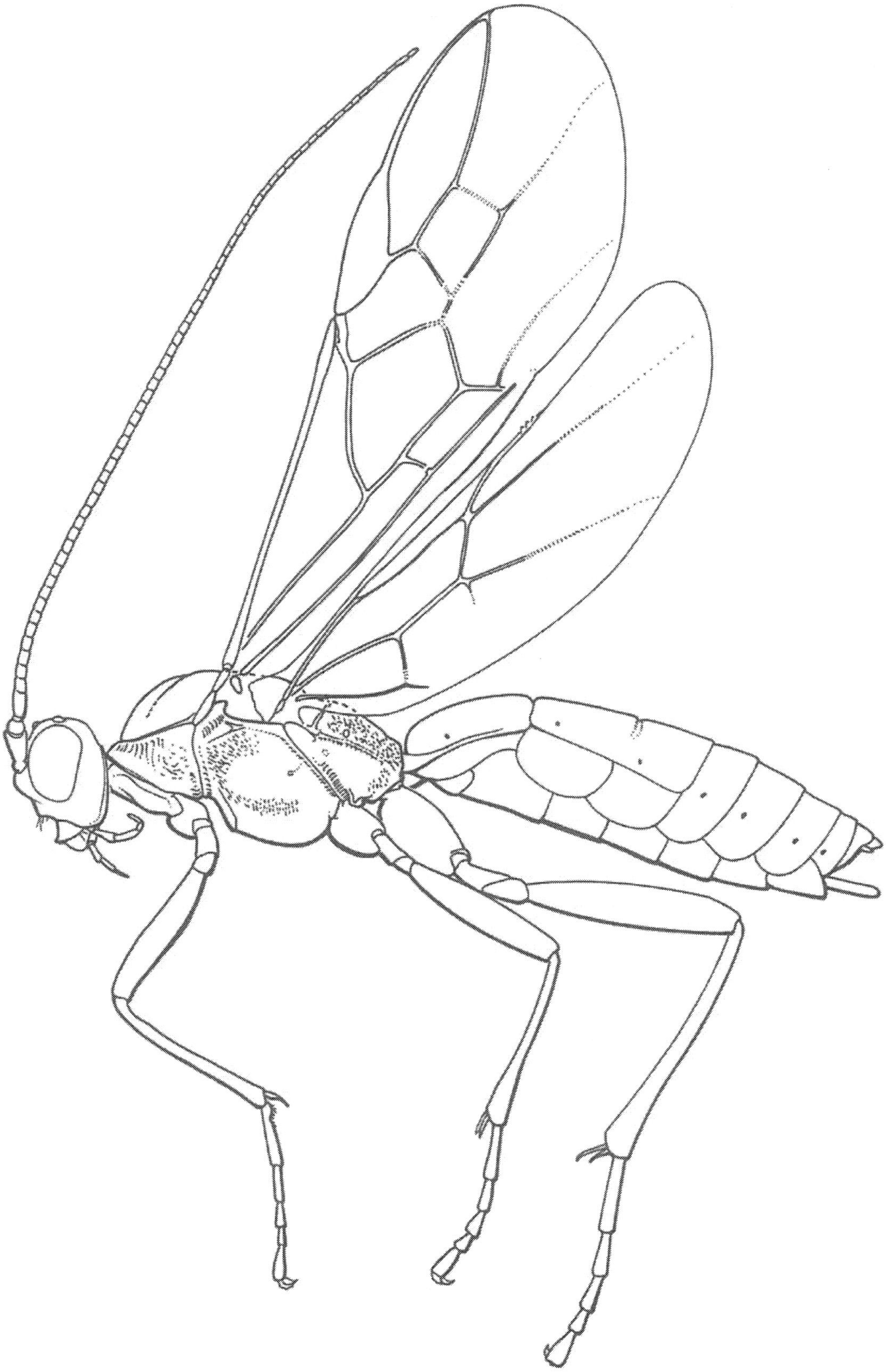

SUBFAMILY ROGADINAE

S. R. SHAW

Department of Plant, Soil, and Insect Sciences, University of Wyoming, Laramie, Wyoming 82071-3354, U.S.A.

INTRODUCTION. The Rogadinae is a very large, cosmopolitan subfamily, well represented in the New World by several hundred species. As defined here, this subfamily includes all those species known to mummify the host larva (= Tribe Rogadini of other authors). A number of small, related groups that are sometimes placed as tribes in the Rogadinae (e.g. Rhysipolini) are treated in the chapter on Hormiinae. Subfamily classification is discussed there and in the Introduction. Although rogadines, especially *Aleiodes* Wesmael and *Rogas* Nees, are very common insects, there is very little revisionary work available pertaining to the New World fauna. The Nearctic species of *Stiropius* Cameron were revised by Whitfield (1988) and the Nearctic species of *Aleiodes* are currently being revised by Marsh and S. Shaw. Van Achterberg (1991) studied the Afrotropical and Western Palaearctic rogadines and treated *Petalodes* Wesmael as a junior synonym of *Aleiodes*. S. Shaw (1993) studied the species of *Eucystomastax* Brues in the New World, treating *Eucystomastax* as a subgenus of *Aleiodes*.

Of the many braconids attacking Lepidoptera, the Rogadinae are unique in their habit of mummifying the remains of the host caterpillar, thus this group is easy to recognize when reared from host lepidopterans. The host mummy is diagnostic for many rogadine species, so associated host remains should always be preserved with reared wasps. Preservation of the mummified host larva is also valuable because it permits future verification of the host identity, it preserves the parasitoid's larval exuvium, and it may be the only clue to the primary parasitoid involved if a hyperparasitoid emerges.

BIOLOGY. Rogadinae are koinobiont endoparasitoids, mostly of exposed-feeding macrolepidopteran larvae (M. Shaw, 1983). However, the *Stiropius* group (*Stiropius*, *Choreborogas* Whitfield, *Polystenidea* Viereck) parasitizes leaf-mining lyonetiid and gracillariid larvae (Whitfield, 1988, 1990). The vast majority of rogadine species are solitary parasitoids, but a few gregarious species are known (e.g. *Aleiodes stigmator* (Say)). Pupation is internal, within the shrunken and mummified remains of the host caterpillar. Species of *Rogas*, for which rearing records exist, parasitize Limacodidae, Zygaenidae, Lycaenidae, and Riodinidae (the hosts of most *Rogas* species are not known). In *Rogas*, the host mummy is usually not glued down by the parasitoid larva, and the emergence hole is ragged and irregular (Smith et al., 1955 as *Pelecystoma* Wesmael). Species of *Aleiodes* parasitize a wide array of macrolepidopteran hosts; however, the Noctuoidea, Geometroidea, and Sphingoidea (as defined by Hodges et al., 1983) are most commonly utilized (Muesebeck, 1960; M. Shaw, 1983). In *Aleiodes*, the host mummy is often glued down by the parasitoid larva through a hole chewed in the bottom of the host's prothorax, consequently the host mummies are commonly found stuck to a leaf or twig. Emerging *Aleiodes* adults normally cut an even, less jagged (as compared with *Rogas*), and more circular emergence hole at the posterior end of the mummifiedcaterpillar. Most rogadine adults are nocturnally active, and are among the most common braconids attracted to lights at night. More complete reviews of rogadine biology are provided by M. Shaw (1983, 1994), M. Shaw and Huddleston (1991) and S. Shaw (1993, 1995).

With the exception of M. Shaw's (1981) study of *Clinocentrus gracilipes* (Thompson), the physiological mechanism underlying the mummification process has not yet been determined, and would be an interesting topic for further investigation. The mummy appears to have an outer larval cuticle, supported by an underlying pupal cuticle, which on the inside is lined lightly by silk produced by the pupating parasitoid larva. However, the parasitoid' s cocoon is often extremely thin, andit appears that the main support for the mummy comes from the cuticular layer produced by the host larva. Since in many species the mummy is produced during a penultimate larval instar, it appears that the parasitoid larva in some way manipulates the normal physiology of the host to induce a premature metamorphosis and production of pupal cuticle. Thus the mummification process might be due to the destruction of the host's corpora allatum (eliminating the source of juvenile hormone), although this hypothesis has not been tested. It is, however, supported by the observation that early feeding by the rogadine larva is localized in the prothoracic region, where the "glue hole" is formed to attach the mummy to the substrate. Excellent observations on European species are provided by M. Shaw (1983).

BIOLOGICAL CONTROL POTENTIAL. Rogadines are probably very important for the natural control of many lepidopterans, however, not many have been studied for biological control purposes. *Aleiodes indiscretus* (Reardon) was introduced into North America from India for suppression of the gypsy moth, but it is not known to be established (Reardon et al., 1973). Several other native *Aleiodes* species are parasitoids of important forest pest species, and several *Aleiodes* are parasitoids of cutworm-type larvae, and may be or importance in agroecosystems. *Aleiodes vaughani* (Muesebeck) and *A. laphygmae* (Viereck) are among species currently being studied as biocontrol agents against noctuid larvae in Central America (R. Cave, pers. comm.).

COMMON GENERA. Species of *Aleiodes* and *Rogas* are the most frequently collected rogadines, partly because of their greater species richness, and partly because they include the most species with large body sizes. There is a definite shift in the proportional species richness of these two genera in the New World along a latitudinal gradient (S. Shaw, 1993). *Aleiodes* species comprise about 95% of the rogadine fauna in cool, temperate parts of North America, and *Rogas* is rather rare. In the Neotropics, *Rogas* is far more abundant and species rich, comprising more than 50% of the rogadine fauna in tropical moist forest habitats. Species of *Cystomastax* Szépligeti, although not commonly collected, are worth noting because of their large body size. *Cystomastax* appear to be mimics of ophionine ichneumonids, and are probably the largest Neotropical braconids that are attracted to lights.

DISTRIBUTION. The Rogadinae occur throughout the New World in virtually every habitat where Lepidoptera feed on foliage. Species of *Stiropius* and *Choreborogas* are numerous in the Neotropics, but can be easily overlooked because of their minute size and nocturnal activity (Whitfield, 1988, 1990). The larger species of *Rogas* are particularly diverse in tropical America, and, together with the less commonly encountered *Cystomastax*, form a very conspicuous element of the Neotropical braconid fauna. Species of *Aleiodes* are also common in the Neotropics, but they are relatively far more diverse in temperate parts of North America.

DISTINGUISHING FEATURES. The biological traits of endoparasitism and mummification of the host larva support the probable monophyly of the Rogadinae *s.s.* (Whitfield, 1992). A morphological synapomorphy for diagnosing the Rogadinae *s.s.* is the antero-lateral dorsal carinae of the petiole, which converge and meet near the anterior end of the segment, and usually continue some distance posteriorly as a single median carina.

KEY TO THE NEW WORLD GENERA OF THE SUBFAMILY ROGADINAE s.s.

It is difficult to use a single character to separate all the species of *Rogas* from all the species of *Aleiodes*. Couplet 7 is thus complex. However, individuals that lack the marginal fringe of setae at the apex of the hind tibia can be taken directly to couplet 9 from couplet 7.

1. Hind wing 1M longer than M+CU (Figs 2-4). Body size minute (1-3mm). Metasomal terga 2+3 carapace-like, and together with terga 1 and 4 forming a "pseudocarapace" that conceals most of the abdomen in dorsal view (Fig. 12); surface sculpturing finely pitted or granular (Fig. 12) (*Stiropius* group of Whitfield, 1990) 2

\- Hind wing 1M shorter than M+CU (Figs 5-8, 10-11); or body size larger than 3mm; or metasomal terga 1 through 4 not carapace-like, or surface sculpturing not granular, striate, or rugose) 4

Distribution: U.S.A. and Canada to Argentina, including Caribbean; 17 described New World species, many undescribed. Biology: parasitoids of Lyonetiidae and Gracillariidae. References: Whitfield (1988, key to North and Central American species), van Achterberg (1995). Includes *Viridipyge* Whitfield, synonymized by van Achterberg (1995).

2(1). Fore wing r-m present (Figs 1, 2) .... ***Stiropius*** Cameron

\- Fore wing r-m absent (Figs 3, 4) 3

3(2). Malar space short, less than half eye height (Fig. 13); malar suture present between compound eye and mandible (Fig. 13). Fore wing stigma usually elongated distally, with r arising basad its midpoint (Fig. 3). Hind femora often swollen or otherwise modified (Fig. 14). Metasomal tergum 4 without a semicircular subbasal groove or depression .... ***Choreborogas*** Whitfield

Distribution: Texas and Mexico to Peru; 8 described New World species, several undescribed. Biology: parasitoids of Lyonetiidae. References: Whitfield (1990, description and key to species), van Achterberg (1995).

\- Malar space long, about equal to eye height; malar suture absent. Fore wing stigma not elongated distally, r arising at or near midpoint of stigma (Fig. 4). Hind femora not swollen or otherwise modified. Metasomal tergum 4 with a semicircular subbasal groove or depression .... ***Polystenidea*** Viereck

Distribution: widespread, but more diverse in the Nearctic; small genus; 2 described New World species, several undescribed. Biology: parasitoids of Lyonetiidae. Reference: van Achterberg (1995).

4(1). Tarsomeres 2-4 of front and middle legs extremely reduced and compact: tarsomeres 2-4 each shorter than wide, fifth tarsomere longer than tarsomeres 2-4 combined. Relatively uncommon. Wings: Fig. 5 .... ***Yelicones*** Cameron

Distribution: widespread; uncommon; 5 described New World species, several un- described. Biology: parasitoids of Pyralidae. Reference: Quicke and Kruft (1995, key to North American species), Quicke et al. (1996, key to Central American species).

\- Tarsomeres 2-4 of front and middle legs normal, unmodified: tarsomeres 2-4 each longer than wide, fifth tarsomere not as long as tarsomeres 2-4 combined. Many common species 5

5(4). Suture between fused metasomal terga 2+3 very weak or absent, never coarsely sculptured. Ovipositor always as long as middle tibia or longer. Relatively uncommon. Wings: Fig. 6 .... ***Clinocentrus*** Haliday

Distribution: widespread; uncommon; 3 described New World species, several un- described. Biology: parasitoids of web-inhabiting and other concealed microlepidoptera, especially Tortricidae. Reference: Marsh (1979).

\- Suture between fused metasomal terga 2+3 distinct, often coarsely sculptured (Fig. 24). Ovipositor variable in length, but usually shorter than middle tibia, and often as short as fifth tarsomere. Many common species 6

6(5). Hind tibial spurs long, strongly curved, smooth, and mostly devoid of setae (Fig. 17). Tarsal claws simple, with a rounded basal lobe but never with a blunt lobe, blunt tooth, or distinct pectin. Eyes extremely large, with malar space shorter than mandible width and shortest distance between eye and ocelli less than ocellar width (Fig. 18). Males with third and fourth segments of maxillary palp strongly expanded and flattened (Fig. 19). Very rare. Wings: Fig. 7 .... ***Macrostomion*** Szépligeti

Distribution: very rare Indo-Australian genus that also occurs in Costa Rica and South America; several previous literature references to this genus in the New World are incorrect (S. Shaw, 1993), but at least one undescribed species does occur. Biology: unknown. Reference: van Achterberg (1991).

\- Hind tibial spurs either relatively straight, or if curved then always distinctly setose (especially outer side) (Fig. 16). Tarsal claws either with a blunt basal lobe, blunt tooth, or a distinct pectin (at high magnification) (Fig. 15). Eyes often not so extremely large: either malar space longer than mandible width or distance between eye and ocelli more than ocellar width (but usually not both). Males with maxillary palps usually normal, not expanded; if modified (rare) then thickened or globose rather than strongly flattened. Many common species 7

7(6). Tarsal claw with a large, blunt basal lobe or tooth, with or without basal pectin (Fig. 15). Inner margin of hind tibia with a distinct fringe of flattened setae (Fig. 16). Sternaulus of mesopleuron usually foveate. Median carina of propodeum diverging posteriorly into an irregular, rugose areola (Figs 20, 22). Fore wing m-cu arising near 2RS, thus (RS+M)b short (Figs 8, 9); 2RS diverging from r-m posteriorly, forming a rather wide and roughly trapezoidal second submarginal cell (Figs 8, 9) 8

Tarsal claw without a large, blunt basal lobe or tooth (Fig. 21); basal portion of tarsal claw rounded, with or without a pectin of sharp accessory spines of variable size (usually these can be seen under high power with light microscopy) (Fig. 21). Inner margin of hind tibia without a distinct fringe of flattened setae (98°/o of species), or (more rarely) fringe present. Sternaulus of mesopleuron not foveate, either smooth or indicated only by shallow rugose sculpture. Median carina of propodeum complete to end of segment, or (more rarely) absent, but never diverging postero-medially into an areola. Fore wing m-cu arising distinctly basad 2RS, thus (RS+M)b long (Figs 10, 11); 2RS nearly parallel with r-m, forming a roughly rectangular or subquadrate second submarginal cell (Figs 10, 11) 9

8(7). Membrane of fore wing first submarginal cell evenly setose, without a distinct windowlike clear patch (Fig. 8). Petiole relatively broad, at most 2.75 times longer than apical width (Fig. 22). Body size variable but usually 10 mm or smaller .... ***Rogas*** Nees von Esenbeck

Distribution: U.S.A. to Argentina, including Caribbean; very common in the Neo- tropical region, very scarce in the Nearctic region; probably 100-150 New World species, most of which are undescribed. Biology: parasitoids of sluggish and slug- like macrolepidoptera including Limacodidae, Zygaenidae, Lycaenidae, and Riodinidae. Reference: van Achterberg (1991). No revisions for New World.

\- Membrane of fore wing first submarginal cell with a distinct window-like clear patch that is devoid of setae (Fig. 9). Petiole extremely long and narrow 2.75 to 4 times longer than apical width (Fig. 23). Large species, 9-14 mm in length .... ***Cystomastax*** Szepligeti

Distribution: Neotropical; relatively uncommon; 6 described species, in need of revision. Biology: unknown, Reference: Shenefelt (1975).

9(7). Metasomal terga 1-4 forming a strong, rigid, often coarsely pitted carapace that conceals the remainder of the abdomen from above (Fig. 25). Wings: Fig. 11 .... ***Tetrasphaeropyx*** Ashmead

Distribution: western U.S.A. and Mexico; uncommon; one described species, at least 7 undescribed. In need of revision; possibly a derived offshoot of *Aleiodes*. Biology: unknown. Reference: Marsh (1979).

\- Metasomal terga 1-4 not forming a strong, rigid, coarsely pitted carapace that conceals the remainder of the abdomen from above (Fig. 24); metasomal tergum 5 (and often 6 or 7) usually visible in dorsal view; metasomal sculpture variable, often striate or rugose 10

10(9). Face strongly convex; hind femur short and strongly compressed .... ***Bulborogas*** van Achterberg

Distribution: Neotropical; rare; 2 described species. Biology: unknown. Reference: van Achterberg (1995).

\- Face not strongly convex; hind femur long and not strongly compressed. Wings: Fig. 10 .... ***Aleiodes*** Wesmael

Distribution: Cosmopolitan; very common; probably 150-200 species in New World. Biology: parasitoids of various macrolepidoptera, especially Noctuoidea, Geometroidea, Arctioidea, and Sphingoidea (M. Shaw, 1994). References: Marsh (1979, as *Rogas*), van Achterberg (1991), S. Shaw et al. (1997, key to species-groups). Nearctic species currently being revised by Marsh and S. Shaw.

REFERENCES

Achterberg, C. van 1991. Revision of the genera of the Afrotropical and W. Palaearctic Rogadinae Foerster (Hymenoptera: Braconidae). Zoologische Verhandelingen Leiden 273: 1-102.

Achterberg, C. van 1995. Generic revision of the subfamily Betylobraconinae (Hymenoptera: Braconidae) and other groups with modified fore tarsus. Zoologische Verhandelingen Leiden 298: 1-242.

Hodges, R. W., Dominick, T., Davis, D. R., Ferguson, D. C., Franclemont, J. G., Munroe, E. G., and Powell, J. A. 1983. Check List of the Lepidoptera of America North of Mexico. E. W. Classey and Wedge Entomological Research Foundation, London. 284 pp.

Marsh, P. M. 1979. Family Braconidae. Pp. 144-295. In: Krombein, K. V., Hurd, P. D., Smith, D.R., and Burks, B. D. (Eds.). Catalog of Hymenoptera in America North of Mexico. Smithsonian Institution Press, Washington, D.C.

Muesebeck, C. F. W. 1960. New reared Neotropical species of *Rogas* Nees (Hymenoptera: Braconidae). Entomological News 71: 257-261.

Quicke, D. L J. and Kruft, R. A. 1995. Species of *Yelicones* (Hymenoptera: Braconidae: Rogadinae) in North America with descriptions of two new species. Annals of the Entomological Society of America 88: 129-138.

Quicke, D. L J., Ch.ishti, M. J. K., and Basibuyuk, H. H. 1996 A revision of the *Yelicones* species (Hymenoptera: Braconidae: Rogadinae) from Central America, with descriptions of sixteen new species. Zoologische MededeJingen Leiden 70: 17-61.

Reardon, R. C., Statler, M. W., and Mcl.ane, W. H. 1973. Rearing techniques and biology of five gypsy moth parasites. Environmental Entomology 2: 124-127.

Shaw, M. R. 1981. Delayed inhibition of host development by the non-paralyzing venoms of parasitic wasps. Journal of Invertebrate Pathology 37: 215-221.

Shaw, M. R. 1983. On evolution of endoparasitism: The biology of some genera of Rogadinae (Braconidae). Contributions of the American Entomological Institute 20: 307-328.

Shaw, M. R. 1994. Parasitoid host ranges. Pp. 111-144. In: Hawkins, B. A. and Sheehan, W. (Eds.). Parasitoid Community Ecology. Oxford University Press, Oxford.

Shaw, M. R. and Huddleston, T. 1991. Classification and biology of braconid wasps (Hymenoptera: Braconidae). Handbooks for the Identification of British Insects 7(1l): l-126.

Shaw, S. R. 1993. Systematic status of *Eucystomastax* Brues and characterization of the Neotropical *Aleiodes melanopterus* (Erichson) species complex (Hymenoptera: Braconidae: Rogadinae: Rogadini). Journal of Hymenoptera Research 2: 1-11.

Shaw, S. R. 1995. The Braconidae. Pp. 431-464. In: Hanson, P. and Gauld, I. (Eds.). The Hymenoptera of Costa Rica. Oxford University Press, Oxford.

Shaw, S. R., Marsh, P. M., and Fortier, J. C. 1997. Revision of North American *Aleiodes* Wesmael (Part 1): the *pulchripes* Wesmael species-group in the New World (Hymenoptera: Braconidae, Rogadinae). Journal of Hymenoptera Research 6(1): 10-35.

Shenefelt, R. D. 1975. Pars 12. Braconidae 8, Exothecinae Rogadinae. Pp. 1115-1262. In: Vecht, J. van der and Shenefelt, R. D. (Eds.), Hymenopterorum Catalogus (nova edito). Dr. W. Junk, The Hague.

Smith, 0. J., Diboll, A. G., and Rosenberger, J. H. 1955. Laboratory studies of *Pelecystoma harrisinae* (Ashmead), an adventive braconid parasite of the western grape leaf skeletonizer. Annals of the Entomological Society of America 48: 232-237.

Whitfield, J. B. 1988. Revision of the Nearctic species of the genus *Stiropius*. Cameron (=*Bucculairiptex* Auct.) with the description of a new related genus (Hymenoptera: Braconidae). Systematic Entomology 13: 373-385.

Whitfield, J. B. 1989. The limits and composition of the rogadine tribes. Unpublished lecture outline distributed at the national meeting of the Entomological Society of America, Lexington, Kentucky.

Whitfield, J. B. 1990. Phylogenetic review of the *Stiropius* group of genera (Hymenoptera: Braconidae, Rogadinae) with description of a new Neotropical genus. Proceedings of the Entomological Society of Washington 92: 36-43.

Whitfield, J. B. 1992. The polyphyletic origin of endoparasitism in the cyclostome lineages of the Braconidae (Hymenoptera). Systematic Entomology 17: 273-286.

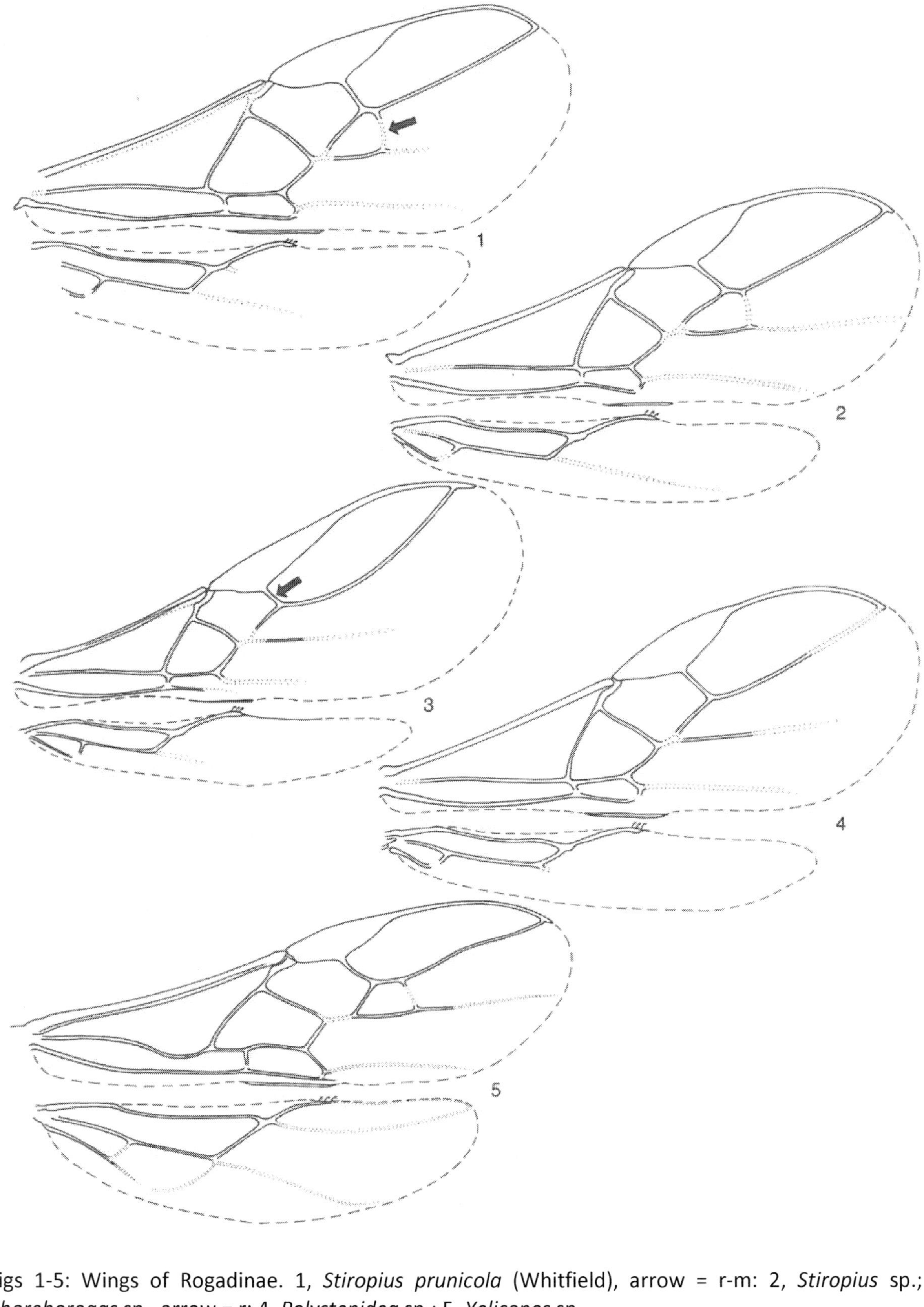

Figs 1-5: Wings of Rogadinae. 1, *Stiropius prunicola* (Whitfield), arrow = r-m: 2, *Stiropius* sp.; 3, *Choreborogas* sp., arrow = r: 4, *Polystenidea* sp.; 5, *Yelicones* sp.

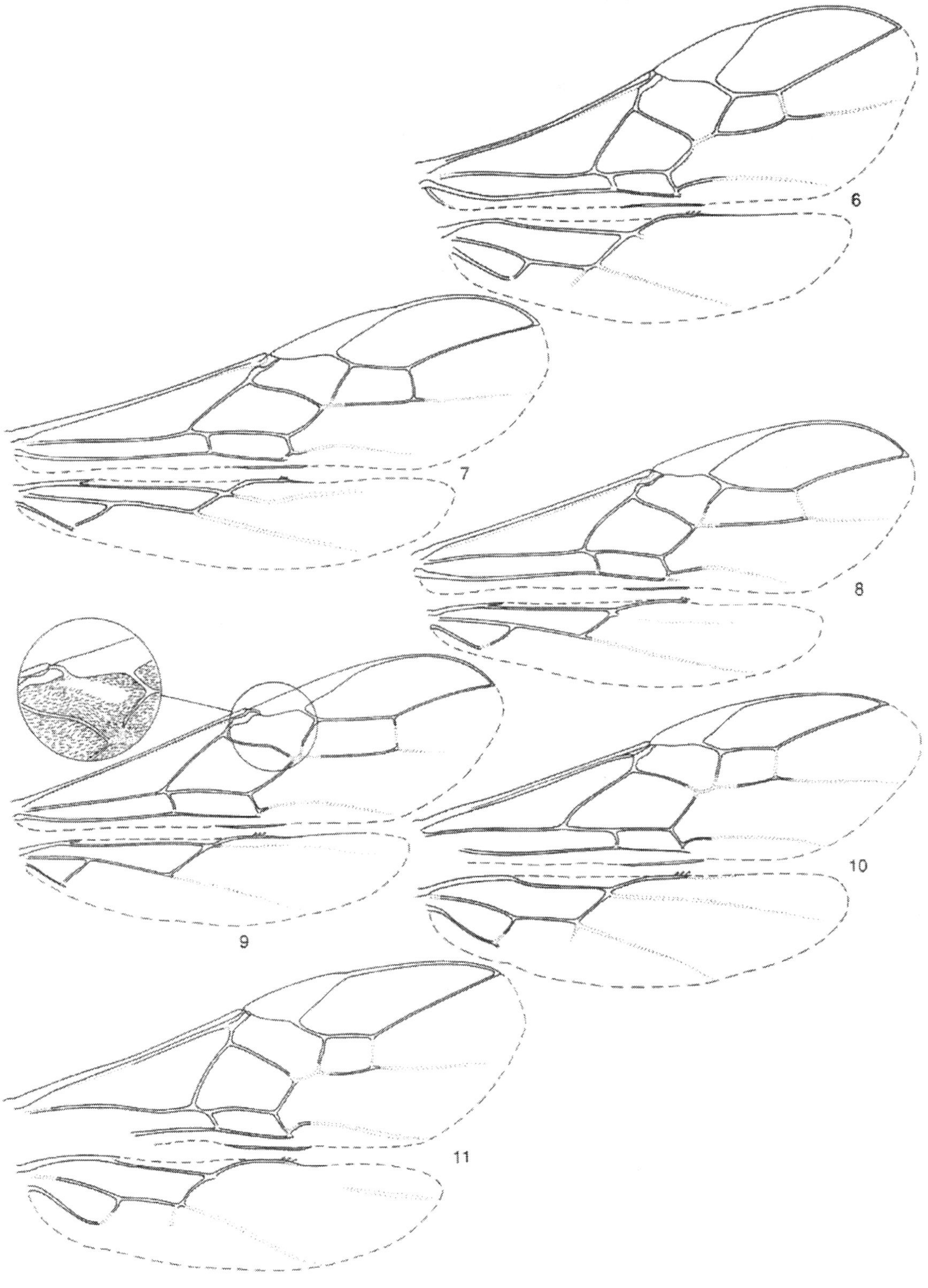

Figs 6-11: Wings of Rogadinae, 6, *Clinocentrus* sp.; 7, *Macrostomion* sp.; 8, *Rogas* sp.; 9, *Cystomastax* sp.; 10, *Aleiodes terminalis* Cresson; 11, *Tetrasphaeropyx* sp.

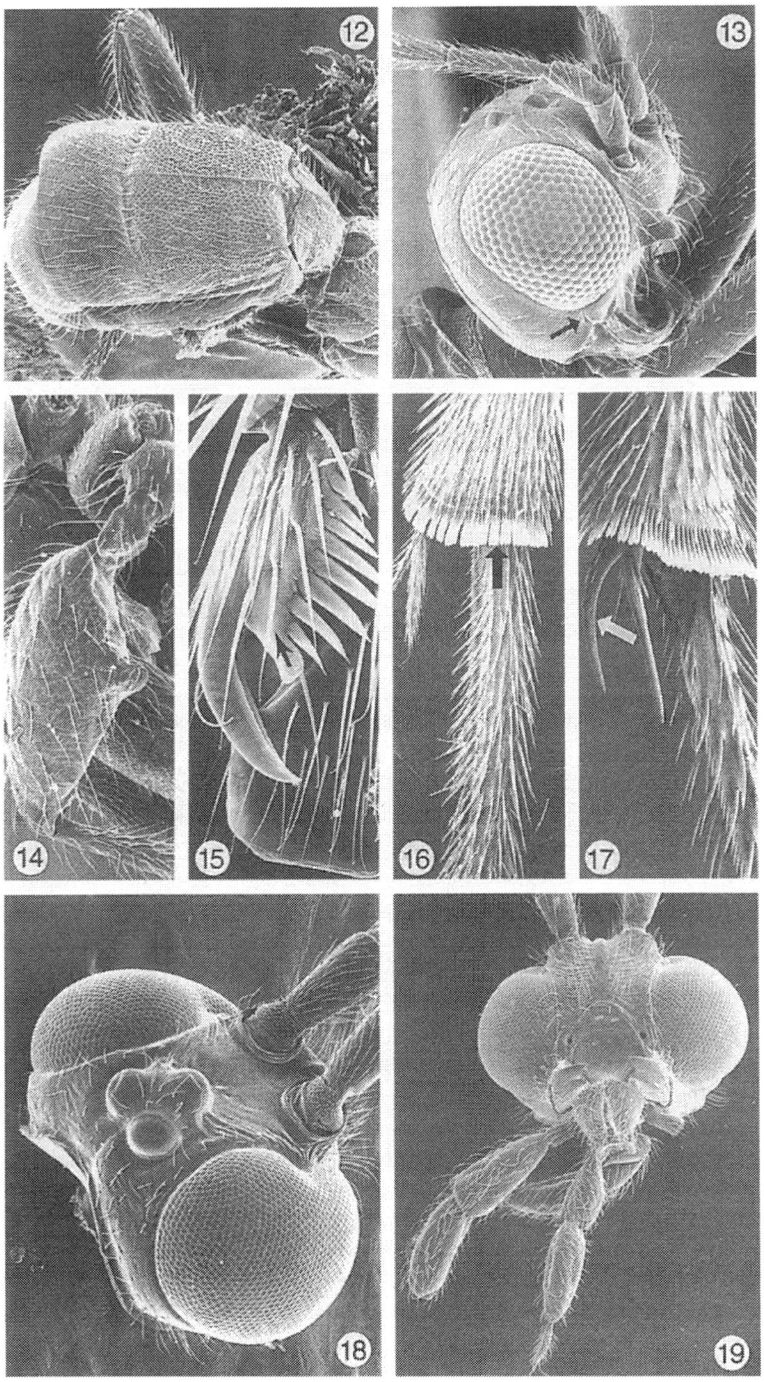

Figs 12-19: Rogadinae. 12, metasoma of *Stiropius* sp., dorsal view; 13, head of *Choreborogas* sp., lateral view, arrow = malar suture; 14, hind femur of *Choreborogas* sp.; 15, tarsal claw of *Rogas fusciceps* Cresson, arrow = blunt basal lobe bearing pectin; 16, apex of inner side of hind tibia of *Rogas fusciceps* Cresson, arrow = fringe of flattened setae; 17, apex of inner side of hind tibia of *Macrostomion* sp., arrow = curved, relatively bare tibial spur; 18, head of *Macrostomion* sp., dorsal view showing large eyes; 19, head of male *Macrostomion* sp., frontal view, showing enlarged maxillary palps.

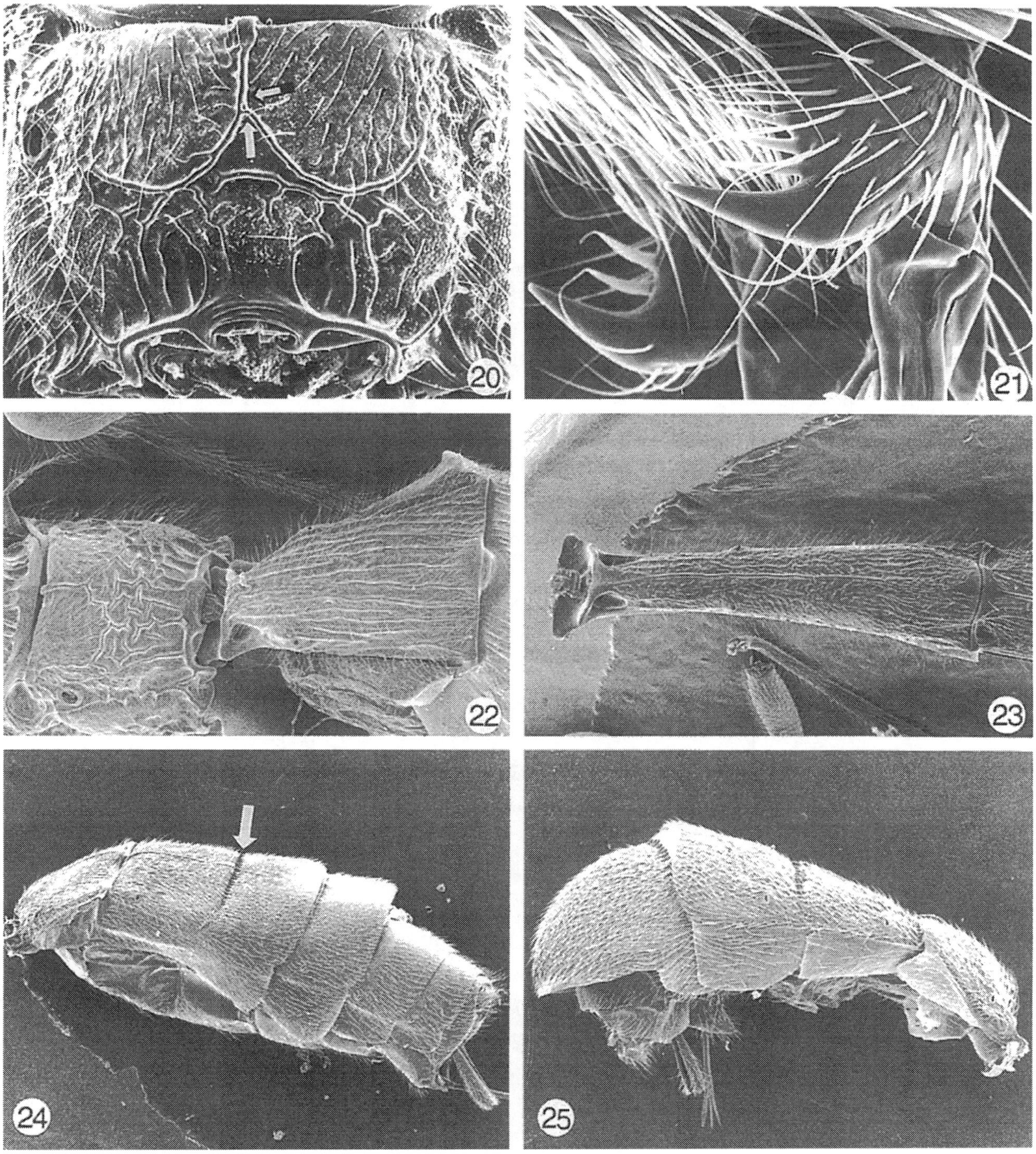

Figs 20-25: Rogadinae. 20, propodeum of *Rogas* sp., arrows = median carina diverging near middle; 21, tarsal claws of *Aleiodes melanopterus* (Erichson) showing pectin not on a basal lobe; 22, propodeum and petiole of *Rogas* sp., dorsal view; 23, petiole of *Cystomastax* sp., dorsal view; 24, metasoma of *Aleiodes* sp., lateral view, arrow = coarsely sculptured suture between terga 2 and 3; 25, metasoma of *Tetrasphaeropyx* sp., lateral view.

ROGADINAE UPDATE

S.R. Shaw
Department of Ecosystem Science and Management, University of Wyoming, Laramie, Wyoming 82071.
braconid@uwyo.edu
D.L.J. Quicke
Department of Biology, Faculty of Life Sciences, Chulalongkorn University, Bangkok, Thailand.
d.quicke@email.com

The Rogadinae chapter of the manual continues to work relatively well for the identification of the common New World genera of the traditional Rogadinae, though it now also includes the two New World genera of Betylobraconinae plus another recently described genus in that group.

The subfamily is now formally split into five tribes - Aleiodini, Betylobraconini, Clinocentrini, Rogadini, Stiropiini (corresponding to the *Stiropius* group of the 1997 key) and Yeliconini (Quicke and Butcher, 2015).

Betylobraconinae (*partim* - specifically Betylobraconini) are now treated as a tribe of Rogadinae based on morphological and molecular evidence (Butcher and Quicke, 2015). The Betylobraconini was until recently known only from the Old World, but a 'Gondwanan' representative was recently described (Quicke and Butcher, 2015). The tribe Facitorini of the original Betylobraconinae have also been transferred to the Rogadini where they are currently treated as a subtribe (Facitorina) within the Yeliconini (Belokobylskij *et al.*, 2008).

New World species formerly classified as *Rogas* were reassigned to *Triraphis* Ruthe (e.g. Zaldívar-Riverón *et al.*, 2008); true *Rogas* is an Old World genus that probably does not occur in the New World.

Specimens keying to the predominantly Old World genus *Colastomion* in Che and He (1997) have also been seen from Neotropical material.

A highly distinctive new genus, *Pseudoyelicones* van Achterberg, Penteado-Dias & Quicke, was described by van Achterberg *et al.* (1997) and additional species described by Areekul and Quicke (2004). It will key to *Yelicones* in the 1997 key, to which it is closely allied, but differs in having the hind wing marginal cell basally expanded with the wing membrane strongly corrugated. It also differs in having a simple rather than strongly pectinate claws.

The composition of the cosmopolitan genus *Aleiodes* has changed somewhat. *Tetrasphaeropyx*, which was treated as a separate genus from *Aleiodes* in the 1997 key, is now treated as the *pilosus* species group within *Aleiodes* (Fortier 2006a). The genus *Heterogamus* was treated as a synonym of *Aleiodes*, often separated as the *A. dispar* species-group, but they are now considered to be separate genera and only anciently related. Both are cosmopolitan genera with similar morphology but are abundantly distinct especially on the basis of DNA sequence data, but they are also morphologically fairly easily separable (Butcher *et al.*, 2012; van Achterberg and Shaw, 2016). *Aleiodes* is by far the larger of the two genera.

Braet and van Achterberg (2011) described two new small, *Aleiodes*-like species from French Guiana in their new genus *Athacryvac* Braet & van Achterberg. Molecular data at the time suggested they were potentially a sister-group of *Aleiodes*. However, whilst they are derived quite basally *Athacryvac* appears, in the light of the discovery of additional species, as well as additional morphological and molecular evidence, to be nested within *Aleiodes* and accordingly Shimbori *et al.* (2016) treat it as a subgenus of *Aleiodes*.

The North American species of *Aleiodes* have been extensively studied and revised in a series of papers by Marsh and Shaw (1998, 1999, 2001, 2003), Shaw (2006), Shaw and Marsh (2004), Fortier (2009), and Shaw *et al.* (1997, 1998a, 1998b, 2006, 2012). These papers follow the species-group arrangement proposed by Shaw *et al.* (1997) and cover all the North American species-groups of *Aleiodes*, with the exception of the *circumscriptus* and *gastritor* groups, which are still being revised. However, some of the most common species of the *circumscriptus* and *gastritor* groups are covered in Shaw (2006). The *seriatus* species-group as treated in the New World literature (Marsh and Shaw 1998) appear on the basis of molecular data not be closely related to the West Palaearctic *A. seriatus* and may be a polyphyletic assemblage (van Achterberg and Shaw, 2016).

Other recent papers have greatly expanded our knowledge of Neotropical *Aleiodes* species and their host records (Shimbori *et al.*, 2016), especially from Ecuador (Townsend and Shaw, 2009a; Shimbori and Shaw, 2014) and Brazil (Shimbori *et al.*, 2015).

Several recent studies have provided high-quality biological data and demonstrated the utility of distinctive host caterpillar-mummy forms for identifying *Aleiodes* species (Shaw, 2006; Townsend and Shaw, 2009a; Shimbori and Shaw, 2014). Shaw (2006) provides color photos of the mummified host-remains of all common *Aleiodes* in the eastern USA.

Annotated list of recent papers on New World Rogadinae

Achterberg C van, Penteado-Dias AM. 1995. Six new species of the *Aleiodes dispar* group (Hymenoptera: Braconidae: Rogadinae). *Zoologische Mededelingen, Leiden* 69: 1-18.
The Neotropical species described here belong to *Heterogamus*.

Achterberg C van, Shaw MR. 2016. Revision of the western Palaearctic species of *Aleiodes* Wesmael (Hymenoptera, Braconidae, Rogadinae). Part 1: Introduction, key to species groups, outlying distinctive species, and revisionary notes on some further species. *ZooKeys* 639: 1-164.
A major work on the genus *Aleiodes* from the West Palaearctic that defines relevant species-groups from there but importantly notes that the '*seriatus*' group from the New World is not monophyletic, being based on a single homoplastic character that is likely adaptive.

Achterberg C van, Penteado Dias AM, Quicke DLJ. 1997. *Pseudoyelicones* (Hymenoptera: Braconidae: Rogadinae), a new genus from Brazil and Costa Rica. *Zoologische Mededelingen, Leiden* 71: 1-8.
Description of a new genus of Yeliconini.

Areekul B, Quicke DLJ. 2004. Two new species of *Pseudoyelicones* (Braconidae: Rogadinae) from Costa Rica. *Journal of Hymenoptera Research* 13: 207-213.
Descriptions of additional species.

Areekul B, Quicke, DLJ. 2006. Systematics of the parasitic wasp genus *Yelicones* Cameron (Hymenoptera: Braconidae: Rogadinae) and revisions of the genus from the New World. *Systematics and Biodiversity* 4(2): 1-122.
Revised the genus *Yelicones* for the New World, providing an identification key for 85 species, of which 63 were newly described.

Belokobylskij SA. 1995. Revision of the Palaearctic species of the genus *Clinocentrus* (Hymenoptera, Braconidae). *Journal of Natural History* 29(3): 803-836.
The Palearctic species of *Clinocentrus* were revised; this work might be used to recognize some North American species that have Holarctic distributions.

Belokobylskij SA, Zaldívar-Riverón A, Maetô K, Sáez AG. 2008. Asian Betylobraconinae (Hymenoptera, Braconidae), with description of a new genus and phylogenetic affinities of the tribe Facitorini. *Insect Systematics and Evolution* 39: 133-154.

Molecular phylogenetic study which led to transfer of Facitorini from Betylobraconinae to Rogadinae as a subtribe within Yeliconini.

Braet Y, Achterberg C van. 2011. *Athacryvac* gen. n. from French Guiana (Hymenoptera: Braconidae: Rogadinae), with description of two new species. *Tijdschrift voor Entomologie* 154: 283-290.
Braet and van Achterberg (2011) described a new genus, *Athacryvac*, from French Guiana; however, further study has concluded that *Athacryvac* should be treated as a subgenus of *Aleiodes* (Shimbori *et al.*, 2016a).

Butcher BA, Quicke DLJ. 2015. A remarkable new genus and species of Rogadinae (Hymenoptera: Braconidae) of uncertain tribal placement, from Papua New Guinea, resembling Betylobraconini *stat. nov. Journal of Natural History* 49: 2045-2054.
Subfamily Betylobraconinae was synonymized under Rogadinae and treated as a separate tribe Betylobraconini within Rogadinae.

Butcher BA, Smith MA, Sharkey MJ, Quicke DLJ. 2012. A turbo-taxonomic study of Thai *Aleiodes (Aleiodes)* and *Aleiodes (Arcaleiodes)* (Hymenoptera: Braconidae: Rogadinae) based largely on COI bar-coded specimens, with rapid descriptions of 179 new species. *Zootaxa* 3457: 1-232.
Although dealing with the Thai fauna, provides illustration of how to separate *Heterogamus* from *Aleiodes*, and illustrated the homoplasy of characters used to recognise species groups.

Chen X, He J. 1997. Revision of the subfamily Rogadinae (Hymenoptera: Braconidae) from China. *Zoologische Verhandelingen* 308: 1-187.
Key allows separation of *Colastomion* from *Macrostomion*.

Fortier JC. 2006a.The first host association for *Aleiodes pilosus* (=*Tetrasphaeropyx* Ashmead) (Hymenoptera: Braconidae: Rogadinae). *Entomological News* 117: 457-459.
The species of *Aleiodes pilosus* species group (often as the subgenus *Tetrasphaeropyx*) have been studied in a series of papers by Fortier (2006a, 2006b, 2007a, 2007b, 2008, 2009) and Fortier and Sherman (2008).

Fortier JC. 2006b.Twelve new species and a new combination of the *Aleiodes pilosus* species-group (= *Tetrasphaeropyx* Ashmead) (Hymenoptera, Braconidae, Rogadinae) in North America: Part 1. *Entomological News* 117: 465-484.

Fortier JC. 2007a. Eight new species and a key to species of the *Aleiodes pilosus* species-group (=*Tetrasphaeropyx* Ashmead) (Hymenoptera, Braconidae, Rogadinae) in North America, Part 2. *Entomological News* 118: 11-30.

Fortier JC. 2007b. The *Aleiodes pilosus* Group (=*Tetrasphaeropyx* Ashmead) species of Canada, including seven new species and a key (Hymenoptera: Braconidae, Rogadinae). *Transactions of the American Entomological Society* 133: 1-20.

Fortier JC. 2008. The first host association for *Aleiodes shawi* (Hymenoptera: Braconidae: Rogadinae). *Entomological News* 119: 214-216.

Fortier JC. 2009. A revision of the *Tetrasphaeropyx* Ashmead Lineage of the Genus *Aleiodes* Wesmael (Hymenoptera: Braconidae: Rogadinae). *Zootaxa* 2256: 1-126.
Provides a comprehensive revision of *Aleiodes* subgenus *Tetrasphaeropyx*, treating 72 species of which 43 were new species (in addition to the 28 newly described species in the aforementioned series of papers).

Fortier JC, Sherman C. 2008. A new *Aleiodes* (*Tetrasphaeropyx*) (=*A. pilosus* group), with a host record from Missouri, USA, biological remarks, and previous type depository corrections. *Entomological News* 119: 445-453.

Fortier JC, Shaw, SR. 1999. Cladistics of the *Aleiodes* lineage of the subfamily Rogadinae (Hymenoptera: Braconidae). *Journal of Hymenoptera Research* 8: 204-237.
Reconstructed the phylogeny of *Aleiodes* species and concluded that "genus" *Tetrasphaeropyx* is a derived offshoot from within *Aleiodes*, as predicted by Shaw (1997), and *Tetrasphaeropyx* is treated as a subgenus of *Aleiodes* by Fortier (2009). This opinion was corroborated by the results of Zaldívar-Riverón *et al.* (2008).

Marsh PM, Shaw SR. 1998. Revision of North American *Aleiodes* (part 3): the *seriatus* species-group (Hymenoptera: Braconidae: Rogadinae). *Proceedings of the Entomological Society of Washington* 100(3): 395-408.
Provided a key to the North American species of the *seriatus* species-group, with descriptions of new species. A more extensive study of the Neotropical species is currently in progress in collaboration with Dr. Eduardo Mitio Shimbori. It estimates that there may be as many as 200 or more new species of the *seriatus* species-group in the neotropics.

Marsh PM, Shaw SR. 1999. Revision of North American *Aleiodes* (part 5): the *melanopterus* (Erichson) species-group in North America (Hymenoptera: Braconidae: Rogadinae). *Journal of Hymenoptera* Research 8: 98-108.
A key to the North American species of the *melanopterus* species-group is provided, along with descriptions of three new species.

Marsh PM, Shaw SR. 2001. Revision of North American *Aleiodes* (part 6): the *gasterator* (Jurine) and *unipunctator* (Thunberg) species groups (Hymenoptera: Braconidae: Rogadinae*). Proceedings of the Entomological Society of Washington* 103(3): 291-307.
Keys for the identification of the *gasterator* and *unipunctator* species-groups in North America are given, along with descriptions of six new species.

Marsh PM, Shaw SR. 2003. Revision of North American *Aleiodes* (part 7): the *compressor* Herrich-Schaeffer, *ufei* (Walley), *gressetti* (Muesebeck) and *procerus* Wesmael species-groups (Hymenoptera: Braconidae: Rogadinae). *Proceedings of the Entomological Society of Washington* 105(3): 698-707.
Six North American species in four species-groups are treated, along with the descriptions of three new species.

Martinez JJ. 2009. Two new species of *Clinocentrus* Haliday (Hymenoptera: Braconidae) from Argentina, with notes on South American species. *Journal of Natural History* 43: 43-44, 2667-2676.
Described two new *Clinocentrus* species from Argentina and provided notes on other species in South America.

Quicke DLJ, Butcher BA. 2015. Description of a new Betylobraconini-like parasitoid wasp genus and species (Hymenoptera: Braconidae: Rogadinae) from Chile. *Zootaxa* 4021: 459-466.
A new genus, *Gondwanocentrus* from Chile, the only genuine member of the Betylobraconini from the New World, is described and illustrated.

Quicke DLJ, Smith MA, Miller SE, Hrcek J, Butcher B. 2012. *Colastomion* Baker (Braconidae, Rogadinae): nine new species from Papua New Guinea reared from Crambidae. *Journal of Hymenoptera Research* 28: 85-121.

Shaw MR. 2002. A new species of *Macrostomion* Szepligeti (Hymenoptera: Braconidae: Rogadinae) from Papua New Guinea, with notes on the biology of the genus. *Zoologische Mededelingen, Leiden* 76: 133-140. ISSN 0024-0672.
Described a new species of *Macrostomion* from New Guinea and provided notes on the biology of the genus.

Shaw SR. 2006. *Aleiodes* wasps of eastern forests: a guide to parasitoids and associated dead caterpillars. *Forest Health Technology Enterprise Team (FHTET), FHTET*-2006-08, 126 pp.
Provided a field guide to the *Aleiodes* species of eastern North America, as well as color images of their associated host caterpillar mummies.

Shaw SR, Marsh, PM. 2004. Two new Eastern North American species of the *Aleiodes coxalis* (Spinola) species-group (Hymenoptera: Braconidae, Rogadinae) reared from Geometridae. *Zootaxa* 656: 1-10.
Two new species from the *coxalis* species-group, both parasitoids of Geometridae caterpillars, are described from eastern North America.

Shaw SR, Marsh PM, Fortier J. 1997. Revision of North American Aleiodes (Part 1): the *pulchripes* species-group in the New World. *Journal of Hymenoptera Research* 6: 10-35.
Provided an identification key for the species-groups of North American *Aleiodes*, a key to New World species of the *pulchripes* species-group, and descriptions of six new species.

Shaw SR, Marsh PM, Fortier J. 1998a. Revision of North American *Aleiodes* (Part 2): the *apicalis* species-group in the New World. *Journal of Hymenoptera Research* 7: 62-73.
Provided an identification key for the seven known species of the apicalis species-group in the New World, with the description of one new species.

Shaw SR, Marsh PM, Fortier J. 1998b. Revision of North American *Aleiodes* (Part 4): the *albitibia* and *praetor* species-groups in the New World. *Proceedings of the Entomological Society of Washington* 100: 553-565.
Provided an identification key for the New World species of the *praetor* species-group with the description of one new species. The *albitibia* species-group is defined and reviewed.

Shaw SR, Marsh PM, Fortier J. 2006. Revision of North American *Aleiodes* (Part 8): the *coxalis* species-group. *Zootaxa* 1314, 1-30.
An identification key for the Nearctic species of the *coxalis* species-group is provided, with the descriptions of nine new species.

Shaw SR, Marsh PM, Talluto MA. 2012. Revision of North American *Aleiodes* (Part 9): the *pallidator* (Thunberg) species-group with description of two new species (Hymenoptera: Braconidae, Rogadinae). *Zootaxa* 3608(3): 204-212.
An identification key for the five known North American species of the *pallidator* species-group is provided, with the descriptions of two new species.

Shimbori EM, Shaw SR. 2014. Twenty-four new species of *Aleiodes* Wesmael from the eastern Andes of Ecuador with associated biological information (Hymenoptera, Braconidae, Rogadinae). Zookeys 7402: 1-81.
Described 24 new *Aleiodes* from Ecuador with detailed host data.

Shimbori EM, Shaw SR, Penteado-Dias AM. 2015. Revision of New World *Aleiodes* Wesmael species of the *compressor* Herrich-Schaeffer species-group (Hymenoptera, Braconidae, Rogadinae). *The Canadian Entomologist* 147: 665-673.

New world species of the *compressor* species-group are revised, and two new species from Brazil are described.

Shimbori EM, Shaw SR, Ventura de Almeida LF, Penteado-Dias AM. 2016. Eleven new species of *Athacryvac* Braet & van Achterberg from the Neotropical Region (Hymenoptera, Braconidae, Rogadinae). *Zootaxa* 4138 (1): 83-117.

Reclassified *Athacryvac* as a subgenus of *Aleiodes*, provides descriptions of 11 new species, and gives an identification key for the Neotropical species of the subgenus.

Shimbori EM, Gessner C da Silva Souza, Penteado-Dias A, Shaw SR. 2017. A revision of the genus *Andesipolis* (Hymenoptera: Braconidae: Mesostoinae) and redefinition of the subfamily Mesostoinae. *Zootaxa* 4216 (2): 101-152.

Revised *Andesipolis*, described new South American species, and recommended that the genus be classified as a member of the Mesostoinae.

Spencer L, Whitfield JB. 1999. Revision of the Nearctic species of *Rhysipolis* Förster (Hymenoptera: Braconidae). *Transactions of the American Entomological Society* 125(3): 295-324.

Provided a revision of the Nearctic *Rhysipolis* with a key to known species.

Townsend A, Shaw SR. 2009. Nine new species of *Aleiodes* reared from caterpillars in the northeastern Andes of Ecuador (Hymenoptera: Braconidae: Rogadinae). *Journal of Insect Science* 9: Article 33, pages 1-21.

Described nine new *Aleiodes* from Ecuador with detailed host data.

Valerio AA, Shaw SR. 2015. Thirteen new Costa Rican species belonging to the genus *Trirhaphis* Ruthe (Hymenoptera: Braconidae, Rogadinae) with their host records. *Zootaxa* 3904 (4): 501-540.

Thirteen new species of *Trirhaphis* from Costa Rica and their host associations are described. Many other new species of *Trirhaphis* from New World tropical lowland forests still remain to be described.

Zaldívar-Riverón A, Shaw MR, Sáez AG, Mori M, Belokobylskij SA, Shaw SR, Quicke, DLJ. 2008. Evolution of the parasitic wasp subfamily Rogadinae (Braconidae): phylogeny and evolution of lepidopteran host ranges and mummy characteristics. *BMC (Biomed Central) Evolutionary Biology* 8(329) 1-20.

Investigated the phylogenetic history of the subfamily Rogadinae, revised their higher-level classification, and assessed the evolution of rogadine host ranges and host-mummy types. The first host record for a *Cystomastax* species is given.

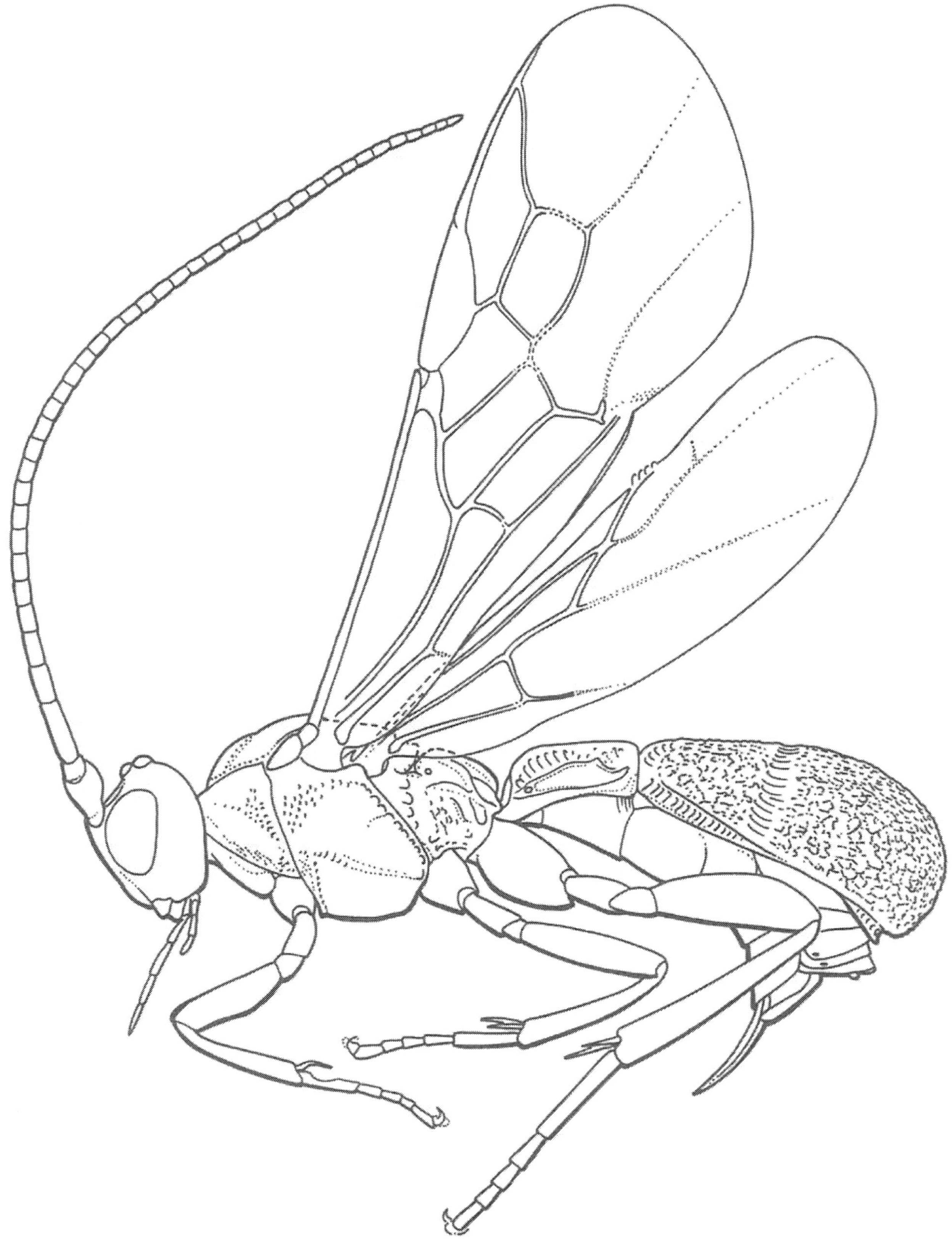

SUBFAMILY SIGALPHINAE

M. J. SHARKEY

Department of Entomology, University of Kentucky, S-225 Agricultural Science Building North, Lexington, Kentucky 40546, U.S.A.

INTRODUCTION. The subfamily Sigalphinae contains eight genera, only two of which are found in the New World. Van Achterberg and Austin (1992) revised and keyed the World genera (except *Pselaphanus* Szépligeti) and all known species. Sharkey and Janzen (1995) described two new species of Sigalphus Latreille, provided a key to World species, and detailed the biology of a species from Costa Rica. Shenefelt (1970, 1973) cataloged the World species of Sigalphinae.

BIOLOGY. Members of the Sigalphinae are koinobiont endoparasites of larval Noctuidae (Lepidoptera). Cushman (1913) observed first instar larvae being attacked by *Sigalphus bicolor* (Cresson) though no adults were successfully reared. Van Achterberg and Austin's (1992) hypothesis that sigalphines may be egg-larval parasites because they have a carapace-like metasoma appears somewhat contradicted by this evidence. *Sigalphus bicolor* is a gregarious, multivoltine parasitoid of *Acronicta clarescens* Cuenée, and parasitoid cocoons are spun within the host cocoon (Cushman, 1913). *Sigalphus romeroi* Sharkey from Costa Rica and *S. irrorator* (Fabricius) from the Palearctic are solitary endoparasitoids of larval Noctuidae (Sharkey and Janzen, 1995). Shaw and Huddleston (1991) reviewed the biology of the subfamily.

BIOLOGICAL CONTROL POTENTIAL. Because there are less than 20 species worldwide (van Achterberg and Austin 1992), and most of these are quite rare, it is doubtful that any members are important in the natural control of pest species of Lepidoptera. None has been used in a classical biological control program.

COMMON GENERA. Both New World genera are rare. *Sigalphus* is represented by two known New World species (Sharkey and Janzen, 1995) and *Pselaphanus* is represented by one species.

DISTRIBUTION. Cosmopolitan. In the New World *S. bicolor* is found in eastern U.S.A. and extreme southeastern Canada. *Sigalphus romeroi* has only been recorded from the Pacific deciduous forest in Costa Rica. *Pselaphanus trogoides* Szépligeti has been recorded from tropical regions of Brazil, Venezuela, and Surinam.

DISTINGUISHING FEATURES. Sigalphines can be recognized by the following combination of characters: hind wing with vein CUb present (Figs 1, 2); fore wing vein M+CU tubular; fore wing second submarginal cell longer than wide; petiole with a pair of longitudinal carinae; metasoma not laterally compressed.

KEY TO THE NEW WORLD GENERA OF THE SUBFAMILY SIGALPHINAE

1. Metasomal terga 1-3 forming carapace covering entire metasoma, tergum 4 not visible dorsally (see habitus). Wings: Fig. 1 .... ***Sigalphus*** Latreille

\- Metasomal terga 1-3 not forming carapace covering entire metasoma, tergum 4 visible dorsally. Wings: Fig. 2 .... ***Pselaphanus*** Szépligeti

REFERENCES

Achterberg, C. van and Austin, A. D. 1992. Revision of the genera of the subfamily Sigalphinae (Hymenoptera: Braconidae), including a revision of the Australian species. Zoologische Verhandelingen Leiden 280: 1-44.

Cushman, R. A. 1913. Biological notes on a few rare or little known parasitic Hymenoptera. Proceedings of the Entomological Society of Washington 15: 153-161.

Shaw, M. R. and Huddleston, T. 1991. Classification and biology of braconid wasps (Hymenoptera: Braconidae). Handbooks for the Identification of British Insects 7(11): 1-126.

Sharkey, M. J. and Janzen, D. 1995. Review of the world species of *Sigalphus* (Hymenoptera: Braconidae: Sigalphinae) and biology of *Sigalphus romeroi*, new species. Journal of Hymenoptera Research 4: 99-109.

Shenefelt, R. D. 1970. Pars 2. Braconidae 2. Helconinae, Calyptinae, Mimagathidinae, Triaspinae. Pp. 177-306. In: Ferrière, Ch. and Vecht, J. van der (Eds.). Hymenopterorum Catalogus (nova editio). Or. W. Junk, The Hague.

Shenefelt, R. D. 1973. Pars 10. Braconidae 6, Cheloninae. Pp. 813-936. In: Vecht, J. van der and Shenefelt, R. D. (Eds.). Hymenopterorum Catalogus (nova editio). Dr. W. Junk, The Hague.

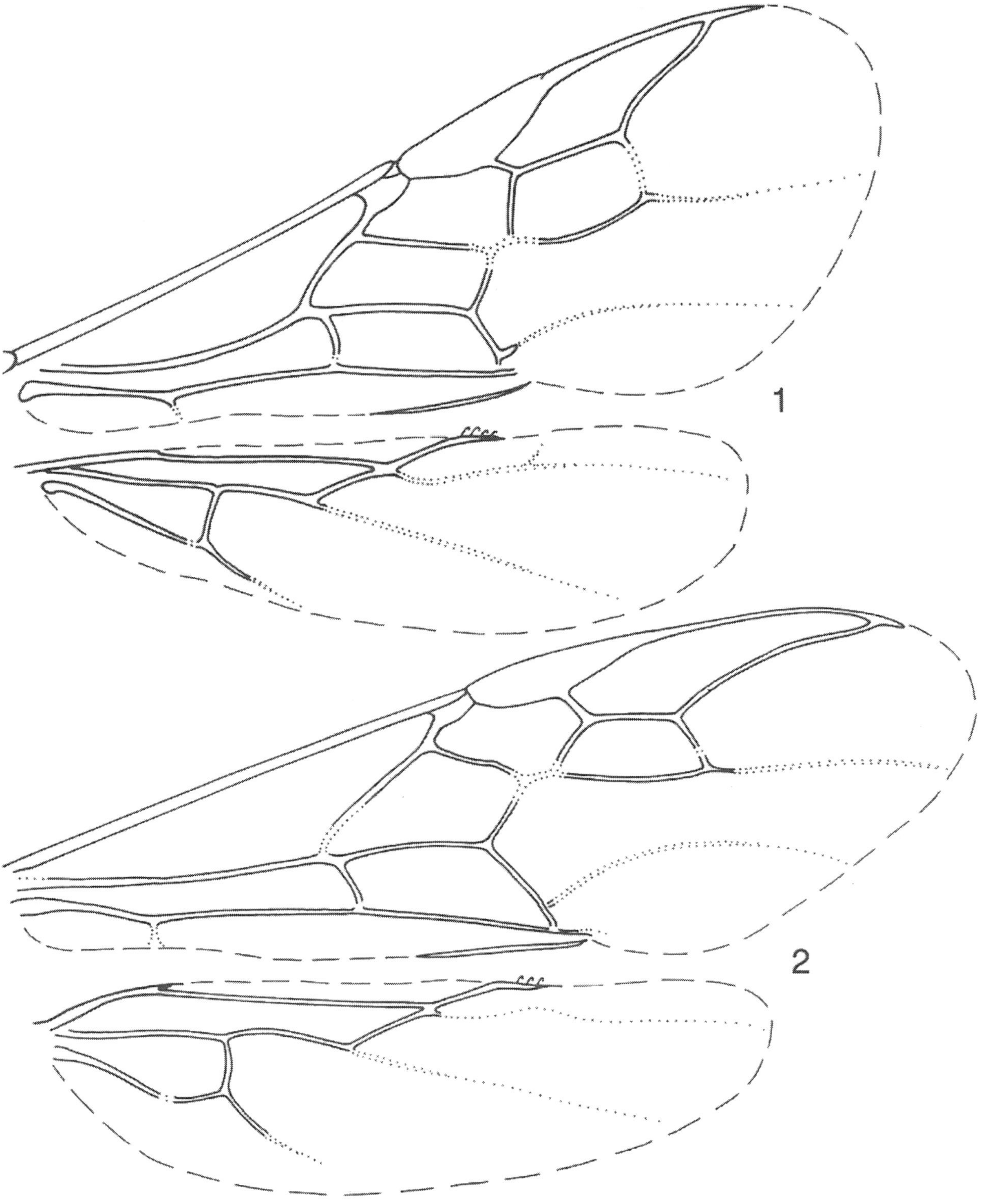

Figs 1-2: Sigalphinae. 1, Wings of *Sigalphus bicolor* (Cresson); 2, wings of *Pselaphanus trogoides* Szépligeti.

SIGALPHINAE UPDATE

M.J. Sharkey
Department of Entomology, University of Kentucky, Lexington, Kentucky 40546-0091, U.S.A. msharkey@uky.edu

Penteado-Dias (1996) reported on the discovery of the genus *Acampsis* (*Acampsis braziliensis*) in Brazil, the first report of the genus from the New World. *Acampsis* will key to *Sigalphus*. The two can be separated as follows: the third tergum is curved ventrad in *Sigalphus* such that it is visable in posterior view, whereas in *Acampsis* it is comparatively flat and not visable in posterior view.

Van Achterberg and Riedel (2009) described a new genus, *Paphanus*, from Paraguay and included a key to separate it from the closely related *Pselaphanus*. Penteado-Dias will describe a new species of *Paphanus*, occurring for the first time in Brazil (in prep.). *Paphanus* will run to *Pselaphanus* in the key to sigalphiine genera in the Manual, and it can be easily separated from the latter by the presence of a prepectal carina which is lacking in *Pselaphanus* and by a completely sculptured third median tergite (smooth in the posterior half in *Pselaphanus*).

Sharkey (2004) reported the first occurrence of *Minanga* in the New World (Mexico). Sharkey and Braet (2012) described a new species of *Minanga* from French Guiana and presented a key to the world species.

Quicke *et al.* (2008) presented a phylogenetic analysis of the world genera and demonstrated that *Pselaphanus* is sister to the remaining Sigalphinae.

Rrecent papers on New World Sigalphinae

Achterberg C van, Riedel M. 2009. *Paphanus drechseli gen. nov & spec. nov.* (Hymenoptera: Braconidae: Sigalphinae: Pselaphanini) from Paraguay. *Zoologische Mededelingen (Leiden)* 83(4): 799-804.

Penteado-Dias AM. 1996. A new species of *Acampsis* Wesmael (Hymenoptera, Braconidae, Sigalphinae) from Brazil. *Revista Brasileira de Zoologia* 13(2): 331-333.

Quicke DLJ, Sharkey MJ, Luarenne NM, Dowling A. 2008. A preliminary molecular phylogeny of the Sigalphinae (Hymenoptera: Braconidae), including *Pselaphanus* Szépligeti, based on 28S rDNA, with descriptions of new Afrotropical and Madagascan *Minanga* and *Malasigalphus* species. *Journal of Natural History* 42(43-44): 2703-2719.

Sharkey MJ. 2004. Afrotropical-North American disjunct distribution of *Minanga* (Hymenoptera: Braconidae) with the description of a new species and first record for the new world. *Annals of the Entomological Society of America* 97: 1198-1203.

Sharkey MJ, Braet Y. 2012. New species of the rare genera *Dentigaster* Zettel, 1990 (Hymenoptera, Braconidae, Cheloninae) and *Minanga* Cameron, 1906 (Sigalphinae) from French Guiana. *Journal of Hymenoptera Research* 25: 93-102.

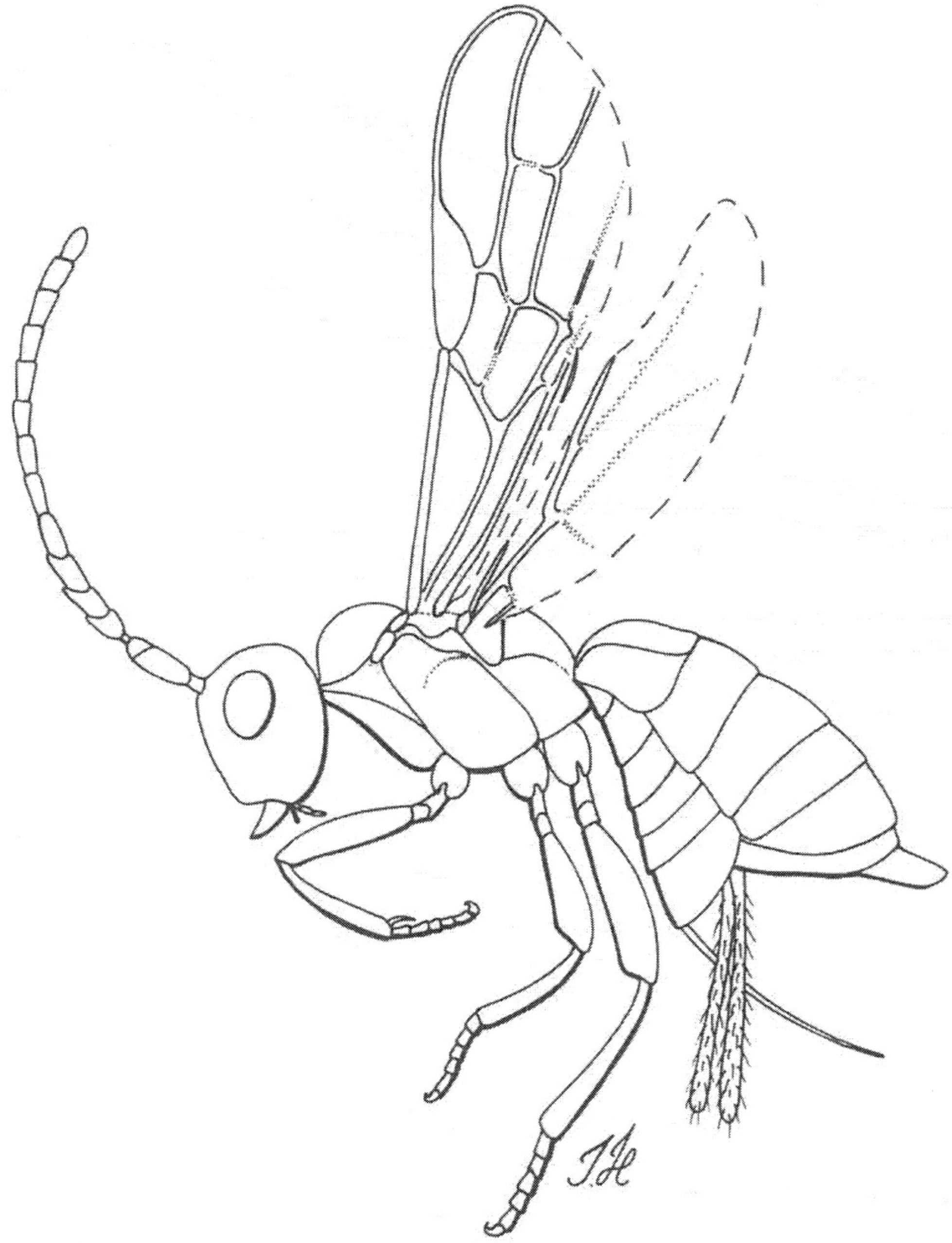
J.H

SUBFAMILY YPSISTOCERINAE

D. L. J. QUICKE

De Department of Biology, Faculty of Life Sciences, Chulalongkorn University, Bangkok, Thailand.

C. VAN ACHTERBERG

Afdeling Entomologie (Hymenoptera), Nationaal Natuurhistorisch Museum, Postbus 9517,
2300 RA Leiden, The Netherlands

INTRODUCTION. The Ypsistocerinae is a very small and little-known subfamily of braconids. Only three genera have been discovered (Cushman, 1923; Brues, 1923; van Achterberg, 1995). The group has long been given independent subfamily status (Cushman, 1923; also see Shenefelt, 1978). Recent evidence (Quicke et al., 1992a, b) strongly suggests that these aberrant wasps are no more than derived members of the Doryctinae, though Sharkey (1993) has regarded them as belonging to the Rogadinae s. 1. Sexual dimorphism is pronounced with the males having considerably reduced wings with very thickened veins (Fig. 3).

BIOLOGY. All known species in two of the three genera, viz. *Ypsistocerus manni* Cushman, *Y. vestigialis* Cushman and *Termitobracon emersoni* Brues, have been recorded from nests of *Nasutitermes* species (Cushman, 1923; Brues, 1923). Whether ypsistocerines are actually parasitoids of the termites themselves rather than of some group of inquilines, is at present unknown. Brues (1923), however, reported dissecting many termites from a nest yielding many ypsistocerines without finding any trace of parasitoid larvae, suggesting that parasitism of inquilines is more likely. The biology of the. only known species of the third genus, *Embobracon brevistigmus* van Achterberg, is unknown.

BIOLOGICAL CONTROL POTENTIAL. Unknown, but probably none.

COMMON GENERA. The subfamily is only known from three, rarely collected genera. Given their biology, however, they may be far more widespread than museum collections suggest.

DISTRIBUTION. The Ypsistocerinae are known only from the Neotropical Region.

DISTINGUISHING FEATURES. The three known genera are rather dissimilar from one another, and it is not possible to give any single easy diagnostic characters though in all of them hind wing vein SC+ R is largely obliterated (Figs 1, 2, 4). The sternites and the base of the ovipositor sheath are also somewhat retracted relative to the apical tergites (Figs 5, 7, 9). Recent work on ovipositor structure (Quicke et al., 1992a) and internal anatomy (Quicke et al., 1992b) suggests that ypsistocerines might be regarded as aberrant members of the Doryctinae, and accordingly, van Achterberg (1995) has treated them as a tribe (Ypsistocerini) of that subfamily.

KEY TO THE NEW WORLD GENERA OF THE SUBFAMILY YPSISTOCERINAE·

1. Maxillary palp with 5 segments (Fig. 5); labial palp with 4 segments (Fig. 5). Propodeum strongly sculptured (Fig. 6). Wings: Fig. 1 .... ***Embobracon*** van Achterberg

   Distribution: Panama; one described species. Biology: unknown. Reference: van Achterberg (1995).

\- Maxillary palp with 3 or fewer segments; labial palp with 2 or fewer segments. Propodeum smooth, without strong sculpture 2

2. Apical margin of first 3 flagellomeres strongly oblique (Figs 7, 8). Fore wing 2RS of female present (Fig. 2). Metasomal terga smooth. Maxillary palp with 3 segments; labial palp with 2 segments (Fig. 7) .... ***Termitobracon*** Brues

Distribution: Panama and Guyana; one described species. Biology: reared from nests of *Nasutitermes* (Brues, 1923). Reference: Brues (1923), van Achterberg (1995).

\- Apical margin of first 3 flagellomeres truncate (Fig. 10). Fore wing 2RS of female absent (Fig. 4). Metasomal terga largely coriaceous (Fig. 9). Maxillary palp with 2 segments (Fig. 9); labial palp absent (Fig. 9) .... ***Ypsistocerus*** Cushman

Distribution: Bolivia; 2 described species. Biology: collected from nests of *Nasutitermes* (Cushman, 1923). Reference: Cushman (1923), van Achterberg (1995).

REFERENCES

Achterberg, C van 1995. Generic revision of the subfamily Betylobraconinae (Hymenoptera; Braconidae) and other groups with modified fore tarsus. Zoologische Verhandelingen Leiden 298: 1-242.

Brues, C. T. 1923. *Termitobracon*, a termitophilous braconid from British Guiana. Zoologica 3: 427-432.

Cushman, R. A. 1923. A new subfamily of Braconidae (Hym.) from termite nests. Proceedings of the Entomological Society of Washington 25: 54-55.

Quicke, D. L. J., Ficken, L. C., and Fitton, M. G. 1992a. New diagnostic ovipositor characters for doryctine wasps (Hymenoptera, Braconidae), Journal of Natural History 26: 1035-1046.

Quicke, D. L. J., Tunstead, J., Falco, J. V., and Marsh, P. M. 1992b. Venom gland and reservoir morphology in the Doryctinae and related braconid wasps (Insecta, Hymenoptera, Braconidae), Zoologica Scripta 21: 403-416.

Sharkey, M. J. 1993. Family Braconidae. Pp. 362-395. In: Goulet, H. and Huber, J. T. (Eds.). Hymenoptera of the World: An Identification Guide to Families. Agriculture Canada, Ottawa.

Shenefelt, R. D. 1978. Pars 15. Braconidae 10, Braconinae, Gnathobraconinae, Mesostoinae, Pseudodicrogeniinae, Telengainae, Ypsistocerinae plus Braconidae in general, major groups, unplaced genera and species. Pp. 1425-1865. In: Achterberg, C. van and Shenefelt, R D. (Eds.). Hymenopterorum Catalogus (nova editio). Dr. W. Junk, The Hague.

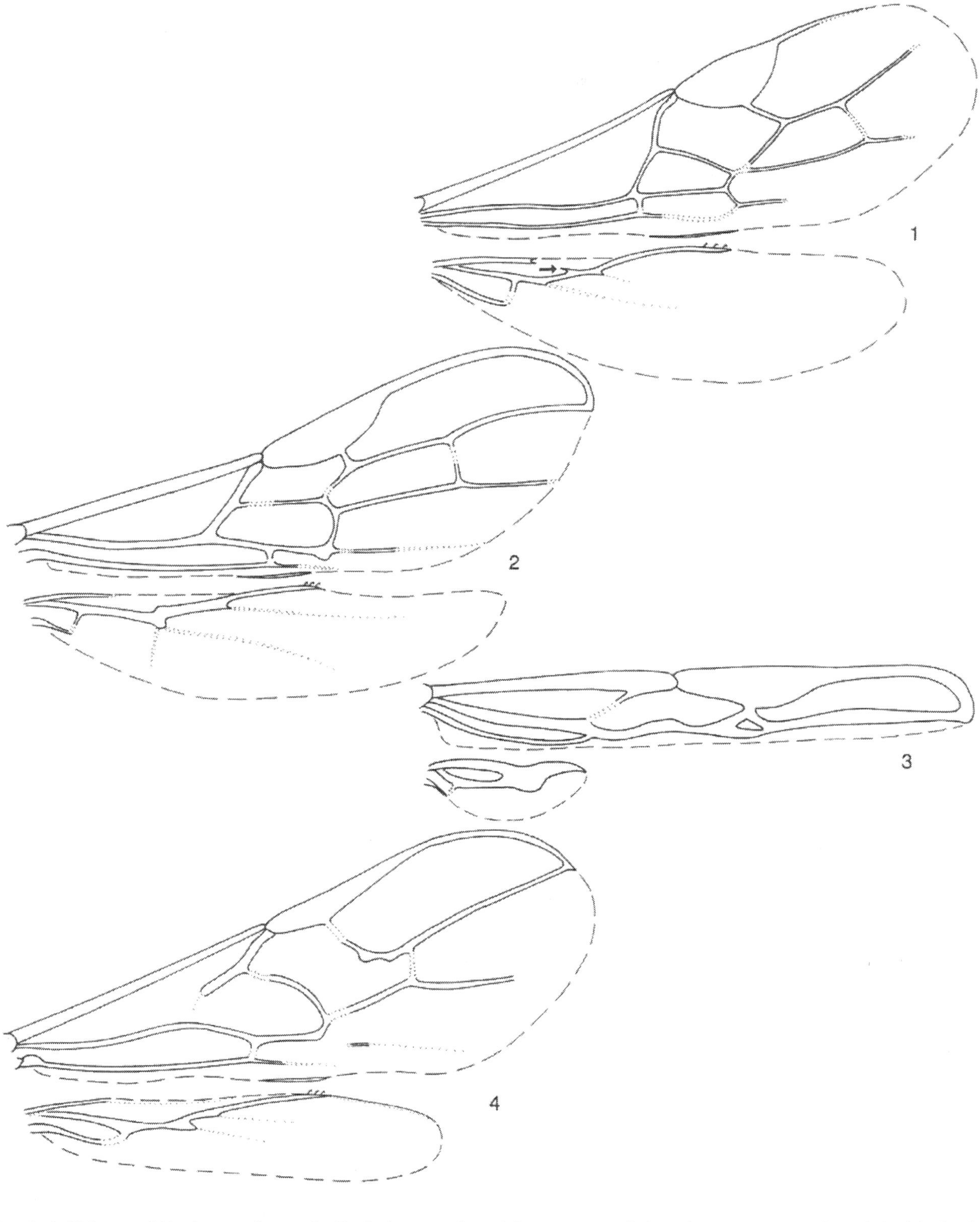

Figs 1-4: Wings of Ypsistocerinae. 1, *Embobracon brevisitgmus* van Achterberg, arrow = break in hind wing SC+R; 2, *Termiiobracon emersoni* Brues, female; 3, *Termitobracon emersoni* Brues, male; 4, *Ypsisiocerus manni* Cushman.

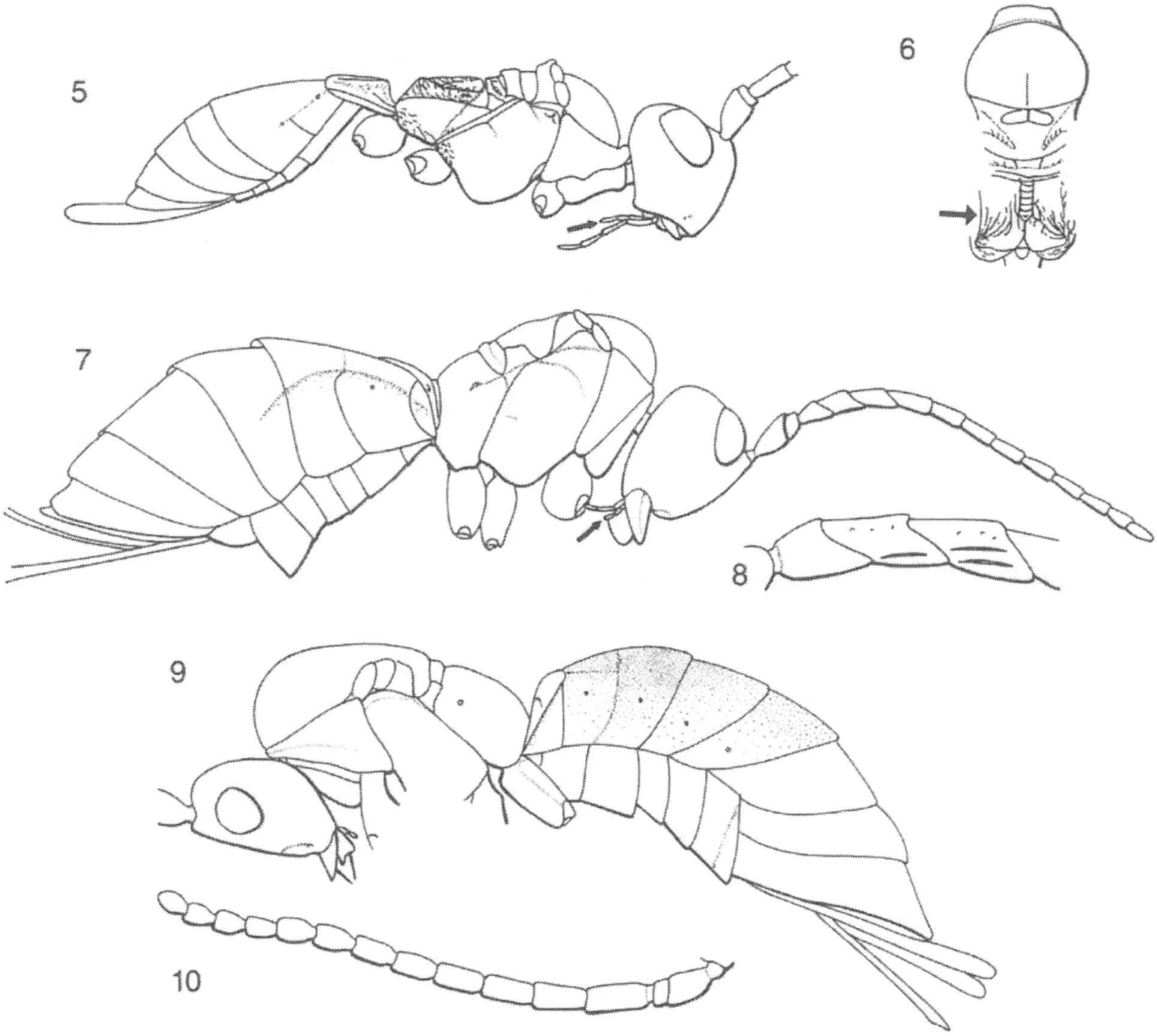

Figs 5-10: Ypsistocerinae, 5, habitus of *Embobracon brevisitgmus* van Achterberg, arrow = labial palp; 6, mesosoma of *Embobracon brevistigmus* van Achterberg, arrow = propodeum; 7, habitus of *Termitobracon emersoni* Brues, arrow = maxillary palp: 8, basal 3 flagellomeres of *Termitobracon emersoni* Brues showing oblique apical margins; 9, habitus of *Ypsistocerus manni* Cushman; 10, antenna of *Ypsistocerus manni* Cushman. All figures van Achterberg (1995).

YPSISTOCERINAE UPDATE

D.L.J. Quicke
Department of Biology, Faculty of Life Sciences, Chulalongkorn University, Bangkok, Thailand, d.quicke@email.com

This group is now regarded as being a tribe within the Doryctinae based on morphological evidence (Belokobylskij *et al.*, 2004); no material has been available for DNA sequencing as yet. No new species have been described in this group since the 1997 Manual, but the two named species of *Ypsistocerus* have been synonymized (Kistner *et al.*, 2000).

Quicke and Butcher (2015) described what they believed at the time was a new genus, *Preembobracon*, related to *Embobracon* van Achterberg, but according to Alejandro Zaldívar-Riverón (pers comm.), that genus was actually *Labania* Hedqvist, a previously described doryctine and probably not closely related to *Embobracon*.

Kistner *et al.* (2000) provided a few more observations on the interactions between *Termitobracon* and *Ypsistocerus* and the termites in whose nests they live - aggression was rarely seen, wasps were not especially attacked by the termites though they often avoided contact.

Annotated list of recent papers on New World Ypsistocerini

Belokobylskij SA, Zaldívar-Riverón A, Quicke DLJ. 2004. Phylogeny of the parasitic wasp genera of the subfamily Doryctinae (Hymenoptera: Braconidae) from morphological evidence. *Zoological Journal of the Linnean* Society 142: 369-404.

Kistner DH, Jacobson HR, Elliott DE. 2000. Some interesting Braconidae that live with termites and their behavior (Hymenoptera; Isoptera: Termitidae, Nasutitermitinae). *Sociobiology* 36: 79-82 and 257-264.
The male of *Ypsistocerus manni* is described for the first time. *Ypsistocerus vestigialis* is synonymized with *Y. manni.*

Quicke DLJ, Butcher BA. 2015. *Preembobracon gen. nov.* (Hymenoptera: Braconidae: Doryctinae: Ypsistocerini: Embobraconina) from Brazil. *Zootaxa* 4000**:** 275-280.

Index of the generic names including Subgenera and Synonyms

Index of Generic Names Including Subgenera and Synonyms

Listed below are all New World genera mentioned in this manual with indication of the page(s) where they appear in the keys. In addition, subgeneric names are listed as well as most relevant synonyms. The authors are not in total agreement with the synonymies listed here, but this is how the names have been treated in this manual. Because many people have been using the previous North American manual to genera (Marsh, et al., 1987), this list should help them place a name that they were using before, but which may not be in usage now.

Appendix

Alphabetical listing, by subfamily, of New World genera cited in the keys

SUBFAMILY ADELIINAE

Adelius Haliday, 1833. Ent. Mag. 1: 262.

Paradelius de Saeger, 1942. Rev. Zool. Botan. Afr. 36: 313.

SUBFAMILY AGATHIDINAE

Aerophilus Szépligeti, 1902. Term. Fuez. 25: 73.

Agathirsia Westwood, 1882. Tijdschr. Ent. 25: 20.

Agathis Latrielle, 1804. Nouv. Dict. Hist. Nat. Deterville (Ed. 1) 24: 173.

Alabagrus Enderlein, 1920 (1918). Arch. Naturgesch. 84A(11): 203.

Bassus Fabricus, 1804. Syst. Piezat., p. 93.

Coccygidium de Saussure, 1892. In: Grandidier, Hist. Phys. Nat. Polit. Madagascar 20: P1. 15, Fig. 27.

Crassomicrodus Ashmead, 1900. Proc. U.S. Natl. Mus. 23: 128.

Cremnops Foerster, 1862. Verh. naturh. Ver. preuss. Rheinl. 19: 246.

Dichelosus Szépligeti, 1902. Term. Fuez. 25: 71.

Earinus Wesmael, 1837. Nouv. Mém. Acad. Brux. 10: 8.

Labagathis Enderlein, 1920 (1918). Arch. Naturgesch. 84A(11): 168.

Liopisa Enderlein, 1920 (1918). Arch. Naturgesch. 84A(11): 182.

Marjoriella Sharkey, 1983. Contrib. Am. Ent. Inst. 20: 94.

Mesocoelus Schulz, 1911. Zool. Ann. 4: 88.

Pharpa Sharkey, 1986. Can. Ent. 118: 1232.

Plesiocoelus van Achterberg, 1990. Zool. Meded. Leiden 64: 44.

Sesioctonus Viereck, 1912. Smithson. Misc. Collns. 59(5): 1.

Trachagathis Viereck, 1913. Proc. U.S. Natl. Mus. 46: 366.

Zacremnops Sharkey and Wharton, 1985. Can. Ent. 117: 599.

Zamicrodus Viereck, 1912. Proc. U.S. Natl. Mus. 42: 630.

SUBFAMILY ALYSIINAE

Alloea Haliday, 1833. Ent. Mag. 1: 265.

Alysia Latreille, 1804. Nouv. Dict. Hist. Nat. Deterville (Ed. 1). 24: 173.

Alysiasta Wharton, 1980. Univ. Calif. Publ. Ent. 88: 22.

Anisocyrta Foerster, 1862. Verh. naturh. Ver. preuss. Rheinl. 19: 268.

Aphaereta Foerster, 1862. Verh. naturh. Ver. preuss. Rheinl. 19: 264.

Apronopa van Achterberg, 1980. Ent. Bericht. 40: 75.

Asobara Foerster, 1862. Verh. naturh. Ver. preuss. Rheinl. 19: 267.

Aspilota Foerster, 1862. Verh. naturh. Ver. preuss. Rheinl. 19: 268.

Chaenusa Haliday, 1839. Hym. Brit. Alysia, fasc. 2: 19.

Chorebus Haliday, 1833. Ent. Mag. 1: 264.

Coelinius Nees, 1818. Nova Acta Acad. Leop. Carol. 9: 301.

Coloneura Foerster, 1862. Verh. naturh. Ver. preuss. Rheinl. 19: 276.

Cratospila Foerster, 1862. Verh. naturh. Ver. preuss. Rheinl. 19: 265.

Dacnusa Haliday, 1833. Ent. Mag. 1: 264.

Dapsilarthra Foerster, 1862. Verh. naturh. Ver. preuss. Rheinl. 19: 267.

Dapsildiastema Wharton, 1994. Proc. Ent. Soc. Wash. 96: 641.

Dinostigma Fischer, 1966. Ann. naturh. Mus. Wein 69: 182.

Dinotrema Foerster, 1862. Verh. naturh. Ver. preuss. Rheinl. 19: 268.

Epimicta Foerster, 1862. Verh. naturh. Ver. preuss. Rheinl. 19: 274.

Exotela Foerster, 1862. Verh. naturh. Ver. preuss. Rheinl. 19: 274.

Glyphogaster Muesebeck and Walkley, 1951. Agric. Monogr. 2: 148.

Gnathopleura Wharton, 1980. Univ. Calif. Publ. Ent. 88: 38.

Hoplitalysia Ashmead, 1900. Proc. U.S. Natl. Mus. 23: 105.

Idiasta Foerster, 1862. Verh. naturh. Ver. preuss. Rheinl. 19: 265.

Ilatha Fischer, 1975. Studia Ent. 18: 139.

Laotris Nixon, 1943. Ent. Mon. Mag. 79: 30.

Mesocrina Foerster, 1862. Verh. naturh. Ver. preuss. Rheinl. 19: 266.

Microcrasis Fischer, 1975. Studia Ent. 18: 148.

Oenonogastra Ashmead, 1900. Proc. U.S. Natl. Mus. 23: 105.

Orthostigma Ratzeburg, 1844. Ichn. Forstins. 1: 53.

Pentapleura Foerster, 1862. Verh. naturh. Ver. preuss. Rheinl. 19: 264.

Phaenocarpa Foerster, 1862. Verh. naturh. Ver. preuss. Rheinl. 19: 267.

Phasmalysia Tobias, 1971. Trudy Vses. Ent. Obshch. 54: 261.

Phasmidiasta Wharton, 1980. Univ. Calif. Publ. Ent. 88: 63.

Pneumosema Fischer, 1966. Ann. naturh. Mus. Wein 69: 207.

Symphanes Foerster, 1862. Verh. naturh. Ver. preuss. Rheinl. 19: 264.

Symphya Foerster, 1862. Verh. naturh. Ver. preuss. Rheinl. 19: 273.
Syncrasis Foerster, 1862. Verh. naturh. Ver. preuss. Rheinl. 19: 264.
Synelix Foerster, 1862. Verh. naturh. Ver. preuss. Rheinl. 19: 276.
Tanycarpa Foerster, 1862. Verh. naturh. Ver. preuss. Rheinl. 19: 265.
Vachterbergia Wharton, 1994. Proc. Ent. Soc. Wash. 96: 658.

SUBFAMILY APHIDIINAE

Acanthocaudus Smith, 1944. Ohio St. Univ. Contrib. Zool. Ent. 6: 96.
Adialytus Foerster, 1862. Verh. naturh. Ver. preuss. Rheinl. 19: 249.
Aphidius Nees, 1818. Nova Acta Acad. Leop. Carol. 9: 302.
Binodoxys Mackauer, 1960. Beitr. Ent. 10: 141.
Boreogalba Mackauer, 1962. Can. Ent. 94: 1107.
Diaeretiella Starý, 1960. Cas. Czsl. Spol. Ent. 57: 242.
Diaeretus Foerster, 1862. Verh. naturh. Ver. preuss. Rheinl. 19: 249.
Ephedrus Haliday, 1833. Ent. Mag. 1: 261.
Euaphidius Mackauer, 1961. Beitr. Ent. 11: 110.
Harkeria Cameron, 1900. Ann. Mag. Nat. Hist. (7)6: 537.
Lysaphidus Smith, 1944. Ohio St. Univ. Contrib. Zool. Ent. 6: 72.
Lysiphlebus Foerster, 1862. Verh. naturh. Ver. preuss. Rheinl. 19: 248.
Monoctonus Haliday, 1833. Ent. Mag. 1: 261.
Pauesia Quilis Pérez, 1931. Eos, Madrid 7: 67.
Praon Haliday, 1833. Ent. Mag. 1: 261.
Pseudephedrus Starý, 1972. Ann. Zool. Warszawa 29: 317.
Pseudopraon Starý, 1975. Acta Ent. Bohemosl. 72: 249.
Quadrictonus Starý and Remaudière, 1982. Ann. Soc. Ent. Fr. 18: 118.
Toxares Haliday, 1840. In: Westwood, Introd. Mod. Class. Insects 2 (Synopsis): 65.
Trioxys Haliday, 1833. Ent. Mag. 1: 261.
Xenostigmus Smith, 1944. Ohio St. Univ. Contrib. Zool. Ent. 6: 36.

SUBFAMILY APOZYGINAE

Apozyx Mason, 1978. Proc. Ent. Soc. Wash. 80: 606.

SUBFAMILY BETYLOBRACONINAE

Conobregma van Achterberg, 1995. Zool. Verh. Leiden 298: 35.
Jannya van Achterberg, 1995. Zool. Verh. Leiden 298: 43.

SUBFAMILY BLACINAE

Apoblacus van Achterberg, 1976. Tijdschr. Ent. 118: 245.
Blacozona van Achterberg, 1988. Zool. Verh. Leiden 249: 22.
Blacus Nees, 1818. Nova Acta Acad. Leop. Carol. 9: 306.
Chalarope van Achterberg, 1988. Zool. Verh. Leiden 249: 35.
Dyscoletes Haliday, 1840. In: Westwood, Introd. Mod. Class. Insects 2 (Synopsis): 62.
Glyptoblacus van Achterberg, 1995. Zool. Meded. Leiden 69: 303.
Grypokeros van Achterberg, 1988. Zool. Verh. Leiden 249: 38.
Mesoxiphium van Achterberg, 1976. Tijdschr. Ent. 118: 179.
Stegnocella van Achterberg, 1976. Tijdschr. Ent. 118: 177.

SUBFAMILY BRACONINAE

Alienoclypeus Shenefelt, 1978. In: Mason, Can. Ent. 110: 765.
Atanycolus Foerster, 1862. Verh. naturh. Ver. preuss. Rheinl. 19: 238.
Bracon Fabricius, 1804. Syst. Piezat., p. 102.
Calobracon Szépligeti, 1902. Term. Fuez. 25: 39.
Cervellus Szépligeti, 1904. Genera Insect. 22: 253.
Chartobracon van Achterberg, 1983. Ent. Scand. 14: 69.
Coeloides Wesmael, 1838. Nouv. Mém. Acad. Brux. 11: 59.
Compsobracon Ashmead, 1900. Proc. U.S. Natl. Mus. 23: 138.
Compsobraconoides Quicke, 1989. Can. Ent. 121: 341.
Cyanopterus Haliday, 1835. Ent. Mag. 3: 22.
Cyclaulacidea Quicke and Delobel, 1995. Ent. Mon. Mag. 131: 218.
Cyclaulax Cameron, 1911. Timehri (3)1: 312.
Digonogastra Viereck, 1912. Proc. U.S. Natl. Mus. 42: 621.
Gnathobracon Costa, 1864 (1862). Annuar. R. Mus. Zool. R. Univ. Napoli 2: 70.
Gracilibracon Quicke, 1995. Ent. Mon. Mag. 131: 223.
Habrobracon Ashmead, 1895. In: Johnson, Ent. News 6: 324.
Hemibracon Szépligeti, 1906. Ann. Hist.-Nat. Mus. Natl. Hung. 4: 558.
Lapicida Quicke, 1989. Can. Ent. 121: 339.
Lasiophorus Haliday, 1838. Ent. Mag. 5: 213.
Leptobracon Szépligeti, 1901. Term. Fuez. 24: 360.
Megabracon Szépligeti, 1906. Ann. Hist.-Nat. Mus. Natl. Hung. 4: 580.
Megacoeloides Quicke, 1989. Ent. Mon. Mag. 125: 9.
Myosoma Brullé, 1846. In: Lepeletier, Hist. Nat. Insectes, Hym. 4: 450.
Myosomatoides Quicke, 1994. J. Hym. Res. 3: 228.
Palabracon Quicke, 1988. Bull. Ent. Res. 78: 15.
Pheloura van Achterberg, 1989. Ent. Bericht. 49: 105.
Pigeria van Achterberg, 1985. Zool. Meded. Leiden 59: 168.
Sacirema Quicke, 1995. Ent. Mon. Mag. 131: 226.
Vipio Latreille, 1804. Nouv. Dict. Hist. Nat. Deterville (Ed. 1) 24: 173.
Vipiomorpha Tobias, 1962. Zool. Zh. 41: 1193.

SUBFAMILY CARDIOCHILINAE

Cardiochiles Nees, 1818. Nova Acta Acad. Leop. Carol. 9: 307.

Hansonia Dangerfield, 1996. Proc. Ent. Soc. Wash. 98: 592.

Heteropteron Brullé, 1846. In: Lepeletier, Hist. Nat. Insectes, Hym. 4: 472.

Neocardiochiles Szépligeti, 1908. Ann. Hist.-Nat. Mus. Natl. Hung. 6: 423.

Schoenlandella Cameron, 1904. Rec. Albany Mus. 1: 169.

Toxoneuron Say, 1836. Boston J. Nat. Hist. 1: 258.

Wesmaelella Spinola, 1853 (1851). Mem. Accad. Sci. Torino (2) 13: 32.

SUBFAMILY CENOCOELIINAE

Aulacodes Cresson, 1865. Proc. Ent. Soc. Philad. 4: 8.

Capitonius Brullé, 1846. In: Lepeletier, Hist. Nat. Insectes, Hym. 4: 544.

Cenocoelius Haliday, 1840. In: Westwood, Introd. Mod. Class. Insects 2 (Synopsis): 62.

Evaniomorpha Szépligeti, 1901. Term. Fuez. 24: 356.

Foenomorpha Szépligeti, 1904. Genera Insect. 22: 8.

SUBFAMILY CHELONINAE

Ascogaster Wesmael, 1835. Nouv. Mém. Acad. Brux. 9: 226.

Chelonus Panzer, 1806. Krit. Revis. Insektenf. Deutschlands, 2: 99.

Dentigaster Zettel, 1990. Ann. naturh. Mus. Wien 91: 158.

Fischeriella Zettel, 1990. Ann. naturh. Mus. Wien 91: 181.

Leptochelonus Zettel, 1990. Ann. naturh. Mus. Wien 91: 157.

Leptodrepana Shaw, 1983. Entomography 2: 37.

Phanerotoma Wesmael, 1838. Nouv. Mém. Acad. Brux. 11: 165.

Phanerotomoides Zettel, 1990. Ann. naturh. Mus. Wien 91: 182.

Pseudophanerotoma Zettel, 1990. Ann. naturh. Mus. Wien 91: 164.

SUBFAMILY DIRRHOPINAE

Dirrhope Foerster, 1861. Verh. naturh. Ver. preuss. Rheinl. 18: 79.

SUBFAMILY DORYCTINAE

Acanthorhogas Szépligeti, 1906. Annls. Hist.-Nat. Mus. Natl. Hung. 4: 609.

Achterbergia Marsh, 1993. Contrib. Amer. Ent. Inst. 28: 4.

Acrophasmus Enderlein, 1912. Arch. Naturgesch. 78A(2): 16.

Aivalykus Nixon, 1938. Proc. R. Ent. Soc. Lond. 7: 152.

Allorhogas Gahan, 1912. Proc. Ent. Soc. Wash. 14: 3.

Aphelopsia Marsh, 1993. Contrib. Amer. Ent. Inst. 28: 5.

Aptenobracon Marsh, 1965. Ann. Ent. Soc. Amer. 58: 675.

Araucania Marsh, 1993. Contrib. Amer. Ent. Inst. 28: 6.

Atopodoryctes Marsh, 1993. Contrib. Amer. Ent. Inst. 28: 8.

Binaria Brullé, 1846. In: Lepeletier, Hist. Nat. Insectes, Hym. 4: 470.

Bohartiellus Marsh, 1983. Pan-Pac. Ent. 59: 148.

Caingangia Marsh, 1993. Contrib. Amer. Ent. Inst. 28: 9.

Callihormius Ashmead, 1900. Proc. U.S. Natl. Mus. 23: 148.

Coiba Marsh, 1993. Contrib. Amer. Ent. Inst. 28: 11.

Curtisella Spinola, 1853 (1851). Mem. Accad. Sci. Torino 13: 30.

Cyphodoryctes Marsh, 1997. Proc. Ent. Soc. Wash. 99: 586.

Dendrosoter Wesmael, 1838. Nouv. Mém. Acad. Brux. 11: 137.

Donquickeia Marsh, 1997. Proc. Ent. Soc. Wash. 99: 586.

Doryctes Haliday, 1836. Ent. Mag. 4: 43.

Ecphyloides Marsh, 1993. Contrib. Amer. Ent. Inst. 28: 14.

Ecphylus Foerster, 1862. Verh. naturh. Ver. preuss. Rheinl. 19: 237.

Evaniodes Szépligeti, 1901. Term. Fuez. 24: 353.

Exontsira Belokobyl'skij, 1982. Ent. Obozr. 61: 602.

Fritziella Marsh, 1993. Contrib. Amer. Ent. Inst. 28: 15.

Glyptocolastes Ashmead, 1900. Proc. U.S. Natl. Mus. 23: 142.

Guaygata Marsh, 1993. Contrib. Amer. Ent. Inst. 28: 16.

Gymnobracon Szépligeti, 1902. Term. Fuez. 25: 43.

Hecabolus Curtis, 1834. Brit. Ent. 11: 507.

Heerz Marsh, 1993. Contrib. Amer. Ent. Inst. 28: 17.

Heterospilus Haliday, 1836. Ent. Mag. 4: 46.

Histeromeroides Marsh, 1993. Contrib. Amer. Ent. Inst. 28: 19.

Holcobracon Cameron, 1905. Spolia Zeylan. 3: 89.

Hybodoryctes Szépligeti, 1906. Ann. Hist.-Nat. Mus. Natl. Hung. 4: 604

Janzenia Marsh, 1993. Contrib. Amer. Ent. Inst. 28: 20.

Jarra Marsh and Austin, 1994. Bull. Ent. Res. 84: 157.

Johnsonius Marsh, 1993. Contrib. Amer. Ent. Inst. 28: 21.

Labania Hedqvist, 1963. Ent. Tidsk. 84: 36.

Leluthia Cameron, 1887. Biologia Cent.-Am., Hym. 1: 392.

Liobracon Szépligeti, 1901. Term. Fuez. 24: 361.

Masonius Marsh, 1993. Contrib. Amer. Ent. Inst. 28: 23.

Megaloproctus Schulz, 1906. Spolia Hymenopterologica, p. 138.

Micrommatus Marsh, 1993. Contrib. Amer. Ent. Inst. 28: 24.

Mixtec Marsh, 1993. Contrib. Amer. Ent. Inst. 28: 25.

Monarea Szépligeti, 1904. Genera Insect. 22: 68.

Monolexis Foerster, 1862. Verh. Naturh. Ver. Preuss. Rheinl. 19: 237.

Mononeuron Fischer, 1981. Zeitschr. Arbeit. Oesterr. Ent. 33: 47.

Nervellius Roman, 1924. Ark. Zool. 16: 5.

Notiospathius Matthews and Marsh, 1973. J. Wash. Acad. Sci. 63: 73.
Odontobracon Cameron, 1887. Biologia Cent.-Am., Hym. 1: 384.
Ontsira Cameron, 1900. Mem. Proc. Manchester Lit. Phil. Soc. 44: 89.
Osmophila Szépligeti, 1902. Term. Fuez. 25: 51.
Pambolidea Ashmead, 1900. Proc. U.S. Natl. Mus. 23: 147.
Panama Marsh, 1993. Contrib. Amer. Ent. Inst. 28: 26.
Parallorhogas Marsh, 1993. Contrib. Amer. Ent. Inst. 28: 27.
Pariodes Fischer, 1981. Polskie Pismo Ent. 51: 94.
Pedinotus Szépligeti, 1902. Term. Fuez. 25: 56.
Percnobracon Kieffer, 1910. In: Kieffer and Joergensen, Zentbl. Bakt. Parasit. 27: 431.
Percnobraconoides Marsh, 1989. Pan-Pac. Ent. 65: 64.
Pioscelus Muesebeck and Walkley, 1951. Agric. Monogr. 2: 180.
Psenobolus Reinhard, 1885. In: Mayr, Verh. Zool.-Bot. Ges. Wien 35: 246.
Ptesimogaster Marsh, 1965. Ann. Ent. Soc. Amer. 58: 691.
Rhaconotus Ruthe, 1854. Stettin. ent. Ztg. 15: 349.
Rhoptrocentroides Marsh, 1993. Contrib. Amer. Ent. Inst. 28: 31.
Rhoptrocentrus Marshall, 1897. In: André, Spec. Hym. Eur. Alg. 5: 99.
Semirhytus Szépligeti, 1902. Term. Fuez. 25: 55.
Sericobracon Shaw, 1985. Psyche 92: 506
Sharkeyella Marsh, 1993. Contrib. Amer. Ent. Inst. 28: 32.
Shawius Marsh, 1993. Contrib. Amer. Ent. Inst. 28: 33.
Spathius Nees, 1818. Nova Acta Acad. Leop. Carol. 9: 301.
Stenocorse Marsh, 1968. Proc. Ent. Soc. Wash. 70: 110.
Syngaster Brullé, 1846. In: Lepeletier, Hist. Nat. Insects, Hym. 4: 454.
Tarasco Marsh, 1993. Contrib. Amer. Ent. Inst. 28: 34.
Trigonophasmus Enderlein, 1912. Arch. Naturgesch. 78A(2): 14.
Tripteria Enderlein, 1912. Arch. Naturgesch. 78A(2): 26.
Verae Marsh, 1993. Contrib. Amer. Ent. Inst. 28: 35.
Waitaca Marsh, 1993. Contrib. Amer. Ent. Inst. 28: 36.
Whartonius Marsh, 1993. Contrib. Amer. Ent. Inst. 28: 37.
Whitfieldiellus Marsh, 1997. Proc. Ent. Soc. Wash. 99: 586.

SUBFAMILY EUPHORINAE

Allurus Foerster, 1862. Verh. naturh. Ver. preuss. Rheinl. 19: 254.
Aridelus Marshall, 1887. Trans. R. Ent. Soc. Lond. 1887: 66.
Betelgeuse Shaw, 1989. Psyche 95: 290.
Bracteodes de Saeger, 1946. Explor. Parc. Natn. Albert Miss. G. F. de Witte 50: 189.
Centistes Haliday, 1835 (1834). Ent. Mag. 2: 462.
Centistina Enderlein, 1912. Arch. Naturgesch. 78A(2): 40.
Centistoides van Achterberg, 1992. Zool. Meded. Leiden 66: 345.
Chrysopophthorus Goidanich, 1948. Boll. Inst. Ent. Univ. Bologna 17: 83.
Cosmophorus Ratzeburg, 1848. Ichn. Forstins. 2: 71.
Cryptoxilos Viereck, 1911. Proc. U.S. Natl. Mus. 40: 180.
Dinocampus Foerster, 1862. Verh. naturh. Ver. preuss. Rheinl. 19: 252.
Ecclitura Kokujev, 1902. Ent. Obozr. 2: 5.
Euphoriella Ashmead, 1900. Proc. U.S. Natl. Mus. 23: 116.
Falcosyntretus Tobias, 1965. Ent. Obozr. 44: 843.
Holdawayella Loan, 1967. Ann. Ent. Soc. Amer. 60: 241.
Lecythodella Enderlein, 1912. Arch. Naturgesch. 78A(2): 41.
Leiophron Nees, 1818. Nova Acta Acad. Leop. Carol. 9: 303.
Litostolus van Achterberg, 1985. Zool. Meded. Leiden 59: 354.
Loxocephalus Foerster, 1862. Verh. naturh. Ver. preuss. Rheinl. 19: 252.
Marshiella Shaw, 1985. Entomography 3: 329.
Microctonus Wesmael, 1835. Nouv. Mém. Acad. Brux. 9: 54.
Orionis Shaw, 1987. Syst. Ent. 12: 106.
Perilitus Nees, 1818. Nova Acta Acad. Leop. Carol. 9: 302.
Peristenus Foerster, 1862. Verh. naturh. Ver. preuss. Rheinl. 19: 251.
Plynops Shaw, 1996. J. Hym. Res. 5: 167.
Pygostolus Haliday, 1833. Ent. Mag. 1: 263.
Ropalophorus Curtis, 1837. Guide Brit. Insects, p. 118.
Spathicopis van Achterberg, 1977. Ent. Bericht. 37: 27.
Streblocera Westwood, 1833. Phil. Mag. Jour. Sci. 3: 342.
Syntretus Foerster, 1862. Verh. naturh. Ver. preuss. Rheinl. 19: 251.
Townesilitus Haeselbarth and Loan, 1983. Contrib. Amer. Ent. Inst. 20: 384.
Wesmaelia Foerster, 1862. Verh. naturh. Ver. preuss. Rheinl. 19: 251.

SUBFAMILY GNAMPTODONTINAE

Gnamptodon Haliday, 1833. Ent. Mag. 1: 265.
Pseudognaptodon Fischer, 1964. Polskie Pismo Ent. 34: 207.

SUBFAMILY HELCONINAE

Aliolus Say, 1836. Boston J. Nat. Hist. 1: 259.
Baeacis Foerster, 1878. Verh. Naturh. Ver. Preuss. Rheinl. 35: 70.
Diospilus Haliday, 1833. Ent. Mag. 1: 262.
Eubazus Nees 1814 (1812). Mag. Ges. nat. Fr. Berl. 6: 214.
Eumacrocentrus Ashmead, 1900. Proc. U.S. Natl. Mus. 23: 120.

Helcon Nees, 1814 (1812). Mag. Ges. nat. Fr. Berl. 6: 216.
Nealiolus Mason, 1974. Proc. Ent. Soc. Wash. 76: 241.
Schizoprymnus Foerster, 1862. Verh. naturh. Ver. preuss. Rheinl. 19: 242.
Taphaeus Wesmael, 1835. Nouv. Mém. Acad. Brux. 9: 189.
Topaldios Papp, 1995. Acta Zool. Acad. Sci. Hung. 41: 107.
Triaspis Haliday, 1835. Ent. Mag. 3: 124.
Urosigalphus Ashmead, 1888. Proc. U.S. Natl. Mus. 11: 637.
Vadum Mason, 1987. Proc. Ent. Soc. Wash. 89: 325.
Wroughtonia Cameron, 1899. Mem. Proc. Manchester Lit. Phil. Soc. 43: 56.

SUBFAMILY HISTEROMERINAE
Histeromerus Wesmael, 1838. Nouv. Mém. Acad. Brux. 11: 63.

SUBFAMILY HOMOLOBINAE
Charmon Haliday, 1833. Ent. Mag. 1: 262.
Charmontia van Achterberg, 1979. Tijdschr. Ent. 122: 262.
Exasticolus van Achterberg, 1979. Tijdschr. Ent. 122: 271.
Homolobus Foerster, 1862. Verh. naturh. Ver. preuss. Rheinl. 19: 256.

SUBFAMILY HORMIINAE
Acrisis Foerster, 1862. Verh. naturh. Ver. preuss. Rheinl. 19: 236.
Allobracon Gahan, 1915. Proc. U.S. Natl. Mus. 49: 94.
Aspilodemon Fischer, 1966. Entomophaga 11: 162.
Cantharoctonus Viereck, 1912. Proc. U.S. Natl. Mus. 42: 617.
Chremylus Haliday, 1833. Ent. Mag. 1: 266.
Colastes Haliday, 1833. Ent. Mag. 1: 266.
Dolopsidea Hincks, 1944. Proc. R. Ent. Soc. Lond. 13: 31.
Hormiopius Blanchard, 1962. Rev. Invest. Agr. 16: 45.
Hormius Nees, 1818. Nova Acta Acad. Leop. Carol. 9: 305.
Hydrangeocola Brèthes, 1927. Rev. Chil. Hist. Nat. 31: 195.
Lysitermoides van Achterberg, 1995. Zool. Verh. Leiden 298: 112.
Lysitermus Foerster, 1862. Verh. naturh. Ver. preuss. Rheinl. 19: 236.
Monitoriella Hedqvist, 1963. Ent. Tidsk. 84: 44.
Neurocrassus Snoflák, 1945. Ent. Listy 8: 26.
Oncophanes Foerster, 1862. Verh. naturh. Ver. preuss. Rheinl. 19: 241.
Pambolus Haliday, 1836. Ent. Mag. 4: 49.
Parahormius Nixon, 1940. Ann. Mag. Nat. Hist. (11)5: 473.
Proacrisis Tobias, 1983. Trudy Vses. Ent. Obsch. 65: 161.
Pseudobathystomus Belokobyl'skij, 1986. Ent. Obozr. 45: 781.
Rhysipolis Foerster, 1862. Verh. naturh. Ver. preuss. Rheinl. 19: 235.
Rhyssalus Haliday, 1833. Ent. Mag. 1: 266.
Shawiana van Achterberg, 1983. Zool. Meded. Leiden 57: 341.
Xenarcha Foerster, 1862. Verh. naturh. Ver. preuss. Rheinl. 19: 235.
Xenosternum Muesebeck, 1935. Ann. Ent. Soc. Amer. 28: 246.

SUBFAMILY ICHNEUTINAE
Hebichneutes Sharkey and Wharton, 1994. J. Nat. Hist. 28: 883.
Helconichia Sharkey and Wharton, 1994. J. Nat. Hist. 28: 886.
Ichneutes Nees, 1816 (1813). Mag. Ges. nat. Fr. Berl. 7: 275.
Lispixys Mason, 1969. Proc. Ent. Soc. Wash. 71: 274.
Masonbeckia Sharkey and Wharton, 1994. J. Nat. Hist. 28: 891.
Muesonia Sharkey and Wharton, 1994. J. Nat. Hist. 28: 898.
Oligoneurus Szépligeti, 1902. Term. Fuez. 25: 77.
Paroligoneurus Muesebeck, 1931. Proc. U.S. Natl. Mus. 79: 7.
Proterops Wesmael, 1835. Nouv. Mém. Acad. Brux. 9: 201.

SUBFAMILY MACROCENTRINAE
Austrozele Roman, 1910. Ent. Tidskr. 31: 113.
Dolichozele Viereck, 1911. Proc. U.S. Natl. Mus. 40: 182.
Hymenochaonia Dalla Torre, 1898. Wein. ent. Ztg. 17: 100.
Macrocentrus Curtis, 1833. Ent. Mag. 1: 187.

SUBFAMILY MASONINAE
Masona van Achterberg, 1995. Zool. Verh. Leiden 298: 97.

SUBFAMILY MENDESELLINAE
Epsilogaster Whitfield and Mason, 1994. Syst. Ent. 19: 72.
Mendesella Whitfield and Mason, 1994. Syst. Ent. 19: 68.

SUBFAMILY METEORIDEINAE
Meteoridea Ashmead, 1900. Proc. U.S. Natl. Mus. 23: 129.

SUBFAMILY METEORINAE
Meteorus Haliday, 1835. Ent. Mag. 3: 24.
Zele Curtis, 1832. Brit. Ent. 9: 415.

SUBFAMILY MICROGASTRINAE
Alphomelon Mason, 1981. Mem. Ent. Soc. Can. 115: 54.
Apanteles Foerster, 1862. Verh. naturh. Ver. preuss. Rheinl. 19: 245.
Austrocotesia Austin and Dangerfield, 1992. Invert. Taxon. 6: 11
Choeras Mason, 1981. Mem. Ent. Soc. Can. 115: 76.
Clarkinella Mason, 1981. Mem. Ent. Soc. Can. 115: 66.

Cotesia Cameron, 1891. Mem. Proc. Manchester Lit. Phil. Soc. 4: 185.
Dasylagon Muesebeck, 1958. Proc. U.S. Natl. Mus. 107: 424.
Deuterixys Mason, 1981. Mem. Ent. Soc. Can. 115: 123.
Diolcogaster Ashmead, 1900. In: Smith, Insects New Jersey, p. 594.
Distatrix Mason, 1981. Mem. Ent. Soc. Can. 115: 93.
Dolichogenidea Viereck, 1911. Proc. U.S. Natl. Mus. 40: 173.
Exix Mason, 1981. Mem. Ent. Soc. Can. 115: 116.
Exoryza Mason, 1981. Mem. Ent. Soc. Can. 115: 40.
Fornicia Brullé, 1846. In: Lepeletier, Hist. Nat. Insectes, Hym. 4: 511.
Glyptapanteles Ashmead, 1904. Proc. U.S. Natl. Mus. 28: 147.
Hygroplitis Thomson, 1895. Opusc. Ent. 20: 2238.
Hypomicrogaster Ashmead, 1898 (1897). Proc. Ent. Soc. Wash. 4: 166.
Iconella Mason, 1981. Mem. Ent. Soc. Can. 115: 74.
Illidops Mason, 1981. Mem. Ent. Soc. Can. 115: 56.
Larissimus Nixon, 1965. Bull. Br. Mus. Nat. Hist., Ent. Suppl. 2: 204.
Lathrapanteles Williams, 1985. Can. J. Zool. 63: 1963.
Microgaster Latreille, 1804. Nouv. Dict. Hist. Nat. Deterville (Ed. 1) 24: 175.
Microplitis Foerster, 1862. Verh. naturh. Ver. preuss. Rheinl. 19: 245.
Papanteles Mason, 1981. Mem. Ent. Soc. Can. 115: 47.
Parapanteles Ashmead, 1900. Proc. U.S. Natl. Mus. 23: 131.
Paroplitis Mason, 1981. Mem. Ent. Soc. Can. 115: 68.
Pelicope Mason, 1981. Mem. Ent. Soc. Can. 115: 57.
Pholetesor Mason, 1981. Mem. Ent. Soc. Can. 115: 37.
Prasmodon Nixon, 1965. Bull. Br. Mus. Nat. Hist., Ent. Suppl. 2: 205.
Promicrogaster Brues and Richardson, 1913. Bull. Amer. Mus. Nat. Hist. 32: 499.
Protapanteles Ashmead, 1898 (1897). Proc. Ent. Soc. Wash. 4: 166.
Protomicroplitis Ashmead, 1898. Proc. Ent. Soc. Wash. 4: 167.
Pseudapanteles Ashmead, 1898 (1897). Proc. Ent. Soc. Wash. 4: 166.
Rasivalva Mason, 1981. Mem. Ent. Soc. Can. 115: 91.
Rhygoplitis Mason, 1981. Mem. Ent. Soc. Can. 115: 81.
Sathon Mason, 1981. Mem. Ent. Soc. Can. 115: 78.
Sendaphne Nixon, 1965. Bull. Brit. Mus. Nat. Hist., Ent. Suppl. 2: 203.
Snellenius Westwood, 1882. Tijdschr. Ent. 25: 19.
Venanides Mason, 1981. Mem. Ent. Soc. Can. 115: 99.
Venanus Mason, 1981. Mem. Ent. Soc. Can. 115: 94.
Wilkinsonellus Mason, 1981. Mem. Ent. Soc. Can. 115: 122.
Xanthapanteles Whitfield, 1995. Proc. Ent. Soc. Wash. 97: 879.
Xanthomicrogaster Cameron, 1911. Timehri (3)1: 324.
Xenogaster Mason, 1981. Mem. Ent. Soc. Can. 115: 71.
Yelicones Cameron, 1887. Biologia Cent.-Am., Hym. 1: 387.

SUBFAMILY SIGALPHINAE

Pselaphanus Szépligeti, 1902. Term. Fuez. 25: 82.
Sigalphus Latrielle, 1802. Hist. Nat. Crust. Insectes 3: 327.

SUBFAMILY MICROTYPINAE

Microtypus Ratzeburg, 1848. Ichn. Forstins. 2: 47.
Neomicrotypus van Achterberg, 1992. Zool. Meded. Leiden 66: 375.

SUBFAMILY MIRACINAE

Mirax Haliday, 1833. Ent. Mag. 1: 263.

SUBFAMILY NEONEURINAE

Elasmosoma Ruthe, 1858. Berl. ent. Zeitschr. 2: 7.
Neoneurus Haliday, 1838. Ent. Mag. 5: 213.

SUBFAMILY OPIINAE

Ademon Haliday, 1833. Ent. Mag. 1: 266.
Biosteres Foerster, 1862. Verh. naturh. Ver. preuss. Rheinl. 19: 259.
Diachasma Foerster, 1862. Verh. naturh. Ver. preuss. Rheinl. 19: 259.
Diachasmimorpha Viereck, 1913. Proc. U.S. Natl. Mus. 44: 641.
Doryctobracon Enderlein, 1920 (1918). Arch. Naturgesch. 84A(11): 144.
Eurytenes Foerster, 1862. Verh. naturh. Ver. preuss. Rheinl. 19: 259.
Exodontiella Wharton, 1977. Pan-Pac. Ent. 53: 298.
Fopius Wharton, 1987. Proc. Ent. Soc. Wash. 89: 68.
Opius Wesmael, 1835. Nouv. Mém. Acad. Brux. 9: 115.
Psyttalia Walker, 1860. Ann. Mag. Nat. Hist. (3) 5: 311.
Utetes Foerster, 1862. Verh. naturh. Ver. preuss. Rheinl. 19: 261.

SUBFAMILY ORGILINAE

Antestrix van Achterberg, 1987. Zool. Verh. Leiden 242: 10.
Bentonia van Achterberg, 1992. Zool. Meded. Leiden 66: 339.
Orgilus Haliday, 1833. Ent. Mag. 1: 262.
Podorgilus van Achterberg, 1994. Zool. Meded. Leiden 68: 173.
Stantonia Ashmead, 1904. Proc. U.S. Natl. Mus. 28: 146.

SUBFAMILY ROGADINAE

Aleiodes Wesmael, 1838. Nouv. Mém. Acad. Brux. 11: 94.
Bulborogas van Achterberg, 1995. Zool. Verh. Leiden 298: 115.
Choreborogas Whitfield, 1990. Proc. Ent. Soc. Wash. 92: 39.
Clinocentrus Haliday, 1833. Ent. Mag. 1: 266.
Cystomastax Szépligeti, 1904. Genera Insect. 22: 81.
Macrostomion Szépligeti, 1900. Term. Fuez. 23: 57.
Polystenidea Viereck, 1911. Proc. U.S. Natl. Mus. 40: 186.
Rogas Nees, 1818. Nova Acta Acad. Leop. Carol. 9: 306.
Stiropius Cameron, 1911. Timehri (3) 1: 329.
Tetrasphaeropyx Ashmead, 1888. Proc. U.S. Natl. Mus. 11: 634.

SUBFAMILY YPSISTOCERINAE

Embobracon van Achterberg, 1995. Zool. Verh. Leiden 298: 49.
Termitobracon Brues, 1923. Zoologica, N.Y. 3: 429.
Ypsistocerus Cushman, 1923. Proc. Ent. Soc. Wash. 25: 54.

PUBLICATIONS FOR SALE BY THE INTERNATIONAL SOCIETY OF HYMENOPTERISTS

SPECIAL PUBLICATIONS

No. 1 Manual of the New World Genera of the Family Braconidae (Hymenoptera), edited by Robert A. Wharton, Paul M. Marsh and Michael J. Sharkey. 439 pp. 1997 .... $20.00

JOURNAL OF HYMENOPTERA RESEARCH

Back issues of the Journal of Hymenoptera Research are available at $31 for volumes 1–5 and $41 for volumes 6 and following.

Prices quoted are U.S. currency; postage is extra except on prepaid orders. All orders should be placed with the Treasurer, International Society of Hymenopterists. For this address, see the inside cover of the latest issue of the Journal of Hymenoptera Research.

Membership. Membership in The International Society of Hymenopterists is open to persons who have demonstrated interest in the science of entomology. Annual dues for members are $35, payable to The International Society of Hymenopterists. Requests for membership should be sent to the Treasurer (for address see inside front cover of the latest issue of the Journal of Hymenoptera Research).

Publications. The *Journal of Hymenoptera Research* is published twice each year by the Society. Subjects include all aspects of the study of Hymenoptera. Members in good standing receive the *Journal* free. Nonmember subscriptions are $60 per year. The *Special Publications of The International Society of Hymenopterists* are published periodically as manuscripts are available. This series is intended for larger comprehensive manuscripts and authors are expected to cover some or all of the publication costs. Manuscripts for both publications should be sent to the Editor (for address see inside front cover of the latest issue of the Journal of Hymenoptera Research).

Made in United States
Troutdale, OR
06/27/2025

32426346R20270